# Lecture Notes in Mathematics

Walter Roth

# Operator-Valued Measures and Integrals for Cone-Valued Functions

Springer

Walter Roth
Department of Mathematics
University of Brunei Darussalam
BE 1410 Gadong
Brunei Darussalam
roth@fos.ubd.edu.bn

ISBN: 978-3-540-87564-2 e-ISBN: 978-3-540-87565-9
DOI: 10.1007/978-3-540-87565-9

Lecture Notes in Mathematics ISSN print edition: 0075-8434
ISSN electronic edition: 1617-9692

Library of Congress Control Number: 2008938191

Mathematics Subject Classification (2000): 28B20, 46A13, 46E40, 46G10



*Cover design*: SPi Publishing Services

Printed on acid-free paper

9 8 7 6 5 4 3 2 1

springer.com

# Preface

The aim of this book is twofold: Firstly, to introduce the developing theory of locally convex cones to a wider audience. This theory generalizes locally convex topological vector spaces and permits many additional and substantially different examples and applications. In the aspects of the theory that have been developed so far, the increase in generality does not lead to any compromises with respect to the depth of its results. The main difference to vector spaces is the presence of infinity-type unbounded elements and the general non-availability of the cancellation law. Some important mathematical models, while close to the structure of vector spaces are of this type. They do not allow subtraction of their elements or multiplication by negative scalars. Examples are certain classes of set-valued or extended real-valued functions that may take infinite values. These arise naturally in integration theory, potential theory and in a variety of other settings and do not form vector spaces. Therefore many results and techniques from classical functional analysis can not be immediately applied. Locally convex cones carry a reflexive and transitive order relation, and their (convex semiuniform) topology is defined using this order structure. The first part of this book contains a review and summary of the aspects of the theory of locally convex cones that have been developed so far, sometimes without detailed proofs, but references to the sources instead. The theory is then developed further, adding some (hopefully) interesting new features.

This leads to the second objective: Locally convex cones are used to provide the setting for a novel approach to integration theory. The generality of their structure allows to deal simultaneously with a wide variety of situations, including extended real-valued, vector-valued, operator-valued and cone-valued measures and functions. Topological limits from the classical theory are replaced by approximations in terms of the order structure of a locally convex cone. The main results include convergence theorems for measures and functions and an integral representation theorem for continuous linear operators on certain cones of functions. The latter establishes that a given operator can be expressed as an integral with respect to some unique measure. This is a very technical result and requires a lengthy proof. It is followed by a comprehensive collection of special cases and applications. Some of these lead to known representation results for compact and weakly compact operators on Banach space-valued functions, but the more general cases are new. The insertions of a special case yield the classical Spectral Representation Theorem for normal linear operators on a complex Hilbert space.

# Contents

# Introduction

Integration theory was originally developed for real-valued functions with respect to real-valued measures. There are a great number of expositions devoted to this, most notably the classical treatises by Lebesgue [116], Carathéodory [30], Radon [158] and Daniell [36]. More recent treatments in the works of Bourbaki [25], Hahn and Rosenthal [80], Halmos [83] and Saks [182] contain excellent historical notes and references on the subject. Vector integration was introduced in the first half of the last century, and exhaustive discussions of the field can for example be found in the works of Dunford and Schwartz [54], [55], [56], Diestel and Uhl [43] or Graves [75].

The aim of this book is to develop a general theory of integration which simultaneously deals with extended real-valued, vector-valued, operator-valued and cone-valued measures and functions. All except the last of these topics have been explored extensively, and integration theory as presented in the available standard texts uses different approaches in each of these cases. Both finitely and countably additive measures have been considered. However, finitely additive measures yield only limited results and are therefore not widely used in analysis. As a consequence only countably additive measures shall be considered in this book.

The order structure of the extended real number system $\overline{\mathbb{R}} = \mathbb{R} \cup \{+\infty\}$ is extensively and indispensably used for the integration of $\overline{\mathbb{R}}$-valued functions with respect to $\overline{\mathbb{R}}$-valued measures. Integrals are defined using suprema and infima in $\overline{\mathbb{R}}$. However, an order structure is generally not available in all of the above-mentioned cases and different techniques need to be applied in integration theory, often replacing suprema and infima with topological limits. The literature on these subjects is of course vast, and the approaches chosen by different authors vary to some degree. Some of the more popular texts shall be cited throughout this book and many more are mentioned in the bibliography.

In very general terms, integration theory deals with measures whose values are in some set, say $\mathcal{L}$, integrating certain functions with values in a second set $\mathcal{P}$, and resulting in integrals in a third set $\mathcal{Q}$. In order to obtain a

W. Roth, *Operator-Valued Measures and Integrals for Cone-Valued Functions*,
Lecture Notes in Mathematics 1964,

meaningful theory, one needs to impose some linear structure on the sets $\mathcal{L}, \mathcal{P}$ and $\mathcal{Q}$ and a bilinear form from $\mathcal{L} \times \mathcal{P}$ into $\mathcal{Q}$ that determines how measures and functions interact with each other. Choosing (topological) vector spaces for $\mathcal{L}, \mathcal{P}$ and $\mathcal{Q}$ does however severely restrict the use of unbounded, that is infinity-type, elements. In classical extended real-valued measure theory, the use of the element $+\infty$ is indeed essential. In more general settings, for example in function spaces, the availability of unbounded and infinite-valued functions is equally desirable. They do however not fit into vector spaces, since subtraction and multiplication by negative scalars may not be available. In order to apply vector space techniques, these infinity-type elements have to be removed only to be (sometimes rather awkwardly) re-adjoined at later stages in building the theory.

A different and, in the context of integration theory, novel approach will therefore be followed in this book: The main idea is to use more general structures, called locally convex cones, which are modeled by cones of convex sets or cones of set-valued functions. The theory of locally convex cones was first developed by the author together with K. Keimel [100] in 1992, mainly in the context of Korovkin-type linear approximation theory. Other than vector spaces, and this signifies the main divergence and generalization in terms of their algebraic structure, cones do not necessarily satisfy the cancellation law, namely the rule that $a+c=b+c$ for given elements $a, b$ and $c$ implies that $a=c$. The validity of this rule would imply that a cone is embeddable into a real vector space, thus leaving out some of the more interesting applications. A simple example for such a cone is the extended real line $\overline{\mathbb{R}}$, endowed with the usual algebraic operations of addition and multiplication by non-negative reals.

Locally convex cones also carry a topological structure which is in some way compatible with the algebraic operations. Continuity of the scalar multiplication is not required in general, as it would exclude some of the most interesting and essential non-vector space examples. However, for invertible elements the topological requirements coincide locally with those for topological vector spaces. In this way, locally convex cones constitute canonical generalizations of locally convex topological vector spaces, still retaining many of their most important properties. In particular, locally convex cones yield a rich duality theory, where the study of the dual cone consisting of all continuous linear $\overline{\mathbb{R}}$-valued functionals offers valuable insight into the given locally convex cone itself. There are powerful Hahn-Banach type extension and separation theorems using sublinear and superlinear functionals which allow infinite values. Locally convex topologies can be generated using dual pairs of cones and a bilinear form. This leads to various notions of weak and polar topologies, including Mackey-Arens type theorems. There are the notions of continuous linear operators between locally convex cones and their adjoints as linear operators between the respective duals. There are analogues of the Uniform Boundedness and Open Mapping theorems. Although in instances considerably more delicate and complicated than their analogues for locally convex topological vector spaces, most of these concepts reduce to their full strength in those special cases. Locally convex cones carry an order relation which is only required to be reflexive and transitive and compatible with the algebraic operations. Equality is such an order, hence cones without an explicit order structure are also included. The topology of a locally convex cone is given through a convex semiuniform structure, but it can alternatively be expressed in terms of its order structure alone, using a subfamily of positive elements called abstract neighborhoods. This often allows for a concise and elegant formulation of topological properties. This latter formalization of locally convex cones shall therefore be used throughout this book. The relationship between topology and order was explored in detail in a seminal work by L. Nachbin [135].

Most importantly for the purposes of this book, and in contrast to vector spaces, locally convex cones allow for the presence of unbounded elements. These infinity-type elements can be thoroughly investigated, classified and grouped into boundedness and topological connectedness components. Modified versions of the cancellation law apply within each of these components. The topological structure of locally convex cones is thus far richer, more complicated (and arguably more interesting) than that of topological vector spaces. Moreover, one shall investigate special types of locally convex cones called locally convex complete lattice cones, which allow suprema and infima of subsets with respect to their order. These suprema and infima are required to be compatible in a rather strong sense with the algebraic operations and with the topological neighborhoods, comparable to the requirements for M-topologies in locally convex topological vector lattices. Cones of $\overline{\mathbb{R}}$-valued functions with their natural operations and pointwise order are the prime examples here. There is a canonical concept for order convergence of nets in a locally convex complete lattice cone, which is generally weaker than convergence in the given locally convex cone topology. This notion will turn out to be suitable for integration theory. Later on one shall establish and make use of the fact that every locally convex topological vector space can be canonically embedded into such a locally convex complete lattice cone. A normed space, for example, can be embedded into the locally convex complete lattice cone of all bounded-below $\overline{\mathbb{R}}$-valued functions on its dual unit ball.

In developing this version of integration theory, locally convex cones shall be chosen for the sets $\mathcal{P}$ and $\mathcal{Q}$ from above, i.e. the ranges for the functions involved and their resulting integrals. Indeed, $\mathcal{Q}$ is required to be a locally convex complete lattice cone in order to allow the formation of suprema and infima, which is essential for the evaluation of integrals of $\mathcal{P}$-valued functions. Correspondingly, the values of the measures will be linear operators from $\mathcal{P}$ into $\mathcal{Q}$ and the convergence of a series of measures - essential for the explanation of countable additivity - will be defined using a variation of strong operator convergence. The same notion of convergence will be used for the formulation of a variety of versions of the standard convergence theorems from integration theory, involving both sequences of functions and sequences of measures.

The generality of this setting permits its application to a large variety of situations. In this way, one shapes a course for a unified approach to extended real-valued, vector-valued and operator-valued measures and the integration of corresponding classes of functions. Positive reals, for example, may be interpreted as linear operators from $\overline{\mathbb{R}}$ into itself. Elements of $\overline{\mathbb{R}}$, on the other hand are linear operators on the positive reals. Vectors may be considered as linear operators from the scalar field into their space, or alternatively, from their dual space into the scalars. Particular attention will be paid to the case in which $\mathcal{Q}$, the range of the integrals, is some cone of linear operators on a locally convex cone or vector space. If this cone or vector space carries some additional structure, let us say a multiplication or an order with lattice properties, then additional conditions on the measures under consideration can guarantee certain desired properties for the resulting integrals. They may be multiplicative operators in the first of the cases mentioned above, or lattice homomorphisms in the second. Among other applications, this type of condition will be used in the final chapter in order to derive a generalized version of the Spectral representation theorem for linear operators on Banach spaces.

In the first chapter of this book, a summary of some known and previously established properties of locally convex cones will be provided, emphasizing essential attributes which turn out to be relevant to integration theory. Details are often referred to [100] and related works. The first sections of Chapter I contain a review of the definitions, basic properties and a set of standard examples for locally convex cones. Many of these can be found in earlier texts. These standard examples serve in particular to illustrate the variety and the wide reach of the theory, and will be used and referred to throughout the book. In order to meet the demands of their utilization in integration theory, some specific properties of locally convex cones will be explored in detail. There is a review of the relative topologies and a thorough investigation of boundedness and connectedness components with respect to these topologies. These were introduced in earlier works by the author ([175], [176]), but some relevant details will be added. Some of the following investigations are new, in particular the parts about locally convex complete lattice cones, and are therefore provided in full detail and with all proofs. Not all of the concepts introduced in this chapter will be used later on, but appear to be of sufficient interest for the development of the general theory of locally convex cones to merit inclusion. Readers who are interested mainly in their application to integration theory may skip some of the details of these sections at a first reading.

Chapter II comprises the general theory of operator-valued measures and integrals of cone-valued functions. For this one considers two locally convex cones $\mathcal{P}$ and $\mathcal{Q}$ (the latter is supposed to be a complete lattice cone) and the cone $\mathfrak{L}(\mathcal{P},\mathcal{Q})$ of linear operators from $\mathcal{P}$ into $\mathcal{Q}$. $\mathfrak{L}(\mathcal{P},\mathcal{Q})$-valued measures are defined on a $\sigma$-field (or a $\sigma$-ring) of a set $X$. These measures are required to be bounded in some sense and countably additive with respect to order convergence in $\mathcal{Q}$, which is generally weaker than the given convergence when a locally convex vector space is embedded into a locally convex complete lattice cone. This is followed by the definition of measurability and integrability for $\mathcal{P}$-valued functions on $X$. The evaluation of their integrals are then elements of $\mathcal{Q}$. The use of locally convex cones as the general setting allows the utilization of their order structure, and hence the concepts and techniques of classical real-valued measure theory. The integral of a function is defined using an approximation in order from below by integrals over step functions. The general convergence theorems for integrals of sequences of integrable functions are also formulated (and valid only) in terms of order convergence, but may be strengthened in special situations. Most of the results in this chapter are new, as is the attempt to use locally convex cones in integration theory. Many of the definitions, statements and proofs turn out to be quite technical and often lengthy, but they are provided in full detail. Unfortunately, the inclusion of unbounded elements in locally convex cones tends to make arguments considerably more delicate and complicated compared to their counterparts for topological vector spaces. Chapter II concludes with an extensive section dealing with examples and special cases. These include the classical settings for extended real-, complex- and vector-valued measure theory that can be found in standard texts such as [25], [43], [55], [83], [178] or [179]. However, the generality of the approach allows a far wider range of applications. Of particular interest is the situation when both locally convex cones $\mathcal{P}$ and $\mathcal{Q}$ carry an additional algebraic or order theoretic structure. This structure then transfers naturally, that is pointwise, to the cone of all $\mathcal{P}$-valued functions on $X$. In these circumstances, some canonical additional conditions on an $\mathfrak{L}(\mathcal{P},\mathcal{Q})$-valued measure can guarantee that integrals of $\mathcal{P}$ -valued functions with respect to such a measure will represent a structure-preserving linear operator from the cone of integrable $\mathcal{P}$-valued functions on $X$ into $\mathcal{Q}$. Further applications include integrals for cone-valued functions with respect to real-valued measures, operator-valued functions with respect to cone-valued measures and positive, real- or complex-valued functions and cone- or operator-valued measures.

Chapter III deals with topological measure theory, that is operator-valued measures on the $\sigma$-field (or $\sigma$-ring) of all Borel subsets (or relatively compact Borel subsets) of a locally compact space $X$. As usual, some regularity properties are required for these measures. The presence of unbounded elements in a locally convex cone $\mathcal{P}$ suggests a slightly modified notion of continuity for $\mathcal{P}$-valued functions in this case, with the result that one considers continuity with respect to the somewhat coarser relative topology of $\mathcal{P}$. These functions yield interesting properties concerning the boundedness components of their range. Using suitable inductive limit-type topologies for these functions one can identify corresponding cones of continuous $\mathcal{P}$-valued functions that vanish at infinity with respect to these topologies. This generalizes the concept of weighted spaces of continuous real-valued functions on a locally compact space which was introduced by Nachbin in [136] and

Prolla in [155]. The main result of this chapter, and perhaps in this book, is an integral representation theorem for continuous linear operators from function cones of this type into a locally convex lattice cone $\mathcal{Q}$. It establishes that the given operator can be expressed as an integral with respect to some unique operator-valued measure. It is of course inspired by the classical Riesz representation theorem for continuous real-valued linear functionals on a function space. Unsurprisingly, the proof of this theorem requires considerable effort and is probably the most technical and surely the longest in the entire book. The rewards are however plentiful. A comprehensive collection of special cases and applications is furnished at the end of this chapter. Some of these lead to known representation results for compact and weakly compact operators on Banach space-valued functions, but the more general cases are new. The results of Chapter II can be applied to derive additional properties for the representing measure if the given operator is of a special type. This applies in particular when the range cones $\mathcal{P}$ and $\mathcal{Q}$ carry an additional algebraic or order theoretic structure. The insertions of a special case yield the classical Spectral representation theorem for normal linear operators on a complex Hilbert space, for example.

It might be interesting, although probably demanding, to try to extend the main theorem of Chapter III into a Choquet-type representation theory. In such an approach a continuous linear operator is given only on a subcone of continuous functions on $X$ and an integral representation is sought using a measure with certain additional properties concerning its support. The techniques for proving Choquet's theorem in the real-valued case rely heavily on the order properties of real-valued functions. Allowing locally convex cones with their rich order structure as the range of the functions involved might therefore present an avenue for the extension of this powerful classical result.

Further investigations into cone-valued functions and operator valued-measures might also focus on generalizations of the Radon-Nikodým theorem. This theorem probes the absolute continuity of a measure with respect to another one and the related fact that the first measure can be expressed through the second one by the use of an integrable density function. Chapter II is concerned with some aspects of these questions, but further investigation would require more detailed studies of special properties of the locally convex cones involved. For vector-valued measures there are well-known relations between the Radon-Nikodým and the Riesz representation theorems. Details about these can be found in the text by Diestel and Uhl [43], for example.

Each chapter of this book concludes with a brief section containing notes and remarks. The bibliography at the end is far from complete. It contains the references and a somewhat arbitrary list of some of the better known publications on integration theory. A more exhaustive bibliography on the subject can be found in the books by Dunford and Schwartz [55], [56], [57] and Diestel and Uhl [43].

The numbering of theorems, corollaries, definitions, examples, etc. is carried out consecutively and takes into account the sections, but not the chapters of the book. Cross-references without further qualification are therefore meant within the same chapter. References to different chapters are pointed out by the addition of the roman numeral for the relevant chapter.

# Chapter I
# Locally Convex Cones

The purpose of this chapter is twofold: Firstly, to provide the tools and the settings for the integration theory which will be developed in Chapters II and III, and secondly, to introduce the theory of locally convex cones to a wider audience. This theory generalizes locally convex topological vector spaces and has (in the author's opinion, quite unsurprisingly) not yet received the attention that it deserves. Locally convex cones permit many more and substantially different examples and applications than locally convex vector spaces. In the aspects of the theory that have been developed so far, the increase in generality leads only to minor, if any at all, compromises with respect to the depth of its results. While some of the methods and arguments employed may at times appear rather technical and indeed counterintuitive, this is largely the consequence of the inclusion of infinity-type unbounded elements and the general non-availability of the cancellation law.

So why is it worth the effort? Endowed with suitable topologies, vector spaces yield rich and well-studied structures. Locally convex topological vector spaces permit an extensive duality theory whose study gives valuable insight into the spaces themselves. Some important mathematical settings, however, while close to the structure of vector spaces do not allow subtraction of their elements or multiplication by negative scalars. Examples are certain classes of functions that may take infinite values or are characterized through inequalities rather than equalities. They arise naturally in integration theory, potential theory and in a variety of other settings. Likewise, families of convex subsets of vector spaces which are of interest in various contexts, do not form vector spaces. If the cancellation law fails, domains of this type can not be embedded into vector spaces in order to apply the results and techniques from classical functional analysis. The inclusion of these and similar examples into an analytical theory merits the investigation of a more general structure. Apart from being useful in this sense, the theory of locally convex cones allows for some interesting and occasionally insightful and elegant mathematics.

W. Roth, *Operator-Valued Measures and Integrals for Cone-Valued Functions*,
Lecture Notes in Mathematics 1964,

The first three sections of this chapter present a review of some of the main concepts of this theory while often referring to [100] and other sources for details and proofs. A brief survey of the subject can also be found in [169]. Section 4 introduces the relative topologies of a locally convex cone and provides definitions and investigations of different types of boundedness and connectedness components. Locally convex lattice cones, quasi-full locally convex cones and cones of linear operators, are studied in Sections 5, 6 and 7, respectively. These will be used extensively in the integration theory of Chapters II and III. Some of the more specialized parts of Sections 4 to 7 are included for reference in the later stages of Chapters II and III and may be skipped at first reading.

## 1. Locally Convex Cones

A *cone* is a set $\mathcal{P}$ endowed with an addition $(a,b) \mapsto a+b$ and a scalar multiplication $(\alpha,a) \mapsto \alpha a$ for real numbers $\alpha \geq 0$. The addition is supposed to be associative and commutative, and there is a neutral element $0 \in \mathcal{P}$. For the scalar multiplication the usual associative and distributive properties hold, that is $\alpha(\beta a) = (\alpha\beta)a$, $(\alpha+\beta)a = \alpha a + \beta a$, $\alpha(a+b) = \alpha a + \alpha b$, $1a = a$ and $0a = 0$ for all $a, b \in \mathcal{P}$ and $\alpha, \beta \geq 0$. The *cancellation law*, stating that $a+c = b+c$ implies $a = b$, however, is not required in general. It holds if and only if the cone $\mathcal{P}$ can be embedded into a real vector space.

An *ordered cone* $\mathcal{P}$ carries a reflexive transitive relation $\leq$ such that $a \leq b$ implies $a+c \leq b+c$ and $\alpha a \leq \alpha b$ for all $a, b, c \in \mathcal{P}$ and $\alpha \geq 0$. Equality on $\mathcal{P}$ is obviously such an order. Note that anti-symmetry is not required for the relation $\leq$.

The theory of locally convex cones as developed in [100] uses order theoretical concepts to introduce a quasiuniform topological structure on an ordered cone. In a first approach, the resulting topological neighborhoods themselves will be considered to be elements of the cone. In this vein, a *full locally convex cone* $(\mathcal{P}, \mathcal{V})$ is an ordered cone $\mathcal{P}$ that contains an *abstract neighborhood system* $\mathcal{V}$, that is a subset of positive elements which is directed downward, closed for addition and multiplication by strictly positive scalars. The elements $v$ of $\mathcal{V}$ define *upper*, resp. *lower neighborhoods* for the elements of $\mathcal{P}$ by

$$v(a) = \{ b \in \mathcal{P} \mid b \leq a + v \} \quad \text{resp.} \quad (a)v = \{ b \in \mathcal{P} \mid a \leq b + v \},$$

Their intersection $v^s(a) = v(a) \cap (a)v$ is the corresponding *symmetric neighborhood* of $a$. These neighborhoods create the *upper, lower* and *symmetric topologies* on $\mathcal{P}$, respectively. All elements of $\mathcal{P}$ are supposed to be *bounded below*, that is for every $a \in \mathcal{P}$ and $v \in \mathcal{V}$ we have $0 \leq a + \lambda v$ for some $\lambda \geq 0$.

Finally, a *locally convex cone* $(\mathcal{P},\mathcal{V})$ is a subcone of a full locally convex cone not necessarily containing the abstract neighborhood system $\mathcal{V}$. Every locally convex ordered topological vector space is a locally convex cone in this sense, as it can be canonically embedded into a full locally convex cone (see Example 1.4(c) below, or Example I.2.7 in [100]).

A subset $\mathcal{V}_0$ of the neighborhood system $\mathcal{V}$ is called a *basis* for $\mathcal{V}$ if for every $v \in \mathcal{V}$ there is $v_0 \in \mathcal{V}_0$ and $\alpha > 0$ such that $\alpha v_0 \leq v$.

An element $a$ of a locally convex cone $(\mathcal{P},\mathcal{V})$ is called *bounded (above)* if for every $v \in \mathcal{V}$ there is $\lambda \geq 0$ such that $a \leq \lambda v$. All invertible elements of $\mathcal{P}$ are bounded. Indeed, if $-a \in \mathcal{P}$ for some $a \in \mathcal{P}$, then given $v \in \mathcal{V}$ there is $\lambda \geq 0$ such that $0 \leq (-a) + \lambda v$ since all elements of $\mathcal{P}$ are required to be bounded below. This yields $a \leq \lambda v$.

For later reference we shall list a few basic properties of locally convex cone topologies. We shall use the following standard notations: A subset $A$ of $\mathcal{P}$ is called

*decreasing* if $b \in A$ whenever $b \leq a$ for some $a \in A$,
*increasing* if $b \in A$ whenever $b \geq a$ for some $a \in A$, or
*order convex* if $b \in A$ whenever $a \leq b \leq c$ for some $a, c \in A$.
*balanced* if $b \in A$ whenever $b = \lambda a$ or $b + \lambda a = 0$
for some $a \in A$ and $0 \leq \lambda \leq 1$.

The last of these definitions is of course derived from corresponding one for real vector spaces, that is the requirement that $\lambda a \in A$ whenever $a \in A$ and $-1 \leq \lambda \leq 1$.

**Proposition 1.1.** *Let $(\mathcal{P},\mathcal{V})$ be a locally convex cone. The upper (lower or symmetric) topology of $\mathcal{P}$ satisfies the following:*

*(i) Every element of $\mathcal{P}$ admits a basis of convex and decreasing (increasing or order convex) neighborhoods. The symmetric neighborhoods in the basis for $0 \in \mathcal{P}$ are also balanced.*
*(ii) The mapping $(a,b) \mapsto a + b : \mathcal{P} \times \mathcal{P} \to \mathcal{P}$ is continuous.*
*(iii) The mapping $(\alpha, a) \mapsto \alpha a : [0,+\infty) \times \mathcal{P} \to \mathcal{P}$ is continuous at all points $(\alpha, a) \in [0,+\infty) \times \mathcal{P}$ such that $a \in \mathcal{P}$ is bounded.*

*Proof.* Clearly, for $a \in \mathcal{P}$ and $v \in \mathcal{V}$ the neighborhoods $v(a)$, $(a)v$ or $v^s(a)$ are convex and decreasing, increasing or order convex, respectively. The symmetric neighborhoods of $0 \in \mathcal{P}$ are also balanced. Indeed, let $v \in \mathcal{V}$ and let $a \in v^s(0)$. Then $a \leq v$ and $0 \leq a + v$. Let $0 \leq \lambda \leq 1$. Then $\lambda a \in v^s(0)$ follows from the convexity of $v^s(0)$ since $\lambda a = \lambda a + (1-\lambda)0$. If on the other hand $b + \lambda a = 0$ for $b \in \mathcal{P}$, then

$$b \leq b + \lambda(a+v) = (b + \lambda a) + v = v \qquad \text{and} \qquad 0 = b + \lambda a \leq b + v$$

Hence $b \in v(0)$ holds in this case as well.

For property (ii), let $a, b \in \mathcal{P}$ and $v \in \mathcal{V}$ and set $u = (1/2)v \in \mathcal{V}$. Then for $c \in u(a)$ and $d \in u(b)$, that is $c \leq a + u$ and $d \leq b + u$ we have $c+d \leq (a+b)+v$, hence $c+d \in v(a+b)$. This shows continuity of the addition with respect to the upper topology. Likewise, $c \in (a)u$ and $d \in (b)u$, that is $a \leq c+u$ and $b \leq d+u'$ implies that $a+b \leq (c+d)+v$, hence $c+d \in (a+b)v$. This yields continuity of the addition with respect to the lower topology. Combining the preceding arguments, we realize that $c \in u^s(a)$ and $d \in u^s(b)$ implies $c+d \in v^s(a+b)$, which proves continuity with respect to the symmetric topology.

For Part (iii) let $(\alpha, a) \in [0,+\infty) \times \mathcal{P}$ for a bounded element $a \in \mathcal{P}$ and let $v \in \mathcal{V}$. There is $\lambda > 0$ such that both $0 \leq a + \lambda v$ and $a \leq \lambda v$. Set $\varepsilon = \min\{1,\ 1/(2\lambda),\ 1/(2\alpha+2)\} > 0$ and $\alpha_0 = \max\{\alpha - \varepsilon,\ 0\}$ and $\alpha_1 = \alpha + \varepsilon$. The interval $[\alpha_0, \alpha_1]$ then is a neighborhood for $\alpha$ in $[0,+\infty)$. For every $\alpha_0 \leq \beta \leq \alpha$ we observe that

$$\alpha a = \beta a + (\alpha - \beta)a \leq \beta a + (\alpha - \beta)\lambda v \leq \beta a + \varepsilon \lambda v \leq \beta a + \frac{1}{2}v$$

and

$$\beta a \leq \beta a + (\alpha - \beta)(a + \lambda v) = \alpha a + (\alpha - \beta)\lambda v \leq \alpha a + \frac{1}{2}v.$$

Likewise, for $\alpha \leq \beta \leq \alpha_1$ we have

$$\alpha a \leq \alpha a + (\beta - \alpha)(a + \lambda v) = \beta a + (\beta - \alpha)\lambda v \leq \alpha a + \frac{1}{2}v$$

and

$$\beta a = \alpha a + (\beta - \alpha)a \leq \alpha a + (\beta - \alpha)\lambda v \leq \beta a + \frac{1}{2}v.$$

Thus

$$\alpha a \leq \beta a + \frac{1}{2}v \qquad \text{and} \qquad \beta a \leq \alpha a + \frac{1}{2}v$$

holds for all $\beta \in [\alpha_0, \alpha_1]$. Now let $u = \varepsilon v \in \mathcal{V}$. Then for every $b \in u(a)$ and every $\beta \in [\alpha_0, \alpha_1]$ we have

$$\begin{aligned}\beta b \leq \beta(a+u) &= \beta a + \varepsilon \beta v \\ &\leq \left(\alpha a + \frac{1}{2}v\right) + \varepsilon(\alpha + \varepsilon)v \leq \alpha a + \frac{1}{2}v + \frac{1}{2}v \leq \alpha a + v\end{aligned}$$

by our construction of $\varepsilon > 0$. Thus $\beta b \in v(\alpha a)$. This shows continuity of the mapping $(\alpha, a) \mapsto \alpha a : [0,+\infty) \times \mathcal{P} \to \mathcal{P}$ at $(\alpha, a)$ with respect to the upper topology of $\mathcal{P}$. Likewise, for $b \in (a)u$ and $\beta \in [\alpha_0, \alpha_1]$ we infer using the above

$$\alpha a \leq \beta a + \frac{1}{2}v \leq \beta(b+u) + \frac{1}{2}v \leq \beta b + \varepsilon(\alpha + \varepsilon)v + \frac{1}{2}v \leq \beta b + v.$$

This yields $\beta b \in (\alpha a)v$ and continuity of the mapping $(\alpha, a) \mapsto \alpha a$: $[0,+\infty) \times \mathcal{P} \to \mathcal{P}$ at $(\alpha, a)$ with respect to the lower topology of $\mathcal{P}$. Combining the preceding arguments, we realize that $b \in u^s(a)$ and $\beta \in [\alpha_0, \alpha_1]$

implies $\beta b \in v^s(\alpha a)$, which proves continuity with respect to the symmetric topology. □

On the subcone $\mathcal{P}_0$ of all invertible elements in a locally convex cone $(\mathcal{P}, V)$ the scalar multiplication can be canonically extended to all real numbers if we set $\alpha a = (-\alpha)(-a)$ for $\alpha < 0$ and $a \in \mathcal{P}_0$. Proposition 1.1 then yields

**Corollary 1.2.** *Let $(\mathcal{P}, \mathcal{V})$ be a locally convex cone and let $\mathcal{P}_0$ be the subcone of all invertible elements of $\mathcal{P}$. The mapping $(\alpha, a) \mapsto \alpha a : \mathbb{R} \times \mathcal{P}_0 \to \mathcal{P}_0$ is continuous with respect to the symmetric topology of $\mathcal{P}$.*

*Proof.* First we observe that $a \in v^s(b)$ if and only if $-a \in v^s(-b)$ for $a, b \in \mathcal{P}_0$ and $v \in \mathcal{V}$. Thus $a_i \to a$ for a net $(a_i)_{i \in \mathcal{I}}$ in $\mathcal{P}_0$ implies that $(-a_i) \to (-a)$. Next suppose that $\alpha_i \to \alpha \in \mathbb{R}$ for $0 \leq \alpha_i \in \mathbb{R}$ and $a_i \to a$ for $a_i, a \in \mathcal{P}_0$. Then $\alpha_i a_i \to \alpha a$ by 1.1(iii) since every invertible element is bounded. Now finally, let $\alpha_i \to \alpha$ in $\mathbb{R}$ and $a_i \to a$ for $a_i, a \in \mathcal{P}_0$. Let $\beta_i = \alpha_i \vee 0$ and $\gamma_i = -(\alpha_i \wedge 0)$. Then $\beta_i, \gamma_i \geq 0$ and $\alpha_i = \beta_i - \gamma_i$. We have $\beta_i a_i \to \beta a$ and $\gamma_i(-a_i) \to \gamma(-a)$, where $\beta = \alpha \vee 0$ and $\gamma = -(\alpha \wedge 0)$, by the preceding. Thus

$$\alpha_i a_i = \beta_i a_i + \gamma_i(-a_i) \;\to\; \beta a + \gamma(-a) = \alpha a,$$

again by 1.1(ii), as claimed. □

**1.3 Locally Convex Cones via Convex Quasiuniform Structures.** As a subcone of a full locally convex cone, a locally convex cone $(\mathcal{P}, \mathcal{V})$ inherits both its order, algebraic structure and neighborhood system from the former. While this approach elegantly permits the use of the order structure of the full cone to describe the topologies of $\mathcal{P}$, it is not always very practical, because for concrete examples such a full cone may be difficult to access. Quite frequently, the topology of a locally convex cone is more visible as a *convex quasiuniform structure* as described in I.5 of [100]. This is a straightforward generalization of the uniform structures that define the topologies of locally convex topological vector spaces. In this vein, a neighborhood is a convex subset $v$ of $\mathcal{P}^2$, where $\mathcal{P}$ is an ordered cone, satisfying the following conditions:

(U1) If $a \leq b$ for $a, b \in \mathcal{P}$, then $(a, b) \in v$.
(U2) If $(a, b) \in \lambda v$ and $(b, c) \in \rho v$ for $a, b, c \in \mathcal{P}$ and $\lambda, \rho > 0$, then $(a, c) \in (\lambda + \rho) v$.
(U3) For every $a \in \mathcal{P}$ there is $\lambda \geq 0$ such that $(0, a) \in \lambda v$.

If a family $\mathcal{V}$ of such neighborhoods fulfills the usual conditions for a quasiuniform structure (see [135]), that is

(U4) For $u, v \in \mathcal{V}$ there is $w \in \mathcal{V}$ such that $w \subset u \cap v$,
(U5) If $v \in \mathcal{V}$ and $\lambda > 0$, then $\lambda v \in \mathcal{V}$,

then a straightforward procedure (see I.5 in [100]) allows the embedding of $\mathcal{P}$ and $\mathcal{V}$ into a full locally convex cone $(\widehat{\mathcal{P}}, \widehat{\mathcal{V}})$ whose neighborhood system $\widehat{\mathcal{V}}$ is generated by the elements of $\mathcal{V}$, and such that $(a,b) \in v$ for $a,b \in \mathcal{P}$ and $v \in \mathcal{V}$ means $a \leq b+v$ in $\widehat{\mathcal{P}}$. Convex quasiuniform structures therefore yield an equivalent approach to locally convex cones.

*Examples 1.4.* (a) In the extended real number system $\overline{\mathbb{R}} = \mathbb{R} \cup \{+\infty\}$ we consider the usual order and algebraic operations, in particular $a+\infty = +\infty$ for all $a \in \overline{\mathbb{R}}$, $\alpha\cdot(+\infty) = +\infty$ for all $\alpha > 0$ and $0\cdot(+\infty) = 0$. Endowed with the neighborhood system $\mathcal{V} = \{\varepsilon \in \mathbb{R} \mid \varepsilon > 0\}$, $\overline{\mathbb{R}}$ is a full locally convex cone. For $a \in \mathbb{R}$ the intervals $(-\infty, a+\varepsilon]$ are the upper and the intervals $[a-\varepsilon, +\infty]$ are the lower neighborhoods, while for $a = +\infty$ the entire cone $\overline{\mathbb{R}}$ is the only upper neighborhood, and $\{+\infty\}$ is open in the lower topology. The symmetric topology is the usual topology on $\mathbb{R}$ with $+\infty$ as an isolated point. It is finer than the usual topology of $\overline{\mathbb{R}}$, where the intervals $[a, +\infty]$ are the neighborhoods of $+\infty$.

(b) For the subcone $\overline{\mathbb{R}}_+ = \{a \in \overline{\mathbb{R}} \mid a \geq 0\}$ of $\overline{\mathbb{R}}$ we may also consider the singleton neighborhood system $\mathcal{V} = \{0\}$. The elements of $\overline{\mathbb{R}}_+$ are obviously bounded below even with respect to the neighborhood $v = 0$, hence $\overline{\mathbb{R}}_+$ is a full locally convex cone. For $a \in \overline{\mathbb{R}}$ the intervals $(-\infty, a]$ and $[a, +\infty]$ are the only upper and lower neighborhoods, respectively. The symmetric topology is the discrete topology on $\overline{\mathbb{R}}_+$.

(c) Let $(E, \leq)$ be a locally convex ordered topological vector space. Recall that equality is an order relation, hence this example will cover locally convex spaces in general. In order to interpret $E$ as a locally convex cone we shall embed it into a larger full cone. This is done in a canonical way: Let $\mathcal{P} = \mathit{Conv}(E)$ be the cone of all non-empty convex subsets of $E$, endowed with the usual addition and multiplication of sets by non-negative scalars, that is $\alpha A = \{\alpha a \mid a \in A\}$ and $A+B = \{a+b \mid a \in A \text{ and } b \in B\}$ for $A, B \in \mathcal{P}$ and $\alpha \geq 0$. We define the order on $\mathcal{P}$ by

$$A \leq B \qquad \text{if} \qquad A \subset {\downarrow}B,$$

where ${\downarrow}B = \{x \in E \mid x \leq b \text{ for some } b \in B\}$ is the *decreasing hull* of the set $B$ in $E$. Note that ${\downarrow}B$ is again a convex subset of $E$. The requirements for an ordered cone are easily checked. The neighborhood system in $\mathcal{P}$ is given by a basis $\mathcal{V} \subset \mathcal{P}$ of convex and balanced neighborhoods of the origin in $E$. That is

$$A \leq B+V \qquad \text{if} \qquad A \subset {\downarrow}(B+V)$$

for $A, B \in \mathcal{P}$ and $V \in \mathcal{V}$. We observe that for every $A \in \mathcal{P}$ and $V \in \mathcal{V}$ there is $\rho > 0$ such that $\rho V \cap A \neq \emptyset$. This yields $0 \in A + \rho V$. Therefore $\{0\} \leq A + \rho V$, and every element $A \in \mathcal{P}$ is indeed bounded below. Thus $(\mathcal{P}, \mathcal{V})$ is a full locally convex cone.

Via the embedding $x \mapsto \{x\} : E \to \mathcal{P}$ of its elements onto singleton subsets, the locally convex ordered topological vector space $E$ itself may be considered as a subcone of $\mathcal{P}$. This embedding preserves the order of $E$, and on its image in $\mathcal{P}$, the upper or lower topologies of $\mathcal{P}$ reflect the order structure of $E$ in the following sense: All upper or lower neighborhoods are decreasing or increasing, respectively, that is for elements $a, b \in E$ and a neighborhood $V \in \mathcal{V}$ we have

$$a \leq b + V \qquad \text{if} \qquad a - b \in \downarrow V.$$

For a linear operator $T : E \to E$ in particular, continuity with respect to either the induced upper or lower topology requires that $T$ is monotone (see Section 2 below). The symmetric topology of $\mathcal{P}$, on the other hand induces a locally convex vector space topology on $E$ in the usual sense. It coincides with the given topology of $E$ if the neighborhoods $V \in \mathcal{V}$ are also order convex, that is if $c \in V$ whenever $a \leq c \leq b$ for $a, b \in V$ and $V \in \mathcal{V}$. If the given order on $E$ is indeed the equality, then the upper, lower and symmetric topologies of $\mathcal{P}$ all coincide on $E$ with the given topology since $a \leq b + V$ for $a, b \in E$ and $V \in \mathcal{V}$ means that $a - b \in V$ in this case, and since the neighborhoods in $\mathcal{V}$ were supposed to be balanced. In this way, every locally convex ordered topological vector space, endowed with a basis $\mathcal{V}$ of balanced, convex and order convex neighborhoods, is a locally convex cone, but not a full cone.

Other subcones of $\mathcal{P}$ that merit further investigation are those of all closed, closed and bounded, or compact convex sets in $\mathcal{P}$, respectively. Note that closed and bounded convex sets satisfy the cancellation law. Details on those and further related examples can be found in [100], I.1.7, I.2.7 and I.2.8.

This example can be further generalized if we replace the vector space $E$ by a locally convex cone.

(d) If $(\mathcal{P}, \mathcal{V})$ is a locally convex cone and if $\mathcal{P}$ is indeed a vector space over $\mathbb{R}$, that is the scalar multiplication in $\mathcal{P}$ is extended to all reals, then all elements of $\mathcal{P}$ are obviously bounded, as boundedness from above for the element $a \in \mathcal{P}$ follows from boundedness from below for the element $-a \in \mathcal{P}$. We have $a \in v(b)$ for $a, b \in \mathcal{P}$ and $v \in \mathcal{V}$ in this case if and only if $a - b \in v(0)$. While the multiplication by negative scalars is in general not continuous with respect to the upper and lower topologies on $\mathcal{P}$, the symmetric topology generated by the neighborhoods of the origin

$$v^s(0) = \{a \in \mathcal{P} \mid a \leq v \ \text{ and } \ -a \leq v\}$$

is a locally convex vector space topology in the usual sense (see Corollary 1.2).

If $\mathcal{P}$ is indeed a vector space over $\mathbb{C}$, then we need to consider the *modular symmetric topology* instead (see Section 2 in [168]). It is generated by the neighborhoods of the origin

$$v^{s_m}(0) = \{a \in \mathcal{P} \mid \gamma a \leq v \ \text{ for all } \ \gamma \in \Gamma\},$$

where $\Gamma = \{\gamma \in \mathbb{C} \mid |\gamma| = 1\}$ denotes the unit circle of $\mathbb{C}$. It is easy to verify that these sets are convex, balanced and absorbing. The modular symmetric topology is therefore a locally convex vector space topology in the usual sense and yields continuity for the multiplication by all scalars in $\mathbb{C}$. Thus endowed with the modular neighborhoods $\mathcal{V}_m = \{v^{sm} \mid v \in \mathcal{V}\}$ and the equality as its order, $(\mathcal{P}, \mathcal{V}_m)$ is again a locally convex cone, and we have $a \le b + v^{sm}$ for $a, b \in \mathcal{P}$ and $v \in \mathcal{V}$ if $\gamma(a-b) \le v$ for all $\gamma \in \Gamma$.

In the sequel, we shall say that a locally convex cone $(\mathcal{P}, \mathcal{V})$ is a locally convex topological vector space over $\mathbb{R}$ or $\mathbb{C}$ if $\mathcal{P}$ is a vector space over $\mathbb{R}$ or $\mathbb{C}$, endowed with the equality as order and a system $\mathcal{V}$ of neighborhoods such that $v(0) = v^{sm}(0)$ holds for all $v \in \mathcal{V}$. The subsets $v(0)$ of $\mathcal{P}$ then are convex, balanced and absorbing, and $\mathcal{P}$ carries its modular symmetric topology.

(e) Let $(\mathcal{P}, \mathcal{V})$ be a locally convex cone, $X$ a set and let $\mathcal{F}(X, \mathcal{P})$ be the cone of all $\mathcal{P}$-valued functions on $X$, endowed with the pointwise operations and order. If $\widehat{\mathcal{P}}$ is a full cone containing both $\mathcal{P}$ and $\mathcal{V}$, then we may identify the elements $v \in \mathcal{V}$ with the constant functions $\hat{v}$ on $X$, that is $x \mapsto v$ for all $x \in X$. Hence $\widehat{\mathcal{V}} = \{\hat{v} \mid v \in \mathcal{V}\}$ is a subset and a neighborhood system for $\mathcal{F}(X, \widehat{\mathcal{P}})$. A function $f \in \mathcal{F}(X, \widehat{\mathcal{P}})$ is uniformly bounded below, if for every $\hat{v} \in \widehat{\mathcal{V}}$ there is $\rho \ge 0$ such that $0 \le f + \rho\hat{v}$. These functions form a full locally convex cone $\big(\mathcal{F}_b(X, \widehat{\mathcal{P}}), \widehat{\mathcal{V}}\big)$, carrying the topology of uniform convergence. As a subcone, $\big(\mathcal{F}_b(X, \mathcal{P}), \widehat{\mathcal{V}}\big)$ is a locally convex cone. Alternatively, a more general neighborhood system $\widehat{\mathcal{V}}$ for $\mathcal{F}(X, \mathcal{P})$ may be created using a family of $\overline{\mathcal{V}}$-valued functions on $X$, where $\overline{\mathcal{V}} = \mathcal{V} \cup \{0, \infty\}$ consists of the neighborhood system $\mathcal{V}$ for $\mathcal{P}$ augmented by $0 \in \mathcal{P}$ and a maximal element $\infty$. (We use $a + \infty = v + \infty = \alpha \cdot \infty = \infty$ and $a \le \infty$ for all $a \in \mathcal{P}$, $v \in \mathcal{V}$ and $\alpha > 0$.) The neighborhoods $\hat{v} \in \widehat{\mathcal{V}}$ are defined for functions $f, g \in \mathcal{F}(X, \mathcal{P})$ as

$$f \le g + \hat{v} \quad \text{if} \quad f(x) \le g(x) + \hat{v}(x) \quad \text{for all} \quad x \in X.$$

In this case we consider the subcone $\mathcal{F}_{\widehat{\mathcal{V}}_b}(X, \mathcal{P})$ of all functions in $\mathcal{F}(X, \mathcal{P})$ that are bounded below relative to the functions in $\widehat{\mathcal{V}}$, that is $f \in \mathcal{F}_{\widehat{\mathcal{V}}_b}(X, \mathcal{P})$ if for every $\hat{v} \in \widehat{\mathcal{V}}$ there is $\lambda \ge 0$ such that $0 \le f + \lambda\hat{v}$. In this way $\big(\mathcal{F}_{\widehat{\mathcal{V}}_b}(X, \mathcal{P}), \widehat{\mathcal{V}}\big)$ forms a locally convex cone. Of particular interest is the case when $\widehat{\mathcal{V}}$ is generated by a suitable family $\mathcal{Y}$ of subsets $Y$ of $X$ and the $\overline{\mathcal{V}}$-valued functions $\hat{v}_Y(x) = v$ for $x \in Y$ and $\hat{v}_Y(x) = \infty$, else, corresponding to some $v \in \mathcal{V}$ and $Y \in \mathcal{Y}$. In this case $\big(\mathcal{F}_{\widehat{\mathcal{V}}_b}(X, \mathcal{P}), \widehat{\mathcal{V}}\big)$ carries the topology of uniform convergence on the sets in $\mathcal{Y}$.

If $X$ is a topological space, then suitable subcones for further investigation are those of continuous functions with respect to any of the given (upper, lower or symmetric) topologies on $\mathcal{P}$. We shall explore different notions of continuity for cone-valued functions and discuss an even wider range of suitable locally convex cone topologies in Chapter III.

Occasionally in applications of this type (see for example the proof of Proposition 5.37 and the construction of the standard lattice completion in 5.57 below) the family $\widehat{\mathcal{V}}$ of $\overline{\mathcal{V}}$-valued functions under consideration is not naturally closed for addition, and including all pointwise sums of the functions in $\widehat{\mathcal{V}}$ might not be desirable. This situation can often be remedied if we consider $\widehat{\mathcal{V}}$ as a system of abstract neighborhoods instead, with a suitably modified addition $\oplus$ for which $\widehat{\mathcal{V}}$ is closed and which is compatible with the scalar multiplication. The neighborhoods $\hat{v} \in \mathcal{F}(X, \mathcal{P})$ are defined as above using associated $\overline{\mathcal{V}}$-valued functions which for simplicity we also denote by $\hat{v}$. The latter amounts to a slight abuse of notation, since for this concept to work we need to allow that the association between neighborhoods and $\overline{\mathcal{V}}$-valued functions is not one-to-one. In order to create a convex quasiuniform structure in the sense of 1.3, hence a locally convex cone $\big(\mathcal{F}_{\widehat{\mathcal{V}}_b}(X, \mathcal{P}), \widehat{\mathcal{V}}\big)$, we require that $\hat{u} \oplus \hat{v} \geq \hat{u} + \hat{v}$ holds for all $\hat{u}, \hat{v} \in \widehat{\mathcal{V}}$, where $+$ stands for the pointwise sum of the associated $\overline{\mathcal{V}}$-valued functions. We shall use this approach in 5.37 and 5.57 below.

(f) For $x \in \overline{\mathbb{R}}$ denote $x^+ = \max\{x, 0\}$ and $x^- = -\min\{x, 0\}$. For $1 \leq p \leq +\infty$ and a sequence $(x_i)_{i\in\mathbb{N}}$ in $\overline{\mathbb{R}}$ let $\|(x_i)\|_p$ denote the usual $l^p$ norm, that is $\|(x_i)\|_p = \big(\sum_{i=1}^{\infty} |x_i|^p\big)^{(1/p)} \in \overline{\mathbb{R}}$ for $p < +\infty$ and $\|(x_i)\|_\infty = \sup\{|x_i| \mid i \in \mathbb{N}\} \in \overline{\mathbb{R}}$. Now let $\overline{l^p}$ be the cone of all sequences $(x_i)_{i\in\mathbb{N}}$ in $\overline{\mathbb{R}}$ such that $\|(x_i^-)\|_p < +\infty$. We use the pointwise order in $\overline{l^p}$ and the neighborhood system $\mathcal{V}_p = \{\rho v_p \mid \rho > 0\}$, where

$$(x_i)_{i\in\mathbb{N}} \leq (y_i)_{i\in\mathbb{N}} + \rho v_p$$

means that $\|(x_i - y_i)^+\|_p \leq \rho$. (In this expression the $l^p$ norm is evaluated only over the indexes $i \in \mathbb{N}$ for which $y_i < +\infty$.) It can be easily verified that $(\overline{l^p}, \mathcal{V}_p)$ is a locally convex cone. In fact $(\overline{l^p}, \mathcal{V}_p)$ can be embedded into a full cone following a procedure analogous to that in 1.4(c). The case for $p = +\infty$ is of course already covered by Example 1.4(e).

## 2. Continuous Linear Operators and Functionals

For cones $\mathcal{P}$ and $\mathcal{Q}$ a mapping $T : \mathcal{P} \to \mathcal{Q}$ is called a *linear operator* if

$$T(a + b) = T(a) + T(b) \qquad \text{and} \qquad T(\alpha a) = \alpha T(a)$$

holds for all $a, b \in \mathcal{P}$ and $\alpha \geq 0$. If both $\mathcal{P}$ and $\mathcal{Q}$ are indeed vector spaces over $\mathbb{R}$, then $0 = T(a - a) = T(a) + T(-a)$ implies that such an operator is linear over $\mathbb{R}$. If both $\mathcal{P}$ and $\mathcal{Q}$ are ordered, then $T$ is called *monotone*, if $a \leq b$ implies $T(a) \leq T(b)$. If both $(\mathcal{P}, \mathcal{V})$ and $(\mathcal{Q}, \mathcal{W})$ are locally convex cones, the operator $T$ is called *(uniformly) continuous* if for every $w \in \mathcal{W}$ one can find $v \in \mathcal{V}$ such that $T(a) \leq T(b) + w$ whenever $a \leq b + v$ for $a, b \in \mathcal{P}$. A family $\mathfrak{T}$ of linear operators is called *equicontinuous* if the above condition holds for every $w \in \mathcal{W}$ with the same $v \in \mathcal{V}$ for all $T \in \mathfrak{T}$.

Uniform continuity is not just continuity. It is immediate from the definition that it implies and combines continuity for the operator $T : \mathcal{P} \to \mathcal{Q}$ with respect to the upper, lower and symmetric topologies on $\mathcal{P}$ and $\mathcal{Q}$, respectively.

A *linear functional* on $\mathcal{P}$ is a linear operator $\mu : \mathcal{P} \to \overline{\mathbb{R}}$. The *dual cone* $\mathcal{P}^*$ of a locally convex cone $(\mathcal{P}, \mathcal{V})$ consists of all continuous linear functionals on $\mathcal{P}$ and is the union of all *polars* $v^\circ$ of neighborhoods $v \in \mathcal{V}$, where $\mu \in v^\circ$ means that $\mu(a) \le \mu(b) + 1$, whenever $a \le b + v$ for $a, b \in \mathcal{P}$. Continuity implies that a linear functional $\mu$ is monotone, and for a full cone $\mathcal{P}$ it requires just that $\mu(v) \le 1$ holds for some $v \in \mathcal{V}$ in addition. Continuous linear functionals can take only finite values on bounded elements. Indeed, let $\mu \in v^\circ$ for some $v \in \mathcal{V}$ and let $a \in \mathcal{P}$ be a bounded element. Then $a \le \lambda v$ for some $\lambda \ge 0$, hence $\mu(a) \le \lambda$ as claimed. We endow $\mathcal{P}^*$ with the canonical algebraic operations and the topology $w(\mathcal{P}^*, \mathcal{P})$ of pointwise convergence on the elements of $\mathcal{P}$, considered as functions on $\mathcal{P}^*$ with values in $\overline{\mathbb{R}}$ with its usual topology. As in locally convex topological vector spaces, the polar $v^\circ$ of a neighborhood $v \in \mathcal{V}$ is seen to be $w(\mathcal{P}^*, \mathcal{P})$-compact and convex ( [100], Theorem II.2.4).

*Examples 2.1.* Revisiting the preceding Examples 1.4 we observe the following:

(a) The dual cone $\overline{\mathbb{R}}^*$ of $\overline{\mathbb{R}}$ (see 1.4(a)) consists of all positive reals (via the usual multiplication), and the singular functional $\bar{0}$ such that $\bar{0}(a) = 0$ for all $a \in \mathbb{R}$ and $\bar{0}(+\infty) = +\infty$.

(b) Likewise, in 1.4(b), the continuous linear functionals on $\overline{\mathbb{R}}_+$, endowed with the neighborhood system $\mathcal{V} = \{0\}$, are the positive reals together with $\bar{0}$, but further include the element $+\infty$, acting as $+\infty(0) = 0$ and $+\infty(a) = +\infty$ for all $0 \ne a \in \overline{\mathbb{R}}_+$. This functional is obviously contained in the polar of the neighborhood $0 \in \mathcal{V}$.

(c) If both $(\mathcal{P}, \mathcal{V})$ and $(\mathcal{Q}, \mathcal{W})$ are locally convex cones and ordered vector spaces over $\mathbb{K} = \mathbb{R}$ or $\mathbb{K} = \mathbb{C}$, let us also consider the modular symmetric topologies on $\mathcal{P}$ and $\mathcal{Q}$ which are defined by the modular symmetric neighborhoods $v^{s_m}$ and $w^{s_m}$ corresponding to the given neighborhoods $v \in \mathcal{V}$ and $w \in \mathcal{W}$, respectively. Recall from 1.4(d) that $a \le b + v^{s_m}$ for $a, b \in \mathcal{P}$ and $v \in \mathcal{V}$ means that $\gamma(a - b) \le v$ for all $\gamma \in \mathbb{K}$ such that $|\gamma| = 1$. The modular topologies were seen to be locally convex vector space topologies. If a linear operator $T : \mathcal{P} \to \mathcal{Q}$ is continuous with respect to the given locally convex cone topologies and indeed linear over $\mathbb{K}$, then it is straightforward to verify that $T$ is also continuous with respect to the respective modular symmetric topologies of $\mathcal{P}$ and $\mathcal{Q}$. The converse does not hold true in general. For $\mathcal{Q} = \overline{\mathbb{R}}$, however, that is for linear functionals, we have the following: If $\mathcal{P}^*$ denotes the given dual of $\mathcal{P}$, and if $\mathcal{P}_m^*$ denotes the dual of $\mathcal{P}$ if endowed with the modular symmetric neighborhoods, then $\mathcal{P}^* \subset \mathcal{P}_m^*$ since the latter topology is finer that the given one. According to Theorem 3.3 in [168], for every linear functional $\mu \in \mathcal{P}_m^*$ there are $\mu_i \in \mathcal{P}^*$ for $i = 1, 2$ in the real or $i = 1, 2, 3, 4$ in the complex case such that

$$\mu(a) = \mu_1(a) + \mu_2(-a) \quad \text{or} \quad \mu(a) = \mu_1(a) + \mu_2(-a) + \mu_3(ia) + \mu_4(-ia)$$

for all $a \in \mathcal{P}$, respectively.

(d) Let $(\mathcal{P}, \mathcal{V})$ be a locally convex vector space over $\mathbb{K}$, that is a locally convex cone which is a vector space over $\mathbb{K}$ and carries the modular symmetric topology. The functionals in the dual cone $\mathcal{P}^*$ of $\mathcal{P}$ are real-valued, but there exists a canonical correspondence between the dual cone $\mathcal{P}^*$ and the usual dual space $\mathcal{P}^*_{\mathbb{K}}$ of $\mathcal{P}$ as a locally convex topological vector space. $\mathcal{P}^*_{\mathbb{K}}$ consists of all $\mathbb{K}$-valued continuous $\mathbb{K}$-linear functionals on $\mathcal{P}$. In the real case this correspondence is obvious, as $\mathcal{P}^*$ and $\mathcal{P}^*_{\mathbb{K}}$ coincide. (If both $a, -a \in \mathcal{P}$, then $\mu(a) + \mu(-a) = 0$ for every $\mu \in \mathcal{P}^*$, hence $\mu$ is linear over $\mathbb{R}$.) In the complex case there is an established correspondence between $\mathcal{P}^*$ and $\mathcal{P}^*_{\mathbb{K}}$ : The real part $\mu$ of every continuous complex linear functional $\mu_{\mathbb{K}}$ on $\mathcal{P}$ is in $\mathcal{P}^*$ and, conversely, for every $\mu \in \mathcal{P}^*$, the mapping $a \mapsto \mu(a) - i\,\mu(ia)$ defines a continuous complex linear functional $\mu_{\mathbb{K}} \in \mathcal{P}^*_{\mathbb{K}}$. $\mathcal{P}^*_{\mathbb{K}}$ is again a vector space over $\mathbb{K}$, and for $\mu_{\mathbb{K}}$ and $\alpha \in \mathbb{K}$ the respective projections $\mu$ and $(\alpha\mu)$ into $\mathcal{P}^*$ of the functionals $\mu_{\mathbb{K}}$ and $\alpha\mu_{\mathbb{K}}$ relate as

$$(\alpha\mu)(a) = \Re\big((\alpha\mu_{\mathbb{K}})(a)\big) = \Re\big(\mu_{\mathbb{K}}(\alpha a)\big) = \mu(\alpha a).$$

The above formula effectively extends the multiplication by non-negative reals in $\mathcal{P}^*$ to all scalars in $\mathbb{K}$ in such a way that the mapping $\mu_{\mathbb{K}} \to \mu : \mathcal{P}^*_{\mathbb{K}} \to \mathcal{P}^*$ becomes a vector space isomorphism. Similarly, every element $\varphi_{\mathbb{K}}$ of the (algebraic) second vector space dual $\mathcal{P}^{**}_{\mathbb{K}}$ of $\mathcal{P}_{\mathbb{K}}$ corresponds to a real-linear functional $\varphi$ on the dual cone $\mathcal{P}^*$ by

$$\varphi(\mu) = \Re\big(\varphi_{\mathbb{K}}(\mu_{\mathbb{K}})\big)$$

for $\mu \in \mathcal{P}^*$. On the other hand, every functional $\varphi$ on $\mathcal{P}^*$ that is linear with respect to the non-negative reals corresponds to a $\mathbb{K}$-valued linear functional $\varphi_{\mathbb{K}} \in \mathcal{P}^{**}_{\mathbb{K}}$ on $\mathcal{P}^*_{\mathbb{K}}$ by

$$\varphi_{\mathbb{K}}(\mu_{\mathbb{K}}) = \varphi(\mu) \qquad \text{or} \qquad \varphi_{\mathbb{K}}(\mu_{\mathbb{K}}) = \varphi(\mu) - i\varphi(i\mu)$$

for $\mu_{\mathbb{K}} \in \mathcal{P}^*_{\mathbb{K}}$ in the real or complex case, respectively. Here we use the above defined extension of the scalar multiplication in $\mathcal{P}^*$. $\mathbb{K}$-linearity for $\varphi_{\mathbb{K}}$ is easily checked. Indeed, additivity is obvious for $\varphi_{\mathbb{K}}$. For compatibility with the scalar multiplication, the real case follows from $\varphi(\mu) + \varphi\big((-1)\mu\big) = \varphi(0) = 0$, hence $\varphi\big((-1)\mu\big) = -\varphi(\mu)$. In the complex case we calculate for $\mu_{\mathbb{K}} \in \mathcal{P}^*_{\mathbb{K}}$ and $\alpha = x + iy \in \mathbb{C}$

$$\begin{aligned} \varphi_{\mathbb{K}}\big((x+iy)\mu_{\mathbb{K}}\big) &= \varphi\big((x+iy)\mu\big) - i\varphi\big((x+iy)i\mu\big) \\ &= (x+iy)\varphi(\mu) + (y-ix)\varphi(i\mu) \\ &= (x+iy)\big(\varphi(\mu) - i\varphi(i\mu)\big) \\ &= (x+iy)\varphi_{\mathbb{K}}(\mu_{\mathbb{K}}). \end{aligned}$$

Thus there is also a canonical correspondence between $\mathcal{P}^{**}$, the cone of all real-valued linear functionals on $\mathcal{P}^*$ (see 7.3(i) below) and the second vector space dual $\mathcal{P}_{\mathbb{K}}^{**}$ of $\mathcal{P}_{\mathbb{K}}$.

(e) In 1.4(c) and (e) on the other hand, due to the generality of the settings, a complete description for the respective dual cones is not immediately available. We may, however, identify some of their elements: In 1.4(c), let $\mu$ be a continuous monotone linear function on the locally convex ordered topological vector space $(E, \leq)$. Then the mapping

$$A \mapsto \sup\{\mu(a) \mid a \in A\} : Conv(E) \to \overline{\mathbb{R}}$$

is seen to be an element of $Conv(E)^*$.

(f) In 1.4(e), if $\mu \in v^\circ \subset \mathcal{P}^*$ for some $v \in \mathcal{V}$, and if $\hat{v}(x) \leq v$ for some $v \in \widehat{\mathcal{V}}$ and $x \in X$, then the mapping $\mu_x : \mathcal{F}_{\widehat{\mathcal{V}}_b}(X, \mathcal{P}) \to \overline{\mathbb{R}}$ such that

$$\mu_x(f) = \mu\big(f(x)\big) \qquad \text{for all} \quad f \in \mathcal{F}_{\widehat{\mathcal{V}}_b}(X, \mathcal{P})$$

is a continuous linear functional on $\mathcal{F}_{\widehat{\mathcal{V}}_b}(X, \mathcal{P})$; more precisely $\mu_x \in \hat{v}^\circ$.

(g) In 1.4(g), for $p < +\infty$ the dual cone of $\overline{l^p}$ consists of all sequences $(y_i)_{i \in \mathbb{N}}$ such that $y_i \geq 0$ for all $i \in \mathbb{N}$ and $\|(y_i)\|_q < +\infty$, where $q$ is the conjugate index of $p$.

**2.2 Embeddings.** We have intuitively used the term embedding before. Let us now establish a precise definition: Let $(\mathcal{P}, \mathcal{V})$ and $(\mathcal{Q}, \mathcal{W})$ be locally convex cones. A linear operator $\Phi : \mathcal{P} \to \mathcal{Q}$ is called an *embedding* of $(\mathcal{P}, \mathcal{V})$ into $(\mathcal{Q}, \mathcal{W})$ if it can be extended to a mapping $\Phi : (\mathcal{P} \cup \mathcal{V}) \to (\mathcal{Q} \cup \mathcal{W})$ such that $\Phi(\mathcal{V}) = \mathcal{W}$ and

$$a \leq b + v \qquad \text{holds if and only if} \qquad \Phi(a) \leq \Phi(b) + \Phi(w)$$

for all $a, b \in \mathcal{P}$ and $v \in \mathcal{V}$.

This condition implies that $\Phi$ is continuous, and in case that $\Phi$ is one-to-one, that the inverse operator $\Phi^{-1} : \Phi(\mathcal{P}) \to \mathcal{P}$ is also continuous. It is easily verified that the composition of two embeddings is again an embedding in this sense. Embeddings are meant to preserve not just the topological structure, but also the particular neighborhood system of a locally convex cone.

**Lemma 2.3.** *Let $(\mathcal{P}, \mathcal{V})$ and $(\mathcal{Q}, \mathcal{W})$ be locally convex cones and let $\Phi : \mathcal{P} \to \mathcal{Q}$ be an embedding of $(\mathcal{P}, \mathcal{V})$ into $(\mathcal{Q}, \mathcal{W})$. If the symmetric topology of $\mathcal{P}$ is Hausdorff, then $\Phi$ is one-to-one.*

*Proof.* Under the assumptions of the Lemma, suppose that $\Phi(a) = \Phi(b)$ holds for $a, b \in \mathcal{P}$. Then $a \leq b + v$ and $b \leq a + v$, hence $a \in v^s(b)$ for all $v \in \mathcal{V}$ follows from 2.2. This yields $a = b$ since the symmetric topology of $\mathcal{P}$ is supposed to be Hausdorff. □

An embedding $\Phi$ of $(\mathcal{P},\mathcal{V})$ into $(\mathcal{Q},\mathcal{W})$ is called an *isomorphism* if the mapping $\Phi:(\mathcal{P}\cup\mathcal{V})\to(\mathcal{Q}\cup\mathcal{W})$ is invertible. Then $\Phi^{-1}$ is an embedding of $(\mathcal{Q},\mathcal{W})$ into $(\mathcal{P},\mathcal{W})$.

Hahn-Banach type extension and separation theorems for linear functionals are most important for the development of a powerful duality theory for locally convex cones. We shall mention a few results from [100] and [172]. A *sublinear functional* on a cone $\mathcal{P}$ is a mapping $p:\mathcal{P}\to\overline{\mathbb{R}}$ such that

$$p(\alpha a)=\alpha p(a)\qquad\text{and}\qquad p(a+b)\leq p(a)+p(b)$$

holds for all $a,b\in\mathcal{P}$ and $\alpha\geq 0$. Likewise, an *extended superlinear functional* on $\mathcal{P}$ is a mapping $q:\mathcal{P}\to\underline{\overline{\mathbb{R}}}=\mathbb{R}\cup\{+\infty,-\infty\}$ such that

$$q(\alpha a)=\alpha q(a)\qquad\text{and}\qquad q(a+b)\geq q(a)+q(b)$$

holds for all $a,b\in\mathcal{P}$ and $\alpha\geq 0$. (We set $\alpha+(-\infty)=-\infty$ for all $\alpha\in\overline{\mathbb{R}}$, $\alpha\cdot(-\infty)=-\infty$ for all $\alpha>0$ and $0\cdot(-\infty)=0$ in this context.) We cite Theorem 3.1 from [172]:

**Sandwich Theorem 2.4.** *Let $(\mathcal{P},\mathcal{V})$ be a locally convex cone, and let $v\in\mathcal{V}$. For a sublinear functional $p:\mathcal{P}\to\overline{\mathbb{R}}$ and an extended superlinear functional $q:\mathcal{P}\to\underline{\overline{\mathbb{R}}}$ there exists a linear functional $\mu\in v^{\circ}$ such that $q\leq\mu\leq p$ if and only if $q(a)\leq p(b)+1$ holds whenever $a\leq b+v$ for $a,b\in\mathcal{P}$.*

This theorem is the basic tool for the development of a duality theory for locally convex cones. It leads to a variety of Hahn-Banach type extension and separation results, the most general ones being Theorems 4.1 and 4.4 in [172]. For future use we shall quote both of these:

**Extension Theorem 2.5.** *Let $(\mathcal{P},\mathcal{V})$ be a locally convex cone, $C$ and $D$ non-empty convex subsets of $\mathcal{P}$, and let $v\in\mathcal{V}$. Let $p:\mathcal{P}\to\overline{\mathbb{R}}$ be a sublinear and $q:\mathcal{P}\to\underline{\overline{\mathbb{R}}}$ an extended superlinear functional. For a convex function $f:C\to\overline{\mathbb{R}}$ and a concave function $g:D\to\underline{\overline{\mathbb{R}}}$ there exists a monotone linear functional $\mu\in v^{\circ}$ such that*

$$q\leq\mu\leq p,\qquad g\leq\mu\quad\text{on } D\qquad\text{and}\qquad \mu\leq f\quad\text{on } C$$

*if and only if*

$$q(a)+\rho g(d)\leq p(b)+\sigma f(c)+1\qquad\text{whenever}\qquad a+\rho d\leq b+\sigma c+v$$

*for $a,b\in\mathcal{P}$, $c\in C$, $d\in D$ and $\rho,\sigma\geq 0$.*

In the context of this theorem (Theorem 4.1 in [172]), an $\overline{\mathbb{R}}$-valued function $f$ defined on a convex subset $\mathcal{C}$ of an ordered cone $\mathcal{P}$ is called *convex* if

$$f\big(\lambda c_1+(1-\lambda)c_2\big)\ \leq\ \lambda f(c_1)+(1-\lambda)f(c_2)$$

holds for all $c_1, c_2 \in \mathcal{C}$ and $\lambda \in [0,1]$. Likewise, $f : \mathcal{C} \to \overline{\mathbb{R}}$ is called *concave* if

$$f(\lambda c_1 + (1-\lambda)c_2) \;\geq\; \lambda f(c_1) + (1-\lambda) f(c_2)$$

holds for all $c_1, c_2 \in \mathcal{C}$ and $\lambda \in [0,1]$. An *affine* function $f : \mathcal{C} \to \overline{\mathbb{R}}$ is both convex and concave.

The generality of Theorem 2.5 leads to a wide variety of applications and special cases. An extension theorem in the true meaning of the words can be obtained by identifying the convex sets $C$ and $D$ and the functions $f$ and $g$. For the following (still very general) corollary we shall also leave out (by setting them equal to $+\infty$ and $-\infty$ outside $0 \in \mathcal{P}$, respectively) the functionals $p$ and $q$.

**Corollary 2.6.** *Let $(\mathcal{P}, \mathcal{V})$ be a locally convex cone, $C$ a non-empty convex subsets of $\mathcal{P}$, and let $v \in \mathcal{V}$. For an affine function $f : C \to \overline{\mathbb{R}}$ there exists a monotone linear functional $\mu \in v^\circ$ such that $\mu = f$ on $C$ if and only if*

$$\rho f(d) \leq \sigma f(c) + 1 \qquad \textit{whenever} \qquad \rho d \leq \sigma c + v$$

*for $c, d \in C$, and $\rho, \sigma \geq 0$.*

If $C$ is indeed a subcone of $\mathcal{P}$, that is $(C, \mathcal{V})$ is a locally convex subcone of $(\mathcal{P}, \mathcal{V})$, then the condition of Corollary 2.6 reduces to: $f(0) = 0$ and $f(d) \leq f(c) + 1$ holds whenever $d \leq c + v$ for $c, d \in C$. But this means that the affine function $f$ is indeed a linear functional on $\mathcal{C}$ and contained in the polar of the neighborhood $v \in V$. This observation leads to the following most frequently used consequence of Theorem 2.5 (see also Theorem II.2.9 in [100]).

**Corollary 2.7.** *Let $(\mathcal{N}, \mathcal{V})$ be a subcone of the locally convex cone $(\mathcal{P}, \mathcal{V})$. Every continuous linear functional on $\mathcal{N}$ can be extended to a continuous linear functional on $\mathcal{P}$; more precisely: For every $\mu \in v^\circ_{\mathcal{N}}$ there is $\hat{\mu} \in v^\circ_{\mathcal{P}}$ such that $\hat{\mu}$ coincides with $\mu$ on $\mathcal{N}$.*

Theorem 4.4 in [172] deals with the separation of convex sets by continuous linear functionals, a result that can also be obtained by special insertions in Theorem 2.5.

**Separation Theorem 2.8.** *Let $(\mathcal{P}, \mathcal{V})$ be a locally convex cone, $C$ and $D$ non-empty convex subsets of $\mathcal{P}$, and let $v \in \mathcal{V}$. For $\alpha \in \mathbb{R}$ there exists a monotone linear functional $\mu \in v^\circ$ such that*

$$\mu(c) \leq \alpha \leq \mu(d) \qquad \textit{for all } c \in C \textit{ and } d \in D$$

*if and only if*

$$\alpha\rho \leq \alpha\sigma + 1 \qquad \textit{whenever} \qquad \rho d \leq \sigma c + v$$

*for $c \in C$, $d \in D$ and $\rho, \sigma \geq 0$.*

In a special case, this leads to a separation result for points and closed convex sets (see Corollary 4.6 in [172]):

**Corollary 2.9.** *Let* $A$ *be a non-empty convex subset of a locally convex cone* $(\mathcal{P},\mathcal{V})$ *such that* $0 \in A$.

*(i) If* $A$ *is closed with respect to the lower topology on* $\mathcal{P}$, *then for every element* $b \notin A$ *in* $\mathcal{P}$ *there exists a monotone linear functional* $\mu \in \mathcal{P}^*$ *such that*

$$\mu(a) \le 1 \le \mu(b) \qquad \text{for all } a \in A$$

*and indeed* $1 < \mu(b)$ *if* $b$ *is bounded above.*

*(ii) If* $A$ *is closed with respect to the upper topology on* $\mathcal{P}$, *then for every element* $b \notin A$ *in* $\mathcal{P}$ *there exists a monotone linear functional* $\mu \in \mathcal{P}^*$ *such that*

$$\mu(b) < -1 \le \mu(a) \qquad \text{for all } a \in A.$$

In view of the corresponding separation results for locally convex topological vector spaces, Corollary 2.9 is not entirely satisfying, in particular since it requires that $0 \in A$. A stronger and more suitable separation statement will be derived in Section 4 (Theorem 4.30). It will make use of the relative topologies of a locally convex cone which are to be introduced below.

We shall quote and make use of another result from [172]. The Range Theorem (Theorem 5.1 in [172]) describes the scope of all linear functionals whose existence is guaranteed by the Sandwich Theorem. It is a powerful and indeed non-trivial consequence even in the special case of vector spaces, where its formulation can however be considerably simplified.

**Range Theorem 2.10.** *Let* $(\mathcal{P},\mathcal{V})$ *be a locally convex cone. Let* $p$ *and* $q$ *be sublinear and extended superlinear functionals on* $\mathcal{P}$ *and suppose that there is at least one linear functional* $\mu \in \mathcal{P}^*$ *satisfying* $q \le \mu \le p$. *Then for all* $a \in \mathcal{P}$

$$\sup_{\substack{\mu\in\mathcal{P}^*\\ q\le\mu\le p}} \mu(a) = \sup_{v\in\mathcal{V}} \inf \{p(b) - q(c) \mid b,c \in \mathcal{P},\ q(c) \in \mathbb{R},\ a + c \le b + v\},$$

*and for all* $a \in \mathcal{P}$ *such that* $\mu(a)$ *is finite for at least one* $\mu \in \mathcal{P}^*$ *satisfying* $q \le \mu \le p$

$$\sup_{\substack{\mu\in\mathcal{P}^*\\ q\le\mu\le p}} \mu(a) = \inf_{v\in\mathcal{V}} \sup \{q(c) - p(b) \mid b,c \in \mathcal{P},\ p(b) \in \mathbb{R},\ c \le a + b + v\}.$$

## 3. Weak Local and Global Preorders

In addition to the given order $\le$ on a locally convex cone, we shall frequently use the *weak (global) preorder* $\preccurlyeq$ (for details, see [175] and Section 4 below) which is slightly weaker then the given order and defined for $a, b \in \mathcal{P}$ by

$$a \preccurlyeq b \qquad \text{if} \qquad a \leq \gamma b + \varepsilon v$$

for all $v \in \mathcal{V}$ and $\varepsilon > 0$ with some $1 \leq \gamma \leq 1 + \varepsilon$. This order represents a closure of the given order with respect to the linear and topological structures of $\mathcal{P}$. It is obviously coarser than the given order, that is $a \leq b$ implies $a \preccurlyeq b$ for $a, b \in \mathcal{P}$. In the preceding Examples 1.4(a) and (b), however, both orders coincide. In 1.4(e) this depends on the order in $\mathcal{P}$ and the neighborhood-valued functions in $\hat{\mathcal{V}}$. If $\mathcal{P} = \overline{\mathbb{R}}$ and if for every $x \in X$ there is $\hat{v} \in \widehat{\mathcal{V}}$ such that $\hat{v}(x) < +\infty$, then the given and the weak preorder coincide. In 1.4(c), on the other hand, we have $A \preccurlyeq B$ if $A \subset \overline{\downarrow B}$, where $\overline{\downarrow B}$ denotes the topological closure in $E$ of the decreasing hull $\downarrow B$ of $B$. Note that $\overline{\downarrow B}$ is again a convex subset of $E$. In 1.4(d), that is the case of a vector space $\mathcal{P}$ over $\mathbb{R}$ or $\mathbb{C}$, the weak preorder is given by $a \preccurlyeq b$ if $a - b \in v(0)$ for all $v \in \mathcal{V}$. In this way $(\mathcal{P}, \preccurlyeq)$ becomes a locally convex ordered topological vector space in the usual sense if endowed with the (modular) symmetric topology resulting from the neighborhood system.

The weak preorder on $\mathcal{P}$ is again compatible with the algebraic operations, as Lemma 4.1 below will imply. In Corollary 4.31 below (see also Theorem 3.1 in [175]) we shall establish that the weak preorder on a locally convex cone $\mathcal{P}$ is entirely determined by its dual cone $\mathcal{P}^*$, that is $a \preccurlyeq b$ holds for $a, b \in \mathcal{P}$ if and only if $\mu(a) \leq \mu(b)$ for all $\mu \in \mathcal{P}^*$. The weak preorder may also be used in a full cone containing $\mathcal{P}$ and $\mathcal{V}$. Consequently, the respective relation involving the neighborhoods in $\mathcal{V}$ is defined for elements $a, b \in \mathcal{P}$ and $v \in \mathcal{V}$ as

$$a \preccurlyeq b + v \qquad \text{if} \qquad a \leq \gamma(b + v) + \varepsilon u$$

for all $u \in \mathcal{V}$ and $\varepsilon > 0$ with some $1 \leq \gamma \leq 1 + \varepsilon$. This condition can be slightly simplified:

**Lemma 3.1.** *Let $a, b \in \mathcal{P}$ and $v \in \mathcal{V}$. We have $a \preccurlyeq b + v$ if and only if for every $\varepsilon > 0$ there is $1 \leq \gamma \leq 1 + \varepsilon$ such that $a \leq \gamma b + (1 + \varepsilon)v$.*

*Proof.* Let $a, b \in \mathcal{P}$ and $v \in \mathcal{V}$. Suppose that $a \preccurlyeq b + v$ and let $\varepsilon > 0$. According to the preceding definition of the weak preorder involving neighborhoods, for $u = v$ and $\varepsilon/2$ in place of $\varepsilon$, there is $1 \leq \gamma \leq 1 + \varepsilon/2$ such that $a \leq \gamma(b+v) + (\varepsilon/2)v \leq \gamma b + \varepsilon v$. For the reverse implication suppose that the condition of the Lemma holds, and let $u \in \mathcal{V}$ and $\varepsilon > 0$. There is $\lambda \geq 0$ such that $0 \leq b + \lambda u$. Choose $0 < \delta \leq \varepsilon/\lambda$. Then there is $1 \leq \gamma \leq 1 + \delta$ such that

$$\begin{aligned} a &\leq \gamma b + (1 + \delta)v \\ &\leq \gamma b + (1 + \delta)v + (1 + \delta - \gamma)(b + \lambda u) \\ &\leq (1 + \delta)(b + v) + \delta \lambda u \\ &\leq (1 + \delta)(b + v) + \varepsilon u. \end{aligned}$$

This shows $a \preccurlyeq b + v$. □

Endowed with the weak preorder $(\mathcal{P},\mathcal{V})$ forms again a locally convex cone. For details we refer to [175]. In Corollary 4.34 below (see also Theorem 3.2 in [175]) we shall demonstrate that for $a,b\in\mathcal{P}$ and a neighborhood $v\in\mathcal{V}$, we have $a\preccurlyeq b+v$ if and only if $\mu(a)\leq\mu(b)+1$ holds for all $\mu\in v^{\circ}$. The neighborhoods with respect to the weak preorder in $\mathcal{P}$ are therefore entirely determined by their polars.

Given a neighborhood $v\in\mathcal{V}$ the *weak local preorder* (see [175]) $\preccurlyeq_v$ on $\mathcal{P}$ is the weak (global) preorder with respect to the neighborhood subsystem $\mathcal{V}_v=\{\alpha v\mid\alpha>0\}$. That is, for $a,b\in\mathcal{P}$ we have

$$a\preccurlyeq_v b\qquad\text{if}\qquad a\leq\gamma b+\varepsilon v$$

for all $\varepsilon>0$ with some $1\leq\gamma\leq1+\varepsilon$. Corollary 4.31 below (see also Theorem 3.1 in [175]) states that $a\preccurlyeq_v b$ if and only if $\mu(a)\leq\mu(b)$ holds for all $\mu\in v^{\circ}$.

**Lemma 3.2.** *Let* $a,b\in\mathcal{P}$.

*(a)* $a\preccurlyeq b$ *if and only if for every* $v\in\mathcal{V}$ *and* $\varepsilon>0$ *there is* $1\leq\gamma\leq1+\varepsilon$ *such that* $a\preccurlyeq\gamma b+\varepsilon v$.

*(b)* $a\preccurlyeq_v b$ *for* $v\in\mathcal{V}$ *if and only if for every* $\varepsilon>0$ *there is* $1\leq\gamma\leq1+\varepsilon$ *such that* $a\preccurlyeq\gamma b+\varepsilon v$.

*Proof.* Part (a) follows from Part (b) as $a\preccurlyeq b$ holds if and only if $a\preccurlyeq_v b$ for all $v\in\mathcal{V}$. For Part (b) let $a,b\in\mathcal{P}$ and $v\in\mathcal{V}$ such that the second condition in (b) holds. Given $\varepsilon>0$ set $\delta=\min\{\varepsilon/3,1\}$. Then $a\preccurlyeq\gamma b+\delta v$ holds with some $1\leq\gamma\leq1+\delta$. We infer from Lemma 3.1 that there is $1\leq\gamma'\leq1+\delta$ such that $a\leq(\gamma'\gamma)b+(1+\delta)\delta v$. Since $(1+\delta)\delta\leq\varepsilon$ and $1\leq\gamma'\gamma\leq(1+\delta)^2\leq1+\varepsilon$, and since $\varepsilon>0$ was arbitrarily chosen, we conclude that $a\preccurlyeq_v b$. The reverse implication is trivial since $a\leq\gamma b+\varepsilon v$ implies that $a\preccurlyeq\gamma b+\varepsilon v$. □

Lemma 3.2 shows in particular that the second iteration of the weak preorder, that is the second weak preorder generated by the first one does indeed coincide with the first one.

We observe that for a linear operator $T$ between locally convex cones $(\mathcal{P},\mathcal{V})$ and $(\mathcal{Q},\mathcal{W})$, continuity with respect to the given orders implies continuity and monotonicity with respect to the respective weak preorders on $\mathcal{P}$ and $\mathcal{Q}$. Indeed, suppose that for $v\in\mathcal{V}$ and $w\in\mathcal{W}$ we have $T(a)\leq T(b)+w$ whenever $a\leq b+v$ for $a,b\in\mathcal{P}$. Let $a\preccurlyeq b+v$ and let $\varepsilon>0$. According to Lemma 3.1 there is $1\leq\gamma\leq1+\varepsilon$ such that $a\leq\gamma b+(1+\varepsilon)v$. Thus $T(a)\leq\gamma T(b)+(1+\varepsilon)w$. Since $\varepsilon>0$ was arbitrarily chosen, we conclude that $T(a)\preccurlyeq T(b)+w$, thus establishing our claim.

The weak preorder may also be used to establish a representation for a locally convex cone $(\mathcal{P},\mathcal{V})$ as a cone of continuous $\overline{\mathbb{R}}$-valued functions on some topological space and as a cone of convex subsets of some locally convex topological vector space, respectively. We shall cite Theorem 4.1 from [175]. Recall the definition of an embedding from 2.2.

**Theorem 3.3.** *Every locally convex cone $(\mathcal{P},\mathcal{V})$ can be embedded with respect to its weak preorder into*

*(i) a locally convex cone of continuous $\overline{\mathbb{R}}$-valued functions on some topological space $X$, endowed with the pointwise order and operations and the topology of uniform convergence on a family of compact subsets of $X$.*

*(ii) a locally convex cone of convex subsets of a locally convex topological vector space, endowed with the usual addition and multiplication by scalars, the set inclusion as order and the neighborhoods inherited from the vector space.*

## 4. Boundedness and the Relative Topologies

While all elements of a locally convex cone are bounded below by definition, they need not to be bounded above. Given a neighborhood $v \in \mathcal{V}$, an element $a$ of a locally convex cone $(\mathcal{P},\mathcal{V})$ is called *v-bounded (above)* (see [100], I.2.3) if there is $\lambda \geq 0$ such that $a \leq \lambda v$. The subset $\mathcal{B}_v \subset \mathcal{P}$ of all $v$-bounded elements is a subcone and even a face of $\mathcal{P}$. Correspondingly, by $\mathcal{B} = \bigcap_{v\in\mathcal{V}} \mathcal{B}_v$ we denote the subcone (and face) of all bounded elements of $\mathcal{P}$ (see Section 1 and Proposition 4.11 below). All invertible elements of $\mathcal{P}$ were seen to be bounded, and continuous linear functionals take only finite values on bounded elements (see Section 2).

The presence of unbounded elements constitutes a significant difference between locally convex cones and locally convex topological vector spaces. It tends to make matters more interesting, but also considerably more complicated. If, for example, the element $a \in \mathcal{P}$ is not bounded, then the mapping $\alpha \mapsto \alpha a : [0,+\infty) \to \mathcal{P}$, is not necessarily continuous if we consider the usual topology of $[0,+\infty)$ and any of the given (upper, lower or symmetric) topologies on $\mathcal{P}$ (see Proposition 1.1(iii)). Hence these topologies appear to be rather restrictive. For similar reasons, our upcoming definition of measurability for $\mathcal{P}$-valued functions in Chapter II would turn out to be very limiting if applied to the given topologies of a locally convex cone. We shall therefore introduce slightly coarser neighborhoods on $\mathcal{P}$ which take unbounded elements suitably into account. Given a neighborhood $v \in \mathcal{V}$ and $\varepsilon > 0$, we define the corresponding *upper* and *lower relative neighborhoods* $v_\varepsilon(a)$ and $(a)v_\varepsilon$ for an element $a \in \mathcal{P}$ by

$$\begin{aligned} v_\varepsilon(a) &= \{\, b \in \mathcal{P} \mid b \leq \gamma a + \varepsilon v \quad \text{for some} \quad 1 \leq \gamma \leq 1+\varepsilon \,\} \\ (a)v_\varepsilon &= \{\, b \in \mathcal{P} \mid a \leq \gamma b + \varepsilon v \quad \text{for some} \quad 1 \leq \gamma \leq 1+\varepsilon \,\}. \end{aligned}$$

Their intersection $v_\varepsilon^s(a) = v_\varepsilon(a) \cap (a)v_\varepsilon$ is the corresponding *symmetric relative neighborhood*. These are of course convex subsets of $\mathcal{P}$. Note that for a positive element $a \in \mathcal{P}$ the above expressions somewhat simplify. Since $\gamma a \leq (1+\varepsilon)a$ in this case, we have $v_\varepsilon(a) = \{\, b \in \mathcal{P} \mid b \leq (1+\varepsilon)a + \varepsilon v\}$ and

$(a)v_\varepsilon = \{ b \in \mathcal{P} \mid a \le (1+\varepsilon)b + \varepsilon v\}$. We shall frequently use the following observations:

**Lemma 4.1.** *Let* $a, b, c, a_i, b_i \in \mathcal{P}$, $v \in \mathcal{V}$, $\lambda \ge 0$ *and* $\varepsilon, \delta > 0$.

*(a) If* $a \in v_\varepsilon(b)$ *and* $b \in v_\delta(c)$, *then* $a \in v_{(\varepsilon+\delta+\varepsilon\delta)}(c)$.
*(b) If* $a \in v_\varepsilon(b)$ *and* $0 \le b + \lambda v$, *then* $a \le (1+\varepsilon)b + \varepsilon(1+\lambda)v$.
*(c) If* $a \in v_\varepsilon(b)$ *and* $0 \le a + \lambda v$, *then* $a \le (1+\varepsilon)b + \varepsilon(1+\lambda+\varepsilon)v$ *and* $0 \le b + (\lambda+\varepsilon)v$.
*(d) If* $a_i \in v_\varepsilon(b_i)$ *and if* $0 \le b_i + \lambda v$ *for* $i = 1, \ldots, n$, *then* $(a_1 + \ldots + a_n) \in v_{\varepsilon n(1+\lambda)}(b_1 + \ldots + b_n)$.

*Proof.* For (a), let $a \in v_\varepsilon(b)$ and $b \in v_\delta(c)$, that is $a \le \gamma b + \varepsilon v$ and $b \le \lambda c + \delta v$ for some $1 \le \gamma \le 1+\varepsilon$ and $1 \le \lambda \le 1+\delta$. Then $a \le \gamma\lambda c + (\gamma\delta+\varepsilon)v$. As

$$\gamma\delta + \varepsilon \le (1+\varepsilon)\delta + \varepsilon = \varepsilon + \delta + \varepsilon\delta$$

and

$$1 \le \gamma\lambda \le (1+\varepsilon)(1+\delta) = 1 + \varepsilon + \delta + \varepsilon\delta,$$

we have $a \in v_{(\varepsilon+\delta+\varepsilon\delta)}(c)$. For (b), let $a \in v_\varepsilon(b)$, that is $a \le \gamma b + \varepsilon v$ for some $1 \le \gamma \le 1+\varepsilon$. If $0 \le b + \lambda v$, then

$$a \le \gamma b + \varepsilon v + (1+\varepsilon-\gamma)(b+\lambda v) \le (1+\varepsilon)b + (\varepsilon + \varepsilon\lambda)v.$$

For (c), let $a \in v_\varepsilon(b)$ and $\lambda \ge 0$ such that $0 \le a + \lambda v$. Then $a \le \gamma b + \varepsilon v$ with some $1 \le \gamma \le 1+\varepsilon$, hence $0 \le \gamma b + (\varepsilon+\lambda)v$, and indeed $0 \le b + \frac{\varepsilon+\lambda}{\gamma}v \le b + (\varepsilon+\lambda)v$. Part (b) yields $a \le (1+\varepsilon)b + \varepsilon(1+\lambda+\varepsilon)v$. For (d), let $a_i \in v_\varepsilon(b_i)$ and $0 \le b_i + \lambda v$. Then $a_i \le (1+\varepsilon)b_i + \varepsilon(1+\lambda)v$ by Part (b). This yields

$$a_1 + \ldots + a_n \le (1+\varepsilon)(b_1 + \ldots + b_n) + n\varepsilon(1+\lambda)v,$$

hence our claim. □

Property 4.1(a) implies in particular that $v_\varepsilon(a) \subset v_{3\varepsilon}(c)$ whenever $a \in v_\varepsilon(b)$ and $b \in v_\varepsilon(c)$ for $a, b, c \in \mathcal{P}$ and $0 < \varepsilon \le 1$. Similar statements as in Lemma 4.1 hold for the lower and for the symmetric relative neighborhoods.

For elements $a, b \in \mathcal{P}$ the weak local and global preorders on $\mathcal{P}$ as defined in Section 3 can be recovered as

$$a \preccurlyeq_v b \qquad \text{if} \qquad a \in v_\varepsilon(b)$$

for some $v \in \mathcal{V}$ and all $\varepsilon > 0$, and

$$a \preccurlyeq b \qquad \text{if} \qquad a \in v_\varepsilon(b)$$

for all $\varepsilon > 0$ and $v \in \mathcal{V}$. Lemma 4.1(d) implies that these orders are compatible with the algebraic operations in $\mathcal{P}$.

For varying $v \in \mathcal{V}$ and $\varepsilon > 0$ the neighborhoods $v_\varepsilon(\cdot)$, $(\cdot)v_\varepsilon$ and $v_\varepsilon^s(\cdot)$ create the *upper, lower* and *symmetric relative topologies* on $\mathcal{P}$, respectively.

We notice that $a \leq b + v$ for $a, b \in \mathcal{P}$ and $v \in \mathcal{V}$ implies that $a \preccurlyeq b + v$, and for a given $\varepsilon > 0$, with $\delta = \min\{1, \varepsilon/2\}$, we notice that $a \preccurlyeq b + \delta v$ implies $a \leq \gamma b + (1+\delta)\delta v \leq \gamma b + \varepsilon v$ for some $1 \leq \gamma \leq 1 + \delta$ (see Lemma 3.1), hence $b \in v_\varepsilon(a)$. This observation demonstrates that the given upper, lower and symmetric topologies on $\mathcal{P}$ are finer than those induced by the same neighborhood system using the weak preorder, and that in turn these topologies are finer than the above defined relative topologies.

However, while the relative neighborhoods form convex subsets of $\mathcal{P}$, they do in general not create a locally convex cone topology. Indeed, the sets $\{(a,b) \mid a \in v_\varepsilon(b)\}$ are not necessarily convex in $\mathcal{P}^2$, hence do not establish a convex semiuniform structure on $\mathcal{P}$ in the sense of 1.3.

For later reference we shall list some further properties of the relative topologies and use the earlier introduced standard notations for subsets of $\mathcal{P}$ (see 1.1).

**Proposition 4.2.** *Let $(\mathcal{P}, \mathcal{V})$ be a locally convex cone. The upper (lower or symmetric) relative topology of $\mathcal{P}$ is coarser than the given upper (lower or symmetric) topology and satisfies the following:*

*(i) Every element of $\mathcal{P}$ admits a basis of convex and decreasing (increasing or order convex) neighborhoods. The symmetric relative neighborhoods in the basis for $0 \in \mathcal{P}$ are also balanced.*
*(ii) The mapping $(a,b) \mapsto a + b : \mathcal{P} \times \mathcal{P} \to \mathcal{P}$ is continuous.*
*(iii) The mapping $(\alpha, a) \mapsto \alpha a : [0, +\infty) \times \mathcal{P} \to \mathcal{P}$ is continuous at all points $(\alpha, a) \in [0, +\infty) \times \mathcal{P}$ such that either $\alpha > 0$ or $a \in \mathcal{P}$ is bounded.*
*(iv) For bounded elements of $\mathcal{P}$ the neighborhoods in the upper (lower or symmetric) relative topology are equivalent to the neighborhoods in the given upper (lower or symmetric) topology.*

*Proof.* We observed before that the relative topologies are coarser than the given topologies on $\mathcal{P}$. Clearly, for $a \in \mathcal{P}$, $v \in \mathcal{V}$ and $\varepsilon > 0$ the relative neighborhoods $v_\varepsilon(a)$, $(a)v_\varepsilon$ or $v_\varepsilon^s(a)$ are convex and decreasing, increasing or order convex, respectively. The symmetric relative neighborhoods of $0 \in \mathcal{P}$ are also balanced. Indeed, let $v \in \mathcal{V}$ and $\varepsilon > 0$ and let $a \in v_\varepsilon^s(0)$. Then $a \leq \varepsilon v$ and $0 \leq \gamma a + \varepsilon v$ for some $1 \leq \gamma \leq 1 + \varepsilon$. Thus $0 \leq a + (\varepsilon/\gamma)v \leq a + \varepsilon v$. Let $0 \leq \lambda \leq 1$. Then $\lambda a \in v_\varepsilon^s(0)$ follows from the convexity of $v_\varepsilon^s(0)$ since $\lambda a = \lambda a + (1-\lambda)0$. If on the other hand $b + \lambda a = 0$ for $b \in \mathcal{P}$, then

$$b \leq b + \lambda(a + \varepsilon v) \leq \varepsilon v \qquad \text{and} \qquad 0 = b + \lambda a \leq b + \varepsilon v$$

Hence $b \in v_\varepsilon(0)$ holds in this case as well.

For property (ii), let $a, b \in \mathcal{P}$, $v \in \mathcal{V}$ and $\varepsilon > 0$. There is $\lambda \geq 0$ such that $0 \leq a + \lambda v$ and $0 \leq b + \lambda v$. Choose $\delta = \varepsilon/(2\lambda + 4)$. Then for $c \in v_\delta(a)$ and $d \in v_\delta(b)$ we have $c + d \in v_{2\delta(1+\lambda)}(a+b) = v_\varepsilon(a+b)$ by Lemma 4.1(d). This shows continuity of the addition with respect to the upper relative topology. Next with the same choice for $\delta$, let $c \in (a)v_\delta$ and $d \in (b)v_\delta$, that is $a \in v_\delta(c)$ and $b \in v_\delta(d)$. Then we have $a \le \gamma c + \delta v$ for some $1 \le \gamma \le 1+\delta$, hence $0 \le \gamma c + (\lambda+\delta)v$ and $0 \le c+(\lambda+1)v$. Likewise, $0 \le d+(\lambda+1)v$. Now 4.1(d) yields $a+b \in v_{2\delta(2+\lambda)}(c+d) \subset v_\varepsilon(c+d)$. Thus $c+d \in (a+b)v_\varepsilon$. This shows continuity of the addition with respect to the lower relative topology. Combining the preceding arguments, we realize that $c \in v_\delta^s(a)$ and $d \in v_\delta^s(b)$ yields $c+d \in v_\varepsilon^s(a+b)$, which proves continuity with respect to the symmetric relative topology.

Next we shall argue Part (iv): Let $a \in \mathcal{P}$ be a bounded element, let $v \in \mathcal{V}$ and $\varepsilon > 0$. There is $\lambda \ge 0$ such that $a \le \lambda v$. We shall verify that

$$(\varepsilon v)(a) \subset v_\varepsilon(a) \subset (\rho v)(a) \qquad \text{and} \qquad (a)(\varepsilon v) \subset (a)v_\varepsilon(a) \subset (a)(\rho v).$$

with $\rho = \varepsilon(1+\lambda)$. Indeed, the inclusions $(\varepsilon v)(a) \subset v_\varepsilon(a)$ and $(\varepsilon v)(a) \subset v_\varepsilon(a)$ are obvious. Moreover, for $b \in v_\varepsilon(a)$ we have $b \le \gamma a + \varepsilon v$ with some $1 \le \gamma \le 1+\varepsilon$. Then $\gamma a = a + (\gamma-1)a \le a + \varepsilon\lambda v$ implies that $b \le a + \varepsilon(1+\lambda)v = a + \rho v$, hence $b \in (\rho v)(a)$. For $b \in (a)v_\varepsilon$ on the other hand, we have $a \le \gamma b + \varepsilon v$ with $1 \le \gamma \le 1+\varepsilon$. Then $\gamma a \le a + \varepsilon\lambda v$ implies $\gamma a \le \gamma b + \varepsilon(1+\lambda)v = \gamma b + \rho v$, hence $a \le b + (\rho/\gamma)v \le b + \rho v$, and therefore $b \in (a)(\rho v)$.

For the first case in Part (iii) let $(\alpha, a) \in (0,+\infty) \times \mathcal{P}$. For $v \in \mathcal{V}$ and $\varepsilon > 0$ let $\lambda \ge 0$ such that $0 \le a + \lambda v$. For $0 < \delta < \min\{1, \varepsilon/3, \varepsilon/\big(2\alpha(1+\lambda)\big)\}$ we consider the neighborhoods $u_\delta(\alpha) = \big[\alpha/(1+\delta), \alpha(1+\delta)\big]$ of $\alpha$ in $[0,+\infty)$ and $v_\delta(a)$ of $a$ in $\mathcal{P}$. For every $b \in v_\delta(a)$ we have $b \le (1+\delta)a + \delta(1+\lambda)v$ by 4.1(b). For $\beta \in u_\delta(\alpha)$ we set $\gamma = \beta(1+\delta)/\alpha$ and estimate

$$\beta b \le \beta(1+\delta)a + \beta\delta(1+\lambda)v = \gamma(\alpha a) + \beta\delta(1+\lambda)v.$$

Now $\alpha/(1+\delta) \le \beta \le \alpha(1+\delta)$ and our choice for $\delta$ implies $1 \le \gamma \le (1+\delta)^2 \le 1+\varepsilon$ as well as $\beta\delta(1+\lambda) \le \alpha(1+\delta)\delta(1+\lambda) \le 2\alpha\delta(1+\lambda) \le \varepsilon$. Thus $\beta b \in v_\varepsilon(\alpha a)$. This shows continuity for the scalar multiplication at $(\alpha, a)$ with respect to the upper relative topology. For the lower topology, with the same choice for $\delta$, let $b \in (a)v_\delta$ and $\beta \in u_\delta(\alpha)$. Then $a \le (1+\delta)b + \delta(2+\lambda)v$ by 4.1(c). We set $\gamma = \alpha(1+\delta)/\beta$ and obtain

$$\alpha a \le \alpha(1+\delta)b + \alpha\delta(2+\lambda)v = \gamma(\beta b) + \alpha\delta(2+\lambda)v.$$

We verify $1 \le \gamma \le 1+\varepsilon$ and $\alpha\delta(2+\lambda) \le \varepsilon$ and infer that $\alpha a \in (\beta b)v_\varepsilon$, hence $\beta b \in (\alpha a)v_\varepsilon$. This shows continuity with respect to the lower relative topology. The combination of both arguments yields continuity with respect to the symmetric relative topology.

The second case of Part (iii), that is the continuity of the scalar multiplication at $(\alpha, a) \in [0,+\infty) \times \mathcal{P}$ for a bounded element $a \in \mathcal{P}$, follows directly from Part (iv) and from Part (iii) of Proposition 1.1. Indeed, the given and the relative upper (lower or symmetric) topologies coincide locally at $a \in \mathcal{P}$ by (iv), thus continuity with respect to any of the given topologies which was established in Proposition 1.1(iii) implies continuity with respect to the corresponding relative topology. □

For $\mathcal{P} = \overline{\mathbb{R}}$, in particular, Part (iv) of the preceding proposition implies that the given and the relative topologies coincide on all reals. They also coincide on the element $+\infty$, thus everywhere, as can be easily verified (for details on this, see Example 4.37(a) below).

Part (iv) together with Corollary 1.2 also yields:

**Corollary 4.3.** *Let $(\mathcal{P}, \mathcal{V})$ be a locally convex cone and let $\mathcal{P}_0$ be the subcone of all invertible elements of $\mathcal{P}$. The mapping $(\alpha, a) \mapsto \alpha a : \mathbb{R} \times \mathcal{P}_0 \to \mathcal{P}_0$ is continuous with respect to the symmetric relative topology of $\mathcal{P}$.*

We observe that the given upper (lower or symmetric) topologies do of course satisfy the properties listed in Proposition 4.2 with the exception of 4.2(iii). More precisely, we take note:

**Proposition 4.4.** *Let $(\mathcal{P}, \mathcal{V})$ be a locally convex cone. The upper (or lower) relative topology is the finest topology on $\mathcal{P}$ which is coarser than the given upper (or lower) topology and satisfies property* (iii) *from Proposition 4.2.*

*Proof.* Let $\tau$ be any topology on $\mathcal{P}$ which is finer than the upper (or lower) topology and satisfies property (iii) from Proposition 4.2. Let $a \in \mathcal{P}$ and let $U(a)$ be a neighborhood in $\tau$ for $a$. We shall show that $U(a)$ contains some upper (or lower) relative neighborhood of $a$. The mapping $(\alpha, b) \mapsto \alpha b : [0, +\infty) \times \mathcal{P} \to \mathcal{P}$ is continuous with respect to $\tau$ at the point $(1, a)$. Thus there is a neighborhood $V(a)$ in $\tau$ and $0 < \varepsilon \leq 1$ such that $\beta b \in U(a)$ for all $b \in V(a)$ and $\beta \in [1-\varepsilon, 1+\varepsilon]$. Moreover, since $\tau$ is coarser than the upper (or lower) topology of $\mathcal{P}$ there is $v \in \mathcal{V}$ such that $v(a) \subset V(a)$ $\big($or $(a)v \subset V(a)\big)$. In the case of the upper topology, then for every $c \in v_\varepsilon(a)$ we have $c \leq \gamma a + \varepsilon v$ for some $1 \leq \gamma \leq 1+\varepsilon$. Thus $d \leq a + (\varepsilon/\gamma)v \leq a + v$ for $d = (1/\gamma)c$. We infer that $d \in v(a) \subset V(a)$, hence $c = \gamma d \in U(a)$ since $\gamma \in [1-\varepsilon, 1+\varepsilon]$. This shows $v_\varepsilon(a) \subset U(a)$. Likewise, in the case of the lower topology, for $c \in (a)v_\varepsilon$ we have $a \leq \gamma c + \varepsilon v$ for some $1 \leq \gamma \leq 1+\varepsilon$, hence $d = \gamma c \in (a)(v) \subset V(a)$. This yields $c = (1/\gamma)d \in U(a)$ since $(1/\gamma) \in [1-\varepsilon, 1+\varepsilon]$. We conclude that $(a)v_\varepsilon \subset U(a)$ in this case. □

**Proposition 4.5.** *Let $(\mathcal{P}, \mathcal{V})$ and $(\mathcal{Q}, \mathcal{W})$ be locally convex cones. A continuous linear operator $T : \mathcal{P} \to \mathcal{Q}$ is also continuous if both $\mathcal{P}$ and $\mathcal{Q}$ are endowed with either their respective upper, lower or symmetric relative topologies.*

*Proof.* Let $T : \mathcal{P} \to \mathcal{Q}$ be a continuous linear operator. Given $w \in \mathcal{W}$, there is $v \in \mathcal{V}$ such that $a \leq b+v$ implies $T(a) \leq T(b)+w$ for elements $a, b \in \mathcal{P}$. Thus $a \in v_\varepsilon(b)$, that is $a \leq \gamma b + \varepsilon v$ with some $1 \leq \gamma \leq 1+\varepsilon$, implies $T(a) \leq \gamma T(b) + \varepsilon w$, hence $T(a) \in w_\varepsilon\big(T(b)\big)$. A similar argument shows continuity with respect to either the lower or symmetric relative topologies of $\mathcal{P}$ and $\mathcal{Q}$. □

For $\mathcal{Q} = \overline{\mathbb{R}}$, in particular, we remarked earlier (see also Example 4.37(a) below) that the given and the relative topologies coincide. A linear functional $\mu \in \mathcal{P}^*$ is therefore also continuous if we endow $\mathcal{P}$ with either of its relative and $\overline{\mathbb{R}}$ with the corresponding given topology.

We shall also use the *(upper, lower, symmetric) relative $v$-topologies* on $\mathcal{P}$, generated by the relative neighborhoods for a fixed $v \in \mathcal{V}$. The symmetric relative $v$-topology, in particular, is induced by the pseudometric

$$d_v(a,b) = \inf\left\{1, \sqrt{\varepsilon} \mid a \in v^s_\varepsilon(b)\right\}.$$

The properties of a pseudometric (see Section 2.1 in [198]) are readily checked for this expression: We obviously have $d_v(a,b) \geq 0$, $d_v(a,a) = 0$ and $d_v(a,b) = d_v(b,a)$ for $a,b \in \mathcal{P}$. The triangular inequality, namely $d_v(a,c) \leq d_v(a,b) + d_v(b,c)$ for $a,b,c \in \mathcal{P}$, holds trivially true if either $d_v(a,b) = 1$ or $d_v(b,c) = 1$. Otherwise, if $d_v(a,b) < \varepsilon < 1$ and $d_v(b,c) < \delta < 1$, then $a \in v^s_{\varepsilon^2}(b)$ and $b \in v^s_{\delta^2}(c)$ implies by Lemma 4.1(a) that $a \in v^s_\rho(c)$, where $\rho = \varepsilon^2 + \delta^2 + \varepsilon^2\delta^2 \leq (\varepsilon+\delta)^2$. Thus $d_v(a,c) \leq \varepsilon + \delta$, hence the triangular inequality holds. As a consequence of the availability of a pseudometric for the symmetric relative $v$-topology, arbitrary subsets of separable subsets of $\mathcal{P}$ remain separable (see 16G in [198]). We shall use this fact in Chapter II.

The (upper, lower, symmetric) relative topologies on $\mathcal{P}$ are the common refinements of all (upper, lower, symmetric) relative $v$-topologies.

**4.6 The Weak Topology $\sigma(\mathcal{P},\mathcal{P}^*)$.** The *weak topology* $\sigma(\mathcal{P},\mathcal{P}^*)$ on a locally convex cone $(\mathcal{P},\mathcal{V})$ is generated by its dual cone in the following way: For an element $a \in \mathcal{P}$ an upper neighborhood $\mathcal{V}_\Upsilon(a)$, corresponding to a finite subset $\Upsilon = \{\mu_1,\ldots,\mu_n\}$ of $\mathcal{P}^*$, is given by

$$\mathcal{V}_\Upsilon(a) = \left\{b \in \mathcal{P} \mid \mu_i(b) \leq \mu_i(a) + 1 \quad \text{for all } \mu_i \in \Upsilon\right\}.$$

Endowed with these neighborhoods, $\mathcal{P}$ forms again a locally convex cone (see Section II.3 in [100]). We are mostly interested in the resulting symmetric topology $\sigma(\mathcal{P},\mathcal{P}^*)$ which is generated by the symmetric neighborhoods

$$\mathcal{V}^s_\Upsilon(a) = \left\{ b \in \mathcal{P} \;\middle|\; \begin{array}{rl} |\mu_i(b) - \mu_i(a)| \leq 1, & \text{if } \mu_i(a) < +\infty \\ \mu_i(b) = +\infty, & \text{if } \mu_i(a) = +\infty \end{array} \right\}$$

In this way *weak convergence* for a net $(a_i)_{i\in\mathcal{I}}$ in $(\mathcal{P},\mathcal{V})$ means that $\big(\mu(a_i)\big)_{i\in\mathcal{I}}$ converges towards $\mu(a)$ in $\overline{\mathbb{R}}$ (with respect to the symmetric locally convex cone topology of $\overline{\mathbb{R}}$) for every continuous linear functional $\mu \in \mathcal{P}^*$.

While the relative topologies of a locally convex cone are generally coarser than the given ones, we observe from the preceding definition that the relative upper, lower and symmetric weak topologies do indeed coincide with the given upper, lower and symmetric weak topologies on $\mathcal{P}$.

**Lemma 4.7.** *The weak topology $\sigma(\mathcal{P},\mathcal{P}^*)$ on a locally convex cone $(\mathcal{P},\mathcal{V})$ is coarser than the symmetric relative topology.*

*Proof.* For this, let $a \in \mathcal{P}$, let $\Upsilon$ be a finite subset of $\mathcal{P}^*$ and consider the weak neighborhood $\mathcal{V}^s_\Upsilon(a)$ from above. Choose $v \in \mathcal{V}$ such that $\mu_i \in v^\circ$ for all $i = 1,\ldots,n$. We shall show that for a suitable $\varepsilon > 0$ the symmetric neighborhood $v^s_\varepsilon(a)$ is contained in $\mathcal{V}^s_\Upsilon(a)$. Indeed, let $b \in v^s_\varepsilon(a)$, that is

$$b \le \gamma a + \varepsilon v \qquad \text{and} \qquad a \le \gamma' b + \varepsilon v$$

for some $1 \le \gamma, \gamma' \le 1+\varepsilon$. Thus

$$\mu(b) \le \gamma\mu(a) + \varepsilon \qquad \text{and} \qquad \mu(a) \le \gamma'\mu(b) + \varepsilon$$

for all $\mu \in \Upsilon$. If $\mu(a) = +\infty$, then $\mu(a) = +\infty$. Moreover, $\varepsilon > 0$ may be chosen such that the above implies $|\mu(b) - \mu(a)| \le 1$ for all $\mu \in \Upsilon$ such that $\mu(a) < +\infty$. This shows $b \in \mathcal{V}^s_\Upsilon(a)$. $\square$

**Proposition 4.8.** *Let $(\mathcal{P},\mathcal{V})$ be a locally convex cone. The following statements are equivalent:*

*(i) The symmetric relative topology on $\mathcal{P}$ is Hausdorff.*
*(ii) The weak topology on $\mathcal{P}$ is Hausdorff.*
*(iii) The weak preorder on $\mathcal{P}$ is antisymmetric.*

*Proof.* Clearly, (ii) implies (i), since the symmetric relative topology is finer than $\sigma(\mathcal{P},\mathcal{P}^*)$. If $a \preccurlyeq b$ and $b \preccurlyeq a$ for $a,b \in \mathcal{P}$, then $a \in v^s_\varepsilon(b)$ for all $v \in \mathcal{V}$ and $\varepsilon > 0$. If the symmetric relative topology is Hausdorff, then this implies $a = b$. Thus (i) implies (iii). If the weak preorder is antisymmetric, then for distinct elements $a,b \in \mathcal{P}$ we have either $a \not\preccurlyeq b$ or $b \not\preccurlyeq a$, thus $a \not\preccurlyeq b+v$ or $b \not\preccurlyeq a+v$ for some $v \in \mathcal{V}$ by Lemma 3.2. Then there exists a linear functional $\mu \in v^\circ$ such that $\mu(a) > \mu(b) + 1$ or $\mu(b) > \mu(a) + 1$, respectively (see Section 3 and Corollary 4.34 below). The weak neighborhoods $\mathcal{V}^s_{\{(1/3)\mu\}}(a)$ and $\mathcal{V}^s_{\{(1/3)\mu\}}(b)$ are therefore seen to be disjoint. Thus (iii) implies (ii) as well. $\square$

**4.9 Boundedness Components.** For an element $a \in \mathcal{P}$ we define the *upper* and *lower boundedness components* of $a$ as

$$\mathcal{B}(a) = \bigcap_{v\in\mathcal{V}} \bigcup_{\varepsilon>0} v_\varepsilon(a) \qquad \text{and} \qquad (a)\mathcal{B} = \bigcap_{v\in\mathcal{V}} \bigcup_{\varepsilon>0} (a)v_\varepsilon,$$

respectively. The elements of $\mathcal{B}(a)$ are called *bounded above relative to* $a$. Correspondingly, the elements of $(a)\mathcal{B}$ are called *bounded below relative to* $a$. By the definition of a locally convex cone we have $0 \in \mathcal{B}(a)$ for all $a \in \mathcal{P}$, and $\mathcal{B}(0) = \mathcal{B}$ consists of all bounded elements of $\mathcal{P}$. We shall first list a few basic properties of the upper boundedness components.

**Proposition 4.10.** *Let $a, b, \in \mathcal{P}$. The following are equivalent:*

*(i)* $b \in \mathcal{B}(a)$.
*(ii)* $\mathcal{B}(b) \subset \mathcal{B}(a)$.
*(iii) For every* $v \in \mathcal{V}$ *there are* $\alpha, \beta \geq 0$ *such that* $b \leq \alpha a + \beta v$.
*(iv) The mapping*

$$\alpha \mapsto a + \alpha b \;:\; [0, +\infty) \to \mathcal{P}$$

*is continuous with respect to the symmetric relative topology of* $\mathcal{P}$.
*(v) For all* $\mu \in \mathcal{P}^*$, $\mu(a) < +\infty$ *implies* $\mu(b) < +\infty$.

*Proof.* Let $a, b \in \mathcal{P}$. We shall first establish the equivalence of (i), (ii) and (iii): Suppose that $b \in \mathcal{B}(a)$ and let $c \in \mathcal{B}(b)$. Then for every $v \in \mathcal{V}$ there are $\varepsilon, \delta > 0$ such that $c \in v_{\varepsilon}(b)$ and $b \in v_{\delta}(a)$. Following Lemma 4.1(a), this implies $c \in v_{(\varepsilon+\delta+\varepsilon\delta)}(a)$. We conclude that $c \in \mathcal{B}(a)$, hence $\mathcal{B}(b) \subset \mathcal{B}(a)$, and (i) implies (ii). If $\mathcal{B}(b) \subset \mathcal{B}(a)$, then $b \in \mathcal{B}(a)$ since $b \in \mathcal{B}(b)$ trivially holds. Thus for every $v \in \mathcal{V}$ there is $\varepsilon > 0$ such that $b \in v_{\varepsilon}(a)$, that is $b \leq \alpha a + \beta v$ for some $\alpha, \beta \geq 0$. Therefore (ii) implies (iii). If, on the other hand, for some $v \in \mathcal{V}$ we have $b \leq \alpha a + \beta v$ for $\alpha, \beta \geq 0$, we choose $\lambda \geq 0$ such that $0 \leq a + \lambda v$. Then

$$b \leq (\alpha a + \beta v) + (a + \lambda v) = (1 + \alpha)a + (\beta + \lambda)v,$$

hence $b \in v_{\varepsilon}(a)$ for every $\varepsilon > \max\{\alpha, \beta + \lambda\}$. If this argument can be made for all $v \in \mathcal{V}$, then we have $b \in \mathcal{B}(a)$, hence (iii) implies (i) as well, and the Conditions (i), (ii) and (iii) are seen to be equivalent.

Next we shall verify that (iii) implies (iv): Following Proposition 4.2(iii), for any choice of $b \in \mathcal{P}$ the mapping $\alpha \mapsto \alpha b$ is continuous with respect to the symmetric relative topology of $\mathcal{P}$ on the open interval $(0, +\infty)$. Likewise, of course, is the constant mapping $\alpha \mapsto a$. Thus by the continuity of the addition in $\mathcal{P}$ (see Proposition 4.2(ii)), the mapping $f : [0, +\infty) \to \mathcal{P}$ such that

$$f(\alpha) = a + \alpha b$$

is also continuous on $(0, +\infty)$. In case that (iii) holds, we shall verify continuity at $\alpha = 0$ as well: Given $v \in \mathcal{V}$ and $\varepsilon > 0$ there is $\lambda > 0$ such that $0 \leq b + \lambda v$, and by (iii) there are $\gamma, \rho \geq 0$ such that $b \leq \gamma a + \rho v$. Then for $\delta = \min\{\varepsilon/\gamma, \varepsilon/\rho, \varepsilon/\lambda\}$ and all $\alpha \in [0, \delta)$ we have

$$a + \alpha b \leq a + \alpha(\gamma a + \rho v) \leq (1 + \alpha\gamma)a + \alpha\rho v.$$

Since our choice of $\delta$ guarantees that both $\alpha\gamma \leq \varepsilon$ and $\alpha\rho \leq \varepsilon$, we infer that $f(\alpha) \in v_{\varepsilon}(f(0))$. Similarly, one observes that

$$a \leq a + \alpha(b + \lambda v) \leq (a + \alpha b) + \alpha\lambda v$$

holds for all $\alpha \geq 0$. If indeed $\alpha \in [0, \delta)$, then our choice for $\delta$ guarantees that $\alpha\lambda \leq \varepsilon$. This shows $f(0) \in v_{\varepsilon}(f(\alpha))$, that is $f(\alpha) \in (f(0))v_{\varepsilon}$, and together with the above, $f(\alpha) \in v_\varepsilon^s(f(0))$ for all $\alpha \in [0,\delta)$. We infer continuity for the function $f$ at $\alpha = 0$ with respect to the symmetric relative topology of $\mathcal{P}$.

Next suppose that (iv) holds. Then for every linear functional $\mu \in \mathcal{P}^*$ the mapping $\varphi : [0,+\infty) \to \overline{\mathbb{R}}$ such that

$$\varphi(\alpha) = \mu(a + \alpha b) = \mu(a) + \alpha\mu(b)$$

is also continuous at $\alpha = 0$ (see the remark after Proposition 4.5) if we consider $\overline{\mathbb{R}}$ in its symmetric topology, for which $= \infty$ is an isolated point (see Example 4.37(a) below). Therefore $\mu(b)$ is finite whenever $\varphi(0) = \mu(a)$ is finite, and we infer that (iv) implies (v).

Finally, suppose that Condition (iii) fails for the element $b$. Given a neighborhood $v \in \mathcal{V}$, we define a corresponding functional $\mu_v$ on $\mathcal{P}$ setting $\mu_v(c) = 0$ for all $c \in \mathcal{P}$ such that $c \le \alpha a + \beta v$ for some $\alpha, \beta \ge 0$, and $\mu_v(c) = +\infty$, else. It is straightforward to check that $\mu_v$ is linear. Indeed, if $\mu_v(c) = \mu_v(d) = 0$, that is $c \le \alpha a + \beta v$ and $c \le \gamma + \delta v$ for some $\alpha, \beta, \gamma, \delta \ge 0$, then $c+d \le (\alpha+\gamma)a + (\beta+\delta)v$, hence $\mu_v(c+d) = 0$ as well. If, on the other hand, $\mu_v(c+d) = 0$, that is $c+d \le \alpha a+\beta v$ for some $\alpha, \beta \ge 0$, we choose $\lambda \ge 0$ such that $0 \le d+\lambda v$ and have $c \le c+d+\lambda v \le \alpha a+(\beta+\lambda)v$. This shows $\mu_v(c) = 0$. Similarly, one verifies that $\mu_v(d) = 0$. Moreover, we realize that $\mu_v$ is an element of the polar $v^\circ$ of $v$, as for $c \le d + v$, we have $\mu_v(c) = 0$ whenever $\mu_v(d) = 0$, hence $\mu_v(c) \le \mu_v(d) + 1$ holds in any case. Using this construction, we proceed with our argument: If (iii) fails for $b$, then there is a neighborhood $v \in \mathcal{V}$ such that $b \not\le \alpha a + \beta v$ for all choices of $\alpha, \beta \ge 0$, hence $\mu_v(b) = +\infty$, while $\mu_v(a) = 0$. Thus Condition (v) does not hold either. This in turn shows that (v) implies (iii) and completes our argument. □

**Proposition 4.11.** *Let $a, b, c \in \mathcal{P}$. Then*

*(a) $\mathcal{B}(a)$ is a subcone of $\mathcal{P}$, and $\mathcal{B} \subset \mathcal{B}(a)$.*
*(b) $\mathcal{B}(a)$ is a face in $\mathcal{P}$, that is $b + c \in \mathcal{B}(a)$ implies both $b, c \in \mathcal{B}(a)$.*
*(c) $\mathcal{B}(\alpha a) = \mathcal{B}(a)$ for $\alpha > 0$, and $\mathcal{B}(a) + \mathcal{B}(b) \subset \mathcal{B}(a+b)$.*
*(d) $\mathcal{B}(a)$ is closed in $\mathcal{P}$ with respect to the lower relative topology of $\mathcal{P}$.*

*Proof.* Part (a) is obvious from Proposition 4.10(iii), since $b \le \alpha a + \beta v$ and $c \le \gamma a + \delta v$ for $v \in \mathcal{V}$ and $\alpha, \beta, \gamma, \delta \ge 0$ implies that $b + c \le (\alpha + \gamma)a + (\beta + \delta)v$ and $\lambda b \le \lambda\alpha a + \lambda\beta v$ for $\lambda \ge 0$. Moreover, since $0 \in \mathcal{B}(a)$, Proposition 4.10(ii) yields that $\mathcal{B} = \mathcal{B}(0) \subset \mathcal{B}(a)$.

For (b), let $b+c \in \mathcal{B}(a)$, that is, given $v \in \mathcal{V}$, we have $b+c \le \alpha a+\beta v$ for some $\alpha, \beta \ge 0$. Because all elements of a locally convex cone are bounded below, there is $\lambda \ge 0$ such that $0 \le c + \lambda v$. Thus $b \le b + c + \lambda v \le \alpha a + (\beta + \lambda)v$. Hence $b \in \mathcal{B}(a)$. Similarly, one verifies that $c \in \mathcal{B}(a)$.

The first statement of (c) is obvious from 4.10(iii). For the second statement, let $c \in \mathcal{B}(a)$, $d \in \mathcal{B}(b)$ and $v \in \mathcal{V}$. Then $c \le \alpha a+\beta v$ and $d \le \gamma b+\delta v$ for some $\alpha, \beta, \gamma, \delta \ge 0$. Let $\lambda \ge 0$ such that both $0 \le a+\lambda v$ and $0 \le b+\lambda v$.

In case that $\alpha \le \gamma$, this yields $c \le \alpha a + \beta v + (\gamma - \alpha)(a + \lambda v) = \gamma a + \rho v$, where $\rho = \beta + (\gamma - \alpha)\lambda$. Thus $c + d \le \gamma(a + b) + (\rho + \delta)v$. In case that $\alpha > \gamma$, a similar argument leads to $c + d \le \alpha(a + b) + (\rho' + \beta)v$, where $\rho' = \delta + (\alpha - \gamma)\lambda$. This verifies $c + d \in \mathcal{B}(a + b)$.

Finally, for Part (c), we remarked before that a linear functional $\mu \in \mathcal{P}^*$ is a continuous mapping from $\mathcal{P}$ into $\overline{\mathbb{R}}$ if we endow $\mathcal{P}$ with either its upper, lower or symmetric relative topology, and $\overline{\mathbb{R}}$ with either its given upper, lower or symmetric topology, respectively. We shall use this observation for the functionals $\mu_v \in \mathcal{P}^*$ for $v \in \mathcal{V}$, that we constructed in the argument for the implication (v) $\Rightarrow$ (iii) in the proof of Proposition 4.10, that is $\mu_v(c) = 0$ if $c \le \alpha a + \beta v$ for some $\alpha, \beta \ge 0$, and $\mu(c) = +\infty$, else. Because $\mathbb{R}$ is a closed subset of $\overline{\mathbb{R}}$ in the lower topology of $\overline{\mathbb{R}}$ (see Example 1.4(a)), its inverse image $\mu_v^{-1}(\mathbb{R})$ under $\mu_v$ is closed in the lower relative topology of $\mathcal{P}$. We have $\mathcal{B}(a) = \bigcap_{v \in \mathcal{V}} \mu_v^{-1}(\mathbb{R})$ by Proposition 4.10(v). Thus $\mathcal{B}(a)$ is indeed closed in the lower relative topology of $\mathcal{P}$. □

We proceed to identify the corresponding properties of the lower boundedness components.

**Proposition 4.12.** *Let $a, b, \in \mathcal{P}$. The following are equivalent:*

*(i) $b \in (a)\mathcal{B}$.*
*(ii) $a \in \mathcal{B}(b)$.*
*(iii) $\mathcal{B}(a) \subset \mathcal{B}(b)$.*
*(iv) $(b)\mathcal{B} \subset (a)\mathcal{B}$.*
*(v) For every $v \in \mathcal{V}$ there are $\alpha, \beta > 0$ such that $\alpha a \le b + \beta v$.*
*(vi) The mapping*

$$\alpha \mapsto \alpha a + b \; : \; [0, +\infty) \to \mathcal{P}$$

*is continuous with respect to the symmetric relative topology of $\mathcal{P}$.*
*(vii) For all $\mu \in \mathcal{P}^*$, $\mu(a) = +\infty$ implies $\mu(b) = +\infty$.*

*Proof.* Let $a, b \in \mathcal{P}$. First we observe that $b \in (a)\mathcal{B}$ holds for $a, b \in \mathcal{P}$ if and only if for every $v \in \mathcal{V}$ there is $\varepsilon > 0$ such that $b \in (a)v_\varepsilon$, that is $a \in v_\varepsilon(b)$. The latter means that $a \in \mathcal{B}(b)$. Hence (i) and (ii) are indeed equivalent.

The equivalence of (ii) and (iii) follows from the corresponding one in Proposition 4.10: We have $b \in (a)\mathcal{B}$ if and only if $a \in \mathcal{B}(b)$ by the preceding argument, and the latter holds if and only if $\mathcal{B}(a) \subset \mathcal{B}(b)$ by 4.10.

Now suppose that $\mathcal{B}(a) \subset \mathcal{B}(b)$ holds and let $c \in (b)\mathcal{B}$. Then $b \in \mathcal{B}(c)$, hence $\mathcal{B}(a) \subset \mathcal{B}(b) \subset \mathcal{B}(c)$ by 4.10(ii). Thus $a \in \mathcal{B}(c)$, hence $c \in (a)\mathcal{B}$. This shows $(b)\mathcal{B} \subset (a)\mathcal{B}$. For the converse suppose that $(b)\mathcal{B} \subset (a)\mathcal{B}$. This implies $b \in (a)\mathcal{B}$, hence $a \in \mathcal{B}(b)$ and $\mathcal{B}(a) \subset \mathcal{B}(b)$ by 4.10(ii). Therefore (iii) and (iv) are also equivalent.

Next suppose that for every $v \in \mathcal{V}$ there are $\alpha, \beta > 0$ such that $\alpha a \le b + \beta v$. Then $a \le (1/\alpha)b + (\beta/\alpha)v$, hence $a \in \mathcal{B}(b)$ by 4.10(iii). For the converse, let $a \in \mathcal{B}(b)$, that is $a \in v_\varepsilon(b)$ for every $v \in \mathcal{V}$ with some $\varepsilon > 0$. This yields

$a \le \gamma b + \varepsilon v$ for some $1 \le \gamma \le 1+\varepsilon$, hence $(1/\gamma)a \le b + (\varepsilon/\gamma)v$, as claimed. We infer that (ii) and (v) are also equivalent. The remaining parts of this proof require only little effort if we use the already established equivalence of (i) and (ii) and the corresponding results for the upper boundedness components in Proposition 4.10:

The equivalence of (ii) and (vi) follows from the equivalence of (i) and (iv) in Proposition 4.10. The equivalence of Conditions (i) and (v) from 4.10, on the other hand, yields that $a \in \mathcal{B}(b)$ if and only if $\mu(b) < +\infty$ implies $\mu(a) < +\infty$ for every $\mu \in \mathcal{P}^*$. But the latter is equivalent to the formulation of Condition (vii) in the present proposition. □

**Proposition 4.13.** *Let* $a, b, c \in \mathcal{P}$. *Then*

*(a) If* $b \in (a)\mathcal{B}$ *and* $c \in \mathcal{P}$, *then* $\beta b + c \in (a)\mathcal{B}$ *for all* $\beta > 0$.
*(b)* $(\alpha a)\mathcal{B} = (a)\mathcal{B}$ *for* $\alpha > 0$, *and* $(a+b)\mathcal{B} = (a)\mathcal{B} \cap (b)\mathcal{B}$.
*(c)* $(a)\mathcal{B}$ *is closed in* $\mathcal{P}$ *with respect to the upper relative topology of* $\mathcal{P}$.

*Proof.* For Part (a), let $b \in (a)\mathcal{B}$, that is $a \in \mathcal{B}(b)$, let $c \in \mathcal{P}$ and $\beta > 0$. 4.11(c) shows that $a \in \mathcal{B}(\beta b)$, hence $a \in \mathcal{B}(\beta b) + \mathcal{B}(c) \subset \mathcal{B}(\beta b + c)$. Thus $\beta b + c \in (a)\mathcal{B}$.

The first part of (b) is obvious from 4.12(v). For the second part let $c \in (a+b)\mathcal{B}$. Then $a+b \in \mathcal{B}(c)$, hence both $a \in \mathcal{B}(c)$ and $b \in \mathcal{B}(c)$, since $\mathcal{B}(c)$ is a face in $\mathcal{P}$ by Proposition 4.11(b). Thus $c \in (a)\mathcal{B} \cap (b)\mathcal{B}$. This argument is indeed reversible: If $c \in (a)\mathcal{B} \cap (b)\mathcal{B}$, then both $a \in \mathcal{B}(c)$ and $b \in \mathcal{B}(c)$. This implies $a+b \in \mathcal{B}(c)$, since $\mathcal{B}(c)$ is a subcone of $\mathcal{P}$ $\big($see 4.11(a)$\big)$. Thus $c \in (a+b)\mathcal{B}$.

For Part (c) we recall that the singleton set $\{+\infty\}$ is closed in the upper topology of $\overline{\mathbb{R}}$, hence its inverse image $\mu^{-1}(\{+\infty\})$ under any linear functional $\mu \in \mathcal{P}^*$ is closed with respect to the upper relative topology of $\mathcal{P}$. Following Proposition 4.12(vii), $(a)\mathcal{B}$ is the intersection of the sets $\mu^{-1}(\{+\infty\})$ for all $\mu \in \mathcal{P}^*$ such that $\mu(a) = +\infty$, hence $(a)\mathcal{B}$ is indeed closed for the upper relative topology. □

The sets

$$\mathcal{B}^s(a) = \mathcal{B}(a) \cap (a)\mathcal{B}$$

are called the *symmetric boundedness components* of $\mathcal{P}$. The elements of $\mathcal{B}^s(a)$ are called *bounded relative to* $a$. The symmetric boundedness components are of particular interest, since they will provide a natural partition of a locally convex cone into boundedness equivalence classes. Before establishing this feature, we shall list a few properties of the symmetric boundedness components:

**Proposition 4.14.** *Let* $a, b, \in \mathcal{P}$. *The following are equivalent:*

*(i)* $b \in \mathcal{B}^s(a)$.
*(ii)* $a \in \mathcal{B}^s(b)$.
*(iii)* $\mathcal{B}(b) = \mathcal{B}(a)$.

*(iv)* $(b)\mathcal{B} = (a)\mathcal{B}$.
*(v)* $\mathcal{B}^s(b) = \mathcal{B}^s(a)$.
*(vi) For every* $v \in \mathcal{V}$ *there are* $\alpha, \beta \geq 0$ *such that both*

$$b \leq \alpha a + \beta v \qquad \text{and} \qquad a \leq \alpha b + \beta v.$$

*(vii) The mapping*

$$\alpha \mapsto \alpha a + (1-\alpha) b \; : \; [0,1] \to \mathcal{P}$$

*is continuous with respect to the symmetric relative topology of* $\mathcal{P}$.
*(viii) For all* $\mu \in \mathcal{P}^*$, $\mu(a) = +\infty$ *if and only if* $\mu(b) = +\infty$.

*Proof.* Let $a, b \in \mathcal{P}$. If $b \in \mathcal{B}^s(a) = \mathcal{B}(a) \cap (a)\mathcal{B}$, then $b \in \mathcal{B}(a)$ and $a \in \mathcal{B}(b)$. This implies $\mathcal{B}(a) = \mathcal{B}(b)$ by 4.10(ii). On the other hand, if $\mathcal{B}(a) = \mathcal{B}(b)$, then $b \in \mathcal{B}(a)$ and $a \in \mathcal{B}(b)$, hence $b \in \mathcal{B}(a) \cap (a)\mathcal{B}$. This yields the equivalence of (i) and (iii).

The equivalence of (iii) and (iv) follows immediately from 4.12(iii) and (iv).

Conditions (iii) and (iv) are symmetric in $a$ and $b$ and therefore also equivalent to (ii).

Conditions (iii) and (iv) imply (v), which in turn obviously renders (i), since $\mathcal{B}^s(b) = \mathcal{B}^s(a)$ implies $b \in \mathcal{B}^s(b) = \mathcal{B}^s(a)$.

Clearly, (vi) implies (i), since by Proposition 4.10(iii) it yields $b \in \mathcal{B}(a)$ and $a \in \mathcal{B}(b)$, hence $b \in \mathcal{B}^s(a)$. On the other hand, if $b \in \mathcal{B}^s(a)$, then $b \in \mathcal{B}(a)$ and $a \in \mathcal{B}(b)$, and by 4.10(iii), given $v \in \mathcal{V}$, there are $\alpha', \alpha'', \beta', \beta'' \geq 0$ such that

$$b \leq \alpha' a + \beta' v \qquad \text{and} \qquad a \leq \alpha'' b + \beta'' v.$$

There is $\lambda \geq 0$ such that both $0 \leq a + \lambda v$ and $0 \leq b + \lambda v$. Set $\alpha = \max\{\alpha', \alpha''\}$ and $\beta = \max\{\, \beta' + (\alpha - \alpha')\lambda,\ \beta'' + (\alpha - \alpha'')\lambda \,\}$. Then

$$\begin{aligned} b &\leq (\alpha' a + \beta' v) + (\alpha - \alpha')(a + \lambda v) \\ &\leq \alpha a + \big(\beta' + (\alpha - \alpha')\lambda\big) v \leq \alpha a + \beta v, \end{aligned}$$

and, likewise,

$$\begin{aligned} a &\leq (\alpha'' a + \beta'' v) + (\alpha - \alpha'')(a + \lambda v) \\ &\leq \alpha a + \big(\beta'' + (\alpha - \alpha'')\lambda\big) v \leq \alpha a + \beta v. \end{aligned}$$

Therefore (i) implies (vi) as well.

Condition (viii) of this proposition is the combination of the corresponding conditions in Propositions 4.10 and 4.12 and therefore also equivalent to Conditions (i) to (vi). All left to show is that (vii) is equivalent to the rest. First let us verify that (vii) implies (viii). If the mapping

$$\alpha \mapsto \alpha a + (1-\alpha) b \; : \; [0,1] \to \mathcal{P}$$

is continuous with respect to the symmetric relative topology of $\mathcal{P}$, then for every linear functional $\mu \in \mathcal{P}^*$ the mapping $\varphi : [0,1] \to \overline{\mathbb{R}}$ such that

$$\varphi(\alpha) = \mu\big(\alpha a + (1-\alpha)b\big) = \alpha\mu(a) + (1-\alpha)\mu(b)$$

is also continuous (see the remark after Proposition 4.5) if we consider $\overline{\mathbb{R}}$ in its symmetric topology, for which $= \infty$ is an isolated point. Therefore $\varphi(0) = \mu(b)$ is finite if and only if $\varphi(1) = \mu(a)$ is finite. Hence (vii) implies (viii). Finally, we shall demonstrate how the other conditions imply (vii). Following Proposition 4.2(iii), for any choice of $a, b \in \mathcal{P}$ the mappings $\alpha \mapsto \alpha a$ and $\alpha \mapsto (1-\alpha)b$ are continuous with respect to the symmetric relative topology of $\mathcal{P}$ on the open intervals $(0, +\infty)$ and $(-\infty, 1)$, respectively. Thus by Proposition 4.2(ii), that is the continuity of the addition in $\mathcal{P}$, the mapping $f : [0,1] \to \mathcal{P}$ such that

$$f(\alpha) = \alpha a + (1-\alpha)b$$

is continuous on the interval $(0,1)$. In case that $b \in \mathcal{B}^s(a)$, we shall verify continuity at the endpoints $\alpha = 0$ and $\alpha = 1$ as well: Proposition 4.12(vi), if applied to the element $(1/2)b \in (a)\mathcal{B}$, states that the mapping

$$\alpha \mapsto \alpha a + \frac{1}{2}b \; : \; \big[0, +\infty\big) \to \mathcal{P}$$

is continuous at $\alpha = 0$. The mapping

$$\alpha \mapsto \left(\frac{1}{2} - \alpha\right) b \; : \; \left(-\infty, \frac{1}{2}\right] \to \mathcal{P},$$

on the other hand, is continuous at $0$ by 4.2(iii). Thus the sum of these mappings, that is the function $f$, is also continuous at $0$. A similar argument holds for $\alpha = 1$. Following Propositions 4.10(iv) and 4.2(iii), respectively, the mappings

$$\alpha \mapsto \frac{1}{2}a + (1-\alpha)b \; : \; \big(-\infty, 1\big]$$

and

$$\alpha \mapsto \left(\alpha - \frac{1}{2}\right) a \; : \; \left[\frac{1}{2}, +\infty\right) \to \mathcal{P}$$

are continuous at $\alpha = 1$. So is their sum, the function $f$. This concludes our argument. □

**Proposition 4.15.** *Let* $a, b, c \in \mathcal{P}$. *Then*

*(a) If* $b, c \in \mathcal{B}^s(a)$, *then* $\beta b + \gamma c \in \mathcal{B}^s(a)$ *for all* $\beta, \gamma > 0$.
*(b)* $\mathcal{B}^s(\alpha a) = \mathcal{B}^s(a)$ *for* $\alpha > 0$, *and* $\mathcal{B}^s(a+b) \supset \mathcal{B}^s(a) \cap \mathcal{B}^s(b)$.
*(c)* $\mathcal{B}^s(a)$ *is closed in* $\mathcal{P}$ *with respect to the symmetric relative topology of* $\mathcal{P}$.

*Proof.* Part (a) follows directly from Propositions 4.11(a) and 4.13(a). The first part of (b) follows from the first parts of 4.11(c) and 4.13(b). The same sources yield the second part of (b) as well, since the relations $(a+b)\mathcal{B} = (a)\mathcal{B} \cap (b)\mathcal{B}$ and $\mathcal{B}(a+b) \supset \mathcal{B}(a) + \mathcal{B}(b) \supset \mathcal{B}(a) \cap \mathcal{B}(b)$ imply that

$$\mathcal{B}^s(a+b) = \mathcal{B}(a+b) \cap (a+b)\mathcal{B} \supset \mathcal{B}(a) \cap \mathcal{B}(b) \cap (a)\mathcal{B} \cap (b)\mathcal{B} = \mathcal{B}^s(a) \cap \mathcal{B}^s(b).$$

Finally, by Propositions 4.11(d) and 4.13(c), the sets $\mathcal{B}(a)$ and $(a)\mathcal{B}$ are closed in the lower and upper relative topologies of $\mathcal{P}$, respectively. Consequently, both of these sets as well as their intersection, that is $\mathcal{B}^s(a)$, are also closed in the symmetric relative topology of $\mathcal{P}$, which is finer than both the upper and the lower relative topologies. □

**Proposition 4.16.** *The symmetric boundedness components satisfy a version of the cancellation law, that is $a+c \preccurlyeq b+c$ for elements $a, b$ and $c$ of the same boundedness component implies that $a \preccurlyeq b$.*

*Proof.* Suppose that the elements $a, b, c \in \mathcal{P}$ are bounded relative to each other and that $a + c \preccurlyeq b + c$. Given $v \in \mathcal{V}$ there is $\lambda \geq 0$ such that $0 \leq c + \lambda v$. Thus $a + (c + \lambda v) \preccurlyeq b + (c + \lambda v)$. As we observed before, $(\mathcal{P}, \mathcal{V})$ endowed with the weak preorder $\preccurlyeq$ forms again a locally convex cone. Following Lemma I.4.2 in [100], if applied to this order and the positive element $(a + \lambda v)$ of a full cone containing $\mathcal{P}$, the above implies $a \preccurlyeq b + \varepsilon(c + \lambda v)$ for all $\varepsilon > 0$. By our assumption, there are $\alpha, \beta \geq 0$ such that $c \leq \alpha b + \beta v$. Now combining the above yields

$$a \preccurlyeq b + \varepsilon\big(\alpha b + (\beta + \lambda)v\big) = (1 + \varepsilon\alpha)b + \varepsilon(\beta + \lambda)v$$

for all $\varepsilon > 0$. This shows $a \preccurlyeq_v b$ by our definition of the weak local preorder in Section 3. Finally, because $a \preccurlyeq_v b$ holds for all $v \in \mathcal{V}$, we infer that $a \preccurlyeq b$. □

**Proposition 4.17.** *The symmetric boundedness components furnish a partition of $\mathcal{P}$ into disjoint convex subsets that are closed and connected in the symmetric relative topology.*

*Proof.* Proposition 4.15(a) implies that the symmetric boundedness components are convex subset of $\mathcal{P}$. They are closed in the symmetric relative topology by 4.15(c). Moreover, the equivalence of (i) and (v) in Proposition 4.14 shows that any two symmetric boundedness components of $\mathcal{P}$ either coincide or are disjoint. For connectedness, let $a \in \mathcal{P}$, and let $b, c \in \mathcal{B}^s(a)$. Then $\mathcal{B}^s(a) = \mathcal{B}^s(b) = \mathcal{B}^s(b)$ by Proposition 4.14(v), and by the equivalent condition in 4.14(vii), the mapping $f : [0,1] \to \mathcal{B}^s(a)$ such that

$$f(\alpha) = \alpha b + (1 - \alpha)c$$

is continuous with respect to the symmetric relative topology of $\mathcal{P}$. As $f(0) = c$ and $f(1) = b$, this shows that $\mathcal{B}^s(a)$ is pathwise connected, hence connected in the symmetric relative topology of $\mathcal{P}$ (see Theorem 27.2 in [198]). $\square$

We shall also consider the local boundedness components of a locally convex cone $\mathcal{P}$ that arise if we endow $\mathcal{P}$ with the neighborhood subsystem $\mathcal{V}_v = \{\alpha v \mid \alpha > 0\}$ consisting of the multiples of a single neighborhood $v \in \mathcal{V}$. For an element $a \in \mathcal{P}$ and a neighborhood $v \in \mathcal{V}$, we define the *(local) upper, lower* and *symmetric $v$-boundedness components* of $a$ as

$$\mathcal{B}_v(a) = \bigcup_{\varepsilon>0} v_\varepsilon(a), \quad (a)\mathcal{B}_v = \bigcup_{\varepsilon>0} (a)v_\varepsilon, \quad \text{and} \quad \mathcal{B}_v^s(a) = \mathcal{B}_v(a) \cap (a)\mathcal{B}_v,$$

respectively. The elements of $\mathcal{B}_v(a)$ are called *$v$-bounded above relative to $a$.* $\mathcal{B}_v(0) = \mathcal{B}_v$ consists of all $v$-bounded elements of $\mathcal{P}$. The global boundedness components may be recovered as

$$\mathcal{B}(a) = \bigcap_{v\in\mathcal{V}} \mathcal{B}_v(a), \qquad (a)\mathcal{B} = \bigcap_{v\in\mathcal{V}} (a)\mathcal{B}_v \qquad \text{and} \qquad \mathcal{B}^s(a) = \bigcap_{v\in\mathcal{V}} \mathcal{B}_v^s(a),$$

respectively. Obviously, the statements of Propositions 4.10 to 4.17 apply also to the local boundedness components, since we may replace the given neighborhood system $\mathcal{V}$ by the subsystem $\mathcal{V}_v$ and consider the locally convex cone $(\mathcal{P}, \mathcal{V}_v)$ for this purpose. The cancellation law in Proposition 4.16 holds with the weak local preorder $\preccurlyeq_v$ in this case. The dual cone $\mathcal{P}^*$ of $(\mathcal{P}, \mathcal{V}_v)$ consists only of the multiples of the functionals in $v^\circ$, and the relative topologies of $\mathcal{P}$ are the relative $v$-topologies.

The main benefit in considering the local boundedness components as compared to the global ones, is the following: We shall proceed to verify that the disjoint partition of $\mathcal{P}$ into symmetric local boundedness components provides indeed a topological partition as well.

**Proposition 4.18.** *Let $a \in \mathcal{P}$ and $v \in \mathcal{V}$.*

*(a) $\mathcal{B}_v(a)$ is open in $\mathcal{P}$ with respect to the upper, closed with respect to the lower and both open and closed with respect to the symmetric relative $v$-topology of $\mathcal{P}$.*
*(b) $(a)\mathcal{B}_v$ is closed in $\mathcal{P}$ with respect to the upper, open with respect to the lower and both open and closed with respect to the symmetric relative $v$-topology of $\mathcal{P}$.*

*Proof.* Let $a \in \mathcal{P}$ and $v \in \mathcal{V}$ Proposition 4.11(d) states that $\mathcal{B}_v(a)$ is closed in the lower relative $v$-topology of $\mathcal{P}$. Let $b \in \mathcal{B}_v(a)$, that is $b \leq \alpha b + \beta v$ for some $\alpha, \beta \geq 0$, and let $v_\varepsilon(b)$ be a lower neighborhood of $b$. Then for $c \in v_\varepsilon(b)$ we have $c \leq \gamma b + \varepsilon v$ with some $1 \leq \gamma \leq 1+\varepsilon$, and therefore $c \leq (\alpha\gamma)a + (\beta\gamma + \varepsilon)v$. This shows $c \in \mathcal{B}_v(a)$, hence $v_\varepsilon(b) \subset \mathcal{B}_v(a)$, and $\mathcal{B}_v(a)$ is seen to be open in the lower relative $v$-topology of $\mathcal{P}$. Moreover, because the symmetric relative $v$-topology is the common refinement of the upper and lower topologies, $\mathcal{B}_v(a)$ is indeed both open and closed in this topology. This completes Part (a).

The argument for Part (b) is similar: Proposition 4.13(c) states that $(a)\mathcal{B}_v$ is closed in the upper relative $v$-topology of $\mathcal{P}$. Let $b \in (a)\mathcal{B}_v$, that is $\alpha a \leq b + \beta v$ for some $\alpha, \beta > 0$, and let $(b)v_\varepsilon$ be a lower neighborhood of $b$. Then for $c \in (b)v_\varepsilon$ we have $b \leq \gamma c + \varepsilon v$ with some $1 \leq \gamma \leq 1 + \varepsilon$, and therefore $\alpha a \leq \gamma c + (\varepsilon + \delta)v$, hence $(\alpha/\gamma)a \leq c + (\varepsilon + \delta)/\gamma v$. This shows $c \in (a)\mathcal{B}_v$, hence $(b)v_\varepsilon \subset (a)\mathcal{B}_v$, and $(a)\mathcal{B}_v$ is seen to be open in the lower relative $v$-topology of $\mathcal{P}$. Hence $(a)\mathcal{B}_v$ is both open and closed in the symmetric relative $v$-topology. □

Propositions 4.18 and 4.17 now yield a topological and algebraic partition of a locally convex cone into local boundedness components.

**Proposition 4.19.** *For every neighborhood $v \in \mathcal{V}$, the symmetric $v$-boundedness components furnish a partition of $\mathcal{P}$ into disjoint convex subsets that are open, closed and connected in the symmetric relative $v$-topology.*

A subset of $\mathcal{P}$ that is open or closed in any of the relative $v$-topologies is of course also open or closed in the corresponding (global) relative topology of $\mathcal{P}$. The same statement does however not hold for connectedness.

**4.20 Connectedness.** Topological vector spaces are connected and all of their elements are bounded. This does not hold for locally convex cones in general. However, Propositions 4.17 and 4.19 suggest relations between the boundedness and the connectedness components of a locally convex cone. Let us recall some of the relevant concepts from topology: The *quasi-component of a point* $x$ in a topological space $X$ is the intersection of all closed and open subsets of $X$ which contain $x$. The quasi-components constitute a decomposition of $X$ into pairwise disjoint and closed subsets (see VIII.26 in [198] or VI.1 in [59]). The *component of a point* $x \in X$, on the other hand is the largest connected subset of $X$ which contains the point $x$. The components are subsets of the quasi-components and constitute a decomposition of $X$ into pairwise disjoint, connected and closed subsets. A topological space is locally connected, if each of its points has a basis of connected neighborhoods. In locally connected spaces the quasi-components and components coincide and are both open and closed (see Corollary 27.10 in [198]).

**Proposition 4.21.** *Let $(\mathcal{P}, \mathcal{V})$ be a locally convex cone.*

*(a) In the symmetric relative topology of $\mathcal{P}$ the components, quasi-components and the symmetric boundedness components coincide.*
*(b) For every neighborhood $v \in \mathcal{V}$ and the symmetric relative $v$-topology, $\mathcal{P}$ is locally connected and the components, quasi-components and the symmetric $v$-boundedness components coincide.*

*Proof.* (a) For an element $a \in \mathcal{P}$ Proposition 4.17 implies that $\mathcal{B}^s(a)$ is contained in its (connectedness) component. On the other hand, $\mathcal{B}^s(a)$ is the intersection of the sets $\mathcal{B}_v^s(a)$ for all $v \in \mathcal{V}$, all of which are open and closed in the respective symmetric relative $v$-topologies, hence in the symmetric relative topology of $\mathcal{P}$ by Proposition 4.19. This shows that the quasi-component of $a$ is contained in $\mathcal{B}^s(a)$. Hence these three components coincide.

For Part (b) let $v \in \mathcal{V}$ and $a \in \mathcal{P}$. The $v$-boundedness component $\mathcal{B}_v^s(a)$ of $a$ contains all the neighborhoods $v_\varepsilon^s(a)$ for $\varepsilon > 0$. Convexity then guarantees (see the corresponding argument in the proof of Proposition 4.17) that these neighborhoods are pathwise connected in the symmetric relative $v$-topology, hence $\mathcal{P}$ is locally connected. The components, quasi-components and the symmetric $v$-boundedness components of $\mathcal{P}$ coincide by Part (a) if we endow $\mathcal{P}$ with the neighborhood subsystem $\mathcal{V}_v = \{\alpha v \mid \alpha > 0\}$. □

**Proposition 4.22.** *A locally convex cone $(\mathcal{P}, \mathcal{V})$ is locally connected in its symmetric relative topology if and only if every point $a \in \mathcal{P}$ has a basis of symmetric relative neighborhoods that are contained in $\mathcal{B}^s(a)$.*

*Proof.* Let $a \in \mathcal{P}$. The argument in the proof of Proposition 4.17 shows that every convex subset of $\mathcal{B}^s(a)$ is pathwise connected, hence connected in the symmetric relative topology. On the other hand, every connected subset of $\mathcal{P}$ containing the element $a$ is a subset of $\mathcal{B}^s(a)$, the component of $a$ by 4.21(a). Because the symmetric relative neighborhoods of $a$ are convex, our claim follows. □

**4.23 Locally Convex Cones with Uniform Boundedness Components.** We shall say that a locally convex cone $(\mathcal{P}, \mathcal{V})$ has *uniform boundedness components* if the boundedness components of $\mathcal{P}$ for all neighborhoods coincide, that is if $\mathcal{B}_v^s(a) = \mathcal{B}^s(a)$ for all $v \in \mathcal{V}$ and $a \in \mathcal{P}$. Locally convex topological vector spaces are obviously of this type as all their elements are bounded with respect to every neighborhood. Also, any locally convex cone whose neighborhood system consists of the multiples of a single neighborhood, has uniform boundedness components. Proposition 4.22 yields that a locally convex cone with uniform boundedness components is locally connected. Its global boundedness components are both open and closed in the each of the symmetric relative $v$-topologies (Proposition 4.19), hence also in the (global) symmetric relative topology.

Similar and related notions of boundedness components in locally convex cones had previously been established in [170] and [176].

**4.24 Bounded Subsets.** We shall also use notions of boundedness for subsets corresponding to those for elements of a locally convex cone $(\mathcal{P}, \mathcal{V})$. A subset $A$ of $\mathcal{P}$ is called

(i) *bounded below* if for every $v \in \mathcal{V}$ there is $\lambda \geq 0$ such that $0 \leq a + \lambda v$ for all $a \in A$;

(ii) *bounded above* if for every $v \in \mathcal{V}$ there is $\lambda \geq 0$ such that $a \leq \lambda v$ for all $a \in A$;

(iii) *bounded* if it is both bounded below and above;
(iv) *bounded above relative to* $b \in \mathcal{P}$ if for every $v \in \mathcal{V}$ there are $\lambda, \rho \geq 0$ such that $a \leq \rho b + \lambda v$ for all $a \in A$;
(v) *relatively bounded above* if it is bounded above relative to some element of $\mathcal{P}$; and
(vi) *relatively bounded* if it is both bounded below and relatively bounded above, that is if there is $b \in \mathcal{P}$ such that for every $v \in \mathcal{V}$ there are $\lambda, \rho \geq 0$ such that $0 \leq a + \lambda v$ and $a \leq \rho b + \lambda v$ for all $a \in A$.

All these notions do of course coincide in a locally convex topological vector space. Similar concepts may be used to define local boundedness, that is boundedness relative to a specific neighborhood $v \in \mathcal{V}$, for subsets of $\mathcal{P}$.

Note that a continuous linear operator $T : \mathcal{P} \to \mathcal{Q}$, where both $(\mathcal{P}, \mathcal{V})$ and $(\mathcal{Q}, \mathcal{W})$ are locally convex cones, maps bounded subsets of one of the above types in $\mathcal{P}$ into bounded subsets of the same type in $\mathcal{Q}$.

A Uniform-Boundedness-type theorem from [172] allows relative boundedness for subsets of a locally convex cone $\mathcal{P}$ to be characterized in terms of its dual cone $\mathcal{P}^*$.

**Proposition 4.25.** *Let $A$ be a subset of a locally convex cone $(\mathcal{P}, \mathcal{V})$, and let $b \in \mathcal{P}$. If for every linear functional $\mu \in \mathcal{P}^*$ such that $\mu(b) < +\infty$ the set $\mu(A)$ is bounded in $\overline{\mathbb{R}}$, then $A$ is bounded above relative to $b$.*

*Proof.* Let $A$ be a subset of $\mathcal{P}$ which is not bounded above relative to the element $b \in \mathcal{P}$. Then there is $v \in \mathcal{V}$ such that the condition in 4.24(iv) does not hold for this neighborhood. We define a monotone sublinear functional $p : \mathcal{P} \to \overline{\mathbb{R}}$ by

$$p(a) = \inf\{\lambda + \rho \mid \lambda, \rho \geq 0, \;\; a \leq \rho b + \lambda v\}$$

and observe that: (i) Let $c \leq d + v$ for $c, d \in \mathcal{P}$. Then $d \leq \rho b + \lambda v$ for $\lambda, \rho \geq 0$ implies that $c \leq \rho b + (\lambda + 1)v$. Thus $p(c) \leq p(d) + 1$, and the functional $p$ is seen to be continuous with respect to $v$ in the sense of Theorem 3.4 in [172]; (ii) $p$ is unbounded on $A$. Assume to the contrary that there is $M > 0$ such that $p(a) < M$ for all $a \in A$. Let $\sigma \geq 0$ such that $0 \leq b + \sigma v$. Then for every $a \in A$ there are $\lambda, \rho \geq 0$ such that $a \leq \rho b + \lambda v$ and $\lambda + \rho \leq M$. Then

$$a \leq (\rho b + \lambda v) + (M - \rho)(b + \sigma v) \leq M b + M(1 + \sigma) v,$$

contradiction our assumption that $A \subset \mathcal{P}$ is not bounded above relative to $b$. Now Theorem 3.4 from [172] yields the existence of a continuous linear functional $\mu \in v^\circ$ such that $\mu(c) \leq p(c)$ for all $c \in \mathcal{P}$, that is $\mu(b) \leq 1$ in particular, and such that $\mu$ is unbounded on the set $A$. □

Similar notions of boundedness will be used for nets in a locally convex cone, that is a net $(a_i)_{i \in \mathcal{I}}$ in $\mathcal{P}$ will be called *bounded (below, above, relative to an element,...)* if the corresponding requirements 4.24(i) to (vi) hold for the set $\{a_i \mid i \geq i_0\}$ for some $i_0 \in \mathcal{I}$.

**4.26 Closed Convex Sets.** We shall proceed making some observations regarding subsets of a locally convex cone which are closed either with respect to the lower or the upper relative topology.

**Lemma 4.27.** *Let $(\mathcal{P}, \mathcal{V})$ be a locally convex cone. Every subset of $\mathcal{P}$ that is closed with respect to the lower (or the upper) relative topology is decreasing (or increasing) with respect to the weak preorder.*

*Proof.* Indeed, suppose that $A \subset \mathcal{P}$ is closed with respect to the lower relative topology and let $b \preccurlyeq a$ for some $b \in \mathcal{P}$ and $a \in A$. Then $b \in v_\varepsilon(a)$, thus $a \in (b)v_\varepsilon$ and $(b)v_\varepsilon \cap A \neq \emptyset$ for all $v \in \mathcal{V}$ and $\varepsilon > 0$. Thus $b$ is in the closure of $A$ with respect to the lower relative topology which coincides with $A$. Similarly one argues for a subset of $\mathcal{P}$ which is closed with respect to the upper relative topology. □

For a subset $A$ of a locally convex cone $(\mathcal{P}, \mathcal{V})$ we denote by $\overline{A}^{(l)}$ and $\overline{A}^{(u)}$ its closure with respect to the lower and the upper relative topology of $\mathcal{P}$, respectively.

**Proposition 4.28.** *Let $A$ be a subset of a locally convex cone $(\mathcal{P}, \mathcal{V})$.*

*(a) The set $\overline{A}^{(l)}$ consists of all elements $b \in \mathcal{P}$ such that for every $v \in \mathcal{V}$ and $\varepsilon > 0$ there is some $a \in A$ such that $b \in v_\varepsilon(a)$.*
*(b) The set $\overline{A}^{(l)}$ is convex whenever $A$ is convex.*
*(c) The set $\overline{A}^{(l)}$ is bounded above whenever $A$ is bounded above.*

*Proof.* (a) We have $b \in \overline{A}^{(l)}$ if and only if $(b)v_\varepsilon \cap A \neq \emptyset$ for all $v \in \mathcal{V}$ and $\varepsilon > 0$, that is if there is $a \in A$ such that $b \in v_\varepsilon(a)$. For Part (b) suppose that $A$ is convex and let $b, b' \in \overline{A}^{(l)}$ and $b'' = \alpha b + (1-\alpha)b'$ for some $0 \leq \alpha \leq 1$. Given $v \in \mathcal{V}$ and $\varepsilon > 0$, by Part (i) there are $a, a' \in A$ such that $b \in v_\varepsilon(a)$ and $b' \in v_\varepsilon(a')$. Then $b \leq \gamma a + \varepsilon v$ and $b' \leq \gamma' a' + \varepsilon v$ for some $1 \leq \gamma, \gamma' \leq 1 + \varepsilon$. Set $\gamma'' = (\alpha\gamma + (1-\alpha)\gamma')$. Then

$$a'' = \frac{\alpha\gamma}{\gamma''} a + \frac{(1-\alpha)\gamma'}{\gamma''} a' \in A$$

and $b'' \leq \gamma'' a'' + (1+\varepsilon)v$. Since $1 \leq \gamma'' \leq 1 + \varepsilon$, this demonstrates that $b'' \in \overline{A}^{(l)}$, and therefore this set is also convex. For Part (c) suppose that $A$ is bounded above in the sense of 4.24(ii). Let $v \in \mathcal{V}$ and suppose that there is $\lambda \geq 0$ such that $a \leq \lambda v$ for all $a \in A$. Then for every $b \in \overline{A}^{(l)}$ there is $a \in A$ such that $b \in v_1(a)$. This means $b \leq \gamma a + v$ for some $1 \leq \gamma \leq 2$, hence $b \leq (\gamma\lambda + 1)v \leq (2\lambda + 1)v$. Our claim follows. □

In a similar way one proves corresponding statements for the closure of a set with respect to the upper relative topology.

**Proposition 4.29.** *Let $A$ be a subset of a locally convex cone $(\mathcal{P}, \mathcal{V})$.*

*(a) The set $\overline{A}^{(u)}$ consists of all elements $b \in \mathcal{P}$ such that for every $v \in \mathcal{V}$ and $\varepsilon > 0$ there is some $a \in A$ such that $a \in v_\varepsilon(b)$.*
*(b) The set $\overline{A}^{(u)}$ is convex whenever $A$ is convex.*
*(c) The set $\overline{A}^{(u)}$ bounded below whenever $A$ is bounded below.*

Proposition 4.28(a) implies in particular that for a singleton set $\{a\}$ we have $b \in \overline{\{a\}}^{(l)}$ if and only if $b \in v_\varepsilon(a)$ for all $v \in \mathcal{V}$ and $\varepsilon > 0$, that is $b \preccurlyeq a$. Thus

$$\overline{\{a\}}^{(l)} = \{b \in \mathcal{P} \mid b \preccurlyeq a\}.$$

Likewise, Proposition 4.29(a) yields

$$\overline{\{a\}}^{(u)} = \{b \in \mathcal{P} \mid a \preccurlyeq b\}.$$

**Theorem 4.30.** *Let $A$ be a convex subset of a locally convex cone $(\mathcal{P}, \mathcal{V})$ and let $b \in \mathcal{P}$. Then*

*(a) $b \in \overline{A}^{(l)}$ if and only if $\mu(b) \le \sup\{\mu(a) \mid a \in A\}$ for all $\mu \in \mathcal{P}^*$.*
*(b) $b \in \overline{A}^{(u)}$ if and only if $\mu(b) \ge \inf\{\mu(a) \mid a \in A\}$ for all $\mu \in \mathcal{P}^*$.*

*Proof.* Let $A \subset \mathcal{P}$ be convex and let $b \in \mathcal{P}$. We may assume that $A \neq \emptyset$, because for $A = \emptyset$ our claim is trivial. (As usual, we set $\inf \emptyset = +\infty$ and $\sup \emptyset = -\infty$ and use the fact that for every $a \in \mathcal{P}$ there is some $\mu \in \mathcal{P}^*$ such that $\mu(a) < +\infty$.) For Part (a), let $b \in \overline{A}^{(l)}$ and let $\mu \in \mathcal{P}^*$, that is $\mu \in v^\circ$ for some $v \in \mathcal{V}$. Given $\varepsilon > 0$, according to Proposition 4.28(a) there is $a \in A$ and $1 \le \gamma \le 1+\varepsilon$ such that $b \le \gamma a + \varepsilon v$, hence $\mu(b) \le \gamma\mu(a) + \varepsilon \le \gamma \sup\{\mu(a) \mid a \in A\} + \varepsilon$. This shows $\mu(a) \le \sup\{\mu(a) \mid a \in A\}$. The proof of the converse implication will however require some advanced Hahn-Banach type arguments that had been established in the [172] and quoted earlier in Section 2: For a fixed number $\beta \in \mathbb{R}$ consider the sublinear functional $p$ on $\mathcal{P}$ defined for $x \in \mathcal{P}$ as

$$p(x) = \inf\{\lambda\beta \mid x = \lambda a \quad \text{for some} \quad a \in A \quad \text{and} \quad \lambda \ge 0\},$$

together with the extended superlinear functional $q(0) = 0$ and $q(x) = -\infty$ for $x \neq 0$. Following Theorem 2.4 (a quote of Theorem 3.1 in [172]) there is a linear functional $\mu \in \mathcal{P}^*$ such that $q \le \mu \le p$ if and only if we can find a neighborhood $v \in \mathcal{V}$ such that $q(x) \le p(y) + 1$ whenever $x \le y + v$ for $x, y \in \mathcal{P}$; that is in our particular case $0 \le \lambda\beta + 1$ whenever $0 \le \lambda a + v$ for some $a \in A$ and $\lambda \ge 0$. For this we shall have to distinguish two cases: (i) If for every $v \in \mathcal{V}$ there is $a \in A$ such that $0 \le a + v$, then we have to require that $\beta \ge 0$. (ii) If there is $v \in \mathcal{V}$ such that $0 \not\le a + v$ for all $a \in A$, then for $\varepsilon > 0$ the condition $0 \le \lambda a + \varepsilon v$ can hold only for $\lambda < \varepsilon$.

Thus we can choose any $\beta = -(1/\varepsilon)$ for the neighborhood $\varepsilon v \in \mathcal{V}$. In other words, in case (ii) for every $\beta \in \mathbb{R}$ we can find a neighborhood in $V$ such that the above condition is satisfied. Now we shall use Theorem 2.10 (a quote of Theorem 4.23 in [172]), which describes the range of all linear functionals $\mu \in \mathcal{P}^*$ such that $q \le \mu \le p$ on a fixed element $b \in \mathcal{P}$. It states that if there is at least one such linear functional $\mu$, then

$$\sup_{\substack{\mu \in \mathcal{P}^* \\ q \le \mu \le p}} \mu(b) \;=\; \sup_{v \in \mathcal{V}} \inf \{p(x) - q(y) \mid x, y \in \mathcal{P},\ q(y) \in \mathbb{R},\ b + y \le x + v\}.$$

With the particular insertions for $p$ and $q$ from above we need to consider only the choice of $y = 0$ and obtain

$$\sup_{\substack{\mu \in \mathcal{P}^* \\ q \le \mu \le p}} \mu(b) \;=\; \sup_{v \in \mathcal{V}} \inf \{\lambda\beta \mid \lambda \ge 0,\ b \le \lambda a + v \quad \text{for some} \quad a \in A\}.$$

Now let us assume that $\mu(b) \le \sup\{\mu(a) \mid a \in A\}$ holds for all $\mu \in \mathcal{P}^*$. As $q \le \mu \le p$ implies that $\sup\{\mu(a) \mid a \in A\} \le \beta$, this yields

$$\sup_{v \in \mathcal{V}} \inf \{\lambda\beta \mid \lambda \ge 0,\ b \le \lambda a + v \quad \text{for some} \quad a \in A\} \;\le\; \beta$$

for all admissible values of $\beta$. We shall use 4.28(a) to derive $b \in \overline{A}^{(l)}$ from this. Let $v \in \mathcal{V}$ and $\varepsilon > 0$. We choose $\beta = 1$ in the above and observe that there is $a \in A$ and $\lambda \ge 0$ such that

$$b \le \lambda a + \frac{\varepsilon}{2} v \qquad \text{and} \qquad \lambda \le 1 + \varepsilon.$$

If $1 \le \lambda$, this satisfies the criterion in 4.28(a). Otherwise we proceed distinguishing the above cases: In case (i) there is $a' \in A$ such that $0 \le a' + (\varepsilon/2)v$. Thus

$$b \le \lambda a + \frac{\varepsilon}{2} v + (1-\lambda)\left(a' + \frac{\varepsilon}{2} v\right) \le a'' + \varepsilon v$$

with $a'' = \lambda a + (1-\lambda)a' \in A$, satisfying the requirement from 4.28(a). In case (ii) we may use the above inequality for $\beta = -1$ as well. There is $\rho > 0$ such that $0 \le b + \rho v$. Set $\delta = \min\{1/2,\ \varepsilon/(4\rho + 2\varepsilon)\}$. We find $a' \in A$ and $\lambda' \ge 0$ such that

$$b \le \lambda' a' + \frac{\varepsilon}{2} v \qquad \text{and} \qquad -\lambda' \le -1 + \delta,$$

that is $\lambda' \ge 1 - \delta$. Next we choose $0 \le \alpha \le 1$ such that $1 - \delta \le \lambda'' \le 1$ holds for $\lambda'' = \alpha\lambda + (1-\alpha)\lambda'$. (Recall that we are considering the case that $\lambda < 1$, therefore such a choice of $\alpha$ is possible.) Then

$$b \le \alpha\left(\lambda a + \frac{\varepsilon}{2} v\right) + (1-\alpha)\left(\lambda' a' + \frac{\varepsilon}{2} v\right) \le \lambda'' a'' + \frac{\varepsilon}{2} v$$

with

$$a'' = \frac{\alpha\lambda}{\lambda''}a + \frac{(1-\alpha)\lambda'}{\lambda''}a' \in A.$$

From $0 \leq b + \rho v$ we infer that $0 \leq \lambda'' a'' + (\rho + \varepsilon/2)v$. Our assumption that $\delta \leq 1/2$ guarantees $1/2 \leq \lambda'' \leq 1$ and $1 - \lambda'' \leq \delta$. Using this we infer

$$\begin{aligned} 0 &\leq (1-\lambda'')a'' + \frac{(1-\lambda'')(2\rho+\varepsilon)}{2\lambda''}v \\ &\leq (1-\lambda'')a'' + \delta(2\rho+\varepsilon)v \\ &\leq (1-\lambda'')a'' + \frac{\varepsilon}{2}v \end{aligned}$$

since $\delta \leq \varepsilon/(4\rho + 2\varepsilon)$. Now combining the above yields

$$b \leq \lambda'' a'' + \frac{\varepsilon}{2}v + \left((1-\lambda'')a'' + \frac{\varepsilon}{2}v\right) \leq a'' + \varepsilon v,$$

again satisfying the requirement from 4.28(a). We conclude that $b \in \overline{A}^{(l)}$, as claimed.

The argument for Part (b) of the Theorem follows similar lines, but is sufficiently different from the preceding one to be presented here too: If $b \in \overline{A}^{(u)}$ and if $\mu \in \mathcal{P}^*$, then a similar argument than before using Proposition 4.29(a) yields $\mu(a) \geq \inf\{\mu(a) \mid a \in A\}$. For the converse implication we will again employ Theorem 2.10. For fixed numbers $0 \leq \alpha \in \overline{\mathbb{R}}$ and $\beta \in \mathbb{R}$ consider the sublinear functional $p$ on $\mathcal{P}$ defined for $x \in \mathcal{P}$ as $p(x) = \rho\alpha$ if $x = \rho b$ and $p(x) = +\infty$ else, together with the extended superlinear functional

$$q(x) = \sup\{\lambda\beta \mid x = \lambda a \quad \text{for some} \quad a \in A \quad \text{and} \quad \lambda \geq 0\}.$$

There is $\mu \in \mathcal{P}^*$ such that $q \leq \mu \leq p$ if and only if there is $v \in \mathcal{V}$ such that $q(x) \leq p(y) + 1$ whenever $x \leq y + v$ for $x, y \in \mathcal{P}$; that is in our particular case $\lambda\beta \leq \rho\alpha + 1$ whenever $\lambda a \leq \rho b + v$ for some $a \in A$ and $\lambda, \rho \geq 0$. For this we shall again have to distinguish two cases:

(i) If for every $v \in \mathcal{V}$ and $\varepsilon > 0$ there is $a \in A$ and $0 \leq \delta \leq \varepsilon$ such that $a \leq \delta b + v$, then we have to require that $\beta \leq 0$.

(ii) If there are $v \in \mathcal{V}$ and $\varepsilon > 0$ such that $a \not\leq \delta b + v$ for all $a \in A$ and $0 \leq \delta \leq \varepsilon$, then the above condition holds for this neighborhood $v$ with any $\beta \in \mathbb{R}$, provided that $\alpha \geq 1/\varepsilon$. Indeed, assume that $\lambda a \leq \rho b + v$ for some $a \in A$ and $\lambda, \rho \geq 0$, but $\lambda\beta > \rho\alpha + 1$. Then

$$a \leq \frac{\rho}{\lambda}b + \frac{1}{\lambda}v \leq \rho b + v.$$

This shows $\rho/\lambda > \varepsilon$, hence

$$\rho > \varepsilon\lambda > \varepsilon\rho\alpha + \varepsilon \geq \rho + \varepsilon,$$

a contradiction. In order to apply Theorem 2.10 we need to guarantee that $\mu(b) < +\infty$ for at least one $\mu \in \mathcal{P}^*$ satisfying $q \le \mu \le p$. Our insertions for $p$ and $q$ imply that $\beta \le \inf\{\mu(a) \mid a \in a\}$ and $\mu(b) \le \alpha$. We shall use $\alpha = +\infty$ in 2.10, but the preceding discussion involving different choices for $\alpha$ demonstrates that there is $\mu \in \mathcal{P}^*$ such that $q \le \mu \le p$ and $\mu(b) < +\infty$ in case (i), for any choice of $\beta \le 0$ and in case (ii) for any choice of $\beta \in \mathbb{R}$. Thus we may use Theorem 2.10 for

$$\inf_{\substack{\mu \in \mathcal{P}^* \\ q \le \mu \le p}} \mu(b) \;=\; \inf_{v \in \mathcal{V}} \sup\,\{q(x) - p(y) \mid x, y \in \mathcal{P},\; p(y) \in \mathbb{R},\; x \le b + y + v\},$$

With the particular insertions for $p$ and $q$ from above we need to consider only the choice of $y = 0$ and obtain

$$\inf_{\substack{\mu \in \mathcal{P}^* \\ q \le \mu \le p}} \mu(b) \;=\; \inf_{v \in \mathcal{V}} \sup\,\{\lambda\beta \mid \lambda \ge 0,\; \lambda a \le b + v \quad \text{for some} \quad a \in A\}.$$

Now let us assume that $\mu(b) \ge \inf\{\mu(a) \mid a \in A\}$ holds for all $\mu \in \mathcal{P}^*$. This yields

$$\inf_{v \in \mathcal{V}} \sup\,\{\lambda\beta \mid \lambda \ge 0,\; \lambda a \le b + v \quad \text{for some} \quad a \in A\} \;\ge\; \beta$$

for all admissible values of $\beta$. We shall use 4.29(a) to derive $b \in \overline{A}^{(u)}$ from this. Let $v \in \mathcal{V}$ and $\varepsilon > 0$. There is $\rho > 0$ such that $0 \le b + \rho v$. We choose $\beta = -1$ in the above and observe that there is $a \in A$ and $\lambda \ge 0$ such that

$$\lambda a \le b + \frac{\varepsilon}{2} v \quad \text{and} \quad -\lambda \ge -1 - \frac{\varepsilon}{2\rho} \quad \text{that is} \quad \lambda \le 1 + \frac{\varepsilon}{2\rho}.$$

If $1 \le \lambda + 1 + (\varepsilon/2\rho)$, we proceed as follows:

$$\lambda a \le b + \frac{\varepsilon}{2} v + (\lambda - 1)(b + \rho v) \le \lambda b + \varepsilon v,$$

since

$$\frac{\varepsilon}{2} + (\lambda - 1)\rho \le \frac{\varepsilon}{2} + \frac{\varepsilon}{2\rho}\rho = \varepsilon.$$

Thus

$$a \le b + \frac{\varepsilon}{\lambda} v \le b + \varepsilon v,$$

demonstrating that $a \in v_\varepsilon(b)$ as required in 4.29(b). Otherwise, that is if $\lambda < 1$, we continue to distinguish the above cases: In case (i) we set $\delta = \varepsilon/(2 - 2\lambda)$ and according to this case can find $a' \in A$ such that $a' \le \delta' b + \delta v$ for some $0 \le \delta' \le \delta$. Thus

$$a'' = \lambda a + (1 - \lambda) a' \le \Big((1 + (1 - \lambda)\delta')\Big) b + \Big(\frac{\varepsilon}{2} + (1 - \lambda)\delta\Big)\, v.$$

Since $a'' \in A$, and since $(1 - \lambda)\delta = \varepsilon/2$ and

$$1 \le 1 + (1-\lambda)\delta' \le 1 + (1-\lambda)\delta \le 1 + (\varepsilon/2),$$

this shows $a'' \in v_\varepsilon(b)$ as required in 4.29(b). In case (ii) we may use the above inequality for $\beta = +1$ as well. For $\sigma = \max\{1/2,\, 1/(1+\varepsilon)\} < 1$ we find $a' \in A$ and $\lambda' \ge 0$ such that

$$\lambda' a' \le b + \frac{\varepsilon}{2} v \qquad \text{and} \qquad \lambda' \ge \sigma.$$

We can choose $0 \le \alpha \le 1$ such that $\sigma \le \lambda'' \le 1$ holds for $\lambda'' = \alpha\lambda + (1-\alpha)\lambda'$. (Recall that we are considering the case that $\lambda < 1$, therefore such a choice of $\alpha$ is possible.) With

$$a'' = \frac{\alpha\lambda}{\lambda''} a + \frac{(1-\alpha)\lambda'}{\lambda''} a' \in A$$

we have

$$\lambda'' a'' \le b + \frac{\varepsilon}{2} v,$$

hence

$$a'' \le \frac{1}{\lambda''} b + + \frac{\varepsilon}{2\lambda''} v.$$

Because $1 \le 1/\lambda'' \le 1/\sigma \le 1+\varepsilon$, and because $\varepsilon/(2\lambda'') \le \varepsilon/2\sigma \le \varepsilon$ we infer that $a'' \in v_\varepsilon(b)$, again satisfying the requirement from 4.29(a). We conclude that $b \in \overline{A}^{(u)}$, as claimed. □

Theorem 4.30 is a generalization of Theorem 3.1 in [175] as the following corollary will show.

**Corollary 4.31.** *Let $(\mathcal{P}, \mathcal{V})$ be a locally convex cone. Then $a \preccurlyeq b$ holds for $a, b \in \mathcal{P}$ if and only if $\mu(a) \le \mu(b)$ for all $\mu \in \mathcal{P}^*$.*

*Proof.* Let $a, b \in \mathcal{P}$. We have $a \preccurlyeq b$ if and only if $a \in \overline{\{b\}}^{(l)}$, and if and only if $b \in \overline{\{a\}}^{(u)}$. By Theorem 4.30, Parts (a) and (b), each of these statements holds if and only if $\mu(a) \le \mu(b)$ for all $\mu \in \mathcal{P}^*$. □

We proceed to define neighborhoods for subsets of a locally convex cone $(\mathcal{P}, \mathcal{V})$. For a subset $A \subset \mathcal{P}$, a neighborhood $v \in \mathcal{V}$ we define *upper* and *lower relative neighborhoods* as subsets of $\mathcal{P}$ by

$$v(A) = \left\{ b \in \mathcal{P} \;\middle|\; \begin{array}{l} \text{for every } \varepsilon > 0 \text{ there is } a \in A \text{ and } 1 \le \gamma \le 1+\varepsilon \\ \text{such that } b \le \gamma a + (1+\varepsilon)v \end{array} \right\}$$

and

$$(A)v = \left\{ b \in \mathcal{P} \;\middle|\; \begin{array}{l} \text{for every } \varepsilon > 0 \text{ there is } a \in A \text{ and } 1 \le \gamma \le 1+\varepsilon \\ \text{such that } a \le \gamma b + (1+\varepsilon)v \end{array} \right\}.$$

Note that this notation is consistent with the one earlier introduced for elements of $\mathcal{P}$, as we have $a \preccurlyeq b+v$ if and only if $a \in v(\{b\})$ (see Lemma 3.1) and if and only if $b \in (\{a\})v$.

**Lemma 4.32.** *Let $A$ be a subset of a locally convex cone $(\mathcal{P},\mathcal{V})$ and let $v \in \mathcal{V}$.*

*(a) The upper neighborhood $v(A)$ is closed in $\mathcal{P}$ with respect to the lower relative topology.*
*(b) The lower neighborhood $(A)v$ is closed in $\mathcal{P}$ with respect to the upper relative topology.*
*(c) If $A$ is convex, then both $v(A)$ and $(A)v$ are convex.*

*Proof.* Let $A \subset \mathcal{P}$ and let $b \in \overline{v(A)}^{(l)}$. Given $\varepsilon > 0$, set $\delta = \min\{1, \varepsilon/4\}$. According to 4.28(a) there is $c \in v(A)$ such that $b \in v_\delta(c)$, that is $b \leq \gamma c + \delta v$ with some $1 \leq \gamma \leq 1+\delta$. Moreover, we have $c \leq \gamma' a + (1+\delta)v$ for some $a \in A$ and $1 \leq \gamma' \leq 1+\delta$. Thus

$$b \leq (\gamma\gamma')a + \big(\gamma(1+\delta)+\delta\big)v \leq \gamma\gamma' a + \big((1+\delta)^2+\delta\big)v.$$

Since both $1 \leq \gamma\gamma' \leq (1+\delta)^2 \leq 1+\varepsilon$ and $(1+\delta)^2+\delta \leq 1+\varepsilon$, we infer that $b \in v(A)$. Similarly one verifies Part (b) of the Lemma. For Part (c) suppose that $A$ is convex and let $b, b' \in v(A)$ and $b'' = \alpha b + (1-\alpha)b'$ for some $0 \leq \alpha \leq 1$. Given $v \in \mathcal{V}$ and $\varepsilon > 0$ there are $a, a' \in A$ such that $b \leq \gamma a + (1+\varepsilon)v$ and $b' \leq \gamma' a' + (1+\varepsilon)v$ for some $1 \leq \gamma, \gamma' \leq 1+\varepsilon$. Set $\gamma'' = (\alpha\gamma + (1-\alpha)\gamma')$. Then

$$a'' = \frac{\alpha\gamma}{\gamma''}a + \frac{(1-\alpha)\gamma'}{\gamma''}a' \in A$$

and $b'' \leq \gamma'' a'' + (1+\varepsilon)v$. Since $1 \leq \gamma'' \leq 1+\varepsilon$, this demonstrates that $b'' \in v(A)$, and therefore this set is also convex. Similarly one argues for the lower neighborhood $(A)v$. □

**Theorem 4.33.** *Let $A$ be a convex subset of a locally convex cone $(\mathcal{P},\mathcal{V})$, let $v \in \mathcal{V}$ and $b \in \mathcal{P}$. Then*

*(a) $b \in v(A)$ if and only if $\mu(b) \leq \sup\{\mu(a) \mid a \in A\} + 1$ for all $\mu \in v^\circ$.*
*(b) $b \in (A)v$ if and only if $\mu(b) \geq \inf\{\mu(a) \mid a \in A\} - 1$ for all $\mu \in v^\circ$.*

*Proof.* We may again assume that $A \neq \emptyset$. For Part (a), suppose that $b \in v(A)$ and let $\mu \in v^\circ$. Given $\varepsilon \geq 0$ there is $a \in A$ such that $b \leq \gamma a + (1+\varepsilon)v$, hence

$$\mu(b) \leq \gamma\,\mu(a) + (1+\varepsilon) \leq \gamma \sup\{\mu(a) \mid a \in A\} + (1+\varepsilon)$$

for some $1 \leq \gamma \leq 1+\varepsilon$. This shows $\mu(b) \leq \sup\{\mu(a) \mid a \in A\} + 1$ since $\varepsilon > 0$ was arbitrarily chosen. In order to prove the converse implication we consider a full locally convex cone $(\widehat{\mathcal{P}},\mathcal{V})$ containing both $\mathcal{P}$ and the neighborhood system $\mathcal{V}$. Then $A \subset \widehat{\mathcal{P}}$. The lower neighborhood $\hat{v}(A)$ formed in $\widehat{\mathcal{P}}$ is larger than $v(A)$ formed in $\mathcal{P}$, but we have $v(A) = \hat{v}(A) \cap \mathcal{P}$. Thus if $b \notin v(A)$ for $b \in \mathcal{P}$ we also have $b \notin \hat{v}(A)$. Since $\hat{v}(A)$ is a convex subset of $\widehat{\mathcal{P}}$ and closed with respect to the lower relative topology, according to Theorem 4.30 there is $\mu \in \widehat{\mathcal{P}}^*$ such that $\mu(b) > \sup\{\mu(c) \mid c \in v(A)\}$. This implies in particular that $\sup\{\mu(c) \mid c \in v(A)\}$ is finite, and since $a+v \in v(A)$ whenever $a \in A$ we have $\mu(a+v) = \mu(a)+\mu(v) < +\infty$, hence $\mu(v) < +\infty$. If $\mu(v) = 0$, then $\lambda\mu \in v^\circ$ for all $\lambda \geq 0$, and we may choose $\lambda$ such that

$$(\lambda\mu)(b) > \sup\{(\lambda\mu)(c) \mid c \in v(A)\} + 1 \geq \sup\{(\lambda\mu)(a) \mid a \in A\} + 1.$$

If $\mu(v) > 0$, we set $\lambda = 1/\mu(v)$ and have again $\lambda\mu \in v^\circ$. Then for every $a \in A$ we have $a+v \in v(A)$, hence

$$(\lambda\mu)(b) > (\lambda\mu)(a+v) = (\lambda\mu)(a) + 1$$

and therefore

$$(\lambda\mu)(b) > \sup\{(\lambda\mu)(a) \mid a \in A\} + 1.$$

Since the restriction of the functional $\lambda\mu \in \widehat{\mathcal{P}}^*$ to $\mathcal{P}$ is an element of $v^\circ \subset \mathcal{P}^*$, this proves our claim for Part (a). The argument for Part (b) uses the easily verified fact that $b \notin (A)v$ implies that $b+v \notin \overline{A}^{(u)}$. Indeed, if $b+v \in \overline{A}^{(u)}$, then by 4.29(b) for every $\varepsilon > 0$ there is $a \in A$ such that $a \leq \gamma(b+v) + (\varepsilon/2)v$ with some $1 \leq \gamma \leq 1+(\varepsilon/2)$. This shows

$$a \leq \gamma b + \big(\gamma + (\varepsilon/2)\big)v \leq a \leq \gamma b + (1+\varepsilon)v,$$

hence $b \in (A)v$. The remainder of the argument is similar to that in Part (a). □

Theorem 4.33 is a generalization of Theorem 3.2 in [175] as the following corollary will show.

**Corollary 4.34.** *Let $(\mathcal{P}, \mathcal{V})$ be a locally convex cone. Then $a \preccurlyeq b+v$ holds for $a, b \in \mathcal{P}$ and $v \in \mathcal{V}$ if and only if $\mu(a) \leq \mu(b)+1$ for all $\mu \in v^\circ$.*

*Proof.* Let $a, b \in \mathcal{P}$ and $v \in \mathcal{V}$ We have $a \preccurlyeq b+v$ if and only if $a \in v(\{b\})$ and if and only if $b \in (\{a\})v$. By Theorem 4.33, Parts (a) and (b), each of these statements holds if and only if $\mu(a) \leq \mu(b)+1$ for all $\mu \in \mathcal{P}^*$. □

**Corollary 4.35.** *Let $(\mathcal{P}, \mathcal{V})$ be a locally convex cone. Let $a, b \in \mathcal{P}$ and $v \in \mathcal{V}$ such that the element $a$ is $v$-bounded. Then $a \preccurlyeq b+v$ holds if and only if $\mu(a) \leq \mu(b)+1$ for all extreme points $\mu$ of $v^\circ$.*

*Proof.* All left to show is the following: Let $a, b \in \mathcal{P}$ and $v \in \mathcal{V}$ such that $a$ is $v$-bounded. Then $\mu(a) < +\infty$ for all $\mu \in v^\circ$. Thus the function

$$\mu \mapsto \big(\mu(b) - \mu(a)\big) \; : \; v^\circ \to \overline{\mathbb{R}}$$

is affine and continuous with respect to the topology $w(\mathcal{P}^*, \mathcal{P})$ (see Section 2). According to Lemmas II.4.4 and II.4.5 in [100] this function attains its minimum value at some extreme point of $v^\circ$. If $a \not\preccurlyeq b + v$, then according to Corollary 4.34 this minimum value is less than $-1$. Our claim follows. □

*Remark 4.36.* The following counterexample will demonstrate that the statement of Theorem 4.30 does in general not hold true for convex subsets $A \subset \mathcal{P}$ which are closed for the given lower or upper topologies rather than for the (coarser) upper or lower relative topologies. For this, let $(\mathcal{P}, \mathcal{V})$ be the locally convex cone of all continuous real-valued and bounded below functions on the interval $[0, +\infty)$, endowed with the positive constant functions $v > 0$ as its neighborhood system $\mathcal{V}$. (see Example 1.4(e)). Let the subset $A \in \mathcal{P}$ consist of all functions in $g \in \mathcal{P}$ with the following properties: (i) $g(x) \leq x$ for all $x \in [0, +\infty)$, and (ii) there is $M \geq 0$ and $\alpha < 1$ such that $g(x) \leq \alpha x$ for all $x \in [M, +\infty)$. We claim that $A$ is closed in the given lower topology. Indeed, let $f \in \mathcal{P}$ be in the closure of $A$. Then, given $v > 0$ there is $g \in (f)v \cap A$, that is $f \leq g + v$, hence $f(x) \leq g(x) + v \leq x + v$ for all $x \in [0, +\infty)$. Thus $f(x) \leq x$ for all $x \in [0, +\infty)$, since $v > 0$ was arbitrarily chosen, hence (i) holds for $f$. For (ii) let $g \in A$ such that $f(x) \leq g(x) + 1$ for all $x \in [0, +\infty)$, and let $M \geq 0$ and $\alpha < 1$ such that $g(x) \leq \alpha x$ for all $x \in [M, +\infty)$. Choose $N = \max\{M, 2/(1-\alpha)\}$. Then for all $x \in [N, +\infty)$ we have $2/(1-\alpha) \leq x$, hence $1 \leq (1-\alpha)x/2$ and

$$f(x) \leq g(x) + 1 \leq \alpha x + \frac{(1-\alpha)}{2} x \leq \frac{(1+\alpha)}{2} x.$$

Since $(1+\alpha)/2 < 1$, this shows $f \in A$, confirming that $A$ is closed with respect to the lower topology. The set $A \subset \mathcal{P}$ is however not closed with respect to the coarser lower relative topology as the function $f(x) = x$ is contained in $\overline{A}^{(l)}$. Indeed, given $v > 0$ and $\varepsilon > 0$, set $\alpha = 1/(1+\varepsilon) < 1$. Then

$$f(x) = x = (1+\varepsilon)\alpha x \leq (1+\varepsilon)\alpha x + \varepsilon v \qquad \text{for all} \quad x \in [0, +\infty).$$

This shows $g \in (f)v_\varepsilon \cap A \neq \emptyset$, where $g(x) = \alpha x$. We therefore have $\mu(f) \leq \sup\{\mu(g) \mid g \in A\}$ for all $\mu \in \mathcal{P}^*$, but $f \notin A$.

*Examples 4.37.* (a) Let $\mathcal{P} = \overline{\mathbb{R}}$, endowed with the neighborhood system $\mathcal{V} = \{\varepsilon \in \mathbb{R} \mid \varepsilon > 0\}$ (see Example 1.4(a)). For the neighborhood $v = 1$ and $\varepsilon > 0$ the relative neighborhoods of an element $a \in \overline{\mathbb{R}}$ are

$$v_\varepsilon(a) = \big(-\infty, (1+\varepsilon)a + \varepsilon\big] \qquad \text{or} \qquad v_\varepsilon(a) = \big(-\infty, a + \varepsilon\big]$$

if $a \geq 0$ or if $a < 0$, respectively. Thus

$$(a)v_\varepsilon = \left[\tfrac{a-\varepsilon}{1+\varepsilon}, +\infty\right] \qquad \text{or} \qquad (a)v_\varepsilon = \left[a-\varepsilon, +\infty\right]$$

if $a \geq \varepsilon$ or if $a < \varepsilon$, respectively. This yields

$$v_\varepsilon^s(a) = \left[\tfrac{a-\varepsilon}{1+\varepsilon}, (1+\varepsilon)a + \varepsilon\right], \qquad v_\varepsilon^s(a) = \left[a-\varepsilon, (1+\varepsilon)a+\varepsilon\right],$$

or

$$v_\varepsilon^s(a) = \left[a-\varepsilon, a+\varepsilon\right]$$

if $a \geq \varepsilon$, if $0 \leq a < \varepsilon$, or if $a < 0$, respectively. The upper, lower and symmetric relative topologies of $\overline{\mathbb{R}}$ therefore coincide with the corresponding given topologies. (see 1.4(a)). The symmetric relative topology, in particular, is the usual topology on $\mathbb{R}$ with $+\infty$ as an isolated point.

(b) Let $\mathcal{P} = \overline{\mathbb{R}}_+ = \{a \in \overline{\mathbb{R}} \mid a \geq 0\}$, endowed with the neighborhood system $\mathcal{V} = \{0\}$ (see Example 1.4(b)). For the only neighborhood $v = 0 \in \mathcal{V}$ and $\varepsilon > 0$ the relative neighborhoods of an element $a \in \overline{\mathbb{R}}_+$ are

$$v_\varepsilon(a) = \left[0, (1+\varepsilon)a\right], \quad (a)v_\varepsilon = \left[\tfrac{a}{1+\varepsilon}, +\infty\right] \quad \text{and} \quad v_\varepsilon^s(a) = \left[\tfrac{a}{1+\varepsilon}, (1+\varepsilon)a\right].$$

The symmetric relative topology therefore coincides with the Euclidean topology on $(0,+\infty)$, but renders $0 \in \mathcal{P}$ and $+\infty \in \mathcal{P}$ as isolated points. Recall from Example 1.4(b) that the symmetric given topology on $\overline{\mathbb{R}}_+$, in contrast, is the discrete topology. For the boundedness components of $\overline{\mathbb{R}}_+$ we have

$$\mathcal{B}(a) = [0,+\infty), \qquad (a)\mathcal{B} = (0,+\infty] \quad \text{and} \quad \mathcal{B}^s(a) = (0,+\infty)$$

for $a \in (0,+\infty)$,

$$\mathcal{B}(0) = \{0\}, \qquad (0)\mathcal{B} = [0,+\infty] \quad \text{and} \quad \mathcal{B}^s(0) = \{0\},$$

and

$$\mathcal{B}(+\infty) = [0,+\infty], \qquad (+\infty)\mathcal{B} = \{\infty\} \quad \text{and} \quad \mathcal{B}^s(+\infty) = \{\infty\}.$$

As claimed, the symmetric boundedness components furnish a partition of $\mathcal{P} = \overline{\mathbb{R}}_+$ into disjoint subsets that are both open and closed in the symmetric relative topology.

(c) Let us consider Example 1.4(e) with the special insertions for $\mathcal{P} = \overline{\mathbb{R}}$ and the neighborhood system $\widehat{\mathcal{V}}$ generated by a family $\mathcal{Y}$ of subsets of the domain $X$ as elaborated in 1.4(e). Recall that $\widehat{\mathcal{V}}$ is spanned by the $\overline{\mathbb{R}}$-valued functions $\hat{v}_Y \in \widehat{\mathcal{V}}$, corresponding to some $Y \in \mathcal{Y}$, and such that $\hat{v}_Y(x) = 1$ for $x \in Y$ and $\hat{v}_Y(x) = +\infty$, else. Thus $\left(\mathcal{F}_{\widehat{\mathcal{V}}_b}(X,\overline{\mathbb{R}}), \widehat{\mathcal{V}}\right)$ is the locally convex cone of all bounded below (on the sets in $\mathcal{Y}$) $\overline{\mathbb{R}}$-valued functions on $X$, carrying the topology of uniform convergence on the sets in $\mathcal{Y}$. For a function $f \in \mathcal{F}_{\widehat{\mathcal{V}}_b}(X,\overline{\mathbb{R}})$ and a neighborhood $\hat{v}_Y \in \widehat{\mathcal{V}}$, the $\hat{v}_Y$-boundedness component $\mathcal{B}^s_{\hat{v}_Y}(f)$ consists of all $g \in \mathcal{F}_{\widehat{\mathcal{V}}_b}(X, \overline{\mathbb{R}})$ such that

$$\alpha f(x) - \beta \le g(x) \le \gamma f(x) + \delta$$

holds with some constants $\alpha, \beta, \gamma, \delta > 0$ for all $x \in Y$. Thus, obviously, $(\hat{v}_Y)^s_\varepsilon(g) \subset \mathcal{B}^s_{\hat{v}_Y}(f)$ for all $\varepsilon > 0$ whenever $g \in \mathcal{B}^s_{\hat{v}_Y}(f)$. This observation confirms that the component $\mathcal{B}^s_{\hat{v}_Y}(f)$ is both open and closed in the symmetric relative $\hat{v}_Y$-topology, which is the topology of uniform convergence on $Y$. Yet the (global) boundedness component $\mathcal{B}^s(f) = \bigcap_{Y \in \mathcal{Y}} \mathcal{B}^s_{\hat{v}_Y}(f)$ is in general only closed in the symmetric relative topology, which is the topology of uniform convergence on all sets $Y \in \mathcal{Y}$. However, if the set $X$ itself is contained in $\mathcal{Y}$, then the multiples of the neighborhood $\hat{v}_X$ form already a basis for $\widehat{\mathcal{V}}$, and the $\hat{v}_X$-boundedness components coincide with the global ones. Following Proposition 4.22, $\mathcal{F}_{\widehat{\mathcal{V}}_b}(X, \overline{\mathbb{R}})$ is locally connected in this case. Its boundedness components therefore coincide with the components and quasi-components in the symmetric relative topology (Proposition 4.21) and are both open and closed. If, for another special case, $\mathcal{Y}$ consists of all finite subsets of $X$, then for $Y \in \mathcal{Y}$ the above condition yields that two functions $f, g \in \mathcal{F}_{\widehat{\mathcal{V}}_b}(X, \overline{\mathbb{R}})$ are contained in the same $\hat{v}_Y$-boundedness component if and only if they take the value $+\infty$ at exactly the same points of $Y$. The symmetric relative $\hat{v}_Y$-topology is the topology of pointwise convergence on the set $Y$ in this case. Correspondingly, the global boundedness components consist of functions that take the value $+\infty$ at exactly the same points of $X$, and the symmetric relative topology is the topology of pointwise convergence on $X$. If $X$ itself is an infinite set, then the global boundedness components are seen to be closed but not open in this topology.

(d) Let $(\mathcal{P}, \mathcal{V})$ be a locally convex cone and let $\mathcal{Q}$ be the family of all non-empty convex subsets of $\mathcal{P}$ which are closed with respect to the lower relative topology. (See 4.26 to 4.35 before.) If we use the standard multiplication for sets by non-negative scalars and a slightly modified addition, that is

$$A \oplus B = \overline{(A+B)}^{(l)} \qquad \text{for} \quad A, B \in \mathcal{Q},$$

then $\mathcal{Q}$ becomes a cone. Indeed, since the set $A + B$ is obviously again convex, so is its closure with respect to the lower relative topology by Proposition 4.28(b). The neutral element of $\mathcal{Q}$ is given by $\overline{\{0\}}^{(l)} = \{b \in \mathcal{P} \mid b \preccurlyeq 0\}$. We use the set inclusion as the order on $\mathcal{Q}$ and define neighborhoods corresponding to those in $\mathcal{P}$: We set

$$A \le B \oplus v \qquad \text{if} \quad A \subset v(B)$$

for $A, B \in \mathcal{Q}$ and $v \in \mathcal{V}$, that is if for every $a \in A$ and $\varepsilon > 0$ there is $b \in B$ such that $a \le \gamma b + (1+\varepsilon)v$ for some $1 \le \gamma \le 1+\varepsilon$. First we observe that for every $A \in \mathcal{Q}$ and a fixed element $a \in A$ $v \in \mathcal{V}$ there is $\lambda \ge 0$ such that $0 \le a + \lambda v$. Since $\overline{\{0\}}^{(l)} = \{b \in \mathcal{P} \mid b \preccurlyeq 0\}$, this yields $\overline{\{0\}}^{(l)} \le A \oplus (\lambda+1)v$. Indeed, for every $b \preccurlyeq 0$, we have $b \le v$, hence

$b \le a+(\lambda+1)v$. Thus every element $A \in \mathcal{Q}$ is seen to be bounded below and $(\mathcal{Q},\mathcal{V})$ satisfies the requirements for a locally convex cone. Next we observe that the weak preorder on $(\mathcal{Q},\mathcal{V})$ coincides with the given order. Indeed, suppose that $A \preccurlyeq B$, and let $a \in A$. Given $v \in \mathcal{V}$ and $\varepsilon > 0$ we set $\delta = \min\{\varepsilon/3, 1\}$ and have $A \le \gamma B \oplus \delta v$ for some $1 \le \gamma \le 1+\delta$. According to Lemma 3.1 there is $1 \le \gamma' \le 1+\delta$ such that $a \le (\gamma'\gamma)b + (1+\delta)\delta v$ for some $b \in B$. Since $(1+\delta)\delta \le \varepsilon$, this yields $a \le (\gamma'\gamma)b + \varepsilon v$, and since $1 \le \gamma\gamma' \le (1+\delta)^2 \le 1+\varepsilon$, we have $a \in v_\varepsilon(b)$ and infer from (i) that $a \in \overline{B}^{(l)} = B$, hence $A \le B$. Therefore $A \preccurlyeq B$ holds if and only if $A \le B$. A similar argument shows that $A \preccurlyeq B \oplus v$ holds for $A, B \in \mathcal{Q}$ and $v \in \mathcal{V}$ if and only if $A \le B \oplus v$. An element $A \in \mathcal{Q}$ is bounded above in $\mathcal{Q}$ if for every $v \in \mathcal{V}$ there is $\lambda \ge 0$ such that $A \le \lambda v$, that is $a \le (\lambda+1)v$ holds for all $a \in A$, that is if the set $A \subset \mathcal{P}$ is bounded above in $\mathcal{P}$ in the sense of 4.25(ii).

(e) Similarly, but less intuitively we may consider the family $\mathcal{Q}$ of all convex subsets of a locally convex cone $\mathcal{P}$ which are closed with respect to the upper relative topology and bounded below in the sense of 4.25(i). (See 4.26 to 4.35 before.) We use the standard multiplication for sets by non-negative scalars and the addition

$$A \oplus B = \overline{(A+B)}^{(u)} \qquad \text{for} \quad A, B \in \mathcal{Q}.$$

Since the sum of two bounded below convex subsets of $\mathcal{P}$ is obviously again bounded below and convex, Proposition 4.29(b) and (c) guarantees that the set $A \oplus B$ is indeed also bounded below and convex. Thus $\mathcal{Q}$ is a cone with the neutral element $\overline{\{0\}}^{(u)} = \{b \in \mathcal{P} \mid 0 \preccurlyeq b\}$. In this example we use the inverse set inclusion as the order on $\mathcal{Q}$, that is

$$A \le B \qquad \text{if} \quad B \subset A$$

and define neighborhoods corresponding to those in $\mathcal{P}$ by

$$A \le B \oplus v \qquad \text{if} \quad B \subset (A)v$$

for $A, B \in \mathcal{Q}$ and $v \in \mathcal{V}$, that is if for every $b \in B$, and $\varepsilon > 0$ there is $a \in A$ such that $a \le \gamma b + (1+\varepsilon)v$ for some $1 \le \gamma \le 1+\varepsilon$. Because for every $A \in \mathcal{Q}$ and $v \in \mathcal{V}$ there is $\lambda > 0$ such that $0 \le a + \lambda v$ for all $a \in A$, we have $\overline{\{0\}}^{(u)} \le A \oplus \lambda v$, and every element $A \in \mathcal{Q}$ is bounded below. Hence $(\mathcal{Q},\mathcal{V})$ is a locally convex cone. A similar argument than in (d) yields that $(\mathcal{Q},\mathcal{V})$ carries its weak preorder. Note that other than in (d) the empty set is a member of $\mathcal{Q}$, indeed its maximal element. We set $A \oplus \emptyset = \emptyset$, $\alpha \cdot \emptyset = \emptyset$ and $0 \cdot \emptyset = \overline{\{0\}}^{(u)}$ for all $A \in \mathcal{Q}$ and $\alpha > 0$. An element $A \in \mathcal{Q}$ is bounded above in $\mathcal{Q}$ if for every $v \in \mathcal{V}$ there is $\lambda \ge 0$ such that $A \le \lambda v$, that is there is $a \in A$ such that $a \le \lambda v$.

Note that in both Examples (d) and (e) the given locally convex cone $\mathcal{P}$ may be considered as a subcone of $\mathcal{Q}$ via the embedding $a \mapsto \overline{\{a\}}$. This is an embedding of $(\mathcal{P},\mathcal{V})$ into $(\mathcal{Q},\mathcal{V})$ in the sense of 2.2, provided that $\mathcal{P}$ is endowed with the weak preorder, that is $\overline{\{a\}} \leq \overline{\{b\}} + v$ holds if and only if $a \preccurlyeq b+v$ for $a,b \in \mathcal{P}$ and $v \in \mathcal{V}$ (see also 2.2(iii)).

*Remarks 4.38.* (a) As a consequence of the last observation in 4.37(a) and of Proposition 4.5 we infer that a continuous linear functional $\mu$ on a locally convex cone $(\mathcal{P},\mathcal{V})$ is still continuous if we endow $\mathcal{P}$ with its relative topologies. More precisely: Let $\mu \in v^{\circ}$, that is the polar of some neighborhood $v \in \mathcal{V}$. Then $\mu$ is a continuous linear operator from $(\mathcal{P},\mathcal{V}_0)$ to $\overline{\mathbb{R}}$, where $\mathcal{V}_0$ consists of the multiples of the single neighborhood $v$. As shown in 4.37(a), the relative topologies on $\overline{\mathbb{R}}$ coincide with the given ones as described in Example 1.4(a). Thus according to 4.5, the functional $\mu$ is also continuous if we endow $\mathcal{P}$ with either the upper, lower or symmetric relative $v$-topology and, correspondingly, $\overline{\mathbb{R}}$ with its given upper, lower or symmetric topology.

(b) We noted earlier that for a locally convex cone $(\mathcal{P},\mathcal{V})$ the mapping

$$(\alpha,a) \mapsto \alpha a \; : \; [0,+\infty) \times \mathcal{P} \to \mathcal{P},$$

is generally not continuous with respect to any of the given topologies of $\mathbb{R}$ and $\mathcal{P}$. However, if we endow $\mathcal{P}$ with either of the relative topologies, this mapping is continuous at all points $(\alpha,a) \in [0,+\infty) \times \mathcal{P}$ such that either $\alpha > 0$ or $a \in \mathcal{P}$ is bounded. This was established in Proposition 4.2(iii). Now using 4.37(b) we realize that this mapping is continuous at all points of $[0,+\infty) \times \mathcal{P}$ if we consider the symmetric relative topology of $\overline{\mathbb{R}}_+$ (see 4.37(b)) and any of the relative topologies on $\mathcal{P}$ instead. Indeed, the symmetric relative topology of $\overline{\mathbb{R}}_+$ coincides with the usual topology of $\mathbb{R}$ on $(0,+\infty)$, hence continuity at all points $(\alpha,a) \in [0,+\infty) \times \mathcal{P}$ such that $\alpha > 0$ follows from 4.2(iii). Continuity at the points $(0,a)$ for all $a \in \mathcal{P}$, on the other hand is obvious, since $0$ is an isolated element in the symmetric relative topology of $\overline{\mathbb{R}}_+$.

## 5. Locally Convex Lattice Cones

Our upcoming integration theory for cone-valued functions in Chapter II deals with locally convex cones that contain suprema and infima for sufficiently many of their subsets. Let us recall the classical concepts: A *topological vector lattice* is a vector lattice and a locally convex topological vector space $E$ over $\mathbb{R}$ that possesses a neighborhood base of solid sets. (See for example Chapter V.7 in [185], also [132] or [184]. Recall that a subset $A$ of $E$ is called *solid* if $b \in A$ whenever $|b| \leq |a|$ for $b \in E$ and $a \in A$.) Some of the following definitions and results are adaptations of the corresponding classical ones. The presence of unbounded elements and the general unavailability of negatives in locally convex cones, however, requires a more delicate approach.

**5.1 Locally Convex Lattice Cones.** We shall say that a locally convex cone $(\mathcal{P},\mathcal{V})$ is a *locally convex* $\vee$*-semilattice cone* if its order is antisymmetric and if for any two elements $a,b\in\mathcal{P}$ their supremum $a\vee b$ exists in $\mathcal{P}$ and if

($\vee$1) *$(a+c)\vee(b+c)=a\vee b+c$ holds for all $a,b,c\in\mathcal{P}$.*
($\vee$2) *$a\le c+v$ and $b\le c+w$ for $a,b,c\in\mathcal{P}$ and $v,w\in\mathcal{V}$ implies that $a\vee b\le c+(v+w)$.*

Likewise, $(\mathcal{P},\mathcal{V})$ is a *locally convex* $\wedge$*-semilattice cone* if its order is antisymmetric and if for any two elements $a,b\in\mathcal{P}$ their infimum $a\wedge b$ exists in $\mathcal{P}$ and if

($\wedge$1) *$(a+c)\wedge(b+c)=a\wedge b+c$ holds for all $a,b,c\in\mathcal{P}$.*
($\wedge$2) *$c\le a+v$ and $c\le b+w$ for $a,b,c\in\mathcal{P}$ and $v,w\in\mathcal{V}$ implies that $c\le a\wedge b+(v+w)$.*

If both sets of the above conditions hold, then $(\mathcal{P},\mathcal{V})$ is called a *locally convex lattice cone.* In case that $(\mathcal{P},\mathcal{V})$ is indeed a locally convex topological vector space, the existence of suprema implies the existence of infima and vice versa, as $a\wedge b=-\big((-a)\vee(-b)\big)$. Conditions ($\vee$1) and ($\vee$2) then are equivalent to ($\wedge$1) and ($\wedge$2) and consistent with the above mentioned definition of a topological vector lattice. Indeed, $a\le c+v$ and $b\le c+w$ means that $a\le c+s$ $b\le c+t$ in this case, for some elements $s$ and $t$ of the neighborhoods $v$ and $w$, respectively. Because these neighborhoods are supposed to be solid, we have $s\vee 0\le v$ and $t\vee 0\le w$ as well. Now $a\le c+s\vee 0+t\vee 0$ and $b\le c+s\vee 0+t\vee 0$ implies

$$a\vee b\le c+s\vee 0+t\vee 0\le c+(v+w)$$

as required in ($\vee$1).

**Proposition 5.2.** *Let $(\mathcal{P},\mathcal{V})$ be a locally convex $\vee$- (or $\wedge$-) semilattice cone. The lattice operation $(a,b)\mapsto a\vee b$ (or $(a,b)\mapsto a\wedge b$) is a continuous mapping from $\mathcal{P}\times\mathcal{P}$ to $\mathcal{P}$ if $\mathcal{P}$ is endowed with the symmetric relative topology.*

*Proof.* Suppose that $(\mathcal{P},\mathcal{V})$ is a locally convex $\vee$-semilattice cone, and let $a\in v_\varepsilon(b)$ and $c\in v_\varepsilon(d)$ for $a,b,c,d\in\mathcal{P}$, $v\in\mathcal{V}$ and $\varepsilon>0$. There is $\lambda\ge 0$ such that both $0\le b+\lambda v$ and $0\le d+\lambda v$. Then $a\le(1+\varepsilon)b+\varepsilon(1+\lambda)v$ and $c\le(1+\varepsilon)d+\varepsilon(1+\lambda)v$ by Lemma 4.1(b). Thus

$$a\le(1+\varepsilon)(b\vee d)+\varepsilon(1+\lambda)v \qquad\text{and}\qquad c\le(1+\varepsilon)(b\vee d)+\varepsilon(1+\lambda)v,$$

hence

$$a\vee c\le(1+\varepsilon)(b\vee d)+2\varepsilon(1+\lambda)v$$

by ($\vee$2). This shows $a\vee c\in v_{(2\varepsilon(1+\lambda))}(b\vee d)$. Similarly, using 4.1(c) one verifies that $a\in(b)v_\varepsilon$ and $c\in(d)v_\varepsilon$ implies $a\vee c\in(b\vee d)v_{(2\varepsilon(1+\lambda+\varepsilon))}(b\vee d)$.

Combining these observations for both the upper and lower relative neighborhoods then demonstrates that $a \in v^s_\varepsilon(b)$ and $c \in v^s_\varepsilon(d)$ implies $a \vee c \in v^s_{(2\varepsilon(1+\lambda+\varepsilon))}(b \vee d)$, hence our claim. A similar argument yields our claim for locally convex $\wedge$-semilattice cones. □

**Proposition 5.3.** *Let* $(\mathcal{P}, \mathcal{V})$ *be a locally convex lattice cone. Then* $a + b = a \vee b + a \wedge b$ *for all* $a, b \in \mathcal{P}$.

*Proof.* We observe that

$$a + b \leq \inf \{a + a \vee b \, , \, b + a \vee b\} = a \wedge b + a \vee b$$

by $(\wedge 1)$, and by $(\vee 1)$

$$a \vee b + a \wedge b = \sup \{a + a \wedge b \, , \, b + a \wedge b\} \leq a + b.$$

As the order of $\mathcal{P}$ is supposed to by antisymmetric, this yields our claim. □

Proposition 5.3 implies in particular that $a = a \vee 0 + a \wedge 0$ for all elements $a$ of a locally convex lattice cone.

Examples of locally convex lattice cones include classical topological vector lattices and the cones $\overline{\mathbb{R}}$ and $\overline{\mathbb{R}}_+$ from Examples 1.4(a) and (b). If $(\mathcal{P}, \mathcal{V})$ is a locally convex $\vee$- or $\wedge$-semilattice cone, if $X$ is a set, and if $\widehat{\mathcal{V}}$ is a neighborhood system consisting of $(\mathcal{V} \cup \{\infty\})$-valued functions on $X$, then the locally convex cone $\big(\mathcal{F}_{\widehat{\mathcal{V}}_b}(X, \mathcal{P}), \hat{\mathcal{V}}\big)$ of $\mathcal{P}$-valued functions from Example 1.4(e) is also a semilattice cone of the same type. Suprema and infima are formed pointwise in this case. The cones $(\overline{l^p}, \mathcal{V}_p)$ from 1.4(f) are locally convex lattice cones. The locally convex cone of all non-empty convex subsets of some locally convex topological vector space $E$ (see Example 1.4(c)) is antisymmetrically ordered by set inclusion (we assume that equality is the order in $E$) and indeed a $\vee$-semilattice cone. The supremum of two convex subsets of $E$ is the convex hull of their union while infima, that is intersections, do not always exist. Requirements $(\vee 1)$ and $(\vee 2)$ are readily checked.

**5.4 Locally Convex Complete Lattice Cones.** Later in this text, in particular when developing our integration theory, we shall consider substantially stronger properties concerning the lattice operations of a locally convex cone. We shall require the existence of suprema and infima for bounded and bounded below subsets, respectively. This assumption corresponds to the notion of order completeness for ordered vector spaces. Moreover, the upper or lower neighborhoods are supposed to be closed for suprema or infima of their subsets, respectively. This requirement corresponds to the properties of M-topologies in locally convex vector lattices.

We shall say that a locally convex cone $(\mathcal{P}, \mathcal{V})$ is a *locally convex* $\bigvee$-*semilattice cone* if $\mathcal{P}$ carries the weak preorder (that is the given order coincides with the weak preorder for the elements and the neighborhoods in $\mathcal{P}$), this order is antisymmetric and if

($\bigvee$1) *Every non-empty subset* $A \subset \mathcal{P}$ *has a supremum* $\sup A \in \mathcal{P}$ *and* $\sup(A+b) = \sup A + b$ *holds for all* $b \in \mathcal{P}$.

($\bigvee$2) *Let* $\emptyset \neq A \subset \mathcal{P}$, $b \in \mathcal{P}$ *and* $v \in \mathcal{V}$. *If* $a \leq b + v$ *for all* $a \in A$, *then* $\sup A \leq b + v$.

In particular, every $\bigvee$-semilattice cone $\mathcal{P}$ contains a largest element, that is $+\infty = \sup \mathcal{P}$, which can be adjoined as a maximal element to any locally convex cone with the convention that $a + \infty = +\infty$, $\alpha \cdot (+\infty) = +\infty$, $0 \cdot (+\infty) = 0$ and that $a \leq +\infty$ for all $a \in \mathcal{P}$ and $\alpha > 0$. Likewise, $(\mathcal{P}, \mathcal{V})$ is said to be a *locally convex* $\bigwedge$*-semilattice cone* if $\mathcal{P}$ carries the weak preorder, this order is antisymmetric and if

($\bigwedge$1) *Every subset* $A \subset \mathcal{P}$ *that is bounded below has an infimum* $\inf A \in \mathcal{P}$ *and* $\inf(A+b) = \inf A + b$ *holds for all* $b \in \mathcal{P}$.

($\bigwedge$2) *Let* $A \subset \mathcal{P}$ *be bounded below,* $b \in \mathcal{P}$ *and* $v \in \mathcal{V}$. *If* $b \leq a + v$ *for all* $a \in A$, *then* $b \leq \inf A + v$.

These requirements are obviously stronger then the corresponding ones in 4.23, so every locally convex $\bigvee$- (or $\bigwedge$-) semilattice cone is also a $\vee$- (or $\wedge$-) semilattice cone. The assumptions ($\bigvee$2) and ($\bigwedge$2) signify that the upper or lower neighborhoods in $\mathcal{P}$ are closed for suprema or infima of their subsets, respectively. If $(\mathcal{P}, \mathcal{V})$ is a full cone, then ($\bigvee$2) is evident, and ($\bigwedge$2) follows from ($\bigwedge$1). Recall from our convention in 4.24(i) that a subset $A$ of $\mathcal{P}$ is said to be *bounded below* if for every $v \in \mathcal{V}$ there is $\lambda \geq 0$ such that $0 \leq a + \lambda v$ for all $a \in A$. This condition does in general not imply the existence of a lower bound in $\mathcal{P}$. However, if $A$ has a lower bound $b \in \mathcal{P}$, that is $b \leq a$ for all $a \in A$, then $A$ is bounded below in the above sense. Indeed, for every $v \in \mathcal{V}$ there is $\lambda \geq 0$ such that $0 \leq b + \lambda v$, hence $0 \leq a + \lambda v$ holds for all $a \in A$. Note that the empty set $\emptyset \subset \mathcal{P}$ is bounded below, and we have $\inf \emptyset = +\infty$ (see the remark above).

Combining both of the above notions, we shall say that a locally convex cone $(\mathcal{P}, \mathcal{V})$ is a *locally convex complete lattice cone* if $\mathcal{P}$ is both a $\bigvee$-semilattice cone and a $\bigwedge$-semilattice cone.

Corresponding to a family $\{A_i\}_{i \in \mathcal{I}}$ of non-empty subsets of a locally convex $\bigvee$-semilattice cone $\mathcal{P}$ we denote the subset

$$\bigvee_{i \in \mathcal{I}} A_i = \left\{ \bigvee_{i \in \mathcal{I}} a_i \;\middle|\; (a_i)_{i \in \mathcal{I}} \in \prod_{i \in \mathcal{I}} A_i, \right\} \subset \mathcal{P}.$$

**Lemma 5.5.** *Let* $(\mathcal{P}, \mathcal{V})$ *be a locally convex* $\bigvee$*-semilattice cone. Let* $A, B$ *and* $\{A_i\}_{i \in \mathcal{I}}$ *be non-empty subsets of* $\mathcal{P}$. *Then*

*(a)* $\sup(A+B) = \sup A + \sup B$.

*(b)* $\sup\left(\bigcup_{i \in \mathcal{I}} A_i\right) = \sup\left(\bigvee_{i \in \mathcal{I}} A_i\right) = \sup_{i \in \mathcal{I}} \{ \sup A_i \mid i \in \mathcal{I} \}$.

*Proof.* For Part (b) we observe that for every $a \in \bigcup_{i \in \mathcal{I}} A_i$ there is $(a_i)_{i \in \mathcal{I}} \in \prod_{i \in \mathcal{I}} A_i$ such that $a$ is one of the projections of $(a_i)_{i \in \mathcal{I}}$ onto the factor spaces $A_i$. This yields $a \le \bigvee_{i\in\mathcal{I}} a_i$, hence

$$\sup\Big(\bigcup_{i\in\mathcal{I}} A_i\Big) \le \sup\Big(\bigvee_{i\in\mathcal{I}} A_i\Big).$$

For every $(a_i)_{i\in\mathcal{I}} \in \prod_{i\in\mathcal{I}} A_i$ on the other hand, we have $a_i \le \sup A_i$ for all $i \in \mathcal{I}$, hence $\bigvee_{i\in\mathcal{I}} a_i \le \sup_{i\in\mathcal{I}} \{\sup A_i \mid i \in \mathcal{I}\}$ and

$$\sup\Big(\bigvee_{i\in\mathcal{I}} A_i\Big) \le \sup_{i\in\mathcal{I}} \{\sup A_i \mid i \in \mathcal{I}\}.$$

Finally, since $\sup A_i \le \sup\Big(\bigcup_{i\in\mathcal{I}} A_i\Big)$ holds for all $i \in \mathcal{I}$, we infer that

$$\sup_{i\in\mathcal{I}} \{\sup A_i \mid i \in \mathcal{I}\} \le \sup\Big(\bigcup_{i\in\mathcal{I}} A_i\Big).$$

Our claim in Part (b) now follows from the requirement that the order in $\mathcal{P}$ is antisymmetric.

For Part (a) we argue as follows: If $A$ and $B$ are non-empty subsets of $\mathcal{P}$, then we use Part (b) and $(\bigvee 1)$ for

$$\begin{aligned}
\sup(A+B) &= \sup\Big(\bigcup_{b\in B}(A+b)\Big)\\
&= \sup_{b\in B}\{\sup(A+b) \mid b \in B\}\\
&= \sup_{b\in B}\{\sup A + b \mid b \in B\}\\
&= \sup A + \sup B. \quad \square
\end{aligned}$$

Similarly, for a family $\{A_i\}_{i\in\mathcal{I}}$ of subsets of a locally convex $\bigwedge$-semilattice cone such that $\bigcup_{i\in\mathcal{I}} A_i$ is bounded below in $\mathcal{P}$ we denote

$$\bigwedge_{i\in\mathcal{I}} A_i = \Big\{\bigwedge_{i\in\mathcal{I}} a_i \;\Big|\; (a_i)_{i\in\mathcal{I}} \in \prod_{i\in\mathcal{I}} A_i,\Big\} \subset \mathcal{P}$$

and obtain in analogy to Lemma 5.5:

**Lemma 5.6.** *Let $(\mathcal{P},\mathcal{V})$ be a locally convex $\bigwedge$-semilattice cone. Let $A, B$ and $\{A_i\}_{i\in\mathcal{I}}$ be bounded below subsets of $\mathcal{P}$ and suppose that $\bigcup_{i\in\mathcal{I}} A_i$ is also bounded below. Then*

*(a)* $\inf(A+B) = \inf A + \inf B$.
*(b)* $\inf\Big(\bigcup_{i\in\mathcal{I}} A_i\Big) = \inf\Big(\bigwedge_{i\in\mathcal{I}} A_i\Big) = \inf_{i\in\mathcal{I}}\{\inf A_i \mid i \in \mathcal{I}\}$.

*Remarks and Examples 5.7.* (a) Every locally convex $\bigwedge$-semilattice cone $\mathcal{P}$ contains also suprema for all of its non-empty subsets. Indeed, the set of all upper bounds for a non-empty subset $A$ of $\mathcal{P}$ is bounded below, and its infimum is the supremum of $A$ in $\mathcal{P}$. Requirement $(\bigvee 1)$ does however not necessarily follow (see Example (e) below). Likewise, every locally convex $\bigvee$-semilattice cone has infima for subsets with lower bounds in $\mathcal{P}$. (Recall the before mentioned subtle difference between "bounded below" and "having a lower bound".) But again, requirement $(\bigwedge 1)$ does not follow (see (d) below).

(b) The locally convex cones $\overline{\mathbb{R}}$ and $\overline{\mathbb{R}}_+$ (Examples 1.4(a) and (b)) are of course complete lattices.

(c) If $(\mathcal{P},\mathcal{V})$ is a locally convex $\bigvee$-semilattice (or $\bigwedge$-semilattice) lattice cone, if $X$ is a set, and if $\widehat{\mathcal{V}}$ is a neighborhood system consisting of $(\mathcal{V}\cup\{\infty\})$-valued functions (see Example 1.4(e)), then the locally convex cone $(\mathcal{F}_{\widehat{\mathcal{V}}_b}(X,\mathcal{P}),\hat{\mathcal{V}})$ of $\mathcal{P}$-valued functions from 1.4(e) is also a locally convex $\bigvee$-semilattice (or $\bigwedge$-semilattice) lattice cone, provided that for every $x \in X$ and $v \in \mathcal{V}$ there is $\hat{v} \in \widehat{\mathcal{V}}$ such that $\hat{v}(x) \le v$. (Using Lemma 3.2, this condition guarantees that $(\mathcal{F}_{\widehat{\mathcal{V}}_b}(X,\mathcal{P}),\hat{\mathcal{V}})$ carries its weak preorder.) Suprema and infima are formed pointwise. For $\mathcal{P} = \overline{\mathbb{R}}$ in particular, $(\mathcal{F}_{\widehat{\mathcal{V}}_b}(X,\overline{\mathbb{R}}),\hat{\mathcal{V}})$ is a locally convex complete lattice cone, provided that for each $x \in X$ there is $\hat{v} \in \widehat{\mathcal{V}}$ such that $\hat{v}(x) < +\infty$.

(d) Example 4.37(d) yields a locally convex $\bigvee$-semilattice cone. The cone $(Q,\mathcal{V})$ of all non-empty closed (with respect to the lower relative topology) convex subsets of a locally convex cone $(\mathcal{P},\mathcal{V})$ is ordered by set inclusion and carries the weak preorder which is antisymmetric (see 4.37(d)). For a non-empty family $\mathcal{A} \subset \mathcal{Q}$ its supremum is given by

$$\sup \mathcal{A} = \overline{conv\Big(\bigcup_{A\in\mathcal{A}} A\Big)}^{(l)},$$

where $conv(C)$ denotes the convex hull of a set $C \subset \mathcal{P}$, and the closure is meant with respect to the lower relative topology of $\mathcal{P}$. Condition $(\bigvee 1)$ can be readily checked: Let $B \in \mathcal{P}$. Clearly $A \oplus B \subset \sup\mathcal{A} \oplus B$ for all $A \in \mathcal{A}$, hence $\sup\{A \oplus B \mid A \in \mathcal{A}\} \le \sup\mathcal{A} \oplus B$. For the converse inequality let $c \in \sup\mathcal{A} \oplus B = \overline{(conv(\bigcup_{A\in\mathcal{A}} A) + B)}^{(l)}$. Then for every lower relative neighborhood $(c)v_\varepsilon$ there is $d \in (c)v_\varepsilon \cap \big(conv(\bigcup_{A\in\mathcal{A}} A) + B\big)$. This means $d = \sum_{i=1}^n \alpha_i a_i + b$ for some $a_i \in A_i \in \mathcal{A}$, $b \in B$ and $0 \le \alpha_i$ such that $\sum_{i=1}^n \alpha_i = 1$. Thus $d = \sum_{i=1}^n \alpha_i(a_i + b) \in \sup\{A \oplus B \mid A \in \mathcal{A}\}$. This implies $c \in \sup\{A \oplus B \mid A \in \mathcal{A}\}$ as well, since this set is closed in the lower topology. Our claim follows.

(e) A similar argument shows that Example 4.37(e) yields a locally convex $\bigwedge$-semilattice cone. In this case $\mathcal{Q}$ consists of all bounded below closed (with respect to the upper relative topology) convex subsets of $\mathcal{P}$ and is ordered by the inverse set inclusion. For a bounded below family $\mathcal{A} \subset \mathcal{Q}$ its infimum is given by

$$\inf \mathcal{A} = \overline{conv\Big(\bigcup_{A\in\mathcal{A}} A\Big)}^{(u)},$$

where the closure is meant with respect to the upper relative topology of $\mathcal{P}$. It is easily checked that for such a bounded below family $\mathcal{A} \subset \mathfrak{Q}$ the convex hull of its union is again a bounded below subset of $\mathcal{P}$, hence by Proposition 4.28 also the closure of the latter with respect to the upper relative topology.

(f) Let $X$ be a topological space, and let $\mathcal{P}$ be the cone of all $\overline{\mathbb{R}}$-valued lower semicontinuous functions on $X$, where $\overline{\mathbb{R}}$ is endowed with the usual, that is the one-point compactification topology. $\mathcal{P}$ is endowed with the pointwise operations and order and neighborhoods $v \in \mathcal{V}$ for $\mathcal{P}$ are given by the strictly positive constant functions. Because the pointwise infimum of any two functions as well as the pointwise supremum of any non-empty family of functions in $\mathcal{P}$ is again an $\overline{\mathbb{R}}$-valued and lower semicontinuous function, $(\mathcal{P}, \mathcal{V})$ forms a locally convex lattice as well as a $\bigvee$-semilattice cone, however in general not a locally convex complete lattice cone. Similarly, the cone of all $\overline{\mathbb{R}}$-valued bounded below upper semicontinuous functions on $X$ forms a locally convex lattice and $\bigwedge$-semilattice cone.

**5.8 Zero Components.** Throughout the following we shall assume that $(\mathcal{P}, \mathcal{V})$ is a locally convex $\bigwedge$-semilattice cone. We define the *zero component* of an element $a$ of a locally convex $\bigwedge$-semilattice cone $\mathcal{P}$ by

$$\mathfrak{O}(a) = \inf\left\{b \geq 0 \mid a \in \mathcal{B}(b)\right\}.$$

This expression is well defined, and $\mathfrak{O}(a) \geq 0$ for all $a \in \mathcal{P}$. Recall from Proposition 4.10 that $a \in \mathcal{B}(b)$ if and only if for every $v \in \mathcal{V}$ there are $\alpha, \beta \geq 0$ such that $a \leq \alpha b + \beta v$. If $(a)\mathcal{B}$ does not contain a positive element, then $\mathfrak{O}(a) = \inf \emptyset = +\infty \in \mathcal{P}$.

The introduction of zero components is especially useful for the investigation of variations of the cancellation law in $\bigwedge$-semilattice cones.

**Proposition 5.9.** *Let $(\mathcal{P}, \mathcal{V})$ be a locally convex $\bigwedge$-semilattice cone, and let $a, b, c \in \mathcal{P}$ and $v \in \mathcal{V}$. If $a + c \preccurlyeq_v b + c$, then $a \preccurlyeq_v b + \mathfrak{O}(c)$.*

*Proof.* Let $a, b, c \in \mathcal{P}$ and $v \in \mathcal{V}$ and suppose that $a + c \preccurlyeq_v b + c$. As we observed before, the weak local preorder $\preccurlyeq_v$ is compatible with the algebraic operations in $\mathcal{P}$. Following Lemma I.4.1 in [100] if applied to this order, the above implies $a + \rho c \preccurlyeq b + \rho c$ for all $\rho > 0$. There is $\lambda > 0$ such that both $0 \leq b + \lambda v$ and $0 \leq c + \lambda v$. Thus $0 \leq (b + \rho c) + 2\lambda v$ for all $0 < \rho \leq 1$. Next we recall that $a + \rho c \preccurlyeq_v b + \rho c$ means that $a + \rho c \in v_\varepsilon(b + \rho c)$ for all $\varepsilon > 0$. Using Lemma 4.1(b) we infer that

$$a + \rho c \leq (1 + \varepsilon)(b + \rho c) + \varepsilon(1 + 2\lambda)v$$

holds for all $\varepsilon > 0$ and $0 < \rho \leq 1$. Thus

$$a \leq a + \rho(c + \lambda v) \leq (1 + \varepsilon)(b + \rho c) + (\varepsilon + 2\varepsilon\lambda + \rho\lambda)v.$$

Let $d \geq 0$ such that $c \in \mathcal{B}(d)$. Then $c \leq \alpha d + \beta v$ holds for some $\alpha, \beta \geq 0$. Consequently, for all $\rho > 0$ such that $\rho \leq \max\{(\varepsilon/\lambda), (1/\alpha), (2\varepsilon/\beta)\}$ we have

$$\rho c \leq (\rho\alpha)d + (\rho\beta)v \leq d + 2\varepsilon v$$

since $d \geq 0$, and

$$(\varepsilon + 2\varepsilon\lambda + \rho\lambda)v \leq 2\varepsilon(1+\lambda)v,$$

hence

$$a \leq (1+\varepsilon)(b+\rho c) + 2\varepsilon(1+\lambda)v \leq (1+\varepsilon)(b+d) + 2\varepsilon(2+\lambda)v.$$

Now we may use rules $(\bigwedge 1)$ and $(\bigwedge 2)$ and take the infimum over the right-hand side of this inequality with respect to all $d \geq 0$ such that $c \in \mathcal{B}(d)$. This yields

$$a \leq (1+\varepsilon)\big(b + \mathfrak{O}(c)\big) + 2\varepsilon(2+\lambda)v.$$

This last inequality holds true for all $\varepsilon > 0$ and therefore demonstrates

$$a \preccurlyeq_v b + \mathfrak{O}(c). \qquad \square$$

**Proposition 5.10.** *Let $(\mathcal{P}, \mathcal{V})$ be a locally convex $\bigwedge$-semilattice cone, and let $a, b, c \in \mathcal{P}$.*

*(a) If $a + c \leq b + c$, then $a \leq b + \mathfrak{O}(c)$.*
*(b) If $a \in \mathcal{B}(b)$, then $\mathfrak{O}(a) \leq \mathfrak{O}(b)$*
*(c) If $a$ is bounded, then $\mathfrak{O}(a) = 0$.*

*Proof.* Let $a, b, c \in \mathcal{P}$. Recall that $a \preccurlyeq b$, that is $a \leq b$ in the case of a completely ordered locally convex cone which is supposed to carry its weak global preorder, means that $a \preccurlyeq_v b$ holds for all $v \in \mathcal{V}$. This yields Part (a) as an immediate consequence of 5.9. For Part (b) suppose that $a \in \mathcal{B}(b)$. Then for every $c \geq 0$ such that $b \in \mathcal{B}(c)$ we have $\mathcal{B}(b) \subset \mathcal{B}(c)$ by 4.10(ii), hence $a \in \mathcal{B}(c)$ as well. This yields $\mathfrak{O}(a) \leq \mathfrak{O}(b)$. Part (c) follows from Part (b) with $b = 0$. $\square$

**Proposition 5.11.** *Let $(\mathcal{P}, \mathcal{V})$ be a locally convex $\bigwedge$-semilattice cone, and let $a, b, \in \mathcal{P}$. Then*

*(a) $\mathfrak{O}(a+b) = \mathfrak{O}(a) + \mathfrak{O}(b)$.*
*(b) $\mathfrak{O}(\alpha a) = \alpha\mathfrak{O}(a) = \mathfrak{O}(a)$ for all $\alpha > 0$.*
*(c) If $\alpha a = a$ for all $\alpha > 0$, then $\mathfrak{O}(a) = a$.*

*Proof.* Let $a, b, \in \mathcal{P}$. Part (b) is obvious since for every $\alpha > 0$ and every $c \in \mathcal{P}$ we have $\alpha a \in \mathcal{B}(c)$ if and only if $a \in \mathcal{B}(c)$ by 4.11(a). For Part (a) let $a \in \mathcal{B}(c)$ and $b \in \mathcal{B}(d)$ for $c, d \geq 0$. Then $a + b \in \mathcal{B}(c+d)$ by 4.11(c). This shows $\mathfrak{O}(a+b) \leq \mathfrak{O}(a) + \mathfrak{O}(b)$. For the converse, given $v \in \mathcal{V}$ there is $\lambda \geq 0$ such that $0 \leq b + \lambda v$. Hence $a \leq (a+b) + \lambda v$, and we infer that $a \in \mathcal{B}(a+b)$. Thus $\mathfrak{O}(a) \leq \mathfrak{O}(a+b)$ by 5.10(b), and likewise $\mathfrak{O}(b) \leq \mathfrak{O}(a+b)$. This yields

$$\mathfrak{O}(a) + \mathfrak{O}(b) \leq 2\mathfrak{O}(a+b) = \mathfrak{O}(a+b).$$

For Part (c) let $a \in \mathcal{P}$ such that $\alpha a = a$ for all $\alpha \geq 0$. For every $v \in \mathcal{V}$ there is $\lambda > 0$ such that $0 \leq a + \lambda v$, hence $0 \leq (1/\lambda)a + v = a + v$. This shows $0 \leq a$, since $\mathcal{P}$ carries the weak preorder. Thus $\mathfrak{O}(a) \leq a$. If on the other hand $a \in \mathcal{B}(c)$ for some $c \geq 0$, then there are $\alpha, \beta \geq 0$ such that $a \leq \alpha c + \beta v$. Since $\varepsilon \alpha c \leq c$ for all $0 < \varepsilon \leq 1/(\alpha+1)$, this implies

$$a = \varepsilon a \leq \varepsilon \alpha c + \varepsilon \beta v \leq c + \varepsilon \beta v$$

for all such $\varepsilon$. This yields $a \leq c$ since $\mathcal{P}$ carries the weak preorder, and we also have $a \leq \mathfrak{O}(a)$. □

Proposition 5.11(b) implies in particular that a linear functional $\mu \in \mathcal{P}^*$ can attain only the values $0$ or $+\infty$ at a zero component.

Some additional properties can be derived if $\mathcal{P}$ contains also suprema of its elements, that is if $(\mathcal{P}, \mathcal{V})$ is also a locally convex lattice or indeed a locally convex complete lattice cone (see Example 5.7(f)).

**Lemma 5.12.** *Suppose $(\mathcal{P}, \mathcal{V})$ is a locally convex lattice and $\bigwedge$-semilattice cone. Then the zero component of an element $a \in \mathcal{P}$ can be alternatively expressed as*

$$\mathfrak{O}(a) = \inf_{\varepsilon > 0} \big\{ \varepsilon \, (a \vee 0) \big\}.$$

*Proof.* Let $a \in \mathcal{P}$. Then $0 \leq a \vee 0$ and $a \leq a \vee b$. Thus $a \in \mathcal{B}(a \vee b)$. This implies $a \in \mathcal{B}\big(\varepsilon\,(a \vee 0)\big)$ for all $\varepsilon > 0$ by 4.11(c). Hence $\inf \big\{ b \geq 0 \mid a \in \mathcal{B}(b) \big\} \leq \inf_{\varepsilon > 0} \big\{ \varepsilon\,(a \vee 0) \big\}$. For the converse inequality let $b \geq 0$ such that $a \in \mathcal{B}(b)$. Given $v \in \mathcal{V}$ and $\varepsilon > 0$ there are $\alpha, \beta \geq 0$ such that $a \leq \alpha b + \beta v$ (see 4.10(iii)). Condition ($\vee$2) then yields $a \vee 0 \leq \alpha b + 2\beta v$. Thus for $0 < \delta \leq \min\{\frac{\varepsilon}{2\beta+1}, \frac{1}{\alpha+1}\}$ we have

$$\delta(a \vee 0) \leq \delta \alpha b + 2\delta\beta v \leq b + \varepsilon v,$$

since $b \geq 0$ and $\delta\alpha \leq 1$ implies $(\delta\alpha)b \leq b$. This shows $\inf_{\varepsilon > 0} \big\{ \varepsilon\,(a \vee 0) \big\} \leq b + \varepsilon v$, hence

$$\inf_{\varepsilon > 0} \big\{ \varepsilon\,(a \vee 0) \big\} \leq \inf \big\{ b \geq 0 \mid a \in \mathcal{B}(b) \big\} + \varepsilon v$$

by $(\bigwedge 2)$. Because this holds for all $v \in \mathcal{V}$ and for all $\varepsilon > 0$, and because $\mathcal{P}$ carries the weak preorder, we conclude that

$$\inf_{\varepsilon > 0} \big\{ \varepsilon\,(a \vee 0) \big\} \leq \inf \big\{ b \geq 0 \mid a \in \mathcal{B}(b) \big\}.$$

□

**Proposition 5.13.** *Suppose $(\mathcal{P}, \mathcal{V})$ is a locally convex lattice and $\bigwedge$-semilattice cone. Let $a, b, c \in \mathcal{P}$ and $v \in \mathcal{V}$.*

*(a) If $a \in \mathcal{B}_v(b)$, then $\mathfrak{O}(a) \preccurlyeq_v \mathfrak{O}(b)$ and $b + \mathfrak{O}(a) \preccurlyeq_v b$.*
*(b) If $a$ is $v$-bounded, then $\mathfrak{O}(a) \preccurlyeq_v 0$.*

*Proof.* Let $a, b, c \in \mathcal{P}$ and $v \in \mathcal{V}$. For Part (a) , suppose that $a \in \mathcal{B}_v(b)$. There are $\alpha, \beta > 0$ such that $a \leq \alpha b + \beta v$ and $\lambda \geq 0$ such that $0 \leq a + \lambda v$, hence also $0 \leq a \wedge 0 + \lambda v$ by $(\bigwedge 2)$. Then

$$\begin{aligned} b + \mathfrak{O}(a) &\leq b + \varepsilon\,(a \vee 0) \\ &\leq b + \varepsilon\,(a \vee 0) + \varepsilon\big((a \wedge 0) + \lambda v\big) = b + \varepsilon a + \varepsilon \lambda v \\ &\leq (1 + \varepsilon\alpha) b + \varepsilon(\beta + \lambda) v. \end{aligned}$$

for all $\varepsilon > 0$ by Lemma 5.12. This shows $b + \mathfrak{O}(a) \preccurlyeq_v b$. Furthermore, using the cancellation rule from Proposition 5.9 for the element $b$ in $\mathfrak{O}(a) + b \preccurlyeq_v 0 + b$ yields $\mathfrak{O}(a) \preccurlyeq_v \mathfrak{O}(b)$ as claimed. Part (b) follows from Part (a) with $b = 0$. □

**Proposition 5.14.** *Suppose* $(\mathcal{P}, \mathcal{V})$ *is a locally convex lattice and* $\bigwedge$-*semilattice cone. Then* $b + \mathfrak{O}(a) = b$ *holds for all* $a, b \in \mathcal{P}$ *whenever* $a \in \mathcal{B}(b)$.

*Proof.* Let $a, b \in \mathcal{P}$ such that $a \in \mathcal{B}(b)$. Then $a \in \mathcal{B}_v(b)$, hence $b + \mathfrak{O}(a) \preccurlyeq_v b$ by Proposition 5.13, for all $v \in \mathcal{V}$. Thus $b + \mathfrak{O}(a) \preccurlyeq b$. Since $\mathcal{P}$ carries the weak preorder which is supposed to be antisymmetric, and since $b \preccurlyeq b + \mathfrak{O}(a)$ is evident, our claim follows. □

**Proposition 5.15.** *Let* $(\mathcal{P}, \mathcal{V})$ *be a locally convex complete lattice cone, and let* $A, B$ *be non-empty subsets of* $\mathcal{P}$. *Then*

(a) $\inf(A \vee B) = \inf A \vee \inf B$ *if both* $A$ *and* $B$ *are bounded below.*
(b) $\sup(A \wedge B) \leq \sup A \wedge \sup B \leq \sup(A \wedge B) + \mathfrak{O}\left(\sup(A \vee B)\right)$.

*Proof.* We first observe that

$$\inf A \vee \inf B \leq a \vee b \qquad \text{and} \qquad a \wedge b \leq \sup A \wedge \sup B$$

holds for all $a \in A$ and $b \in B$. Thus

$$\inf A \vee \inf B \leq \inf(A \vee B) \qquad \text{and} \qquad \sup(A \wedge B) \leq \sup B \wedge \sup A.$$

For Part (a) we assume that both sets $A$ and $B$ are bounded below and use Proposition 5.3 for

$$\begin{aligned} \inf(A \vee B) + \inf(A \wedge B) &\leq \inf\{a \vee b + a \wedge b \mid a \in a,\ b \in B\} \\ &= \inf(A + B) \\ &= \inf A + \inf B \\ &= \inf A \vee \inf B + \inf A \wedge \inf B. \end{aligned}$$

As $\inf(A \wedge B) = \inf A \wedge \inf B$, the cancellation law in Proposition 5.10(a) yields

$$\inf(A \vee B) \leq \inf A\ \vee\ \inf B + \mathfrak{O}\big(\inf(A \wedge B)\big).$$

Similarly, one obtains

$$\sup A \wedge \sup B \leq \sup(A \wedge B) + \mathfrak{O}\left(\sup(A \vee B)\right),$$

that is Part (b). Finally, as $\inf(A \wedge B) = \inf A \wedge \inf B \leq \inf A \vee \inf B$, Proposition 5.14 shows

$$\inf A \vee \inf B + \mathfrak{O}\big(\inf A \wedge \inf B\big) = \inf A \vee \inf B.$$

This completes our proof of Part (a). □

We proceed to refine the cancellation rules in Proposition 5.10 further. Let $(\mathcal{P}, \mathcal{V})$ be a locally convex lattice and $\bigwedge$-semilattice cone. We define the *zero component* of an element $a \in \mathcal{P}$ *relative to* $b \in \mathcal{P}$ by

$$\mathfrak{O}(a \searrow b) = \inf\big\{c \geq 0 \mid c + \mathfrak{O}(b) \geq \mathfrak{O}(a)\big\}.$$

Obviously, $\mathfrak{O}(a \searrow 0) = \mathfrak{O}(a)$. Also, $\mathfrak{O}(\alpha a \searrow \beta b) = \alpha\mathfrak{O}(a \searrow b) = \mathfrak{O}(a \searrow b)$ holds for all $\alpha, \beta > 0$.

**Proposition 5.16.** *Let* $(\mathcal{P}, \mathcal{V})$ *be a locally convex lattice and* $\bigwedge$*-semilattice cone, and let* $a, b, c \in \mathcal{P}$.

*(a)* $0 \leq \mathfrak{O}(a \searrow b) \leq \mathfrak{O}(a) \leq \mathfrak{O}(a \searrow b) + \mathfrak{O}(b)$.
*(b) If* $a + c \leq b + c$, *then* $a \leq b + \mathfrak{O}(c \searrow b)$.
*(c) If* $a \in \mathcal{B}(b)$, *then* $\mathfrak{O}(a \searrow b) = 0$ *and* $b + \mathfrak{O}(c) = b + \mathfrak{O}(c \searrow a)$.

*Proof.* Part (a) follows directly from the definition of $\mathfrak{O}(a \searrow b)$ together with $(\bigwedge 2)$. For (b) we recall that $a + c \leq b + c$ implies $a \leq b + \mathfrak{O}(c)$ by 5.10(a). As $\mathfrak{O}(c) \leq \mathfrak{O}(c \searrow b) + \mathfrak{O}(b)$ by Part (a) and $b + \mathfrak{O}(b) = b$ by 5.14, our claim follows. For (c), let $a \in \mathcal{B}(b)$. Then $\mathfrak{O}(a) \leq \mathfrak{O}(b)$ by 5.10(b), and we may use $c = 0$ in the definition of $\mathfrak{O}(a \searrow b)$. Thus indeed $\mathfrak{O}(a \searrow b) = 0$. Furthermore, we have

$$b + \mathfrak{O}(c) \leq b + \big(\mathfrak{O}(c \searrow a) + \mathfrak{O}(a)\big) = \big(b + \mathfrak{O}(a)\big) + \mathfrak{O}(c \searrow a) = b + \mathfrak{O}(c \searrow a)$$

by Part (a) and 5.14. □

*Examples 5.17.* (a) If $\mathcal{P} = \overline{\mathbb{R}}$ or $\mathcal{P} = \overline{\mathbb{R}}_+$ (see 1.4(a) and 1.4(b)), then $\mathfrak{O}(a) = 0$ for all $a < +\infty$, and $\mathfrak{O}(+\infty) = +\infty$.

(b) Consider $\big(\mathcal{F}_{\widehat{\mathcal{V}}_b}(X, \mathcal{P}), \widehat{\mathcal{V}}\big)$, where $(\mathcal{P}, \mathcal{V})$ is a locally convex $\bigwedge$-semilattice cone, $X$ a set, and $\widehat{\mathcal{V}}$ is a neighborhood system consisting of $(\mathcal{V} \cup \{\infty\})$-valued functions (see Example 1.4(e)) such that for every $x \in X$ and $v \in \mathcal{V}$ there is $\hat{v} \in \widehat{\mathcal{V}}$ such that $\hat{v}(x) \leq v$ (see 5.7(c)). For $f \in \mathcal{F}_{\widehat{\mathcal{V}}_b}(X, \mathcal{P})$ then $\mathfrak{O}(f)$ is the mapping $x \mapsto \mathfrak{O}\big(f(x)\big)$. For $\mathcal{P} = \overline{\mathbb{R}}$, in particular, the zero component of an $\overline{\mathbb{R}}$-valued function $f \in \mathcal{F}_{\widehat{\mathcal{V}}_b}(X, \overline{\mathbb{R}})$ is the mapping $\mathfrak{O}(f)(x) = 0$ if $f(x) < +\infty$, and $\mathfrak{O}(f)(x) = +\infty$ else. The same observation applies to the second part of Example 5.7(f), that is the cone of $\overline{\mathbb{R}}$-valued bounded below upper semicontinuous functions on a topological space with the positive constants as neighborhoods. This was seen to be an example of a locally convex lattice and $\bigwedge$-semilattice cone.

(c) Let us consider Example 4.37(e) (see also 5.7(e)), that is the locally convex $\bigwedge$-semilattice cone $(\mathcal{Q}, \mathcal{V})$ of all convex subsets locally convex cone $(\mathcal{P}, \mathcal{V})$, which are bounded below and closed with respect to the upper relative topology. Recall that the order in $\mathcal{Q}$ is the inverse set inclusion and the neighborhoods are given by $A \le B \oplus v$ for $A, B \in \mathcal{Q}$ and $v \in \mathcal{V}$, if for every $b \in B$, and $\varepsilon > 0$ there is $a \in A$ such that $a \le \gamma b + (1+\varepsilon)v$ for some $1 \le \gamma \le 1+\varepsilon$. The closed convex subsets (including the empty set) of $\overline{\{0\}}^{(u)} = \{b \in \mathcal{P} \mid 0 \preccurlyeq b\}$ are the positive elements in $\mathcal{Q}$. We claim that for an element $A \in \mathcal{Q}$ we have

$$\mathfrak{D}(A) = \{b \succcurlyeq 0 \mid \mathcal{B}_v(b) \cap A \neq \emptyset \quad \text{for all} \quad v \in \mathcal{V}\}.$$

We shall argue for this using the following steps: Let $B$ denote the set on the right-hand side of the above equation.

(i) The set $B \subset \mathcal{P}$ is convex. Indeed, let $b_1, b_2 \in B$, $0 \le \lambda_1, \lambda_2 \le 1$ such that $\lambda_1 + \lambda_2 = 1$ and $b = \lambda_1 b_1 + \lambda_2 b_2$. Given $v \in \mathcal{V}$ there are $a_1 \in \mathcal{B}_v(b_1) \cap A$ and $a_2 \in \mathcal{B}_v(b_2) \cap A$. Set $a = \lambda_1 a_1 + \lambda_2 a_2 \in A$ and choose $\alpha_1, \alpha_2, \beta, \rho \ge 0$ such that

$$a_1 \le \alpha_1 b_1 + \beta v, \qquad a_2 \le \alpha_2 b_2 + \beta v, \qquad 0 \le b_1 + \rho v \quad \text{and} \quad 0 \le b_2 + \rho v.$$

Setting $\alpha = \max\{\alpha_1, \alpha_2\}$ we have

$$a_1 \le \big(\alpha_1 b_1 + \beta v\big) + (\alpha - \alpha_1)(b_1 + \rho v) + \alpha_1 \rho v = \alpha b_1 + (\beta + \alpha\rho)v$$

and, likewise

$$a_2 \le \big(\alpha_2 b_1 + \beta v\big) + (\alpha - \alpha_2)(b_2 + \rho v) + \alpha_2 \rho v = \alpha b_2 + (\beta + \alpha\rho)v.$$

Thus

$$a \le \lambda_1\big(\alpha b_1 + (\beta + \alpha\rho)v\big) + \lambda_2\big(\alpha b_1 + (\beta + \alpha\rho)v\big) = \alpha b + (\beta + \alpha\rho)v.$$

We infer that $a \in \mathcal{B}_v(b) \cap A$, hence $\mathcal{B}_v(b) \cap A \neq \emptyset$. Since this holds for all $v \in \mathcal{V}$ and since $b \succcurlyeq 0$ is evident from $b_1, b_2 \succcurlyeq 0$, we conclude that $b \in B$.

(ii) The set $B \subset \mathcal{P}$ is closed with respect to the upper topology. Indeed, let $c \in \overline{B}^{(u)}$ and let $v \in \mathcal{V}$. There is $b \in v_1(c) \cap B$, that is $b \le \gamma c + v$ for some $1 \le \gamma \le 2$. There is $a \in \mathcal{B}_v(b) \cap A$, that is $a \le \alpha b + \beta v$ for some $\alpha, \beta \ge 0$. Combining these yields $a \le \alpha\gamma c + (\alpha + \beta)v$. This shows $\mathcal{B}_v(c) \cap A \neq \emptyset$ for all $v \in \mathcal{V}$. Furthermore, since $B \subset \overline{\{0\}}^{(u)} = \{b \in \mathcal{P} \mid 0 \preccurlyeq b\}$ which is closed with respect to the upper relative topology, we have $c \in \overline{\{0\}}^{(u)}$ as well, hence $c \succcurlyeq 0$. Together with the above this yields $c \in B$. Since $B \subset \mathcal{P}$ is obviously bounded below (we have $0 \leq b + v$ for all $b \in \mathcal{P}$), we conclude from (i) and (ii) that $B \in \mathcal{Q}$.

(iii) We have $A \in \mathcal{B}\big(\overline{\{b\}}^{(u)}\big)$ for all $b \in B$. Indeed, let $v \in \mathcal{V}$. Given $b \in B$ there is some $a \in \mathcal{B}_v(b) \cap A$, that is there are $\alpha, \beta, \lambda \geq 0$ such that $a \leq \alpha b + \beta v$ and $0 \leq b + \lambda v$. Then for every $c \in \overline{\{b\}}^{(u)}$, that is $b \preccurlyeq c$, we have $b \in v_1(c)$, hence $b \leq 2c + (2 + \lambda)v$ (see Lemma 4.1(c) with $\varepsilon = 1$). This yields $a \leq 2\alpha c + (2\alpha + \lambda\alpha + \beta)v$ and $A \leq 2\alpha\,\overline{\{b\}}^{(u)} \oplus (2\alpha + \lambda\alpha + \beta)v$, hence $A \in \mathcal{B}\big(\overline{\{b\}}^{(u)}\big)$. Consequently,

$$\mathfrak{O}(A) \leq \inf\big\{\overline{\{b\}}^{(u)} \mid b \in B\big\} = \overline{conv\big(\bigcup_{b \in B} \overline{\{b\}}^{(u)}\big)}^{(u)} = B.$$

(iv) On the other hand, let $C \in \mathcal{Q}$ such that $C \geq 0$, that is $C \subset \overline{\{0\}}^{(u)}$, and $A \in \mathcal{B}(C)$. Let $c \in C$. Given $v \in \mathcal{V}$ there are $\alpha, \beta \geq 0$ such that $A \leq \alpha C \oplus \beta v$. According to our definition of the neighborhoods in $\mathcal{Q}$ (see 4.37(e)), for $\varepsilon = 1$ we find $a \in A$ such that $a \leq \gamma(\alpha c) + 2(\beta v)$ with some $1 \leq \gamma \leq 2$. This yields $\mathcal{B}_v(b) \cap A \neq \emptyset$ for all $v \in \mathcal{V}$, hence $c \in B$ since $c \succcurlyeq 0$. Thus

$$C = \overline{conv\big(\bigcup_{c \in C} \overline{\{c\}}^{(u)}\big)}^{(u)} \subset B.$$

This shows $\mathfrak{O}(A) \subset B$, that is $\mathfrak{O}(A) \geq B$, and our claim follows.

In particular, we have $\mathfrak{O}(A) = \overline{\{0\}}^{(u)}$ if and only if $\mathcal{B}_v(0) \cap A \neq \emptyset$ for all $v \in \mathcal{V}$, that is if and only if for every $v \in \mathcal{V}$ there are $a \in A$ and $\lambda \geq 0$ such that $a \leq \lambda v$, that is if and only if the element $A \in \mathcal{Q}$ is bounded above (see 4.37(e)).

For a concrete example let $\mathcal{P}$ be the cone of all real-valued bounded below continuous functions on the open interval $(0, 1)$, endowed with the positive constants as neighborhoods (see 1.4(e)) and let $\mathcal{Q}$ be as before. Consider the subset

$$C = \left\{ f \in \mathcal{P} \;\middle|\; f(x) \geq \frac{1}{x} - 2 \text{ for all } x \in (0,1) \right\}.$$

This set is convex, bounded below and closed with respect to the upper relative topology, hence $C \in \mathcal{Q}$. For a function $g \geq 0$ in $\mathcal{P}$, we have $\mathcal{B}(g) \cap C \neq \emptyset$ if and only if there are $\alpha, \beta \geq 0$ such that $1/x \leq \alpha g(x) + \beta$ for all $x \in (0, 1)$, that is if and only if the inferior limit of $xg(x)$ at $0$ is greater than 0. Thus

$$\mathfrak{O}(C) = \{g \in \mathcal{P} \,|\, g \geq 0 \text{ and } \varliminf_{x \to 0} xg(x) > 0\}.$$

Now according to the cancellation rule in Proposition 5.10(a), if $A, B \in \mathcal{Q}$ such that $A + C \leq B + C$, that is $B + C \subset A + C$, then $A \leq B + \mathfrak{O}(c)$, that is $B + \mathfrak{O}(C) \subset A$.

**5.18 Order Convergence.** We proceed to define order convergence for nets in a locally convex complete lattice cone $(\mathcal{P},\mathcal{V})$. A net $(a_i)_{i\in\mathcal{I}}$ in $\mathcal{P}$ is called *bounded below* if there is $i_0\in\mathcal{I}$ such that the set $\{a_i \mid i\geq i_0\}$ is bounded below in the sense of 4.24(i). We define the superior and inferior limits of a bounded below net $(a_i)_{i\in\mathcal{I}}$ in $\mathcal{P}$ by

$$\underline{\lim}_{i\in\mathcal{I}}\, a_i = \sup_{i\in\mathcal{I}}\left(\inf_{k\geq i} a_k\right) \qquad \text{and} \qquad \overline{\lim}_{i\in\mathcal{I}}\, a_k = \inf_{i\in\mathcal{I}}\left(\sup_{k\geq i} a_k\right).$$

Because the order of $\mathcal{P}$ is supposed to be antisymmetric, both limits are uniquely defined. Obviously, $\underline{\lim}_{i\in\mathcal{I}}\, a_i \leq \overline{\lim}_{i\in\mathcal{I}}\, a_i$. If $\underline{\lim}_{i\in\mathcal{I}}\, a_i$ and $\overline{\lim}_{i\in\mathcal{I}}\, a_i$ coincide, we shall denote their common value by $\lim_{i\in\mathcal{I}} a_i$ and say that the net $(a_i)_{i\in\mathcal{I}}$ is *order convergent*. Obviously, every increasing or decreasing bounded below net is order convergent in this sense, converging towards the supremum or the infimum of the set of its elements, respectively.

**Lemma 5.19.** *Let $(\mathcal{P},\mathcal{V})$ be a locally convex complete lattice cone, and let $(a_i)_{i\in\mathcal{I}}$ and $(b_i)_{i\in\mathcal{I}}$ be bounded below nets in $\mathcal{P}$. Then*

$$\underline{\lim}_{i\in\mathcal{I}}\, a_i+\underline{\lim}_{i\in\mathcal{I}}\, b_i \leq \underline{\lim}_{i\in\mathcal{I}}(a_i+b_i) \leq \overline{\lim}_{i\in\mathcal{I}}\, a_i+\underline{\lim}_{i\in\mathcal{I}}\, b_i \leq \overline{\lim}_{i\in\mathcal{I}}(a_i+b_i) \leq \overline{\lim}_{i\in\mathcal{I}}\, a_i+\overline{\lim}_{i\in\mathcal{I}}\, b_i.$$

*Proof.* For any bounded below net $(c_i)_{i\in\mathcal{I}}$ in $\mathcal{P}$, for $i\in\mathcal{I}$, let

$$s_i^{(c)} = \inf_{k\geq i} c_k \qquad \text{and} \qquad S_i^{(c)} = \sup_{k\geq i} c_k.$$

The nets $(s_i^{(c)})_{i\in\mathcal{I}}$ and $(S_i^{(c)})_{i\in\mathcal{I}}$ are increasing and decreasing, respectively, and

$$\underline{\lim}_{i\in\mathcal{I}}\, c_i = \sup_{i\in\mathcal{I}} s_i^{(c)} \qquad \text{and} \qquad \overline{\lim}_{i\in\mathcal{I}}\, c_i = \inf_{i\in\mathcal{I}} S_i^{(c)}.$$

Now, using the nets $(a_i)_{i\in\mathcal{I}}$, $(b_i)_{i\in\mathcal{I}}$ and $(a_i+b_i)_{i\in\mathcal{I}}$ in place of $(c_i)_{i\in\mathcal{I}}$ we observe that

$$s_i^{(a+b)} \geq s_i^{(a)}+s_i^{(b)} \qquad \text{and} \qquad S_i^{(a+b)} \leq S_i^{(a)}+S_i^{(b)}$$

for all $i\in\mathcal{I}$. For every $k\in\mathcal{I}$ we have by $(\bigvee 1)$

$$s_k^{(a)}+\sup_{i\in\mathcal{I}} s_i^{(b)} = \sup_{i\in\mathcal{I}}(s_k^{(a)}+s_i^{(b)}) \leq \sup_{l\in\mathcal{I}}(s_l^{(a)}+s_l^{(b)}),$$

as $s_k^{(a)}+s_i^{(b)} \leq s_l^{(a)}+s_l^{(b)}$ whenever $i,k\leq l$. This shows

$$\underline{\lim}_{i\in\mathcal{I}}\, a_i+\underline{\lim}_{i\in\mathcal{I}}\, b_i = \sup_{k\in\mathcal{I}} s_k^{(a)}+\sup_{i\in\mathcal{I}} s_k^{(b)} \leq \sup_{l\in\mathcal{I}}(s_l^{(a)}+s_l^{(b)}) = \underline{\lim}_{i\in\mathcal{I}}(a_i+b_i),$$

the first part of our claim. A similar argument using the decreasing nets $(S_i^{(c)})_{i\in\mathcal{I}}$ yields

$$\overline{\lim_{i\in\mathcal{I}}}\, a_i + \overline{\lim_{i\in\mathcal{I}}}\, b_i = \inf_{k\in\mathcal{I}} S_k^{(a)} + \inf_{i\in\mathcal{I}} S_k^{(b)} \geq \inf_{l\in\mathcal{I}} (S_l^{(a)} + S_l^{(b)}) = \overline{\lim_{i\in\mathcal{I}}}(a_i + b_i).$$

Finally, for all $i, l \in \mathcal{I}$ and $j \geq i, l$ we have

$$s_i^{(a+b)} = \inf_{k\geq i}(a_k + b_k) \leq \inf_{k\geq j}(S_l^{(a)} + b_k) = S_l^{(a)} + \inf_{k\geq j} b_k \leq S_l^{(a)} + \underline{\lim}_{i\in\mathcal{I}}\, b_i,$$

hence

$$\underline{\lim}_{i\in\mathcal{I}}(a_i + b_i) = \sup_{i\in\mathcal{I}} s_i^{(a+b)} \leq \inf_{l\in\mathcal{I}} S_l^{(a)} + \underline{\lim}_{i\in\mathcal{I}}\, b_i = \overline{\lim_{i\in\mathcal{I}}}\, a_i + \underline{\lim}_{i\in\mathcal{I}}\, b_i.$$

A similar argument shows that

$$\overline{\lim_{i\in\mathcal{I}}}\, a_i + \underline{\lim}_{i\in\mathcal{I}}\, b_i \leq \overline{\lim_{i\in\mathcal{I}}}(a_i + b_i).$$

□

Note that Lemma 5.19 implies in particular that

$$\underline{\lim}_{i\in\mathcal{I}}(a + b_i) = a + \underline{\lim}_{i\in\mathcal{I}}\, b_i \qquad \text{and} \qquad \overline{\lim_{i\in\mathcal{I}}}(a + b_i) = a + \overline{\lim_{i\in\mathcal{I}}}\, b_i$$

holds for $a \in \mathcal{P}$ and a bounded below net $(b_i)_{i\in\mathcal{I}}$. We shall use Conditions $(\bigvee 2)$ and $(\bigwedge 2)$ for a comparison of the inferior and superior limits of nets:

**Lemma 5.20.** *Let $(\mathcal{P}, \mathcal{V})$ be a locally convex complete lattice cone, let $(a_i)_{i\in\mathcal{I}}$ and $(b_j)_{j\in\mathcal{J}}$ be nets in $\mathcal{P}$, and let $v \in \mathcal{V}$.*

*(a) If for every $i_0 \in \mathcal{I}$ there is $j_0 \in \mathcal{J}$ such that for every $j \geq j_0$ there is $i \geq i_0$ such that $a_i \leq b_j + v$, then $\underline{\lim}_{i\in\mathcal{I}}\, a_i \leq \underline{\lim}_{j\in\mathcal{J}}\, b_j + v$.*
*(b) If for every $j_0 \in \mathcal{J}$ there is $i_0 \in \mathcal{I}$ such that for every $i \geq i_0$ there is $j \geq j_0$ such that $a_i \leq b_j + v$, then $\overline{\lim}_{i\in\mathcal{I}}\, a_i \leq \overline{\lim}_{j\in\mathcal{J}}\, b_j + v$.*
*(c) If $\mathcal{I} = \mathcal{J}$ and if there is $i_0 \in \mathcal{I}$ such that $a_i \leq b_i + v$ for all $i \geq i_0$, then $\underline{\lim}_{i\in\mathcal{I}}\, a_i \leq \underline{\lim}_{i\in\mathcal{I}}\, b_i + v$ and $\overline{\lim}_{i\in\mathcal{I}}\, a_i \leq \overline{\lim}_{i\in\mathcal{I}}\, b_i + v$.*
*(d) If $(a_{i_l})_{l\in\mathcal{L}}$ is a subnet of $(a_i)_{i\in\mathcal{I}}$, then $\underline{\lim}_{i\in\mathcal{I}}\, a_i \leq \underline{\lim}_{l\in\mathcal{L}}\, a_{i_l}$ and $\overline{\lim}_{l\in\mathcal{L}}\, a_{i_l} \leq \overline{\lim}_{i\in\mathcal{I}}\, a_i$.*

*Proof.* (a) Given $i_0 \in \mathcal{I}$ choose $j_0 \in \mathcal{J}$ as in the assumption of Part (a). Then $\inf_{i\geq i_0} a_i \leq b_j + v$ for all $j \geq j_0$, hence

$$\inf_{i\geq i_0} a_i \leq \inf_{j\geq j_0} b_j + v \leq \underline{\lim}_{j\in\mathcal{J}}\, b_j + v$$

by $(\bigwedge 2)$. Thus by $(\bigvee 2)$ we have $\underline{\lim}_{i\in\mathcal{I}}\, a_i \leq \underline{\lim}_{j\in\mathcal{J}}\, b_j + v$ as well. The argument for Part (b) is similar. The assumptions for Part (c) yield those for Parts (a) and (b) with $j_0 = i_0$ and $j = i$. Part (d) follows from (a) and (b) if we set $\mathcal{J} = \mathcal{L}$ and $b_l = a_{i_j}$. □

Lemma 5.20(c) yields in particular that $\lim_{i\in\mathcal{I}} a_i \le \lim_{i\in\mathcal{I}} b_i$ for order convergent nets $(a_i)_{i\in\mathcal{I}}$ and $(b_i)_{i\in\mathcal{I}}$ whenever $a_i \le b_i$ for all $i \in \mathcal{I}$. Part (d) implies that every subnet of an order convergent net is again order convergent with the same limit.

**Lemma 5.21.** *Let $(\mathcal{P},\mathcal{V})$ be a locally convex complete lattice cone. Let $(a_i)_{i\in\mathcal{I}}$ be a bounded below net in $\mathcal{P}$, and let $(\alpha_i)_{i\in\mathcal{I}}$ be a bounded net of non-negative reals such that $\lim_{i\in\mathcal{I}} \alpha_i > 0$. Then*

$$\Big(\lim_{i\in\mathcal{I}} \alpha_i\Big)\Big(\underline{\lim}_{i\in\mathcal{I}} a_i\Big) \le \underline{\lim}_{i\in\mathcal{I}}(\alpha_i a_i) \le \overline{\lim}_{i\in\mathcal{I}}(\alpha_i a_i) \le \Big(\lim_{i\in\mathcal{I}} \alpha_i\Big)\Big(\overline{\lim}_{i\in\mathcal{I}} a_i\Big).$$

*Proof.* Obviously the net $(\alpha_i a_i)_{i\in\mathcal{I}}$ is also bounded below in $\mathcal{P}$. We set $\alpha = \lim_{i\in\mathcal{I}} \alpha_i > 0$. Given $v \in \mathcal{V}$ there is $\lambda \ge 0$ such that $0 \le a_i + \lambda v$ for all $i \in \mathcal{I}$. For $\varepsilon > 0$ we set $\gamma = 1 + \varepsilon$ and find $i_0 \in \mathcal{I}$ such that $(1/\gamma)\alpha \le \alpha_i \le \gamma\alpha$ for all $i \ge i_0$. Thus

$$\alpha_i a_i + \frac{\alpha}{\gamma}\lambda v \le \alpha_i(a_i + \lambda v) \le \gamma\alpha(a_i + \lambda v),$$

hence, using the cancellation law for positive elements (see Lemma I.4.2 in [100])

$$\alpha_i a_i \le \gamma\alpha a_i + \alpha\lambda\Big(\gamma - \frac{1}{\gamma}\Big)v + \varepsilon v \le \gamma\alpha a_i + \varepsilon(2\alpha\lambda + 1)v.$$

Using Lemma 5.20(c) we infer that

$$\overline{\lim}_{i\in\mathcal{I}} \alpha_i a_i \le \gamma\Big(\alpha\, \overline{\lim}_{i\in\mathcal{I}} a_i\Big) + \varepsilon(2\alpha\lambda + 1)v.$$

Since the latter holds for all $\varepsilon > 0$ and since $\mathcal{P}$ carries the weak preorder, we conclude that

$$\overline{\lim}_{i\in\mathcal{I}} \alpha_i a_i \le \alpha\, \overline{\lim}_{i\in\mathcal{I}} a_i.$$

The first part of the inequality in our claim follows in a similar fashion. □

**Proposition 5.22.** *Let $(\mathcal{P},\mathcal{V})$ be a locally convex complete lattice cone. Let $(a_i)_{i\in\mathcal{I}}$ and $(b_i)_{i\in\mathcal{I}}$ be order convergent nets in $\mathcal{P}$, and let $(\alpha_i)_{i\in\mathcal{I}}$ be a bounded net of non-negative reals such that $\lim_{i\in\mathcal{I}} \alpha_i > 0$. Then*

$$\lim_{i\in\mathcal{I}}(a_i + b_i) = \lim_{i\in\mathcal{I}} a_i + \lim_{i\in\mathcal{I}} b_i \qquad \textit{and} \qquad \lim_{i\in\mathcal{I}}(\alpha_i a_i) = \Big(\lim_{i\in\mathcal{I}} \alpha_i\Big)\Big(\lim_{i\in\mathcal{I}} a_i\Big).$$

The latter is an obvious consequence of our previous results 5.19 and 5.21. Note that the requirement that $\lim_{i\in\mathcal{I}} \alpha_i > 0$ can not be omitted if the elements of the net $(a_i)_{i\in\mathcal{I}}$ are not bounded in $\mathcal{P}$ : In the locally convex complete lattice cone $\overline{\mathbb{R}}$ choose $a_n = +\infty$ and $\alpha_n = (1/n)$. Then $\lim_{n\to\infty}(\alpha_n a_n) = +\infty$, but $\big(\lim_{n\to\infty} \alpha_n\big)\big(\lim_{n\to\infty} a_n\big) = 0$.

The following will provide a useful criterion for the convergence of a given net.

**Proposition 5.23.** *Let $(\mathcal{P},\mathcal{V})$ be a locally convex complete lattice cone, and let $(a_i)_{i\in\mathcal{I}}$ be a bounded below net in $\mathcal{P}$. If for every $v\in\mathcal{V}$ there is a convergent net $(b_i)_{i\in\mathcal{I}}$ in $\mathcal{P}$ such that $(a_i+b_i)_{i\in\mathcal{I}}$ is convergent and the limit of $(b_i)_{i\in\mathcal{I}}$ is $v$-bounded, then the net $(a_i)_{i\in\mathcal{I}}$ is also convergent.*

*Proof.* Let $(a_i)_{i\in\mathcal{I}}$ be a net in $\mathcal{P}$ and for $v\in\mathcal{V}$ let $(b_i)_{i\in\mathcal{I}}$ be as stated. We use Lemma 5.19 for

$$\overline{\lim_{i\in\mathcal{I}}}\, a_i + \lim_{i\in\mathcal{I}} b_i \le \lim_{i\in\mathcal{I}}(a_i+b_i) \le \underline{\lim_{i\in\mathcal{I}}}\, a_i + \lim_{i\in\mathcal{I}} b_i.$$

As $b=\lim_{i\in\mathcal{I}} b_i$ is $v$-bounded, following Proposition 5.13(b) we have $\mathfrak{O}(b)\le\varepsilon v$ for all $\varepsilon>0$, hence

$$\overline{\lim_{i\in\mathcal{I}}}\, a_i \le \underline{\lim_{i\in\mathcal{I}}}\, a_i + \varepsilon v.$$

by Proposition 5.10(a). Because this holds for all $v\in\mathcal{V}$ and $\varepsilon>0$ and because $\mathcal{P}$ is a complete lattice cone, we infer that

$$\overline{\lim_{i\in\mathcal{I}}}\, a_i \le \underline{\lim_{i\in\mathcal{I}}}\, a_i$$

holds as claimed. □

**Proposition 5.24.** *Let $(\mathcal{P},\mathcal{V})$ be a locally convex complete lattice cone, and let $(a_i)_{i\in\mathcal{I}}$ be a bounded below net in $\mathcal{P}$. Then*

$$\underline{\lim_{i\in\mathcal{I}}}\, \mathfrak{O}(a_i) \le \mathfrak{O}\big(\underline{\lim_{i\in\mathcal{I}}}\, a_i\big) \qquad \textit{and} \qquad \overline{\lim_{i\in\mathcal{I}}}\, \mathfrak{O}(a_i) \le \mathfrak{O}\big(\overline{\lim_{i\in\mathcal{I}}}\, a_i\big).$$

*Proof.* Let $(a_i)_{i\in\mathcal{I}}$ be a bounded below net, let $v\in\mathcal{V}$ and $\lambda\ge 0$ such that $0\le a_i+\lambda v$ for all $i\ge i_0\in\mathcal{I}$. Then $\mathfrak{O}(a_i)\le\varepsilon(a_i+\lambda v)$ for all $i\ge i_0$ and $\varepsilon>0$, hence

$$\underline{\lim_{i\in\mathcal{I}}}\, \mathfrak{O}(a_i) \le \varepsilon\, \underline{\lim_{i\in\mathcal{I}}}\, a_i + \varepsilon\lambda v \le \varepsilon \sup\Big\{ \underline{\lim_{i\in\mathcal{I}}}\, a_i\,,\, 0\Big\} + \varepsilon\lambda v$$

by 5.20(a). Taking the infimum over all $\varepsilon>0$ on the right-hand side we obtain

$$\underline{\lim_{i\in\mathcal{I}}}\, \mathfrak{O}(a_i) \le \mathfrak{O}\Big(\underline{\lim_{i\in\mathcal{I}}}\, a_i\Big) + \mathfrak{O}(\lambda v) \le \mathfrak{O}\Big(\underline{\lim_{i\in\mathcal{I}}}\, a_i\Big) + v.$$

Because this last inequality holds for all $v\in\mathcal{V}$ and because $\mathcal{P}$ carries the weak preorder, we conclude that

$$\underline{\lim_{i\in\mathcal{I}}}\, \mathfrak{O}(a_i) \le \mathfrak{O}\Big(\underline{\lim_{i\in\mathcal{I}}}\, a_i\Big)$$

holds as claimed. A similar argument demonstrates the same inequality for the superior limits. □

A simple example can show that equality does in general not hold in the expressions of Proposition 5.24: The locally convex cone $\mathcal{P} = \overline{\mathbb{R}}$ is a complete lattice. Order convergence in $\overline{\mathbb{R}}$ means convergence with respect to its usual one-point compactification topology, which at the point $+\infty$ differs from the symmetric topology of $\overline{\mathbb{R}}$ as a locally convex cone. For each $n \in \mathbb{N}$ let $a_n = n \in \overline{\mathbb{R}}$. Then $\lim_{n\to\infty} a_n = +\infty$ with respect to order convergence (but not with respect to the symmetric topology). We therefore have $\mathfrak{O}(a_n) = 0$ for all $n \in \mathbb{N}$, but $\mathfrak{O}\big(\lim_{n\to\infty} a_n\big) = +\infty$.

We proceed to investigate continuity of the lattice operations with respect to order convergence (c.f. Proposition 5.2).

**Proposition 5.25.** *Let $(\mathcal{P},\mathcal{V})$ be a locally convex complete lattice cone and let $(a_i)_{i\in\mathcal{I}}$ and $(b_i)_{i\in\mathcal{I}}$ be convergent nets in $\mathcal{P}$. Then*

*(a)* $\lim_{i\in\mathcal{I}}(a_i \vee b_i) = \Big(\lim_{i\in\mathcal{I}} a_i\Big) \vee \Big(\lim_{i\in\mathcal{I}} b_i\Big)$.

*(b)* $\overline{\lim}_{i\in\mathcal{I}}(a_i \wedge b_i) \le \Big(\lim_{i\in\mathcal{I}} a_i\Big) \wedge \Big(\lim_{i\in\mathcal{I}} b_i\Big) \le \underline{\lim}_{i\in\mathcal{I}}(a_i \wedge b_i) + \mathfrak{O}\Big(\lim_{i\in\mathcal{I}}(a_i \vee b_i)\Big)$.

*Proof.* (a) Let $(a_i)_{i\in\mathcal{I}}$ and $(b_i)_{i\in\mathcal{I}}$ be convergent nets. Then

$$\overline{\lim}_{i\in\mathcal{I}}(a_i \vee b_i) = \inf_{i\in\mathcal{I}}\Big(\sup_{l\ge i}(a_l \vee b_l)\Big) \le \inf_{i\in\mathcal{I}}\Big(\big(\sup_{l\ge i} a_l\big) \vee \big(\sup_{j\ge i} b_j\big)\Big).$$

Because for any choice of $i,k \in \mathcal{I}$ and any $p \in \mathcal{I}$ such that both $i \le p$ and $k \le p$ we have

$$\big(\sup_{l\ge p} a_l\big) \vee \big(\sup_{j\ge p} b_j\big) \le \big(\sup_{l\ge i} a_l\big) \vee \big(\sup_{j\ge k} b_j\big),$$

we realize that

$$\inf_{i\in\mathcal{I}}\Big(\big(\sup_{l\ge i} a_l\big) \vee \big(\sup_{j\ge i} b_j\big)\Big) \le \inf_{i,k\in\mathcal{I}}\Big(\big(\sup_{l\ge i} a_l\big) \vee \big(\sup_{j\ge k} b_j\big)\Big).$$

Now we use Proposition 5.15(a) for

$$\inf_{i,k\in\mathcal{I}}\Big(\big(\sup_{l\ge i} a_l\big) \vee \big(\sup_{j\ge k} b_j\big)\Big) = \inf_{i\in\mathcal{I}}\Big(\sup_{l\ge i} a_l\Big) \vee \inf_{k\in\mathcal{I}}\Big(\sup_{j\ge k} b_j\Big) = \Big(\overline{\lim}_{i\in\mathcal{I}} a_i\Big) \vee \Big(\overline{\lim}_{i\in\mathcal{I}} b_i\Big).$$

Both nets $(a_i)_{i\in\mathcal{I}}$ and $(b_i)_{i\in\mathcal{I}}$ are supposed to be convergent. So we have

$$\Big(\overline{\lim}_{i\in\mathcal{I}} a_i\Big) \vee \Big(\overline{\lim}_{i\in\mathcal{I}} b_i\Big) = \Big(\underline{\lim}_{i\in\mathcal{I}} a_i\Big) \vee \Big(\underline{\lim}_{i\in\mathcal{I}} b_i\Big) \le \underline{\lim}_{i\in\mathcal{I}}(a_i \vee b_i).$$

Summarizing, the above yields

$$\overline{\lim}_{i\in\mathcal{I}}(a_i \vee b_i) \le \Big(\lim_{i\in\mathcal{I}} a_i\Big) \vee \Big(\lim_{i\in\mathcal{I}} b_i\Big) \le \underline{\lim}_{i\in\mathcal{I}}(a_i \vee b_i)$$

as claimed in Part (a). Similarly, one verifies Part (b): The inequality

$$\overline{\lim_{i\in\mathcal{I}}}(a_i \wedge b_i) \le \Big(\overline{\lim_{i\in\mathcal{I}}}\, a_i\Big) \wedge \Big(\overline{\lim_{i\in\mathcal{I}}}\, b_i\Big) = \Big(\lim_{i\in\mathcal{I}} a_i\Big) \wedge \Big(\lim_{i\in\mathcal{I}} b_i\Big)$$

is obvious. Next we use Part (a), Proposition 5.3 and the limit rules from Lemma 5.17 for

$$\begin{aligned}
\Big(\lim_{i\in\mathcal{I}} a_i\Big)\wedge\Big(\lim_{i\in\mathcal{I}} b_i\Big)+\lim_{i\in\mathcal{I}}(a_i \vee b_i) &= \Big(\lim_{i\in\mathcal{I}} a_i\Big) \wedge \Big(\lim_{i\in\mathcal{I}} b_i\Big) + \Big(\lim_{i\in\mathcal{I}} a_i\Big)\vee\Big(\lim_{i\in\mathcal{I}} b_i\Big)\\
&= \lim_{i\in\mathcal{I}} a_i + \lim_{i\in\mathcal{I}} b_i = \lim_{i\in\mathcal{I}}(a_i + b_i)\\
&= \lim_{i\in\mathcal{I}}(a_i \wedge b_i + a_i \vee b_i)\\
&\le \underline{\lim_{i\in\mathcal{I}}}(a_i \wedge b_i) + \overline{\lim_{i\in\mathcal{I}}}(a_i \vee b_i)\\
&= \underline{\lim_{i\in\mathcal{I}}}(a_i \wedge b_i) + \lim_{i\in\mathcal{I}}(a_i \vee b_i).
\end{aligned}$$

Now the cancellation rule from Lemma 5.9(a) yields the remaining part of (b). □

**5.26 Series.** A series $\sum_{i=1}^{\infty} a_i$ with terms $a_i$ in a locally convex complete lattice cone $(\mathcal{P}, \mathcal{V})$ is said to be convergent with limit $s \in \mathcal{P}$ if the sequence $s_n = \sum_{i=1}^{n} a_i$ of its partial sums is order convergent to $s$. We write $\sum_{i=1}^{\infty} a_i = s$ in this case. Convergence of a series requires in particular that the sequence of its partial sums is bounded below (see 5.18).

**Proposition 5.27.** *Let $(\mathcal{P}, \mathcal{V})$ be a locally convex complete lattice cone and let $a_i, b_i \in \mathcal{P}$ for $i \in \mathbb{N}$. If the series $\sum_{i=1}^{\infty} a_i$ is convergent and if $a_i \le b_i$ for all $i \in \mathbb{N}$, then the series $\sum_{i=1}^{\infty} b_i$ is also convergent.*

*Proof.* Let $a_i, b_i \in \mathcal{P}$ such that $a_i \le b_i$ for all $i \in \mathbb{N}$. Let $s_n = \sum_{i=1}^{n} a_i$ and $r_n = \sum_{i=1}^{n} b_i$ be the partial sums of the series $\sum_{i=1}^{\infty} a_i$ and $\sum_{i=1}^{\infty} b_i$, and let $s = \sum_{i=1}^{\infty} a_i$. Then $s_n \le r_n$ for all $n \in \mathbb{N}$, hence $s \le \underline{\lim}_{n\to\infty} r_n$. For $m \ge n$ we have

$$r_n + s_m = r_n + s_n + \sum_{i=n+1}^{m} a_i \le r_n + s_n + \sum_{i=n+1}^{m} b_i = r_m + s_n.$$

For a fixed $n \in \mathbb{N}$ and $m \to \infty$ this leads to

$$r_n + s = r_n + \underline{\lim_{m\to\infty}}\, s_m = \underline{\lim_{m\to\infty}}\,(r_n + s_m) \le \underline{\lim_{m\to\infty}}\,(s_n + r_m) = \underline{\lim_{m\to\infty}}\, r_m + s_n.$$

Now we let $n \to \infty$ and obtain

$$\begin{aligned}
\overline{\lim_{n\to\infty}}\, r_n + s = \overline{\lim_{n\to\infty}}\,(r_n + s) &\le \overline{\lim_{n\to\infty}}\left(\underline{\lim_{m\to\infty}}\, r_m + s_n\right)\\
&= \underline{\lim_{m\to\infty}}\, r_m + \overline{\lim_{n\to\infty}}\, s_n = \underline{\lim_{m\to\infty}}\, r_m + s.
\end{aligned}$$

The cancellation law from Proposition 5.10(a) now yields

$$\overline{\lim_{n\to\infty}}\, r_n \leq \underline{\lim_{n\to\infty}}\, r_n + \mathfrak{O}(s).$$

But $s \leq \underline{\lim}_{n\to\infty}\, r_n$, as we observed before, and therefore $\underline{\lim}_{n\to\infty}\, r_n + \mathfrak{O}(s) = \underline{\lim}_{n\to\infty}\, r_n$ by Proposition 5.19. This yields

$$\overline{\lim_{n\to\infty}}\, r_n \leq \underline{\lim_{n\to\infty}}\, r_n,$$

hence convergence of the sequence $(r_n)_{n\in\mathbb{N}}$, that is the partial sums of the series $\sum_{i=1}^{\infty} b_i$. $\quad\square$

We shall say that a series $\sum_{i=1}^{\infty} A_i$ of non-empty subsets $A_i$ of a locally convex complete lattice cone $\mathcal{P}$ is *convergent* if the series $\sum_{i=1}^{\infty} \inf A_i$ converges in $\mathcal{P}$. In this case, all series $\sum_{i=1}^{\infty} a_i$, for any choice of elements $a_i \in A_i$, are convergent by Proposition 5.27, and we shall denote the set of all limits of these series by $\sum_{i=1}^{\infty} A_i$.

**Proposition 5.28.** *Let $(\mathcal{P}, \mathcal{V})$ be a locally convex complete lattice cone and let $\sum_{i=1}^{\infty} A_i$ be a convergent series of non-empty subsets of $\mathcal{P}$. Then*

(a) $\sum_{i=1}^{\infty} \sup A_i = \sup\{\sum_{i=1}^{\infty} A_i\}$.
(b) $\sum_{i=1}^{\infty} \inf A_i \leq \inf\{\sum_{i=1}^{\infty} A_i\} \leq \sum_{i=1}^{\infty} \inf A_i + \mathfrak{O}\left(\inf\{\sum_{i=1}^{\infty} A_i\}\right)$.

*Proof.* Let $\sum_{i=1}^{\infty} A_i$ be a convergent series of non-empty subsets of $\mathcal{P}$. We shall consider Parts (a) and (b) simultaneously. Let $S_i = \sup A_i$ and $s_i = \inf A_i$ for all $i \in \mathbb{N}$. By our assumption on the series $\sum_{i=1}^{\infty} A_i$ of sets, and following Proposition 5.27, all the series $\sum_{i=1}^{\infty} s_i$, $\sum_{i=1}^{\infty} S_i$ and $\sum_{i=1}^{\infty} a_i$ for any choice of $a_i \in A_i$ are convergent in $\mathcal{P}$. Moreover, for any choice of elements $a_i \in A_i$, for $i \in \mathbb{N}$, as $\sum_{i=1}^{n} s_i \leq \sum_{i=1}^{n} a_i \leq \sum_{i=1}^{n} S_i$ holds for all $n \in \mathbb{N}$, we have

$$\sum_{i=1}^{\infty} s_i \leq \sum_{i=1}^{\infty} a_i \leq \sum_{i=1}^{\infty} S_i,$$

and therefore

$$\sum_{i=1}^{\infty} s_i \leq \inf\left\{\sum_{i=1}^{\infty} A_i\right\} \leq \sup\left\{\sum_{i=1}^{\infty} A_i\right\} \leq \sum_{i=1}^{\infty} S_i.$$

Thus all left to show is that

$$\sum_{i=1}^{\infty} S_i \leq \sup\left\{\sum_{i=1}^{\infty} A_i\right\}$$

and

$$\inf\left\{\sum_{i=1}^{\infty} A_i\right\} \leq \sum_{i=1}^{\infty} s_i + \mathfrak{O}\left(\inf\left\{\sum_{i=1}^{\infty} A_i\right\}\right).$$

For this, let us fix $n \in \mathbb{N}$ and choose arbitrary elements $a_i, b_i \in A_i$. We set $c_i = b_i$ for $i = 1, \ldots, n$ and $c_i = a_i$, else. Then, obviously, for every $m \geq n$ we have

$$\sum_{i=1}^{n} b_i + \sum_{i=1}^{m} a_i = \sum_{i=1}^{m} c_i + \sum_{i=1}^{n} a_i.$$

We let $m$ tend to infinity and obtain

$$\sum_{i=1}^{n} b_i + \sum_{i=1}^{\infty} a_i = \sum_{i=1}^{\infty} c_i + \sum_{i=1}^{n} a_i.$$

As $c_i \in A_i$ for all $i \in \mathbb{N}$, this yields

$$\inf\left\{\sum_{i=1}^{\infty} A_i\right\} + \sum_{i=1}^{n} a_i \leq \sum_{i=1}^{n} b_i + \sum_{i=1}^{\infty} a_i \leq \sup\left\{\sum_{i=1}^{\infty} A_i\right\} + \sum_{i=1}^{n} a_i.$$

As $\sup\{\sum_{i=1}^{n} A_i\} = \sum_{i=1}^{n} S_i$ and $\inf\{\sum_{i=1}^{n} A_i\} = \sum_{i=1}^{n} s_i$ by Lemma 5.6(a), variation of the elements $b_1, \ldots, b_n$ yields

$$\inf\left\{\sum_{i=1}^{\infty} A_i\right\} + \sum_{i=1}^{n} a_i \leq \sum_{i=1}^{n} s_i + \sum_{i=1}^{\infty} a_i$$

and

$$\sum_{i=1}^{n} S_i + \sum_{i=1}^{\infty} a_i \leq \sup\left\{\sum_{i=1}^{\infty} A_i\right\} + \sum_{i=1}^{n} a_i.$$

Now we let $n$ tend to infinity and infer that

$$\inf\left\{\sum_{i=1}^{\infty} A_i\right\} + \sum_{i=1}^{\infty} a_i \leq \sum_{i=1}^{\infty} s_i + \sum_{i=1}^{\infty} a_i$$

and

$$\sum_{i=1}^{\infty} S_i + \sum_{i=1}^{\infty} a_i \leq \sup\left\{\sum_{i=1}^{\infty} A_i\right\} + \sum_{i=1}^{\infty} a_i.$$

Finally, we take the infimum over all choices for the elements $a_i \in A_i$ in this last pair of inequalities and obtain

$$\inf\left\{\sum_{i=1}^{\infty} A_i\right\} + \inf\left\{\sum_{i=1}^{\infty} A_i\right\} \leq \sum_{i=1}^{\infty} s_i + \inf\left\{\sum_{i=1}^{\infty} A_i\right\}$$

and

$$\sum_{i=1}^{\infty} S_i + \inf\left\{\sum_{i=1}^{\infty} A_i\right\} \leq \sup\left\{\sum_{i=1}^{\infty} A_i\right\} + \inf\left\{\sum_{i=1}^{\infty} A_i\right\}.$$

Now the cancellation rule in Proposition 5.10(a) yields

$$\inf\left\{\sum_{i=1}^{\infty} A_i\right\} \le \sum_{i=1}^{\infty} s_i + \mathfrak{O}\left(\inf\left\{\sum_{i=1}^{\infty} A_i\right\}\right)$$

and

$$\sum_{i=1}^{\infty} S_i \le \sup\left\{\sum_{i=1}^{\infty} A_i\right\} + \mathfrak{O}\left(\inf\left\{\sum_{i=1}^{\infty} A_i\right\}\right).$$

As $\inf\{\sum_{i=1}^{\infty} A_i\} \le \sup\{\sum_{i=1}^{\infty} A_i\}$, Proposition 5.14 yields

$$\sup\left\{\sum_{i=1}^{\infty} A_i\right\} + \mathfrak{O}\left(\inf\left\{\sum_{i=1}^{\infty} A_i\right\}\right) = \sup\left\{\sum_{i=1}^{\infty} A_i\right\}.$$

This demonstrates our claim. □

**5.29 Order Continuous Linear Operators and Functionals.** Let $(\mathcal{P},\mathcal{V})$ and $(\mathcal{Q},\mathcal{W})$ be locally convex complete lattice cones. We shall say that a continuous linear operator $T:\mathcal{P}\to\mathcal{Q}$ is *order continuous* if it is continuous with respect to order convergence, that is if

$$T\Big(\lim_{i\in\mathcal{I}} a_i\Big) = \lim_{i\in\mathcal{I}} T(a_i)$$

holds for every order convergent net $(a_i)_{i\in\mathcal{I}}$ in $\mathcal{P}$. The limits refer to order convergence in $\mathcal{P}$ and $\mathcal{Q}$, respectively. Sums and non-negative multiples of order continuous linear operators are again order continuous. We are particularly interested in *order continuous linear functionals* in $\mathcal{P}^*$, that is order continuous linear operators from $\mathcal{P}$ into the locally convex complete lattice cone $\overline{\mathbb{R}}$. They form a subcone of $\mathcal{P}^*$. For every bounded below net $(a_i)_{i\in\mathcal{I}}$ in $\mathcal{P}$ and every order continuous linear operator $T:\mathcal{P}\to\mathcal{Q}$ we have

$$T\Big(\underline{\lim}_{i\in\mathcal{I}}\, a_i\Big) = T\Big(\lim_{i\in\mathcal{I}}\inf_{k\ge i} a_k\Big) = \lim_{i\in\mathcal{I}} T\Big(\inf_{k\ge i} a_k\Big) \le \lim_{i\in\mathcal{I}}\inf_{k\ge i} T(a_k) = \underline{\lim}_{i\in\mathcal{I}}\, T(a_i)$$

and, likewise

$$T\Big(\overline{\lim}_{i\in I}\, a_i\Big) = T\Big(\lim_{i\in\mathcal{I}}\sup_{k\ge i} a_k\Big) = \lim_{i\in\mathcal{I}} T\Big(\sup_{k\ge i} a_k\Big) \ge \lim_{i\in\mathcal{I}}\sup_{k\ge i} T(a_k) = \overline{\lim}_{i\in\mathcal{I}}\, T(a_i),$$

that is

$$T\Big(\underline{\lim}_{i\in\mathcal{I}}\, a_i\Big) \le \underline{\lim}_{i\in\mathcal{I}}\, T(a_i) \le \overline{\lim}_{i\in\mathcal{I}}\, T(a_i) \le T\Big(\overline{\lim}_{i\in\mathcal{I}}\, a_i\Big).$$

**5.30 Lattice Homomorphisms.** Let both $(\mathcal{P},\mathcal{V})$ and $(\mathcal{Q},\mathcal{W})$ be locally convex $\vee$- (or $\wedge$-)semilattice cones. A continuous linear operator $T:\mathcal{P}\to\mathcal{Q}$ is called a $\vee$- (*or* $\wedge$-)*semilattice homomorphism* if it is compatible with the lattice operations in $\mathcal{P}$ and $\mathcal{Q}$, that is if

$$T(a\vee b) = T(a)\vee T(b) \qquad \big(\text{or} \quad T(a\wedge b) = T(a)\wedge T(b)\big)$$

holds for all $a, b \in \mathcal{P}$. If both $(\mathcal{P}, \mathcal{V})$ and $(\mathcal{Q}, \mathcal{W})$ are locally convex lattice cones and $T : \mathcal{P} \to \mathcal{Q}$ is both a $\vee$- and a $\wedge$-semilattice homomorphism, then $T$ is called a *lattice homomorphism*. Non-negative multiples of lattice homomorphism are again lattice homomorphisms, but sums are generally not.

Linear operators that are both order continuous and lattice homomorphisms are of particular interest. Suppose that both $(\mathcal{P}, \mathcal{V})$ and $(\mathcal{Q}, \mathcal{W})$ are locally convex complete lattice cones. A continuous linear operator $T : \mathcal{P} \to \mathcal{Q}$ is an order continuous lattice homomorphism if and only if

$$T(\sup A) = \sup T(A) \qquad \text{and} \qquad T(\inf B) = \inf T(B)$$

holds for all non-empty subsets $A$ and bounded below subsets $B$ of $\mathcal{P}$, that is if and only if $T$ preserves that lattice operations of $\mathcal{P}$ and $\mathcal{Q}$. Indeed, $\sup A$ or $\inf B$ is the limit with respect to order convergence of the net of suprema or infima of finite subsets of $A$ or $B$, respectively. Since an order continuous lattice homomorphism $T : \mathcal{P} \to \mathcal{Q}$ preserves finite suprema and infima as well as order convergence, we conclude that $T$ preserves infinite suprema and infima as well. Conversely, if $T$ preserves the suprema and infima of subsets of $\mathcal{P}$, then we have

$$T\Big(\underline{\lim}_{i \in \mathcal{I}} a_i\Big) = \underline{\lim}_{i \in \mathcal{I}} T(a_i) \qquad \text{and} \qquad T\Big(\overline{\lim}_{i \in \mathcal{I}} a_i\Big) = \overline{\lim}_{i \in \mathcal{I}} T(a_i)$$

for every bounded below net $(a_i)_{i \in \mathcal{I}}$ in $\mathcal{P}$. Thus $T$ maps order convergent nets in $\mathcal{P}$ into order convergent nets in $\mathcal{Q}$ and is therefore an order continuous lattice homomorphism.

*Examples 5.31.* (a) Theorem II.6.7 in [100] states that for every neighborhood $v \in \mathcal{V}$ in an M-type locally convex $\vee$- (or $\wedge$-)semilattice cone $(\mathcal{P}, \mathcal{V})$ all the extreme points of its polar $v^\circ \subset \mathcal{P}^*$ are $\vee$- (or $\wedge$-)semilattice homomorphisms from $\mathcal{P}$ into $\overline{\mathbb{R}}$.

(b) Let $(\mathcal{P}, \mathcal{V})$ be a locally convex cone with dual $\mathcal{P}^*$ and let $(\mathcal{Q}, \mathcal{V})$ be the cone of all non-empty convex subsets of $\mathcal{P}$ which are closed with respect to the lower topology (see Example 4.37(d)). In 5.7(d) we showed that $(\mathcal{Q}, \mathcal{V})$ is a locally convex $\bigvee$-semilattice cone ordered by the set inclusion. There is a natural embedding $\mu \mapsto \tilde{\mu} : \mathcal{P}^* \to \mathcal{Q}^*$, where

$$\tilde{\mu}(A) = \sup\{\mu(a) \mid a \in A\}$$

for $\mu \in \mathcal{P}^*$ and $A \in \mathcal{Q}$. Indeed, if $\mu \in v^\circ$ for some $v \in \mathcal{V}$, then $A \leq B \oplus v$ for $A, B \in \mathcal{Q}$ means that for every $a \in A$ and $\varepsilon \geq 0$ there is $b \in B$ such that $a \leq \gamma b + (1+\varepsilon)v$ (see 4.37(e)) with some $1 \leq \gamma \leq 1+\varepsilon$. This yields

$$\mu(a) \leq \gamma\mu(b) + (1+\varepsilon) \leq \gamma\tilde{\mu}(b) + (1+\varepsilon)$$

for all $\varepsilon > 0$, hence $\mu(a) \leq \tilde{\mu}(B) + 1$. We infer $\hat{\mu}(A) \leq \tilde{\mu}(B) + 1$, and conclude that $\tilde{\mu} \in v^\circ \subset \mathcal{Q}^*$. Moreover, $\tilde{\mu}$ is a $\vee$-semilattice homomorphism even with respect to arbitrary suprema in $\mathcal{Q}$: Let $\mathcal{A}$ be a subset of $\mathcal{Q}$ and let $c$ be an element of $conv(\bigcup_{A \in \mathcal{A}} A)$, the convex hull of the union of all elements of $\mathcal{A}$. Then $c = \sum_{i=1}^{n} \alpha_i a_i$ for some $a_i \in A_i \in \mathcal{A}$ and $\alpha_i \geq 0$ such that $\sum_{i=1}^{n} \alpha_i = 1$. Thus

$$\mu(c) = \sum_{i=1}^{n} \alpha_i \mu(a_i) \leq \sum_{i=1}^{n} \alpha_i \tilde{\mu}(A_i) \leq \sup_{A \in \mathcal{A}} \tilde{\mu}(A).$$

Since the functional $\mu : \mathcal{P} \to \overline{\mathbb{R}}$ is also continuous with respect to the lower relative topology on $\mathcal{P}$, we conclude that

$$\begin{aligned} \tilde{\mu}(\sup \mathcal{A}) &= \sup \left\{ \mu(a) \;\middle|\; a \in \overline{conv(\bigcup_{A \in \mathcal{A}} A)}^{(l)} \right\} \\ &= \sup \left\{ \mu(a) \;\middle|\; a \in conv(\bigcup_{A \in \mathcal{A}} A) \right\} \leq \sup_{A \in \mathcal{A}} \tilde{\mu}(A). \end{aligned}$$

The converse inequality is obvious.

(c) Similarly one argues for the locally convex cone $(\mathcal{Q}, \mathcal{V})$ of all bounded below convex subsets of $\mathcal{P}$ which are closed with respect to the upper topology (see Examples 4.37(e) and 5.7(e)). In this case $(\mathcal{Q}, \mathcal{V})$ is a locally convex $\bigwedge$-semilattice cone, ordered by the inverse set inclusion. There is a natural embedding $\mu \mapsto \tilde{\mu} : \mathcal{P}^* \to \mathcal{Q}^*$, where

$$\tilde{\mu}(A) = \inf\{\mu(a) \mid a \in A\}$$

for $\mu \in \mathcal{P}^*$ and $A \in \mathcal{Q}$. As similar argument as in (b) shows that

$$\tilde{\mu}(\inf \mathcal{A}) = \inf_{A \in \mathcal{A}} \tilde{\mu}(A)$$

holds for every bounded below family of sets $\mathcal{A} \subset \mathcal{Q}$.

(d) Let $(\mathcal{P}, \mathcal{V})$ be a locally convex $\vee$- (or $\wedge$-)semilattice cone, X a set, and consider the locally convex cone $\big(\mathcal{F}_{\widehat{\mathcal{V}}_b}(X, \mathcal{P}), \widehat{\mathcal{V}}\big)$ of $\mathcal{P}$-valued functions on $X$, endowed with the neighborhood system $\widehat{\mathcal{V}}$ consisting of $(\mathcal{V} \cup \{\infty\})$ -valued functions. (Example 1.4(e)). This was seen to be again a locally convex $\vee$- (or $\wedge$-)semilattice cone, provided that for every $x \in X$ and $v \in \mathcal{V}$ there is $\hat{v} \in \widehat{\mathcal{V}}$ such that $\hat{v}(x) \leq v$ (see 5.7(c)). For $\mu \in v^\circ \subset \mathcal{P}^*$, a neighborhood $\hat{v} \in \widehat{\mathcal{V}}$ and $x \in X$ such that $\hat{v}(x) \leq v$, the mapping $\mu_x : \mathcal{F}_{\widehat{\mathcal{V}}_b}(X, \mathcal{P}) \to \overline{\mathbb{R}}$ such that $\mu_x(f) = \mu\big(f(x)\big)$ for all $f \in \mathcal{F}_{\widehat{\mathcal{V}}_b}(X, \mathcal{P})$ is a continuous linear functional on $\mathcal{F}_{\widehat{\mathcal{V}}_b}(X, \mathcal{P})$ (see 2.1(f)), more precisely: an element of $\hat{v}^\circ$. Moreover, if $\mu$ is a $\vee$- (or $\wedge$-)semilattice homomorphism for $\mathcal{P}$, then $\mu_x$ is a semilattice homomorphism of the same type for $\mathcal{F}_{\widehat{\mathcal{V}}_b}(X, \mathcal{P})$.

**5.32 Functionals Supporting the Separation Property.** Corollary 4.34 (see also the Separation Theorem 3.2 in [175]) guarantees that in a locally convex cone the neighborhoods with respect to the weak preorder are completely determined by their polars, that is $a \preccurlyeq b+v$ holds for $a,b \in \mathcal{P}$ and $v \in \mathcal{V}$ if and only if $\mu(a) \leq \mu(b)+1$ for all $\mu \in v^\circ$. In this vein, for a locally convex cone $(\mathcal{P},\mathcal{V})$ we shall say that a subset $\Upsilon$ of $\mathcal{P}^*$ *supports the separation property for* $\mathcal{P}$ if for $a,b \in \mathcal{P}$ and $v \in \mathcal{V}$ such that $a \not\preccurlyeq b+v$ there is $\alpha \geq 0$ and $\mu \in \Upsilon \cap (\alpha v^\circ)$ such that $\mu(a) > \mu(b)+\alpha$. This property implies in particular that the functionals in $\Upsilon$ determine the weak preorder of $\mathcal{P}$, that is $a \preccurlyeq b$ holds for $a,b \in \mathcal{P}$ if and only if $\mu(a) \leq \mu(b)$ for all $\mu \in \Upsilon$. Indeed, the latter implies that $a \preccurlyeq b+v$ for all $v \in \mathcal{V}$, which by Lemma 3.2(a) yields $a \preccurlyeq b$.

*Examples 5.33.* (a) In Examples 1(a) and (b), that is for $\mathcal{P}=\overline{\mathbb{R}}$ or $\mathcal{P}=\overline{\mathbb{R}}_+$ the dual cone contains all positive reals, and the set $\Upsilon=\{1\}$ supports the separation property.

(b) If $\mathcal{V}$ consists of the multiples of a single neighborhood $v$, then we may choose

$$\Upsilon = \{\mu \in \mathcal{P}^* \mid \psi_v(\mu)=0 \ \text{ or } \ \psi_v(\mu)=1\},$$

where $\psi_v(\mu) = \inf\{\alpha \geq 0 \mid \mu \in \alpha v^\circ\}$. (In case that $v \in \mathcal{P}$, we have $\psi_v(\mu)=\mu(v)$.) Indeed, if $a \not\leq b+(\rho v)$ for $a,b \in \mathcal{P}$ and $\rho v \in \mathcal{P}$, then by Corollary 4.34 there is $\mu \in (\rho v)^\circ = (1/\rho)v^\circ$ such that $\mu(a) > \mu(b)+1$. This implies $\psi_v(\mu) \leq 1/\rho$. If $\psi_v(\mu)=0$, then $\mu \in \Upsilon \cap (\rho v)^\circ$ as required. Otherwise, we set $\alpha = 1/\psi_v(\mu) > 0$ and $\nu = \alpha\mu \in \Upsilon$ and observe that both $\nu \in \alpha(\rho v)^\circ$ and $\nu(a) > \nu(b)+\alpha$, again satisfying the requirement.

If in addition all elements of $\mathcal{P}$ are bounded, that is for example, if $\mathcal{P}$ is normed vector space, then according to Corollary 4.35 we may further reduce the size of $\Upsilon$ and choose

$$\Upsilon = \mathrm{Ex}(v^\circ),$$

that is the set of all extreme points of the $w(\mathcal{P}^*,\mathcal{P})$-compact convex set $v^\circ$. (Obviously, $\psi_v(\mu)=1$ holds for every $\mu \in \mathrm{Ex}(v^\circ)$.)

(c) More generally, a locally convex cone $(\mathcal{P},\mathcal{V})$ is said to be *tightly covered by its bounded elements* (see II.2.13 in [100]) if for all $a,b \in \mathcal{P}$ and $v \in \mathcal{V}$ such that $a \not\preccurlyeq b+v$ there is some bounded element $a' \in \mathcal{P}$ such that $a' \preccurlyeq a$ and $a' \not\preccurlyeq b+v$. In this case, if $\mathcal{V}_0$ is a subcollection of $\mathcal{V}$ such that every $v \in \mathcal{V}$ is a multiple of some $v_0 \in \mathcal{V}_0$, then according to Corollary II.4.7 in [100] the set

$$\Upsilon = \bigcup_{v_0 \in \mathcal{V}_0} \mathrm{Ex}(v_0^\circ)$$

supports the separation property for $\mathcal{P}$.

(d) Let $(\mathcal{P},\mathcal{V})$ be a locally convex cone with dual $\mathcal{P}^*$ and let $(\mathcal{Q},\mathcal{V})$ be the cone of all non-empty convex subsets of $\mathcal{P}$ which are closed with respect to the lower topology (see Example 4.37(d)). For every $\mu \in \mathcal{P}^*$ the formula

$$\tilde{\mu}(A) = \sup\{\mu(a) \mid a \in A\} \qquad \text{for} \quad A \in \mathcal{Q}$$

defines an element $\tilde{\mu} \in \mathcal{Q}^*$, more precisely, $\tilde{\mu} \in v^\circ$ whenever $\mu \in v^\circ$ (see Example 5.31(b) before). Now Theorem 4.33 guarantees that the set

$$\Upsilon = \{\tilde{\mu} \mid \mu \in \mathcal{P}^*\} \subset \mathcal{Q}^*$$

supports the separation property for $\mathcal{Q}$. Indeed, if $A \not\leq B \oplus v$ for $A, B \in \mathcal{Q}$ and $v \in \mathcal{V}$, then there is $a \in A$ such that $a \notin v(B)$ (see 4.37(d)). Following Theorem 4.33(a) then there is $\mu \in v^\circ$, hence $\tilde{\mu} \in v^\circ$ such that

$$\mu(a) > \sup\{\mu(b) \mid b \in B\} + 1 = \tilde{\mu}(B) + 1.$$

Thus $\tilde{\mu}(A) = \sup\{\mu(a) \mid a \in A\} > \tilde{\mu}(B) + 1$.

(e) Similarly, if $(\mathcal{Q}, \mathcal{V})$ is the locally convex cone of all bounded below convex subsets of $\mathcal{P}$ which are closed with respect to the upper topology (see Example 4.37(e)), then for every $\mu \in \mathcal{P}^*$ the formula

$$\tilde{\mu}(A) = \inf\{\mu(a) \mid a \in A\} \qquad \text{for} \quad A \in \mathcal{Q}$$

defines an element $\tilde{\mu} \in \mathcal{Q}^*$, more precisely, $\tilde{\mu} \in v^\circ$ whenever $\mu \in v^\circ$ (see 5.31(c)). Then

$$\Upsilon = \{\tilde{\mu} \mid \mu \in \mathcal{P}^*\} \subset \mathcal{Q}^*$$

supports the separation property for $\mathcal{Q}$. Indeed, if $A \not\leq B \oplus v$ for $A, B \in \mathcal{Q}$ and $v \in \mathcal{V}$, then there is $b \in B$ such that $b \notin (A)v$ (see 4.37(e)). Following Theorem 4.33(b) then there is $\mu \in v^\circ$, hence $\tilde{\mu} \in v^\circ$ such that $\mu(b) < \inf\{\mu(b) \mid b \in B\} - 1 = \tilde{\mu}(A) - 1$. Thus $\tilde{\mu}(B) = \inf\{\mu(b) \mid b \in B\} < \tilde{\mu}(A) - 1$, that is $\tilde{\mu}(A) > \tilde{\mu}(B) + 1$.

(f) Let $(\mathcal{P}, \mathcal{V})$ be a locally convex cone, $X$ a set, and consider the locally convex cone $\big(\mathcal{F}_{\widehat{\mathcal{V}}_b}(X, \mathcal{P}), \widehat{\mathcal{V}}\big)$ of $\mathcal{P}$-valued functions on $X$, where the neighborhood system $\widehat{\mathcal{V}}$ is generated by a family of $(\mathcal{V} \cup \{\infty\})$-valued functions on $X$ as elaborated in Example 1.4(e). For every $\mu \in v\circ \subset \mathcal{P}^*$ for $v \in \mathcal{V}$, and $x \in X$ such that $\hat{v}(x) \leq v$ for $\hat{v} \in \widehat{\mathcal{V}}$, the formula

$$\mu_x(f) = \mu\big(f(x)\big) \qquad \text{for} \quad f \in \mathcal{F}_{\widehat{\mathcal{V}}_b}(X, \mathcal{P})$$

defines a continuous linear functional on $\mathcal{F}_{\widehat{\mathcal{V}}_b}(X, \mathcal{P})$ (see 2.1(f)), more precisely: We have $\mu_x \in \hat{v}^\circ$. Let us denote by $X_0$ the subset of all $x \in X$ such that $\hat{v}(x) \neq \infty$ for at least one $\hat{v} \in \widehat{\mathcal{V}}$. If $\Upsilon \subset \mathcal{P}^*$ supports the separation property for $\mathcal{P}$, then

$$\widehat{\Upsilon} = \{\mu_x \mid \mu \in \Upsilon, \ x \in X_0\} \subset \mathcal{F}_{\widehat{\mathcal{V}}_b}(X, \mathcal{P})^*$$

supports the separation property for $\mathcal{F}_{\widehat{\mathcal{V}}_b}(X, \mathcal{P})$. Indeed, if $f \not\leq g + \hat{v}$ for $f, g \in \mathcal{F}_{\widehat{\mathcal{V}}_b}(X, \mathcal{P})$ and $\hat{v} \in \widehat{\mathcal{V}}$, then there is $x \in X$ such that $f(x) \not\leq g(x) + \hat{v}(x)$. This implies $\hat{v}(x) \neq \infty$, hence $v = \hat{v}(x) \in \mathcal{V}$. Following our assumption there is $\alpha \geq 0$ and $\mu \in \Upsilon \cap (\alpha v^\circ)$ such that $\mu\big(f(x)\big) > \mu\big(g(x)\big) + \alpha$.

We therefore have $\mu_x \in \widehat{\Upsilon} \cap (\alpha\hat{v}^\circ)$ and $\mu_x(f) > \mu_x(g) + \alpha$, as required. In case that $\mathcal{P} = \overline{\mathbb{R}}$ or $\mathcal{P} = \overline{\mathbb{R}}_+$ we may choose $\Upsilon = \{1\}$ (see 5.33(a)). Then $\widehat{\Upsilon}$ consists of all point evaluations at the points $x \in X$ such that $\hat{v}(x) < +\infty$ for at least one of the $\overline{\mathbb{R}}_+$-valued neighborhood functions $\hat{v} \in \widehat{\mathcal{V}}$.

The presence of suitable subsets of $\mathcal{P}^*$ supporting the separation property permits a strengthening of certain statements for the general case. The following Propositions 5.34 and 5.35 will improve on Proposition 5.15(b) and Propositions 5.25(b) and 5.28(b) under these circumstances. Recall that a subset $A$ of an ordered cone $\mathcal{P}$ is said to be *directed upward* (or *downward*) if for $a, b \in A$ there is $c \in A$ such that both $a \le c$ and $b \le c$ (or $c \le a$ and $c \le b$.)

**Proposition 5.34.** *Let $(\mathcal{P}, \mathcal{V})$ be a locally convex complete lattice cone, and suppose that the order continuous lattice homomorphisms support the separation property for $\mathcal{P}$. Then*

*(a)* $\sup(A \wedge B) = \sup A \wedge \sup B$ *for non-empty subsets $A, B$ of $\mathcal{P}$.*

*(b)* $\lim_{i \in \mathcal{I}} (a_i \wedge b_i) = \left(\lim_{i \in \mathcal{I}} a_i\right) \wedge \left(\lim_{i \in \mathcal{I}} b_i\right)$
*for convergent nets $(a_i)_{i \in \mathcal{I}}$ and $(b_i)_{i \in \mathcal{I}}$ in $\mathcal{P}$.*

*Proof.* Let $\Upsilon$ be the subset of all order continuous lattice homomorphisms in $\mathcal{P}^*$ and suppose that $\Upsilon$ supports the separation property for $\mathcal{P}$. For Part (a), let $A, B$ be non-empty subsets of $\mathcal{P}$. In Proposition 5.15(b) we already demonstrated $\sup(A \wedge B) \le \sup A \wedge \sup B$. For the converse inequality, it suffices to verify that

$$\mu(\sup A \wedge \sup B) \le \mu\big(\sup(A \wedge B)\big)$$

holds for all $\mu \in \Upsilon$ (see 5.32). For this, let $\mu \in \Upsilon$. Then

$$\mu(\sup A \wedge \sup B) = \mu(\sup A) \wedge \mu(\sup B) = \sup\big(\mu(A)\big) \wedge \sup\big(\mu(B)\big),$$

since $\mu$ is an order continuous lattice homomorphism. We may assume that $\mu(\sup A) \le \mu(\sup B)$. Then for every $a \in A$ and $\varepsilon > 0$ there is $b \in B$ such that $\mu(a) \le \mu(b) + \varepsilon$. Thus also $\mu(a) \le \mu(a \wedge b) + \varepsilon$. This shows

$$\sup\big(\mu(A)\big) \wedge \sup\big(\mu(B)\big) = \sup\big(\mu(A)\big) \le \mu\big(\sup(A \wedge B)\big) + \varepsilon$$

and verifies our claim.

For Part (b), let $(a_i)_{i \in \mathcal{I}}$ and $(b_i)_{i \in \mathcal{I}}$ be convergent nets in $\mathcal{P}$. In the light of 5.25(b) and our assumption on $\Upsilon$ it suffices to verify that

$$\mu\left(\left(\lim_{i \in \mathcal{I}} a_i\right) \wedge \left(\lim_{i \in \mathcal{I}} b_i\right)\right) \le \mu\left(\underline{\lim}_{i \in \mathcal{I}} (a_i \wedge b_i)\right),$$

that is

$$\left(\lim_{i\in\mathcal{I}}\mu(a_i)\right)\wedge\left(\lim_{i\in\mathcal{I}}\mu(b_i)\right)\le\underline{\lim}_{i\in\mathcal{I}}\left(\mu(a_i)\wedge\mu(b_i)\right)$$

holds for all $\mu\in\Upsilon$. For this, given a functional $\mu\in\Upsilon$, we may assume that $\lim_{i\in\mathcal{I}}\mu(a_i)\le\lim_{i\in\mathcal{I}}\mu(b_i)$. Then for every $\varepsilon>0$ there is $i_0\in\mathcal{I}$ such that $\mu(a_i)\le\mu(b_i)+\varepsilon$ for all $i\ge i_0$. This implies $\mu(a_i)\le\mu(a_i)\wedge\mu(b_i)+\varepsilon$. Thus $\lim_{i\in\mathcal{I}}\mu(a_i)\le\underline{\lim}_{i\in\mathcal{I}}\left(\mu(a_i)\wedge\mu(b_i)\right)+\varepsilon$. This yields our claim. □

**Proposition 5.35.** *Let $(\mathcal{P},\mathcal{V})$ be a locally convex complete lattice cone. If the order continuous lattice homomorphisms (or the order continuous linear functionals) support the separation property for $\mathcal{P}$, then*

$$\sum_{i=1}^{\infty}\inf A_i=\inf\left\{\sum_{i=1}^{\infty}A_i\right\}$$

*for every convergent series $\sum_{i=1}^{\infty}A_i$ of non-empty (or non-empty directed downward) subsets of $\mathcal{P}$.*

*Proof.* Let $\Upsilon$ be the subset of all order continuous lattice homomorphisms (or order continuous linear functionals) in $\mathcal{P}^*$ and suppose that $\Upsilon$ supports the separation property for $\mathcal{P}$. In the second case we assume in addition that the sets $A_i\subset\mathcal{P}$ are directed downward. Thus, in each of the cases for $\Upsilon$ we have

$$\mu(\inf A_i)=\inf\{\mu(A_i)\}$$

for all $i\in\mathbb{N}$ and $\mu\in\Upsilon$. The order continuity of the functionals $\mu$ then yields

$$\mu\left(\sum_{i=1}^{\infty}\inf A_i\right)=\sum_{i=1}^{\infty}\mu(\inf A_i)=\sum_{i=1}^{\infty}\inf\{\mu(A_i)\}.$$

Likewise, since the sets $\sum_{i=1}^{\infty}A_i$ are seen to be directed downward in the second case for $\Upsilon$, we have

$$\begin{aligned}\mu&\left(\inf\left\{\sum_{i=1}^{\infty}A_i\right\}\right)\\&=\inf\left\{\mu\left(\sum_{i=1}^{\infty}a_i\right)\,\middle|\,a_i\in A_i\right\}=\inf\left\{\sum_{i=1}^{\infty}\mu(a_i)\,\middle|\,a_i\in A_i\right\}.\end{aligned}$$

Given $\mu\in\Upsilon$ we choose $a_i\in A_i$ such that $\mu(a_i)\le\inf\{\mu(A_i)\}+2^{-i}$. Then

$$\inf\left\{\sum_{i=1}^{\infty}\mu(a_i)\,\middle|\,a_i\in A_i\right\}\le\sum_{i=1}^{\infty}\mu(a_i)\le\sum_{i=1}^{\infty}\inf\{\mu(A_i)\}+1,$$

hence

$$\mu\left(\inf\left\{\sum_{i=1}^{\infty} A_i\right\}\right) \leq \mu\left(\sum_{i=1}^{\infty} \inf A_i\right) + 1.$$

Because $\Upsilon$ supports the separation property for $\mathcal{P}$, this shows

$$\inf\left\{\sum_{i=1}^{\infty} A_i\right\} \leq \sum_{i=1}^{\infty} \inf A_i + v$$

for all $v \in \mathcal{V}$, hence

$$\inf\left\{\sum_{i=1}^{\infty} A_i\right\} \leq \sum_{i=1}^{\infty} \inf A_i,$$

since $\mathcal{P}$ carries its weak preorder. The reverse inequality was established in Proposition 5.28(b). □

We shall say that a subcone $\mathcal{N}$ of $\mathcal{P}$ is a *locally convex lattice subcone of* $(\mathcal{P},\mathcal{V})$ if $a \vee b \in \mathcal{N}$ and $a \wedge b \in \mathcal{N}$ whenever $a, b \in \mathcal{N}$. Likewise, $\mathcal{N}$ is a *locally convex complete lattice subcone of* $(\mathcal{P},\mathcal{V})$ if $\sup A \in \mathcal{N}$ and $\inf B \in \mathcal{N}$ whenever $A, B \subset \mathcal{N}$, $A$ is not empty and $B$ is bounded below. The suprema and infima are taken in $\mathcal{P}$.

A family $\mathfrak{A}$ of subsets of $\mathcal{P}$ will be called *sup-bounded below* if the set $\{\sup A \mid A \in \mathfrak{A}\}$ is bounded below in $\mathcal{P}$. This implies in particular that $\emptyset \notin \mathfrak{A}$ and that $\inf\{\sup A \mid A \in \mathfrak{A}\}$ exists in $\mathcal{P}$.

**Proposition 5.36.** *Let $(\mathcal{P},\mathcal{V})$ be a locally convex complete lattice cone, and suppose that the order continuous lattice homomorphisms support the separation property for $\mathcal{P}$. Let $\mathcal{N}$ be a subcone of $\mathcal{P}$. The smallest locally convex complete lattice subcone of $\mathcal{P}$ that contains $\mathcal{N}$ consists of all elements $a \in \mathcal{P}$ which can be expressed in the following way:*

$$a = \inf\{\sup A \mid A \in \mathfrak{A}\},$$

*where $\mathfrak{A}$ is a sup-bounded below family of subsets of $\mathcal{N}$.*

*Proof.* Let $(\mathcal{P},\mathcal{V})$ be a locally convex complete lattice cone. Corresponding to a sup-bounded below family $\mathfrak{A}$ of non-empty subsets of $\mathcal{P}$ let us define the element $a_{\mathfrak{A}} \in \mathcal{P}$ by

$$a_{\mathfrak{A}} = \inf\{\sup A \mid A \in \mathfrak{A}\}.$$

For families $\mathfrak{A}$ and $\mathfrak{B}$ of this type and $\alpha \geq 0$ we denote $\alpha\mathfrak{A} = \{\alpha A \mid A \in \mathfrak{A}\}$ and $\mathfrak{A} + \mathfrak{B} = \{A + B \mid A \in \mathfrak{A},\ B \in \mathfrak{B}\}$. It is evident from Lemmas 5.5 and 5.6 that these are again sup-bounded below families of subsets of $\mathcal{P}$. We also use 5.5 and 5.6 for the following observations:

(i) $\alpha a_{\mathfrak{A}} = \alpha \inf\{\sup A \mid A \in \mathfrak{A}\} = \inf\{\sup \alpha A \mid A \in \mathfrak{A}\} = a_{\alpha A}.$

$$\text{(ii)} \qquad \begin{aligned} a_{\mathfrak{A}} + a_{\mathfrak{B}} &= \inf\big\{ \sup A \mid A \in \mathfrak{A}\big\} + \inf\big\{ \sup B \mid B \in \mathfrak{B}\big\} \\ &= \inf\big\{ \sup A + \sup B \mid A \in \mathfrak{A},\ B \in \mathfrak{B}\big\} \\ &= \inf\big\{ \sup(A+B) \mid A \in \mathfrak{A},\ B \in \mathfrak{B}\big\} \\ &= \inf\big\{ \sup C \mid C \in (\mathfrak{A}+\mathfrak{B})\big\} \\ &= a_{(\mathfrak{A}+\mathfrak{B})}. \end{aligned}$$

Let $\{\mathfrak{A}_i\}_{i\in\mathcal{I}}$ be a collection of sup-bounded families $\mathfrak{A}_i$ of subsets of $\mathcal{P}$. In a first instance, suppose that this collection is not empty, and let $\mathfrak{A} = \{\cup_{i\in\mathcal{I}} A_i \mid (A_i)_{i\in\mathcal{I}} \in \prod_{i\in\mathcal{I}} \mathfrak{A}_i\}$, that is the elements $A$ of $\mathfrak{A}$ are all unions of the type $A = \cup_{i\in\mathcal{I}} A_i$, where $A_i \in \mathfrak{A}_i$. (The Axiom of Choice is required for this construction.) This family $\mathfrak{A}$ is also sup-bounded below. Indeed, given $v \in \mathcal{V}$ and a fixed $k \in \mathcal{I}$ there is $\lambda \geq 0$ such that $0 \leq \sup A_k + \lambda v$ for all $A_k \in \mathfrak{A}_k$. Thus for every $A \in \mathfrak{A}$ we have $A_k \subset A$ for some $A_k \in \mathfrak{A}_k$, hence $0 \leq \sup A_k + \lambda v \leq \sup A + \lambda v$. We claim that

$$\text{(iii)} \qquad \sup_{i\in\mathcal{I}} a_{\mathfrak{A}_i} = \sup_{i\in\mathcal{I}} \inf\{\sup A_i \mid A_i \in \mathfrak{A}_i\} = \inf\{\sup A \mid A \in \mathfrak{A}\} = a_{\mathfrak{A}}.$$

Indeed, for every fixed $i \in \mathcal{I}$ and every $A \in \mathfrak{A}$ there is some $A_i \in \mathfrak{A}_i$ such that $A_i \subset A$. This shows $\inf\{\sup A_i \mid A_i \in \mathfrak{A}_i\} \leq \sup A$ for all $A \in \mathfrak{A}$, hence $\inf\{\sup A_i \mid A_i \in \mathfrak{A}_i\} \leq \inf\{\sup A \mid A \in \mathfrak{A}\}$ holds for all $i \in \mathcal{I}$. This yields

$$\sup_{i\in\mathcal{I}} \inf\{\sup A_i \mid A_i \in \mathfrak{A}_i\} \leq \inf\{\sup A \mid A \in \mathfrak{A}\}.$$

For the converse inequality we will have to use the fact that the lattice operations are formed in a locally convex complete lattice cone for which the order continuous lattice homomorphisms support the separation property, that is it suffices to verify that

$$\mu\big(\inf\{\sup A \mid A \in \mathfrak{A}\}\big) \leq \mu\big(\sup_{i\in\mathcal{I}} \inf\{\sup A_i \mid A_i \in \mathfrak{A}_i\}\big)$$

holds for every order continuous lattice homomorphism $\mu \in \mathcal{P}^*$. For this assume to the contrary that there is

$$\rho < \mu\big(\inf\{\sup A \mid A \in \mathfrak{A}\}\big) = \inf\{\sup \mu(A) \mid A \in \mathfrak{A}\},$$

and that

$$\mu\big(\sup_{i\in\mathcal{I}} \inf\{\sup A_i \mid A_i \in \mathfrak{A}_i\}\big) = \sup_{i\in\mathcal{I}} \inf\{\sup \mu(A_i) \mid A_i \in \mathfrak{A}_i\} < \rho$$

holds for some order continuous lattice homomorphism $\mu \in \mathcal{P}^*$. This means $\inf\{\sup \mu(A_i) \mid A_i \in \mathfrak{A}_i\} < \rho$ for all $i \in \mathcal{I}$, hence $\sup \mu(A_i) < \rho$ for some $A_i \in \mathfrak{A}_i$. We use these sets $A_i$ for $A = \cup_{i\in\mathcal{I}} A_i \in \mathfrak{A}$. We have $\sup \mu(A) = \sup\{\mu(A_i) \mid i \in \mathcal{I}\} \leq \rho$, contradicting the assumption that $\rho < \sup \mu(A)$ holds for all $A \in \mathfrak{A}$. This yields our claim.

In a second instance, suppose that the set $\{a_{\mathfrak{A}_i}\}_{i\in\mathcal{I}}$ is bounded below in $\mathcal{P}$. Then the family $\mathfrak{A} = \cup_{i\in\mathcal{I}} \mathfrak{A}_i$ is also sup-bounded below. Indeed, given $v \in \mathcal{V}$

there is $\lambda \geq 0$ such that $0 \leq a_{\mathfrak{A}_i} + \lambda v$ for all $i \in \mathcal{I}$. Thus for every $A \in \mathfrak{A}$ we have $A \in \mathfrak{A}_i$ for some $i \in \mathcal{I}$ and therefore $0 \leq a_{\mathfrak{A}_i} + \lambda v \leq \sup A + \lambda v$. Now we infer that

$$\text{(iv)} \qquad \begin{aligned} \inf_{i\in\mathcal{I}} a_{\mathfrak{A}_i} &= \inf_{i\in\mathcal{I}} \big\{ \inf\{\sup A_i \mid A_i \in \mathfrak{A}_i\}\big\} \\ &= \inf\big\{ \sup A \mid A \in \cup_{i\in\mathcal{I}}\mathfrak{A}_i\big\} \\ &= a_{\mathfrak{A}} \end{aligned}$$

Now let $\mathcal{N}$ be a subcone of $\mathcal{P}$ and denote by $\widehat{\mathcal{N}}$ the subset of $\mathcal{P}$ consisting of all elements $a_{\mathfrak{A}}$, where $\mathfrak{A}$ is an sup-bounded below family of subsets of $\mathcal{N}$. The preceding arguments (i) to (iv) yield that $\widehat{\mathcal{N}}$ is a locally convex lattice subcone of $\mathcal{P}$. Since $a_{\mathfrak{A}} = a$ for every $a \in \mathcal{N}$ with $\mathfrak{A} = \big\{\{a\}\big\}$, we have $\mathcal{N} \subset \widehat{\mathcal{N}}$. On the other hand, every locally convex lattice subcone of $\mathcal{P}$ that contains $\mathcal{N}$, necessarily contains also all elements $a_{\mathfrak{A}} \in \mathcal{P}$ of this type. Thus $\widehat{\mathcal{N}}$ is indeed the smallest locally convex complete lattice subcone of $\mathcal{P}$ that contains $\mathcal{N}$. □

For the following recall the notations from Example 1.4(e). We observed before that $\big(\mathcal{F}_{\widehat{\mathcal{V}}_b}(X,\overline{\mathbb{R}}),\widehat{\mathcal{V}}\big)$ is a locally convex complete lattice cone for any choice of the set $X$ and the neighborhood system $\widehat{\mathcal{V}}$, provided that for every $x \in X$ there is $\hat{v} \in \widehat{\mathcal{V}}$ such that $\hat{v}(x) < +\infty$ $\big($see 5.7(c)$\big)$. The point evaluations at the points $x \in X$ are order continuous lattice homomorphisms and according to Example 5.33(f) support the separation property for $\mathcal{F}_{\widehat{\mathcal{V}}_b}(X,\overline{\mathbb{R}})$. Thus for every locally convex complete lattice subcone of $\big(\mathcal{F}_{\widehat{\mathcal{V}}_b}(X,\overline{\mathbb{R}}),\widehat{\mathcal{V}}\big)$ the order continuous lattice homomorphisms support the separation property. For an inverse implication recall the definition of an embedding in 2.2:

**Proposition 5.37.** *Let $(\mathcal{P},\mathcal{V})$ be a locally convex complete lattice cone. If the set $\Upsilon$ of all order continuous lattice homomorphisms in $\mathcal{P}^*$ supports the separation property, then $(\mathcal{P},\mathcal{V})$ can be embedded into the locally convex complete lattice of $\big(\mathcal{F}_{\widehat{\mathcal{V}}_b}(\Upsilon,\overline{\mathbb{R}}),\widehat{\mathcal{V}}\big)$, endowed with a suitable system $\widehat{\mathcal{V}}$ of $\overline{\mathbb{R}}_+$ -valued neighborhood functions. This embedding is one-to-one and preserves the lattice operations.*

*Proof.* Let $\Upsilon$ be the set of order continuous lattice homomorphisms in $\mathcal{P}^*$. Recall that $\alpha\mu \in \Upsilon$ whenever $\mu \in \Upsilon$ and $\alpha \geq 0$. With every element $a \in \mathcal{P}$ we associate the function $\varphi_a : \Upsilon \to \overline{\mathbb{R}}$ such that

$$\varphi_a(\mu) = \mu(a) \qquad \text{for all} \quad \mu \in \Upsilon.$$

The mapping $\Phi : \mathcal{P} \to \mathcal{F}(\Upsilon,\overline{\mathbb{R}})$ such that $\Phi(a) = \varphi_a$ for all $a \in \mathcal{P}$ is obviously linear, monotone, and since $\mathcal{P}$ carries the weak preorder which is supposed to be antisymmetric, $\Phi$ is also one-to one. Since the elements of $\Upsilon$ are all order continuous lattice homomorphisms in $\mathcal{P}^*$, for every subset $A$ of $\mathcal{P}$ and $\mu \in \Upsilon$ we have

$$\mu(\sup A) = \sup\{\mu(a) \mid a \in A\} = \sup\{\varphi_a(\mu) \mid a \in A\} = \big(\sup \varphi_a\big)(\mu).$$

Thus $\Phi(\sup A) = \sup \Phi(A)$. Likewise, we have $\Phi(\inf B) = \inf \Phi(B)$ for every bounded below subset $B$ of $\mathcal{P}$. Recall that the lattice operations are carried out pointwise in $\mathcal{F}(\Upsilon, \overline{\mathbb{R}}.)$ Corresponding to the neighborhoods $v \in \mathcal{V}$ we consider the $\overline{\mathbb{R}}$-valued functions $\psi_v$ on $\Upsilon$ such that

$$\psi_v(\mu) = \inf\{\alpha > 0 \mid \mu \in \alpha v^\circ\}$$

for all $\mu \in \mathcal{P}^*$. As usual, we set $\inf \emptyset = +\infty$, but observe that for every $\mu \in \Upsilon$ there is $v \in \mathcal{V}$ such that $\psi_v(\mu) < +\infty$. Note that $\psi_v = \varphi_v$ in case that $v \in \mathcal{P}$. We also note that the family of all functions $\psi_v$ for $v \in \mathcal{V}$ is not necessarily closed for the pointwise addition of its functions. For this reason we refer to the last remark in Example 1.4(e) relating to the construction of a locally convex cone of cone-valued functions. For $\mathcal{F}(\Upsilon, \overline{\mathbb{R}})$ we use the abstract neighborhood system $\widehat{V} = \{\hat{v} \mid v \in \mathcal{V}\}$ with the addition $\oplus$ and multiplication by scalars carried over by the corresponding operations in $\mathcal{V}$, that is $\hat{u} \oplus \hat{v} = \widehat{u+v}$ and $\alpha\hat{v} = \widehat{\alpha v}$ for $u, v \in \mathcal{V}$ and $\alpha > 0$. The neighborhood system $\widehat{\mathcal{V}}$ corresponds to the family $\{\psi_v \mid v \in \mathcal{V}\}$ of $\overline{\mathbb{R}}_+$-valued neighborhood functions which define the neighborhoods $\hat{v} \in \widehat{\mathcal{V}}$ for $\mathcal{F}(\Upsilon, \overline{\mathbb{R}})$ by

$$f \le g + \hat{v} \qquad \text{if} \qquad f(\mu) \le g(\mu) + \psi_v(\mu) \quad \text{for all} \quad \mu \in \Upsilon$$

(see 1.4(e)) for functions $f, g \in \mathcal{F}(\Upsilon, \overline{\mathbb{R}})$. As required, we have $\psi_{(\alpha v)} = \alpha\psi_v$ and $\psi_{(u+v)} \ge \psi_u + \psi_v$ for all $u, v \in \mathcal{V}$ and $\alpha > 0$. The first of these claims is obvious. For the second one, let $\mu \in \Upsilon$ and let $\sigma < \psi_u(\mu)$ and $\rho < \psi_v(\mu)$. Since both $\mu \notin \sigma u^\circ$ and $\mu \notin \rho v^\circ$, there are $a, b, c, d \in \mathcal{P}$ such that $a \le b+u$ and $\mu(a) > \mu(b) + \sigma$ as well as $c \le d + v$ and $\mu(c) > \mu(d) + \sigma$. Then from $(a+c) \le (b+d)+(u+w)$ and $\mu(a+c) > \mu(b+d)+(\sigma+\rho)$ we conclude that $\mu \notin (\sigma + \rho)(u + v)^\circ$. This shows $\psi_{(u+v)}(\mu) \ge (\sigma + \rho)$, yielding our claim. Thus $\big(\mathcal{F}_{\widehat{\mathcal{V}}_b}(\Upsilon, \overline{\mathbb{R}}), \widehat{\mathcal{V}}\big)$ is a locally convex lattice cone in the sense of 1.4(e). Moreover, since $\psi_v(\mu) < +\infty$ holds for all $\mu \in \mathcal{P}^*$ with some $\hat{v} \in \widehat{\mathcal{V}}$, we established in Example 5.7 that $\big(\mathcal{F}_{\widehat{\mathcal{V}}_b}(\mathcal{P}^*, \overline{\mathbb{R}}), \widehat{\mathcal{V}}\big)$ is indeed a locally convex complete lattice cone. We claim that $\Phi(\mathcal{P})$ is contained in $\big(\mathcal{F}_{\widehat{\mathcal{V}}_b}(\Upsilon, \overline{\mathbb{R}}), \widehat{\mathcal{V}}\big)$ and that

$$a \le b + v \qquad \text{if and only if} \qquad \Phi(a) \le \Phi(b) + \hat{v}$$

holds for $a, b \in \mathcal{P}$ and $v \in \mathcal{V}$. Indeed, suppose that $a \le b + v$. Then for every $\mu \in \Upsilon$ and $\alpha > 0$ such that $\mu \in \alpha v^\circ$ we have $\mu(a) \le \mu(b) + \alpha$, that is $\varphi_a(\mu) \le \varphi_b(\mu) + \psi_v(\mu)$, hence $\Phi(a) \le \Phi(b) + \psi_v$. Conversely, if $a \not\le b + v$, then following our assumption that $\Upsilon$ supports the separation property, there is $\mu \in v^\circ \cap \Upsilon$ such that $\mu(a) > \mu(b) + 1$. The former implies $\psi_v(\mu) \le 1$, hence $\varphi_a(\mu) > \varphi_b(\mu) + \psi_v(\mu)$ and therefore $\Phi(a) \not\le \Phi(b) + \psi_v$. We infer in particular that the functions $\Phi(a) \in \mathcal{F}(\Upsilon, \overline{\mathbb{R}})$ are bounded below relative to the neighborhoods in $\widehat{\mathcal{V}}$. Indeed, given $a \in \mathcal{P}$ and $v \in \mathcal{V}$ there is $\lambda \geq 0$ such that $0 \leq a + \lambda v$, hence $0 \leq \Phi(a) + \lambda \hat{v}$. Therefore the element $\Phi(a)$ is contained in $\mathcal{F}_{\widehat{\mathcal{V}}_b}(\Upsilon, \overline{\mathbb{R}})$ as claimed.

Finally we establish that the linear operator $\Phi : \mathcal{P} \to \mathcal{F}_{\widehat{\mathcal{V}}_b}(\Upsilon, \overline{\mathbb{R}})$ is an embedding in the sense of 2.2 of the locally convex complete lattice cone $(\mathcal{P}, \mathcal{V})$ into $\big(\mathcal{F}_{\widehat{\mathcal{V}}_b}(\Upsilon, \overline{\mathbb{R}}), \widehat{\mathcal{V}}\big)$. Indeed, we set $\Phi(v) = \hat{v}$ for $v \in \mathcal{V}$ towards the extension

$$\Phi : (\mathcal{P} \cup \mathcal{V}) :\to \big(\mathcal{F}_{\widehat{\mathcal{V}}_b}(\Upsilon, \overline{\mathbb{R}}) \cup \widehat{\mathcal{V}}\big).$$

Then $\Phi(\mathcal{V}) = \widehat{\mathcal{V}}$, and by the above $a \leq b + v$ holds for $a, b \in \mathcal{P}$ and $v \in \mathcal{V}$ if and only if $\Phi(a) \leq \Phi(b) + \Phi(v)$, as required in 2.2. Moreover, since the (weak pre-)order of the locally convex complete lattice cone $\mathcal{P}$ is antisymmetric, its symmetric topology is Hausdorff by Proposition 4.8. Lemma 2.3 therefore yields that the operator $\Phi : \mathcal{P} \to \mathcal{F}_{\widehat{\mathcal{V}}_b}(\Upsilon, \overline{\mathbb{R}})$ is one-to-one, as claimed. □

We shall demonstrate in 5.57 below that every locally convex cone $(\mathcal{P}, \mathcal{V})$ can be canonically embedded into a locally convex complete lattice cone for which the set of order continuous lattice homomorphisms in $\mathcal{P}^*$ supports the separation property.

**5.38 Almost Order Convergent Nets.** The concept of order convergence can in some cases be meaningfully extended to nets that are not necessarily bounded below. We shall say that a net $(a_i)_{i \in \mathcal{I}}$ in a locally convex complete lattice cone $(\mathcal{P}, \mathcal{V})$ is *almost order convergent towards* $a \in \mathcal{P}$ if for every $k \in \mathcal{I}$ the net $(a_i \vee a_k)_{i \in \mathcal{I}}$ is order convergent and if

$$\lim_{k \in \mathcal{I}} \Big( \lim_{i \in \mathcal{I}} (a_i \vee a_k) \Big) = a.$$

The net $(a_i \vee a_k)_{i \in \mathcal{I}}$ is of course bounded below for any choice of $k \in \mathcal{I}$. Indeed, given $v \in \mathcal{V}$ there is $\lambda \geq 0$ such that $0 \leq a_k + \lambda v \leq (a_i \vee a_k) + \lambda v$ for all $i \in \mathcal{I}$.

**Lemma 5.39.** *Let $(\mathcal{P}, \mathcal{V})$ be a locally convex complete lattice cone. A bounded below net $(a_i)_{i \in \mathcal{I}}$ in $\mathcal{P}$ is order convergent if and only if it is almost order convergent with the same limit.*

*Proof.* Let $(a_i)_{i \in \mathcal{I}}$ be a bounded below net in $\mathcal{P}$. If $(a_i)_{i \in \mathcal{I}}$ is order convergent and $\lim_{i \in \mathcal{I}} a_i = a$, then

$$\lim_{i \in \mathcal{I}} (a_i \vee a_k) = \Big( \lim_{i \in \mathcal{I}} a_i \Big) \vee a_k = a \vee a_k$$

for all $k \in \mathcal{I}$ by Proposition 5.25(a). Therefore

$$\lim_{k \in \mathcal{I}} \Big( \lim_{i \in \mathcal{I}} (a_i \vee a_k) \Big) = \lim_{k \in \mathcal{I}} (a \vee a_k) = a \vee \Big( \lim_{k \in \mathcal{I}} a_k \Big) = a$$

again by 5.25(a), and we infer that $(a_i)_{i \in \mathcal{I}}$ is almost order convergent towards $a$. On the other hand, for every $b \in \mathcal{P}$ we have

$$\begin{aligned}\underline{\lim}_{i\in\mathcal{I}}(a_i \vee b) &= \sup_{i\in\mathcal{I}}\Big(\inf_{j\geq i}(a_j \vee b)\Big) = \sup_{i\in\mathcal{I}}\Big(\big(\inf_{j\geq i} a_j\big)\vee b)\Big) \\ &= \Big(\sup_{i\in\mathcal{I}}\big(\inf_{j\geq i} a_j\big)\Big)\vee b = \Big(\underline{\lim}_{i\in\mathcal{I}} a_i\Big)\vee b\end{aligned}$$

by Proposition 5.15 and Lemma 5.5. Similarly one realizes that

$$\begin{aligned}\overline{\lim}_{i\in\mathcal{I}}(a_i \vee a_k) &= \inf_{i\in\mathcal{I}}\Big(\sup_{j\geq i}(a_j \vee b)\Big) = \inf_{i\in\mathcal{I}}\Big(\big(\sup_{j\geq i} a_j\big)\vee b)\Big) \\ &= \Big(\inf_{i\in\mathcal{I}}\big(\sup_{j\geq i} a_j\big)\Big)\vee b = \Big(\overline{\lim}_{i\in\mathcal{I}} a_i\Big)\vee b\end{aligned}$$

If the net $(a_i)_{i\in\mathcal{I}}$ is almost order convergent towards $a\in\mathcal{P}$, this yields

$$\lim_{i\in\mathcal{I}}(a_i \vee a_k) = \Big(\underline{\lim}_{i\in\mathcal{I}} a_i\Big)\vee a_k = \Big(\overline{\lim}_{i\in\mathcal{I}} a_i\Big)\vee a_k$$

for all $k\in\mathcal{I}$. Thus, again using the above

$$\begin{aligned} a &= \lim_{k\in\mathcal{I}}\Big(\lim_{i\in\mathcal{I}}(a_i\vee a_k)\Big) \\ &= \underline{\lim}_{k\in\mathcal{I}}\bigg(\Big(\underline{\lim}_{i\in\mathcal{I}} a_i\Big)\vee a_k\bigg) \\ &= \Big(\underline{\lim}_{i\in\mathcal{I}} a_i\Big)\vee\Big(\underline{\lim}_{k\in\mathcal{I}} a_k\Big) = \Big(\underline{\lim}_{i\in\mathcal{I}} a_i\Big),\end{aligned}$$

as well as

$$\begin{aligned} a &= \lim_{k\in\mathcal{I}}\Big(\lim_{i\in\mathcal{I}}(a_i\vee a_k)\Big) \\ &= \overline{\lim}_{k\in\mathcal{I}}\bigg(\Big(\overline{\lim}_{i\in\mathcal{I}} a_i\Big)\vee a_k\bigg) \\ &= \Big(\overline{\lim}_{i\in\mathcal{I}} a_i\Big)\vee\Big(\overline{\lim}_{k\in\mathcal{I}} a_k\Big) = \Big(\overline{\lim}_{i\in\mathcal{I}} a_i\Big).\end{aligned}$$

This yields $\lim_{i\in\mathcal{I}} a_i = a$. □

*Examples 5.40.* Let $\mathcal{P}$ be the cone of all bounded below $\overline{\mathbb{R}}$-valued functions on $[0,+\infty)$, endowed with the pointwise operations and order, and the positive constant functions $v>0$ as its neighborhood system $\mathcal{V}$ (see Example 1.4(e)). $(\mathcal{P},\mathcal{V})$ is a locally convex complete lattice cone, and order convergence in $\mathcal{P}$ implies pointwise convergence on $[0,+\infty)$ for the functions involved. Pointwise convergence, on the other hand does not require that a net in $\mathcal{P}$ is bounded below and therefore does not always imply order convergence. Let us illustrate this in a simple example: For $n\in\mathbb{N}$ let $f_n\in\mathcal{P}$ such that $f_n(x)=-n$ for $0<x\leq 1/n$, and $f_n(x)=0$ else. The sequence $(f_n)_{n\in\mathbb{N}}$ converges pointwise to $0\in\mathcal{P}$, but it is not bounded below in $\mathcal{P}$ and therefore not order convergent. However, for every $m\in\mathbb{N}$ the sequence

$(f_n \vee f_m)_{n\in\mathbb{N}}$ is bounded below and converges pointwise, hence in order towards $0 \in \mathcal{P}$. We infer that $(f_n)_{n\in\mathbb{N}}$ is almost order convergent towards $0 \in \mathcal{P}$. In fact, it can be easily verified that pointwise convergence coincides with almost order convergence in this example (see Proposition 5.51 below).

We proceed probing different patterns of convergence in a locally convex complete lattice cone $(\mathcal{P},\mathcal{V})$. For a net $(a_i)_{i\in\mathcal{I}}$ in $\mathcal{P}$, convergence with respect to the symmetric relative topology of $\mathcal{P}$ towards $a \in \mathcal{P}$ means that for every $v \in \mathcal{V}$ and $\varepsilon > 0$ there is $i_0 \in I$ such that $a_i \in v_\varepsilon^s(a)$ for all $i \geq i_0$. $(a_i)_{i\in\mathcal{I}}$ is a *Cauchy net* if for every $v \in \mathcal{V}$ and $\varepsilon > 0$ there is $i_0 \in I$ such that $a_i \in v_\varepsilon(a_k)$ for all $i, k \geq i_0$. Obviously, convergence implies that $(a_i)_{i\in\mathcal{I}}$ is a Cauchy net. The converse, that is topological completeness holds also true:

**Proposition 5.41.** *Every locally convex complete lattice cone is complete with respect to the symmetric relative topology.*

*Proof.* Suppose that $(a_i)_{i\in\mathcal{I}}$ is a Cauchy net in $\mathcal{P}$. We shall first demonstrate that $(a_i)_{i\in\mathcal{I}}$ is order convergent. Let $v \in \mathcal{V}$ and $0 < \varepsilon \leq 1$. There is $i_0 \in \mathcal{I}$ such that $a_i \in v_\varepsilon(a_k)$ for all $i, k \geq i_0$. Choose $\lambda \geq 0$ such that $0 \leq a_{i_0} + \lambda v$. Following Lemma 4.1(b) and (c) this implies

$$a_i \leq (1+\varepsilon)a_{i_0} + \varepsilon(1+\lambda)v \qquad \text{and} \qquad a_{i_0} \leq (1+\varepsilon)a_i + \varepsilon(2+\lambda)v$$

for all $i > i_0$. This shows in particular that $(a_i)_{i\in\mathcal{I}}$ is bounded below and also that

$$a_i \leq (1+\varepsilon)^2 a_k + 3\varepsilon(2+\lambda)v$$

for all $i, k \geq i_0$. This shows

$$\overline{\lim_{i\in\mathcal{I}}}\, a_i \leq (1+\varepsilon)^2 \,\underline{\lim_{k\in I}}\, a_k + 3\varepsilon(2+\lambda)v.$$

As this holds for all $v \in \mathcal{V}$ and $0 < \varepsilon \leq 1$, and as $\mathcal{P}$ carries the weak preorder which is supposed to be antisymmetric, we infer that $\overline{\lim}_{i\in\mathcal{I}} a_i = \underline{\lim}_{k\in I} a_k$, hence order convergence towards an element $a \in \mathcal{P}$. Moreover, the above shows that

$$a_i \leq (1+\varepsilon^2)a + 3\varepsilon(2+\lambda)v \qquad \text{and} \qquad a \leq (1+\varepsilon^2)a_i + 3\varepsilon(2+\lambda)v$$

holds for all $i \geq i_0$. Thus the net $(a_i)_{i\in\mathcal{I}}$ converges to $a$ in the symmetric relative topology as well. □

In fact, we just verified that every Cauchy net, hence every convergent net in the symmetric relative topology of $(\mathcal{P},\mathcal{V})$ is indeed order convergent with the same limit. We shall formulate this as a separate proposition:

**Proposition 5.42.** *Let $(\mathcal{P},\mathcal{V})$ be a locally convex complete lattice cone. Convergence of a net $(a_i)_{i\in\mathcal{I}}$ in $\mathcal{P}$ towards $a \in \mathcal{P}$ in the symmetric relative topology implies order convergence towards $a$.*

While convergence in the symmetric relative topology implies order convergence, the converse is not necessarily true, as a simple example can show: In the locally convex complete lattice cone $\overline{\mathbb{R}}$ order convergence means convergence in the usual (one-point compactification) topology of $\overline{\mathbb{R}}$ which for the element $+\infty$ does not coincide with the symmetric relative topology of $\overline{\mathbb{R}}$. The sequence $(n)_{n\in\mathbb{N}}$, for example, is order convergent towards $+\infty \in \overline{\mathbb{R}}$, but does not converge in the symmetric relative topology, as $+\infty$ is an isolated point in this topology.

**5.43 Order Topology.** While order convergence in a locally convex complete lattice cone $(\mathcal{P}, \mathcal{V})$ does not necessarily correspond to a topology on $\mathcal{P}$ in the sense that order and topological convergence for nets coincide (see 1.1.9 in [132]), there is a finest topology $\mathcal{O}(\mathcal{P})$ on $\mathcal{P}$ with the following properties:

(OT1) *Every very element of $\mathcal{P}$ admits a basis of both convex and order convex neighborhoods. The neighborhoods in the basis for $0 \in \mathcal{P}$ are also balanced.*

(OT2) *The mappings $(a,b) \mapsto a+b$, $(a,b) \mapsto a \vee b$ and $(a,b) \mapsto a \wedge b$ from $\mathcal{P}^2$ into $\mathcal{P}$ are continuous.*

(OT3) *The mapping $(\alpha, a) \mapsto \alpha a : [0,+\infty) \times \mathcal{P} \to \mathcal{P}$ is continuous at all points $(\alpha, a) \in [0,+\infty) \times \mathcal{P}$ such that either $\alpha > 0$ or $a \in \mathcal{P}$ is bounded.*

(OT4) *All almost order convergent nets in $\mathcal{P}$ are topologically convergent with the same limit.*

Indeed, let $\mathfrak{T}$ be the family of all topologies on $\mathcal{P}$ with these properties. These topologies need not be Hausdorff. Therefore $\mathfrak{T}$ is not empty as it contains the discrete topology. Let $\mathcal{O}(\mathcal{P})$ be the supremum of this family in the lattice of topologies on $\mathcal{P}$. A neighborhood basis in $\mathcal{O}(\mathcal{P})$ for a point $a \in \mathcal{P}$ is generated by the intersections of finitely many neighborhoods for $a$ taken from topologies in $\mathfrak{T}$. This shows that $\mathcal{O}(\mathcal{P})$ again satisfies (OT1) to (OT4), hence is the finest topology with these properties. We shall call $\mathcal{O}(\mathcal{P})$ the *(strong) order topology* on $\mathcal{P}$. Note that $\mathcal{O}(\mathcal{P})$ is not necessarily a locally convex cone topology. For $\mathcal{P} = \overline{\mathbb{R}}$, for example, the order topology is the usual topology of $\overline{\mathbb{R}}$ where $+\infty$ is not an isolated point.

In Proposition 4.2 we verified that the symmetric relative topology of $\mathcal{P}$ satisfies (OT1), (OT2) and (OT3), however it does not meet (OT4) in general.

**Proposition 5.44.** *Let $(\mathcal{P}, \mathcal{V})$ be a locally convex complete lattice cone. The order topology $\mathcal{O}(\mathcal{P})$ on $\mathcal{P}$ is coarser than the symmetric relative topology.*

*Proof.* We observed in Proposition 5.42 that convergence for a net in the symmetric relative topology implies order convergence, hence convergence in $\mathcal{O}(\mathcal{P})$. Since the closure in any topology of a given subset $A$ of $\mathcal{P}$ can be described as the set of all limit points of convergent nets in this subset, Proposition 5.42 implies that the closure of $A$ with respect to the symmetric relative topology is contained in the closure of $A$ with respect to $\mathcal{O}(\mathcal{P})$. We infer that $\mathcal{O}(\mathcal{P})$ is generally coarser than the symmetric relative topology. □

**Lemma 5.45.** *Let $(\mathcal{P},\mathcal{V})$ be a locally convex complete lattice cone and let $\mathcal{P}_0$ be the subcone of all invertible elements of $\mathcal{P}$. The mapping $(\alpha,a)\mapsto \alpha a:\mathbb{R}\times\mathcal{P}_0\to\mathcal{P}_0$ is continuous with respect to the order topology $\mathcal{O}(\mathcal{P})$.*

*Proof.* We shall make this argument in several short steps: First suppose that $a_i\to 0$ for $a_i\in\mathcal{P}_0$ in any topology satisfying (OT1) to (OT4). Given a neighborhood $U$ in the basis for $0$ there is $i_0$ such that $a_i\in U$ for all $i\geq i_0$. This implies $-a_i\in U$ as well since $U$ is supposed to be balanced by (OT1). Thus $(-a_i)\to 0$. Next suppose that $a_i\to a$ for $a_i,a\in\mathcal{P}_0$. Then $\big(a_i+(-a)\big)\to 0$ by (OT2), hence $\big((-a_i)+a\big)\to 0$ by the preceding step, and $(-a_i)\to(-a)$ by (OT2). In a third step, suppose that $\alpha_i\to\alpha\in\mathbb{R}$ for $0\leq\alpha_i\in\mathbb{R}$ and $a_i\to a$ for $a_i,a\in\mathcal{P}_0$. Then $\alpha_i a_i\to\alpha a$ by (OT3) since every invertible element is bounded. Now in the fourth and final step of our argument, let $\alpha_i\to\alpha$ in $\mathbb{R}$ and $a_i\to a$ for $a_i,a\in\mathcal{P}_0$. Let $\beta_i=\alpha_i\vee 0$ and $\gamma_i=-(\alpha_i\wedge 0)$. Then $\beta_i,\gamma_i\geq 0$ and $\alpha_i=\beta_i-\gamma_i$. We have $\beta_i a_i\to\beta a$ and $\gamma_i(-a_i)\to\gamma(-a)$, where $\beta=\alpha\vee 0$ and $\gamma=-(\alpha\wedge 0)$, by the second and third steps of our argument. Thus

$$\alpha_i a_i=\beta_i a_i+\gamma_i(-a_i)\ \to\ \beta a+\gamma(-a)=\alpha a,$$

again by (OT2), as claimed. □

**Proposition 5.46.** *Let $(\mathcal{P},\mathcal{V})$ and $(\mathcal{Q},\mathcal{W})$ be a locally convex complete lattice cones. An order continuous lattice homomorphism $T:\mathcal{P}\to\mathcal{Q}$ is also continuous with respect to the respective order topologies $\mathcal{O}(\mathcal{P})$ and $\mathcal{O}(\mathcal{Q})$.*

*Proof.* Let $T:\mathcal{P}\to\mathcal{Q}$ be an order continuous lattice homomorphism, consider the order topology $\mathcal{O}(\mathcal{Q})$ on $\mathcal{Q}$ and let $\tau$ be the initial topology induced on $\mathcal{P}$ by $T$, that is $\tau$ is the coarsest topology on $\mathcal{P}$ for which the mapping $T:\mathcal{P}\to\mathcal{Q}$ is continuous. The sets in $\tau$ then are just the inverse images under $T$ of the sets in $\mathcal{O}(\mathcal{Q})$. It is straightforward to verify that $\tau$ satisfies the requirements (OT1) to (OT4): For $a\in\mathcal{P}$ the element $T(a)\in\mathcal{Q}$ admits a basis of neighborhoods in $\mathcal{O}(\mathcal{Q})$ satisfying (OT1). Their inverse images under $T$ have the same properties and form a neighborhood basis for $a$ in $\tau$. Next suppose that $a_i\to a$ and $b_i\to b$ in $\tau$. Then $T(a_i)\to T(a)$ and $T(b_i)\to T(b)$ in $\mathcal{O}(\mathcal{Q})$. Thus $T(a_i)+T(b_i)\to T(a)+T(b)=T(a+b)$ since $\mathcal{O}(\mathcal{Q})$ satisfies (OT2). Because every neighborhood of $a+b$ in $\tau$ is the inverse image under $T$ of a neighborhood of $T(a+b)$, this shows that $(a_i+b_i)\to(a+b)$ in $\tau$. Similarly one verifies the continuity of the mappings $(a,b)\mapsto a\vee b$, $(a,b)\mapsto a\wedge b$ and $(\alpha,a)\mapsto\alpha a$ with respect to $\tau$. For (OT4) let $(a_i)_{i\in\mathcal{I}}$ be an almost order convergent net in $\mathcal{P}$ with limit $a\in\mathcal{P}$. Then

$$T(a)=T\left(\lim_{k\in\mathcal{I}}\left(\lim_{i\in\mathcal{I}}(a_i\vee a_k)\right)\right)=\lim_{k\in\mathcal{I}}\left(\lim_{i\in\mathcal{I}}\Big(T(a_i)\vee T(a_k)\Big)\right)$$

since $T$ is an order continuous lattice homomorphism. The net $\big(T(a_i)\big)_{i\in\mathcal{I}}$ is therefore almost order convergent with limit $T(a)$ in $\mathcal{Q}$. As $\mathcal{O}(\mathcal{Q})$ satisfies (OT4), this implies $T(a_i) \to T(a)$ in $\mathcal{O}(\mathcal{Q})$, and therefore $a_i \to a$ in $\tau$, since the neighborhoods of $a$ in $\tau$ are inverse images under $T$ of neighborhoods of $T(a)$ in $\mathcal{O}(\mathcal{Q})$. Summarizing, we have verified that the topology $\tau$ on $\mathcal{P}$ satisfies conditions (OT1) to (OT4) and is therefore coarser then the order topology $\mathcal{O}(\mathcal{P})$. Hence the operator $T : \mathcal{P} \to \mathcal{Q}$ is also continuous if we endow $\mathcal{P}$ with $\mathcal{O}(\mathcal{P})$. □

**Proposition 5.47.** *Let $(\mathcal{P},\mathcal{V})$ be a locally convex complete lattice cone and let $\mathcal{N}$ be a subcone of $\mathcal{P}$. Then the closure $\overline{\mathcal{N}}$ of $\mathcal{N}$ with respect to $\mathcal{O}(\mathcal{P})$ is again a subcone of $\mathcal{P}$. If $\mathcal{N}$ is a lattice subcone of $\mathcal{P}$, then $\overline{\mathcal{N}}$ is a complete lattice subcone of $\mathcal{P}$.*

*Proof.* The first part of the claim follows directly from (OT2) and (OT3). For the second part suppose that $\mathcal{N}$ is a lattice subcone of $\mathcal{P}$ and let $a,b \in \mathcal{N}$. There are nets $(a_i)_{i\in\mathcal{I}}$ and $(b_j)_{j\in\mathcal{J}}$ in $\mathcal{N}$ converging in the order topology towards $a$ and $b$, respectively. Then the net $(a_i \vee b_j)_{(i,j)\in\mathcal{I}\times\mathcal{J}}$ in $\mathcal{N}$ converges to $a \vee b$ by (OT3). Thus $a \vee b \in \overline{\mathcal{N}}$. Similarly one shows that $a \wedge b \in \overline{\mathcal{N}}$, hence $\overline{\mathcal{N}}$ is also a lattice subcone of $\mathcal{P}$. Now let $A$ be a non-empty subset of $\overline{\mathcal{N}}$. For every finite subset $i = \{a_1,\ldots,a_n\}$ of $A$ set $a_i = a_1 \vee \ldots \vee a_n \in \overline{\mathcal{N}}$. Then $\sup A = \lim_{i\in\mathcal{I}} a_i$, where $\mathcal{I}$ is the collection of all finite subsets of $A$, ordered by set inclusion. This shows $\sup A \in \overline{\mathcal{N}}$ by (OT4). Similarly one shows that $\inf B \in \overline{\mathcal{N}}$ whenever $B$ is a bounded below subset of $\overline{\mathcal{N}}$. Thus $\overline{\mathcal{N}}$ is indeed a complete lattice subcone of $\mathcal{P}$. □

**Proposition 5.48.** *Let $(\mathcal{P},\mathcal{V})$ be a locally convex complete lattice cone and let $\mathcal{N}$ be a complete lattice subcone of $\mathcal{P}$. The restriction of $\mathcal{O}(\mathcal{P})$ to $\mathcal{N}$ is coarser than the order topology $\mathcal{O}(\mathcal{N})$ of $\mathcal{N}$.*

*Proof.* This follows from the easily verifiable fact that the restriction of $\mathcal{O}(\mathcal{P})$ to the complete lattice subcone $\mathcal{N}$ satisfies the requirements (OT1) to (OT4). □

**5.49 Weak Order Convergence.** *Weak order convergence* for a net $(a_i)_{i\in\mathcal{I}}$ in a locally convex complete lattice cone $(\mathcal{P},\mathcal{V})$ means that $\big(\mu(a_i)\big)_{i\in\mathcal{I}}$ converges towards $\mu(a)$ in $\overline{\mathbb{R}}$ (with respect to order convergence) for every order continuous lattice homomorphism $\mu \in \mathcal{P}^*$. This notion of convergence results from the *weak order topology* $o(\mathcal{P},\mathcal{P}^*)$ on $\mathcal{P}$ which is generated by the (both convex and order convex) neighborhoods

$$\mathcal{V}^o_{\Upsilon}(a) = \left\{ b \in \mathcal{P} \;\middle|\; \begin{array}{rl} |\mu_i(b) - \mu_i(a)| \le 1\,, & \text{if} \quad \mu_i(a) < +\infty \\ \mu_i(b) \ge 1\,, & \text{if} \quad \mu_i(a) = +\infty \end{array} \right\},$$

for an element $a \in \mathcal{P}$, corresponding to a finite set $\Upsilon = \{\mu_1,\ldots,\mu_n\}$ of order continuous lattice homomorphisms in $\mathcal{P}^*$. Like the order topology $\mathcal{O}(\mathcal{P})$, this is in general not a locally convex cone topology.

**Proposition 5.50.** *Let $(\mathcal{P},\mathcal{V})$ be a locally convex complete lattice cone. The weak order topology $o(\mathcal{P},\mathcal{P}^*)$ on $\mathcal{P}$ is coarser than the order topology $\mathcal{O}(\mathcal{P})$ and also coarser than the weak topology $\sigma(\mathcal{P},\mathcal{P}^*)$.*

*Proof.* Requirements (OT1) to (OT4) from 5.40 are readily checked for the weak order topology: (OT1) and the first part of (OT2) are self evident. The second part of (OT2) follows from the easily verified fact that $\mu(a'\vee b')\le\mu(a\vee b)+1$ holds whenever $\mu(a')\le\mu(a)+1$ and $\mu(b')\le\mu(b)+1$ for elements $a,a',b,b'\in\mathcal{P}$ and an order continuous lattice homomorphism $\mu\in\mathcal{P}^*$. Similarly one argues for the third part of (OT2). For (OT4) let $(a_i)_{i\in\mathcal{I}}$ be an almost order convergent net in $\mathcal{P}$ with limit $a\in\mathcal{P}$. Then

$$\mu(a)=\mu\left(\lim_{k\in\mathcal{I}}\left(\lim_{i\in\mathcal{I}}(a_i\vee a_k)\right)\right)=\lim_{k\in\mathcal{I}}\left(\lim_{i\in\mathcal{I}}\left(\mu(a_i)\vee\mu(a_k)\right)\right)$$

for every order continuous lattice homomorphism $\mu\in\mathcal{P}^*$. The limit on the right-hand side is taken with respect to the usual (that is the order) topology of $\overline{\mathbb{R}}$. The net $(a_i)_{i\in\mathcal{I}}$ is therefore also convergent with respect to the weak order topology. We infer that $o(\mathcal{P},\mathcal{P}^*)$ is generally coarser than the order topology $\mathcal{O}(\mathcal{P})$. The second statement of Proposition 5.50 follows immediately from a comparison of the respective neighborhoods in 4.6 and in 5.49: For $a\in\mathcal{P}$ and a finite set $\Upsilon=\{\mu_1,\ldots,\mu_n\}$ of order continuous lattice homomorphisms in $\mathcal{P}^*$ we have $\mathcal{V}^s_\Upsilon(a)\subset\mathcal{V}^o_\Upsilon(a)$. Thus $\sigma(\mathcal{P},\mathcal{P}^*)$ is indeed finer than $o(\mathcal{P},\mathcal{P}^*)$. □

**Proposition 5.51.** *Let $(\mathcal{P},\mathcal{V})$ be a locally convex complete lattice cone, and suppose that the order continuous lattice homomorphisms support the separation property for $\mathcal{P}$. Then the order and the weak order topologies coincide on $\mathcal{P}$ and are Hausdorff. A net in $\mathcal{P}$ is convergent in the (weak) order topology if and only if it is almost order convergent.*

*Proof.* Let $(\mathcal{P},\mathcal{V})$ be a locally convex complete lattice cone such that the order continuous lattice homomorphisms support the separation property for $\mathcal{P}$. Let us fist argue that the weak order topology is Hausdorff. Indeed, for distinct elements $a,b\in\mathcal{P}$ we have either $a\not\le b$ or $b\not\le a$, since the order of $\mathcal{P}$ is supposed to be antisymmetric. Thus $a\not\le b+v$ or $b\not\le a+v$ for some $v\in\mathcal{V}$ by Lemma 3.2. Then there exists an order continuous linear functional $\mu\in v^\circ$ such that $\mu(a)>\mu(b)+1$ or $\mu(b)>\mu(a)+1$, respectively. For suitable $\varepsilon,\delta>0$ then the neighborhoods $\mathcal{V}^o_{\{\varepsilon\mu\}}(a)$ and $\mathcal{V}^o_{\{\delta\mu\}}(b)$ are seen to be disjoint. Next we shall verify the last statement of our claim: Let $(a_i)_{i\in\mathcal{I}}$ be a net in $\mathcal{P}$. If $(a_i)_{i\in\mathcal{I}}$ is almost order convergent, then it is convergent with the same limit in $\mathcal{O}(\mathcal{P})$ by (OT4), hence weakly order convergent since the weak order topology is coarser than $\mathcal{O}(\mathcal{P})$. For the converse suppose that $(a_i)_{i\in\mathcal{I}}$ is weakly order convergent toward $a\in\mathcal{P}$. Then for every $b\in\mathcal{P}$ and every order continuous lattice homomorphism $\mu\in\mathcal{P}^*$ we have

$$\begin{aligned}\mu\Big(\overline{\lim_{i\in\mathcal{I}}}(a_i\vee b)\Big) &= \overline{\lim_{i\in\mathcal{I}}}\,\big(\mu(a_i)\vee\mu(b)\big)\\ &= \Big(\overline{\lim_{i\in\mathcal{I}}}\,\mu(a_i)\Big)\vee\mu(b)\\ &= \mu(a)\vee\mu(b)\\ &= \mu(a\vee b).\end{aligned}$$

This shows $\overline{\lim}_{i\in\mathcal{I}}(a_i\vee b)=(a\vee b)$ since the weak order topology was seen to be Hausdorff. Similarly one verifies that $\underline{\lim}_{i\in\mathcal{I}}(a_i\vee b)=(a\vee b)$, hence

$$\lim_{i\in\mathcal{I}}(a_i\vee b)=(a\vee b).$$

For $b=a_k$ in particular, this renders $\lim_{i\in\mathcal{I}}(a_i\vee a_k)=(a\vee a_k)$ for every $k\in\mathcal{I}$. Repeating this argument with $b=a$ and $a_k$ in place of $a_i$ then yields

$$\lim_{k\in\mathcal{I}}(a\vee a_k)=(a\vee a)=a.$$

We thus verified that the net $(a_i)_{i\in\mathcal{I}}$ is almost order convergent towards $a\in\mathcal{P}$. This completes our argument for convergent nets and also implies the first part of our claim. Indeed, since the closed sets in any given topology can be described in terms of limits of convergence nets alone, having the same notion of convergence for nets means that the topologies involved coincide. □

**Proposition 5.52.** *Let $(\mathcal{P},\mathcal{V})$ be a locally convex complete lattice cone such that the order continuous lattice homomorphisms support the separation property for $\mathcal{P}$, and let $\mathcal{N}$ be a complete lattice subcone of $\mathcal{P}$. Then $\mathcal{N}$ is closed in $\mathcal{O}(\mathcal{P})$. The order topology $\mathcal{O}(\mathcal{N})$ of $\mathcal{N}$ coincides with the restriction of $\mathcal{O}(\mathcal{P})$ to $\mathcal{N}$.*

*Proof.* Let $(\mathcal{P},\mathcal{V})$ be a locally convex complete lattice cone and let $\mathcal{N}$ be a complete lattice subcone of $\mathcal{P}$. Because the restriction to $\mathcal{N}$ of an order continuous lattice homomorphism on $\mathcal{P}$ is an order continuous lattice homomorphism on $\mathcal{N}$, under the assumptions of the Proposition these functionals support the separation property for both $\mathcal{P}$ and $\mathcal{N}$. The conclusions of Proposition 5.51 therefore apply to both of these cones. Let $(a_i)_{i\in\mathcal{I}}$ be a net in $\mathcal{N}$. We observe the following: If $(a_i)_{i\in\mathcal{I}}$ is almost order convergent as a net in $\mathcal{N}$ with limit $a\in\mathcal{N}$, then it is also almost order convergent as a net in $\mathcal{P}$ with the same limit. Conversely, if $(a_i)_{i\in\mathcal{I}}$ is almost order convergent in as a net in $\mathcal{P}$ with limit $a\in\mathcal{P}$, then $a\in\mathcal{N}$, and $(a_i)_{i\in\mathcal{I}}$ is also almost order convergent as a net in $\mathcal{N}$ with the same limit. This is an immediate consequence of the fact that the subcone $\mathcal{N}$ contains the infima and suprema of its sets as elements, hence the limits of its order convergent nets. Now both of our claims follow, since the convergent nets in the order topologies of $\mathcal{P}$ and of $\mathcal{N}$ coincide with the almost order convergent nets in $\mathcal{P}$ and $\mathcal{N}$, respectively. □

In Proposition 5.37 we established that every locally convex lattice cone can be represented as a as a cone of $\overline{\mathbb{R}}$-valued functions on some set $X$. The preceding considerations now allow us to identify the weak and strong order topologies as the topology of pointwise convergence in this representation.

**Proposition 5.53.** *Let $(\mathcal{P}, \widehat{\mathcal{V}})$ be a complete lattice subcone of $\big(\mathcal{F}_{\widehat{\mathcal{V}}_b}(X, \overline{\mathbb{R}}), \widehat{\mathcal{V}}\big)$ for some set $X$ and a neighborhood system $\widehat{\mathcal{V}}$ consisting of $\overline{\mathbb{R}}$-valued functions such that $\hat{v}(x) < +\infty$ for every $x \in X$ with some $\hat{v} \in \widehat{\mathcal{V}}$. Then the order topology, the weak order topology and the topology of pointwise convergence on $X$ (with respect to the usual topology of $\overline{\mathbb{R}}$) all coincide on $\mathcal{P}$ and are Hausdorff.*

*Proof.* Under the assumptions of the Proposition, the order continuous lattice homomorphisms support the separation property for $\big(\mathcal{F}_{\widehat{\mathcal{V}}_b}(X, \overline{\mathbb{R}}), \widehat{\mathcal{V}}\big)$ $\big($see 5.33(f)$\big)$, hence also for the complete lattice subcone $(\mathcal{P}, \widehat{\mathcal{V}})$. The coincidence of the order and the weak order topology was established in Proposition 5.51. Since for every $x \in X$ the point evaluation $f \mapsto f(x)$ is an order continuous lattice homomorphism on $\mathcal{P}$ $\big($see 5.31(d)$\big)$, week order convergence for a net in $\mathcal{P}$ implies pointwise convergence on $X$. A pointwise convergent net, on the other hand is seen to be almost order convergent and therefore convergent in the order topology. The three notions of convergence, hence the respective topologies therefore coincide. □

**5.54 Extension of Linear Operators.** A short inspection of the Hahn-Banach type extension results for linear functionals in [172] (see also Section 2) shows that they are still valid if the range $\overline{\mathbb{R}}$ for the functionals is replaced by some locally convex cone $(\mathcal{Q}, \mathcal{W})$, provided that

(i) $(\mathcal{Q}, \mathcal{W})$ is full and a complete lattice cone,
(ii) all elements of $\mathcal{Q}$, with the exception of the element $+\infty = \sup \mathcal{Q}$, are invertible,
(iii) the neighborhood system $\mathcal{W}$ consists of all (strictly) positive multiples of a single neighborhood $w \in \mathcal{W}$.

Requirement (ii) means of course that $\mathcal{Q}$ is a Dedekind complete Riesz space with an adjoint maximal element $+\infty$. Results about the extension of monotone linear operators between vector spaces and Dedekind complete Riesz spaces are due to Kantorovič $\big($[96] and [98]$\big)$ and can for example be found in Section 1.5 of [132]. Without furnishing the details of this, we reformulate Corollary 4.1 in [172] (see also Corollary 2.7 before).

**Theorem 5.55.** *Let $(\mathcal{N}, \mathcal{V})$ be a subcone of the locally convex cone $(\mathcal{P}, \mathcal{V})$. Suppose that $(\mathcal{Q}, \mathcal{W})$ is a full locally convex complete lattice cone, that all elements of $\mathcal{Q}$ other than $+\infty$ are invertible, and that $\mathcal{W} = \{\alpha w \mid \alpha > 0\}$ for some $w \in \mathcal{W}$. Then every continuous linear operator $T : \mathcal{N} \to \mathcal{Q}$ can be extended to a continuous linear operator $\overline{T} : \mathcal{P} \to \mathcal{Q}$.*

Unfortunately, a similar result is not generally available if the locally convex complete lattice cone $(\mathcal{Q},\mathcal{W})$ does not meet the stringent additional requirements of Theorem 5.55. However, we have the following:

**Theorem 5.56.** *Let $\mathcal{N}$ be a subcone of the locally convex cone $(\mathcal{P},\mathcal{V})$ and let $(\mathcal{Q},\mathcal{W})$ be a locally convex complete lattice cone. Every continuous linear operator $T:\mathcal{N}\to\mathcal{Q}$ can be uniquely extended to $\overline{\mathcal{N}}$, the closure of $\mathcal{N}$ in $\mathcal{P}$ with respect to the symmetric relative topology.*

*Proof.* Let $T:\mathcal{N}\to\mathcal{Q}$ be a continuous linear operator and let $a\in\overline{\mathcal{N}}$. There is a net $(a_i)_{i\in\mathcal{I}}$ in $\mathcal{N}$ converging to $a$ in the symmetric relative topology. Given $w\in\mathcal{W}$ and $\varepsilon>0$ there is $v\in\mathcal{V}$ such that $T(b)\le T(c)+w$ whenever $b\le c+v$ for $b,c\in\mathcal{N}$. Because $(a_i)_{i\in\mathcal{I}}$ is a Cauchy net in $\mathcal{N}$, there is $i_0\in\mathcal{I}$ such that $a_i\in v_\varepsilon(a_k)$ for all $i,k\ge i_0$. This implies $T(a_i)\in w_\varepsilon\big(T(a_k)\big)$ for all $i,k\ge i_0$, hence $\big(T(a_i)\big)_{i\in\mathcal{I}}$ is a Cauchy net in $\mathcal{Q}$ as well. Proposition 5.41 shows that this net converges in $\mathcal{Q}$. Moreover, if $(b_j)_{j\in\mathcal{J}}$ is a second net in $\mathcal{N}$ converging toward the same element $a$, given $w\in\mathcal{W}$ and $\varepsilon>0$ we choose $v\in\mathcal{V}$ as above and find $i_0\in\mathcal{I}$ and $j_0\in\mathcal{J}$ such that both $a_i\in v_\varepsilon(b_j)$ and $b_j\in v_\varepsilon(a_i)$, hence $T(a_i)\in w_\varepsilon\big(T(b_j)\big)$ and $T(b_j)\in v_\varepsilon\big(T(a_i)\big)$, for all $i\ge i_0$ and $j\ge j_0$. This shows that both nets $\big(T(a_i)\big)_{i\in\mathcal{I}}$ and $\big(T(b_j)\big)_{j\in\mathcal{J}}$ have the same limit in $\mathcal{Q}$ which we denote $\overline{T}(a)$. It is now straightforward to verify that this procedure results in a bounded linear extension $\overline{T}:\overline{\mathcal{N}}\to\mathcal{Q}$ of the operator $T$. Uniqueness of this extension is obvious. □

**5.57 The Standard Lattice Completion of a Locally Convex Cone.** We proceed to establish that every locally convex cone $(\mathcal{P},\mathcal{V})$ can be canonically embedded into a locally convex complete lattice cone. For this, we use a representation for $(\mathcal{P},\mathcal{V})$ as a cone of $\overline{\mathbb{R}}$-valued functions on its dual cone $\mathcal{P}^*$, in analogy to the construction that we employed in the proof of Proposition 5.37: With the element $a\in\mathcal{P}$ we associate the $\overline{\mathbb{R}}$-valued function $\varphi_a$ on $\mathcal{P}^*$ such that

$$\varphi_a(\mu)=\mu(a)\qquad\text{for all}\quad \mu\in\mathcal{P}^*.$$

The mapping $\Phi:\mathcal{P}\to\mathcal{F}(\mathcal{P}^*,\overline{\mathbb{R}})$ such that $\Phi(a)=\varphi_a$ for all $a\in\mathcal{P}$ is linear and monotone. Corresponding to the neighborhoods $v\in\mathcal{V}$ we consider the $\overline{\mathbb{R}}$-valued functions $\Phi(v)=\psi_v$ on $\mathcal{P}^*$ such that

$$\psi_v(\mu)=\inf\{\alpha>0\mid \mu\in\alpha v^\circ\}$$

for all $\mu\in\mathcal{P}^*$ and denote $\widehat{\mathcal{V}}=\{\hat{v}\mid v\in\mathcal{V}\}$. The neighborhoods $\hat{v}\in\widehat{\mathcal{V}}$ are defined for $\mathcal{F}(\mathcal{P}^*,\overline{\mathbb{R}})$ by

$$f\le g+\hat{v}\qquad\text{if}\qquad f(\mu)\le g(\mu)+\psi_v(\mu)\quad\text{for all}\quad \mu\in\mathcal{P}^*$$

(see 1.4(e)) for functions $f,g \in \mathcal{F}(\mathcal{P}^*,\overline{\mathbb{R}})$ and $\hat{v} \in \widehat{\mathcal{V}}$. We have $\psi_{(\alpha v)} = \alpha\psi_v$ and $\psi_{(u+v)} \geq \psi_u + \psi_v$ for all $u,v \in \mathcal{V}$ and $\alpha > 0$ (see the proof of 5.37 for details). Thus $\big(\mathcal{F}_{\widehat{\mathcal{V}}_b}(\mathcal{P}^*,\overline{\mathbb{R}}),\widehat{\mathcal{V}}\big)$ is a locally convex cone in the sense of 1.4(e), and a complete lattice since $\psi_v(\mu) < +\infty$ holds for all $\mu \in \mathcal{P}^*$ with some $\hat{v} \in \widehat{\mathcal{V}}$. We claim that $\Phi(\mathcal{P}) \subset \mathcal{F}_{\widehat{\mathcal{V}}_b}(\mathcal{P}^*,\overline{\mathbb{R}})$ and that

$$a \preccurlyeq b + v \qquad \text{if and only if} \qquad \varphi_a \leq \varphi_b + \psi_v.$$

holds for $a,b \in \mathcal{P}$ and $v \in \mathcal{V}$. Indeed, suppose that $a \preccurlyeq b + v$. Then for every $\mu \in \mathcal{P}^*$ and $\alpha > 0$ such that $\mu \in \alpha v^\circ$ we have $\mu(a) \leq \mu(b) + \alpha$, that is $\varphi_a(\mu) \leq \varphi_b(\mu) + \psi_v(\mu)$, hence $\varphi_a \leq \varphi_b + \psi_v$. Conversely, if $a \not\preccurlyeq b + v$, then following Corollary 4.34 there is $\mu \in v^\circ \subset \mathcal{P}^*$ such that $\mu(a) > \mu(b) + 1$. The former implies $\psi_v(\mu) \leq 1$, hence $\varphi_a(\mu) > \varphi_b(\mu) + \psi_v(\mu)$ and therefore $\varphi_a \not\leq \varphi_b + \psi_v$. We infer in particular that the functions $\Phi(a) = \varphi_a$ are contained in $\mathcal{F}_{\widehat{\mathcal{V}}_b}(\mathcal{P}^*,\overline{\mathbb{R}})$ for all $a \in \mathcal{P}$. Indeed, given $a \in \mathcal{P}$ and $v \in \mathcal{V}$ there is $\lambda \geq 0$ such that $0 \preccurlyeq a + \lambda v$, hence $0 \leq \Phi(a) + \lambda\hat{v}$. Therefore the element $\Phi(a)$ is contained in $\mathcal{F}_{\widehat{\mathcal{V}}_b}(\mathcal{P}^*,\overline{\mathbb{R}})$ as claimed. Finally we establish that the linear operator $\Phi : \mathcal{P} \to \mathcal{F}_{\widehat{\mathcal{V}}_b}(\mathcal{P}^*,\overline{\mathbb{R}})$ is an embedding in the sense of 2.2 of the locally convex cone $(\mathcal{P},\mathcal{V})$ into $\big(\mathcal{F}_{\widehat{\mathcal{V}}_b}(\mathcal{P}^*,\overline{\mathbb{R}}),\widehat{\mathcal{V}}\big)$, provided that we consider $(\mathcal{P},\mathcal{V})$ in its weak preorder. Indeed, we set $\Phi(v) = \hat{v}$ for $v \in \mathcal{V}$ towards the extension

$$\Phi : (\mathcal{P} \cup \mathcal{V}) :\to \big(\mathcal{F}_{\widehat{\mathcal{V}}_b}(\Upsilon,\overline{\mathbb{R}}) \cup \widehat{\mathcal{V}}\big).$$

Then $\Phi(\mathcal{V}) = \widehat{\mathcal{V}}$, and by the above $a \preccurlyeq b + v$ holds for $a,b \in \mathcal{P}$ and $v \in \mathcal{V}$ if and only if $\Phi(a) \leq \Phi(b) + \Phi(v)$, as required in 2.2.

Finally, we denote by $\widehat{\mathcal{P}}$ the smallest locally convex complete lattice subcone of $\mathcal{F}_{\widehat{\mathcal{V}}_b}(\mathcal{P}^*,\overline{\mathbb{R}})$ that contains the embedding $\Phi(\mathcal{P})$ of $\mathcal{P}$. proposition 5.36 specifies that $\widehat{\mathcal{P}}$ consists of all functions in $\mathcal{F}_{\widehat{\mathcal{V}}_b}(\mathcal{P}^*,\overline{\mathbb{R}})$ that can be expressed in the following way:

$$\varphi_{\mathfrak{A}} = \inf\big\{\sup \Phi(A) \mid A \in \mathfrak{A}\big\}.$$

where $\mathfrak{A}$ is family of subsets of $\mathcal{P}$ such that $\Phi(\mathfrak{A}) = \{\Phi(A) \mid A \in \mathfrak{A}\}$ is supbounded below in $\mathcal{F}_{\widehat{\mathcal{V}}_b}(\mathcal{P}^*,\overline{\mathbb{R}})$ (see 5.36). We call $(\widehat{\mathcal{P}},\widehat{\mathcal{V}})$ the *standard lattice completion* of the locally convex cone $(\mathcal{P},\mathcal{V})$. According to Proposition 5.52, the subcone $\widehat{\mathcal{P}}$ of $\mathcal{F}_{\widehat{\mathcal{V}}_b}(\mathcal{P}^*,\overline{\mathbb{R}})$ is closed in the order topology $\mathcal{O}\big(\mathcal{F}_{\widehat{\mathcal{V}}_b}(\mathcal{P}^*,\overline{\mathbb{R}})\big)$, and $\mathcal{O}(\widehat{\mathcal{P}})$ coincides with the restriction of $\mathcal{O}\big(\mathcal{F}_{\widehat{\mathcal{V}}_b}(\mathcal{P}^*,\overline{\mathbb{R}})\big)$ to $\widehat{\mathcal{P}}$. According to Proposition 5.53 the order topology, the weak order topology and the topology of pointwise convergence on $X$ all coincide on $\widehat{\mathcal{P}}$ and are Hausdorff. Moreover, the restriction of this topology to the subcone $\Phi(\mathcal{P})$ of $\widehat{\mathcal{P}}$ is generally coarser than the image of the weak topology $\sigma(\mathcal{P},\mathcal{P}^*)$ (see 4.6) on $\Phi(\mathcal{P})$. Indeed, while the domain of the functions $\varphi \in \widehat{\mathcal{P}}$ is the dual cone $\mathcal{P}^*$ of $\mathcal{P}$, pointwise convergence for a net $(\varphi_{a_i})_{i\in\mathcal{I}}$ of $\overline{\mathbb{R}}$-valued functions in

$\Phi(\mathcal{P})$ is treated differently from weak convergence for the corresponding net $(a_i)_{i\in\mathcal{I}}$ in $\mathcal{P}$ if the function value $+\infty \in \overline{\mathbb{R}}$ is involved. The order topology of $\overline{\mathbb{R}}$, which is used for pointwise convergence of the functions is coarser than the given (locally convex cone) topology of $\overline{\mathbb{R}}$ at this point (see 4.6 and 5.40). However, if all elements of $\mathcal{P}$ are bounded, that is for example in the case of a vector space, then continuous linear functionals take only finite values on $\mathcal{P}$, hence the elements of $\Phi(\mathcal{P})$ take only finite values as functions on $\mathcal{P}^*$. In this case the order topology, the weak order topology, the weak topology and the topology of pointwise convergence all coincide on $\Phi(\mathcal{P})$.

The embedding of a locally convex cone $(\mathcal{P},\mathcal{V})$ into some locally convex complete lattice cone is of course not unique. However, the standard lattice completion $(\widehat{\mathcal{P}},\widehat{\mathcal{V}})$ of $(\mathcal{P},\mathcal{V})$ is distinguished by the fact that every continuous linear operator from $\mathcal{P}$ into some locally convex complete lattice cone $(\mathcal{Q},\mathcal{W})$ can be extended to an order continuous lattice homomorphism from $(\widehat{\mathcal{P}},\widehat{\mathcal{V}})$ into $(\mathcal{Q},\mathcal{W})$. More precisely:

**Proposition 5.58.** *Let $(\mathcal{P},\mathcal{V})$ be a locally convex cone, and let $\Phi$ be the canonical embedding of $\mathcal{P}$ into its standard lattice completion $\widehat{\mathcal{P}}$. Let $(\mathcal{Q},\mathcal{W})$ be a locally convex complete lattice cone such that the order continuous lattice homomorphisms support the separation property for $\mathcal{Q}$. For every continuous linear operator $T:\mathcal{P}\to\mathcal{Q}$ there exists an order continuous lattice homomorphism $\widehat{T}:\widehat{\mathcal{P}}\to\mathcal{Q}$ such that $T=\widehat{T}\circ\Phi$. Moreover, if $v\in\mathcal{V}$ and $w\in\mathcal{W}$ are such that $a\le b+v$ implies $T(a)\le T(b)+w$ for $a,b\in\mathcal{P}$, then $\varphi\le\psi+\Phi(v)$ implies $\widehat{T}(\varphi)\le\widehat{T}(\psi)+w$ for $\varphi,\psi\in\widehat{\mathcal{P}}$.*

*Proof.* Let $(\mathcal{P},\mathcal{V})$, $(\mathcal{Q},\mathcal{W})$ and $T:\mathcal{P}\to\mathcal{Q}$ be as stated. The *adjoint operator* $T^*:\mathcal{Q}^*\to\mathcal{P}^*$ is defined as follows (see II.2.15 in [100]): For any $\nu\in\mathcal{Q}^*$ define the linear functional $T^*(\nu)$ on $\mathcal{P}$ by $T^*(\nu)(a)=\nu\big(T(a)\big)$ for all $a\in\mathcal{P}$. More precisely: If $\nu\in w^\circ$ for some $w\in\mathcal{W}$ and if $v\in\mathcal{V}$ is such that $T(a)\le T(b)+w$ whenever $a\le b+v$ for $a,b\in\mathcal{P}$, then $T^*(\nu)\in v^\circ$. Indeed, $a\le b+v$ for $a,b\in\mathcal{P}$ implies that

$$T^*\big(\nu\big)(a)=\nu\big(T(a)\big)\le\nu\big(T(b)\big)+1=T^*\big(\nu\big)(b)+1.$$

Now let $\mathfrak{A}$ be a family of subsets of $\mathcal{P}$ such that $\Phi(\mathfrak{A})=\{\Phi(A)\mid A\in\mathfrak{A}\}$ is sup-bounded below in $\mathcal{F}_{\widehat{\mathcal{V}}_b}(\mathcal{P}^*,\overline{\mathbb{R}})$. We claim that the family $T(\mathfrak{A})=\{T(A)\mid A\in\mathfrak{A}\}$ is sup-bounded below in $\mathcal{Q}$: Indeed, given $w\in W$ there is $v\in\mathcal{V}$ such that $T^*(w^\circ)\subset v^\circ$. There is $\lambda\ge 0$ such that $0\le\sup\Phi(A)+\lambda v$ for all $A\in\mathfrak{A}$. This means $\mu\big(\sup\Phi(A)\big)\ge-\lambda$ for all $\mu\in v^\circ$. For an order continuous lattice homomorphism $\nu\in w^\circ$ set $\mu=T^*(\nu)\in v^\circ$. Then for $A\in\mathfrak{A}$ we have

$$\nu\big(\sup T(A)\big)=\sup\nu\big((T(A)\big)=\sup\mu(A)=\mu\big(\sup\Phi(A)\big)\ge-\lambda.$$

This shows $0\le\sup T(A)+\lambda v$, since the order continuous lattice homomorphisms are supposed to support the separation property for $\mathcal{Q}$. Our claim has therefore been verified. Now consider the elements of $\varphi_{\mathfrak{A}} \in \widehat{\mathcal{P}}$ and $\widetilde{\varphi}_{\mathfrak{A}} \in \mathcal{Q}$ defined as

$$\varphi_{\mathfrak{A}} = \inf\left\{ \sup \Phi(A) \mid A \in \mathfrak{A}\right\} \qquad \text{and} \qquad \widetilde{\varphi}_{\mathfrak{A}} = \inf\left\{ \sup T(A) \mid A \in \mathfrak{A}\right\},$$

where $\Phi$ denotes the canonical embedding of $\mathcal{P}$ into $\widehat{\mathcal{P}}$. For every $\mu \in \mathcal{P}^*$, that is the domain of the functions in $\widehat{\mathcal{P}}$, and for every order continuous lattice homomorphism $\nu \in \mathcal{Q}^*$ we calculate

$$\varphi_{\mathfrak{A}}(\mu) = \inf\left\{ \sup \Phi(A) \mid A \in \mathfrak{A}\right\}(\mu) = \inf\left\{ \sup \mu(A) \mid A \in \mathfrak{A}\right\}$$

and

$$\begin{aligned}\nu(\widetilde{\varphi}_{\mathfrak{A}}) &= \sup\left\{ \inf \nu\big(T(A)\big) \mid A \in \mathfrak{A}\right\} = \sup\left\{ \inf T^*(\nu)\big(A\big) \mid A \in \mathfrak{A}\right\} \\ &= \varphi_{\mathfrak{A}}\big((T^*(\nu)\big).\end{aligned}$$

Thus, if $w \in \mathcal{W}$, if $v \in \mathcal{V}$ is such that $T(a) \le T(b) + w$ whenever $a \le b + v$ for $a, b \in \mathcal{P}$, and if $\varphi_{\mathfrak{A}} \le \varphi_{\mathfrak{B}} + \Phi(v)$ for such families $\mathfrak{A}$ and $\mathfrak{B}$ of subsets of $\mathcal{P}$, then

$$\nu(\widetilde{\varphi}_{\mathfrak{A}}) = \varphi_{\mathfrak{A}}\big((T^*(\nu)\big) \le \varphi_{\mathfrak{B}}\big((T^*(\nu)\big) + \psi_v\big(T^*(\nu)\big) \le \nu(\widetilde{\varphi}_{\mathfrak{B}}) + 1$$

holds for all order continuous lattice homomorphisms $\nu \in w^\circ$, since $T^*(\nu) \in v^\circ$ implies that $\psi_v\big(T^*(\nu)\big) \le 1$. This shows

$$\widetilde{\varphi}_{\mathfrak{A}} \le \widetilde{\varphi}_{\mathfrak{B}} + w,$$

since these functionals are supposed to support the separation property for $\mathcal{Q}$. In particular, we infer that $\widetilde{\varphi}_{\mathfrak{A}} = \widetilde{\varphi}_{\mathfrak{B}}$ whenever $\varphi_{\mathfrak{A}} = \varphi_{\mathfrak{B}}$. This follows from the fact that both cones $\widehat{\mathcal{P}}$ and $\mathcal{Q}$ carry their respective weak preorders, which are supposed to be antisymmetric. We are now prepared to define the operator $\widehat{T} : \widehat{\mathcal{P}} \to \mathcal{Q}$. For a family $\mathfrak{A}$ of subsets of $\mathcal{P}$ such that $\Phi(\mathfrak{A}) = \{\Phi(A) \mid A \in \mathfrak{A}\}$ is sup-bounded below in $\mathcal{F}_{\widehat{\mathcal{V}}_b}(\mathcal{P}^*, \overline{\mathbb{R}})$ and the corresponding element $\varphi_{\mathfrak{A}} \in \widehat{\mathcal{P}}$ we set

$$\widehat{T}\big(\varphi_{\mathfrak{A}}\big) = \widetilde{\varphi}_{\mathfrak{A}}$$

and observe the following:

(i) The operator $\widehat{T}$ is linear. Indeed, we note that $\Phi(\alpha\mathfrak{A}) = \alpha\Phi(\mathfrak{A})$ and $T(\alpha\mathfrak{A}) = \alpha T(\mathfrak{A})$ as well as $\Phi(\mathfrak{A}+\mathfrak{B}) = \Phi(\mathfrak{A})+\Phi(\mathfrak{B})$ and $T(\mathfrak{A}+\mathfrak{B}) = T(\mathfrak{A})+T(\mathfrak{B})$ holds for any such families $\mathfrak{A}$ and $\mathfrak{B}$ of subsets of $\mathcal{P}$ and $\alpha \ge 0$. Then the arguments in Parts (i) and (ii) of the proof for Proposition 5.36 yield that

$$\varphi_{\mathfrak{A}} + \varphi_{\mathfrak{B}} = \varphi_{(\mathfrak{A}+\mathfrak{B})} \qquad \text{and} \qquad \widetilde{\varphi}_{\mathfrak{A}} + \widetilde{\varphi}_{\mathfrak{B}} = \widetilde{\varphi}_{(\mathfrak{A}+\mathfrak{B})}.$$

Thus

$$\widehat{T}(\varphi_{\mathfrak{A}}+\varphi_{\mathfrak{B}})=\widehat{T}(\varphi_{(\mathfrak{A}+\mathfrak{B})})=\widetilde{\varphi}_{(\mathfrak{A}+\mathfrak{B})}=\widetilde{\varphi}_{\mathfrak{A}}+\widetilde{\varphi}_{\mathfrak{B}}=\widehat{T}(\varphi_{\mathfrak{A}})+\widehat{T}(\varphi_{\mathfrak{B}}).$$

Likewise, $\alpha\varphi_{\mathfrak{A}}=\varphi_{\alpha\mathfrak{A}}$ and $\alpha\widetilde{\varphi}_{\mathfrak{A}}=\widetilde{\varphi}_{\alpha\mathfrak{A}}$ yields $\widehat{T}(\alpha\varphi_{\mathfrak{A}})=\alpha\widehat{T}(\varphi_{\mathfrak{A}})$ for all $\alpha\geq 0$.

(ii) We observed before that, given $w\in\mathcal{W}$ and $v\in\mathcal{V}$ such that $a\leq b+v$ for $a,b\in\mathcal{P}$ implies $T(a)\leq T(b)+w$, then $\varphi_{\mathfrak{A}}\leq\varphi_{\mathfrak{B}}+\Phi(v)$ for $\varphi_{\mathfrak{A}},\varphi_{\mathfrak{B}}\in\widehat{\mathcal{P}}$ implies

$$\widehat{T}(\varphi_{\mathfrak{A}})=\widetilde{\varphi}_{\mathfrak{A}}\leq\widetilde{\varphi}_{\mathfrak{B}}+w=\widehat{T}(\varphi_{\mathfrak{B}})+w.$$

This entails continuity for the operator $\widehat{T}$.

(iii) Let $a\in\mathcal{P}$ and set $\mathfrak{A}=\big\{\{a\}\big\}$. Then $\varphi_{\mathfrak{A}}=\Phi(a)\in\widehat{\mathcal{P}}$ and $\widetilde{\varphi}_{\mathfrak{A}}=T(a)$. Thus

$$\big(\widehat{T}\circ\Phi\big)(a)=\widehat{T}(\varphi_{\mathfrak{A}})=\widetilde{\varphi}_{\mathfrak{A}}=T(a).$$

This shows $T=\widehat{T}\circ\Phi$.

(iv) Let $\{\mathfrak{A}_i\}_{i\in\mathcal{I}}$ be a collection of such families $\mathfrak{A}_i$ of subsets of $\mathcal{P}$. In a first instance, suppose that this collection is not empty, and let $\mathfrak{A}=\{\cup_{i\in\mathcal{I}}A_i\mid(A_i)_{i\in\mathcal{I}}\in\prod_{i\in\mathcal{I}}\mathfrak{A}_i\}$, that is the elements $A$ of $\mathfrak{A}$ are all unions of the type $A=\cup_{i\in\mathcal{I}}A_i$, where $A_i\in\mathfrak{A}_i$. Then $\Phi(\mathfrak{A})=\{\cup_{i\in\mathcal{I}}\Phi(A_i)\mid(A_i)_{i\in\mathcal{I}}\in\prod_{i\in\mathcal{I}}\mathfrak{A}_i\}$ and $T(\mathfrak{A})=\{\cup_{i\in\mathcal{I}}T(A_i)\mid(A_i)_{i\in\mathcal{I}}\in\prod_{i\in\mathcal{I}}\mathfrak{A}_i\}$. Therefore Part (iii) in the proof for Proposition 5.36 yields

$$\sup_{i\in\mathcal{I}}\varphi_{\mathfrak{A}_i}=\varphi_{\mathfrak{A}}\qquad\text{and}\qquad\sup_{i\in\mathcal{I}}\widetilde{\varphi}_{\mathfrak{A}_i}=\widetilde{\varphi}_{\mathfrak{A}}.$$

This shows

$$\widehat{T}\big(\sup_{i\in\mathcal{I}}\varphi_{\mathfrak{A}_i}\big)=\sup_{i\in\mathcal{I}}\widehat{T}\big(\varphi_{\mathfrak{A}_i}\big).$$

In a second instance, suppose that the set $\{a_{\mathfrak{A}_i}\}_{i\in\mathcal{I}}$ is bounded below in $\mathcal{P}$, and let $\mathfrak{A}=\cup_{i\in\mathcal{I}}\mathfrak{A}_i$. Then $\Phi(\mathfrak{A})=\{\cup_{i\in\mathcal{I}}\Phi(\mathfrak{A}_i)\mid i\in\mathcal{I}\}$ and $T(\mathfrak{A})=\{\cup_{i\in\mathcal{I}}T(A_i)\mid i\in\mathcal{I}\}$. Part (iv) in the proof for Proposition 5.36 yields

$$\inf_{i\in\mathcal{I}}\varphi_{\mathfrak{A}_i}=\varphi_{\mathfrak{A}}\qquad\text{and}\qquad\inf_{i\in\mathcal{I}}\widetilde{\varphi}_{\mathfrak{A}_i}=\widetilde{\varphi}_{\mathfrak{A}}.$$

Thus

$$\widehat{T}\big(\inf_{i\in\mathcal{I}}\varphi_{\mathfrak{A}_i}\big)=\inf_{i\in\mathcal{I}}\widehat{T}\big(\varphi_{\mathfrak{A}_i}\big)$$

holds as well. The operator $\widehat{T}:\widehat{\mathcal{P}}\to\mathcal{Q}$ is therefore an order continuous lattice homomorphism. □

For $\mathcal{Q}=\overline{\mathbb{R}}$ in particular, Proposition 5.38 states that for $v\in\mathcal{V}$ and every linear functional $\mu\in v^\circ$ on $\mathcal{P}$ there is an order continuous lattice homomorphism $\hat{\mu}\in\big(\Phi(v)\big)^\circ$ on $\widehat{\mathcal{P}}$ such that $\mu=\hat{\mu}\circ\Phi$. We proceed to demonstrate that the standard lattice completion $(\widehat{\mathcal{P}},\widehat{\mathcal{V}})$ of a locally convex cone $(\mathcal{P},\mathcal{V})$ is indeed the only (up to embedding) locally convex complete lattice cone which contains an embedding of $\mathcal{P}$ and satisfies this property.

**Proposition 5.59.** *Let $(\mathcal{P},\mathcal{V})$ be a locally convex cone, let $(\widetilde{\mathcal{P}},\widetilde{\mathcal{V}})$ be a locally convex complete lattice cone such that the order continuous lattice homomorphisms support the separation property for $\widetilde{\mathcal{P}}$. Suppose that there is an embedding $\Psi:\mathcal{P}\to\widetilde{\mathcal{P}}$ with respect to the weak preorder of $\mathcal{P}$ and that for every $v\in\mathcal{V}$ and every linear functional $\mu\in v^\circ$ on $\mathcal{P}$ there is an order continuous lattice homomorphism $\tilde{\mu}\in\big(\Psi(v)\big)^\circ$ on $\widetilde{\mathcal{P}}$ such that $\mu=\tilde{\mu}\circ\Psi$. Then there exists an embedding $\widehat{\Psi}:\widehat{\mathcal{P}}\to\widetilde{\mathcal{P}}$, where $(\widehat{\mathcal{P}},\widehat{\mathcal{V}})$ denotes the standard lattice completion of $(\mathcal{P},\mathcal{V})$. This embedding preserves the lattice operations for $\widehat{\mathcal{P}}$ and $\widetilde{\mathcal{P}}$.*

*Proof.* We shall use the notations from the proof of the preceding proposition, in particular we denote by $\Phi:\mathcal{P}\to\widehat{\mathcal{P}}$ the canonical embedding of $(\mathcal{P},\mathcal{V})$ into its standard lattice completion $(\widehat{\mathcal{P}},\widehat{\mathcal{V}})$. Now suppose that the linear operator $\Psi:\mathcal{P}\to\widetilde{\mathcal{P}}$ is also an embedding in the sense of 2.2, that is there is an extension

$$\Psi:(\mathcal{P}\cup\mathcal{V})\to(\widetilde{\mathcal{P}}\cup\widetilde{\mathcal{V}})$$

with the required properties. According to Proposition 5.58 there exists an order continuous lattice homomorphism $\widehat{\Psi}:\widehat{\mathcal{P}}\to\widetilde{\mathcal{P}}$ such that $\Psi=\widehat{\Psi}\circ\Phi$ and

$$\varphi_{\mathfrak{A}}\leq\varphi_{\mathfrak{B}}+\Phi(v)\qquad\text{implies that}\qquad\widehat{\Psi}(\varphi_{\mathfrak{A}})\leq\widehat{\Psi}(\varphi_{\mathfrak{B}})+\Psi(v)$$

for $\varphi_{\mathfrak{A}},\varphi_{\mathfrak{B}}\in\widehat{\mathcal{P}}$ and $v\in\mathcal{V}$. As before we abbreviate $\hat{v}=\Phi(v)\in\widehat{\mathcal{V}}$ for $v\in\mathcal{V}$ and use this notation for the extension

$$\widehat{\Psi}:(\widehat{\mathcal{P}}\cup\widehat{\mathcal{V}})\to(\widetilde{\mathcal{P}}\cup\widetilde{\mathcal{V}})$$

setting $\widehat{\Psi}(\hat{v})=\Psi(v)$ for all $v\in\mathcal{V}$. Clearly $\widehat{\Psi}(\widehat{\mathcal{V}})=\Psi(\mathcal{V})=\widetilde{\mathcal{V}}$, and rewriting the above yields that

$$\varphi_{\mathfrak{A}}\leq\varphi_{\mathfrak{B}}+\hat{v}\qquad\text{implies that}\qquad\widehat{\Psi}(\varphi_{\mathfrak{A}})\leq\widehat{\Psi}(\varphi_{\mathfrak{B}})+\widehat{\Psi}(\hat{v})$$

for $\varphi_{\mathfrak{A}},\varphi_{\mathfrak{B}}\in\widehat{\mathcal{P}}$ and $\hat{v}\in\widehat{\mathcal{V}}$. All left to verify for $\widehat{\Psi}:\widehat{\mathcal{P}}\to\widetilde{\mathcal{P}}$ to be an embedding is the reverse implication in the above. For this, suppose that $\widehat{\Psi}(\varphi_{\mathfrak{A}})\leq\widehat{\Psi}(\varphi_{\mathfrak{B}})+\widehat{\Psi}(\hat{v})$ and let $\mu\in v^\circ$. Following our assumption there is an order continuous lattice homomorphism $\tilde{\mu}\in\big(\Psi(v)\big)^\circ$ on $\widetilde{\mathcal{P}}$ such that $\mu=\tilde{\mu}\circ\Psi$. Then

$$\tilde{\mu}\big(\widehat{\Psi}(\varphi_{\mathfrak{A}})\big)\leq\tilde{\mu}\big(\widehat{\Psi}(\varphi_{\mathfrak{B}})\big)+1.$$

On the other hand, we have

$$\begin{aligned}\varphi_{\mathfrak{A}}(\mu) &= \inf\left\{\sup \mu(A) \mid A \in \mathfrak{A}\right\}\\ &= \inf\left\{\sup \tilde{\mu}(\Psi(A)) \mid A \in \mathfrak{A}\right\}\\ &= \tilde{\mu}\left(\inf\left\{\sup \Psi(A) \mid A \in \mathfrak{A}\right\}\right)\end{aligned}$$

and

$$\begin{aligned}\inf\left\{\sup \Psi(A) \mid A \in \mathfrak{A}\right\} &= \inf\left\{\sup \widehat{\Psi}(\Phi(A)) \mid A \in \mathfrak{A}\right\}\\ &= \widehat{\Psi}\left(\inf\left\{\sup \Phi(A) \mid A \in \mathfrak{A}\right\}\right)\\ &= \widehat{\Psi}(\varphi_{\mathfrak{A}})\end{aligned}$$

since $\Psi = \widehat{\Psi} \circ \Phi$ by 5.58 and since $\widehat{\Psi} : \widehat{\mathcal{P}} \to \widetilde{\mathcal{P}}$ is an order continuous lattice homomorphism. Combining the above yields $\varphi_{\mathfrak{A}}(\mu) = \tilde{\mu}\big(\widehat{\Psi}(\varphi_{\mathfrak{A}})\big)$ and, likewise $\varphi_{\mathfrak{B}}(\mu) = \tilde{\mu}\big(\widehat{\Psi}(\varphi_{\mathfrak{B}})\big)$. Therefore

$$\varphi_{\mathfrak{A}}(\mu) \leq \varphi_{\mathfrak{B}}(\mu) + 1$$

holds for all $\mu \in v^{\circ}$. Form this we infer that $\varphi_{\mathfrak{A}}(\mu) \leq \varphi_{\mathfrak{B}}(\mu) + \psi_v(\mu)$ holds for all $\mu \in \mathcal{P}^*$, and conclude that $\varphi_{\mathfrak{A}} \leq \varphi_{\mathfrak{B}} + \hat{v}$, as claimed. □

*Remarks 5.60.* (a) We shall make extensive use of the standard lattice completion $(\widehat{\mathcal{P}}, \widehat{\mathcal{V}})$ of a locally convex cone $(\mathcal{P}, \mathcal{V})$ in the integration theory for cone-valued functions in Chapters II and III. However, many of the results will refer only to the order closure of the embedding of $\mathcal{P}$ into $\widehat{\mathcal{P}}$. It is therefore useful to observe that the elements of this order closure can be interpreted as elements of some second dual of $\mathcal{P}$. Indeed, let $\varphi \in \widehat{\mathcal{P}}$ be an element of this closure. Since convergence in the order topology of $\widehat{\mathcal{P}}$ coincides with pointwise convergence on $\mathcal{P}^*$, there is a net $(a_i)_{i\in\mathcal{I}}$ in $\mathcal{P}$ such that the functions $\varphi_{a_i} \in \widehat{\mathcal{P}}$ from above converge pointwise to $\varphi$. Thus for $\mu, \nu \in \mathcal{P}^*$ we have

$$\varphi(\mu+\nu) = \lim_{i\in\mathcal{I}} \varphi_{a_i}(\mu+\nu) = \lim_{i\in\mathcal{I}} \varphi_{a_i}(\mu) + \lim_{i\in\mathcal{I}} \varphi_{a_i}(\nu) = \varphi(\mu) + \varphi(\nu)$$

by (OT2). Since $\varphi(\alpha\mu) = \alpha\varphi(\mu)$ holds for all $\varphi \in \widehat{\mathcal{P}}$ and $\mu \in \mathcal{P}^*$ and $\alpha \geq 0$, the function $\varphi$ is an $\overline{\mathbb{R}}$-valued linear functional on $\mathcal{P}^*$, that is an element of $\mathcal{P}^{**}$, the dual cone of $\mathcal{P}^*$ under its finest locally convex topology which renders all linear functionals on $\mathcal{P}^*$ continuous (see 7.3(i) below). Moreover, as an element of $\widehat{\mathcal{P}}$, the functional $\varphi$ is bounded below on all polars of neighborhoods in $\mathcal{V}$.

(b) If $(\mathcal{P}, \mathcal{V})$ is a locally convex vector space over $\mathbb{K} = \mathbb{R}$ or $\mathbb{K} = \mathbb{C}$ in its symmetric (modular) topology (see Example 1.4(d)), that is a locally convex topological vector space, then the dual cone $\mathcal{P}^*$ of $\mathcal{P}$ consist of the real parts $\mu$ of all continuous $\mathbb{K}$-linear functionals $\mu_{\mathbb{K}}$ in the vector space dual $\mathcal{P}^*_{\mathbb{K}}$ of $\mathcal{P}$ (see 2.1(d)). Similarly, in 2.1(d) we established a correspondence between real-valued linear (with respect to multiplication by non-negative scalars) functionals $\varphi$ on $\mathcal{P}^*$ and $\mathbb{K}$-valued $\mathbb{K}$-linear functionals $\varphi_{\mathbb{K}}$ on $\mathcal{P}^*_{\mathbb{K}}$. This correspondence is given by

$$\varphi(\mu) = \mathfrak{Re}\big(\varphi_{\mathbb{K}}(\mu_{\mathbb{K}})\big)$$

for all $\mu \in \mathcal{P}^*$, and

$$\varphi_{\mathbb{K}}(\mu_{\mathbb{K}}) = \varphi(\mu) \qquad \text{or} \qquad \varphi_{\mathbb{K}}(\mu_{\mathbb{K}}) = \big(\varphi(\mu) - i\,\varphi(i\mu)\big)$$

for all $\mu_{\mathbb{B}} \in \mathcal{P}^*_{\mathbb{K}}$ in the real or complex case, respectively. If the real-linear functional $\varphi$ on $\mathcal{P}^*$ is contained in $\widehat{\mathcal{P}}$, that is for example if $\varphi$ is contained in the order closure of the embedding of $\mathcal{P}$ (see Part (a) before) into $\widehat{\mathcal{P}}$, then $\varphi$ is bounded below on the polars $v^\circ \subset \mathcal{P}^*$ of all neighborhoods $v \in \mathcal{V}$. Therefore the corresponding $\mathbb{K}$-linear functional $\varphi_{\mathbb{K}}$ in the second vector space dual $\mathcal{P}^{**}_{\mathbb{K}}$ of $\mathcal{P}$ is also bounded on all polars of neighborhoods in $\mathcal{V}$. In the case of a normed space $(\mathcal{P}, \|\ \|)$, for example, the latter implies that $\varphi_{\mathbb{K}}$ is an element of the (strong) second vector space dual of $\mathcal{P}$.

(c) For a concrete example to (b) let $\mathcal{P} = \mathbb{K}$ endowed with the Euclidean topology, that is the neighborhood system $\mathcal{V} = \{\varepsilon\mathbb{B} \mid \varepsilon > 0\}$, where $\mathbb{B}$ is the unit ball in $\mathbb{K}$. The vector space dual $\mathcal{P}_{\mathbb{K}}$ of $\mathbb{K}$ then is of course $\mathbb{K}$ itself, which corresponds to the dual cone $\mathcal{P}^*$ of $\mathbb{K}$ as a locally convex cone as elaborated in 2.1(d), that is every $z \in \mathbb{K}$ defines a real-linear functional in $\mathcal{P}^*$ via

$$a \mapsto \mathfrak{Re}(za) : \mathbb{K} \to \mathbb{R}.$$

On the other hand, every real-valued linear functional $\varphi$ on $\mathcal{P}^* = \mathbb{K}$ corresponds to an element $z \in \mathbb{K}$, that is the second vector space dual of $\mathbb{K}$, by

$$z = \varphi(1) \qquad \text{or} \qquad z = \varphi(1) - i\varphi(i)$$

in the real or complex case, respectively.

(d) If under the assumptions of (b), $(\mathcal{Q}, \mathcal{W})$ is a second locally convex vector space over $\mathbb{K}$, then we shall say that a linear operator $T : \mathcal{Q} \to \widehat{\mathcal{P}}$ is $\mathbb{K}$-*linear* if

(i) $T(f)(\mu + \nu) = T(f)(\mu) + T(f)(\nu)$ and
(ii) $T(\alpha f)(\mu) = T(f)(\alpha\mu)$

holds for all $f \in \mathcal{Q}$, $\mu, \nu \in \mathcal{P}^*$ and $\alpha \in \mathbb{K}$. In this case $T$ corresponds to a $\mathbb{K}$-linear operator $\widetilde{T} : \mathcal{Q} \to \mathcal{P}^{**}_{\mathbb{K}}$.

**5.61 Simplified Standard Lattice Completion.** It is often preferable to realize a lattice completion of a locally convex cone $(\mathcal{P}, \mathcal{V})$ as a cone of $\overline{\mathbb{R}}$-valued functions on a suitable subset of $\mathcal{P}^*$ rather than on the whole of $\mathcal{P}^*$. For this we use a subset $\Upsilon$ of $\mathcal{P}^*$ which supports the separation property for $\mathcal{P}$ in the sense of 5.32. (Following Corollary 4.34 this holds true for $\Upsilon = \mathcal{P}^*$). Let us denote by $\widehat{\mathcal{P}}_\Upsilon$ and $\widehat{\mathcal{V}}_\Upsilon$ the restrictions to $\Upsilon$ of the functions in $\widehat{\mathcal{P}}$ and of the associated neighborhood functions in $\widehat{\mathcal{V}}$. Then

$(\widehat{\mathcal{P}}_\Upsilon, \widehat{\mathcal{V}}_\Upsilon)$ is again a full locally convex complete lattice cone. Consider the restriction map $\widehat{\Psi} : \widehat{\mathcal{P}} \to \widehat{\mathcal{P}}_\Upsilon$ and its composition $\Psi = \widehat{\Psi} \circ \Phi$ with the canonical embedding $\Phi$ of $\mathcal{P}$ into $\widehat{\mathcal{P}}$. We claim that $\Psi : \mathcal{P} \to \widehat{\mathcal{P}}_\Upsilon$ is an embedding of $\mathcal{P}$ into $\widehat{\mathcal{P}}_\Upsilon$ if we consider $\mathcal{P}$ in its weak preorder. Indeed, if $a \preccurlyeq b+\psi_v$ for $a, b \in \mathcal{P}$ and $v \in \mathcal{V}$, then $\widehat{\Psi}(\varphi_a) \le \widehat{\Psi}(\varphi_b)+\psi_v$ holds as well in $\widehat{\mathcal{P}}_\Upsilon$, that is $\Psi(a) \le \Psi(b) + \Psi(v)$. (We use the earlier notations.) Conversely, if $a \not\preccurlyeq b+v$, then by our assumption there is $\alpha \ge 0$ and $\mu \in \Upsilon \cap \alpha v^\circ$ such that $\mu(a) > \mu(b)+\alpha$. Then $\psi_v(\mu) \le \alpha$, hence $\varphi_a(\mu) > \varphi_b(\mu)+\psi_v(\mu)$. This shows $\widehat{\Psi}(\varphi_a) \not\le \widehat{\Psi}(\varphi_b)+\psi_v$, that is $\Psi(a) \not\le \Psi(b) + \Psi(v)$. Thus $\Phi : \mathcal{P} \to \widehat{\mathcal{P}}_\Upsilon$ is an embedding in the sense of 2.2 as claimed. Since the lattice operations are performed pointwise, we have $\widehat{\Psi}(\sup A) = \sup\big(\widehat{\Psi}(A)\big)$ for every non-empty subset $A$ of $\widehat{\mathcal{P}}$ and $\widehat{\Psi}(\inf A) = \inf\big(\widehat{\Psi}(A)\big)$ for every non-empty bounded below subset $A$ of $\widehat{\mathcal{P}}$. The operator

$$\widehat{\Psi} : \widehat{\mathcal{P}} \to \widehat{\mathcal{P}}_\Upsilon$$

is therefore a surjective order continuous lattice homomorphism, but not necessarily an embedding. Because the operator $\widehat{\Psi}$ is also continuous with respect to the order topologies on $\widehat{\mathcal{P}}$ and $\widehat{\mathcal{P}}_\Upsilon$ (Proposition 5.46), the image under $\widehat{\Psi}$ of the order closure of $\Phi(\mathcal{P})$ in $\widehat{\mathcal{P}}$ is contained in the order closure of $\Psi(\mathcal{P})$ in $\widehat{\mathcal{P}}_\Upsilon$. According to the preceding Proposition 5.59, $\widehat{\Psi}$ is an embedding and indeed an isomorphism if and only if for every $v \in \mathcal{V}$ and every linear functional $\mu \in v^\circ$ on $\mathcal{P}$ there is an order continuous lattice homomorphism $\tilde{\mu} \in \big(\Psi(v)\big)^\circ$ on $\widehat{\mathcal{P}}_\Upsilon$ such that $\mu = \tilde{\mu} \circ \Psi$. This condition is satisfied if for every $\mu \in \mathcal{P}^*$ there is $\nu \in \Upsilon$ and $\alpha \ge 0$ such that $\mu = \alpha\nu$. In this case the conclusion of Proposition 5.58 applies to $\widehat{\mathcal{P}}_\Upsilon$ as it does to $\widehat{\mathcal{P}}$.

We shall at times us such a simplified standard lattice completion $(\widehat{\mathcal{P}}_\Upsilon, \widehat{\mathcal{V}}_\Upsilon)$ and the order continuous lattice homomorphism $\Psi : \widehat{\mathcal{P}} \to \widehat{\mathcal{P}}_\Upsilon$ in order to represent and visualize results that were obtained in the standard lattice completion $(\widehat{\mathcal{P}}, \widehat{\mathcal{V}})$ of a locally convex cone $(\mathcal{P}, \mathcal{V})$.

*Examples 5.62.* In the preceding Examples 5.33 we investigated a range of locally convex cones $(\mathcal{P}, \mathcal{V})$ and identified subsets $\Upsilon$ of the dual cone which support the separation property. All of these choices are suitable for the construction of a simplified standard lattice completion $(\widehat{\mathcal{P}}_\Upsilon, \widehat{\mathcal{V}}_\Upsilon)$. Let us elaborate on the most important of these situations.

(a) For $\mathcal{P} = \mathbb{R}$ with the usual order and neighborhood system $\mathcal{V} = \{\varepsilon \in \mathbb{R} \mid \varepsilon > 0\}$ the dual cone is $\mathcal{P}^* = \{\alpha \in \mathbb{R} \mid \alpha \ge 0\}$. Therefore the standard lattice completion $\widehat{\mathbb{R}}$ of $\mathbb{R}$ is the cone of all linear $\overline{\mathbb{R}}$-valued functions on $\mathcal{P}^*$. This can be visualized more easily if we choose the subset $\Upsilon = \{1\}$ of $\mathcal{P}^*$ for the above simplified construction, since $\widehat{\mathcal{P}}_\Upsilon$ then coincides with $\overline{\mathbb{R}}$. Following the remark in 5.61, we realize that the standard lattice completion of $\mathbb{R}$ is isomorphic to $\overline{\mathbb{R}}$.

(b) If $(\mathcal{P}, \| \|)$ is a normed vector space $\big($see 5.33(b)$\big)$, then we may choose the dual unit sphere for $\Upsilon \subset \mathcal{P}^*$. According to our preceding remark, the lattice completion $(\widehat{\mathcal{P}}_\Upsilon, \widehat{\mathcal{V}}_\Upsilon)$ then is isomorphic to the standard lattice completion $(\widehat{\mathcal{P}}, \widehat{\mathcal{V}})$. Alternatively, we may choose $\Upsilon = Ex(\mathbb{B})$, that is the set of all extreme points of the dual unit ball $\mathbb{B}$ in $\mathcal{P}^*$. However, the conclusion of Proposition 5.58 does not generally apply to $\widehat{\mathcal{P}}_\Upsilon$ for the latter choice of $\Upsilon$. In both cases the lattice completion $\widehat{\mathcal{P}}_\Upsilon$ of $\mathcal{P}$ consists of a cone of $\overline{\mathbb{R}}$-valued bounded below functions on $\Upsilon$, endowed with the topology of uniform convergence.

(c) For a special case of 5.33(f) consider the locally convex cone $\big(\mathcal{F}_{b_\mathcal{Y}}(X, \overline{\mathbb{R}}), \mathcal{V}_\mathcal{Y}\big)$ of $\overline{\mathbb{R}}$-valued functions on a set $X$ endowed with the topology of uniform convergence on the sets in a family $\mathcal{Y}$ of subsets of $X$ $\big($see 1.4(e)$\big)$. Then

$$\Upsilon = \left\{ \varepsilon_x \;\middle|\; x \in \bigcup_{Y \in \mathcal{Y}} Y \right\} \subset \mathcal{F}_{b_\mathcal{Y}}(X, \overline{\mathbb{R}})^*,$$

where $\varepsilon_x$ denotes the point evaluation at $x \in X$, supports the separation property for $\mathcal{F}_{b_\mathcal{Y}}(X, \mathcal{P})$ $\big($see 5.33(f)$\big)$. The corresponding lattice completion $\widehat{\mathcal{F}_{b_\mathcal{Y}}(X, \overline{\mathbb{R}})}_\Upsilon$ of $\mathcal{F}_{b_\mathcal{Y}}(X, \overline{\mathbb{R}})$ then consists of $\overline{\mathbb{R}}$-valued bounded below functions on $\Upsilon$, endowed with the topology of uniform convergence.

(d) For a special case of (c) let $X$ be a compact set and let $\mathcal{P} = \mathcal{C}(X)$ be the space of all continuous real-valued functions on $X$, endowed with the pointwise operations and order. The neighborhood system $\mathcal{V}$ consisting of all positive constants generates the topology of uniform convergence. The set $\Upsilon$ of all point evaluations $\varepsilon_x$ for $x \in X$ supports the separation property, and the lattice completion $(\widehat{\mathcal{P}}, \widehat{\mathcal{V}})$ of $(\mathcal{P}, \mathcal{V})$ can be realized as a cone of $\overline{\mathbb{R}}$-valued functions on $X$.

(e) In Section 7 below we shall provide another example, that is cones $\mathfrak{H}(\mathcal{N}, \mathcal{M})$ of linear operators from a cone $\mathcal{N}$ into a second cone $\mathcal{M}$, endowed with suitable locally convex cone topologies, where the canonical choice for $\Upsilon$ for a lattice completion $\widehat{\mathfrak{H}}(\mathcal{N}, \mathcal{M})$ is a proper subset rather than the whole dual cone of $\mathfrak{H}(\mathcal{N}, \mathcal{M})$.

(f) Let $\mathcal{P} = \mathbb{K}$, endowed with the Euclidean topology, that is the neighborhood system $\mathcal{V} = \{\varepsilon\mathbb{B} \mid \varepsilon > 0\}$, where $\mathbb{B}$ is the unit ball in $\mathbb{K}$ (see the preceding Remark 5.60(c)). The vector space dual $\mathcal{P}_\mathbb{K}$ of $\mathbb{K}$ then is $\mathbb{K}$ itself which corresponds to the dual cone $\mathcal{P}^*$ of $\mathbb{K}$ as a locally convex cone as elaborated in 2.1(d) and in 5.60(c).

For the construction of a simplified standard lattice completion $\widehat{\mathbb{K}}_\Upsilon$ of $\mathbb{K}$ we choose $\Upsilon = \Gamma$, the unit circle in $\mathbb{K}$. It is straightforward to verify that $\widehat{\mathbb{K}}_\Upsilon$ consists of all bounded below $\overline{\mathbb{R}}$-valued functions on $\Gamma$, endowed with the (strictly) positive constants as neighborhoods. A function $\varphi \in \widehat{\mathbb{K}}_\Upsilon$ can be canonically extended to a real-linear functional on all of $\mathcal{P}^* = \mathbb{K}$ if and only if it takes only finite values in $\mathbb{R}$ and if

$$\sum_{i=1}^{n} \alpha_i \varphi(\gamma_i) = 0$$

holds whenever $\sum_{i=1}^{n} \alpha_i \gamma_i = 0$ for $\alpha_i \in \mathbb{R}$ and $\gamma_i \in \Gamma$. In the real case, that is for $\mathbb{K} = \mathbb{R}$, the latter requires just that

$$\varphi(-1) = -\varphi(1).$$

If the above condition holds, then the corresponding $\mathbb{K}$-linear functional $\varphi_{\mathbb{K}}$ in the second vector space dual of $\mathbb{K}$, that is $\mathbb{K}$ itself, is represented by the number

$$\varphi_{\mathbb{K}} = \varphi(1) \in \mathbb{R} \qquad \text{or} \qquad \varphi_{\mathbb{K}} = \varphi(1) - i\varphi(i) \in \mathbb{C}$$

in the real or in the complex case, respectively.

## 6. Quasi-Full Locally Convex Cones

In Section 1, a locally convex cone $(\mathcal{P}, \mathcal{V})$ was defined to be a subcone of a full locally convex cone, inheriting both the order and the algebraic structure from the latter. Using only the convex quasiuniform structure of $\mathcal{P}$ (see I.3), a procedure described in Chapter I.5 of [100] allows to recover such a full locally convex cone containing $\mathcal{P}$. However this construction is rather unwieldy and far from unique. In situations like in our upcoming measure and integration theory we shall require more immediate access to a canonically constructed full locally convex cone, containing the given cone of interest. This will be possible for a restricted class of locally convex cones which we shall define and describe in the following.

**6.1 Quasi-Full Locally Convex Cones.** In a locally convex cone $(\mathcal{P}, \mathcal{V})$ the scalar multiples and sums for neighborhoods in $\mathcal{V}$ are not necessarily reflected in the corresponding operations for their upper, lower or symmetric neighborhoods as subsets of $\mathcal{P}$. In general we only have

$$\alpha\, v(a) = \big(\alpha v\big)(\alpha a) \qquad \text{and} \qquad u(a) + v(b) \subset \big(u + v\big)(a + b)$$

for $u, v \in \mathcal{V}$, $a, b \in \mathcal{P}$ and $\alpha > 0$, as well as similar relations for the lower and symmetric neighborhoods. Stronger links for the addition are however desirable in some cases. In this vein, we shall say that a locally convex cone $(\mathcal{P}, \mathcal{V})$ is *quasi-full* if for $a, b \in \mathcal{P}$ and $u, v \in \mathcal{V}$

(QF1) $a \le b + v$ *for* $a, b \in \mathcal{P}$ *and* $v \in \mathcal{V}$ *if and only if* $a \le b + s$ *for some* $s \in \mathcal{P}$ *such that* $s \le v$, and

(QF2) $a \le u + v$ *for* $a \in \mathcal{P}$ *and* $u, v \in \mathcal{V}$ *if and only if* $a \le s + t$ *for some* $s, t \in \mathcal{P}$ *such that* $s \le u$ *and* $t \le v$.

These conditions can be reformulated as

$$v(a) = v(0) + a \qquad \text{and} \qquad (u+v)(0) = \downarrow\big(u(0) + v(0)\big),$$

where $\downarrow A = \{b \in \mathcal{P} \mid b \le a \text{ for some } a \in A\}$ denotes the decreasing hull of a subset $A$ of $\mathcal{P}$. Indeed, the first statement is clearly equivalent to (QF1), whereas $\downarrow\big(u(0)+v(0)\big) \subset (u+v)(0)$ always holds as the latter set is decreasing. The reverse inclusion is equivalent to (QF2).

Obviously, every full cone, that is every locally convex cone that contains its neighborhoods as elements, is quasi-full. Most importantly, every ordered locally convex topological vector space $(\mathcal{P}, \mathcal{V})$, where $\mathcal{V}$ denotes a basis of balanced convex neighborhoods of the origin, is seen to be a quasi-full locally convex cone in this sense. Recall from Example 1.4(c) that the cone topologies on $\mathcal{P}$ are defined for elements $a, b \in \mathcal{P}$ and $V \in \mathcal{V}$ by

$$a \le b + V \qquad \text{if} \qquad a - b \le s \quad \text{for some} \quad s \in V.$$

(QF1) is evident, since $s \in V$ implies $s \le V$. For (QF2), let $a \le (U + V)$ for $a \in \mathcal{P}$ and $U, V \in \mathcal{V}$. Then $a \le s + t$ for some $s \in U$ and $t \in V$, since the addition in $\mathcal{V}$ is the usual addition for subsets of $\mathcal{P}$. As $s \le U$ and $t \le V$, this yields (QF2). Recall that equality is a possible choice for the order on $\mathcal{P}$.

In fact, quasi-full locally convex cones are close to locally convex topological vector spaces in the sense that the neighborhoods of every element $a \in \mathcal{P}$ are already determined by the neighborhoods of the element $0 \in \mathcal{P}$. The sum of two neighborhoods in $\mathcal{V}$ coincides with the usual sum of the corresponding subsets of $\mathcal{P}$, that is $u(a) + v(b) = (u+v)(a+b)$ and $(a)u + (b)v = (a+b)(u+v)$ for $u, v \in \mathcal{V}$ and $a, b \in \mathcal{P}$.

Another advantage of quasi-full locally convex cones is that for complete lattice structures in the sense of Section 5.4, Condition $(\bigvee 2)$ transfers from zero-neighborhoods to general ones and from individual neighborhoods to their sums, thus needs to be checked only for a subsystem of zero-neighborhoods that span the entire neighborhood system. Indeed, suppose that $(\mathcal{P}, \mathcal{V})$ is a quasi-full locally convex cone that contains suprema of non-empty sets, and that Condition $(\bigvee 2)$ holds with $b = 0$ and the neighborhoods $u$ and $v$ in $\mathcal{V}$, that is for a non-empty subset $A \subset \mathcal{P}$, $a \le v$ for all $a \in A$ implies $\sup A \le v$, and $a \le u$ for all $a \in A$ implies $\sup A \le u$. Now if there is $b \in \mathcal{P}$ such that $a \le b + (u+v)$ for all $a \in A$, then $a \le b + s_a + t_a$ for some $s_a \le u$ and $t_a \le v$. Then $s = \sup_{a \in A} s_a \le u$ and $t = \sup_{a \in A} t_a \le v$ by our assumption on $u$ and $v$. This shows that $a \le b+(s+t)$ for all $a \in A$, hence $\sup A \le b+(s+t) \le b+(u+v)$, as claimed. If $(\mathcal{P}, \mathcal{V})$ contains both suprema of non-empty and infima of bounded below subsets and satisfies $(\bigwedge 1)$, then Condition $(\bigvee 2)$ for some neighborhood $v \in \mathcal{V}$ implies $(\bigwedge 2)$ for the same $v$. Indeed, suppose that there is $b \in \mathcal{P}$ such that $b \le a + v$ holds for all elements $a$ of some bounded below subset

$A \subset \mathcal{P}$. Then $b \le a + t_a$ for some $t_a \le v$, hence $b \le a + t$ for all $a \in A$, where $t = \sup_{a\in A} t_a \le v$. This yields $b \le \inf(A+t) = \inf A + t \le \inf A + v$ by $(\bigwedge 1)$, hence our claim.

**6.2 The Standard Full Extension of a Quasi-Full Cone.** We shall construct a canonical embedding of a quasi-full locally convex cone $(\mathcal{P}, \mathcal{V})$ into a full locally convex cone in the following manner. Let

$$\mathcal{P}_{\mathcal{V}} = \{a \oplus v \mid a \in \mathcal{P},\ v \in \mathcal{V} \cup \{0\}\}.$$

We use the obvious algebraic operations on $\mathcal{P}_{\mathcal{V}}$, that is

$$(a \oplus v) + (b \oplus u) = (a+b) \oplus (v+u) \qquad \text{and} \qquad \alpha(a \oplus v) = (\alpha a \oplus \alpha v)$$

for $a, b \in \mathcal{P}$, $u, v \in \mathcal{V} \cup \{0\}$ and $\alpha \ge 0$. The order on $\mathcal{P}_{\mathcal{V}}$ is defined as

$$a \oplus v \le b \oplus u$$

if $c \le a + v$ implies that $c \le b + u$ for all $c \in \mathcal{P}$. This order relation is reflexive, and transitive, as for $a, b, c \in \mathcal{P}$ and $u, v, w \in \mathcal{V} \cup \{0\}$ such that $a \oplus v \le b \oplus u$ and $b \oplus u \le c \oplus w$, for every $d \in \mathcal{P}$ such that $d \le a + v$, we have $d \le b + u$, hence $d \le c + w$. Thus $a \oplus v \le c \oplus w$. Similarly, one verifies compatibility with the algebraic operations: Compatibility with the multiplication by positive scalars is obvious; for compatibility with the addition, let $(a \oplus v), (b \oplus u), (c \oplus w) \in \mathcal{P}_{\mathcal{V}}$ such that $a \oplus v \le b \oplus u$. If $d \le (a + c) + (v + w)$, then $d \le (a + c) + s$ for some $s \le v + w$ by (QF1), and $s \le s' + s''$ for some $s' \le v$ and $s'' \le w$ by (QF2.) Hence $d \le (a+s') + (c+s'')$. Because $a + s' \le a + v$ implies that $a + s' \le b + u$, we infer that $d \le (b+c)+(u+w)$. This shows $(a+c)\oplus(v+w) \le (b+c)\oplus(u+w)$. The embedding

$$a \mapsto a \oplus 0 \ :\ \mathcal{P} \to \mathcal{P}_{\mathcal{V}}$$

therefore preserves the algebraic operations and the order of $\mathcal{P}$, since $a \le b$ holds for elements $a, b \in \mathcal{P}$ if and only $a \oplus 0 \le b \oplus 0$ holds in $\mathcal{P}_{\mathcal{V}}$. Moreover, for a neighborhood $v \in \mathcal{V}$ and $a, b \in \mathcal{P}$ we have $a \le b + v$ in $\mathcal{P}$ if and only if $a \oplus 0 \le (b \oplus 0) + (0 \oplus v) = b \oplus v$ holds in $\mathcal{P}_{\mathcal{V}}$. We may therefore identify the neighborhoods $v \in \mathcal{V}$ with the elements $0 \oplus v$ in $\mathcal{P}_{\mathcal{V}}$. In this way $\mathcal{V}$ is embedded into $\mathcal{P}_{\mathcal{V}}$ as well, and $(\mathcal{P}_{\mathcal{V}}, \mathcal{V})$ becomes a full locally convex cone, containing $(\mathcal{P}, \mathcal{V})$ as a subcone. If a certain neighborhood $v \in \mathcal{V}$ is already contained in the given cone $\mathcal{P}$, then the above definition of the order in $\mathcal{P}_{\mathcal{V}}$ yields that both $v \oplus 0 \le 0 \oplus v$ and $0 \oplus v \le v \oplus 0$. The elements $v \oplus 0$ and $0 \oplus v$ are therefore equivalent with respect to the canonical equivalence relation defined by the order on $\mathcal{P}_{\mathcal{V}}$. Thus for a full cone $\mathcal{P}$, this extension $\mathcal{P}_{\mathcal{V}}$ yields only elements that in terms of the order relation are equivalent to existing ones in $\mathcal{P}$. We shall call $(\mathcal{P}_{\mathcal{V}}, \mathcal{V})$ the *standard full extension* of the locally convex cone $(\mathcal{P}, \mathcal{V})$.

**Theorem 6.3.** *Let $(\mathcal{P},\mathcal{V})$ be a quasi-full locally convex cone, and let $(\mathcal{Q},\mathcal{W})$ be a locally convex complete lattice cone. Every continuous linear operator $T:\mathcal{P}\to\mathcal{Q}$ can be extended to a continuous linear operator $\overline{T}:\mathcal{P}_{\mathcal{V}}\to\mathcal{Q}$.*

*Proof.* Let $(\mathcal{P},\mathcal{V})$ be quasi-full, $(\mathcal{Q},\mathcal{W})$ a complete lattice cone, and let $T:\mathcal{P}\to\mathcal{Q}$ be a continuous linear operator. Recall from Section 3 that a continuous linear operator between locally convex cones is monotone with respect to the respective weak preorders. Because $\mathcal{Q}$ carries its weak preorder, this implies monotonicity with respect to the given orders of $\mathcal{P}$ and $\mathcal{Q}$ as well. For an element $a\oplus v\in\mathcal{P}_{\mathcal{V}}$ we define

$$\overline{T}(a\oplus v)=\sup\big\{T(b)\mid b\in\mathcal{P},\ b\leq a+v\big\}\in\mathcal{Q}.$$

Let us first check linearity: Clearly $\overline{T}(\alpha a\oplus\alpha v)=\alpha\overline{T}(a\oplus v)$ for $\alpha\geq 0$. For additivity, let $(a\oplus v),(b\oplus u)\in\mathcal{P}_{\mathcal{V}}$. Using Lemma 5.5(a), we infer

$$\begin{aligned}\overline{T}(a\oplus v)+\overline{T}(b\oplus u)&=\sup\big\{T(c)\mid c\in\mathcal{P},\ c\leq a+v\big\}\\&\qquad+\sup\big\{T(d)\mid d\in\mathcal{P},\ d\leq b+u\big\}\\&=\sup\big\{T(c+d)\mid c,d\in\mathcal{P},\ \ c\leq a+v,\ \ d\leq b+u\big\}\\&\leq\sup\big\{T(e)\mid e\in\mathcal{P},\ \ e\leq(a+b)+(v+u)\big\}\\&=\overline{T}\big((a\oplus v)+(b\oplus u)\big).\end{aligned}$$

If on the other hand $c\leq(a+b)+(v+u)$ for $c\in\mathcal{P}$, then $c\leq c'+c''$ for some $c',c''\in\mathcal{P}$ such that $c'\leq a+v$ and $c''\leq b+v$. by (QF1) and (QF2). Thus

$$T(c)\leq T(c')+T(c'')\leq\overline{T}(a\oplus v)+\overline{T}(b\oplus u).$$

Taking the supremum over all such elements $c\leq(a+b)+(v+u)$ on the left-hand side yields

$$\overline{T}\big((a\oplus v)+(b\oplus u)\big)\leq\overline{T}(a\oplus v)+\overline{T}(b\oplus u).$$

Next we observe that the operator $\overline{T}$ is monotone. Indeed, let $a\oplus v\leq b\oplus u$ for $(a\oplus v),(b\oplus u)\in\mathcal{P}_{\mathcal{V}}$, and let $c\in\mathcal{P}$ such that $c\leq a+v$. Then $c\leq b+u$ by our definition of the order in $\mathcal{P}_{\mathcal{V}}$. This shows $T(c)\leq\overline{T}(b\oplus u)$. Taking the supremum over all such elements $c\leq a+v$ on the left-hand side yields $\overline{T}(a\oplus v)\leq\overline{T}(b\oplus u)$. Finally, for every $w\in\mathcal{W}$ there is $v\in\mathcal{V}$ such that $a\leq b+v$ implies that $T(a)\leq T(b)+w$ for all $a,b\in\mathcal{P}$. Thus

$$\overline{T}(0\oplus v)=\sup\big\{T(s)\mid s\in\mathcal{P},\ s\leq v\big\}\leq w.$$

As $(\mathcal{P}_{\mathcal{V}},\mathcal{V})$ is a full locally convex cone, this demonstrates the continuity of the monotone linear operator $\overline{T}:\mathcal{P}_{\mathcal{V}}\to\mathcal{Q}$. All left to verify is that $\overline{T}$ is indeed an extension of $T$ if we consider $\mathcal{P}$ as a subcone of $\mathcal{P}_{\mathcal{V}}$ via its canonical embedding $a \mapsto (a \oplus 0)$. But this is obvious, as for $a \in \mathcal{P}$ we have

$$\overline{T}(a \oplus 0) = \sup\{T(b) \mid b \in \mathcal{P},\ b \leq a\} = T(a).$$

□

*Remarks 6.4.* Let $(\mathcal{P}, \mathcal{V})$ be a (not necessarily quasi-full) locally convex cone satisfying the following condition:

(QF*) *For every* $a \in \mathcal{P}$ *and* $v \in \mathcal{V}$ *there is* $s \in \mathcal{P}$ *such that* $s \leq v$ *and* $\lambda \geq 0$ *such that* $0 \leq a + \lambda s$.

In this case we may remodel $\mathcal{P}$ into a quasi-full locally convex cone if we define an alternative neighborhood system $\mathfrak{V}$ consisting of all families $(r_v)_{v \in \mathcal{V}}$, where $r_v$ is a non-negative real and $r_v > 0$ for at least one $v \in \mathcal{V}$ and $r_v = 0$ else. Endowed with componentwise defined algebraic operations and order $\mathfrak{V}_0 = \mathfrak{V} \cup 0$ is an ordered cone. Let $\widetilde{\mathcal{P}} = \mathcal{P} \oplus \mathfrak{V}_0$ be the direct sum of $\mathcal{P}$ and $\mathcal{V}_0$. We define the order on $\widetilde{\mathcal{P}}$ in the following way: We set $a \oplus r \leq b \oplus s$ for elements $a, b \in \mathcal{P}$ and $r, s \in \mathcal{V}_0$ if $r \leq s$ and if there are elements $c_1, \ldots, c_n \in \mathcal{P}$ such that $a \leq b + (c_1 + \ldots + c_n)$ and $c_i \leq (s_{v_i} - r_{v_i})v_i$ for distinct elements $v_1, \ldots, v_n \in \mathcal{V}$. In this way, $(\widetilde{\mathcal{P}}, \mathfrak{V})$ becomes a full locally convex cone. Condition (QF*) in particular guarantees that its elements are bounded below. The neighborhoods $u \in \mathcal{V}$ may be identified with the elements $r(u) \in \mathfrak{V}$ such that $r(u)_u = 1$ and $r(u)_v = 0$ else. As a subcone of $(\widetilde{\mathcal{P}}, \mathfrak{V})$, the locally convex cone $(\mathcal{P}, \mathfrak{V})$ is seen to be quasi-full. (Conditions (QF1) and (QF2) from 6.1 are implied by our definition of the neighborhoods in $\mathfrak{V}$.) Because $a \leq b \oplus r(v)$ implies $a \leq b + v$ for $a, b \in \mathcal{P}$ and $v \in \mathcal{V}$, the (upper, lower, symmetric) topologies induced on $\mathcal{P}$ by $\mathfrak{V}$ are generally finer than the given ones. The dual cone $\mathcal{P}^*_{\mathfrak{V}}$ of $\mathcal{P}$ under this new locally convex topology is therefore larger than the given dual cone $\mathcal{P}^*$. The polar $r^*$ of a neighborhood $r \in \mathfrak{V}$ consists of all monotone $\overline{\mathbb{R}}$-valued linear functionals $\mu$ on $\mathcal{P}$ satisfying $\mu(c) \leq 1$ for all $c \in \mathcal{P}$ such that $c \leq r$.

## 7. Cones of Linear Operators

Endowed with the canonical (pointwise) algebraic operations, the linear operators between two cones $\mathcal{N}$ and $\mathcal{M}$ form again a cone $L(\mathcal{N}, \mathcal{M})$. We may introduce neighborhoods for $L(\mathcal{N}, \mathcal{M})$ in the following way (for a similar construction in the case of vector spaces see III.3 in [185]): Let $\mathcal{W}$ be a neighborhood system and let $\leq$ be an order for $\mathcal{M}$ such that $(\mathcal{M}, \mathcal{W})$ is a locally convex cone. Let $\mathfrak{Z}$ be a family of subsets of $\mathcal{N}$, directed upward by set inclusion. For every $Z \in \mathfrak{Z}$ and $w \in W$ we define a neighborhood $V_{(Z,w)}$, setting $S \leq T + V_{(Z,w)}$ for linear operators $S, T \in L(\mathcal{N}, \mathcal{M})$ if

$$S(a) \leq T(a) + w \quad \text{for all} \quad a \in Z.$$

The collection $\mathfrak{V}_{(\mathfrak{Z},\mathcal{W})} = \{V_{(Z,w)} \mid Z \in \mathfrak{Z},\ w \in \mathcal{W}\}$ of these neighborhoods defines a convex quasiuniform structure on a subcone $\mathfrak{H}(\mathcal{N},\mathcal{M})$ of $L(\mathcal{N},\mathcal{M})$ in the sense of I.5.3 in [100] provided that its elements are bounded below, that is if for each $T \in \mathfrak{H}(\mathcal{N},\mathcal{M})$ and $Z \in \mathfrak{Z}$ and $w \in \mathcal{W}$ there is $\lambda \geq 0$ such that

$$0 \leq T(a) + \lambda w \quad \text{for all} \quad a \in Z.$$

It is elaborated in I.5.3 [100] how such a convex quasiuniform structure can be used to construct an abstract neighborhood system $\mathfrak{V}$ for the cone $\mathfrak{H}(\mathcal{N},\mathcal{M})$, turning $(\mathfrak{H}(\mathcal{N},\mathcal{M}),\mathfrak{V})$ into a locally convex cone in such a way that the neighborhoods in $\mathfrak{V}_{(\mathfrak{Z},\mathcal{W})}$ form a basis for the neighborhood system $\mathfrak{V}$. In fact, all that needs to be done is to define suitable sums of the elements of $\mathfrak{V}_{(\mathfrak{Z},\mathcal{W})}$ and thus create a cone that can be adjoined to $\mathfrak{H}(\mathcal{N},\mathcal{M})$. The induced order for $\mathfrak{H}(\mathcal{N},\mathcal{M})$ is given by $S \leq T$ for operators $S,T \in \mathfrak{H}(\mathcal{N},\mathcal{M})$ if

$$S(a) \leq T(a) + w \qquad \text{for all} \qquad a \in \bigcup_{Z \in \mathfrak{Z}} Z \quad \text{and} \quad w \in \mathcal{W}.$$

Alternatively, if Condition (QF*) from 6.4 holds for the neighborhoods in $\mathfrak{V}_{\mathfrak{Z}}$ (with the order from above), then we may use the procedure from 6.4 in order to turn $\mathfrak{H}(\mathcal{N},\mathcal{M})$ into a quasi-full locally convex cone $(\mathfrak{H}(\mathcal{N},\mathcal{M}),\widetilde{\mathfrak{V}})$. As elaborated in 6.4, the topologies induced by $\widetilde{\mathfrak{V}}$ are generally finer than those resulting from $\mathfrak{V}$.

*Remark 7.1.* The standard lattice completion $\widehat{\mathfrak{H}}(\mathcal{N},\mathcal{M})$ of $\mathfrak{H}(\mathcal{N},\mathcal{M})$ (see 5.57) leads to a rather unwieldy setting in this case. It consists of $\overline{\mathbb{R}}$-valued functions defined on the dual cone $\mathfrak{H}(\mathcal{N},\mathcal{M})^*$ which is difficult to approach and depends on the particular topology of $\mathfrak{H}(\mathcal{N},\mathcal{M})$, that is the choice for the family $\mathfrak{Z}$ of subsets of $\mathcal{N}$. It is therefore preferable to employ a simplified lattice completion $\widehat{\mathfrak{H}}(\mathcal{N},\mathcal{M})_{\Upsilon}$ in the sense of 5.61 for which we shall use the subset $\Upsilon = \left(\bigcup_{Z \in \mathfrak{Z}} Z\right) \times M^*$ of $\mathfrak{H}(\mathcal{N},\mathcal{M})^*$, consisting of all continuous linear functionals whose elements $(a,\mu)$ act as linear functionals on $\mathfrak{H}(\mathcal{N},\mathcal{M})$ as

$$(a,\mu)(T) = \mu(T(a)) \qquad \text{for all} \qquad T \in \mathfrak{H}(\mathcal{N},\mathcal{M}).$$

By our definition of the neighborhoods in $\mathfrak{H}(\mathcal{N},\mathcal{M})$, this set $\Upsilon$ supports the separation property. The locally convex cone $\mathfrak{H}(\mathcal{N},\mathcal{M})$ is therefore embedded into $\widehat{\mathfrak{H}}(\mathcal{N},\mathcal{M})_{\Upsilon}$, which in turn permits a more easily accessible realization of the lattice completion of $\mathfrak{H}(\mathcal{N},\mathcal{M})$. In case that the subcone spanned by the sets $Z \in \mathfrak{Z}$ is all of $\mathcal{N}$, we may interpret the elements of the order closure of $\mathfrak{H}(\mathcal{N},\mathcal{M})$ in its lattice completion $\widehat{\mathfrak{H}}(\mathcal{N},\mathcal{M})$ as linear operators from $\mathcal{N}$ into $\mathcal{M}^{**}$, the second dual of $\mathcal{M}$. Indeed, we observed in 5.60(a) that every element $\varphi \in \widehat{\mathfrak{H}}(\mathcal{N},\mathcal{M})$ in the order closure of $\mathfrak{H}(\mathcal{N},\mathcal{M})$ is a linear functional on $\mathfrak{H}(\mathcal{N},\mathcal{M})^*$. Since $\Upsilon = \mathcal{N} \times \mathcal{M}^* \subset \mathfrak{H}(\mathcal{N},\mathcal{M})^*$ as elaborated above, the function $\varphi : \mathcal{N} \times \mathcal{M}^* \to \overline{\mathbb{R}}$ is linear in both arguments from $\mathcal{N}$ and from $\mathcal{M}^*$. Thus the mapping

$$a \mapsto \varphi_a : \mathcal{N} \to \mathcal{M}^{**},$$

where $\varphi_a(\mu) = \varphi(a, \mu)$ for $\mu \in \mathcal{M}^*$ is indeed a linear operator from $\mathcal{N}$ into $\mathcal{M}^{**}$. Moreover, if both $\mathcal{N}$ and $\mathcal{M}$ are in fact vector spaces over $\mathbb{K} = \mathbb{R}$ or $\mathbb{K} = \mathbb{C}$ and if all operators in $\mathfrak{H}(\mathcal{N}, \mathcal{M})$ are $\mathbb{K}$-linear, then a similar argument shows the every function $\varphi$ in the order closure of $\mathfrak{H}(\mathcal{N}, \mathcal{M})$ in $\widehat{\mathfrak{H}}(\mathcal{N}, \mathcal{M})$ can be interpreted as a $\mathbb{K}$-linear operator from $\mathcal{N}$ into $\mathcal{M}^{**}$.

*Examples 7.2.* (a) If both $(\mathcal{N}, \mathcal{U})$ and $(\mathcal{M}, \mathcal{W})$ are locally convex cones, and if all the sets $Z \in \mathfrak{Z}$ are bounded below in $\mathcal{N}$, then every continuous linear operator from $\mathcal{N}$ in to $\mathcal{M}$ is bounded below with respect to the neighborhoods in $\mathfrak{V}_{(\mathfrak{Z}, \mathcal{W})}$. Thus, if $\mathfrak{H}(\mathcal{N}, \mathcal{M})$ is a cone of continuous linear operators from $\mathcal{N}$ into $\mathcal{M}$, we may consider either of the following:

(i) If $\mathfrak{Z}$ is the family of all bounded below subset of $\mathcal{N}$, we obtain the *uniform operator topology* for $\mathfrak{H}(\mathcal{N}, \mathcal{M})$. We may alternatively choose the families $\mathfrak{Z}$ of all bounded or of all relatively bounded subsets of $\mathcal{P}$ (see 4.24) in this case.

(ii) If $\mathfrak{Z}$ is the family of all finite subsets of $\mathcal{N}$, we obtain the *strong operator topology* for $\mathfrak{H}(\mathcal{N}, \mathcal{M})$.

We shall also consider topologies on $\mathfrak{H}(\mathcal{N}, \mathcal{M})$ that arise if $\mathcal{M}$ is endowed with an alternative weak topology $\sigma(\mathcal{M}, \mathcal{L})$ generated by a third cone $\mathcal{L}$ and a bilinear form on $\mathcal{M} \times \mathcal{L}$ (see II.3 in [100]). In particular:

(iii) If $\mathfrak{Z}$ is the family of all finite subsets of $\mathcal{N}$, and if $\mathcal{M}$ is endowed with the topology $\sigma(\mathcal{M}, \mathcal{M}^*)$, we obtain the *weak operator topology* for $\mathfrak{H}(\mathcal{N}, \mathcal{M})$.

(iv) If $\mathfrak{Z}$ is the family of all finite subsets of $\mathcal{N}$, and if $\mathcal{M} = \mathcal{L}^*$ is the dual cone of some locally convex cone $(\mathcal{L}, \mathcal{V})$, endowed with the topology $\sigma(\mathcal{L}^*, \mathcal{L})$, we obtain the *weak* operator topology* for $\mathfrak{H}(\mathcal{N}, \mathcal{L}^*)$.

(b) If $\mathfrak{Z}$ consists of the set $Z = \mathcal{N}$, then $V = 0$ is the only resulting neighborhood for $L(\mathcal{N}, \mathcal{M})$, and boundedness from below requires that we consider linear operators that take only positive values on $\mathcal{N}$ for the cone $\mathfrak{H}(\mathcal{N}, \mathcal{M})$. The resulting order for operators $S, T \in \mathfrak{H}(\mathcal{N}, \mathcal{M})$ is $S \leq T$ if $S(a) \leq T(a)$ for all $a \in \mathcal{N}$. If on the other hand, $\mathfrak{Z}$ consists of the set $Z = \{0\}$, then $V = \infty$ is the only resulting neighborhood and the indiscrete topology arises for any subcone $\mathfrak{H}(\mathcal{N}, \mathcal{M})$ of $L(\mathcal{N}, \mathcal{M})$.

(c) If $\mathcal{N} = \mathcal{M}$, then $\mathfrak{H}(\mathcal{M}, \mathcal{M}) = \mathbb{R}_+ = \{a \in \mathbb{R} \mid a \geq 0\}$ is an example of a cone of linear operators on $\mathcal{M}$, with the scalar multiplication as its operation. If $(\mathcal{M}, \mathcal{W})$ is a locally convex cone and if $\mathfrak{Z}$ is an upward directed family of bounded below subsets of subsets of $\mathcal{M}$, then the neighborhood $V_{(Z,w)}$ in $\mathbb{R}_+$ corresponding to some $Z \in \mathfrak{Z}$ and $w \in \mathcal{W}$ according to the above is given by $\alpha \leq \beta + V_{(Z,w)}$ for $\alpha, \beta \in \mathbb{R}_+$ if

$$\alpha a \leq \beta a + w \qquad \text{for all} \qquad a \in Z.$$

If all elements of the set $Z$ are bounded in $\mathcal{M}$, then this condition can be interpreted as follows: Let $\delta = \inf\{\lambda \geq 0 \mid 0 \leq a + \lambda w \text{ for all } a \in Z\}$ and $\gamma = \inf\{\lambda \geq 0 \mid a \leq \lambda w \text{ for all } a \in Z\}$. A simple argument using the cancellation rule I.4.2 in [100] then yields that the above is equivalent to

$$\beta \leq \alpha + \frac{1}{\delta} \qquad \text{and} \qquad \alpha \leq \beta + \frac{1}{\gamma}.$$

(We set of course $\frac{1}{0} = +\infty$ and $\frac{1}{+\infty} = 0$ is these expressions.) Thus depending on our choice for $\mathfrak{Z}$, one of the following can emerge as the upper neighborhoods $V_{(Z,w)}(\alpha)$ for an element $\alpha \in \mathbb{R}_+$: The intervals (for $\varepsilon > 0$) (i) $[\alpha - \varepsilon, \alpha + \varepsilon]$, yielding the Euclidean topology with equality as order; (ii) $[0, \alpha + \varepsilon]$, yielding the upper Euclidean topology with the natural order; (iii) $[\alpha - \varepsilon, +\infty)$, yielding the lower Euclidean topology with reverse natural order; (iv) $[\alpha - \varepsilon, \alpha]$, yielding the equality as order; (v) $[0, \alpha]$, yielding the natural order. Note that only in cases (ii) and (v) the resulting locally convex cone $(\mathbb{R}_+, \mathfrak{V})$ is quasi-full.

If $(\mathcal{N}, \mathcal{U})$ is indeed a locally convex topological vector space over $\mathbb{K} = \mathbb{R}$ or $\mathbb{K} = \mathbb{C}$, endowed with its (modular) symmetric topology, then we may also consider $\mathfrak{H}(\mathcal{N}, \mathcal{N}) = \mathbb{K}$. Most useful choices for $\mathfrak{Z}$ will yield the Euclidean neighborhoods $\mathbb{B}_\varepsilon(\alpha) = \{\beta \in \mathbb{K} \mid |\beta - \alpha| \leq \varepsilon\}$ for elements $\alpha$ of $\mathbb{K}$ and the equality as order. Alternatively, there may be a subcone $C$ of negative elements in $\mathbb{K}$ in this case, and the upper neighborhoods are the sets $\mathbb{B}_\varepsilon(\alpha) + C$.

(d) Every locally convex cone $(\mathcal{P}, \mathcal{V})$ can be represented as a locally convex cone of linear operators. Indeed, algebraically, $\mathcal{P}$ coincides with the cone $\mathfrak{H}(\mathbb{R}_+, \mathcal{P})$ of all linear operators from $\mathbb{R}_+ = \{a \in \mathbb{R} \mid a \geq 0\}$ into $\mathcal{P}$ if we identify an element $a \in \mathcal{P}$ with the operator $\alpha \mapsto \alpha a$ in $\mathfrak{H}(\mathbb{R}_+, \mathcal{P})$. The neighborhoods of $\mathcal{P}$ may be recovered for $\mathfrak{H}(\mathbb{R}_+, \mathcal{P})$ if we use the above procedure with $\mathfrak{Z}$ containing only the singleton set $\{1\} \subset \mathbb{R}_+$. We obtain a copy of the locally convex cone $(\mathcal{P}, \mathcal{V})$.

**7.3 Cones of Linear Functionals. The Second Dual.** Let $(\mathcal{P}, \mathcal{V})$ be a locally convex cone. In the general settings of this section we choose $(\mathcal{N}, \mathcal{U}) = (\mathcal{P}, \mathcal{V})$ and $\mathcal{M} = \overline{\mathbb{R}}$ with its usual neighborhood system $\mathcal{W} = \{\varepsilon \in \mathbb{R} \mid \varepsilon > 0\}$ (see 1.4(a)). For the subcone $\mathfrak{H}(\mathcal{P}, \overline{\mathbb{R}})$ of $L(\mathcal{P}, \overline{\mathbb{R}})$ we choose the dual $\mathcal{P}^*$ of $\mathcal{P}$. As in 7.2(a) let $\mathfrak{Z}$ be a family of bounded below subsets of $\mathcal{P}$. In this way, $(\mathcal{P}^*, \mathfrak{V})$ becomes a locally convex cone. Its own dual cone, that is the second dual of $\mathcal{P}$, then is well-defined and depends on the choice for the topology of $\mathcal{P}^*$, that is on the choice for the family $\mathfrak{Z}$ of subsets of $\mathcal{P}$. Considering the particular choices for $\mathfrak{Z}$ as elaborated in 7.2(a) we shall use the following notations for the second dual of a locally convex cone $(\mathcal{P}, \mathcal{V})$ :

(i) $\mathcal{P}^{**}$ denotes the cone of all $\overline{\mathbb{R}}$-valued linear functionals on $\mathcal{P}^*$.
(ii) $\mathcal{P}^{**}_{sl}$, $\mathcal{P}^{**}_{sr}$ and $\mathcal{P}^{**}_{sl}$ denote the dual of $(\mathcal{P}^*, \mathfrak{V})$ if $\mathfrak{Z}$ consists of all (bounded below, relatively bounded) or bounded subsets of $\mathcal{P}$. These

are referred to as the *(lower strong, relative strong)* or *strong second dual* of $\mathcal{P}$.

(iii) $\mathcal{P}_w^{**}$ denotes the *weak second dual* of $\mathcal{P}$, that is the dual of $(\mathcal{P}^*, \mathfrak{V})$ if $\mathfrak{Z}$ consists of all finite subsets of $\mathcal{P}$.

Since every element $a \in \mathcal{P}$ acts as an $\overline{\mathbb{R}}$-valued linear functional $\varphi_a$ on $\mathcal{P}^*$, and since this linear functional is obviously contained in the polar of the neighborhood $V_{(Z,1)}$, where $Z = \{a\} \subset \mathcal{P}$, the given cone $\mathcal{P}$ can be envisioned as a subcone of its second dual $\mathcal{P}_w^{**}$. Indeed, we have

$$\mathcal{P} \subset \mathcal{P}_w^{**} \subset \mathcal{P}_s^{**} \subset \mathcal{P}_{sr}^{**} \subset \mathcal{P}_{sl}^{**} \subset \mathcal{P}^{**}$$

in general. Now let us recall the construction of the standard lattice completion $\widehat{\mathcal{P}}$ of $\mathcal{P}$ from Section 5.57. The elements of $\widehat{\mathcal{P}}$ were realized as $\overline{\mathbb{R}}$-valued functions on $\mathcal{P}^*$, and in Remark 5.60(a) we observed that the elements of the order closure of $\mathcal{P}$ in $\widehat{\mathcal{P}}$ are linear on $\mathcal{P}^*$. This order closure can therefore also be considered as a subcone of $\mathcal{P}^{**}$.

Furthermore, we observe that for every choice of the family $\mathfrak{Z}$ of bounded below subsets of $\mathcal{P}$ the linear functionals in the thus generated second dual $\mathcal{P}_{\mathfrak{Z}}^{**}$ of $\mathcal{P}$, if considered as $\overline{\mathbb{R}}$-valued functions on $\mathcal{P}^*$, are bounded below on the polars of all neighborhoods in $\mathcal{V}$. Indeed, every functional $\varphi$ in the second dual $\mathcal{P}_{\mathfrak{Z}}^{**}$ of $\mathcal{P}$ is contained in the polar of some neighborhood $V_{(Z,\varepsilon)}$ for $Z \in \mathfrak{Z}$ and $\varepsilon > 0$. Given $v \in \mathcal{V}$ there is $\lambda \geq 0$ such that $0 \leq z + \lambda v$ for all $z \in Z$. Then for every $\mu \in v^\circ$ we have $0 \leq \mu(z) + \lambda$ for all $z \in Z$, hence $0 \leq \mu + (\lambda/\varepsilon) V_{(Z,\varepsilon)}$ by the definition of the neighborhood $V_{(Z,\varepsilon)}$. Since $\varphi \in V_{(Z,\varepsilon)}^\circ$, this yields $\varphi(\mu) \geq -(\lambda/\varepsilon)$ for all $\mu \in v^\circ$. Consequently, for every such choice of the family $\mathfrak{Z}$, the resulting second dual $\mathcal{P}_{\mathfrak{Z}}^{**}$ of $\mathcal{P}$ may be considered as a subcone of the locally convex complete lattice cone $\mathcal{F}_{\widehat{\mathcal{V}}_b}(\mathcal{P}^*, \overline{\mathbb{R}})$ from 5.57. Recall that the standard lattice completion $\widehat{\mathcal{P}}$ of $\mathcal{P}$ had been introduced as the smallest locally convex complete lattice subcone of $\mathcal{F}_{\widehat{\mathcal{V}}_b}(\mathcal{P}^*, \overline{\mathbb{R}})$ that contains $\mathcal{P}$ (see 5.57). We have $\mathcal{P} \subset \mathcal{P}_{\mathfrak{Z}}^{**} \subset \mathcal{F}_{\widehat{\mathcal{V}}_b}(\mathcal{P}^*, \overline{\mathbb{R}})$ for any such choice of the family $\mathfrak{Z}$ by the above.

Therefore both the order closure of $\mathcal{P}$ in $\widehat{\mathcal{P}}$ and the second dual $\mathcal{P}_{\mathfrak{Z}}^{**}$ are contained in the intersection of $\mathcal{P}^{**}$ and $F_{\widehat{\mathcal{V}}_b}(\mathcal{P}^*, \overline{\mathbb{R}})$, but it is in general not possible to identify one of these as a subcone of the other.

We can, however, add the following often helpful observation: Let $Z$ be a bounded below subset of $\mathcal{P}$, and suppose that the element $\varphi \in \widehat{\mathcal{P}}$ is in the closure with respect to the order topology of (the embedding of) $Z$ in $\widehat{\mathcal{P}}$. This means that there is a net $(a_i)_{i \in \mathcal{I}}$ in $Z$ converging pointwise as functions on $\mathcal{P}^*$ towards $\varphi$, that is

$$\varphi(\mu) = \lim_{i \in \mathcal{I}} \mu(a_i)$$

for all $\mu \in \mathcal{P}^*$. (Convergence is meant in the usual (order) topology of $\overline{\mathbb{R}}$.) Then the function $\varphi$ is linear on $\mathcal{P}^*$ and $\mu \leq \nu + V_{(Z,1)}$ for elements $\mu, \nu \in \mathcal{P}^*$ implies that $\mu(a_i) \leq \nu(a_i) + 1$ holds for all $i \in \mathcal{I}$, and therefore

$\varphi(\mu) \leq \varphi(\nu)+1$ as well. This shows that $\varphi \in V^{\circ}_{(Z,1)}$. Hence the element $\varphi \in \widehat{\mathcal{P}}$ is contained in the dual cone $\mathcal{P}^{**}_{\mathfrak{Z}}$ of $(\mathcal{P}^*, \mathfrak{V})$ whenever the neighborhood generating family $\mathfrak{Z}$ contains the bounded below set $Z$.

For a locally convex vector space $(\mathcal{P}, \mathcal{V})$ over $\mathbb{K} = \mathbb{R}$ or $\mathbb{K} = \mathbb{C}$ the different notions in (ii) for the strong second dual coincide, and according to 2.1(d) every real-valued (real) linear functional $\varphi \in \mathcal{P}^{**}$ corresponds canonically to a $\mathbb{K}$-valued $\mathbb{K}$-linear functional $\varphi_{\mathbb{K}}$ on $\mathcal{P}^*_{\mathbb{K}}$, that is an element of $\mathcal{P}^{**}_{\mathbb{K}}$, the (algebraic) second vector space dual of $\mathcal{P}$.

These final observations will prove particularly useful in the subsequent chapters when we shall investigate integrals of cone-valued functions.

## 8. Notes and Remarks

The theory of locally convex cones originated in a joint work [100] by the author and K. Keimel in 1992. We were then looking for a suitable setting for the formulation of Korovkin-type approximation theory which deals with certain restricted classes of continuous linear operators on locally convex vector spaces. These may be positive operators on ordered spaces, contractions on normed spaces, multiplicative operators on Banach algebras, etc. Approximation processes are modeled by sequences or nets of operators in such a class. The given restrictions then guarantee convergence towards the identity operator on a large subset of their domain if this property can be checked for a relatively small test set. The use of locally convex cones instead of locally convex vector spaces turns out to be very advantageous in this context, since it allows to formulate all those different restriction properties for the operators in terms of the order structure alone, thus yielding a unifying approach. Subsequently the theory of locally convex cones has been expanded, mostly by the author of this book. Readers interested in further aspects of the subject should in particular familiarize themselves with the Hahn-Banach type extension and separation results that were laid out in [172] and form the foundations for the duality theory of locally convex cones. Ordered cones were earlier studied by various authors, in particular Fuchssteiner and Lusky [63] whose book contains a Hahn-Banach type sandwich theorem for $(\mathbb{R} \cup \{-\infty\})$-valued linear functionals on an ordered cone, a non-topological predecessor to the results from [172]. An in-depth investigation for the relationship between order and topology can be found in the seminal work [135] by Nachbin. The compendia of continuous lattices [68] and [69] by Gierz, Hofmann, Keimel, Lawson, Mislove and Scott contain a detailed analysis of various ways to introduce topologies on lattices.

The weak (global) preorder $\preccurlyeq$ as defined in Section 3 has an earlier analogue in the (global) preorder $\preceq$ which was defined in Section I.3 of [100] for elements $a, b \in \mathcal{P}$ as follows:

$$a \preceq b \qquad \text{if} \qquad a \leq b + v$$

for all $v \in \mathcal{V}$. Clearly $a \leq b$ implies $a \preceq b$ which in turn implies that $a \preccurlyeq b$. In some sense the preorder $\preceq$ can be considered as a topological closure of the given order $\leq$ whereas the weak preorder $\preccurlyeq$ signifies a closure with respect to both topology and the linear structure. Like for the weak preorder there is also a local version $\preceq_v$ of the preorder (see I.3 in [100]) referring to a particular neighborhood $v \in \mathcal{V}$ rather than the whole neighborhood system. Relationships between the different orders of a locally convex cone are investigated in detail in [175]. Since it provides the separation properties from Section 4, the weak preorder turns out to be the most suitable one for our purposes.

An excellent historical account of the extensive literature on ordered topological vector spaces can be found in the classical book by Day [39]. The notions of order convergence and of order topology for complete locally convex lattice cones from Section 5 are also used in ordered vector spaces, but introduced in a slightly different way which does not require a given topological or lattice structure (see Chapter V.6 in [185]). However, on topological vector lattices this notion coincides with ours form Section 5. Topological vector lattices had first been introduced as Banach lattices, and comprehensive treatments can for example be found in the books by Schäfer [184] and [185] and by Meyer-Nieberg [132]. In locally convex vector spaces there are compatibility requirements between the algebraic and the lattice operations as well as the topology. These are reflected in the corresponding requirements of Section 5 for locally convex cones. The strong conditions for locally convex lattice cones mirror those for M-topologies in topological vector lattices. Since under circumstances the latter permit representations as function spaces (see [94]), the result of Proposition 5.37 is not unexpected. Proposition 5.37 gives also the reason for using $\overline{\mathbb{R}}$-valued functions in the standard lattice completion of a locally convex cone. General lattices carrying different orders leading to notions of order convergence and of approximation of elements are thoroughly investigated in [68] and [69].

# Chapter II
# Measures and Integrals. The General Theory

In this chapter we shall develop a general integration theory for cone-valued functions with respect to operator-valued measures. The structure of locally convex cones will allow the use of many of the main concepts of classical measure theory for (extended) real-valued functions. Section 1 introduces measurability for cone-valued functions on a set $X$ with respect to a (weak) $\sigma$-ring of subsets of $X$. This notion does not involve any reference to a particular measure. Bounded operator-valued measures will be defined in Section 3. The introduction of its modulus allows the extension of any given measure to a full locally convex cone containing the given cone and its neighborhood system, thus greatly facilitating the expansion of our concepts. This yields a new understanding of the variation of a measure, not as a separate positive real-valued measure associated with the given one, but as a component of its extension. The development of an integration theory for cone-valued functions with respect to an operator-valued measure follows in Section 4. Section 5 contains the general convergence theorems for sequences of functions and measures, that is variations and adaptations of the dominated convergence theorem. Chapter II concludes with a long list of special cases and examples in Section 6, demonstrating the generality of the approach. These examples include classical real-valued measure theory as well as settings with vector-, cone-, functional- and operator-valued measures and functions.

## 1. Measurable Cone-Valued Functions

Throughout the following let $X$ be a set, $(\mathcal{P},\mathcal{V})$ a locally convex cone with dual $\mathcal{P}^*$. Endowed with the pointwise algebraic operations and order, the $\mathcal{P}$-valued functions on $X$ form again a cone, denoted by $\mathcal{F}(X,\mathcal{P})$. As usual, we say that a function $f \in \mathcal{F}(X,\mathcal{P})$ is *supported by a set* $E \subset X$ if $f(x) = 0$ for all $x \in X \setminus E$. For a positive real-valued function $\varphi$ on $X$ and $f \in \mathcal{F}(X,\mathcal{P})$ we denote by $\varphi_{\otimes} f \in \mathcal{F}(X,\mathcal{P})$ the mapping

W. Roth, *Operator-Valued Measures and Integrals for Cone-Valued Functions*, Lecture Notes in Mathematics 1964,

$$x \mapsto \varphi(x)f(x) : X \to \mathcal{P}.$$

For an element $a$ of $\mathcal{P}$ or of $\mathcal{V}$ we shall also use its symbol to denote the constant function $x \mapsto a$, hence $\varphi \otimes a$ for $x \mapsto \varphi(x)a$.

**1.1 Weak $\sigma$-Rings.** We shall develop our measure and integration theory with respect to a family $\mathfrak{R}$ of subsets of $X$ with the following properties:

(R1) $\emptyset \in \mathfrak{R}$.
(R2) *If* $E_1, E_2 \in \mathfrak{R}$, *then* $E_1 \bigcup E_2 \in \mathfrak{R}$ *and* $E_1 \setminus E_2 \in \mathfrak{R}$.
(R3) *If* $E_n \in \mathfrak{R}$ *for* $n \in \mathbb{N}$ *and* $E_n \subset E$ *for some* $E \in \mathfrak{R}$, *then* $\bigcup_{n \in \mathbb{N}} E_n \in \mathfrak{R}$.

We shall call a family $\mathfrak{R}$ with these properties a *(weak) $\sigma$-ring*. (Condition (R3) is weaker then the usual one for $\sigma$-rings.) As $E_1 \cap E_2 = E_1 \setminus (E_1 \setminus E_2)$, Condition (R2) implies that $E_1 \cap E_2 \in \mathfrak{R}$ whenever $E_1, E_2 \in \mathfrak{R}$. Of course, any $\sigma$-algebra is a $\sigma$-ring in this sense, and a $\sigma$-ring $\mathfrak{R}$ is a $\sigma$-algebra if and only if $X \in \mathfrak{R}$. However, because we shall require boundedness for measures defined on $\mathfrak{R}$, using $\sigma$-algebras from the beginning would impose undue limitations. We may, however, associate with $\mathfrak{R}$ in a canonical way the $\sigma$-algebra

$$\mathfrak{A}_{\mathfrak{R}} = \{A \subset X \mid A \cap E \in \mathfrak{R} \quad \text{for all} \quad E \in \mathfrak{R}\}$$

of *measurable* subsets of $X$. As usual, $\chi_E$ stands for the characteristic (or indicator) function on $X$ of a subset $E \subset X$, and $\mathcal{S}_{\mathfrak{R}}(X, \mathcal{P})$ is the subcone of $\mathcal{F}(X, \mathcal{P})$ of all $\mathcal{P}$-valued step functions supported by $\mathfrak{R}$, that is functions $h = \sum_{i=1}^{n} \chi_{E_i} \otimes a_i$ with $E_i \in \mathfrak{R}$ and $a_i \in \mathcal{P}$. If the sets $E_i$ are pairwise disjoint, then we shall call the above the *standard representation* for the step function $h$. Measurability for vector-valued functions has been introduced in various places (see for example Dunford & Schwartz [55], III.2.10). A suitable adaptation for cone-valued functions needs to consider the presence of unbounded elements in $\mathcal{P}$ and the absence of negatives. We shall therefore employ the relative topologies.

**1.2 Measurable Functions.** We shall say that a function $f \in \mathcal{F}(X, \mathcal{P})$ is *measurable with respect to the $\sigma$-ring* $\mathfrak{R}$ if for every $v \in \mathcal{V}$, with respect to the symmetric relative $v$-topology of $\mathcal{P}$

(M1) $f^{-1}(O) \cap E \in \mathfrak{R}$ *for every open subset* $O$ *of* $\mathcal{P}$ *and every* $E \in \mathfrak{R}$.
(M2) $f(E)$ *is separable in* $\mathcal{P}$ *for every* $E \in \mathfrak{R}$.

Note that Condition (M1) means that $f^{-1}(O) \in \mathfrak{A}_{\mathfrak{R}}$ for all open subsets $O$ of $\mathcal{P}$. Obviously the functions in $\mathcal{S}_{\mathfrak{R}}(X, \mathcal{P})$ are measurable.

**Proposition 1.3.** *A function* $f \in \mathcal{F}(X, \overline{\mathbb{R}})$ *is measurable if and only if it is measurable in the usual sense with respect to the* $\sigma$*-algebra* $\mathfrak{A}_{\mathfrak{R}}$.

*Proof.* Let $f \in \mathcal{F}(X, \overline{\mathbb{R}})$. The neighborhood system for $\overline{\mathbb{R}}$ consists of the positive reals $\varepsilon > 0$. The symmetric relative topology therefore coincides with the usual topology on the elements of $\mathbb{R}$, while $+\infty$ is an isolated point. The range of $f$, hence $f(E)$ for every $E \in \mathfrak{R}$, is separable in any case. Thus for measurability we require that $f^{-1}(O) \in \mathfrak{A}_{\mathfrak{R}}$ for every open subset of $\mathbb{R}$ and also that $f^{-1}(+\infty) \in \mathfrak{A}_{\mathfrak{R}}$. This coincides with the usual definition of measurability. □

**Theorem 1.4.** *A function* $f \in \mathcal{F}(X,\mathcal{P})$ *is measurable if and only if for every* $E \in \mathfrak{R}$, $v \in \mathcal{V}$ *and* $\varepsilon > 0$ *there are sets* $E_n \in \mathfrak{R}$, $n \in \mathbb{N}$, *such that* $\bigcup_{n\in\mathbb{N}} E_n = E$ *and* $f(x) \in v_\varepsilon(f(y))$ *whenever* $x, y \in E_n$ *for some* $n \in \mathbb{N}$.

*Proof.* First assume that the function $f \in \mathcal{F}(X,\mathcal{P})$ is measurable. For $E \in \mathfrak{R}$ and $v \in \mathcal{V}$ let $A = \{a_n \mid n \in \mathbb{N}\}$ be a dense subset (with respect to the symmetric relative $v$-topology) of $f(E)$. For $a \in \mathcal{P}$ and $\varepsilon > 0$ the sets

$$\overset{\circ}{v}_\varepsilon(a) = \bigcup_{0<\varepsilon'<\varepsilon} v_{\varepsilon'}(a) \qquad \text{and} \qquad (a)\overset{\circ}{v}_\varepsilon = \bigcup_{0<\varepsilon'<\varepsilon} (a)v_{\varepsilon'}$$

are open in the upper and lower relative $v$-topologies, respectively, and their intersection $\overset{\circ}{v}{}^s_\varepsilon(a)$ is an open neighborhood of $a$ in the symmetric relative $v$-topology. Set

$$E_n = f^{-1}\big(\overset{\circ}{v}{}^s_{(\varepsilon/3)}(a_n)\big) \cap E \in \mathfrak{R}.$$

Then Lemma I.4.1(a) shows that $f(x) \in v_\varepsilon(f(y))$ whenever $x, y \in E_n$. Furthermore, for every $x \in E$ there is some $a_n \in \overset{\circ}{v}{}^s_{\varepsilon/2}(f(x))$. Thus $f(x) \in \overset{\circ}{v}{}^s_{\varepsilon/2}(a_n) \subset \overset{\circ}{v}{}^s_\varepsilon(a_n)$, hence $x \in E_n$. This shows $\bigcup_{n\in\mathbb{N}} E_n = E$, as required.

For the converse, assume that the above condition holds for the function $f \in \mathcal{F}(X,\mathcal{P})$ and let $E \in \mathfrak{R}$ Then for every $m \in \mathbb{N}$ there are $E_n^m \in \mathfrak{R}$, such that $\bigcup_{n\in\mathbb{N}} E_n^m = E$ and $f(x) \in v_{(1/m)}(f(y))$ whenever $x, y \in E_n^m$ for some $n \in \mathbb{N}$. Choose $a_n^m = f(x_n^m)$ for some $x_n^m \in E_n^m$. Then the set $A = \{a_n^m \mid n, m \in \mathbb{N}\}$ is seen to be dense in $f(E)$, which yields Condition (M1). For (M2), let $O \subset \mathcal{P}$ be open in the symmetric relative $v$-topology. With the sets $E_n^m$ from above set

$$F = \bigcup \{\, E_n^m \mid E_n^m \subset f^{-1}(O) \,\}.$$

Then $F \in \mathfrak{R}$ by (R2) and $F \subset f^{-1}(O) \cap E$. For $x \in f^{-1}(O) \cap E$ on the other hand, there is $m \in \mathbb{N}$ such that $v^s_{(1/m)}(f(x)) \subset O$. We find $x \in E_n^m$ for some $n \in \mathbb{N}$. Then for every $y \in E_n^m$ we have $f(y) \in v_{(1/m)}(f(x)) \subset O$, hence $y \in f^{-1}(O)$. This shows $x \in E_n^m \subset F$, hence $f^{-1}(O) \cap E = F \in \mathfrak{R}$. □

We can indeed assume that the sets $E_n \in \mathfrak{R}$ from Theorem 1.4 are disjoint, since otherwise we may set

$$G_1 = E_1 \qquad \text{and} \qquad G_n = E_n \setminus \bigcup_{i=1}^{n-1} E_i$$

for $n \geq 2$. Then $G_n \in \mathfrak{R}$ and $G_n \subset E_n$. The sets $G_n$ are disjoint and their union equals the union of the sets $E_n$, that is the given set $E \in \mathfrak{R}$.

**Corollary 1.5.** *The measurable functions form a subcone of $\mathcal{F}(X, \mathcal{P})$.*

*Proof.* Clearly, $\alpha f$ is measurable, whenever $f \in \mathcal{F}(X, \mathcal{P})$ is measurable and $\alpha \geq 0$. We proceed to show that for measurable functions $f, g \in \mathcal{F}(X, \mathcal{P})$ their sum $f + g$ is also measurable. In a first step, given $E \in \mathfrak{R}$ and $v \in \mathcal{V}$, using Theorem 1.4 and Lemma I.4.1(c) we can find sets $E_n \in \mathfrak{R}$ and $\lambda_n \geq 0$ such that $\bigcup_{n \in \mathbb{N}} E_n = E$ and $0 \leq f(x) + \lambda_n$ and $0 \leq g(x) + \lambda_n$ whenever $x \in E_n$. Now, given $\varepsilon > 0$ we set $\delta_n = \varepsilon/(2 + 2\lambda_n)$ and again using Theorem 1.4, for every $n \in \mathbb{N}$ we find sets $E_n^m \in \mathfrak{R}$ such that $\bigcup_{m \in \mathbb{N}} E_n^m = E_n$ and $f(x) \in v_{\delta_n}\big(f(y)\big)$ as well as $g(x) \in v_{\delta_n}\big(g(y)\big)$ whenever $x, y \in E_n^m$. Following I.4.1(b) this yields

$$f(x) \leq (1 + \delta) f(y) + \delta(1 + \lambda)v = (1 + \delta) f(z) + (\varepsilon/2)v$$

and, likewise $g(x) \leq (1 + \delta) f(y) + (\varepsilon/2)v$. Thus

$$f(x) + g(x) \leq (1 + \delta)\big(f(y) + g(y)\big) + \varepsilon v$$

and indeed $f(x) + g(x) \in v_\varepsilon\big(f(y) + g(y)\big)$. As $\bigcup_{m,n \in \mathbb{N}} E_n^m = E$, this proves that the function $f + g$ is also measurable. □

**Theorem 1.6.** *For measurable functions $f, g \in \mathcal{F}(X, \mathcal{P})$ and $v \in \mathcal{V}$ the set $\{x \in X \mid f(x) \preccurlyeq_v g(x)\}$ is measurable, that is in $\mathfrak{A}_\mathfrak{R}$.*

*Proof.* Let $f, g \in \mathcal{F}(X, \mathcal{P})$ be measurable, $v \in \mathcal{V}$ and $E \in \mathfrak{R}$. According to Theorem 1.4, for $0 < \varepsilon \leq 1$ there are disjoint sets $E_i \in \mathfrak{R}$ such that $\bigcup_{i \in \mathbb{N}} = E$ and $f(x) \in v_\varepsilon\big(f(y)\big)$ as well as $g(x) \in v_\varepsilon\big(g(y)\big)$ whenever $x, y \in E_i$ for some $i \in \mathbb{N}$. Set

$$F_\varepsilon = \bigcup \big\{E_i \mid f(x_i) \in v_\varepsilon\big(g(x_i)\big) \quad \text{for some} \quad x_i \in E_i\big\} \in \mathfrak{R}.$$

If $x \in F_\varepsilon$, then $f(x) \in v_\varepsilon\big(f(x_i)\big)$, $f(x_i) \in v_\varepsilon\big(g(x_i)\big)$ and $g(x_i) \in v_\varepsilon\big(g(x)\big)$, hence $f(x) \in v_{(7\varepsilon)}\big(g(x)\big)$ by Lemma 2,1(a). This shows

$$\big\{x \in E \mid f(x) \in v_\varepsilon\big(g(x)\big)\big\} \subset F_\varepsilon \subset \big\{x \in E \mid f(x) \in v_{7\varepsilon}\big(g(x)\big)\big\}.$$

Then

$$F = \bigcap_{n \in \mathbb{N}} F_{(\frac{1}{n})} = \{x \in E \mid f(x) \preccurlyeq_v g(x)\} \in \mathfrak{R}$$

as well. As $F = E \bigcap \{x \in X \mid f(x) \preccurlyeq_v g(x)\}$, our claim follows. □

We shall use different patterns of pointwise convergence for sequences of cone-valued functions. For functions $(f_n)_{n \in \mathbb{N}}$ and $f$ in $\mathcal{F}(X, \mathcal{P})$, a subset $F \subset X$ and a neighborhood $v \in \mathcal{V}$ we shall write

$$f_n \overset{v}{\underset{F}{\searrow}} f, \qquad f_n \overset{v}{\underset{F}{\nearrow}} f, \qquad \text{or} \qquad f_n \overset{v}{\underset{F}{\rightarrow}} f$$

if $(f_n)_{n\in\mathbb{N}}$ converges to $f$ pointwise on $F$ with respect to the upper, lower or symmetric relative $v$-topology of $\mathcal{P}$, that is if for every $x \in F$ and $\varepsilon > 0$ there is $n_0 \in \mathbb{N}$ such that $f_n(x) \in v_\varepsilon(f(x))$, $f_n(x) \in (f(x))v_\varepsilon$ or $f_n(x) \in v^s_\varepsilon(f(x))$ for all $n \geq n_0$, respectively. Convergence in this sense for all $v \in \mathcal{V}$ means convergence in the (global) upper, lower or symmetric relative topology of $\mathcal{P}$. We shall denote this by

$$f_n \underset{F}{\searrow} f, \qquad f_n \underset{F}{\nearrow} f, \qquad \text{or} \qquad f_n \underset{F}{\rightleftarrows} f.$$

All the above notions of convergence are compatible with the algebraic operations in $\mathcal{P}$ (see Lemma I.4.1(d)).

**Theorem 1.7.** *If for $f \in \mathcal{F}(X,\mathcal{P})$ and every $E \in \mathfrak{R}$ and $v \in \mathcal{V}$ there is a sequence of measurable functions $f_n \in \mathcal{F}(X,\mathcal{P})$ such that $f_n \overset{v}{\underset{E}{\rightleftarrows}} (f)$, then $f$ is also measurable.*

*Proof.* Let $f \in \mathcal{F}(X,\mathcal{P})$, $E \in \mathfrak{R}$ and $v \in \mathcal{V}$. Suppose that there is a sequence of measurable functions $(f_n)_{n\in\mathbb{N}}$ such that $f_n \overset{v}{\underset{E}{\rightleftarrows}} (f)$. If for all $n \in \mathbb{N}$ the sets $A_n = \{a_n^i \mid i \in \mathbb{N}\}$ are dense (in the symmetric relative $v$-topology) in $f_n(E)$, then $A = \bigcup_{n\in\mathbb{N}} A_n$ is obviously dense in $f(E)$, which is thus seen to be separable. Now let $O \subset \mathcal{P}$ be open in the symmetric relative $v$-topology. For $\varepsilon > 0$ let $U_\varepsilon$ be the topological interior of the set $O_\varepsilon = \{a \in O \mid v^s_\varepsilon(a) \subset O\}$. For $m, n \in \mathbb{N}$ set

$$E_n^m = E \bigcap_{k\geq n} (f_k)^{-1}\big(U_{(1/m)}\big) \in \mathfrak{R}$$

and $F = \bigcup_{m,n\in\mathbb{N}} E_n^m \in \mathfrak{R}$. If $x \in f^{-1}(O) \cap E$, then there is $\varepsilon > 0$ such that $v^s_\varepsilon(f(x)) \subset O$. For $m \geq 7/\varepsilon$ there is $n \in \mathbb{N}$ such that $f_k(x) \in v^s_{(1/m)}(f(x))$ for all $k \geq n$. For any such $k$ let $a \in v^s_{(1/m)}(f_k(x))$. Then

$$v^s_{(1/m)}(a) \subset v^s_{(3/m)}\big(f_k(x)\big) \subset v^s_{(7/m)}\big(f(x)\big) \subset v^s_\varepsilon\big(f(x)\big) \subset O$$

by Lemma I.4.1(a). Thus $a \in O_{(1/m)}$, hence $v^s_{(1/m)}(f_k(x)) \subset O_{(1/m)}$, therefore $f_k(x) \in U_{(1/m)}$ and $x \in E_n^m \subset F$. This shows $f^{-1}(O) \subset F$. For $x \in F$, on the other hand, there are $m, n \in \mathbb{N}$ such that $f_k(x) \in U_{(1/m)} \subset O_{(1/m)}$ for all $k \geq n$. There is such $k$ such that $f(x) \in v^s_{(1/m)}(f_k(x))$, hence $f(x) \in O$. Thus $f^{-1}(O) \cap E = F \in \mathfrak{R}$. □

**Theorem 1.8.** *Let $f \in \mathcal{F}(X,\mathcal{P})$ be measurable.*

*(a) Let $\varphi : X \to \mathbb{R}$. If $\varphi$ is positive and measurable with respect to $\mathfrak{A}_\mathfrak{R}$, then $\varphi{\otimes}f \in \mathcal{F}(X,\mathcal{P})$ is also measurable.*

*(b) Let $\Phi : X \to X$. If $\Phi^{-1}(A) \in \mathfrak{A}_\mathfrak{R}$ for all $A \in \mathfrak{A}_\mathfrak{R}$, and $\Phi(E) \subset F$ for every $E \in \mathfrak{R}$ with some $F \in \mathfrak{R}$, then the function then $f\circ\Phi \in \mathcal{F}(X,\mathcal{P})$ is also measurable.*

(c) *Let $(\mathcal{N},\mathcal{U})$ be a locally convex cone and let $\Psi : \mathcal{P} \to \mathcal{N}$. If for every $u \in \mathcal{U}$ there is $v \in \mathcal{V}$ such that the mapping $\Psi$ is continuous with respect to the symmetric relative $v$- and $u$-topologies of $\mathcal{P}$ and $\mathcal{N}$, then the function $\Psi \circ f \in \mathcal{F}(X,\mathcal{N})$ is also measurable.*
*The assumption on $\Psi$ holds in particular if $\Psi : \mathcal{P} \to \mathcal{N}$ is a continuous linear operator.*

*Proof.* (a) Our claim is obvious if $\varphi$ is a real-valued step function (supported by $\mathfrak{A}_{\mathfrak{R}}$,) since the validity of the criterion from Theorem 1.4 for $\varphi{\otimes}f$ follows straight from its validity for $f$. Generally, there is a sequence $(\psi_n)_{n\in\mathbb{N}}$ of positive real-valued step functions that converges pointwise from below to $\varphi$. All the functions $f_n = \psi_n{\otimes}f$ are measurable by the above. If $\varphi(x) = 0$ for $x \in X$, then $\psi_n(x) = 0$ for all $n \in \mathbb{N}$. Otherwise, for $v \in \mathcal{V}$ and $\varepsilon > 0$ choose $\lambda > 0$ such that $0 \le f(x) + \lambda v$ and set $\gamma = \min\{1+\varepsilon,\ 1+\frac{\varepsilon}{2\lambda\varphi(x)}\}$. There is $n_0 \in \mathbb{N}$ such that $\psi_n(x) \le \varphi(x) \le \gamma\psi_n(x)$, hence

$$\psi_n(x)(f(x)+\lambda v) \le \varphi(x)(f(x)+\lambda v) \le \gamma\psi_n(x)(f(x)+\lambda v)$$

for all $n \ge n_0$. This shows

$$f_n(x) + \lambda\psi_n(x)v \le \varphi(x)f(x) + \lambda\varphi(x)v \le \varphi(x)f(x) + \gamma\lambda\psi_n(x)v,$$

hence $f_n(x) \le \varphi(x)f(x) + \varepsilon v$ by the cancellation law for positive elements (see Lemma I.4.2 in [100]), as $\gamma\lambda\psi_n < \lambda\psi_n(x) + \varepsilon$. Likewise, the above implies

$$\varphi(x)f(x) + \lambda\varphi(x)v \le \gamma f_n(x) + \gamma\lambda\psi_n(x)v \le \gamma f_n(x) + \gamma\lambda\varphi(x)v,$$

and $\varphi(x)f(x) \le \gamma f_n(x) + \varepsilon v$ as well. Thus $f_n(x) \in v^s_\varepsilon(\varphi(x)f(x))$ for all $n \ge n_0$. This shows $f_n \underset{X}{\to} \varphi{\otimes}f$, and by Theorem 1.7 the function $\varphi{\otimes}f$ is seen to be measurable.

For (b), let $f$ and $\Phi : X \to X$ be as stated, let $g = f\circ\Phi$ and $v \in \mathcal{V}$. For $E \in \mathfrak{R}$ we have $\Phi(E) \subset F$ for some $F \in \mathfrak{R}$, hence $g(E) \subset f(F)$ which is separable in the symmetric relative $v$-topology. Secondly, for an open subset $O$ of $\mathcal{P}$ we have $f^{-1}(O) \in \mathfrak{A}_{\mathfrak{R}}$, hence $g^{-1}(O) = \Phi^{-1}(f^{-1}(O)) \in \mathfrak{A}_{\mathfrak{R}}$, and the function $g$ is seen to be measurable.

For Part (c), let $f$ and $\Psi : \mathcal{P} \to \mathcal{N}$ be as stated, let $g = f \circ \Psi$ and $u \in \mathcal{U}$. Let $v \in \mathcal{V}$ be such that $\Psi$ is continuous with respect to the symmetric relative $v$- and $u$-topologies of $\mathcal{P}$ and $\mathcal{N}$. For every $E \in \mathfrak{R}$, the set $f(E)$ is separable with respect to the symmetric relative $v$-topology of $\mathcal{P}$, hence its continuous image $g(E) = \Psi(f(E))$ is separable with respect to the symmetric relative $u$-topology of $\mathcal{N}$. Secondly, for an open subset $O$ of $\mathcal{N}$ its inverse image $\Psi^{-1}(O)$ is open in $\mathcal{P}$, hence $g^{-1}(O) = f^{-1}(\Psi^{-1}(O)) \in \mathfrak{A}_{\mathfrak{R}}$, and the function $g$ is seen to be measurable. For the additional statement in (c), suppose that $\Psi : \mathcal{P} \to \mathcal{N}$ is a continuous linear operator. Given $u \in \mathcal{U}$ there is $v \in \mathcal{V}$ such that $\Psi(a) \le \Psi(b) + u$ holds whenever $a \le b + v$ for

$a, b \in \mathcal{P}$. Thus $a \in v_\varepsilon(b)$, that is $a \leq \gamma b + \varepsilon v$ with some $1 \leq \gamma \leq 1+\varepsilon$ implies $\Psi(a) \leq \gamma \Psi(b) + \varepsilon$, hence $\Psi(a) \in u_\varepsilon\big(\Psi(b)\big)$. Likewise, $a \in v_\varepsilon^s(b)$ implies that $\Psi(a) \in u_\varepsilon^s\big(\Psi(b)\big)$. The function $\Psi$ is therefore continuous with respect to the symmetric relative $v$- and $u$-topologies of $\mathcal{P}$ and $\mathcal{N}$. □

In the literature the terms *weak measurability* or *scalar measurability* are often used for a vector-valued function $f$ if all the scalar-valued functions $\mu \circ f$ for linear functionals $\mu$ in the dual of the range of $f$ are measurable in the usual sense. The following theorem states that measurability in our sense implies scalar measurability. The converse holds true for functions with bounded values and separable ranges.

**Theorem 1.9.** *Let $f \in \mathcal{F}(X, \mathcal{P})$.*

*(a) If $f$ is measurable, then the $\overline{\mathbb{R}}$-valued functions $\mu \circ f$ are measurable for all $\mu \in \mathcal{P}^*$.*
*(b) If the values of $f$ are bounded, $f(E)$ is separable in the symmetric relative $v$-topology for all $E \in \mathfrak{R}$ and $v \in \mathcal{V}$, and the $\overline{\mathbb{R}}$-valued functions $\mu \circ f$ are measurable for all $\mu \in \mathcal{P}^*$, then $f$ is measurable.*

*Proof.* For (a), suppose that the function $f \in \mathcal{F}(X, \mathcal{P})$ is measurable, and let $\mu \in \mathcal{P}^*$, that is $\mu \in v^\circ$ for some $v \in \mathcal{V}$. Recall from Proposition I.4.5 and Example I.4.37(a) that a continuous linear functional $\mu : \mathcal{P} \to \overline{\mathbb{R}}$ is also continuous, if we endow $\mathcal{P}$ with the symmetric relative $v$-topology and $\overline{\mathbb{R}}$ with its given symmetric topology (which of course coincides with its symmetric relative topology). Following Theorem 1.8(c), this shows that the function $\mu \circ f : X \to \overline{\mathbb{R}}$ is measurable whenever the function $f$ is measurable.

Now suppose that the assumptions of Part (b) hold for the function $f \in \mathcal{F}(X, \mathcal{P})$. We shall verify the criterion of Theorem 1.4 for measurability. Recall from Proposition I.4.2(iv) that on the subcone $\mathcal{B}$ of bounded elements of $\mathcal{P}$ the corresponding given and relative topologies coincide. As the values of $f$ are supposed to be bounded, this will greatly facilitate our arguments.

In a first step, let us consider an element $a \in \mathcal{B}$ and neighborhood $v \in \mathcal{V}$. Then $b \notin v(a)$, that is $b \not\leq a + v$, for an element $b \in \mathcal{B}$ implies that $b \not\leq a + v/2$, as indeed otherwise, for $\lambda > 0$ such that $a \leq \lambda v$ and $\varepsilon = \min\{1/9,\, 1/(3\lambda)\}$ there would be $1 \leq \gamma \leq 1+\varepsilon$ such that

$$b \leq \gamma\,(a + v/2) + \varepsilon v = \gamma a + \Big(\frac{\gamma}{2} + \varepsilon\Big)\, v \leq \gamma a + \frac{2}{3} v.$$

Because

$$\gamma a = a + (\gamma - 1) a \leq a + (\gamma - 1)\lambda v \leq a + \varepsilon \lambda v \leq a + \frac{1}{3} v,$$

this yields $a \leq b + v$, contradicting our assumption. Consequently, following Theorem 3.2 in [175] (see also Corollary 4.34 in Chapter I), there is a linear functional $\mu \in v^\circ$ such that

$$\mu(b) > \mu(a) + \frac{1}{2}.$$

Now, in a second step of our argument, consider an element $a \in \mathcal{B}$ and for $v \in \mathcal{V}$ the symmetric neighborhood

$$v^s(a) = v(a) \cap (a)v = \{c \in \mathcal{P} \mid c \leq a + v \quad \text{and} \quad a \leq c + v\}.$$

Given a set $E \in \mathfrak{R}$, let $\{b_i\}_{i \in \mathbb{N}}$ be a countable subset of $f(E) \setminus v^s(a) \subset \mathcal{B}$ that is dense with respect to the (given) symmetric topology. Such a subset exists because on $\mathcal{B}$ the given and the relative topologies of $\mathcal{P}$ coincide. For each $i \in \mathbb{N}$ we have either $b_i \notin v(a)$ or $b_i \notin (a)v$. Accordingly, we may choose linear functionals $\mu_i \in v^\circ$ corresponding to the elements $b_i$ such that either

$$\mu_i(b) > \mu_i(a) + \frac{1}{2} \qquad \text{or} \qquad \mu_i(a) > \mu_i(b) + \frac{1}{2}$$

if $b_i \notin v(a)$ or $b_i \notin (a)v$, respectively. We denote

$$O_i = \left(-\infty\,,\ \mu_i(a) + \frac{1}{4}\right] \qquad \text{or} \qquad O_i = \left[\mu_i(a) - \frac{1}{4}\,,\ +\infty\right]$$

in these respective cases and set $A_i = \mu_i^{-1}(O_i) \subset \mathcal{P}$ and $A = \bigcap_{i \in \mathbb{N}} A_i$. For every $c \in (v/4)^s(a)$, that is $c \leq a + v/4$ and $a \leq c + v/4$ we have $\mu(c) \leq \mu(a) + 1/4$ and $\mu(a) \leq \mu(c) + 1/4$ for all $\mu \in v^\circ$, hence $c \in A_i$ for all $i \in \mathbb{N}$. This shows $v^s(a) \subset A$. We shall proceed to verify that $A \cap f(E) \subset (2v)^s(a)$. For this, consider any element $c \in f(E) \setminus (2v)^s(a)$. First we observe that $v^s(c) \cap v^s(a) = \emptyset$, because the existence of an element $d \in v^s(c) \cap v^s(a)$ would lead to $c \in (2v)^s(a)$, contradicting our choice of $c$. Thus $f(E) \cap (v/4)^s(c) \subset f(E) \setminus v^s(a)$ holds as well, and there is some $b_i \in (v/4)^s(c)$. We have

$$\mu_i(b_i) \leq \mu_i(c) + \frac{1}{4} \qquad \text{and} \qquad \mu_i(c) \leq \mu_i(b) + \frac{1}{4}$$

since $\mu_i \in v^\circ$. Recall that either $b_i \notin v(a)$ or $b_i \notin (a)v$. In the first case, this implies $\mu_i(b_i) > \mu_i(a) + 1/2$, hence $\mu_i(c) > \mu_i(a) + 1/4$, and $c \notin A_i$. In the second case, we have $\mu_i(b_i) < \mu_i(a) - 1/2$, hence $\mu_i(c) < \mu_i(a) - 1/4$ and, likewise $c \notin A_i$. Thus indeed $c \notin A$. Summarizing, we verified that

$$v^s(a) \subset A \qquad \text{and} \qquad A \cap f(E) \subset (2v)^s(a).$$

Let us apply this to an element $a = f(x)$ for some $x \in E$. By our assumption, all the $\overline{\mathbb{R}}$-valued functions $\varphi_i = \mu_i \circ f$ are measurable, hence the sets $F_i = \varphi_i^{-1}(O_i)$ are contained in $\mathfrak{A}_{\mathfrak{R}}$. Likewise,

$$F = \bigcap_{i \in \mathbb{N}} F_i = \bigcap_{i \in \mathbb{N}} f^{-1}\big(\mu_i^{-1}(O_i)\big) = \bigcap_{i \in \mathbb{N}} f^{-1}(A_i) = f^{-1}(A) \in \mathfrak{A}_{\mathfrak{R}}.$$

Thus

$$f^{-1}\big(v^s(a)\big) \subset F \qquad \text{and} \qquad F \cap E \subset f^{-1}\big((2v)^s(a)\big).$$

Now in the third and final step of our argument we shall verify the criterion of Theorem 1.4: For $E \in \mathfrak{R}$ and $v \in \mathcal{V}$ let $\{a_n\}_{n\in\mathbb{N}}$ be a subset of $f(E)$ that is dense with respect to the symmetric relative $v$-topology, hence with respect to the given symmetric $v$-topology. For each element $a_n$ and the neighborhood $u = (\varepsilon/4)v \in \mathcal{V}$ in place of $v$ choose the set $F = F_n \in \mathfrak{A}_{\mathfrak{R}}$ as in the last part of the preceding step and set $E_n = F_n \cap E$. Then

$$f^{-1}\big(u^s(a_n)\big) \cap E \subset E_n \qquad \text{and} \qquad E_n \subset f^{-1}\big((2u)^s(a)\big)$$

holds for all $n \in \mathbb{N}$ by the above. Thus, firstly, for $x, y \in E_n$ we have $f(x), f(y) \in (2u)^s(a_n)$. But this obviously implies that $f(x) \in (4u)\big(f(y)\big) \subset v_\varepsilon\big(f(y)\big)$. Secondly, for any $x \in E$ there is some $a_n$ such that $f(x) \in u^s(a_n)$, that is $x \in f^{-1}\big(u^s(a_n)\big) \cap E \subset E_n$. This demonstrates $\bigcup_{n\in\mathbb{N}} E_n = E$ and completes our argument. □

## 2. Inductive Limit Neighborhoods for Cone-Valued Functions

Let $(\mathcal{P}, \mathcal{V})$ be a locally convex cone. In preparation of our integration theory for cone-valued functions with respect to an operator-valued measure, we shall introduce appropriate neighborhoods for the cone $\mathcal{F}(X, \mathcal{P})$ and corresponding subcones of measurable functions. Our integrals will constitute continuous linear operators on these cones. First, in order to allow greater generally, we shall extend the given neighborhood system of $\mathcal{V}$.

**2.1 Infinity as a Neighborhood.** We shall adjoin the maximal element $\infty$ to the neighborhood system $\mathcal{V}$ such that $a \leq b + \infty$ holds for all $a, b \in \mathcal{P}$. The addition and multiplication by scalars involving this element is defined in a canonical way: We set $v + \infty = \infty$, $0 \cdot \infty = 0$ and $\alpha \cdot \infty = \infty$ for all $v \in \mathcal{V}$ and $\alpha > 0$. The augmented neighborhood system which includes this infinite element and $0 \in \mathcal{P}$ will be denoted by $\overline{\mathcal{V}}$, that is $\overline{\mathcal{V}} = \mathcal{V} \cup \{0, \infty\}$. Obviously, $(\overline{\mathcal{V}}, \mathcal{V})$ is a full locally convex cone.

**2.2 Inductive Limit Neighborhoods.** Let $X$ and $\mathfrak{R}$ be as before, and let $\mathcal{F}\big(X, \overline{\mathcal{V}}\big)$ be the family of $\overline{\mathcal{V}}$-valued functions on $X$, endowed with the pointwise operations and order. For functions $f, g \in \mathcal{F}(X, \mathcal{P})$ and $s \in \mathcal{F}\big(X, \overline{\mathcal{V}}\big)$ we write $f \leq g + s$ if $f(x) \leq g(x) + s(x)$ for all $x \in X$. The addition and multiplication by scalars for functions $s, t \in \mathcal{F}\big(X, \overline{\mathcal{V}}\big)$ is defined pointwise, and $s \leq t$ means that $f \leq g + s$ implies $f \leq g + t$ for $f, g \in \mathcal{F}(X, \mathcal{P})$.

An (*$\mathfrak{R}$-compatible) inductive limit neighborhood* for $\mathcal{F}(X,\mathcal{P})$ is a convex subset $\mathfrak{v}$ of measurable functions in $\mathcal{F}(X,\overline{\mathcal{V}})$ such that for every $E \in \mathfrak{R}$ there is $v_E \in \mathcal{V}$ and $s \in \mathfrak{v}$ such that $\chi_{E}{\otimes}v_E \leq s$. Measurability is meant with respect to $\mathfrak{R}$ and the locally convex cone $(\overline{\mathcal{V}},\mathcal{V})$. For functions $f,g \in \mathcal{F}(X,\mathcal{P})$ and an inductive limit neighborhood $\mathfrak{v}$ we denote

$$f \leq g+\mathfrak{v} \quad \text{if} \quad f \leq g+s, \quad \text{for some} \quad s \in \mathfrak{v}.$$

We define sums and multiples by positive scalars for inductive limit neighborhoods through the addition and multiplication of their elements. A canonical order relation is given by

$$\mathfrak{v} \leq \mathfrak{u} \quad \text{if for every } s \in \mathfrak{v} \text{ there is } t \in \mathfrak{u} \text{ such that } s \leq t.$$

Inductive limit neighborhoods include uniform neighborhoods, consisting of a single constant function $x \mapsto v$; and if $X \in \mathfrak{R}$, that is if $\mathfrak{R}$ is a $\sigma$-algebra, then the uniform neighborhoods form a base for the family of all inductive limit neighborhoods.

**2.3 The Cone $\mathcal{F}_{\mathfrak{R}}(X,\mathcal{P})$.** We shall in the sequel deal with measurable functions in $\mathcal{F}(X,\mathcal{P})$ that can be reached from below by step functions; more precisely: We denote by $\mathcal{F}_{\mathfrak{R}}(X,\mathcal{P})$ the subcone of all measurable functions $f \in \mathcal{F}(X,\mathcal{P})$ such that for every inductive limit neighborhood $\mathfrak{v}$ there is $h \in \mathcal{S}_{\mathfrak{R}}(X,\mathcal{P})$ satisfying $h \leq f+\mathfrak{v}$.

**Lemma 2.4.** *Let $f \in \mathcal{F}_{\mathfrak{R}}(X,\mathcal{P})$.*

*(a) For every inductive limit neighborhood $\mathfrak{v}$ there is $\lambda \geq 0$ such thats $0 \leq f+\lambda\mathfrak{v}$.*
*(b) There is $E \in \mathfrak{R}$ such that $f(x) \geq 0$ for all $x \in X \setminus E$, and for every $v \in \mathcal{V}$ there is $\lambda \geq 0$ such that $0 \leq f+\lambda\,\chi_{E\otimes}v$.*

*Proof.* Let $f \in \mathcal{F}_{\mathfrak{R}}(X,\mathcal{P})$. For (a), given an inductive limit neighborhood $\mathfrak{v}$, there is a step function $h = \sum_{i=1}^{n} \chi_{E_i}{\otimes}a_i \in \mathcal{S}_{\mathfrak{R}}(X,\mathcal{P})$ such that $h \leq f+\mathfrak{v}$, that is $h \leq f+s$ for some $s \in \mathfrak{v}$. We may assume that the sets $E_i \in \mathfrak{R}$ are disjoint and $E = \bigcup_{i=1}^{n} E_i \in \mathfrak{R}$. There is $v \in \mathcal{V}$ such that $\chi_{E\otimes}v \leq \mathfrak{v}$, and in turn there is $\lambda \geq 0$ such that $0 \leq a_i+\lambda v$ for all $i = 1,\ldots,n$. This shows $0 \leq f(x)+s(x)+\lambda v$ for all $x \in E$ and $0 \leq f(x)+s(x)$ for all $x \in X \setminus E$. Thus $0 \leq f+(s+\lambda\,\chi_{E\otimes}v)$, hence indeed $0 \leq f+(1+\lambda)\mathfrak{v}$.

For (b), let the inductive limit neighborhood $\mathfrak{v}$ consist of all $\mathcal{V}$-valued functions that are supported by some set in $\mathfrak{R}$. By (a) there is $\lambda \geq 0$ and a function $s \in \mathfrak{v}$ such that $0 \leq f+\lambda s$. Because $s$ is supported by some set $E \in \mathfrak{R}$, that is $s(x) = 0$ for all $x \in X \setminus E$, we have indeed $f(x) \geq 0$ for all $x \in X \setminus E$. Now let $v \in \mathcal{V}$ and let $\mathfrak{v}$ consist of the single function $x \to v$. Then $0 \leq f+\lambda\mathfrak{v}$ with some $\lambda \geq 0$ by (a). Thus $0 \leq f+\lambda\,\chi_{E\otimes}v$, as claimed. □

The following lemma provides a more straightforward characterization of the functions in $\mathcal{F}_{\mathfrak{R}}(X,\mathcal{P})$, avoiding the use of inductive limit neighborhoods.

**Lemma 2.5.** *A measurable $\mathcal{P}$-valued function $f$ is in $\mathcal{F}_{\mathfrak{R}}(X,\mathcal{P})$ if and only if*

*(i) there is $E \in \mathfrak{R}$ such that $f(x) \geq 0$ for all $x \in X \setminus E$, and*
*(ii) for every $v \in \mathcal{V}$ there is $h \in \mathcal{S}_{\mathfrak{R}}(X,\mathcal{P})$ such that $h \leq f + \chi_{X} \otimes v$.*

*Proof.* If $f \in \mathcal{F}_{\mathfrak{R}}(X,\mathcal{P})$, then (i) follows from Lemma 2.4(b). Statement (ii) follows from the definition of the cone $\mathcal{F}_{\mathfrak{R}}(X,\mathcal{P})$ if we consider the singleton inductive limit neighborhood $\mathfrak{v} = \{\chi_{X} \otimes v\}$. For the converse, suppose that (i) and (ii) hold for the measurable function $f \in \mathcal{F}(X,\mathcal{P})$, and let $\mathfrak{v}$ be an inductive limit neighborhood. For the set $E \in \mathfrak{R}$ from (i) there is $v \in \mathcal{V}$ such that $\chi_{E} \otimes v \leq s$ for some $s \in \mathfrak{v}$. According to (ii), let $h \in \mathcal{S}_{\mathfrak{R}}(X,\mathcal{P})$ such that $h \leq f + \chi_{X} \otimes v$ and set $h' = \chi_{E} \otimes h \in \mathcal{S}_{\mathfrak{R}}(X,\mathcal{P})$. Then $h' \leq f + s$, hence $h' \leq f + \mathfrak{v}$. This shows $f \in \mathcal{F}_{\mathfrak{R}}(X,\mathcal{P})$. □

Note that a function $\chi_{F} \otimes a$ for $F \in \mathfrak{A}_{\mathfrak{R}}$ and $a \in \mathcal{P}$ is contained in $\mathcal{F}_{\mathfrak{R}}(X,\mathcal{P})$ if and only if either $a \geq 0$ or $F \in \mathfrak{R}$.

**Lemma 2.6.** *Let $f \in \mathcal{F}_{\mathfrak{R}}(X,\mathcal{P})$ and let $\varphi$ be a positive real-valued function on $X$, measurable with respect to $\mathfrak{A}_{\mathfrak{R}}$. If either $f$ is positive or if $\varphi$ is bounded, then $\varphi \otimes f \in \mathcal{F}_{\mathfrak{R}}(X,\mathcal{P})$.*

*Proof.* Following Theorem 1.8(a), the function $\varphi \otimes f$ is measurable. If $f$ is positive, then $\varphi \otimes f$ is also positive, hence in $\mathcal{F}_{\mathfrak{R}}(X,\mathcal{P})$. Otherwise, there is $\rho > 0$ such that $0 \leq \varphi(x) \leq \rho$ for all $x \in X$. Given an inductive limit neighborhood $\mathfrak{v}$ there is $h \in \mathcal{S}_{\mathfrak{R}}(X,\mathcal{P})$ such that $h \leq f + (1/2\rho)\mathfrak{v}$. Also, there is $\lambda > 0$ such that $0 \leq f + \lambda\mathfrak{v}$. As $\varphi$ is bounded and measurable, there is a real-valued positive step function $\psi$ on $X$ such that

$$\psi(x) \leq \varphi(x) \leq \psi(x) + \frac{1}{2\lambda}$$

for all $x \in X$. Then $l = \psi \otimes h \in \mathcal{S}_{\mathfrak{R}}(X,\mathcal{P})$, and indeed

$$l \leq \psi \otimes f + \frac{1}{2\rho}\psi \otimes v \leq \psi \otimes f + \frac{1}{2}v + (\varphi - \psi) \otimes (f + \lambda\mathfrak{v}) \leq \varphi \otimes f + \mathfrak{v}.$$

□

Lemma 2.6 implies in particular that $\chi_{F} \otimes f \in \mathcal{F}_{\mathfrak{R}}(X,\mathcal{P})$ whenever $f \in \mathcal{F}_{\mathfrak{R}}(X,\mathcal{P})$ and $F \in \mathfrak{A}_{R}$. Also, if $\varphi$ is a positive real-valued measurable function and $a \in \mathcal{P}$, then $\varphi \otimes a \in \mathcal{F}_{\mathfrak{R}}(X,\mathcal{P})$ if $a \geq 0$, and $(\chi_{E}\varphi) \otimes a = \varphi \otimes (\chi_{E} \otimes a) \in \mathcal{F}_{\mathfrak{R}}(X,\mathcal{P})$ for every $E \in \mathfrak{R}$ in general if the function $\varphi$ is bounded.

A sequence $(f_n)_{n\in\mathbb{N}}$ in $\mathcal{F}_{\mathfrak{R}}(X,\mathcal{P})$ is said to be *bounded below* if for every inductive limit neighborhood $\mathfrak{v}$ there is $\lambda \geq 0$ such that $0 \leq f_n + \lambda\mathfrak{v}$ for all $n \in \mathbb{N}$.

**Theorem 2.7.** *Let* $f \in \mathcal{F}_{\mathfrak{R}}(X,\mathcal{P})$ *and* $E \in \mathfrak{R}$. *For every inductive limit neighborhood* $\mathfrak{v}$, *every* $v \in \mathcal{V}$ *and* $\varepsilon > 0$ *there is* $1 \le \gamma \le 1+\varepsilon$ *and a bounded below sequence* $(h_n)_{n\in\mathbb{N}}$ *of step functions in* $\mathcal{S}_{\mathfrak{R}}(X,\mathcal{P})$ *such that*

*(i)* $h_n \le \gamma f + \mathfrak{v}$ *for all* $n \in \mathbb{N}$.
*(ii) For every* $x \in E$ *there is* $n_0 \in \mathbb{N}$ *such that* $f(x) \le h_n(x) + v$ *for all* $n \ge n_0$.

*Proof.* Let $f \in \mathcal{F}_{\mathfrak{R}}(X,\mathcal{P})$, $E \in \mathfrak{R}$, let $\mathfrak{v}$ be an inductive limit neighborhood, let $v \in \mathcal{V}$ and $\varepsilon > 0$. Following Lemma 2.4(b) we may assume that $f(x) \ge 0$ for all $x \in X \setminus E$. There is $u \in \mathcal{V}$ such that both $u \le v$ and $\chi_{E^{\circledast}} u \le \mathfrak{v}$. Again using 2.4(b) we find $\lambda \ge 0$ such that $0 \le f + \lambda\chi_{E^{\circledast}} u$ We set $\delta = \min\{1, \frac{\varepsilon}{3}, \frac{1}{4(1+\lambda)}\}$ and $\gamma = (1+\delta)^2 \le 1+\varepsilon$. By Theorem 1.4 there is a partition of $E$ into disjoint subsets $E_i \in \mathfrak{R}$, $i \in \mathbb{N}$, such that $f(x) \in u_\delta\big(f(y)\big)$ holds for all $x, y \in E_i$. Thus

$$f(x) \le (1+\delta)f(y) + \delta(1+\lambda)u$$

by Lemma I.4.1(b). We set $a_i = (1+\delta)f(x_i) \in \mathcal{P}$ for some $x_i \in E_i$. Thus for any $x \in E_i$ we have

$$f(x) \le a_i + \delta(1+\lambda)u \le a_i + v$$

and

$$a_i \le (1+\delta)^2 f(x) + \delta(1+\delta)(1+\lambda)u \le \gamma f(x) + \frac{1}{2}u.$$

Thus

$$\sum_{i=1}^{n} \chi_{E_i^{\circledast}} a_i \le \chi_{E^{\circledast}}(\gamma f) + \frac{1}{2}\chi_{E^{\circledast}} u \le \chi_{E^{\circledast}}(\gamma f) + \frac{1}{2}\mathfrak{v}.$$

Furthermore, there is $h_0 \in \mathcal{S}_{\mathfrak{R}}(X,\mathcal{P})$ such that $h_0 \le \gamma f + \frac{1}{2}\mathfrak{v}$, and therefore

$$\chi_{(X\setminus E)^{\circledast}} h_o \le \chi_{(X\setminus E)^{\circledast}}(\gamma f) + \frac{1}{2}\mathfrak{v}$$

holds as well. Now we choose the step functions

$$h_n = \sum_{i=1}^{n} \chi_{E_i^{\circledast}} a_i \; + \; \chi_{(X\setminus E)^{\circledast}} h_0.$$

Adding the above yields indeed

$$h_n \le \gamma f + \mathfrak{v}$$

for all $n \in \mathbb{N}$, hence (i). Part (ii) of our claim follows directly from the above, as

$$f(x) \le h_n(x) + v \qquad \text{for all} \qquad x \in \bigcup_{i=1}^{n} E_i.$$

Finally, given an inductive limit neighborhood $\mathfrak{u}$, there is $\lambda \geq 0$ such that $0 \leq h_0 + \lambda\mathfrak{u}$, that is $0 \leq h_0 + \lambda s$ for some $s \leq \mathfrak{u}$. Also there is $u \in \mathcal{V}$ such that $\chi_{E \otimes} u \in \mathfrak{u}$ and $\rho \geq 0$ such that $0 \leq f + \rho\chi_{E \otimes} u$. The latter implies $0 \leq a_i + \rho u$ for all $i \in \mathbb{N}$, hence

$$0 \leq h_n + \lambda s + \rho\chi_{E \otimes} u \leq h_n + (\lambda + \rho)\mathfrak{u}$$

for all $n \in \mathbb{N}$. The sequence $(h_n)_{n\in\mathbb{N}}$ is therefore bounded below. Finally, if $f \geq 0$, then we may choose $h_0 = 0$, and as all the elements $a_i = (1+\delta)f(x_i)$ are also positive, we realize that $h_n \geq 0$ for all $n \in \mathbb{N}$. □

If $(\mathcal{P}, \mathcal{V})$ is indeed a full locally convex cone, as will frequently occur in the subsequent sections, then the preceding result can obviously be simplified. We shall formulate this in a corollary.

**Corollary 2.8.** *Let $(\mathcal{P}, \mathcal{V})$ be a full locally convex cone. Let $f \in \mathcal{F}_{\mathfrak{R}}(X, \mathcal{P})$ and $E \in \mathfrak{R}$. For every inductive limit neighborhood $\mathfrak{v}$ and $\varepsilon > 0$ there is $1 \leq \gamma \leq 1 + \varepsilon$ and a bounded below sequence $(h_n)_{n\in\mathbb{N}}$ of step functions in $\mathcal{S}_{\mathfrak{R}}(X, \mathcal{P})$ such that*

*(i) $h_n \leq \gamma f + \mathfrak{v}$ for all $n \in \mathbb{N}$.*
*(ii) For every $x \in E$ there is $n_0 \in \mathbb{N}$ such that $f(x) \leq h_n(x)$ for all $n \geq n_0$.*

*Proof.* We choose $v \in \mathcal{V}$ such that $\chi_{E \otimes} v \leq (1/2)\mathfrak{v}_w$ and apply Theorem 2.7 with this $v$, the inductive limit neighborhood $(1/2)\mathfrak{v}_w$ and the given $\varepsilon > 0$. There is a sequence $(h_n)_{n\in\mathbb{N}}$ as in 2.7. The functions $h'_n = h_n + \chi_{E \otimes} v$ then satisfy our claim. □

## 3. Operator-Valued Measures

Let $(\mathcal{P}, \mathcal{V})$ be a quasi-full locally convex cone and let $(\mathcal{Q}, \mathcal{W})$ be a locally convex complete lattice cone (see Sections 5 and 6 in Chapter I). Let $\mathfrak{L}(\mathcal{P}, \mathcal{Q})$ denote the cone of all (uniformly) continuous linear operators from $\mathcal{P}$ to $\mathcal{Q}$. Recall from Section 3 in Chapter I that a continuous linear operator between locally convex cones is monotone with respect to the respective weak preorders. Because $\mathcal{Q}$ carries its weak preorder, this implies monotonicity with respect to the given orders of $\mathcal{P}$ and $\mathcal{Q}$ as well. Let $X$ be a set, $\mathfrak{R}$ a (weak) $\sigma$-ring of subsets of $X$. An $\mathfrak{L}(\mathcal{P}, \mathcal{Q})$-*valued measure* $\theta$ on $\mathfrak{R}$ is a set function

$$E \mapsto \theta_E : \mathfrak{R} \to \mathfrak{L}(\mathcal{P}, \mathcal{Q})$$

such that $\theta(\emptyset) = 0$ and

$$\theta_{(\bigcup_{i\in\mathbb{N}} E_i)} = \sum_{i=1}^{\infty} \theta_{E_i}$$

holds whenever the sets $E_i \in \mathfrak{R}$ are disjoint and $\bigcup_{i=1}^{\infty} E_i \in \mathfrak{R}$. Convergence for the series on the right-hand side is meant in the following way: For every $a \in \mathcal{P}$ the series $\sum_{i=1}^{\infty} \theta_{E_i}(a)$ is order convergent in $\mathcal{Q}$ in the sense of I.5.26. (Recall from Proposition I.5.42 that order convergence is implied by convergence in the symmetric relative topology.)

**Lemma 3.1.** *Let $E \in \mathfrak{R}$ and $a \in \mathcal{P}$.*

*(a) If $E_i \in \mathfrak{R}$ are such that $E_i \subset E_{i+1}$ for all $i \in \mathbb{N}$ and $E = \cup_{i=1}^{\infty} E_i$, then $\theta_E(a) = \lim_{i\to\infty} \theta_{E_i}(a)$.*

*(b) If $E_i \in \mathfrak{R}$ are such that $E \supset E_i \supset E_{i+1}$ for all $i \in \mathbb{N}$, and $\cap_{i=1}^{\infty} E_i = \emptyset$, then $0 \leq \underline{\lim}_{i\to\infty} \theta_{E_i}(a) + \mathfrak{O}(\theta_E(a))$ and $\overline{\lim}_{i\to\infty} \theta_{E_i}(a) \leq \mathfrak{O}(\theta_E(a))$.*

*Proof.* For Part (a), let $F_1 = E_1$ and $F_i = E_i \setminus F_{i-1}$ for $i > 1$. The sets $F_i$ are disjoint, $E_n = \cup_{i=1}^{n} F_i$ and $E = \cup_{i=1}^{\infty} F_i$. From the countable additivity of the measure $\theta$ we infer that $\theta_{E_n}(a) = \sum_{i=1}^{n} \theta_{F_i}(a)$ and $\theta_E(a) = \sum_{i=1}^{\infty} \theta_{F_i}(a)$, hence our claim. For Part (b), let $F_i = E \setminus E_i$ for $i \in \mathbb{N}$. Thus $F_i \subset F_{i+1}$ and $\cup_{i=1}^{\infty} F_i = E$. This shows $\theta_E(a) = \lim_{i\to\infty} \theta_{E_i}(a)$ by Part (a). Furthermore, $\theta_E(a) = \theta_{E_i}(a) + \theta_{F_i}(a)$ holds for all $i \in \mathbb{N}$ by Part (a). Using the limit rules in Lemma I.5.19 we infer that

$$\theta_E(a) \leq \underline{\lim}_{i\to\infty} \theta_{E_i}(a) + \theta_E(a) \leq \overline{\lim}_{i\to\infty} \theta_{E_i}(a) + \theta_E(a) \leq \theta_E(a),$$

hence equality for these terms as $\mathcal{Q}$ carries the weak preorder which is supposed to be antisymmetric. Now the cancellation rule in Proposition I.5.10(a) yields our claim. □

For our upcoming integration theory for $\mathcal{P}$-valued functions with respect to an $\mathfrak{L}(\mathcal{P}, \mathcal{Q})$-valued measure $\theta$ (see Section 4 below) we shall also have to assign values of $\theta$ to the neighborhoods in $\mathcal{P}$. This will be done by the introduction of its modulus $|\theta|$. Recall that we require the locally convex cone $(\mathcal{P}, \mathcal{V})$ to be quasi-full.

**3.2 The Modulus of a Measure.** Throughout the following, let $\theta$ be a fixed $\mathfrak{L}(\mathcal{P}, \mathcal{Q})$-valued measure on $\mathfrak{R}$. For a neighborhood $v \in \mathcal{V}$ and a set $E \in \mathfrak{R}$, *modulus (or semivariation) of* $\theta$ is defined as

$$|\theta|(E, v) = \sup\left\{ \sum_{i=1}^{n} \theta_{E_i}(s_i) \;\middle|\; s_i \in \mathcal{P},\ s_i \leq v,\ E_i \in \mathfrak{R} \text{ disjoint subsets of } E \right\}.$$

The following is obvious from this definition.

**Lemma 3.3.** *Let $v \in \mathcal{V}$ and $E \in \mathfrak{R}$. If $v \in \mathcal{P}$, then $|\theta|(E, v) = \theta_E(v)$.*

*Proof.* Let $E \in \mathfrak{R}$ and $v \in \mathcal{V} \cap \mathcal{P}$. If $s_i \in \mathcal{P}$ such that $s_i \leq v$ and $E_i \in \mathfrak{R}$ are disjoint subsets of $E$ for $i = 1, \ldots, n$, then

$$\sum_{i=1}^{n} \theta_{E_i}(s_i) \leq \sum_{i=1}^{n} \theta_{E_i}(v) = \theta_{(\cup_{i=1}^{n} E_i)}(v) \leq \theta_E(v).$$

Thus $|\theta|(E,v) \leq \theta_E(v)$. The reverse inequality is obvious, as we may choose $E_1 = E$ and $s_1 = v$ in 3.2. □

**Lemma 3.4.** *Let $v \in \mathcal{V}$ and $E \in \mathfrak{R}$. Then*

*(a)* $0 \leq |\theta|(E,v)$ *and* $|\theta|(\emptyset, v) = 0$.
*(b)* $\theta_E(a) \leq \theta_E(b) + |\theta|(E,v)$ *whenever* $a \leq b + v$ *for* $a, b \in \mathcal{P}$.
*(c) If* $E_i \in \mathfrak{R}$ *are disjoint sets such that* $E = \bigcup_{i=1}^{\infty} E_i$, *then* $|\theta|(E,v) = \sum_{i=1}^{\infty} |\theta|(E_i, v)$.

*Proof.* Part (a) is obvious. Part (b) follows as the locally convex cone is supposed to be quasi-full. Indeed, for $a \leq b + v$ there is $s \leq v$ such that $a \leq b + s$. This implies $\theta_E(a) \leq \theta_E(b) + \theta_E(s)$, and as $\theta_E(s) \leq |\theta|(E,v)$, our claim follows immediately from $\bigwedge 1$. For Part (c), let $E = \cup_{i=1}^{\infty} E_i$ for disjoint sets $E_i \in \mathfrak{R}$. Let $F_1, \ldots, F_n \in \mathfrak{R}$ be disjoint subsets of $E$ and $s_k \in \mathcal{P}$ such that $s_k \leq v$ for $k = 1, \ldots, n$. Then

$$\theta_{F_k}(s_k) = \sum_{i=1}^{\infty} \theta_{(F_k \cap E_i)}(s_k)$$

for every $k = 1, \ldots, n$ by the countable additivity of $\theta$, hence

$$\begin{aligned} \sum_{k=1}^{n} \theta_{F_k}(s_k) &= \sum_{k=1}^{n} \left( \sum_{i=1}^{\infty} \theta_{(F_k \cap E_i)}(s_k) \right) \\ &= \sum_{i=1}^{\infty} \left( \sum_{k=1}^{n} \theta_{(F_k \cap E_i)}(s_k) \right) \\ &\leq \sum_{i=1}^{\infty} |\theta|(E_i, v) \end{aligned}$$

by the limit rules established in Section 5 of Chapter I. For the converse inequality, let $n \in \mathbb{N}$ and for each $i = 1, \ldots, n$, let $F_1^i, \ldots, F_{n_i}^i \in \mathfrak{R}$ be disjoint subsets of $E_i$ and $s_1^i, \ldots, s_{n_i}^i \leq v$. Then

$$\sum_{i=1}^{n} \left( \sum_{k=1}^{n_i} \theta_{F_k^i}(s_k^i) \right) \leq |\theta|(E,v),$$

as the sets $F_k^i \subset E$ are pairwise disjoint. Now taking the supremum over all such choices of sets $F_i^k$ yields with $(\bigvee 1)$

$$\sum_{i=1}^{n} |\theta|(E_i, v) \leq |\theta|(E,v), \qquad \text{hence} \qquad \sum_{i=1}^{\infty} |\theta|(E_i, v) \leq |\theta|(E,v),$$

as $n \in \mathbb{N}$ was arbitrary. □

**Lemma 3.5.** *Let* $E \in \mathfrak{R}$, $\alpha > 0$ *and* $u, v \in \mathcal{V}$. *Then*

*(a)* $|\theta|(E, \alpha v) = \alpha|\theta|(E, v)$.
*(b)* $|\theta|(E, u + v) = |\theta|(E, u) + |\theta|(E, v)$.

*Proof.* Part (a) is obvious. For Part (b), let $E_1, \ldots, E_n \in \mathfrak{R}$ be disjoint subsets of $E$ and let $r_i \in \mathcal{P}$ such that $r_i \leq u+v$ for $i = 1, \ldots, n$. According to (QF2) in I.6.1 there are elements $s_i, t_i \in \mathcal{P}$ such that $s_i \leq u$, $t_i \leq v$ and $s_i \leq r_i + t_i$. This shows

$$\sum_{i=1}^{n} \theta_{E_i}(r_i) \leq \sum_{i=1}^{n} \theta_{E_i}(s_i) + \sum_{i=1}^{n} \theta_{E_i}(t_i) \leq |\theta|(E, u) + |\theta|(E, v).$$

As the sets $E_i \in \mathfrak{R}$ and the elements $r_i \leq u+v$ were chosen arbitrarily, this shows $|\theta|(E, u + v) \leq |\theta|(E, u) + |\theta|(E, v)$. For the converse inequality, let $E_1, \ldots, E_n \in \mathfrak{R}$ and $F_1, \ldots, F_m \in \mathfrak{R}$ be two collections of disjoint subsets of $E$. We may assume that $\bigcup_{i=1}^{n} E_i = \bigcup_{k=1}^{m} F_k = E$. Let $s_i \leq u$ and $t_k \leq v$ for $s_i, t_k \in \mathcal{P}$. Then

$$\sum_{i=1}^{n} \theta_{E_i}(s_i) + \sum_{k=1}^{m} \theta_{F_k}(t_k) = \sum_{i=1}^{n} \sum_{k=1}^{m} \theta_{(E_i \cap F_k)}(s_i + t_k) \leq |\theta|(E, u + v)$$

by the above. Taking first the supremum over all choices for the sets $E_i \in \mathfrak{R}$ and the elements $s_i \leq u$ on the left-hand side of this inequality and using $(\bigvee 1)$ yields

$$|\theta|(E, u) + \sum_{k=1}^{m} \theta_{F_k}(t_k) \leq |\theta|(E, v + u).$$

In a second step, we obtain $|\theta|(E, u) + |\theta|(E, v) \leq |\theta|(E, u + v)$ if we repeat this argument for the sets $F_k \in \mathfrak{R}$ and the elements $t_k \leq v$. □

**3.6 Bounded Measures.** Let $(\mathcal{P}, \mathcal{V})$ be a quasi-full locally convex cone and let $(\mathcal{Q}, \mathcal{W})$ be a locally convex complete lattice cone. We shall say that an $\mathfrak{L}(\mathcal{P}, \mathcal{Q})$-valued measure $\theta$ on $\mathfrak{R}$ is $\mathfrak{R}$-*bounded* or *of bounded semivariation on* $\mathfrak{R}$ if

(BV) *For every* $w \in \mathcal{W}$ *and* $E \in \mathfrak{R}$ *there is* $v \in \mathcal{V}$ *such that* $|\theta|(E, v) \leq w$.

In the sequel we shall always assume boundedness in this sense.

*Remarks 3.7.* (a) If $(\mathcal{P}, \mathcal{V})$ is a full locally convex cone, then every $\mathfrak{L}(\mathcal{P}, \mathcal{Q})$-valued measure on $\mathfrak{R}$ is bounded. Indeed, let $E \in \mathfrak{R}$ and $w \in \mathcal{W}$. Because the operator $\theta_E : \mathcal{P} \to \mathcal{Q}$ is supposed to be continuous, there is $v \in \mathcal{V}$ such that $\theta_E(a) \leq \theta_E(b) + w$ whenever $a \leq b + v$ for $a, b \in \mathcal{P}$. Following Lemma 3.3, this shows $|\theta|(E, v) = \theta_E(v) \leq w$ in particular.

(b) Let $\mathcal{P} = \mathbb{K}$ for $\mathbb{K} = \mathbb{R}$ or $\mathbb{K} = \mathbb{C}$, endowed with the equality as order and the usual Euclidean topology; that is $\mathcal{V} = \{\varepsilon\mathbb{B} \mid \varepsilon > 0\}$, where $\mathbb{B}$ is the unit ball in $\mathbb{K}$ and $a \leq b+\varepsilon\mathbb{B}$ means that $a \in b+\varepsilon\mathbb{B}$. Let $\mathcal{Q} = \overline{\mathbb{R}}$. Then $\mathfrak{L}(\mathcal{P}, \mathcal{Q})$ can be identified with $\mathbb{K}$, since every linear operator (functional) from $\mathbb{K}$ to $\overline{\mathbb{R}}$ is given by an element $z \in \mathbb{K}$ via the evaluation $a \mapsto \mathfrak{Re}(za)$ for $a \in \mathbb{K}$. This is therefore the case of a real- or complex-valued measure $\theta$. According to 3.2, its modulus is computed as

$$|\theta|(E, \mathbb{B}) = \sup\left\{ \sum_{i=1}^{n} \theta_{E_i} \cdot s_i \;\middle|\; s_i \in \mathbb{B},\ E_i \in \mathfrak{R} \text{ disjoint subsets of } E \right\}$$

$$= \sup\left\{ \sum_{i=1}^{n} |\theta_{E_i}| \;\middle|\; E_i \in \mathfrak{R} \text{ disjoint subsets of } E \right\},$$

that is the usual *total variation* of the real- or complex-valued measure $\theta$ (see II.1.4 in [55]).

(c) If $(\mathcal{P}, \mathcal{V})$ is a locally convex topological vector space, and $\mathcal{Q} = \overline{\mathbb{R}}$, that is the case of a functional-valued measure, our requirement of boundedness corresponds to Dieudonné's notion of *p-domination* in [44] and to Prolla's of *finite p-semivariation* in [155] (Ch. 5.5) for measures with values in the dual of a locally convex vector space.

(d) If $(\mathcal{N}, \|\ \|)$ is a normed space over $\mathbb{K} = \mathbb{R}$ or $\mathbb{K} = \mathbb{C}$, then every $\mathcal{N}$-valued measure $\theta$ may be considered to be an operator-valued measure in our sense. Indeed, the elements of $\mathcal{N}$ are linear operators from $\mathcal{P} = \mathbb{K}$, endowed with the Euclidean topology, into the standard lattice completion $(\widehat{\mathcal{N}}, \widehat{\mathcal{W}})$ of $\mathcal{N}$ as introduced in I.5.57. The notion of the semivariation of a vector-valued measure as given for example in IV.10.3 in [55] slightly differs from our notion of the modulus, as there it is a real-valued expression (in fact, it is in some sense the norm in $\mathcal{N}$ of our modulus; see Section 8 below), which is however not countably additive in general. We shall consider this example in more detail in Section 6 below.

(e) If $(\mathcal{P}, \mathcal{V})$ is a quasi-full locally convex cone and if we endow the subcone $\mathcal{P}_+$ of its positive elements with the neighborhood system $\widetilde{\mathcal{V}} = \{0\}$, (for this, see also Example I.1.4(b)), then every $\mathfrak{L}(\mathcal{P}, \mathcal{Q})$-valued measure $\theta$ can be canonically extended to an $\mathfrak{L}(\mathcal{P}_+, \mathcal{Q})$-valued measure on the whole $\sigma$-algebra $\mathfrak{A}_{\mathfrak{R}}$ : For every set $F \in \mathfrak{A}_{\mathfrak{R}}$ we define the operator $\theta_F \in \mathfrak{L}(\mathcal{P}_+, \mathcal{Q})$ by

$$\theta_F(a) = \sup\{\theta_E(a) \mid E \subset F,\ E \in \mathfrak{R}\} \in \mathcal{Q}$$

for $a \in \mathcal{P}_+$. Linearity of this operator follows from I.5.22, and continuity is trivial, since $\mathcal{P}_+$ is endowed with the neighborhood system $\widetilde{\mathcal{V}} = \{0\}$. For countable additivity on $\mathfrak{A}_{\mathfrak{R}}$ let $F_n \in \mathfrak{A}_{\mathfrak{R}}$, for $n \in \mathbb{N}$, be disjoint sets and let $F = \bigcup_{n\in\mathbb{N}} F_n$. Let $a \in \mathcal{P}_+$. Then

$$\begin{aligned}\theta_F(a) &= \sup\{\theta_E(a) \mid E \subset F,\ E \in \mathfrak{R}\}\\ &= \sup\left\{\sum_{i=1}^{\infty} \theta_{E\cap F_i}(a) \mid E \subset F,\ E \in \mathfrak{R}\right\} \le \sum_{i=1}^{\infty} \theta_{F_i}(a).\end{aligned}$$

For every $n \in \mathbb{N}$, on the other hand, and $E_i \in \mathfrak{R}$ such that $E_i \subset F_i$ for $i = 1, \ldots, n$, we set $E = \bigcup_{i=1}^n E_i \in \mathfrak{R}$ and have $\sum_{i=1}^n \theta_{E_i}(a) = \theta_E(a) \le \theta_F(a)$. Taking the supremum over all such choices of sets $E_i \in \mathfrak{R}$ yields with Lemma I.5.5(a) that $\sum_{i=1}^n \theta_{F_i}(a) \le \theta_F(a)$. This holds for all $n \in \mathbb{N}$ and therefore yields the reverse inequality $\sum_{i=1}^\infty \theta_{F_i}(a) \le \theta_F(a)$.

**3.8 Extension of a Measure.** We may use the modulus of an $\mathfrak{R}$-bounded $\mathfrak{L}(\mathcal{P}, \mathcal{Q})$-valued measure $\theta$ to define an extension to an $\mathfrak{R}$-bounded $\mathfrak{L}(\mathcal{P}_{\mathcal{V}}, \mathcal{Q})$-valued measure, where $(\mathcal{P}_{\mathcal{V}}, \mathcal{V})$ denotes the standard full extension of the quasi-full locally convex cone $(\mathcal{P}, \mathcal{V})$ as constructed in Section 6.2 of Chapter I, that is

$$\mathcal{P}_{\mathcal{V}} = \big\{a \oplus v \mid a \in \mathcal{P},\ v \in \mathcal{V} \cup \{0\}\big\}.$$

This follows the extension of a continuous linear operator from $\mathcal{P}$ to $\mathcal{Q}$ into a continuous linear operator from $\mathcal{P}_{\mathcal{V}}$ to $\mathcal{Q}$ as elaborated in Theorem I.6.3. For $E \in \mathfrak{R}$ and $a \oplus v \in \mathcal{P}_{\mathcal{V}}$ we set

$$\theta_E(a \oplus v) = \theta_E(a) + |\theta|(E, v).$$

The required properties for a measure are readily checked. Indeed, for a fixed set $E \in \mathfrak{R}$, Lemma 3.5 shows that $\theta_E$ is a linear operator on $\mathcal{P}_{\mathcal{V}}$. In order to verify that this operator is monotone, let $a \oplus v \le b \oplus u$ for $a \oplus v, b \oplus u \in \mathcal{P}_{\mathcal{V}}$. Let $E_1, \ldots, E_n \in \mathfrak{R}$ be disjoint subsets of $E$ and $s_1, \ldots, s_n \le v$. We set $E_0 = E \setminus \bigcup_{i=1}^n E_i$ and $s_0 = 0$. Then $\bigcup_{i=0}^n E_i = E$ and $a + s_i \le b + u$ for all $i = 0, \ldots, n$ by our definition of the order in $\mathcal{P}_{\mathcal{V}}$, hence $a + s_i \le b + t_i$ for some $t_i \le u$ by Condition (QF1) from I.6.1. Thus $\theta_{E_i}(a + s_i) \le \theta_{E_i}(b + t_i)$ for all $i = 0, \ldots, n$ by the monotonicity of the operators $\theta_{E_i}$, hence

$$\begin{aligned}\theta_E(a) + \sum_{i=1}^n \theta_{E_i}(s_i) &= \sum_{i=0}^n \theta_{E_i}(a + s_i)\\ &\le \sum_{i=0}^n \theta_{E_i}(b + t_i)\\ &= \theta_E(b) + \sum_{i=1}^n \theta_{E_i}(t_i)\\ &\le \theta_E(b) + |\theta|(E, u).\end{aligned}$$

Taking the supremum over all such choices of sets $E_i \in \mathfrak{R}$ and elements $s_i \le v$ on the left-hand side of this inequality yields

$$\theta_E(a \oplus v) = \theta_E(a) + |\theta|(E, v) \le \theta_E(b) + |\theta|(E, u) = \theta_E(b \oplus u).$$

Furthermore, given $w \in \mathcal{W}$, by the $\mathfrak{R}$-boundedness of the given measure $\theta$, there is $v \in \mathcal{V}$ such that $\theta_E(0 \oplus v) = |\theta|(E, v) \leq w$. This implies that the linear operator $\theta_E : \mathcal{P}_\mathcal{V} \to \overline{\mathbb{R}}$ is indeed continuous. The countable additivity of the extended measure follows from Lemma 3.4(c). Furthermore, as $(\mathcal{P}_\mathcal{V}, \mathcal{V})$ is a full cone, the extension of $\theta$ remains $\mathfrak{R}$-bounded (see 3.7(a)), that is, $|\theta|(E, 0 \oplus v) = \theta_E(0 \oplus v) = |\theta|(E, v)$ holds for all $E \in \mathfrak{R}$ and $v \in \mathcal{V}$. If on the other hand, $\theta$ is an $\mathfrak{R}$-bounded $\mathfrak{L}(\mathcal{P}_\mathcal{V}, \mathcal{Q})$-valued measure on $\mathfrak{R}$, and if $\theta_0$ denotes its restriction to an $\mathfrak{L}(\mathcal{P}, \mathcal{Q})$-valued measure, then we have $|\theta_0|(E, v) \leq \theta_E(0 \oplus v)$.

This procedure of extending a given $\mathfrak{R}$-bounded measure from a quasi-full to a full cone yields an interesting new understanding of the *(total) variation* of a given measure, not as a separate positive real-valued measure associated with the given one, but as an integral part of its extension. Because this extension, evaluated at the neighborhoods is also $\mathcal{Q}$- and not necessarily positive real-valued, its countable additivity is preserved, thus removing a major inconvenience that arises in the classical approach (see IV.10.3 in [55]). This therefore avoids the need to introduce the separate terms of *variation* and *semivariation* for a measure (see I.2 in [43]).

The extension of a given measure as carried out in 3.8 will turn out to be invaluable in our upcoming integration theory for cone-valued functions with respect to an operator-valued measure. It does in fact justify the use of a full cone for $\mathcal{P}$, that is the range of the concerned functions and the domain of the linear operators resulting from our measures.

**3.9 Composition of Measures and Continuous Linear Operators.** Let $(\mathcal{P}, \mathcal{V})$ and $(\widetilde{\mathcal{P}}, \widetilde{\mathcal{V}})$ be quasi-full locally convex cones, and let $(\mathcal{Q}, \mathcal{W})$ and $(\widetilde{\mathcal{Q}}, \widetilde{\mathcal{W}})$ be locally convex complete lattice cones. For continuous linear operators $S \in \mathfrak{L}(\widetilde{\mathcal{P}}, \mathcal{P})$, $T \in \mathfrak{L}(\mathcal{P}, \mathcal{Q})$ and $U \in \mathfrak{L}(\mathcal{Q}, \widetilde{\mathcal{Q}})$ let $U \circ T \circ S \in \mathfrak{L}(\widetilde{\mathcal{P}}, \widetilde{\mathcal{Q}})$ denote their composition, that is the continuous linear operator

$$l \mapsto U\big(T\big(S(l)\big)\big) \;:\; \widetilde{\mathcal{P}} \to \widetilde{\mathcal{Q}}.$$

It is straightforward to verify that this operator is indeed linear and continuous. We shall use this in the following way: If $\theta$ is an $\mathfrak{L}(\mathcal{P}, \mathcal{Q})$-valued measure on $\mathfrak{R}$, if $S \in \mathfrak{L}(\widetilde{\mathcal{P}}, \mathcal{P})$ and if the operator $U \in \mathfrak{L}(\mathcal{Q}, \widetilde{\mathcal{Q}})$ is order continuous (see I.5.29), then the set function

$$E \mapsto (U \circ \theta_E \circ S) \;:\; \mathfrak{R} \to \mathfrak{L}(\widetilde{\mathcal{P}}, \widetilde{\mathcal{Q}})$$

is an $\mathfrak{L}(\widetilde{\mathcal{P}}, \widetilde{\mathcal{Q}})$-valued measure, called the *composition of $\theta$ with $U$ and $S$* and denoted as $(U \circ \theta \circ S)$. Countable additivity follows from the order continuity of the operator $U$. Indeed, let $E_i \in \mathfrak{R}$ be disjoint sets such that $E = \bigcup_{i=1}^{\infty} E_i \in \mathfrak{R}$. Then for every $l \in \widetilde{\mathcal{P}}$ we have $\theta_E\big(S(l)\big) = \sum_{i=1}^{\infty} \theta_{E_i}\big(S(l)\big)$ by the countable additivity of $\theta$, hence

$$\begin{aligned}(U\circ\theta\circ S)_E(l) &= U\big(\theta_E(S(l))\big)\\ &= U\left(\sum_{i=1}^{\infty}\theta_{E_i}(S(l))\right)\\ &= \sum_{i=1}^{\infty}U\big(\theta_{E_i}(S(l))\big)\\ &= \sum_{i=1}^{\infty}(U\circ\theta\circ S)_{E_i}(l)\end{aligned}$$

by the order continuity of $U$. The modulus of the measure $(U\circ\theta\circ S)$ can be estimated as follows: Let $E\in\mathfrak{R}$, and for $v\in\mathcal{V}$ let $\tilde{v}\in\widetilde{\mathcal{V}}$ such that $S(l)\le S(m)+v$ whenever $l\le m+\tilde{v}$ for $l,m\in\widetilde{\mathcal{P}}$. If $E_1,\ldots,E_n\in\mathfrak{R}$ are disjoint subsets of $E$ and if $l_i\in\widetilde{\mathcal{P}}$ such that $l_i\le\tilde{v}$ for $i=1,\ldots,n$, then

$$\sum_{i=1}^{n}\big(U\circ\theta\circ S\big)_{E_i}(l_i)=U\left(\sum_{i=1}^{n}\theta_{E_i}(S(l_i))\right)\le U\big(|\theta|(E,v)\big).$$

Taking the supremum over all such choices for sets $E_i\in\mathfrak{R}$ and elements $l_i\le\tilde{v}$ yields

$$|U\circ\theta\circ S|(E,\tilde{v})\le U\big(|\theta|(E,v)\big).$$

The $\mathfrak{L}(\widetilde{\mathcal{P}},\widetilde{\mathcal{Q}})$-valued measure $(U\circ\theta\circ S)$ is therefore $\mathfrak{R}$-bounded whenever the $\mathfrak{L}(\mathcal{P},\mathcal{Q})$-valued measure $\theta$ is $\mathfrak{R}$-bounded. Indeed, for $E\in\mathfrak{R}$ and $\tilde{w}\in\widetilde{W}$ there is $w\in\mathcal{W}$ such that $U(s)\le U(t)+\tilde{w}$ whenever $s\le t+\tilde{w}$ for $s,t\in\mathcal{Q}$. There is $v\in\mathcal{V}$ such that $|\theta|(E,v)\le w$, hence $|U\circ\theta\circ S|(E,\tilde{v})\le\tilde{w}$ if $\tilde{v}\in\widetilde{V}$ is chosen as above.

We shall in particular make use of the combination of an $\mathfrak{L}(\mathcal{P},\mathcal{Q})$-valued measure $\theta$ with an order continuous linear functional $\mu\in\mathcal{Q}^*$. (We choose $\widetilde{\mathcal{P}}=\mathcal{P}$ and the identity operator for $S$.) The resulting measure $(\mu\circ\theta)$ is $\mathfrak{L}(\mathcal{P},\overline{\mathbb{R}})$-, that is $\mathcal{P}^*$-valued in this case.

**3.10 Strong Additivity.** Countable additivity of an $\mathfrak{L}(\mathcal{P},\mathcal{Q})$-valued measure $\theta$ is meant with respect to order convergence in the locally convex complete lattice cone $(\mathcal{Q},\mathcal{W})$. Order convergence does in general not imply convergence in the weak or indeed convergence in the symmetric relative topology of $\mathcal{Q}$ (see I.5.42). However, the following result based on a well-known theorem by Pettis (see Theorem IV.10.1 in Dunford & Schwartz, [55]) will show that in special cases some stronger type of convergence is implied.

**Theorem 3.11.** *Let $\theta$ be an $\mathfrak{L}(\mathcal{P},\mathcal{Q})$-valued measure, let $a$ be a bounded element of $\mathcal{P}$ and let $\mathcal{Q}_0$ be the subcone of $\mathcal{Q}$ spanned by the set $\{\theta_E(a)\mid E\in\mathfrak{R}\}$. If every continuous linear functional on $\mathcal{Q}_0$ can be extended to an order continuous linear functional on $\mathcal{Q}$, then for disjoint sets $E_i\in\mathfrak{R}$ such that $\cup_{i=1}^{\infty}E_i\in\mathfrak{R}$ the series*

$$\theta_{(\bigcup_{i\in\mathbb{N}} E_i)}(a) = \sum_{i=1}^{\infty} \theta_{E_i}(a)$$

*converges in the symmetric topology of $\mathcal{Q}$.*

*Proof.* We shall follow the main lines of the arguments in the proof of Pettis' Theorem as presented in [55]. Let $a \in \mathcal{P}$ be a bounded element, and let $\mathcal{Q}_0$ be the subcone of $\mathcal{Q}$ spanned by the set $\{\theta_E(a) \in \mathcal{Q}_0 \mid E \in \mathfrak{R}\}$. As all the operators $\theta_E$ are continuous, the elements of $\mathcal{Q}_0$ are bounded in $\mathcal{Q}$. We may therefore consider $\mathcal{Q}_0$ as a locally convex cone endowed with the symmetric topology generated by the neighborhood system $\mathcal{W}$. Let $\mathcal{Q}_0^{s*}$ be the dual of $\mathcal{Q}_0$ under this topology. According to Proposition II.2.21 in [100], the linear functionals $\mu \in \mathcal{Q}_0^{s*}$ can be expressed as the difference of two elements of the given dual cone (with respect to the given topology) $\mathcal{Q}_0^*$ of $\mathcal{Q}_0$, that is $\mathcal{Q}_0^{s*} = \mathcal{Q}_0^* - \mathcal{Q}_0^*$. As the elements of $\mathcal{Q}_0^*$ were supposed to be order continuous on $\mathcal{Q}_0$, so are the elements of $\mathcal{Q}_0^{s*}$.

Now let us consider a sequence of disjoint sets $E_i \in \mathfrak{R}$ such that $E = \cup_{i=1}^{\infty} E_i \in \mathfrak{R}$. Let $\mathfrak{Z}_0 \subset \mathfrak{R}$ be the set algebra in $E$ generated by the sets $E_i$, and let $\mathfrak{Z} \subset \mathfrak{R}$ be the $\sigma$-algebra in $E$ generated by $\mathfrak{Z}_0$. The algebra $\mathfrak{Z}_0$ is known to be countable (see III.8.4 in [55]). Let $\mathcal{Q}_1$ be the closure (with respect to the symmetric topology) in $\mathcal{Q}_0$ of the subcone that is spanned by the countable set $\{\theta_E(a) \mid E \in \mathfrak{Z}_0\}$.

In a first step, an argument using the separation result from Corollary 4.6 in [172] will demonstrate that $\theta_E(a) \in \mathcal{Q}_1$ for all $E \in \mathfrak{Z}$. For this, assume to the contrary that $\theta_E(a) \notin \mathcal{Q}_1$ for some $E \in \mathfrak{Z}$. Then according to the separation result 4.6 in [172] there is a linear functional $\mu \in \mathcal{Q}_0^{s*}$ such that such that $\mu(\theta_E(a)) \leq -1 \leq \mu(l)$ for all $l \in \mathcal{Q}_1$. As $\mathcal{Q}_1$ is a cone, this implies indeed that $\mu(l) \geq 0$ holds for all $l \in \mathcal{Q}_1$. As the linear functional $\mu \in \mathcal{Q}_0^{s*}$ was seen to be order continuous, $G \mapsto \mu(\theta_G(a)) : \mathfrak{Z} \to \mathbb{R}$ defines a countably additive real-valued measure $(\mu \circ \theta \circ a)$ on $\mathfrak{Z}$. This measure, taking non-negative values on $\mathfrak{Z}_0$ and a negative value on $E \in \mathfrak{R}$ contradicts the uniqueness part of Hahn's extension theorem for measures from an algebra $\mathfrak{Z}_0$ to the $\sigma$-algebra $\mathfrak{Z}$ generated by $\mathfrak{Z}_0$ (see III.5.9 in [55] or 12.2.8 in [178]). Thus $\theta_E(a) \in \mathcal{Q}_1$ as claimed.

Now set $F_n = \bigcup_{i=1}^{n} E_i$ for $n \in \mathbb{N}$, and let us assume that, contrary to our claim, there exists a neighborhood $w \in \mathcal{W}$, and a subsequence $(F_m)_{m\in\mathbb{N}}$ of $(F_n)_{n\in\mathbb{N}}$ such that either

$$\theta_E(a) \not\leq \theta_{F_m}(a) + w \qquad \text{or} \qquad \theta_{F_m}(a) \not\leq \theta_E(a) + w$$

for all $m \in \mathbb{N}$. Then according to Theorem 3.11 in [175] (see also Corollary 4.34 in Chapter I) there are linear functionals $\mu_m \in \mathcal{Q}_0^{s*}$, contained in the polar of the symmetric neighborhood $w$, such that $\mu_m(\theta_E(a)) > \mu_m(\theta_{F_m}(a)) + 1$ for all $m \in \mathbb{N}$. Let $\{l_k \mid k \in \mathbb{N}\}$ be a countable dense (with respect to the symmetric topology) subset of $\mathcal{Q}_1$. Since for every $k \in \mathbb{N}$ the sequence $(\mu_m(l_k))_{m\in\mathbb{N}}$ is bounded in $\mathbb{R}$, we may use a Cantor diagonal procedure to find a subsequence $(\mu_{m_j})_{j\in\mathbb{N}}$ of $(\mu_m)_{m\in\mathbb{N}}$ such that the limit $\lim_{j\to\infty} \mu_{m_j}(l_k)$

exists in $\mathbb{R}$ for all $k \in \mathbb{N}$ : Indeed, there is a subsequence $(\mu_{m_{(1,j)}})_{j\in\mathbb{N}}$ of $(\mu_m)_{m\in\mathbb{N}}$ such that $\lim_{j\to\infty} \mu_{m_{(1,j)}}(l_1)$ exists in $\mathbb{R}$. Then there is a subsequence $(\mu_{m_{(2,j)}})_{m\in\mathbb{N}}$ of $(\mu_{m_{(1,j)}})_{m\in\mathbb{N}}$ such that $\lim_{j\to\infty} \mu_{m_{(2,j)}}(l_2)$ exists, etc. We set $\mu_{m_j} = \mu_{m_{(j,j)}}$ for all $j \in \mathbb{N}$. Then $(\mu_{m_j})_{j\in\mathbb{N}}$ is a subsequence of each of the sequences $(\mu_{m_{(k,j)}})_{m\in\mathbb{N}}$ for $k \in \mathbb{N}$, thus satisfying our requirement. Now a simple argument will show that the limit $\lim_{j\to\infty} \mu_{m_j}(l)$ exists indeed for all $l \in \mathcal{Q}_1$. In fact, given $l \in \mathcal{Q}_1$ and $\varepsilon > 0$ there is some $l_k$ such that both $l \le l_k + \varepsilon w$ and $l_k \le l + \varepsilon w$, hence $|\mu_{m_j}(l) - \mu_{m_j}(l_k)| \le \varepsilon$ for all $j \in \mathbb{N}$. Moreover, there is $j_0 \in \mathbb{N}$ such that $|\mu_{m_{j_1}}(l_k) - \mu_{m_{j_2}}(l_k)| \le \varepsilon$ whenever $j_1, j_2 \ge j_0$. This implies $|\mu_{m_{j_1}}(l_k) - \mu_{m_{j_2}}(l)| \le 3\varepsilon$. Thus the sequence $\big(\mu_{m_j}(l)\big)_{j\in\mathbb{N}}$ is a Cauchy sequence, hence convergent in $\mathbb{R}$. For every $j \in \mathbb{N}$ let $(\mu_{m_j} \circ \theta \circ a)$ denote the real-valued measure $G \mapsto \mu_{m_j}\big(\theta_G(a)\big) : \mathfrak{Z} \to \mathbb{R}$. Then $\lim_{j\to\infty} (\mu_{m_j} \circ \theta \circ a)(G)$ exists for every $G \in \mathfrak{Z}$ by the above, hence following Nikodým's theorem (see Corollary III.7.4 in [55]) the countable additivity of these measures is uniform in $j$. As $E = \bigcup_{j=1}^{\infty} F_j$, this contradicts our assumption that $(\mu_{m_j} \circ \theta \circ a)(E) > (\mu_{m_j} \circ \theta \circ a)(F_j) + 1$ holds for all $j \in \mathbb{N}$. □

This result applies in particular if $(\mathcal{Q}, \mathcal{W})$ is the standard lattice completion of some subcone $\mathcal{Q}_0$ of $\mathcal{Q}$ and if the measure $\theta$ is $\mathfrak{L}(\mathcal{P}, \mathcal{Q}_0)$-valued. In this case, all continuous linear functionals on $\mathcal{Q}_0$ extend to order continuous linear functionals on $\mathcal{Q}$, as required in Theorem 3.11. If all elements of $\mathcal{P}$ are bounded, then countable additivity of an $\mathfrak{L}(\mathcal{P}, \mathcal{Q}_0)$-valued measure implies convergence of the concerned operators with respect to the strong operator topology of $\mathfrak{L}(\mathcal{P}, \mathcal{Q}_0)$ $\big($see I.7.2(ii)$\big)$.

We shall provide a simple example of a measure that is countably additive with respect to order convergence but not with respect the symmetric topology of $\mathcal{Q}$.

*Example 3.12.* Let $\mathcal{P} = \mathbb{R}$ with its usual (Euclidean) topology, and let $\mathcal{Q}$ be the cone of all $\overline{\mathbb{R}}$-valued bounded below functions on the interval $[0, 1]$, endowed with the pointwise algebraic operations and order, and the constant functions $w > 0$ as neighborhoods. Then $(\mathcal{Q}, \mathcal{W})$ is a locally convex complete lattice cone. Let $\mathfrak{R}$ be the $\sigma$-algebra of Borel sets on $X = [0, 1]$. For every $E \in \mathfrak{R}$ let $\theta_E$ be the linear operator in $\mathfrak{L}(\mathcal{P}, \mathcal{Q})$ that maps $\rho \in \mathbb{R}$ into $\rho\chi_E \in \mathcal{Q}$, where $\chi_E$ denotes the characteristic function of the set $E$. Clearly $\theta$ is countably additive with respect to order convergence, but not with respect to uniform convergence, that is convergence with respect to the symmetric topology in $\mathcal{Q}$.

If $(\mathcal{Q}_0, \mathcal{W}_0)$ is locally convex topological vector space over $\mathbb{K} = \mathbb{R}$ or $\mathbb{K} = \mathbb{C}$, then the elements of $\mathcal{Q}_0$ may be considered as continuous linear operators from $\mathcal{P} = \mathbb{K}$, endowed with the Euclidean topology, into the standard lattice completion $(\mathcal{Q}, \mathcal{W})$ of $\mathcal{Q}_0$. (This situation will be explored in greater detail in Example 6.23 below.) A $\mathcal{Q}_0$-valued measure is required to be countably additive with respect to order convergence in $\mathcal{Q}$, that is weak convergence in $\mathcal{Q}_0$. According to Theorem 3.11 (use $a = 1 \in \mathcal{P}$), this implies convergence with respect to the symmetric topology of $\mathcal{Q}$, that is the given topology of $\mathcal{Q}_0$. This result is commonly known as Pettis' theorem.

**Corollary 3.13.** *Let $(\mathcal{P},\mathcal{V})$ be a locally convex topological vector space over $\mathbb{R}$ or $\mathbb{C}$. For a $\mathcal{P}$-valued measure countable additivity with respect to the weak topology implies countable additivity with respect to the given topology of $\mathcal{P}$.*

**3.14 Weak Compactness.** A well-known result due to Bartle, Dunford and Schwartz (see Corollary I.2.7 in [43] or Theorem VI.7.3 in [55]) about the relative weak compactness of the range of a vector-valued measure implies the following for operator-valued measures:

**Theorem 3.15.** *Suppose that $(\mathcal{Q},\mathcal{W})$ is the standard lattice completion of a Banach space $(\mathcal{Q}_0,\|\,\|)$ over $\mathbb{R}$ or $\mathbb{C}$ and that $\theta$ is a bounded $\mathfrak{L}(\mathcal{P},\mathcal{Q}_0)$-valued measure. Then for every $a \in \mathcal{P}$ and every $E \in \mathfrak{R}$ the set*

$$\{\theta_G(a) \mid F \in \mathfrak{R},\ G \subset E\}$$

*is relatively compact in $\mathcal{Q}_0$ with respect to the weak topology $\sigma(\mathcal{Q}_0,\mathcal{Q}_0^*)$.*

*Proof.* Let $a \in \mathcal{P}$ and $E \in \mathfrak{R}$. The family $\mathfrak{R}_E = \{G \in \mathfrak{R},\ G \subset E\}$ is a $\sigma$-algebra on $E$, and the set function

$$G \mapsto \theta_G(a) : \mathfrak{R}_E \to \mathcal{Q}_0$$

is a countably additive $\mathcal{Q}_0$-valued, that is a Banach space-valued measure on $\mathfrak{R}_E$. Our claim then follows directly from Corollary I.2.7 in [43]. □

## 4. Integrals for Cone-Valued Functions

Throughout this section, let $(\mathcal{P},\mathcal{V})$ be a full locally convex cone and let $(\mathcal{Q},\mathcal{W})$ be a locally convex complete lattice cone. Let $\mathfrak{R}$ be a (weak) $\sigma$-ring of subsets of $X$ and $\theta$ an $\mathfrak{L}(\mathcal{P},\mathcal{Q})$-valued measure on $\mathfrak{R}$. The requirement that the locally convex cone $(\mathcal{P},\mathcal{V})$ is full does in fact accommodate quasi-full cones as well. Indeed, in this case we may take advantage of the embedding of a quasi-full cone $(\mathcal{P},\mathcal{V})$ into the full locally convex cone $(\mathcal{P}_{\mathcal{V}},\mathcal{V})$, that is its standard full extension, as elaborated in I.6, and make use of the corresponding extension of an $\mathfrak{R}$-bounded $\mathfrak{L}(\mathcal{P},\mathcal{Q})$-valued measure $\theta$ to an $\mathfrak{L}(\mathcal{P}_{\mathcal{V}},\mathcal{Q})$-valued measure as constructed in Section 3.8; that is, we may set $\theta_E(v) = |\theta|(E,v)$ for every set $E \in \mathfrak{R}$ and every neighborhood $v \in \mathcal{V} \subset \mathcal{P}_{\mathcal{V}}$.

We proceed to define integrals for cone-valued functions with respect to $\theta$. The values of these integrals will be elements of $\mathcal{Q}$. We shall use the cone $\mathcal{F}_{\mathfrak{R}}(X,\mathcal{P})$ of all $\mathcal{P}$-valued measurable functions on $X$ that can be reached from below by $\mathcal{P}$-valued step functions in the sense of Section 2.3. Similarly, $\mathcal{F}_{\mathfrak{R}}(X,\mathcal{V})$ denotes the cone of all measurable $(\mathcal{V}\cup\{0\})$-valued functions on $X$.

In a first step, we shall define integrals for $\mathcal{P}$- and $\mathcal{V}$-valued step functions on $X$, that is functions $s = \sum_{i=1}^{n} \chi_{E_i} \otimes a_i$ for $E_i \in \mathfrak{R}$ and elements $a_i$ in $\mathcal{P}$ or $\mathcal{V}$, respectively. We shall denote the corresponding subcones of $\mathcal{F}(X,\mathcal{P})$ by $\mathcal{S}_{\mathfrak{R}}(X,\mathcal{P})$ and $\mathcal{S}_{\mathfrak{R}}(X,\mathcal{V})$. Note that the functions in $\mathcal{S}_{\mathfrak{R}}(X,\mathcal{V})$ are $(\mathcal{V} \cup \{0\})$-valued. Obviously, any representation $\sum_{i=1}^{n} \chi_{E_i} \otimes a_i$ for a given step function is not unique. To prepare our definition of the integral for functions in $\mathcal{S}_{\mathfrak{R}}(X,\mathcal{P})$ we observe:

**Lemma 4.1.** *Let $E_i, F_k \in \mathfrak{R}$ and $a_i, b_k \in \mathcal{P}$ for $i = 1, \ldots, n$ and $k = 1, \ldots, m$. If $\sum_{i=1}^{n} \chi_{E_i} \otimes a_i \leq \sum_{k=1}^{m} \chi_{F_k} \otimes b_k$, then $\sum_{i=1}^{n} \theta_{E_i}(a_i) \leq \sum_{k=1}^{m} \theta_{F_k}(b_k)$.*

*Proof.* First we shall verify that for any step function $s = \sum_{i=1}^{n} \chi_{E_i} \otimes a_i$ there exists a representation $\sum_{k=1}^{m} \chi_{F_k} \otimes b_k$ such that the sets $F_k \in \mathfrak{R}$ are pairwise disjoint and such that $\sum_{i=1}^{n} \theta_{E_i}(a_i) = \sum_{k=1}^{m} \theta_{F_k}(b_k)$. We shall use induction with respect to $n$. For $n = 1$ there is nothing to prove. Assume that our claim holds true for some $n \geq 1$ and let $s = \sum_{i=1}^{n+1} \chi_{E_i} \otimes a_i$. There are disjoint sets $F_k \in \mathfrak{R}$ such that $\sum_{i=1}^{n} \chi_{E_i} \otimes a_i = \sum_{k=1}^{m} \chi_{F_k} \otimes b_k$ satisfying the above. By adding a suitable term $\chi_F \otimes 0$ to the right-hand of the last equation, we may assume that $E_{n+1} \subset \bigcup_{k=1}^{m} F_k$. Hence

$$
\begin{aligned}
s &= \sum_{k=1}^{m} \chi_{F_k} \otimes b_k + \chi_{E_{n+1}} \otimes a_{n+1} \\
&= \sum_{k=1}^{m} \chi_{(F_k \cap E_{n+1})} \otimes (b_k + a_{n+1}) + \sum_{k=1}^{m} \chi_{(F_k \setminus E_{n+1})} \otimes b_k .
\end{aligned}
$$

The sets in the above representation for $s$ are disjoint, and we have indeed

$$
\begin{aligned}
\sum_{i=1}^{n+1} \theta_{E_i}(a_i) &= \sum_{k=1}^{m} \theta_{F_k}(b_k) + \theta_{E_{n+1}}(a_{n+1}) \\
&= \sum_{k=1}^{m} \theta_{(F_k \cap E_{n+1})}(b_k + a_{n+1}) + \sum_{k=1}^{m} \theta_{(F_k \setminus E_{n+1})}(b_k)
\end{aligned}
$$

as claimed. Thus, to prove our claim in Lemma 4.1, we may assume that both families of sets $E_i$ and $F_k$ are pairwise disjoint and that $\sum_{i=1}^{n} \chi_{E_i} \otimes a_i \leq \sum_{k=1}^{m} \chi_{F_k} \otimes b_k$. By adding suitable terms $\chi_E \otimes 0$ and $\chi_F \otimes 0$ on the left- and right-hand sides of the above inequality, we may assume in addition that $E = \bigcup_{i=1}^{n} E_i = \bigcup_{k=1}^{m} F_k$. Under these assumptions the sets $E_i \cap F_k$ form a disjoint partition of $E$, and we have either $E_i \cap F_k = \emptyset$ or $a_i \leq b_k$. This yields

$$
\sum_{i=1}^{n} \theta_{E_i}(a_i) = \sum_{i=1}^{n} \sum_{k=1}^{m} \theta_{(E_i \cap F_k)}(a_i) \leq \sum_{k=1}^{m} \sum_{i=1}^{n} \theta_{(E_i \cap F_k)}(b_k) = \sum_{k=1}^{m} \theta_{F_k}(b_k),
$$

as claimed. □

**4.2 Integrals for $\mathcal{P}$-Valued Step Functions.** We are now in a position to define the integral for a $\mathcal{P}$-valued step function

$$h = \sum_{i=1}^{n} \chi_{E_i} \otimes a_i \in \mathcal{S}_{\mathfrak{R}}(X, \mathcal{P})$$

over a measurable set $F \in \mathfrak{A}_{\mathfrak{R}}$ with respect to $\theta$ by

$$\int_F h \, d\theta = \sum_{i=1}^{n} \theta_{(E_i \cap F)}(a_i).$$

Lemma 4.1 implies that the sum on the right-hand side is independent of the particular representation for $h$. The integral represents a monotone linear operator from $\mathcal{S}_{\mathfrak{R}}(X, \mathcal{P})$ into $\mathcal{Q}$.

**Lemma 4.3.** *Let $F \in \mathfrak{A}_{\mathfrak{R}}$, let $h, g \in \mathcal{S}_{\mathfrak{R}}(X, \mathcal{P})$ and $\alpha \geq 0$. Then*
*(a) $\int_F (\alpha h) \, d\theta = \alpha \int_F h \, d\theta$.*
*(b) $\int_F (g + h) \, d\theta = \int_F g \, d\theta + \int_F h \, d\theta$.*
*(c) $\int_F g \, d\theta \leq \int_F h \, d\theta$ whenever $g \leq h$.*
*(d) $\int_F g \, d\theta = \int_X (\chi_F \otimes g) \, d\theta$.*

All these properties are obvious from the definition of the integral and from Lemma 4.1.

We shall demonstrate in the following lemma that, if the full cone $(\mathcal{P}, \mathcal{V})$ is in fact the standard full extension $(\mathcal{P}_{0\mathcal{V}}, \mathcal{V})$ of a quasi-full cone $(\mathcal{P}_0, \mathcal{V})$, and if $\theta$ is the canonical extension of an $\mathfrak{R}$-bounded $\mathfrak{L}(\mathcal{P}_0, \mathcal{Q})$-valued measure $\theta_0$, as elaborated in I.6 and 3.8, then the way in which this extension was constructed, guarantees that the integral is already determined by its values on the subcone $\mathcal{S}_{\mathfrak{R}}(X, \mathcal{P}_0)$ of $\mathcal{S}_{\mathfrak{R}}(X, \mathcal{P})$, that is by $\mathcal{P}_0$-valued step functions and the given measure $\theta_0$.

**Lemma 4.4.** *Let $F \in \mathfrak{A}_{\mathfrak{R}}$ and $g \in \mathcal{S}_{\mathfrak{R}}(X, \mathcal{P})$. If $(\mathcal{P}, \mathcal{V})$ is the standard full extension of the quasi-full cone $(\mathcal{P}_0, \mathcal{V})$, and if $\theta$ is the canonical extension of an $\mathfrak{R}$-bounded $\mathfrak{L}(\mathcal{P}_0, \mathcal{Q})$-valued measure $\theta_0$, then*

$$\int_F g \, d\theta = \sup \left\{ \int_F h \, d\theta \;\middle|\; h \in \mathcal{S}_{\mathfrak{R}}(X, \mathcal{P}_0), \quad h \leq g \right\}.$$

*Proof.* Following 4.3(c), we may assume that $F = X$. Let us first recall and reformulate from 3.2 the definition of the modulus of $\theta_0$ for a set $E \in \mathfrak{R}$ and a neighborhood $v \in \mathcal{V}$.

$$\begin{aligned} |\theta_0|(E, v) &= \sup \left\{ \sum_{i=1}^{n} \theta_{E_i}(s_i) \;\middle|\; s_i \in \mathcal{P}_0, \; s_i \leq v, \; E_i \in \mathfrak{R} \text{ disjoint subsets of } E \right\} \\ &= \sup \left\{ \int_X h \, d\theta \;\middle|\; h \in \mathcal{S}_{\mathfrak{R}}(X, \mathcal{P}_0), \; h \leq \chi_E \otimes v \right\}. \end{aligned}$$

Recall that the extension of $\theta_0$ into $\theta$ was constructed by setting $\theta_E(v) = |\theta_0|(E, v)$. Now we consider the case that

$$g = \sum_{i=1}^{n} \chi_{E_i^{\otimes}} v_i \in \mathcal{S}_{\mathfrak{R}}(X, \mathcal{V})$$

is a $\mathcal{V}$-valued function. We compute using Lemma I.5.5(a)

$$\begin{aligned}
\int_X g\,d\theta &= \sum_{i=1}^{n} \theta_{E_i}(v_i) \\
&= \sum_{i=1}^{n} |\theta_0|(E_i, v_i) \\
&= \sup\left\{ \sum_{i=1}^{n} \int_X h_i\,d\theta \;\middle|\; h_i \in \mathcal{S}_{\mathfrak{R}}(X, \mathcal{P}_0),\;\; h_i \le \chi_{E_i \otimes} v \right\} \\
&= \sup\left\{ \int_X h\,d\theta \;\middle|\; h \in \mathcal{S}_{\mathfrak{R}}(X, \mathcal{P}_0),\;\; h \le g \right\},
\end{aligned}$$

as claimed. Now for the general case, let

$$g = \sum_{i=1}^{n} \chi_{E_i^{\otimes}} (a_i + v_i) \in \mathcal{S}_{\mathfrak{R}}(X, \mathcal{P}),$$

for $a_i \in \mathcal{P}_0$ and $v_i \in \mathcal{V}$. Set $g_1 = \sum_{i=1}^{n} \chi_{E_i^{\otimes}} a_i \in \mathcal{S}_{\mathfrak{R}}(X, \mathcal{P}_0)$ and $g_2 = \sum_{i=1}^{n} \chi_{E_i^{\otimes}} v_i \in \mathcal{S}_{\mathfrak{R}}(X, \mathcal{V})$. Then $g = g_1 + g_2$, and the above yields with property $(\bigvee 1)$

$$\begin{aligned}
\int_X g\,d\theta &= \int_X g_1\,d\theta + \int_X g_2\,d\theta \\
&= \sup\left\{ \int_F (g_1 + h)\,d\theta \;\middle|\; h \in \mathcal{S}_{\mathfrak{R}}(X, \mathcal{P}_0),\;\; h \le g_2 \right\} \\
&\le \sup\left\{ \int_F h'\,d\theta \;\middle|\; h' \in \mathcal{S}_{\mathfrak{R}}(X, \mathcal{P}_0),\;\; h' \le g \right\}.
\end{aligned}$$

The converse inequality is obvious from 4.3(c). □

Subsequently, with every neighborhood $w \in \mathcal{W}$ we associate the inductive limit neighborhood $\mathfrak{v}_w$, defined as

$$\mathfrak{v}_w = \left\{ s \in \mathcal{S}_{\mathfrak{R}}(X, \mathcal{V}) \;\middle|\; \int_X s\,d\theta \le w \right\}.$$

(We shall write $\mathfrak{v}_w(\theta)$ if different measures are involved in our considerations.) The boundedness of $\theta$ guarantees that for every $E \in \mathfrak{R}$ there is $v \in \mathcal{V}$ such that $\chi_{E \otimes} v \in \mathfrak{v}_w$. Convexity follows from Lemma 4.3. We have

$\mathfrak{v}_{(\lambda w)} = \lambda \mathfrak{v}_w$ and $\mathfrak{v}_w + \mathfrak{v}_{w'} \leq \mathfrak{v}_{(w+w')}$ for $w, w' \in \mathcal{W}$ and $\lambda > 0$. We proceed to develop the integral over a measurable set $F \in \mathfrak{A}_{\mathfrak{R}}$ for a function $f \in \mathcal{F}_{\mathfrak{R}}(X, \mathcal{P})$ in the following manner: First, for a neighborhood $w \in \mathcal{W}$ we set

$$\int_F^{(w)} f\, d\theta = \sup \left\{ \int_F h\, d\theta \;\middle|\; h \in \mathcal{S}_{\mathfrak{R}}(X, \mathcal{P}),\;\; h \leq f + \mathfrak{v}_w \right\}.$$

We note that in the situation of Lemma 4.4, the integral of a function in $\mathcal{F}_{\mathfrak{R}}(X, \mathcal{P})$ is already determined by $\mathcal{P}_0$-valued step functions alone:

**Lemma 4.5.** *Let $F \in \mathfrak{A}_{\mathfrak{R}}$, $f \in \mathcal{F}_{\mathfrak{R}}(X, \mathcal{P})$ and $w \in \mathcal{W}$. If $(\mathcal{P}, \mathcal{V})$ is the standard full extension of a quasi-full cone $(\mathcal{P}_0, \mathcal{V})$, and if $\theta$ is the canonical extension of an $\mathfrak{L}(\mathcal{P}_0, \mathcal{Q})$-valued measure $\theta_0$, then*

$$\int_F^{(w)} f\, d\theta = \sup \left\{ \int_F h\, d\theta \;\middle|\; h \in \mathcal{S}_{\mathfrak{R}}(X, \mathcal{P}_0),\;\; h \leq f + \mathfrak{v}_w \right\}.$$

We proceed with a simple observation for step functions.

**Lemma 4.6.** *Let $F \in \mathfrak{A}_{\mathfrak{R}}$, $f \in \mathcal{S}_{\mathfrak{R}}(X, \mathcal{P})$ and $w \in \mathcal{W}$. Then*

$$\int_F f\, d\theta \leq \int_F^{(w)} f\, d\theta \leq \int_F f\, d\theta + w.$$

*Proof.* The first part of the inequality is trivial. For the second part, let $h \leq f + \mathfrak{v}_w$ for $h \in \mathcal{S}_{\mathfrak{R}}(X, \mathcal{P})$, that is $h \leq f + s$ for some $\mathcal{V}$-valued step function $s \in \mathfrak{v}_w$. Following Lemma 4.3(b) and (c), this implies

$$\int_F h\, d\theta \leq \int_F f\, d\theta + \int_F s\, d\theta \leq \int_F f\, d\theta + w$$

for each such step function $h \in \mathcal{S}_{\mathfrak{R}}(X, \mathcal{P})$, hence $\int_F^{(w)} f\, d\theta \leq \int_F f\, d\theta + w$ as claimed. □

**Proposition 4.7.** *Let $E \in \mathfrak{R}$ and $f \in \mathcal{F}_{\mathfrak{R}}(X, \mathcal{P})$ and let $(h_n)_{n \in \mathbb{N}}$ be a bounded below sequence of step functions in $\mathcal{S}_{\mathfrak{R}}(X, \mathcal{P})$ such that for every $x \in E$ there is $n_0 \in \mathbb{N}$ such that $f(x) \leq h_n(x)$ for all $n \geq n_0$. Then*

$$\int_E^{(w)} f\, d\theta \leq \underline{\lim}_{n \to \infty} \int_E h_n\, d\theta + w$$

*for every $w \in \mathcal{W}$.*

*Proof.* Let $E \in \mathfrak{R}$, $f \in \mathcal{F}_{\mathfrak{R}}(X, \mathcal{P})$, and let $(h_n)_{n \in \mathbb{N}}$ be a sequence of step functions satisfying our assumptions. For $w \in \mathcal{W}$ let $l \in \mathcal{S}_{\mathfrak{R}}(X, \mathcal{P})$ such that $l \leq f + \mathfrak{v}_w$, that is $l \leq f + s$ for some $s \in \mathfrak{v}_w$. Now we set

$$E_n = \{x \in E \mid l(x) \leq h_m(x) + s(x) \quad \text{for all} \quad m \geq n\}.$$

All the sets $E_n$ are measurable, $E_n \subset E_{n+1}$ and $E = \bigcup_{n\in\mathbb{N}} E_n$ by our assumption. Given any $u \in \mathcal{W}$ there is $v \in \mathcal{V}$ such that $\theta_E(v) \le u$, and as the sequence $(h_n)_{n\in\mathbb{N}}$ is bounded below, there is $\rho \ge 0$ such that $0 \le h_n + \rho\chi_{X}\otimes v$ for all $n \in \mathbb{N}$. Thus

$$\begin{aligned}\chi_{E_n}\otimes l &\le \chi_{E_n}\otimes(h_n + s) + \chi_{(E\setminus E_n)}(h_n + \rho\chi_X\otimes v)\\ &\le \chi_E\otimes h_n + \chi_{E_n}\otimes s + \rho\chi_{(E\setminus E_n)}\otimes v.\end{aligned}$$

Hence by Lemma 4.3, and because

$$\int_X \chi_{E_n} s\, d\theta \le \int_X s\, d\theta \le w, \qquad \text{and} \qquad \int_X \chi_{(E\setminus E_n)}\otimes v = \theta_{(E\setminus E_n)}(v),$$

when taking the integrals over $X$ in the above inequality, we obtain

$$\int_{E_n} l\, d\theta \le \int_E h_n\, d\theta + \rho\theta_{(E\setminus E_n)}(v) + w.$$

Because $E_n \subset E_{n+1}$ and $\bigcup_{n\in\mathbb{N}} E_n = E$, Lemma 3.1(a) yields

$$\theta_{(F\cap E)}(a) = \lim_{n\to\infty} \theta_{(F\cap E_n)}(a)$$

for all $F \in \mathfrak{R}$ and $a \in \mathcal{P}$. Considering the definition of the integral for a step function in 4.2, this renders

$$\lim_{n\to\infty} \int_{E_n} l\, d\theta = \int_E l\, d\theta,$$

and Lemma 3.1(b) yields

$$\overline{\lim_{n\to\infty}}\, \theta_{(E\setminus E_n)}(v) \le \mathfrak{O}(\theta_E(v)) \le \varepsilon' u$$

for all $\varepsilon' \ge 0$. Thus, using the limit rules from Lemma I.5.19, we obtain

$$\int_E l\, d\theta \le \underline{\lim_{n\to\infty}} \int_E h_n\, d\theta + w + \varepsilon' u.$$

Because $u \in \mathcal{W}$ and $\varepsilon' > 0$ were arbitrary, and because $\mathcal{Q}$ carries the weak preorder, this shows

$$\int_E l\, d\theta \le \underline{\lim_{n\to\infty}} \int_E h_n\, d\theta + w.$$

Our claim follows, since the above inequality holds true for all step functions $l \in \mathcal{S}_{\mathfrak{R}}(X, \mathcal{P})$ such that $l \le f + \mathfrak{v}_w$. $\square$

**Corollary 4.8.** *Let* $F \in \mathfrak{A}_{\mathfrak{R}}$, $f \in \mathcal{F}_{\mathfrak{R}}(X, \mathcal{P})$ *and* $u, w \in \mathcal{W}$. *Then*

$$\int_F^{(w)} f\,d\theta \leq \int_F^{(u)} f\,d\theta + w.$$

*Proof.* Let $F \in \mathfrak{A}_{\mathfrak{R}}$, $f \in \mathcal{F}_{\mathfrak{R}}(X, \mathcal{P})$ and $u, w \in \mathcal{W}$. Let $l \in \mathcal{S}_{\mathfrak{R}}(X, \mathcal{P})$ such that $l \leq f + \mathfrak{v}_w$, and according to Lemma 2.4, we choose $E \in \mathfrak{R}$ such that both $h$ is supported by $E$ and such that $f(x) \geq 0$ for all $x \in X \setminus E$. For the set $E \cap F \in \mathfrak{R}$, the inductive limit neighborhood $\mathfrak{v} = \mathfrak{v}_u$ and $\varepsilon \geq 0$, let $(h_n)_{n \in \mathbb{N}}$ be a sequence of step functions in $\mathcal{S}_{\mathfrak{R}}(X, \mathcal{P})$ approaching $f$ as in Corollary 2.8. We may assume that the functions $h_n$ are supported by the set $E \cap F$, since we may otherwise replace them by their product with the characteristic function of this set. Proposition 4.7 yields

$$\int_{(E \cap F)}^{(w)} f\,d\theta \leq \varliminf_{n \to \infty} \int_{(E \cap F)} h_n\,d\theta + w.$$

On the other hand, we have

$$\int_F l\,d\theta = \int_{(E \cap F)} l\,d\theta \leq \int_{(E \cap F)}^{(w)} f\,d\theta,$$

since the function $l$ is supported by $E$. Similarly, for the functions $h_n$ we observe that

$$\int_{(E \cap F)} h_n\,d\theta = \int_F h_n\,d\theta \leq \gamma \int_F^{(u)} f\,d\theta,$$

since $h_n \leq \gamma f + \mathfrak{v}_w$ for all $n \in \mathbb{N}$. Combining all of the above then yields

$$\int_F l\,d\theta \leq \gamma \int_F^{(u)} f\,d\theta + w$$

with some $1 \leq \gamma \leq 1 + \varepsilon$, and indeed

$$\int_F l\,d\theta \leq \int_F^{(u)} f\,d\theta + w,$$

since $\varepsilon > 0$ was chosen independently. Finally, because this last inequality holds true for all $l \in \mathcal{S}_{\mathfrak{R}}(X, \mathcal{P})$ such that $h \leq f + \mathfrak{v}_w$, our claim follows. □

**4.9 Integrals for Functions in $\mathcal{F}_{\mathfrak{R}}(X, \mathcal{P})$.** We may now define the *integral over a set* $F \in \mathfrak{A}_{\mathfrak{R}}$ *for a function* $f \in \mathcal{F}_{\mathfrak{R}}(X, \mathcal{P})$ as

$$\int_F f\,d\theta = \inf_{w \in \mathcal{W}} \int_F^{(w)} f\,d\theta.$$

The above infimum is well-defined and yields an element of the locally convex complete lattice cone $\mathcal{Q}$. Indeed, given any neighborhood $u \in \mathcal{W}$ there is $\lambda \geq 0$ such that $0 \leq f + \lambda \mathfrak{v}_u$. Thus $0 \leq \int_F^{(\lambda u)} f\,d\theta$. According to Corollary 4.8, this yields

$$0 \leq \int_F^{(\lambda u)} f\,d\theta \leq \int_F^{(w)} f\,d\theta + \lambda u.$$

for all $w \in \mathcal{V}$. This demonstrates that the set $\big\{ \int_F^{(w)} f\,d\theta \mid w \in \mathcal{W}\big\}$ is bounded below, and its infimum exists by $(\bigwedge 1)$. Moreover, our earlier observation in Lemma 4.5 justifies that the above definition of the integral is consistent with the preceding one for step functions. Obviously, the integral is monotone, and we shall proceed to verify that it determines a continuous linear operator from $\mathcal{F}_{\mathfrak{R}}(X,\mathcal{P})$ into $\mathcal{Q}$. For Part (a) of the following lemma, recall from Lemma 2.6 that $\chi_F \circledast f \in \mathcal{F}_{\mathfrak{R}}(X,\mathcal{P})$ whenever $f \in \mathcal{F}_{\mathfrak{R}}(X,\mathcal{P})$ and $F \in \mathfrak{A}_{\mathfrak{R}}$. In Part (b) we consider $\mathfrak{R}$ as the index set of a net, directed upward by set inclusion.

**Lemma 4.10.** *Let $f \in \mathcal{F}_{\mathfrak{R}}(X,\mathcal{P})$ and $F \in \mathfrak{A}_{\mathfrak{R}}$. Then*

*(a) $\int_F f\,d\theta = \int_X (\chi_F \circledast f)\,d\theta$.*
*(b) $\int_F f\,d\theta = \lim\limits_{E \in \mathfrak{R}} \int_{(E\cap F)} f\,d\theta$.*

*Proof.* For Part (a) we first note that $\chi_F \circledast f \in \mathcal{F}_{\mathfrak{R}}(X,\mathcal{P})$ (see Lemma 2.6). Let $w \in \mathcal{W}$ and $h_0 \in \mathcal{S}_{\mathfrak{R}}(X,\mathcal{P})$ such that $h_0 \leq f + \mathfrak{v}_w$. We have

$$\int_F^{(w)} f\,d\theta = \sup\left\{ \int_F h\,d\theta \;\middle|\; h \in \mathcal{S}_{\mathfrak{R}}(X,\mathcal{P}),\;\; h \leq f + \mathfrak{v}_w \right\}$$

and

$$\int_X^{(w)} \chi_F \circledast f\,d\theta = \sup\left\{ \int_X h'\,d\theta \;\middle|\; h' \in \mathcal{S}_{\mathfrak{R}}(X,\mathcal{P}),\;\; h' \leq \chi_F \circledast f + \mathfrak{v}_w \right\}$$

First, let $h \in \mathcal{S}_{\mathfrak{R}}(X,\mathcal{P})$ such that $h \leq f + \mathfrak{v}_w$. Then $h' = \chi_F \circledast h \leq \chi_F \circledast f + \mathfrak{v}_w$, and $\int_X h'\,d\theta = \int_F h\,d\theta$ by 4.3(d). This shows

$$\int_F^{(w)} f\,d\theta \leq \int_X^{(w)} \chi_F \circledast f\,d\theta.$$

For the converse inequality, let $h' \in \mathcal{S}_{\mathfrak{R}}(X,\mathcal{P})$ such that $h' \leq \chi_F \circledast f + \mathfrak{v}_w$. Then $\chi_F \circledast h' \leq \chi_F \circledast f + \mathfrak{v}_w$ and $\chi_{(X\setminus F)} \circledast h_0 \leq \chi_{(X\setminus F)} \circledast f + \mathfrak{v}_w$, hence $h = \chi_F \circledast h' + \chi_{(X\setminus F)} \circledast h_0 \leq f + 2\mathfrak{v}_w$, and $\int_F h\,d\theta = \int_F h'\,d\theta$. As $\chi_{(X\setminus F)} \circledast h' \leq \mathfrak{v}_w$, we have $\int_{(X\setminus F)} h'\,d\theta \leq w$, hence $\int_X h'\,d\theta \leq \int_F h\,d\theta + w$. This shows

$$\int_X^{(w)} \chi_F \circledast f\,d\theta \leq \int_F^{(2w)} f\,d\theta + w.$$

Taking the infima over all $w \in \mathcal{W}$ in the above inequality yields Part (a).

For Part (b) it is therefore sufficient to consider the case $F = X$, because the function $f$ may be replaced by its product with the characteristic function $\chi_F$. Let $E_0 \in \mathfrak{R}$ such that $f(x) \geq 0$ for all $x \in X \setminus E_0$. Then $\chi_{E} \otimes f \leq \chi_{E'} \otimes f$ whenever $E_0 \subset E \subset E'$ for $E, E' \in \mathfrak{R}$, hence $\int_E f\,d\theta \leq \int_{E'} f\,d\theta$ by Part (a) and the monotony of the integral. This shows

$$\lim_{E \in \mathfrak{R}} \int_E f\,d\theta = \sup_{E_0 \subset E \in \mathfrak{R}} \int_E f\,d\theta \leq \int_X f\,d\theta.$$

For the converse, let $w \in \mathcal{W}$ and $h \leq f + \mathfrak{v}_w$ for $h \in \mathcal{S}_{\mathfrak{R}}(X, \mathcal{P})$. Because $h$ is supported by a set in $\mathfrak{R}$, there is $E_0 \subset E \in \mathfrak{R}$ such that $\int_X h\,d\theta = \int_E h\,d\theta \leq \int_E^{(w)} f\,d\theta$. Moreover, Corollary 4.8 shows that $\int_E^{(w)} f\,d\theta \leq \int_E f\,d\theta + w$. Thus

$$\int_X h\,d\theta \leq \sup_{E_0 \subset E \in \mathfrak{R}} \int_E f\,d\theta + w.$$

This shows

$$\int_X f\,d\theta \leq \int_X^{(w)} f\,d\theta \leq \sup_{E_0 \subset E \in \mathfrak{R}} \int_E f\,d\theta + w,$$

hence our claim, since $w \in \mathcal{W}$ was arbitrary and $\mathcal{Q}$ carries the weak preorder. □

**4.11 Sets of Measure Zero and Properties Holding Almost Everywhere.** A set $Z \in \mathfrak{A}_{\mathfrak{R}}$ is said to be *of measure zero (with respect to* $\theta$*)* if $\theta_{(E \cap Z)} = 0$ for all $E \in \mathfrak{R}$. The family $\mathfrak{Z}(\theta)$ of all sets of measure zero is obviously closed for set complements and for countable unions. For a subset $F$ of $X$ we shall say that a pointwise defined property of functions on $X$ holds *$\theta$-almost everywhere on* $F$ if it holds on $F \setminus Z$ with some $Z \in \mathfrak{Z}(\theta)$. In particular, we shall use the symbols $\underset{a.e.F}{\leq}$ or $\underset{a.e.F}{=}$ if the relations $\leq$ or $=$ hold $\theta$-almost everywhere on the set $F$, respectively; that is for example, $f \underset{a.e.F}{\leq} g + \mathfrak{v}$ for functions $f, g \in \mathcal{F}(X, \mathcal{P})$ and an inductive limit neighborhood $\mathfrak{v}$ means that $\chi_{(F \setminus Z)} \otimes f \leq \chi_{(F \setminus Z)} \otimes g + \mathfrak{v}$ holds with some $Z \in \mathfrak{Z}(\theta)$. These relations are of course transitive and compatible with the algebraic operations.

As $\theta_{(E \cap Z)} = 0$ holds for all $E \in \mathfrak{R}$ and $Z \in \mathfrak{Z}(\theta)$, we infer that $\theta_E = \theta_{(E \setminus Z)}$. Now Definition 4.2 yields that $\int_F h\,d\theta = \int_{(F \setminus Z)} h\,d\theta$ for all step functions $h \in \mathcal{S}_{\mathfrak{R}}(X, \mathcal{P})$, $F \in \mathfrak{A}_{\mathfrak{R}}$ and $Z \in \mathfrak{Z}(\theta)$. Considering our definition of the integral in 4.9 we observe that this yields $\int_F f\,d\theta = \int_{(F \setminus Z)} f\,d\theta$ for all $f \in \mathcal{F}_{\mathfrak{R}}(X, \mathcal{P})$ as well. Consequently, $f \underset{a.e.F}{\leq} g$ for functions $f, g \in \mathcal{F}_{\mathfrak{R}}(X, \mathcal{P})$ implies that $\chi_{(F \setminus Z)} \otimes f \leq \chi_{(F \setminus Z)} \otimes g$ for some $Z \in \mathfrak{Z}(\theta)$, hence

$$\int_F f\,d\theta = \int_{(F \setminus Z)} f\,d\theta \leq \int_{(F \setminus Z)} g\,d\theta = \int_F g\,d\theta.$$

In particular, any two functions in $\mathcal{F}_{\mathfrak{R}}(X,\mathcal{P})$ that coincide $\theta$-almost everywhere on a set $F \in \mathfrak{A}_{\mathfrak{R}}$ have the same integrals over $F$ with respect to $\theta$.

**4.12 Integrability over a Set $E \in \mathfrak{R}$.** We may now define integrability for cone-valued functions over measurable sets with respect to an operator-valued measure. First, for a set $E \in \mathfrak{R}$ we shall say that a function $f \in \mathcal{F}(X,\mathcal{P})$ is *integrable over $E$ with respect to $\theta$* if for every $w \in \mathcal{W}$ and $\varepsilon > 0$ there are functions $f_{(w,\varepsilon)} \in \mathcal{F}_{\mathfrak{R}}(X,\mathcal{P})$ and $s_{(w,\varepsilon)} \in \mathcal{F}_{\mathfrak{R}}(X,\mathcal{V})$ such that

$$f \underset{a.e.E}{\leq} f_{(w,\varepsilon)} \underset{a.e.E}{\leq} \gamma f + s_{(w,\varepsilon)} \qquad \text{and} \qquad \int_E s_{(w,\varepsilon)}\, d\theta \leq \varepsilon w$$

for some $1 \leq \gamma \leq 1+\varepsilon$. Recall that the functions in $\mathcal{F}_{\mathfrak{R}}(X,\mathcal{V})$ are actually $(\mathcal{V} \cup \{0\})$-valued. However, in the case of Definition 4.12, without loss of generality we may assume that the function $s_{(w,\varepsilon)} \in \mathcal{F}_{\mathfrak{R}}(X,\mathcal{V})$ is indeed $\mathcal{V}$-valued, as we can otherwise replace it by a function $\tilde{s}_{(w,\varepsilon)} = s_{(w,(\varepsilon/2))} + \chi_{X}\otimes v$, where $v \in \mathcal{V}$ is such that $\theta_E(v) \leq (\varepsilon/2)w$, hence $\int_E \tilde{s}_{(w,\varepsilon)}\, d\theta \leq \varepsilon w$.

Consequently, for an integrable function $f \in \mathcal{F}(X,\mathcal{P})$ and a net $(f_{(w,\varepsilon)})_{w\in\mathcal{W}}^{\varepsilon>0}$ a of functions in $\mathcal{F}_{\mathfrak{R}}(X,\mathcal{P})$ satisfying the above, we shall show that the limit

$$\int_E f\, d\theta = \lim_{\substack{\varepsilon > 0 \\ w \in \mathcal{W}}} \int_E f_{(w,\varepsilon)}\, d\theta$$

exists and is independent of the particular choice for the net $(f_{(w,\varepsilon)})_{w\in\mathcal{W}}^{\varepsilon>0}$. (The index set for this net is $\mathcal{W} \times \{\varepsilon > 0\}$ with the reverse componentwise order.) Indeed, given $w \in \mathcal{W}$ and $\varepsilon > 0$, for all $w_1, w_2 \in \mathcal{W}$ such that $w_1, w_2 \leq w$ and $0 < \varepsilon_1, \varepsilon_2 \leq \varepsilon$ we have

$$f_{(w_1,\varepsilon_1)} \leq \gamma f + s_{(w_1,\varepsilon_1)} \leq \gamma_1 f_{(w_2,\varepsilon)} + s_{(w_1,\varepsilon_1)}$$

for some $1 \leq \gamma \leq 1+\varepsilon$, hence

$$\int_E f_{(w_1,\varepsilon_1)} \leq \gamma_1 \int_E f_{(w_2,\varepsilon_2)} + \varepsilon w.$$

Thus $\left(\int_E f_{(w,\varepsilon)}\right)_{w\in\mathcal{W}}^{\varepsilon>0}$ forms a Cauchy net in the symmetric relative topology of $\mathcal{Q}$, hence is convergent by Proposition I.5.41. The preceding argument together with Lemma I.5.20(c) also shows that this limit is independent of the particular choice for the net $(f_{(w,\varepsilon)})_{w\in\mathcal{W}}^{\varepsilon>0}$.

**4.13 Integrability over a Set $F \in \mathfrak{A}_{\mathfrak{R}}$.** Obviously, integrability in the sense of 4.12 for a function $f \in \mathcal{F}_{\mathfrak{R}}(X,\mathcal{P})$ over a set $E \in \mathfrak{R}$ implies integrability over all subsets $G \in \mathfrak{R}$ of $E$. This observation, together with Lemma 4.10 shows that we may consistently define integrability over sets in the $\sigma$-algebra $\mathfrak{A}_{\mathfrak{R}}$ in the following way: We shall say that a function $f \in \mathcal{F}(X,\mathcal{P})$ is *integrable over $F \in \mathfrak{A}_{\mathfrak{R}}$ with respect to $\theta$* if $f$ is integrable over the sets $E \cap F$ for all $E \in \mathfrak{R}$ and if the limit

$$\int_F f\,d\theta = \lim_{E\in\mathfrak{R}} \int_{(E\cap F)} f\,d\theta$$

exists in $\mathcal{Q}$. The set of all functions in $\mathcal{F}(X,\mathcal{P})$ that are integrable over $F$ shall be denoted by $\mathcal{F}_{(F,\theta)}(X,\mathcal{P})$. Lemma 4.10(b) implies that $\mathcal{F}_{\mathfrak{R}}(X,\mathcal{P}) \subset \mathcal{F}_{(F,\theta)}(X,\mathcal{P})$ for every $F \in \mathfrak{A}_R$ and every $\mathfrak{L}(\mathcal{P},\mathcal{Q})$-valued measure $\theta$ on $\mathfrak{R}$.

We may use this definition of integrability also for functions that take the value $\infty \in \overline{\mathcal{V}}$ on a set of measure zero (see Section 2.1).

**Theorem 4.14.** *Let $F \in \mathfrak{A}_{\mathfrak{R}}$. Then $\mathcal{F}_{(F,\theta)}(X,\mathcal{P})$ is a subcone of $\mathcal{F}(X,\mathcal{P})$ containing $\mathcal{F}_{\mathfrak{R}}(X,\mathcal{P})$. More precisely, for $f,g \in \mathcal{F}_{(F,\theta)}(X,\mathcal{P})$ and $0 \le \alpha \in \mathbb{R}$ we have*

*(a)* $\int_F (\alpha f)\,d\theta = \alpha \int_F f\,d\theta$
*(b)* $\int_F (f+g)\,d\theta = \int_F f\,d\theta + \int_F g\,d\theta$
*(c)* $\int_F f\,d\theta \le \int_F g\,d\theta$ *whenever* $f \underset{a.e.F}{\le} g$.

*Proof.* In a first case, let us assume that $f,g \in \mathcal{F}_{\mathfrak{R}}(X,\mathcal{P})$ and that $F = E \in \mathfrak{R}$. Then Part (a) follows trivially from our definition of the integral. For (b), let $w \in \mathcal{W}$, and $h_1 \le f + \mathfrak{v}_w$ and $h_2 \le g + \mathfrak{v}_w$ for $h_1, h_2 \in \mathcal{S}_{\mathfrak{R}}(X,\mathcal{P})$. Then $h_1 + h_2 \le (f+g) + 2\mathfrak{v}_w$, hence

$$\int_E^{(w)} f\,d\theta + \int_E^{(w)} g\,d\theta \le \int_E^{(2w)} (f+g)\,d\theta,$$

and therefore

$$\int_E f\,d\theta + \int_F g\,d\theta \le \int_E (f+g)\,d\theta.$$

For the converse inequality, let $u \in \mathcal{W}$. For the set $E \in \mathfrak{R}$ the inductive limit neighborhood $\mathfrak{v}_u$ and any $\varepsilon > 0$ choose sequences $(h_n)_{n\in\mathbb{N}}$ and $(l_n)_{n\in\mathbb{N}}$ of step functions in $\mathcal{S}_{\mathfrak{R}}(X,\mathcal{P})$ approaching $f$ and $g$ as in Corollary 2.8, respectively. The sequence $(k_n)_{n\in\mathbb{N}}$, where $k_n = h_n + l_n$ then approaches the function $f+g$ with respect to $F$, the inductive limit neighborhood $2\mathfrak{v}_u$ and $\varepsilon$. Thus by Proposition 4.7 we have

$$\begin{aligned}\int_E^{(w)} (f+g)\,d\theta &\le \underline{\lim}_{n\to\infty} \int_E k_n\,d\theta + w \\ &\le \overline{\lim}_{n\to\infty} \int_E h_n\,d\theta + \overline{\lim}_{n\to\infty} \int_E l_n\,d\theta + w \\ &\le \int_E^{(u)} f\,d\theta + \int_E^{(u)} g\,d\theta + w\end{aligned}$$

for all $w \in \mathcal{W}$. This yields

$$\int_E (f+g)\,d\theta \le \int_E^{(u)} f\,d\theta + \int_E^{(u)} g\,d\theta,$$

since $\mathcal{Q}$ is a locally convex complete lattice cone, and indeed

$$\int_E (f+g)\,d\theta \le \int_E f\,d\theta + \int_E f\,d\theta$$

after applying the infima over all $u \in \mathcal{W}$ on the right-hand side and using the rules from Section I.5.

Now in a second case, we still suppose that $F = E \in \mathfrak{R}$, and let $f, g \in \mathcal{F}_{(E,\theta)}(X,\mathcal{P})$. Let $(f_{(w,\varepsilon)})_{w\in\mathcal{W}}^{\varepsilon>0}$ and $(g_{(w,\varepsilon)})_{w\in\mathcal{W}}^{\varepsilon>0}$ be nets of functions in $\mathcal{F}_{\mathfrak{R}}(X,\mathcal{P})$ approaching the functions $f$ and $g$ as in 4.12. Then the nets $(\alpha f_{(w,\varepsilon)})_{w\in\mathcal{W}}^{\varepsilon>0}$ and $(f_{(w,\varepsilon)}+g_{(w,\varepsilon)})_{w\in\mathcal{W}}^{\varepsilon>0}$ approach the functions $\alpha f$ and $f+g$, respectively, and the limit rules from Section 4 yield

$$\int_E \alpha f\,d\theta = \lim_{\substack{\varepsilon>0\\ w\in\mathcal{W}}} \int_E \alpha f_{(w,\varepsilon)}\,d\theta = \alpha \int_E f_{(w,\varepsilon)}\,d\theta = \alpha \int_E f\,d\theta$$

and

$$\begin{aligned}\int_E (f+g)\,d\theta &= \lim_{\substack{\varepsilon>0\\ w\in\mathcal{W}}} \int_E \big(f_{(w,\varepsilon)}+g_{(w,\varepsilon)}\big)\,d\theta \\ &= \lim_{\substack{\varepsilon>0\\ w\in\mathcal{W}}} \int_E f_{(w,\varepsilon)}\,d\theta + \lim_{\substack{\varepsilon>0\\ w\in\mathcal{W}}} \int_E g_{(w,\varepsilon)}\,d\theta \\ &= \int_E f\,d\theta + \int_E g\,d\theta.\end{aligned}$$

For Part (c) in this case, suppose that $f \underset{\text{a.e.}E}{\lesssim} g$ and let $(f_{(w,\varepsilon)})_{w\in\mathcal{W}}^{\varepsilon>0}$ and $(g_{(w,\varepsilon)})_{w\in\mathcal{W}}^{\varepsilon>0}$ be nets in $\mathcal{F}_{\mathfrak{R}}(X,\mathcal{P})$ as before. Then

$$f_{(w,\varepsilon)} \underset{\text{a.e.}E}{\lesssim} \gamma f + s_{(w,\varepsilon)} \underset{\text{a.e.}E}{\lesssim} \gamma g + s_{(w,\varepsilon)} \underset{\text{a.e.}E}{\lesssim} \gamma g_{(w,\varepsilon)} + s_{(w,\varepsilon)},$$

hence

$$\int_E f_{(w,\varepsilon)}\,d\theta \le \gamma \int_E g_{(w,\varepsilon)}\,d\theta + \varepsilon w$$

with some $1 \le \gamma \le 1+\varepsilon$ for all $w \in \mathcal{W}$ and $\varepsilon > 0$. According to the limit rules in Section I.5, this yields

$$\int_E f\,d\theta = \lim_{\substack{\varepsilon>0\\ w\in\mathcal{W}}} \int_E f_{(w,\varepsilon)}\,d\theta \le \lim_{\substack{\varepsilon>0\\ w\in\mathcal{W}}} \int_E g_{(w,\varepsilon)}\,d\theta = \int_E g\,d\theta.$$

For the final and general case, let $F \in \mathfrak{A}_{\mathfrak{R}}$ and $f, g \in \mathcal{F}_{(F,\theta)}(X,\mathcal{P})$. Then the claims of Parts (a),(b) and (c) hold for integrals over all sets $E \cap F$ for $E \in \mathfrak{R}$. The definition of the respective integrals over $F$ together with the limit rules from Lemma I.5.19 yield the validity of these claims for the integrals over $F$ as well. □

Simple examples can show that $F \subset G$ for $F, G \in \mathfrak{A}_{\mathfrak{R}}$ does not necessarily imply that $\mathcal{F}_{(F,\theta)}(X,\mathcal{P}) \subset \mathcal{F}_{(G,\theta)}(X,\mathcal{P})$, but we have the following:

**Proposition 4.15.** *Let $f \in \mathcal{F}(X,\mathcal{P})$ and $F, G \in \mathfrak{A}_{\mathfrak{R}}$*

*(a) If $F, G \in \mathfrak{A}_{\mathfrak{R}}$, then $f$ is integrable over $F \cap G$ if and only if $\chi_{G} \circledast f$ is integrable over $F$, if and only if $\chi_{F} \circledast f$ is integrable over $G$. In this case we have $\int_{(F\cap G)} f\,d\theta = \int_F \chi_{G} \circledast f\,d\theta = \int_G \chi_{F} \circledast f\,d\theta$.*

*(b) If $F$ and $G$ are disjoint and $f$ is integrable over $F$ and $G$, then $f$ is integrable over $F \cup G$ and $\int_{(F\cup G)} f\,d\theta = \int_F f\,d\theta + \int_G f\,d\theta$.*

*(c) If $F \subset G$ and $f$ is integrable over $F, G$ and $G \setminus F$, then $\mathfrak{O}\left(\int_F f\,d\theta\right) \leq \mathfrak{O}\left(\int_G f\,d\theta\right)$.*

*Proof.* For Part (a), in a first step let $E \in \mathfrak{R}$. First we observe from Definition 4.12 that a function $f \in \mathcal{F}(X,\mathcal{P})$ is integrable over $E$ if and only if $\chi_E \circledast f$ is integrable over $E$ and that $\int_E f\,d\theta = \int_E \chi_E \circledast f\,d\theta$. Thus, if for $f \in \mathcal{F}(X,\mathcal{P})$, the function $\chi_E \circledast f$ is integrable over $X$, then by Definition 4.13 the function $\chi_E \circledast f$ and therefore $f$ is integrable over $E$. For the converse, assume that $f \in \mathcal{F}(X,\mathcal{P})$ is integrable over $E$. Let $E' \in \mathfrak{R}$, $w \in \mathcal{W}$ and $\varepsilon > 0$. According to 4.12 there are $f_{(w,\varepsilon)} \in \mathcal{F}_{\mathfrak{R}}(X,\mathcal{P})$ and $s_{(w,\varepsilon)} \in \mathcal{F}_{\mathfrak{R}}(X,\mathcal{V})$ such that $f \underset{a.e.E}{\leq} f_{(w,\varepsilon)} \underset{a.e.E}{\leq} \gamma f + s_{(w,\varepsilon)}$ with some $1 \leq \gamma \leq 1+\varepsilon$ and $\int_E s_{(w,\varepsilon)}\,d\theta \leq \varepsilon w$. Then we have

$$\chi_E \circledast f \underset{a.e.E'}{\leq} \chi_E \circledast f_{(w,\varepsilon)} \underset{a.e.E'}{\leq} \gamma f + \chi_E \circledast s_{(w,\varepsilon)}$$

and

$$\int_{E'} \chi_E \circledast s_{(w,\varepsilon)}\,d\theta \leq \varepsilon w$$

as well. Because the functions $\chi_E \circledast f_{(w,\varepsilon)}$ and $\chi_E \circledast s_{(w,\varepsilon)}$ are also contained in $\mathcal{F}_{\mathfrak{R}}(X,\mathcal{P})$ and $\mathcal{F}_{\mathfrak{R}}(X,\mathcal{V})$, respectively, we conclude that the function $\chi_E \circledast f$ is integrable over $E'$ and that

$$\int_{E'} \chi_E \circledast f\,d\theta = \lim_{\substack{\varepsilon > 0 \\ w \in \mathcal{W}}} \int_{E'} \chi_E \circledast f_{(w,\varepsilon)}\,d\theta = \lim_{\substack{\varepsilon > 0 \\ w \in \mathcal{W}}} \int_X \chi_{(E'\cap E)} \circledast f_{(w,\varepsilon)}\,d\theta.$$

The last equality follows from Lemma 4.10(a). The above holds for all sets $E' \in \mathfrak{R}$, hence using Definition 4.13, we realize that the function $\chi_E \circledast f$ is indeed integrable over $X$ and that

$$\int_X \chi_E \circledast f\,d\theta = \lim_{E' \in \mathfrak{R}} \int_{E'} \chi_E \circledast f\,d\theta = \lim_{\substack{\varepsilon > 0 \\ w \in \mathcal{W}}} \int_X \chi_E \circledast f_{(w,\varepsilon)}\,d\theta$$

$$= \lim_{\substack{\varepsilon > 0 \\ w \in \mathcal{W}}} \int_E f_{(w,\varepsilon)}\,d\theta = \int_E f\,d\theta$$

holds. Thus we have verified that a function $f \in \mathcal{F}(X,\mathcal{P})$ is integrable over a set $E \in \mathfrak{R}$ if and only if $\chi_{E} \otimes f$ is integrable over $X$ and that $\int_E f\, d\theta = \int_X \chi_{E} \otimes f\, d\theta$ in this case. Now in a second step, let $F \in \mathfrak{A}_{\mathfrak{R}}$ and $f \in \mathcal{F}(X,\mathcal{P})$. By the above, the function $f$ is integrable over all sets $E \cap F$ for $E \in \mathfrak{R}$, if and only if all the functions $\chi_{(E\cap F)} \otimes f = \chi_{E} \otimes (\chi_{F} \otimes f)$ are integrable over $X$. In this case

$$\int_{(E\cap F)} f\, d\theta = \int_X \chi_{(E\cap F)} \otimes f\, d\theta = \int_X \chi_{E} \otimes (\chi_{F} \otimes f)\, d\theta = \int_E \chi_{F} \otimes f\, d\theta$$

holds by our first step. According to Definition 4.13 therefore $f$ is integrable over $F$ if and only $\chi_{F} \otimes f$ is integrable over $X$ and

$$\int_F f\, d\theta = \lim_{E \in \mathfrak{R}} \int_{(E\cap F)} f\, d\theta = \lim_{E \in \mathfrak{R}} \int_E \chi_{F} \otimes f\, d\theta = \int_X \chi_{F} \otimes f\, d\theta.$$

In a third and final step for Part (a), let $F, G \in \mathfrak{A}_{\mathfrak{R}}$. From the preceding we conclude that $\chi_{G} \otimes f$ is integrable over $F$ if and only if $\chi_{F} \otimes (\chi_{G} \otimes f) = \chi_{(F\cap G)} \otimes f$ is integrable over $X$, that is $f$ is integrable over $F \cap G$, and all the integrals coincide.

For Part (b), suppose that $F \cap G = \emptyset$ and that $f$ is integrable over both $F$ and $G$. Then both functions $\chi_{F} \otimes f$ and $\chi_{G} \otimes f$ are integrable over $X$ by Part (a), hence $\chi_{(F\cup G)} \otimes f = \chi_{F} \otimes f + \chi_{G} \otimes f$ is also integrable over $X$ by Theorem 4.14(b). Thus $f$ is indeed integrable over $F \cup G$ and

$$\int_{(F\cup G)} f\, d\theta = \int_X \chi_{(F\cup G)} \otimes f = \int_X \chi_{F} \otimes f + \int_X \chi_{G} \otimes f = \int_F f\, d\theta + \int_G f\, d\theta$$

by 4.14(b)

For Part (c), suppose that $F \subset G$ and that $f$ is integrable over $F, G$ and $G \setminus F$. Then

$$\int_G f\, d\theta = \int_F f\, d\theta + \int_{(G\setminus F)} f\, d\theta$$

by Part (b), and

$$\mathfrak{O}\left(\int_F f\, d\theta\right) \leq \mathfrak{O}\left(\int_F f\, d\theta\right) + \mathfrak{O}\left(\int_{(G\setminus F)} f\, d\omega\right) = \mathfrak{O}\left(\int_G f\, d\theta\right).$$

by Proposition I.5.11(a). □

**Proposition 4.16.** *Let $f, g \in \mathcal{F}_{(E,\theta)}(X,\mathcal{P})$ for $E \in \mathfrak{R}$ and let $v \in \mathcal{V}$. If $f(x) \preccurlyeq_v g(x)$ holds $\theta$-almost everywhere on $E$, then $\int_E f\, d\theta \leq \int_E g\, d\theta + \mathfrak{O}(\theta_E(v))$.*

*Proof.* Let $E \in \mathfrak{R}$, let $v \in \mathcal{V}$ and $f, g \in \mathcal{F}_{(E,\theta)}(X,\mathcal{P})$ such that $f(x) \preccurlyeq_v g(x)$ $\theta$-almost everywhere on $E$. In a first case, let us assume in addition that $g \in \mathcal{F}_{\mathfrak{R}}(X,\mathcal{P})$. Lemma 2.4(b) implies that there is $\lambda \geq 0$ such that

$0 \le g(x) + \lambda v$ for all $x \in E$. Recall from Section 2 that $f(x) \preccurlyeq_v g(x)$ means that $f(x) \in v_\varepsilon\big(g(x)\big)$ for all $\varepsilon > 0$. In turn, $f(x) \in v_\varepsilon\big(g(x)\big)$ and $0 \le g(x) + \lambda v$ implies $f(x) \le (1+\varepsilon)g(x) + \varepsilon(1+\lambda)v$ by Lemma I.4.1(b). Thus our assumption yields

$$f \underset{\text{a.e.}E}{\le} (1+\varepsilon)g + \varepsilon(1+\lambda)\chi_{E\otimes v}$$

for all $\varepsilon > 0$. By Theorem 4.14(c), this implies

$$\int_E f\,d\theta \le (1+\varepsilon)\int_E g\,d\theta + \varepsilon(1+\lambda)\theta_E(v).$$

Now we let $\varepsilon$ tend to $0$ in the right-hand side of this expression. Lemma I.5.21 together with the definition of the zero component in I.5.8 leads to

$$\int_E f\,d\theta \le \int_E g\,d\theta + \mathfrak{O}\big(\theta_E(v)\big).$$

Now we may argue the general case: Suppose that $f, g \in \mathcal{F}_{(E,\theta)}(X,\mathcal{P})$, let $w \in \mathcal{W}$ and $\varepsilon > 0$, and for $g$ choose the functions $g_{(w,\varepsilon)} \in \mathcal{F}_{\mathfrak{R}}(X,\mathcal{P})$ and $s_{(w,\varepsilon)} \in \mathcal{F}_{\mathfrak{R}}(X,\mathcal{V})$ as in 4.13, that is $g \underset{\text{a.e.}E}{\le} g_{(w,\varepsilon)} \underset{\text{a.e.}E}{\le} \gamma g + s_{(w,\varepsilon)}$ and $\int_E s_{(w,\varepsilon)}\,d\theta \le \varepsilon w$ for some $1 \le \gamma \le 1+\varepsilon$. Then $f(x) \preccurlyeq_v g_{(w,\varepsilon)}(x)$ holds $\theta$-almost everywhere on $E$, and our first case together with 4.14(c) yields

$$\int_E f\,d\theta \le \int_E g_{(w,\varepsilon)}\,d\theta + \mathfrak{O}\big(\theta_E(v)\big) \le \gamma\int_E g\,d\theta + \mathfrak{O}\big(\theta_E(v)\big) + \varepsilon w.$$

Because $w \in \mathcal{W}$ and $\varepsilon > 0$ were arbitrarily chosen, our claim follows. □

The following Proposition 4.17 is an immediate consequence of 4.16 and strengthens Part (c) of Theorem 4.14(c).

**Proposition 4.17.** *Let $f, g \in \mathcal{F}_{(F,\theta)}(X,\mathcal{P})$ for $F \in \mathfrak{A}_{\mathfrak{R}}$. If $f(x) \preccurlyeq g(x)$ holds $\theta$-almost everywhere on $F$, then $\int_F f\,d\theta \le \int_F g\,d\theta$.*

*Proof.* Let $F \in \mathfrak{A}_{\mathfrak{R}}$ and $f, g \in \mathcal{F}_{(F,\theta)}(X,\mathcal{P})$ such that $f(x) \preccurlyeq g(x)$ holds $\theta$-almost everywhere on $F$. Let $E \in \mathfrak{R}$, $w \in \mathcal{W}$, and choose $v \in \mathcal{V}$ such that $\theta_E(v) \le w$. As $f(x) \preccurlyeq g(x)$ implies $f(x) \preccurlyeq_v g(x)$, Proposition 4.16 yields

$$\int_{(E\cap F)} f\,d\theta \le \int_{(E\cap F)} g\,d\theta + w,$$

hence $\int_{(E\cap F)} f\,d\theta \le \int_{(E\cap F)} g\,d\theta$, since $w \in \mathcal{W}$ was arbitrarily chosen. Now our definition of the integral over a set $F \in \mathfrak{A}_{\mathfrak{R}}$ in 4.13 together with Lemma I.5.20(c) yields our claim. □

**Proposition 4.18.** *Let $f \in \mathcal{F}_{(E,\theta)}(X,\mathcal{P})$ for $E \in \mathfrak{R}$.*

*(a) If $E_n \in \mathfrak{R}$ such that $E_n \subset E_{n+1}$ for all $n \in \mathbb{N}$, and $E = \bigcup_{n\in\mathbb{N}} E_n$, then $\int_E f\,d\theta = \lim_{n\to\infty} \int_{E_n} f\,d\theta$.*

*(b) If $E_n \in \mathfrak{R}$ such that $E \supset E_n \supset E_{n+1}$ for all $n \in \mathbb{N}$, and $\bigcap_{n\in\mathbb{N}} E_n = \emptyset$, then* $0 \le \underline{\lim}_{n\to\infty} \int_{E_n} f\,d\theta \le \overline{\lim}_{n\to\infty} \int_{E_n} f\,d\theta \le \mathfrak{O}\left(\int_E f\,d\theta\right)$.

*Proof.* For Part (a), let $E_n \in \mathfrak{R}$ such that $E_n \subset E_{n+1}$ for all $n \in \mathbb{N}$, and $E = \bigcup_{n\in\mathbb{N}} E_n \in \mathfrak{R}$. We shall first assume that $f \in \mathcal{F}_{\mathfrak{R}}(X,\mathcal{P})$. By Lemma 2.4, for $w \in \mathcal{W}$ there is a neighborhood $v \in \mathcal{V}$ and $\lambda \ge 0$ such that $\theta_E(v) \le w$ and $0 \le f + \lambda\chi_{E\otimes}v$. This implies

$$\chi_{E_n\otimes}f \le \chi_{E_n\otimes}f + \chi_{(E\setminus E_n)\otimes}(f + \lambda\chi_{E\otimes}v) = \chi_{E\otimes}f + \lambda\chi_{(E\setminus E_n)\otimes}v.$$

Thus

$$\int_{E_n} f\,d\theta \le \int_E (f + \lambda\chi_{(E\setminus E_n)\otimes}v)\,d\theta = \int_E f\,d\theta + \lambda\theta_{(E\setminus E_n)}(v).$$

Following Lemma 3.1(b), this yields

$$\overline{\lim_{n\to\infty}} \int_{E_n} f\,d\theta \le \int_E f\,d\theta + \varepsilon w$$

for all $\varepsilon \ge 0$. Now let $h \in \mathcal{S}_{\mathfrak{R}}(X,\mathcal{P})$ be a step function such that $h \le f + \mathfrak{v}_w$, that is $h \le f + s$ for some $s \in \mathcal{S}_{\mathfrak{R}}(X,\mathcal{V})$ such that $\int_X s\,d\theta \le w$. Then $\int_{E_n} h\,d\theta \le \int_{E_n} f\,d\theta + w$ by 4.13(b) and (c), hence

$$\int_E h\,d\theta = \lim_{n\to\infty} \int_{E_n} h\,d\theta \le \underline{\lim}_{n\to\infty} \int_{E_n} f\,d\theta + w.$$

Taking the supremum over all such step functions $h \le f + \mathfrak{v}_w$ yields

$$\int_F^{(w)} f\,d\theta \le \underline{\lim}_{n\to\infty} \int_{E_n} f\,d\theta + w.$$

Combining with the above we infer that

$$\int_E f\,d\theta \le \underline{\lim}_{n\to\infty} \int_{E_n} f\,d\theta + w \le \overline{\lim_{n\to\infty}} \int_{E_n} f\,d\theta + w \le \int_E f\,d\theta + (1+\varepsilon)w.$$

Thus indeed $\int_E f\,d\theta = \lim_{n\to\infty} \int_{E_n} f\,d\theta$, since $w \in \mathcal{W}$ and $\varepsilon > 0$ were arbitrary. Now for the general case, let $f \in \mathcal{F}_{(E,\theta)}(X,\mathcal{P})$. Given $w \in \mathcal{W}$ and $\varepsilon > 0$ choose the functions $f_{(w,\varepsilon)} \in \mathcal{F}_{\mathfrak{R}}(X,\mathcal{P})$ and $s_{(w,\varepsilon)} \in \mathcal{F}_{\mathfrak{R}}(X,\mathcal{V})$ as in Definition 4.12. Then the preceding yields

$$\int_E f\,d\theta \le \int_E f_{(w,\varepsilon)}\,d\theta = \lim_{n\to\infty} \int_{E_n} f_{(w,\varepsilon)}\,d\theta \le \gamma \underline{\lim}_{n\to\infty} \int_{E_n} f\,d\theta + \varepsilon w,$$

and

$$\overline{\lim_{n\to\infty}} \int_{E_n} f\,d\theta \le \overline{\lim_{n\to\infty}} \int_{E_n} f_{(w,\varepsilon)}\,d\theta = \int_E f_{(w,\varepsilon)}\,d\theta \le \gamma \int_E f\,d\theta + \varepsilon w.$$

Our claim from Part (a) follows, since both $w \in \mathcal{W}$ and $\varepsilon > 0$ were arbitrary.

For Part (b), let $E_n \in \mathfrak{R}$ such that $E \supset E_n \supset E_{n+1}$ for all $n \in \mathbb{N}$, and $\bigcap_{n\in\mathbb{N}} E_n = \emptyset$. For the left-hand side of the inequality in (b) we shall again first assume that $f \in \mathcal{F}_{\mathfrak{R}}(X, \mathcal{P})$. Let $w \in \mathcal{W}$. Following Lemma 2.4(b), there is $v \in \mathcal{V}$ and $\lambda \geq 0$ such that $\theta_E(v) \leq w$ and $0 \leq f + \lambda \chi_{E \otimes} v$, hence $0 \leq \chi_{E_n \otimes} f + \lambda \chi_{E_n \otimes} v$. Then

$$0 \leq \int_{E_n} (f + \lambda \chi_{E_n \otimes} v)\, d\theta = \int_{E_n} f\, d\theta + \lambda \theta_{E_n}(v).$$

This yields

$$0 \leq \underline{\lim}_{n\to\infty} \int_{E_n} f\, d\theta + \lambda \mathfrak{O}\big(\theta_E(v)\big)$$

by Lemma 3.1(b). Because $\mathfrak{O}\big(\theta_E(v)\big) \leq \varepsilon w$ for all $\varepsilon > 0$, because $w \in \mathcal{W}$ was arbitrary and $\mathcal{Q}$ carries the weak preorder, we infer that $0 \leq \underline{\lim}_{n\to\infty} \int_{E_n} f\, d\theta$. For the general case, that is $f \in \mathcal{F}_{(E,\theta)}(X, \mathcal{P})$, given $w \in \mathcal{W}$ and $\varepsilon >, 0$ we choose functions $f_{(w,\varepsilon)} \in \mathcal{F}_{\mathfrak{R}}(X, \mathcal{P})$ and $s_{(w,\varepsilon)} \in \mathcal{F}_{\mathfrak{R}}(X, \mathcal{V})$ as in Definition 4.12. Then the preceding yields together with the limit rules from Lemma I.5.19

$$\begin{aligned} 0 \leq \underline{\lim}_{n\to\infty} \int_{E_n} f_{(w,\varepsilon)}\, d\theta &\leq \gamma \underline{\lim}_{n\to\infty} \int_{E_n} f\, d\theta + \overline{\lim}_{n\to\infty} \int_{E_n} s_{(w,\varepsilon)}\, d\theta \\ &\leq \gamma \underline{\lim}_{n\to\infty} \int_{E_n} f\, d\theta + \varepsilon w. \end{aligned}$$

Thus indeed $0 \leq \underline{\lim}_{n\to\infty} \int_{E_n} f\, d\theta$, since $w \subset \mathcal{W}$ and $\varepsilon > 0$ were arbitrarily chosen. For the right-hand side of the inequality in (b), let $G_n = E \setminus E_n$. Then $G_n \subset G_{n+1}$, $E = \cup_{n=1}^{\infty} G_n$ and $E = G_n \cup E_n$. Thus

$$\int_{F_n} f\, d\theta + \int_{E_n} f\, d\theta = \int_E f\, d\theta$$

for all $n \in \mathbb{N}$ by 4.15(b). Part (a) of 4.18 yields $\int_E f\, d\theta = \lim_{n\to\infty} \int_{G_n} f\, d\theta$. Again using the limit rules in Lemma I.5.19 we infer that

$$\begin{aligned} \overline{\lim}_{n\to\infty} \int_{E_n} f\, d\theta + \int_E f\, d\theta &= \overline{\lim}_{n\to\infty} \int_{E_n} f\, d\theta + \lim_{n\to\infty} \int_{G_n} f\, d\theta \\ &\leq \overline{\lim}_{n\to\infty} \left( \int_{E_n} f\, d\theta + \int_{G_n} f\, d\theta \right) = \int_E f\, d\theta. \end{aligned}$$

Now the cancellation rule in Proposition I.5.10(a) yields

$$\overline{\lim}_{n\to\infty} \int_{E_n} f\, d\theta \leq \mathfrak{O}\left( \int_E f\, d\theta \right). \quad \square$$

Given a set $F \in \mathfrak{A}_{\mathfrak{A}}$ we shall denote by $\mathcal{F}_{(|F|,\theta)}(X,\mathcal{P})$ the subcone of all functions in $\mathcal{F}(X,\mathcal{P})$ that are integrable over all complements in $F$ of sets in $\mathfrak{R}$, that is

$$\mathcal{F}_{(|F|,\theta)}(X,\mathcal{P}) = \bigcap_{E\in\mathfrak{R}} \mathcal{F}_{(F\setminus E,\theta)}(X,\mathcal{P}).$$

Using this notion, we obtain:

**Proposition 4.19.** *Let* $f \in \mathcal{F}_{(|F|,\theta)}(X,\mathcal{P})$ *for* $F \in \mathfrak{A}_{\mathfrak{R}}$.
*Then* $0 \leq \underline{\lim}_{E\in\mathfrak{R}} \int_{(F\setminus E)} f\,d\theta \leq \overline{\lim}_{E\in\mathfrak{R}} \int_{(F\setminus E)} f\,d\theta \leq \mathfrak{O}\left(\int_F f\,d\theta\right)$.

*Proof.* For every $E \in \mathfrak{R}$ the function $f$ is integrable over $E \cap F \in \mathfrak{R}$ and $F \setminus E \in \mathfrak{A}_{\mathfrak{R}}$. Thus $\int_{(F\setminus E)} f\,d\theta + \int_{(E\cap F)} f\,d\theta = \int_F f\,d\theta$ by 4.15(b). Taking the limit over all $E \in \mathfrak{R}$ and using the definition of the integral in 4.13 and Lemma I.5.19, we obtain

$$\overline{\lim_{E\in\mathfrak{R}}} \int_{(F\setminus E)} f\,d\theta + \int_F f\,d\theta \leq \int_F f\,d\theta,$$

hence

$$\overline{\lim_{E\in\mathfrak{R}}} \int_{(F\setminus E)} f\,d\theta \leq \mathfrak{O}\left(\int_F f\,d\theta\right)$$

by the cancellation rule Proposition I.5.10(a). For the first part of the inequality in 4.19, we fix $E_0 \in \mathfrak{R}$ and let $E_0 \subset E \in \mathfrak{R}$. Then

$$\int_{(F\setminus E_0)} f\,d\theta = \int_{(F\setminus E)} f\,d\theta + \int_{\left((F\setminus E_0)\cap E\right)} f\,d\theta.$$

Passing to the limits over $E \in \mathfrak{R}$ in this equation and again using I.5.19 and the definition of the integral leads to

$$\int_{(F\setminus E_0)} f\,d\theta \leq \underline{\lim}_{E\in\mathfrak{R}} \int_{(F\setminus E)} f\,d\theta + \int_{(F\setminus E_0)} f\,d\theta.$$

Now passing to the limit over $E_0 \in \mathfrak{R}$, we obtain

$$\underline{\lim}_{E\in\mathfrak{R}} \int_{(F\setminus E)} f\,d\theta \leq \underline{\lim}_{E\in\mathfrak{R}} \int_{(F\setminus E)} f\,d\theta + \underline{\lim}_{E\in\mathfrak{R}} \int_{(F\setminus E)} f\,d\theta.$$

Following Proposition I.5.10(a) and Proposition I.5.14, the latter implies

$$0 \leq \underline{\lim}_{E\in\mathfrak{R}} \int_{(F\setminus E)} f\,d\theta,$$

as claimed. □

# 5. The General Convergence Theorems

We shall proceed to establish a range of general convergence results for sequences of measures and functions and their respective integrals. These results are modeled after the dominated convergence theorem from classical measure theory. However, the presence of unbounded elements and the general absence of negatives will considerably complicate some technical aspects of the approach. First we shall extend some of the concepts of the preceding section from a single measure to families of measures. Subsequently, we shall set up suitable notions for convergence of sequences of measures and functions. Convergence for sequences of integrals will generally refer to order convergence in $\mathcal{Q}$, though in some special cases we will be able to establish stronger convergence with respect to the symmetric topology.

As in the preceding section, let $(\mathcal{P},\mathcal{V})$ be a full locally convex cone and let $(\mathcal{Q},\mathcal{V})$ be a locally convex complete lattice cone. $\mathfrak{R}$ denotes a (weak) $\sigma$-ring of subsets of $X$. We shall consider $\mathfrak{L}(\mathcal{P},\mathcal{Q})$-valued measures on $\mathfrak{R}$.

**5.1 Families of Measures and Properties Holding Almost Everywhere.** In the following we shall simultaneously deal with families of measures, and therefore need to extend our notion of properties holding almost everywhere from 4.11 to this situation: Given a (non-empty) family $\Theta$ of $\mathfrak{L}(\mathcal{P},\mathcal{Q})$-valued measures, we denote by $\mathfrak{Z}(\Theta)$ the collection of all sets $Z\in\mathfrak{A}_{\mathfrak{R}}$ such that $\theta_{(E\cap Z)}=0$ for all $E\in\mathfrak{R}$ and $\theta\in\Theta$. This collection is obviously closed for set complements and for countable unions. Correspondingly, for a subset $F$ of $X$ we shall say that a pointwise defined property of functions on $X$ holds *$\Theta$-almost everywhere on* $F$ if it holds on $F\setminus Z$ with some $Z\in\mathfrak{Z}(\Theta)$. If the concerned family $\Theta$ of measures is clearly identified, for the sake of simplicity we may use the symbols $\underset{a.e.F}{\leq}$ or $\underset{a.e.F}{=}$ if the relations $\leq$ or $=$ hold $\Theta$-almost everywhere on the set $F$, respectively.

**5.2 Equibounded Families of Measures.** A family $\Theta$ of measures on $\mathfrak{R}$ is called *equibounded* if for every $E\in\mathfrak{R}$ and $w\in\mathcal{W}$ there is $v\in\mathcal{V}$ such that $|\theta|(E,v)=\theta_E(v)\leq w$ for all $\theta\in\Theta$.

**5.3 Integrability with Respect to Equibounded Families of Measures.** Likewise, we need to adapt our notation of integrability from Section 4.12 and 4.13. We shall say that a function $f\in\mathcal{F}(X,\mathcal{P})$ is *integrable over a set* $E\in\mathfrak{R}$ *with respect to a family* $\Theta$ of $\mathfrak{L}(\mathcal{P},\mathcal{Q})$-valued measures if $\Theta$ is equibounded and if for every $w\in\mathcal{W}$ and $\varepsilon>0$ there are functions $f_{(w,\varepsilon)}\in\mathcal{F}_{\mathfrak{R}}(X,\mathcal{P})$ and $s_{(w,\varepsilon)}\in\mathcal{F}_{\mathfrak{R}}(X,\mathcal{V})$ such that

$$f\underset{a.e.E}{\leq} f_{(w,\varepsilon)}\underset{a.e.E}{\leq}\gamma f+s_{(w,\varepsilon)}\qquad\text{and}\qquad\int_E s_{(w,\varepsilon)}\,d\theta\leq\varepsilon w$$

for some $1\leq\gamma\leq 1+\varepsilon$ and all $\theta\in\Theta$. The almost-everywhere relation $\underset{a.e.E}{\leq}$ is meant with respect to the family $\Theta$. As in 4.12, we may again assume that the function $s_{(w,\varepsilon)} \in \mathcal{F}_{\mathfrak{R}}(X,\mathcal{V})$ is indeed $\mathcal{V}$- rather than $(\mathcal{V} \cup \{0\})$-valued. Integrability over a set $F \in \mathfrak{A}_{\mathfrak{R}}$ with respect to $\Theta$ then follows as in 4.13: The function $f \in \mathcal{F}(X,\mathcal{P})$ is *integrable over* $F \in \mathfrak{A}_R$ *with respect to* $\Theta$ if $f$ is integrable over the sets $E \cap F$ with respect to $\Theta$ for all $E \in \mathfrak{R}$ and all $\theta \in \Theta$ the limit

$$\int_F f\,d\theta = \lim_{E \in \mathfrak{R}} \int_{(E \cap F)} f\,d\theta$$

exists. The subcone of all these functions $f \in \mathcal{F}(X,\mathcal{P})$ is denoted by $\mathcal{F}_{(F,\Theta)}(X,\mathcal{P})$.

Likewise, $\mathcal{F}_{(|F|,\Theta)}(X,\mathcal{P})$ denotes the subcone of all functions in $\mathcal{F}(X,\mathcal{P})$ that are integrable with respect to $\Theta$ over all complements in $F$ of sets in $\mathfrak{R}$, that is

$$\mathcal{F}_{(|F|,\Theta)}(X,\mathcal{P}) = \bigcap_{E \in \mathfrak{R}} \mathcal{F}_{(F \setminus E,\Theta)}(X,\mathcal{P}).$$

Repeating the argument from Proposition 4.15(a), one can verify that a function $f \in \mathcal{F}(X,\mathcal{P})$ is in $\mathcal{F}_{(F,\Theta)}(X,\mathcal{P})$ or in $\mathcal{F}_{(|F|,\Theta)}(X,\mathcal{P})$ if and only if the function $\chi_{F \circledast} f$ is contained in $\mathcal{F}_{(X,\Theta)}(X,\mathcal{P})$ or in $\mathcal{F}_{(|X|,\Theta)}(X,\mathcal{P})$, respectively.

While integrability with respect to a family of measures obviously implies integrability with respect to every member of this family, the converse is not always true (see Example 5.15 below).

The following results 5.4 to 5.7 are already of interest for integration with respect to a single measure and might therefore have been placed into the preceding section. We shall, however, also refer to the subsequent more general versions which refer to integration with respect to equibounded families of measures.

**Proposition 5.4.** *Let* $\Theta$ *be an equibounded family of measures on* $\mathfrak{R}$. *Let* $E \in \mathfrak{R}$ *and* $f \in \mathcal{F}_{(E,\Theta)}(X,\mathcal{P})$. *For every* $w \in \mathcal{W}$ *there is* $s \in \mathcal{F}_{\mathfrak{R}}(X,\mathcal{V})$ *and* $\lambda \geq 0$ *such that* $0 \underset{a.e.E}{\lesssim} f + s$ *and* $\int_E s\,d\theta \leq \lambda w$ *for all* $\theta \in \Theta$.

*Proof.* Let $E \in \mathfrak{R}$, let $f \in \mathcal{F}_{(E,\Theta)}(X,\mathcal{P})$ and $w \in \mathcal{W}$. According to the definition of integrability in 4.12, for $\varepsilon = 1$ there are $g \in \mathcal{F}_{\mathfrak{R}}(X,\mathcal{P})$ and $s \in \mathcal{F}_{\mathfrak{R}}(X,\mathcal{V})$ such that $f \underset{a.e.E}{\lesssim} g \underset{a.e.E}{\lesssim} \gamma f + s$ for some $1 \leq \gamma \leq 2$ and $\int_E s\,d\theta \leq w$ for all $\theta \in \Theta$. We choose $v \in \mathcal{V}$ such that $\theta_E(v) \leq w$ for all $\theta \in \Theta$. Following Lemma 2.4(b) there is $G \in \mathfrak{R}$ and $\lambda \geq 0$ such that $0 \leq g + \lambda \chi_{G \circledast} v$. The latter implies $0 \underset{a.e.E}{\lesssim} g + \lambda \chi_{E \circledast} v$, hence

$$0 \underset{a.e.E}{\lesssim} \frac{1}{\gamma}(g + \lambda \chi_{E \circledast} v) \underset{a.e.E}{\lesssim} f + \frac{1}{\gamma}(s + \lambda \chi_{E \circledast} v) \underset{a.e.E}{\lesssim} f + (s + \lambda \chi_{E \circledast} v).$$

As $s + \lambda \chi_{E \circledast} v \in \mathcal{F}_{\mathfrak{R}}(X,\mathcal{V})$ and $\int_E (s + \lambda \chi_{E \circledast} v)\,d\theta \leq (1+\lambda) w$ for all $\theta \in \Theta$, our claim follows. $\square$

**5.5 The Locally Convex Cone** $\boldsymbol{(\mathcal{F}_{(F,\Theta)}(X,\mathcal{P}), \mathfrak{V}(F,\Theta))}$. Let $\Theta$ be an equibounded family of measures on $\mathfrak{R}$. Endowed with the order $\underset{a.e.F}{\lesssim}$, that is the given pointwise order $\Theta$-almost everywhere on the set $F \in \mathfrak{A}_{\mathfrak{R}}$, $\mathcal{F}_{(F,\Theta)}(X,\mathcal{P})$ is an ordered cone. We generate a canonical convex quasi-uniform structure (see I.1.3) in the following way: With every $w \in \mathcal{W}$ and $E \in \mathfrak{R}$ we associate the neighborhood $\breve{\mathfrak{v}}_w^E(\Theta)$, defined for functions $f,g \in \mathcal{F}_{(F,\Theta)}(X,\mathcal{P})$ by

$$f \leq g + \breve{\mathfrak{v}}_w^E(\Theta) \qquad \text{if} \qquad f \underset{a.e.E}{\leq} g + s$$

for some $s \in \mathcal{F}_{\mathfrak{R}}(X,\mathcal{V})$ such that $\int_E s\,d\theta \leq w$ for all $\theta \in \Theta$. Let $\mathfrak{V}(F,\Theta)$ denote the neighborhood system generated by the neighborhoods $\breve{\mathfrak{v}}_w^E(\Theta)$ for all $w \subset \mathcal{W}$ and $E \in \mathfrak{R}$ such that $E \subset F$. As $\Theta$ is equibounded, according to Proposition 5.4, for every function $f \in \mathcal{F}_{(F,\Theta)}(X,\mathcal{P})$, every $w \in \mathcal{W}$ and $E \in \mathfrak{R}$ such that $E \subset F$ there is $s \in \mathcal{F}_{\mathfrak{R}}(X,\mathcal{V})$ and $\lambda \geq 0$ such that $0 \underset{a.e.E}{\leq} f + s$ and $\int_E s\,d\theta \leq \lambda w$ holds for all $\theta \in \Theta$. Thus $0 \leq f + \lambda \breve{\mathfrak{v}}_w^E(\Theta)$. All functions in $\mathcal{F}_{(F,\Theta)}(X,\mathcal{P})$ are therefore bounded below with respect to these neighborhoods. In this way, $\big(\mathcal{F}_{(F,\Theta)}(X,\mathcal{P}), \mathfrak{V}(F,\Theta)\big)$ becomes a locally convex cone as elaborated in I.1.3. Theorem 4.14(c) implies that for every $E \in \mathfrak{R}$ such that $E \subset F$ and every $\theta \in \Theta$ the mapping

$$f \mapsto \int_E f\,d\theta \ : \ \mathcal{F}_{(F,\Theta)}(X,\mathcal{P}) \to \mathcal{Q}$$

is a continuous linear operator. Indeed, for $w \in \mathcal{W}$ we have $\int_E f\,d\theta \leq \int_E g\,d\theta + w$ whenever $f \leq g + \breve{\mathfrak{v}}_w^E(\Theta)$ for $f,g \in \mathcal{F}_{(F,\Theta)}(X,\mathcal{P})$.

**5.6 Subcone-Based Integrability.** The following definition of subcone-based integrability is motivated by the fact that in many realizations $(\mathcal{P},\mathcal{V})$ is indeed the standard full extension of some subcone of $\mathcal{P}$, and we might be particularly interested in functions with values in this subcone. Given a subcone $\mathcal{P}_0$ of $\mathcal{P}$ and a neighborhood subsystem $\mathcal{V}_0$ of $\mathcal{V}$, we shall say that a function $f$ in $\mathcal{F}(X,\mathcal{P})$ is *$(\mathcal{P}_0,\mathcal{V}_0)$-based integrable* over a set $E \in \mathfrak{R}$ with respect to an equibounded family $\Theta$ of measures if for every $w \in \mathcal{W}$ and $\varepsilon > 0$ there are functions $f_{(w,\varepsilon)} \in \mathcal{F}_{\mathfrak{R}}(X,\mathcal{P}_0)$ and $s_{(w,\varepsilon)} \in \mathcal{F}_{\mathfrak{R}}(X,\mathcal{V}_0)$ such that

$$f \underset{a.e.E}{\leq} f_{(w,\varepsilon)} + s_{(w,\varepsilon)}, \quad f_{(w,\varepsilon)} \underset{a.e.E}{\leq} \gamma f + s_{(w,\varepsilon)} \quad \text{and} \quad \int_E s_{(w,\varepsilon)}\,d\theta \leq \varepsilon w$$

for some $1 \leq \gamma \leq 1+\varepsilon$ and all $\theta \in \Theta$. The almost-everywhere relation $\underset{a.e.E}{\leq}$ is meant with respect to the family $\Theta$. In this context, $\mathcal{F}_{\mathfrak{R}}(X,\mathcal{P}_0)$ is the subcone of $\mathcal{F}_{\mathfrak{R}}(X,\mathcal{P})$ consisting of all measurable $\mathcal{P}_0$-valued functions such that for every inductive limit neighborhood $\mathfrak{v}$ for $\mathcal{F}(X,\mathcal{P})$ there is a $\mathcal{P}_0$-valued step function $h \in \mathcal{S}_{\mathfrak{R}}(X,\mathcal{P}_0)$ satisfying $h \leq f_{(w,\varepsilon)} + \mathfrak{v}$. Measurability is still defined with respect to the given neighborhood system $\mathcal{V}$ rather than the subsystem $\mathcal{V}_0$. Similarly, $\mathcal{F}_{\mathfrak{R}}(X,\mathcal{V}_0)$ consists of all $\mathcal{V}_0$-valued functions in $\mathcal{F}_{\mathfrak{R}}(X,\mathcal{P})$.

Because $(\mathcal{P}_0, \mathcal{V}_0)$-based integrability over a set $E \in \mathfrak{R}$ implies $(\mathcal{P}_0, \mathcal{V}_0)$-based integrability over all subsets $G \in \mathfrak{R}$ of $E$, $(\mathcal{P}_0, \mathcal{V}_0)$-based integrability over a set $F \in \mathfrak{A}_{\mathfrak{R}}$ may be defined as in 5.3.

Note that a $(\mathcal{P}_0, \mathcal{V}_0)$-based integrable function is not required to be $\mathcal{P}_0$-valued. Obviously, this notion of subcone-based integrability implies integrability based on the given cone $(\mathcal{P}, \mathcal{V})$ in the sense of 4.13 and 5.3, and the $(\mathcal{P}_0, \mathcal{V}_0)$-based integrable functions form a subcone of $\mathcal{F}_{(F,\Theta)}(X, \mathcal{P})$. For $\mathcal{P}_0 = \mathcal{P}$ and $\mathcal{V}_0 = \mathcal{V}$ the definition of $(\mathcal{P}_0, \mathcal{V}_0)$-based integrability coincides of course with the definition of integrability from 5.3: Clearly, integrability in the sense of 5.3 implies $(\mathcal{P}, \mathcal{V})$-based integrability. For the converse, use $\tilde{f}_{(w,\varepsilon)} = f_{(w,\varepsilon)} + s_{(w,\varepsilon)}$ instead of $f_{(w,\varepsilon)}$ in 5.3.

Other than in the classical scenario (see for example [25], [55], [178] and [179]), our definition of integrability does not generally guarantee that an integrable cone-valued function $f \in \mathcal{F}(X, \mathcal{P})$ can be approximated (even with respect to pointwise convergence) by a sequence of step functions whose integrals then converge towards the integral of $f$. However, a combination of Theorem 2.7 with Proposition 4.7 yields some corresponding results.

**Theorem 5.7.** *Let $\Theta$ be an equibounded family of measures on $\mathfrak{R}$. Let $E \in \mathfrak{R}$ and let $f \in \mathcal{F}(X, \mathcal{P})$ be $(\mathcal{P}_0, \mathcal{V}_0)$-based integrable over $E$ with respect to $\Theta$ for a subcone $\mathcal{P}_0$ of $\mathcal{P}$ and a subsystem $\mathcal{V}_0$ of $\mathcal{V}$. For every $w \in \mathcal{W}$ such that $\theta_E(v) \le w$ for some $v \in \mathcal{V}_0$ and all $\theta \in \Theta$, and every $\varepsilon > 0$ there is $s \in \mathcal{F}_{\mathfrak{R}}(X, \mathcal{V}_0)$ such that $\int_E s\, d\theta \le w$ for all $\theta \in \Theta$, $1 \le \gamma \le 1 + \varepsilon$ and a bounded below sequence $(h_n)_{n \in \mathbb{N}}$ of $\mathcal{P}_0$-valued step functions such that:*

*(i) $h_n \underset{a.e.E}{\le} \gamma f + s$ holds for all $n \in \mathbb{N}$.*

*(ii) $\Theta$-almost everywhere on $E$, for $x \in E$ there is $n_0 \in \mathbb{N}$ such that $f(x) \le h_n(x) + s(x)$ for all $n \ge n_0$.*

*(iii) $\int_G f\, d\theta \le \lim_{n \to \infty} \int_G h_n\, d\theta + w$ and $\int_G h_n\, d\theta \le \gamma \int_G f\, d\theta + w$ for all $n \in \mathbb{N}$, all $G \in \mathfrak{R}$ such that $G \subset E$, and all $\theta \in \Theta$.*

*Proof.* Let $f \in \mathcal{F}(X, \mathcal{P})$ be $(\mathcal{P}_0, \mathcal{V}_0)$-based integrable over $E \in \mathfrak{R}$. Given $w \in \mathcal{W}$ and $0 < \varepsilon \le 1$, following our assumption there are $f_{(w,\varepsilon)} \in \mathcal{F}_{\mathfrak{R}}(X, \mathcal{P}_0)$ and $s_{(w,\varepsilon)} \in \mathcal{F}_{\mathfrak{R}}(X, \mathcal{V}_0)$ such that $\int_E s_{(w,\varepsilon)}\, d\theta \le w/4$ for all $\theta \in \Theta$ and $f \underset{a.e.E}{\le} f_{(w,\varepsilon)} + s$ and $f_{(w,\varepsilon)} \underset{a.e.E}{\le} \gamma f + s_{(w,\varepsilon)}$ with some $1 \le \gamma \le 1 + \varepsilon/3$. By our assumption there is $v \in \mathcal{V}_0$ such that $\theta_E(v) \le w/2$ for all $\theta \in \Theta$. We shall apply Theorem 2.7 to the locally convex cone $(\mathcal{P}_0, \mathcal{V}_0)$, the function $f_{(w,\varepsilon)} \in \mathcal{F}_{\mathfrak{R}}(X, \mathcal{P}_0)$, the neighborhood $v \in \mathcal{V}_0$, $\varepsilon/3$ in place of $\varepsilon$, and the inductive limit neighborhood $\mathfrak{v} = \{\chi_{X \otimes} v\}$. For this we observe that the measurability conditions (M1) and (M2) in Section 1 with respect to the neighborhood system $\mathcal{V}$ imply those with respect to the subsystem $\mathcal{V}_0 \subset \mathcal{V}$. There is $1 \le \gamma' \le 1 + \varepsilon/3$ and a bounded below sequence $(h_n)_{n \in \mathbb{N}}$ of $\mathcal{P}_0$-valued step functions such that (i), $h_n(x) \le \gamma' f_{(w,\varepsilon)}(x) + v$ for all $x \in E$ and $n \in \mathbb{N}$, and (ii), for every $x \in E$ there is $n_0 \in \mathbb{N}$ such that $f_{(w,\varepsilon)}(x) \le h_n(x) + v$ for all $n \ge n_0$. This yields

$$h_n \underset{a.e.E}{\leq} (\gamma\gamma')f + (\gamma' s_{(w,\varepsilon)} + \chi_{E\otimes} v)$$

for all $n \in \mathbb{N}$. We set

$$s' = \gamma' s_{(w,\varepsilon)} + \chi_{E\otimes} v \in \mathcal{F}_{\mathfrak{R}}(X, \mathcal{V}_0)$$

and observe that $\int_E s'\,d\theta \leq \gamma' w/4 + w/2 \leq w$ for all $\theta \in \Theta$. Because $1 \leq \gamma\gamma' \leq (1+\varepsilon/3)^2 \leq 1+\varepsilon$, this yields Part (i) of our claim with $s'$ in place of $s$. Part (ii) also follows from the above, since $f_{(w,\varepsilon)}(x) \leq h_n(x) + v$ for $x \in E$ implies that

$$f(x) \leq f_{(w,\varepsilon)}(x) + s_{(w,\varepsilon)}(x) \leq h_n(x) + s_{(w,\varepsilon)}(x) + v \leq h_n(x) + s'(x).$$

For Part (iii) let $G \in \mathfrak{R}$ such that $G \subset E$ and let $\theta \in \Theta$. The second part of (iii) is obvious, since $h_n \underset{a.e.E}{\leq} \gamma f + s'$ implies that $\int_G h_n\,d\theta \leq \gamma \int_G f\,d\theta + w$ for all $n \in \mathbb{N}$. For the first part of (iii), consider the full cone $\mathcal{P}$ and let $h'_n = h_n + \chi_{E\otimes} v \in \mathcal{S}_{\mathfrak{R}}(X, \mathcal{P})$. The sequence $(h'_n)_{n\in\mathbb{N}}$ of step functions approaches $f_{(w,\varepsilon)} \in \mathcal{F}_{\mathfrak{R}}(X, \mathcal{P})$ as required in Proposition 4.7, which therefore yields

$$\int_G f_{(w,\varepsilon)}\,d\theta \leq \int_G^{(w')} f_{(w,\varepsilon)}\,d\theta \leq \lim_{n\to\infty} \int_G h'_n\,d\theta + w'$$

for every $\theta \in \Theta$ and all $w' \in \mathcal{W}$, hence

$$\int_G f_{(w,\varepsilon)}\,d\theta \leq \lim_{n\to\infty} \int_G h'_n\,d\theta = \lim_{n\to\infty} \int_G h_n\,d\theta + \theta_G(v)$$

since $\int_G h'_n\,d\theta = \int_G h_n\,d\theta + \theta_G(v)$. Thus

$$\begin{aligned} \int_G f\,d\theta &\leq \int_G f_{(w,\varepsilon)}\,d\theta + \int_G s_{(w,\varepsilon)}\,d\theta \\ &\leq \lim_{n\to\infty} \int_G h_n\,d\theta + \int_G s_{(w,\varepsilon)}\,d\theta + \theta_G(v) \\ &\leq \lim_{n\to\infty} \int_G h_n + w \end{aligned}$$

since $\int_G s_{(w,\varepsilon)}\,d\theta \leq w/4$ and $\theta_G(v) \leq w/2$. This yields the first inequality in (iii). □

If the family $\Theta$ of measures is equibounded relative to the subsystem $\mathcal{V}_0$ of $\mathcal{V}$, that is if for every $E \in \mathfrak{R}$ and every $w \in \mathcal{W}$ there is $v \in \mathcal{V}_0$ such that $\theta_E(v) \leq w$ for all $\theta \in \Theta$, then the condition on the neighborhood $v \in \mathcal{V}_0$ in Theorem 5.7 is obviously superfluous. Indeed, given $w \in \mathcal{V}$ and any $v \in \mathcal{V}_0$ there is $v' \in \mathcal{V}_0$ as above. Because the neighborhood system $\mathcal{V}_0$ is supposed to be directed downward, there is $v'' \in \mathcal{V}_0$ such that both $v'' \leq v$ and $v'' \leq v'$. Thus $\theta_E(v'') \leq w$ for all $\theta \in \Theta$, and we may apply Theorem 5.7 with $v''$ in place of $v$.

For future use, it is worthwhile to formulate as a corollary the simplifications that occur in Theorem 5.7 if the subcone $(\mathcal{P}_0, \mathcal{V}_0)$ of $(\mathcal{P}, \mathcal{V})$ is indeed a full cone, that is if $\mathcal{V}_0 \subset \mathcal{P}_0$.

**Corollary 5.8.** *Let $\Theta$ be an equibounded family of measures on $\mathfrak{R}$. Let $E \in \mathfrak{R}$ and let $f \in \mathcal{F}(X, \mathcal{P})$ be $(\mathcal{P}_0, \mathcal{V}_0)$-based integrable over $E$ with respect to $\Theta$ for a subcone $\mathcal{P}_0$ of $\mathcal{P}$ and a subsystem $\mathcal{V}_0 \subset \mathcal{P}_0$ of $\mathcal{V}$. For every $w \in \mathcal{W}$ such that $\theta_E(v) \le w$ for some $v \in \mathcal{V}_0$ and all $\theta \in \Theta$, and every $\varepsilon > 0$, there is $s \in \mathcal{F}_{\mathfrak{R}}(X, \mathcal{V}_0)$ such that $\int_E s\, d\theta \le w$ for all $\theta \in \Theta$, $1 \le \gamma \le 1 + \varepsilon$ and a bounded below sequence $(h_n)_{n \in \mathbb{N}}$ of $\mathcal{P}_0$-valued step functions such that:*

*(i) $h_n \underset{a.e.E}{\le} \gamma f + s$ holds for all $n \in \mathbb{N}$.*
*(ii) $\Theta$-almost everywhere on $E$, for $x \in E$ there is $n_0 \in \mathbb{N}$ such that $f(x) \le h_n(x)$ for all $n \ge n_0$.*
*(iii) $\int_G f\, d\theta \le \underline{\lim}_{n \to \infty} \int_G h_n\, d\theta$ and $\int_G h_n\, d\theta \le \gamma \int_G f\, d\theta + w$ for all $n \in \mathbb{N}$, all $G \in \mathfrak{R}$ such that $G \subset E$, and all $\theta \in \Theta$.*

*Proof.* Given a neighborhood $w \in \mathcal{W}$ satisfying the requirement of the corollary, and $0 < \varepsilon \le 1$ we apply Theorem 5.7 with the neighborhood $w/4 \in \mathcal{W}$ instead of $w$. As in the proof of 5.7 we choose $v \in \mathcal{V}$ such that $\theta_E(v) \le w/8$. Let $s \in \mathcal{F}_{\mathfrak{R}}(X, \mathcal{V}_0)$ and the sequence $(h_n)_{n \in \mathbb{N}}$ of $\mathcal{P}_0$-valued step functions as in 5.7. We apply Corollary 2.8 to the full cone $(\mathcal{V}_0, \mathcal{V}_0)$ for the function $s \subset \mathcal{F}_{\mathfrak{R}}(X, \mathcal{V}_0)$ with the inductive limit neighborhood $\mathfrak{v} = \{\chi_{X \circledast} v\}$: There is a bounded below sequence $(s_n)_{n \in \mathbb{N}}$ of $\mathcal{V}_0$-valued step functions satisfying (i) $s_n \le \gamma' s + \chi_{X \circledast} v$ with some $1 \le \gamma' \le 1 + \varepsilon$ and (ii) for every $x \in E$ there is $n_0$ such that $s(x) \le s_n(x)$ for all $n \ge n_0$. The latter implies

$$\int_G s(x)\, d\theta \le \underline{\lim}_{n \to \infty} \int_G s_n(x)\, d\theta$$

for all $G \in \mathfrak{R}$ such that $G \subset E$, and all $\theta \in \Theta$, by Proposition 4.7. Now we set

$$h'_n = h_n + s_n + \chi_{E \circledast} v \in \mathcal{S}_{\mathfrak{R}}(X, \mathcal{P}_0) \qquad \text{and} \qquad s' = 3s + 2\chi_{E \circledast} v \in \mathcal{F}_{\mathfrak{R}}(X, \mathcal{V}_0).$$

These are the functions that we use for Corollary 5.8: We have

$$\int_E s'\, d\theta \le 3(w/4) + 2(w/8) = w,$$

and

$$\begin{aligned} h'_n &\underset{aeE}{\le} (\gamma f + s) + s_n + \chi_{E \circledast} v \\ &\le (\gamma f + s) + (\gamma' s + \chi_{X \circledast} v) + \chi_{E \circledast} v \\ &\le \gamma f + s' \end{aligned}$$

since $1+\gamma' \leq 3$. This implies $\int_G h'_n \, d\theta \leq \gamma \int_G f \, d\theta + w$ for all $n \in \mathbb{N}$, all $G \in \mathfrak{R}$ such that $G \subset E$, and all $\theta \in \Theta$. The first part of (iii) follows from the last inequality in the proof of 5.7, that is

$$\begin{aligned}\int_G f \, d\theta &\leq \underline{\lim}_{n\to\infty} \int_G h_n \, d\theta + \int_G s \, d\theta + \theta_G(v) \\ &\leq \underline{\lim}_{n\to\infty} \int_G h_n \, d\theta + \underline{\lim}_{n\to\infty} \int_G s_n \, d\theta + \theta_G(v) \\ &\leq \underline{\lim}_{n\to\infty} \int_G (h_n + s_n + \chi_{E \circledast v}) \, d\theta \\ &\leq \underline{\lim}_{n\to\infty} \int_G h'_n \, d\theta,\end{aligned}$$

hence our claim. □

For the following recall the definition of the order topology of a locally complete lattice cone from Section I.5.43. We shall also consider integrals of measurable $\overline{\mathcal{V}}$-, that is $\mathcal{V} \cup \{0, \infty\}$-valued functions (see Section 2.1), if they take the value $\infty \in \overline{\mathcal{V}}$ only on a set of measure zero (see 4.12 and 4.13).

**Corollary 5.9.** *Let $\Theta$ be an equibounded family of measures on $\mathfrak{R}$. Let $F \in \mathfrak{A}_{\mathfrak{R}}$ and let $f \in \mathcal{F}(X, \mathcal{P})$ be $(\mathcal{P}_0, \mathcal{V})$-based integrable over $F$ with respect to $\Theta$ for a subcone $\mathcal{P}_0$ of $\mathcal{P}$. Then there is a net $(h_i)_{i \in \mathcal{I}}$ of $\mathcal{P}_0$-valued step functions, a net $(s_i)_{i \in \mathcal{I}}$ of measurable $\overline{\mathcal{V}}$-valued functions and a net $(\gamma_i)_{i \in \mathcal{I}}$ in $\mathbb{R}$ such that for every $\theta \in \Theta$ :*

*(i) For every $x \in F$ there is $i_0 \in \mathcal{I}$ such that $f(x) \leq \gamma_i h_i(x) + s_i(x)$ and $h_i(x) \leq f(x) + s_i(x)$ for all $i \geq i_0$.*
*(ii) $\lim_{i \in \mathcal{I}} \int_F h_i \, d\theta = \int_F f \, d\theta$ in the order topology of $\mathcal{Q}$.*
*(iii) $\lim_{i \in \mathcal{I}} \int_F s_i \, d\theta = 0$ in the symmetric topology of $\mathcal{Q}$.*
*(iv) $\gamma_i \geq 1$ for all $i \in \mathcal{I}$ and $\lim_{i \in \mathcal{I}} \gamma_i = 1$.*

*Consequently, $\int_F f \, d\theta$ is contained in the closure with respect to the order topology of the subcone of $\mathcal{Q}$ spanned by the set $\{\theta_{(E \cap F)}(a) \mid E \in \mathfrak{R},\ a \in \mathcal{P}_0\}$.*

*Proof.* Suppose that the function $f \in \mathcal{F}(X, \mathcal{P})$ is $(\mathcal{P}_0, \mathcal{V})$-based integrable over the set $F \in \mathfrak{A}_{\mathfrak{R}}$ with respect to $\Theta$ and in a first step let $E \in \mathfrak{R}$ be a subset of $F$. For every $w \in \mathcal{W}$ and $\varepsilon > 0$, let $(s_n^{w,\varepsilon})_{n \in \mathbb{N}}$ be a sequence of $\mathcal{P}_0$-valued step functions, $1 \leq \gamma^{w,\varepsilon} \leq 1 + \varepsilon$ and $s^{w,\varepsilon} \in \mathcal{S}_{\mathfrak{R}}(X, \mathcal{V})$ as in Theorem 5.7 with $\varepsilon w$ in place of $w$. According to 5.7 we have $\int_E s^{w,\varepsilon} \, d\theta \leq \varepsilon w$ for all $\theta \in \Theta$, and

$$\int_E f \, d\theta \leq \underline{\lim}_{n\to\infty} \int_E h_n^{w,\varepsilon} \, d\theta + \varepsilon w \quad \text{and} \quad \overline{\lim}_{n\to\infty} \int_E h_n^{w,\varepsilon} \, d\theta \leq \gamma^{w,\varepsilon} \int_E f \, d\theta + \varepsilon w$$

follows from 5.7(iii). That is, for all $\theta \in \Theta$, both

$$\underline{\lim}_{n\to\infty} \int_E h_n^{w,\varepsilon}\, d\theta \qquad \text{and} \qquad \overline{\lim}_{n\to\infty} \int_E h_n^{w,\varepsilon}\, d\theta$$

are elements of the symmetric relative neighborhood $w_\varepsilon^s\big(\int_E f\, d\theta\big)$. Let the index set $\mathcal{J}$ consist of all triples $(w,\varepsilon,\phi)$, where $w \in \mathcal{W}$, $\varepsilon > 0$ and $\phi : \mathcal{W} \times \{\varepsilon > 0\} \to \mathbb{N}$. The set $\mathcal{J}$ is ordered and directed upward by $(w_1,\varepsilon_1,\phi_1) \le (w_2,\varepsilon_2,\phi_2)$ if $w_2 \le w_1$, $\varepsilon_2 \le \varepsilon_1$, and $\phi_1(w,\varepsilon) \le \phi_2(w,\varepsilon)$ for all $w \in \mathcal{W}$ and $\varepsilon > 0$. Note that the index set $\mathcal{J}$ does not depend on the subset $E \in \mathfrak{R}$ of $F$. We set

$$h_j = \chi_{E^\otimes} h^{w,\varepsilon}_{\phi(w,\varepsilon)}$$

for $j = (w,\varepsilon,\phi) \in \mathcal{J}$, as well as

$$s_j = \chi_{E^\otimes} s^{w,\varepsilon} + \chi_{(Z\cup Z_E)^\otimes}\infty \qquad \text{and} \qquad \gamma_j = \gamma^{w,\varepsilon},$$

where $\infty$ is the infinite element of the augmented neighborhood system $\overline{\mathcal{V}}$ (see 2.1), and $Z = X \setminus \big(\bigcup_{E\in\mathfrak{R}} E\big)$. This is a set of $\Theta$-measure zero. Likewise, $Z_E \in \mathfrak{R}$ is a subset of $E$ of $\Theta$-measure zero and such that the conclusions of 5.7(i) and (ii) hold for all $x \in E \setminus Z_E$. (For this, recall that the union of countably many sets of measure zero is again of measure zero.) Therefore, 5.7(i) and (ii) hold for all $x \in E$, not just $\Theta$-almost everywhere if we replace the function $s^{w,\varepsilon}$ by $s_j$. Moreover, since the function $s_j$ takes the value $\infty \in \overline{\mathcal{V}}$ only on a zero set, we infer

$$\int_F s_j\, d\theta = \int_E s^{w,\varepsilon}\, d\theta \le \varepsilon w$$

for all $\theta \in \Theta$. We have $1 \le \gamma_j \le 1+\varepsilon$. Thus

$$\lim_{j\in\mathcal{J}} \int_F s_j\, d\theta = 0 \qquad \text{and} \qquad \lim_{j\in\mathcal{J}} \gamma_j = 1.$$

The first of these limits is taken in the symmetric topology of $\mathcal{Q}$. Next we shall verify that

$$\int_E f\, d\theta = \lim_{j\in\mathcal{J}} \int_F h_j\, d\theta$$

holds for every $\theta \in \Theta$ in the order topology of $\mathcal{Q}$. Indeed, let $\theta \in \Theta$ and let $U$ be a convex and order convex neighborhood of $\int_E f\, d\theta \in \mathcal{Q}$ in the order topology. As the order topology is coarser than the symmetric relative topology of $\mathcal{Q}$ (see Proposition I.5.44), there are $w_0 \in \mathcal{W}$ and $\varepsilon_0 > 0$ such that $U$ is a neighborhood for every point in the symmetric relative neighborhood $w_{0\,\varepsilon_0}^{\,s}(\int_E f\, d\theta)$. Then for each choice of $w \le w_0$ and $\varepsilon \le \varepsilon_0$ we have both

$$\underline{\lim}_{n\to\infty} \int_E h_n^{w,\varepsilon}\, d\theta \in w_{0\,\varepsilon_0}^{\,s}\Big(\int_E f\, d\theta\Big) \quad \text{and} \quad \overline{\lim}_{n\to\infty} \int_E h_n^{w,\varepsilon}\, d\theta \in w_{0\,\varepsilon_0}^{\,s}\Big(\int_E f\, d\theta\Big)$$

by the above. As $U$ is an order topology neighborhood of every element in $v^s_{\varepsilon_0}\left(\int_E f\,d\theta\right)$, there is an integer $\phi_0(w,\varepsilon)\in\mathbb{N}$ such that both

$$\inf_{n\geq\phi_0(w,\varepsilon)}\left\{\int_E h_n^{w,\varepsilon}\,d\theta\right\}\in U \qquad\text{and}\qquad \sup_{n\geq\phi_0(w,\varepsilon)}\left\{\int_E h_n^{w,\varepsilon}\,d\theta\right\}\in U.$$

Now the order convexity of the neighborhood $U$ guarantees that

$$\int_E h_n^{w,\varepsilon}\,d\theta\in U$$

whenever $w\leq w_0$, $\varepsilon\leq\varepsilon_0$ and $n\geq\phi_0(w,\varepsilon)$. We set $j_0=(w_0,\varepsilon_0,\phi_0)\in\mathcal{J}$ and $\phi_0(w,\varepsilon)=1$ if either $w\not\leq w_0$ or $\varepsilon\not\leq\varepsilon_0$. Then the above yields indeed that

$$\int_F h_j\,d\theta=\int_E h^{w,\varepsilon}_{\phi(w,\varepsilon)}\,d\theta\in U$$

for all $j=(w,\varepsilon,\phi)\in\mathcal{J}$ such that $j\geq j_0$, that is $w\leq w_0$, $\varepsilon\leq\varepsilon_0$ and $\phi(w,\varepsilon)\geq\psi_0(w,\varepsilon)$, thus demonstrating our claim. Finally, given $x\in E$, for every $w\in\mathcal{W}$ and $\varepsilon>0$ there is $\phi_0(w,\varepsilon)\in\mathbb{N}$ such that

$$f(x)\leq h_j(x)+s_j(x)\qquad\text{and}\qquad h_j(x)\leq\gamma_j\,f(x)+s_j(x)$$

for all $j=(w,\varepsilon,\phi)\in\mathcal{J}$ such that $\phi(w,\varepsilon)\geq\phi_0(w,\varepsilon)$. If we set $j_0=(w_0,1,\phi_0)$ for any choice of $w_0\in\mathcal{W}$, then the above inequalities hold whenever $j\geq j_0$.

Now in the second step of our construction, for every set $E\in\mathfrak{R}$ we shall construct a net $(h_j^E)_{j\in\mathcal{J}}$ of $(\mathcal{P}_0,\mathcal{V})$-valued step functions as in our first step with respect to the set $E\cap F\in\mathfrak{R}$, that is in particular

$$\int_{(E\cap F)} f\,d\theta=\lim_{j\in\mathcal{J}}\int_F h_j^E\,d\theta$$

for every $\theta\in\Theta$. Similarly, we select the corresponding nets $(s_j^E)_{j\in\mathcal{J}}$ and $(\gamma_j^E)_{j\in\mathcal{J}}$. Now we choose another index set $\mathcal{I}$ consisting of all pairs $(E,\psi)$, where $E\in\mathfrak{R}$ and $\psi:\mathfrak{R}\to\mathcal{J}$, ordered and directed upward by $(E_1,\psi_1)\leq(E_2,\psi_2)$ if $E_1\leq E_2$, and $\psi_1(E)\leq\psi_2(E)$ for all $E\in\mathfrak{R}$. We set $h_i=h^E_{\psi(E)}$ for $i=(E,\psi)\in\mathcal{I}$ and realize that the net $(h_i)_{i\in\mathcal{I}}$ of $(\mathcal{P}_0,\mathcal{V})$-valued step functions satisfies the properties stated in our Corollary. A straightforward diagonal argument similar to the preceding one shows that

$$\int_F f\,d\theta=\lim_{i\in\mathcal{I}}\int_F h_i\,d\theta$$

holds for every $\theta\in\Theta$ in the order topology of $\mathcal{Q}$, as claimed in (ii). Because the integrals of all step functions $h_i$ involved are linear combinations of elements $\theta_{(E\cap F)}(a)$ for $E\in\mathfrak{R}$ and $a\in\mathcal{P}_0$, (ii) does indeed imply that

$\int_F f\,d\theta$ is contained in the closure with respect to the order topology of the subcone of $\mathcal{Q}$ spanned by these elements. Similarly, as claimed in (iii) and in (iv), we verify that

$$\lim_{i\in\mathcal{I}} \int_F s_i\,d\theta = 0 \qquad \text{and} \qquad \lim_{i\in\mathcal{I}} \gamma_i = 1,$$

where the first of these limits is taken in the symmetric topology of $\mathcal{Q}$. For Part (i), let $x \in F$. If $x \notin \bigcup_{E\in\mathfrak{R}} E$, then our claim is trivial, as we have $s_i(x) = \infty \in \overline{\mathcal{V}}$ for all $i \in \mathcal{I}$. Otherwise, there is $E_0 \in \mathfrak{R}$ such that $x \in E_0$. We fix any $w_0 \in \mathcal{W}$ and choose the index $i_0 = (E_0, \psi_0) \in \mathcal{I}$, where $\psi_0 : \mathfrak{R} \to \mathcal{J}$ is the mapping $E \mapsto (w_0, 1, \phi_E) \in \mathcal{J}$. The mapping $\phi_E : \mathcal{W} \times \{\varepsilon > 0\} \to \mathbb{N}$ is chosen as constant $\phi_E(w,\varepsilon) = 1$ if $E_0 \not\subset E$, and otherwise we chose $\phi_E(w,\varepsilon) \in \mathbb{N}$ such that

$$f(x) \le h_j(x) + s_j(x) \qquad \text{and} \qquad h_j(x) \le \gamma_j\, f(x) + s_j(x)$$

for every $j = (w,\varepsilon,\phi) \in \mathcal{J}$ such that $j \ge (w_0, 1, \phi_E)$. This holds for all $E \in \mathfrak{R}$ such that $E_0 \subset E$, hence we infer that

$$f(x) \le h_i(x) + s_i(x) \qquad \text{and} \qquad h_i(x) \le \gamma_i\, f(x) + s_i(x)$$

holds for all $i \ge i_0$. □

It its important to keep in mind that the limit in 5.9(ii) refers to the order topology of $\mathcal{Q}$, not necessarily to order convergence as defined in I.5.18. Because in general the order topology is not known to be Hausdorff, this limit need therefore not be unique.

Corollary 5.9 is of particular interest in case that the locally convex complete lattice cone $(\mathcal{Q},\mathcal{W})$ is indeed the standard completion of some locally convex cone $(\mathcal{Q}_0,\mathcal{W})$ (see I.5.57) and that the measure $\theta$ is indeed $\mathfrak{L}(\mathcal{P},\mathcal{Q}_0)$-valued. The closure of $\mathcal{Q}_0$ in $\mathcal{Q}$ with respect to the order topology was seen to be a subcone of the second dual $\mathcal{Q}_0^{**}$ (see Remark I.5.60(a) and Section I.7.3) in this case, and integrals of functions in $\mathcal{F}(X,\mathcal{P})$ are therefore elements of $\mathcal{Q}_0^{**}$. Moreover, if the full locally convex cone $(\mathcal{P},\mathcal{V})$ is indeed the standard full extension of a quasi-full locally convex cone $(\mathcal{P}_0,\mathcal{V})$ (see I.6.2) and if for all $E \in \mathfrak{R}$ the operator $\theta_E$ maps the elements of $\mathcal{P}_0$ into $\mathcal{Q}_0$, then a similar statement holds for all functions $f \in \mathcal{F}(X,\mathcal{P})$ that are $(\mathcal{P}_0,\mathcal{V})$-based integrable over $F$ (see Proposition 6.7 below).

Because the values of our measures, that is continuous linear operators from $\mathcal{P}$ into $\mathcal{Q}$, may be restricted to linear operators on a subcone $\mathcal{P}_0$ of $\mathcal{P}$, one may raise the question, if and how such a restriction does affect the integrals of functions with values only in this subcone $\mathcal{P}_0$. Let us be precise: Let $\mathcal{P}_0$ be a subcone of $\mathcal{P}$, and let $\mathcal{V}_0 \subset \mathcal{P}_0$ be a neighborhood subsystem of $\mathcal{V}$. If for a given $\mathfrak{L}(\mathcal{P},\mathcal{Q})$-valued measure $\theta$, for all $E \in \mathfrak{R}$, the restrictions of the linear operators $\theta_E$ from the full cone $(\mathcal{P},\mathcal{V})$ to $(\mathcal{Q},\mathcal{W})$ are continuous linear operators from the full cone $(\mathcal{P}_0,\mathcal{V}_0)$ to $(\mathcal{Q},\mathcal{W})$, then, obviously, $\theta$

may also be considered to be an $\mathfrak{L}(\mathcal{P}_0, \mathcal{Q})$-valued measure. This situation requires that for every $E \in \mathfrak{R}$ and $w \in \mathcal{W}$ there is $v \in \mathcal{V}_0$ such that $\theta_E(v) \leq w$.

To avoid confusion, we shall denote this restriction of the measure $\theta$ to $(\mathcal{P}_0, \mathcal{V}_0)$ by $\theta_0$, and by $\mathcal{F}_{(F,\theta_0)}(X, \mathcal{P}_0)$ the cone of all $\mathcal{P}_0$-valued functions that are integrable over a set $F \in \mathfrak{A}_{\mathfrak{R}}$ with respect to $\theta_0$. Similarly, we shall use $\mathcal{F}_{(F,\Theta_0)}(X, \mathcal{P}_0)$ for functions that are integrable with respect to a family of restricted measures. Because our notions of measurability, of being reached from below by step functions and, consequently, of integrability depend on the given neighborhood system as well as on the cone, we shall have to clarify our notions for this situation.

For a $\mathcal{P}_0$-valued function, measurability with respect to $(\mathcal{P}, \mathcal{V})$ obviously implies measurability with respect to $(\mathcal{P}_0, \mathcal{V}_0)$, since $\mathcal{V}_0 \subset \mathcal{V}$ (see Conditions (M1) and (M2) in Section 1.2). The cone $\mathcal{F}_{\mathfrak{R}}(X, \mathcal{P}_0)$ is however not necessarily a subcone of $\mathcal{F}_{\mathfrak{R}}(X, \mathcal{P})$ since the condition for the elements of $\mathcal{F}_{\mathfrak{R}}(X, \mathcal{P}_0)$ of being reached from below (see Section 2.3) involves only inductive neighborhoods that use the neighborhoods in $\mathcal{V}_0 \subset \mathcal{V}$. Positive functions in $\mathcal{F}_{\mathfrak{R}}(X, \mathcal{P}_0)$ are however contained in $\mathcal{F}_{\mathfrak{R}}(X, \mathcal{P})$, since they can be trivially reached from below by the step function $h = 0$. Conversely, every $\mathcal{P}_0$-valued function in $\mathcal{F}_{\mathfrak{R}}(X, \mathcal{P})$ that can be reached from below by $\mathcal{P}_0$-valued step functions is contained in $\mathcal{F}_{\mathfrak{R}}(X, \mathcal{P}_0)$. This implies in particular that $\mathcal{F}_{\mathfrak{R}}(X, \mathcal{V}_0)$ consists of the $\mathcal{V}_0$-valued elements of $\mathcal{F}_{\mathfrak{R}}(X, \mathcal{V})$, that is $\mathcal{F}_{\mathfrak{R}}(X, \mathcal{V}_0) = \mathcal{F}_{\mathfrak{R}}(X, \mathcal{V}) \cap \mathcal{F}(X, \mathcal{V}_0)$.

Furthermore, we note that every set $Z \in \mathfrak{A}_{\mathfrak{R}}$ of measure zero with respect to a measure $\theta$ is also of measure zero with respect to its restriction $\theta_0$. The almost everywhere notion with respect to $\theta$ therefore implies the almost everywhere notion with respect to $\theta_0$. The converse does not necessarily hold true.

**Proposition 5.10.** *Let $\mathcal{P}_0$ be a subcone of $\mathcal{P}$, and let $\mathcal{V}_0 \subset \mathcal{P}_0$ be a neighborhood subsystem of $\mathcal{V}$. Let $\Theta$ be an equibounded family of $\mathfrak{L}(\mathcal{P}, \mathcal{Q})$-valued measures on $\mathfrak{R}$ such that the family $\Theta_0$ of all restrictions of the measures in $\Theta$ to $(\mathcal{P}_0, \mathcal{V}_0)$ is an equibounded family of $\mathfrak{L}(\mathcal{P}_0, \mathcal{Q})$-valued measures. Let $F \in \mathfrak{A}_{\mathfrak{R}}$. If a $\mathcal{P}_0$-valued function $f$ is $(\mathcal{P}_0, \mathcal{V}_0)$-based integrable over $F$ with respect to $\Theta$, then $f \in \mathcal{F}_{(F,\Theta_0)}(X, \mathcal{P}_0)$, and*

$$\int_F f \, d\theta = \int_F f \, d\theta_0$$

*holds for all $\theta \in \Theta$.*

*Proof.* Let $\mathcal{P}_0$ be a subcone of $\mathcal{P}$, and let $\mathcal{V}_0 \subset \mathcal{P}_0$ be a neighborhood subsystem of $\mathcal{V}$. Let $\Theta$ be an equibounded family of $\mathfrak{L}(\mathcal{P}, \mathcal{Q})$-valued measures on $\mathfrak{R}$. The family $\Theta_0$ of all restrictions of the measures in $\Theta$ to $(\mathcal{P}_0, \mathcal{V}_0)$ is an equibounded family of $\mathfrak{L}(\mathcal{P}_0, \mathcal{Q})$-valued measures if and only if for every $E \in \mathfrak{R}$ and $w \in \mathcal{W}$ there is $v \in \mathcal{V}_0$ such that $\theta_E(v) \leq w$ for all $\theta \in \Theta$. By our assumption, $\Theta$ satisfies this requirement. Given a set $F \in \mathfrak{A}_{\mathfrak{R}}$, we shall consider $\mathcal{P}_0$-valued functions as elements of the cones $\mathcal{F}_{(F,\Theta)}(X,\mathcal{P})$ or $\mathcal{F}_{(F,\Theta_0)}(X,\mathcal{P}_0)$, respectively. We shall proceed in several steps:

First we observe that every $\mathcal{P}_0$-valued step function $h=\sum_{i=1}^n \chi_{E_i}\otimes a_i$, for $E_i\in\mathfrak{R}$ and $a_i\in\mathcal{P}_0$, is contained in both $\mathcal{F}_{(F,\Theta)}(X,\mathcal{P})$ and $\mathcal{F}_{(F,\Theta_0)}(X,\mathcal{P}_0)$, and we have $\int_F h\,d\theta=\int_F h\,d\theta_0$ for all $\theta\in\Theta$.

In a second step we consider a neighborhood-valued function $s\in\mathcal{F}_{\mathfrak{R}}(X,\mathcal{V}_0)$. As we remarked before, positivity implies that $s$ is also contained in $\mathcal{F}_{\mathfrak{R}}(X,\mathcal{V})$. Given a neighborhood $w\in\mathcal{W}$, the inductive limit neighborhood $\mathfrak{v}_w$ formed by the neighborhoods in $\mathcal{V}$ contains the corresponding neighborhood $\mathfrak{v}_w^0$ formed by the neighborhoods in $\mathcal{V}_0$ as a subset (see Section 4). Thus

$$\begin{aligned}\int_F^{(w)} s\,d\theta_0 &= \sup\left\{\int_F s\,d\theta \;\middle|\; h\in\mathcal{S}_{\mathfrak{R}}(X,\mathcal{P}_0),\ \ h\le s+\mathfrak{v}_w^0\right\}\\ &\le \sup\left\{\int_F h\,d\theta \;\middle|\; h\in\mathcal{S}_{\mathfrak{R}}(X,\mathcal{P}),\ \ h\le s+\mathfrak{v}_w\right\} = \int_F^{(w)} s\,d\theta.\end{aligned}$$

Taking the infima over all $w\in\mathcal{W}$ on both sides yields

$$\int_F s\,d\theta_0 \le \int_F s\,d\theta.$$

Now in a third step, let $E\in\mathfrak{R}$, and let us consider a $\mathcal{P}_0$-valued function that is $(\mathcal{P}_0,\mathcal{V}_0)$-based integrable over $E$ with respect to $\Theta$. We shall first verify that $f\in\mathcal{F}_{(E,\Theta_0)}(X,\mathcal{P}_0)$. Indeed, the former property requires that for $w\in\mathcal{W}$ and $\varepsilon>0$ there is a $\mathcal{V}_0$-valued function $s_{(w,\varepsilon)}\in\mathcal{F}_{\mathfrak{R}}(X,\mathcal{V})$, and a $\mathcal{P}_0$-valued function $f_{(w,\varepsilon)}\in\mathcal{F}_{\mathfrak{R}}(X,\mathcal{P})$ that can be reached from below by $\mathcal{P}_0$-valued step functions, such that

$$f \underset{a.e.E}{\le} f_{(w,\varepsilon)} \underset{a.e.E}{\le} \gamma f + s_{(w,\varepsilon)}$$

for some $1\le\gamma\le 1+\varepsilon$ and such that $\int_E s_{(w,\varepsilon)}\,d\theta\le\varepsilon w$ holds for all $\theta\in\Theta$. As $\mathcal{F}_{\mathfrak{R}}(X,\mathcal{V}_0)=\mathcal{F}_{\mathfrak{R}}(X,\mathcal{V})\cap\mathcal{F}(X,\mathcal{V}_0)$, we have

$$\int_E s_{(w,\varepsilon)}\,d\theta_0 \le \int_E s_{(w,\varepsilon)}\,d\theta \le \varepsilon w$$

for all $\theta_0\in\Theta_0$ by our first step. As we mentioned before, the almost everywhere relation $\underset{a.e.E}{\le}$ with respect to $\theta$ implies the same relation with respect to $\theta_0$. The function $f$ is therefore indeed integrable over $E$ with respect to the family $\Theta_0$ of the restricted measures, that is $f\in\mathcal{F}_{(E,\Theta_0)}(X,\mathcal{P}_0)$.

Now let $w\in\mathcal{W}$ and $\varepsilon>0$. We shall apply Corollary 5.8 with the family $\Theta$ and the given subcone $(\mathcal{P}_0,\mathcal{V}_0)$ to find a sequence $(h_n)_{n\in\mathbb{N}}$ of $\mathcal{P}_0$-valued step functions as in 5.8. Statements (i) and (iii) refer to the measures $\theta\in\Theta$. However, all functions involved are also contained in $\mathcal{F}_{(E,\Theta_0)}(X,\mathcal{P}_0)$, the integrals with respect to the measures in $\Theta$ and in $\Theta_0$ coincide for the step functions $h_n$, and for the function $s\in\mathcal{F}_{\mathfrak{R}}(X,\mathcal{V}_0)$ in (i) we have

$$\int_F s\,d\theta_0 \le \int_F s\,d\theta \le w$$

by our second step. Property (ii) therefore yields together with Proposition 4.7 that

$$\int_E f\,d\theta_0 \le \varliminf_{n\to\infty} \int_E h_n\,d\theta_0$$

holds for all $\theta_0 \in \Theta_0$. Using this together with our first step and the second part of statement (iii) in 5.8, we obtain

$$\int_E f\,d\theta_0 \le \varliminf_{n\to\infty} \int_E h_n\,d\theta_0 = \varliminf_{n\to\infty} \int_E h_n\,d\theta \le \gamma \int_E f\,d\theta_0 + w$$

and, likewise,

$$\begin{aligned}\int_E f\,d\theta \le \varliminf_{n\to\infty} \int_E h_n\,d\theta &= \varliminf_{n\to\infty} \int_E h_n\,d\theta_0 \\ &\le \gamma \int_E f\,d\theta_0 + \int_E s\,d\theta_0 \le \gamma \int_E f\,d\theta_0 + w\end{aligned}$$

with some $1 \le \gamma \le 1+\varepsilon$. Because $w \in \mathcal{W}$ and $\varepsilon > 0$ were arbitrarily chosen, this yields $\int_E f\,d\theta = \int_E f\,d\theta_0$.

Now for the final step of our argument, let $F \in \mathfrak{A}_{\mathfrak{R}}$, and let $f \in \mathcal{F}(X,\mathcal{P})$ be $\mathcal{P}_0$-valued and $(\mathcal{P}_0,\mathcal{V}_0)$-based integrable over $F$. Then $f$ is $(\mathcal{P}_0,\mathcal{V}_0)$-based integrable over the sets $E \cap F$, for all $E \in \mathfrak{R}$, hence $f \in \mathcal{F}_{(E\cap F,\Theta_0)}(X,\mathcal{P}_0)$ by our second step, and $\int_{(E\cap F)} f\,d\theta = \int_{(E\cap F)} f\,d\theta_0$ by our first step. This shows

$$\int_F f\,d\theta = \lim_{E\in\mathfrak{R}} \int_{(E\cap F)} f\,d\theta = \lim_{E\in\mathfrak{R}} \int_{(E\cap F)} f\,d\theta_0 = \int_F f\,d\theta_0.$$

□

**5.11 Sums, Multiples and Order for Measures.** Let $\theta$ and $\vartheta$ be two $\mathfrak{R}$-bounded $\mathfrak{L}(\mathcal{P},\mathcal{Q})$-valued measures, and let $\alpha \ge 0$. We define the $\mathfrak{L}(\mathcal{P},\mathcal{Q})$-valued measures $\theta+\vartheta$ and $\alpha\theta$ by

$$(\theta+\vartheta)_E(a) = \theta_E(a) + \vartheta_E(a)$$

and

$$(\alpha\theta)_E(a) = \alpha\big(\theta_E(a)\big)$$

for $E \in \mathfrak{R}$ and $a \in \mathcal{P}$. The properties of a measure are readily checked.

Corresponding to a subcone $\mathcal{P}_0$ of $\mathcal{P}$ we define an order relation for measures $\theta$ and $\vartheta$ setting

$$\theta \le_{\mathcal{P}_0} \vartheta \qquad \text{if} \qquad \theta_E(a) \le \vartheta_E(a)$$

holds for all $E \in \mathfrak{R}$ and $a \in \mathcal{P}_0$. We write $\theta \le \vartheta$ for the canonical choice of $\mathcal{P}_0 = \mathcal{P}_+ = \{a \in \mathcal{P} \mid a \ge 0\}$. In this case, for any family $\Theta$ of measures and every set $F \in \mathfrak{R}$, every $\mathcal{P}_+$-valued function $f \in \mathcal{F}_{(F,\Theta)}(X,\mathcal{P})$ is seen to be $(\mathcal{P}_0,\mathcal{V})$-based integrable over $F$ with respect to $\Theta$. Note that $\theta \leq \vartheta$ and $\vartheta \leq \theta$ in this sense implies that $\theta = \vartheta$, that is equality for the positive elements implies equality for all elements of $\mathcal{P}$. Indeed, given $E \in \mathfrak{R}$, $a \in \mathcal{P}$ and $w \in \mathcal{W}$ there is $v \in \mathcal{V}$ such that $\theta_E(v) = \vartheta_E(v) \leq w$. There is $\lambda \geq 0$ such that $0 \leq a+\lambda v$. Thus $\theta_E(a+\lambda v) = \vartheta_E(a+\lambda v)$ and $\theta_E(a) \leq \vartheta_E(a)+w$ by the cancellation rules. This shows $\theta_E(a) \leq \vartheta_E(a)$ and likewise, $\vartheta_E(a) \leq \theta_E(a)$.

**Proposition 5.12.** *Let $\theta$ and $\vartheta$ be $\mathfrak{L}(\mathcal{P},\mathcal{Q})$-valued measures, let $\alpha \geq 0$, $F \in \mathfrak{A}_{\mathfrak{R}}$, and let $\mathcal{P}_0$ be a subcone of $\mathcal{P}$.*

*(a) If $f \in \mathcal{F}_{(F,\{\theta,\vartheta\})}(X,\mathcal{P})$, then $f \in \mathcal{F}_{(F,\theta+\vartheta)}(X,\mathcal{P})$ and $\int_F f\,d(\theta+\vartheta) = \int_F f\,d\theta + \int_F f\,d\vartheta$.*
*(b) If $f \in \mathcal{F}_{(F,\theta)}(X,\mathcal{P})$, then $f \in \mathcal{F}_{(F,\alpha\theta)}(X,\mathcal{P})$ and $\int_F f\,d(\alpha\theta) = \alpha\int_F f\,d\theta$.*
*(c) If $\theta \leq_{\mathcal{P}_0} \vartheta$, then $\int_F f\,d\theta \leq \int_F f\,d\vartheta$ holds for every $f \in \mathcal{F}(X,\mathcal{P})$ that is $(\mathcal{P}_0,\mathcal{V})$-based integrable over $F$ with respect to $\Theta = \{\theta,\vartheta\}$.*

*Proof.* Without loss of generality, we may assume that $F = X$. For Part (a), it is clear from our definition of the sum of two measures that our claim, namely $\int_X h\,d(\theta+\vartheta) = \int_X h\,d\theta + \int_X h\,d\vartheta$ holds for all step functions $h \in \mathcal{S}_{\mathfrak{R}}(X,\mathcal{P})$. Let $\Theta = \{\theta,\vartheta\}$. Every zero set for $\Theta$ is obviously a zero set for $\theta+\vartheta$. We shall first show that every function $f \in \mathcal{F}_{(X,\Theta)}(X,\mathcal{P})$ is integrable over every set $E \in \mathfrak{R}$ with respect to $\theta+\vartheta$. Indeed, given $w \in \mathcal{W}$ and $\varepsilon > 0$, let $w' = w/2$ and let the functions $f_{(w',\varepsilon)} \in \mathcal{F}_{\mathfrak{R}}(X,\mathcal{P})$ and $s_{(w',\varepsilon)} \in \mathcal{F}_{\mathfrak{R}}(X,\mathcal{V})$ be as in Definition 5.3, that is

$$f \underset{a.e.E}{\leq} f_{(w',\varepsilon)} \underset{a.e.E}{\leq} \gamma f + s_{(w',\varepsilon)}$$

for some $1 \leq \gamma \leq 1+\varepsilon$ and $\int_E s_{(w',\varepsilon)}\,d\theta \leq \varepsilon w'$ and $\int_E s_{(w',\varepsilon)}\,d\vartheta \leq \varepsilon w'$. The function $s_{(w',\varepsilon)} \in \mathcal{F}_{\mathfrak{R}}(X,\mathcal{V})$ is integrable with respect to every measure on $\mathfrak{R}$, and for any $u \in \mathcal{W}$ we realize that

$$\begin{aligned}\int_E^{(u)} s_{(w',\varepsilon)}\,d(\theta+\vartheta) &= \sup\left\{\int_E h\,d(\theta+\vartheta) \,\middle|\, h \in \mathcal{S}_{\mathfrak{R}}(X,\mathcal{P}),\ \ h \leq s_{(w',\varepsilon)} + \mathfrak{v}_w\right\}\\ &\leq \int_E^{(u)} s_{(w',\varepsilon)}\,d\theta + \int_E^{(u)} s_{(w',\varepsilon)}\,d\vartheta.\end{aligned}$$

Taking the respective infima over all neighborhoods $u \in \mathcal{W}$ and using Lemma I.5.20(c), we infer that

$$\int_E s_{(w',\varepsilon)}\,d(\theta+\vartheta) \leq \int_E s_{(w',\varepsilon)}\,d\theta + \int_E s_{(w',\varepsilon)}\,d\vartheta \leq \varepsilon w.$$

This shows integrability for $f$ over $E$ with respect to the family $\Theta = \{\theta,\vartheta,\theta+\vartheta\}$ of measures. Next for $\mathcal{P}_0 = \mathcal{P}$ and $\mathcal{V}_0 = \mathcal{V}$, the family $\Theta$ from above, a set $E \in \mathfrak{R}$, a neighborhood $w \in \mathcal{W}$ and $\varepsilon \geq 0$ let $(h_n)_{n\in\mathbb{N}}$

be a sequence of step functions in $\mathcal{S}_{\mathfrak{R}}(X,\mathcal{P})$ approaching the function $f \in \mathcal{F}_{(X,\Theta)}(X,\mathcal{P})$ as in Corollary 5.8. Part (iii) of 5.8 then yields

$$\begin{aligned}\int_E f\,d(\theta+\vartheta) &\le \underline{\lim}_{n\to\infty}\int_E h_n\,d(\theta+\vartheta)\\ &\le \overline{\lim}_{n\to\infty}\int_E h_n\,d\theta + \overline{\lim}_{n\to\infty}\int_E h_n\,d\vartheta\\ &\le \gamma\int_E f\,d\theta + \gamma\int_E f\,d\vartheta + 2w.\end{aligned}$$

And similarly,

$$\begin{aligned}\int_E f\,d\theta + \int_E f\,d\vartheta &\le \underline{\lim}_{n\to\infty}\int_E h_n\,d\theta + \underline{\lim}_{n\to\infty}\int_E h_n\,d\vartheta\\ &\le \underline{\lim}_{n\to\infty}\int_E h_n\,d(\theta+\vartheta)\\ &\le \gamma\int_E f\,d(\theta+\vartheta) + w.\end{aligned}$$

This in turn shows

$$\int_E f\,d(\theta+\vartheta) = \int_E f\,d(\theta+\vartheta),$$

since $w \in \mathcal{W}$ and $\varepsilon > 0$ were arbitrarily chosen. The latter equality holds for all $E \in \mathfrak{R}$, hence or claim follows from the definition of the integral over $X$. Part (b) may be verified in a similar way. For Part (c), let $\theta \le_{\mathcal{P}_0} \vartheta$ for a subcone $\mathcal{P}_0$ of $\mathcal{P}$, and let $f \in \mathcal{F}(X,\mathcal{P})$ be $(\mathcal{P}_0,\mathcal{V})$-based integrable over $X$ with respect to $\Theta = \{\theta,\vartheta\}$. For $E \in \mathfrak{R}$, $w \in \mathcal{W}$ and $\varepsilon > 0$ we choose $v \in \mathcal{V}$ such that both $\theta_E(v) \le w$ and $\vartheta_E(v) \le w$. Let $(h_n)_{n\in\mathbb{N}}$ be a sequence of $\mathcal{P}_0$-valued step functions in $\mathcal{S}_{\mathfrak{R}}(X,\mathcal{P})$ approaching the function $f$ as in Theorem 5.7. We have $\int_E h_n\,d\theta \le \int_E h_n\,d\vartheta$ for all $n \in \mathbb{N}$ since $\theta \le_{\mathcal{P}_0} \vartheta$, hence by Part (iii) of 5.7

$$\int_E f\,d\theta \le \underline{\lim}_{n\to\infty}\int_E h_n\,d\theta + w \le \underline{\lim}_{n\to\infty}\int_E h_n\,d\vartheta + w \le \gamma\int_E f\,d\vartheta + 2w.$$

Thus $\int_E f\,d\theta \le \int_E f\,d\vartheta$, since $w \in \mathcal{W}$ and $\varepsilon > 0$ were arbitrary. Our claim now follows from the definition of the integral over a set $F \in \mathfrak{R}$. □

By the *restriction of a measure* $\theta$ on $\mathfrak{R}$ to a subset $F \in \mathfrak{A}_{\mathfrak{R}}$ we mean the measure $\theta|_F$ on $\mathfrak{R}$, defined as

$$(\theta|_F)_E = \theta_{E\cap F}$$

for all $E \in \mathfrak{R}$. It is immediate from the definition of the integral in Section 4 that $f \in \mathcal{F}_{(X,\theta|_F)}(X,\mathcal{P})$ if and only if $f \in \mathcal{F}_{(F,\theta)}(X,\mathcal{P})$ for a function $f \in \mathcal{F}(X,\mathcal{P})$, and that $\int_X f\,d\theta|_F = \int_F f\,d\theta$ in this case.

**5.13 Convergence of Sequences of Measures.** Let $\theta$ and $(\theta_n)_{n\in\mathbb{N}}$ be $\mathfrak{L}(\mathcal{P},\mathcal{Q})$-valued measures on $\mathfrak{R}$. We shall define lower and upper setwise convergence for measures and denote $\theta_n \nearrow \theta$ or $\theta_n \searrow \theta$ if

(i) $\theta_E(a) \leq \underline{\lim}_{n\to\infty} \theta_{n\,E}(a)$ or $\overline{\lim}_{n\to\infty} \theta_{n\,E}(a) \leq \theta_E(a)$ holds for all $E \in \mathfrak{R}$ and $a \in \mathcal{P}$, respectively.

(ii) There is a set $E_0 \in \mathfrak{R}$ such that $\theta|_{(X\setminus E_0)} \leq_{\mathcal{P}} \theta_n|_{(X\setminus E_0)}$ or $\theta_n|_{(X\setminus E_0)} \leq_{\mathcal{P}} \theta|_{(X\setminus E_0)}$ holds for all $n \in \mathbb{N}$, respectively.

We shall denote $\theta_n \longrightarrow \theta$ if both $\theta_n \nearrow \theta$ and $\theta_n \searrow \theta$.

**Lemma 5.14.** *Let $\Theta = \{\theta_n\}_{n\in\mathbb{N}}$ be equibounded $\mathfrak{L}(\mathcal{P},\mathcal{Q})$-valued measures on $\mathfrak{R}$ such that $\theta_n \nearrow \theta$ for a measure $\theta$. Let $E \in \mathfrak{R}$.*

*(a) If $f \in \mathcal{F}_{(E,\Theta)}(X,\mathcal{P})$, then $f \in \mathcal{F}_{(E,\Theta\cup\{\theta\})}(X,\mathcal{P})$.*
*(b) $\int_E f\,d\theta \leq \underline{\lim}_{n\to\infty} \int_E f\,d\theta_n$ for every $f \in \mathcal{F}_{(E,\Theta)}(X,\mathcal{P})$.*
*(c) $\int_E f\,d\theta = \lim_{n\to\infty} \int_E f\,d\theta_n$ for every invertible function $f$ such that both $f \in \mathcal{F}_{(E,\Theta)}(X,\mathcal{P})$ and $-f \in \mathcal{F}_{(E,\Theta)}(X,\mathcal{P})$.*

*Proof.* Let $\Theta = \{\theta_n\}_{n\in\mathbb{N}}$ be equibounded $\mathfrak{L}(\mathcal{P},\mathcal{Q})$-valued measures on such that $\theta_n \nearrow \theta$ for a measure $\theta$. Let $E \in \mathfrak{R}$. We shall defer the proof of Part (a) since it will use elements of the statement of Part (b). For our proof of Part (b) we shall therefore assume that the function $f$ is integrable over $E$ with respect to the family $\bar{\Theta} = \Theta \cup \{\theta\}$. First, let us consider a step function $h = \sum_{i=1}^n \chi_{E_i} \otimes a_i \in \mathcal{S}_{\mathfrak{R}}(X,\mathcal{P})$. Using Lemma I.5.19 we observe that $\theta_n \nearrow \theta$ implies

$$\int_X h\,d\theta = \sum_{i=1}^m \theta_{E_i}(a_i) \leq \sum_{i=1}^m \left( \underline{\lim}_{n\to\infty} \theta_{n\,E_i}(a_i) \right)$$
$$\leq \underline{\lim}_{n\to\infty} \left( \sum_{i=1}^m \theta_{n\,E_i}(a_i) \right) = \underline{\lim}_{n\to\infty} \int_X h\,d\theta_n.$$

Now let $f \in \mathcal{F}_{(E,\bar{\Theta})}(X,\mathcal{P})$. We shall use Corollary 5.8 with $\mathcal{P}_0 = \mathcal{P}$ and $\mathcal{V}_0 = \mathcal{V}$ in order to establish our claim. Given $w \in \mathcal{W}$ and $\varepsilon > 0$ there is a bounded below sequence $(h_k)_{k\in\mathbb{N}}$ of $\mathcal{P}$-valued step functions such that:

(i) there is $1 \leq \gamma \leq 1+\varepsilon$ and $s \in \mathcal{F}_{\mathfrak{R}}(X,\mathcal{V})$ such that $\int_E s\,d\vartheta \leq w$ for all $\vartheta \in \bar{\Theta}$, and $h_n \underset{a.e.E}{\leq} \gamma f + s$ holds for all $n \in \mathbb{N}$;

(ii) $\bar{\Theta}$-almost everywhere on $E$, for $x \in E$ there is $n_0 \in \mathbb{N}$ such that $f(x) \leq h_n(x)$ for all $n \geq n_0$;

(iii) $\int_E f\,d\vartheta \leq \underline{\lim}_{n\to\infty} \int_E h_n\,d\vartheta$ and $\int_E h_n\,d\vartheta \leq \gamma \int_E f\,d\vartheta + w$ for all $n \in \mathbb{N}$ and $\vartheta \in \bar{\Theta}$.

For every $k \in \mathbb{N}$ then we have by the above

$$\int_E h_k\,d\theta \leq \underline{\lim}_{n\to\infty} \int_X h_k\,d\theta_n \leq \gamma\, \underline{\lim}_{n\to\infty} \int_E f\,d\theta_n + w.$$

Note that this argument implies in particular that the sequence $\left(\int_E f\,d\theta_n\right)_{n\in\mathbb{N}}$ is bounded below. Using (iii), we proceed from this and conclude that

$$\int_E f\,d\theta \leq \gamma \varliminf_{n\to\infty} \int_E f\,d\theta_n + w.$$

Claim (b) follows, since this last inequality holds for all $w \in \mathcal{W}$ and $\varepsilon > 0$. In Part (c) we assume in addition that the negative $-f$ of the function $f$ is also contained in $\mathcal{F}_{(E,\bar{\Theta})}(X,\mathcal{P})$. Then Part (b) yields that both sequences $\left(\int_E f\,d\theta_n\right)_{n\in\mathbb{N}}$ and $\left(\int_E (-f)\,d\theta_n\right)_{n\in\mathbb{N}}$ are bounded below, and

$$\int_E f\,d\theta \leq \varliminf_{n\to\infty} \int_E f\,d\theta_n \qquad \text{and} \qquad \int_E (-f)\,d\theta \leq \varliminf_{n\to\infty} \int_E (-f)\,d\theta_n.$$

Thus

$$\begin{aligned}
\varlimsup_{n\to\infty} \int_E f\,d\theta_n &= \varlimsup_{n\to\infty} \int_E f\,d\theta_n + \int_E (-f)\,d\theta + \int_E f\,d\theta \\
&\leq \varlimsup_{n\to\infty} \int_E f\,d\theta_n + \varliminf_{n\to\infty} \int_E (-f)\,d\theta_n + \int_E f\,d\theta \\
&\leq \varlimsup_{n\to\infty} \int_E \big((-f) + f\big)\,d\theta_n + \int_E f\,d\theta \\
&\leq \int_E f\,d\theta \leq \varliminf_{n\to\infty} \int_E f\,d\theta_n,
\end{aligned}$$

and our claim (c) follows. We shall finally prove Part (a) of the lemma: Let $Z \in \mathfrak{A}_R$ be a zero-set for $\Theta = \{\theta_n\}_{n\in\mathbb{N}}$, that is $\theta_{n\,(E\cap Z)} = 0$ for all $n \in \mathbb{N}$ and $E \in \mathfrak{R}$. As $\theta_n \nearrow \theta$, this implies $\theta_{(E\cap Z)}(a) \leq 0$ for all $a \in \mathcal{P}$, hence $\theta_{(E\cap Z)}(a) = 0$ for all $0 \leq a \in \mathcal{P}$. However, for every $a \in \mathcal{P}$ there is $v \in \mathcal{V}$ such that $0 \leq a + v$. Hence

$$\theta_{(E\cap Z)}(a) = \theta_{(E\cap Z)}(a) + \theta_{(E\cap Z)}(v) = \theta_{(E\cap Z)}(a+v) = 0.$$

Thus $\theta_{(E\cap Z)} = 0$. Every zero-set for $\Theta$ is therefore a zero-set for $\Theta \cup \{\theta\}$ as well. Now let $f \in \mathcal{F}_{(|X|,\Theta)}(X,\mathcal{P})$ and let $E \in \mathfrak{R}$. According to Definition 5.3, for every $w \in \mathcal{W}$ and $\varepsilon > 0$ there are functions $f_{(w,\varepsilon)} \in \mathcal{F}_{\mathfrak{R}}(X,\mathcal{P})$ and $s_{(w,\varepsilon)} \in \mathcal{F}_{\mathfrak{R}}(X,\mathcal{V})$ such that

$$f \underset{a.e.E}{\leq} f_{(w,\varepsilon)} \underset{a.e.E}{\leq} \gamma f + s_{(w,\varepsilon)} \qquad \text{and} \qquad \int_E s_{(w,\varepsilon)}\,d\vartheta \leq \varepsilon w$$

for some $1 \leq \gamma \leq 1+\varepsilon$ and all $\vartheta \in \Theta$. The almost everywhere relations refer to the family $\Theta$ and by the above therefore also to $\Theta \cup \{\theta\}$. Because the function $s_{(w,\varepsilon)} \in \mathcal{F}_{\mathfrak{R}}(X,\mathcal{V})$ is integrable over $E$ with respect to every family of measures on $\mathfrak{R}$, we may use Part (b) of the lemma for

$$\int_E s_{(w,\varepsilon)}\,d\theta \le \overline{\lim_{n\to\infty}} \int_E s_{(w,\varepsilon)}\,d\theta_n \le \varepsilon w.$$

This shows that the function $f$ is indeed integrable over $E$ with respect to the family $\Theta \cup \{\theta\}$. □

*Example 5.15.* The following example will demonstrate that a result corresponding to 5.14(b) for upper convergence of measures is not available in general, that is $\theta_n \searrow \theta$ for measures $\theta_n, \theta$ on $\mathfrak{R}$ does not necessarily imply that $\overline{\lim_{n\to\infty}} \int_E f\,d\theta_n \le \int_E f\,d\theta$ holds for every integrable function $f \in \mathcal{F}(X,\mathcal{P})$. For this, let $X = [0,1]$, let $\mathfrak{R}$ be the $\sigma$-algebra of all Borel sets on $X$, and let $\theta$ be the Lebesgue measure. This may be considered as an $\mathfrak{L}(\mathcal{P},\mathcal{Q})$-valued measure if we set $\mathcal{P} = \mathcal{Q} = \overline{\mathbb{R}}$ (see Examples I.2.1). We define the measures $\theta_n$ as $\theta_n(E) = \sqrt{n}\;\theta\big(E \cap [0,\frac{1}{n}]\big)$ for $E \in \mathfrak{R}$. This yields $\theta_n(E) \le \frac{1}{\sqrt{n}}$ for all $E \in \mathfrak{R}$ and $n \in \mathbb{N}$, hence $\theta_n \longrightarrow 0$, that is the zero measure on $\mathfrak{R}$. Now consider the function $f$ on $X$ defined as $f(x) = \frac{1}{\sqrt{x}}$ for $x > 0$ and $f(0) = 0$. As $f$ is positive and measurable, it is contained in $\mathcal{F}_{\mathfrak{R}}(X,\mathcal{P})$, hence in $\mathcal{F}_{(E,\Theta)}(X,\mathcal{P})$, where $\Theta$ is the equibounded family $\{\theta_n\}_{n\in\mathbb{N}}$. We calculate $\int_X f\,d\theta_n = \sqrt{n}\int_0^{\frac{1}{n}} f\,d\theta = 2$, hence $\int_X f\,d\theta_n \not\to \int_X f\,d0 = 0$, indeed. Note that Part (c) of Lemma 5.14 does not apply in this case. The function $f$ is in fact invertible in $\mathcal{F}(X,\mathcal{P})$ and its inverse $-f$ is integrable with respect to each of the measures $\theta_n$. Indeed, given $\varepsilon > 0$ we may choose $f_\varepsilon(x) = -f(x)$ for $x > \varepsilon$ and $f_\varepsilon(x) = 0$ else. Then $f_\varepsilon$ is bounded below, hence in $\mathcal{F}_{\mathfrak{R}}(X,\mathcal{P})$, and we have $-f \le f_\varepsilon \le -f + s_\varepsilon$, where $s_\varepsilon(x) = 0$ for $x = 0$ or $x \ge \varepsilon$, and $s_\varepsilon(x) = f(x)$ else. (This function $s_\varepsilon$ is $\big(\mathcal{V}\cup\{0\}\big)$-valued as required in Definition 4.12, since the neighborhood system $\mathcal{V}$ of $\overline{\mathbb{R}}$ consists of all strictly positive reals.) For $\varepsilon \le \frac{1}{n}$ we calculate $\int_X s_\varepsilon\,d\theta_n = 2\sqrt{n\varepsilon}$. Thus $-f \in \mathcal{F}_{(E,\theta_n)}(X,\mathcal{P})$ for all $n \in \mathbb{N}$, but $-f$ is not contained in $\mathcal{F}_{(E,\Theta)}(X,\mathcal{P})$ as required in 5.14(c).

**5.16 Residual Components.** Let $(\theta_n)_{n\in\mathbb{N}}$ be an equibounded sequence of measures, let $F \in \mathfrak{A}_{\mathfrak{R}}$ and $f \in \mathcal{F}_{(F,\{\theta_n\})}(X,\mathcal{P})$. We define the *residual component of* $f$ *on* $F$ *with respect to* $(\theta_n)_{n\in\mathbb{N}}$ as follows: Let $\mathfrak{F}$ be the collection of all sequences $(E_n)_{n\in\mathbb{N}}$ of sets in $\mathfrak{R}$ such that $E_n \subset F$, $E_n \supset E_{n+1}$ and $\bigcap_{n\in\mathbb{N}} E_n = \emptyset$. Recall that integrability for a function $f \in \mathcal{F}(X,\mathcal{P})$ over $F$ requires integrability over all subsets $E \in \mathfrak{R}$ of $F$. Thus for $f \in \mathcal{F}_{(F,\{\theta_n\})}(X,\mathcal{P})$ we define

$$\mathfrak{Rs}(\theta_n, F, f) = \sup_{(E_m)\in\mathfrak{F}} \left\{ \lim_{m\to\infty} \left( \overline{\lim_{n\to\infty}} \int_{E_m} f\,d\theta_n \right) \right\}.$$

This appears to be a rather unwieldy expression. It will however turn out to be useful for our continuing investigations.

**Lemma 5.17.** *Let $(\theta_n)_{n\in\mathbb{N}}$ be an equibounded sequence of measures, and let $F \in \mathfrak{A}_{\mathfrak{R}}$. Then*

*(a) $\mathfrak{Rs}(\theta_n, F, f) \geq 0$ for all $f \in \mathcal{F}_{(F,\{\theta_n\})}(X, \mathcal{P})$.*
*(b) If $\theta_n \leq \omega$ for a measure $\omega$ and all $n \in \mathbb{N}$, then*
*$\mathfrak{Rs}(\theta_n, F, f) \leq \mathfrak{O}\left(\int_F f\,d\omega\right)$ for all $f \in \mathcal{F}_{(|F|,\{\theta_n,\omega\})}(X, \mathcal{P})$.*

*Proof.* Part (a) is trivial, as we may choose the stationary sequence $(E_m)_{m\in\mathbb{N}} \in \mathfrak{F}$, where $E_m = \emptyset$ for all $n \in \mathbb{N}$. For Part (b), suppose that $\theta_n \leq \omega$ holds for a measure $\omega$ on $\mathfrak{R}$ and all $n \in \mathbb{N}$, let $\Theta = \{\theta_n, \omega\}_{n\in\mathbb{N}}$, let $f \in \mathcal{F}_{(F,\Theta)}(X, \mathcal{P})$ and $(E_m)_{m\in\mathbb{N}} \in \mathfrak{F}$. For every $w \in \mathcal{W}$ there is by Proposition 5.4 a function $s \in \mathcal{F}_{\mathfrak{R}}(X, \mathcal{V})$ and $\lambda \geq 0$ such that $0 \underset{a.e.E_1}{\leq} f + s$ and $\int_{E_1} s\,d\vartheta \leq \lambda w$ for all $\vartheta \in \Theta$. Because $s \geq 0$, this yields for all $m \in \mathbb{N}$

$$\overline{\lim_{n\to\infty}} \int_{E_m} f\,d\theta_n \leq \overline{\lim_{n\to\infty}} \int_{E_m} (f+s)\,d\theta_n \leq \int_{E_m} (f+s)\,d\omega.$$

Thus by Proposition 4.18(b) and Proposition I.5.11

$$\begin{aligned}
\underline{\lim_{m\to\infty}} \left( \overline{\lim_{n\to\infty}} \int_{E_m} f\,d\theta_n \right) &\leq \underline{\lim_{m\to\infty}} \int_{E_m} (f+s)\,d\omega \\
&\leq \mathfrak{O}\left( \int_{E_1} (f+s)\,d\omega \right) \\
&= \mathfrak{O}\left( \int_{E_1} f\,d\omega \right) + \mathfrak{O}\left( \int_{E_1} s\,d\omega \right) \\
&\leq \mathfrak{O}\left( \int_{E_1} f\,d\omega \right) + w,
\end{aligned}$$

since $\mathfrak{O}\left(\int_{E_1} s\,d\omega\right) \leq \varepsilon w$ for all $\varepsilon > 0$. Because $w \in \mathcal{W}$ was arbitrarily chosen, and because $\mathcal{Q}$ carries the weak preorder, this yields

$$\underline{\lim_{m\to\infty}} \left( \overline{\lim_{n\to\infty}} \int_{E_m} f\,d\theta_n \right) \leq \mathfrak{O}\left( \int_{E_1} f\,d\omega \right).$$

Furthermore, Proposition 4.15(c) states that $\mathfrak{O}\left(\int_{E_1} f\,d\omega\right) \leq \mathfrak{O}\left(\int_F f\,d\omega\right)$. Now combining all of the above, we have indeed

$$\mathfrak{Rs}(\theta_n, F, f) = \sup_{(E_m)\in\mathfrak{F}} \left\{ \underline{\lim_{m\to\infty}} \left( \overline{\lim_{n\to\infty}} \int_{E_m} f\,d\theta_n \right) \right\} \leq \mathfrak{O}\left( \int_F f\,d\omega \right),$$

as claimed. □

Lemma 5.17(b) implies in particular that for a stationary sequence $(\theta_n)_{n\in\mathbb{N}}$ of measures, that is $\theta_n = \theta$ for all $n \in \mathbb{N}$, we have $\mathfrak{Rs}(\theta_n, F, f) \leq \mathfrak{O}\left(\int_F f\,d\theta\right)$ for all $f \in \mathcal{F}_{(|F|,\theta)}(X, \mathcal{P})$. This leads to the following notation:

For a set $F \in \mathfrak{A}_R$, and an equibounded sequence $(\theta_n)_{n\in\mathbb{N}}$ of measures, a measure $\theta$ and a family $\mathfrak{F}$ of functions in $\mathcal{F}_{(F,\{\theta_n,\theta\})}(X,\mathcal{P})$ we shall denote

$$(\theta_n) \overset{F}{\underset{\mathfrak{F}}{\prec}} \theta \qquad \text{if} \qquad \mathfrak{Rs}(\theta_n, F, f) \leq \mathfrak{O}\left(\int_F f\,d\theta\right)$$

holds for all $f \in \mathfrak{F}$. Setwise convergence of the measures $\theta_n$ towards $\theta$, that is $\theta_n \longrightarrow \theta$, does however not necessarily imply that $(\theta_n) \overset{F}{\underset{\{f\}}{\prec}} \theta$ holds for every integrable function $f \in \mathcal{F}_{(|F|,\{\theta_n,\theta\})}(X,\mathcal{P})$, as our preceding Example 5.15 can demonstrate. Indeed, let us calculate the residual component of the function $f$ in 5.15 on the interval $F = [0,1]$ with respect to the given sequence $(\theta_n)_{n\in\mathbb{N}}$ of measures on $[0,1]$. First, let $(E_m)_{m\in\mathbb{N}}$ be a sequence of Borel sets in $[0,1]$ such that $E_m \supset E_{m+1}$ and $\bigcap_{m\in\mathbb{N}} E_m = \emptyset$. Then

$$\int_{E_m} f\,d\theta_n \leq \int_{[0,1]} f\,d\theta_n \leq 2$$

for all $k,l \in \mathbb{N}$. This shows $\mathfrak{Rs}(\theta_n, F, f) \leq 2$. For $E_m = \left[0, \frac{1}{m}\right]$, on the other hand, we have $E_m \supset E_{m+1}$ and $\bigcap_{m\in\mathbb{N}} E_m = \emptyset$, and $\int_{E_m} f\,d\theta_n = 2$ whenever $n \geq m$. Thus $\overline{\lim_{n\to\infty}} \int_{E_m} f\,d\theta_n \geq 2$ for all $m \in \mathbb{N}$, and therefore $\mathfrak{Rs}(\theta_n, F, f) \geq 2$. Together with the above, this yields $\mathfrak{Rs}(\theta_n, F, f) = 2$. But we have $\theta_n \longrightarrow 0$.

A stronger requirement on the integrability of the function $f$ will however avoid such cases.

**5.18 Strongly Integrable Functions.** Let $\Theta$ be a an equibounded family of measures, and let $E \in \mathfrak{R}$. We shall say that a function $f \in \mathcal{F}(X,\mathcal{P})$ is *strongly integrable over* $E$ *with respect to* $\Theta$ if it is integrable over $E$ in the sense of 5.3, and if in addition, for every $w \in \mathcal{W}$ there is a step function $h \in \mathcal{S}_{\mathfrak{R}}(X,\mathcal{P})$ such that $\int_G f\,d\theta \leq \int_G h\,d\theta + w$ and $\int_G h\,d\theta$ is $w$-bounded relative to $\int_G f\,d\theta$ in $\mathcal{Q}$, for all $\theta \in \Theta$ and every subset $G \in \mathfrak{R}$ of $E$. Note that this requirement strengthens the corresponding property from Theorem 5.7 which holds for integrable functions in general.

Similarly, for a set $F \in \mathfrak{A}_{\mathfrak{R}}$, a function $f \in \mathcal{F}(X,\mathcal{P})$ is *strongly integrable over* $F$ *with respect to* $\Theta$ if it is integrable over $F$ in the sense of 5.3 and strongly integrable over the sets $E \cap F$ for all $E \in \mathfrak{R}$. Because strong integrability over a set $E \in \mathfrak{R}$ obviously implies strong integrability over every subset $G \in \mathfrak{R}$ of $E$, this last part of our definition is consistent with the first one.

It is straightforward to verify that the strongly integrable functions form a subcone of $\mathcal{F}_{(F,\Theta)}(X,\mathcal{P})$.

**Lemma 5.19.** *Let* $\Theta = \{\theta, \theta_n\}_{n\in\mathbb{N}}$ *be equibounded measures on* $\mathfrak{R}$ *such that* $\theta_n \searrow \theta$. *Let* $F \in \mathfrak{A}_{\mathfrak{R}}$, *and suppose that the function* $f \in \mathcal{F}_{(|F|,\Theta)}(X,\mathcal{P})$ *is strongly integrable over* $F$ *with respect to* $\Theta$. *Then* $(\theta_n) \overset{F}{\underset{\{f\}}{\prec}} \theta$.

*Proof.* Let $\Theta = \{\theta, \theta_n\}_{n\in\mathbb{N}}$ be equibounded measures such that $\theta_n \searrow \theta$. As in the first step of the proof of Lemma 5.14, one easily verifies that

$$\varlimsup_{n\to\infty} \int_G h\,d\theta_n \le \int_G h\,d\theta$$

holds for every step function $h \in \mathcal{S}_{\mathfrak{R}}(X,\mathcal{P})$ and all $G \in \mathfrak{A}_{\mathfrak{R}}$. Let $F \in \mathfrak{A}_{\mathfrak{R}}$ and suppose that the function $f \in \mathcal{F}_{(F,\Theta)}(X,\mathcal{P})$ is strongly integrable over $F$ with respect to $\Theta$. Let $E_m \in \mathfrak{R}$ for $m \in \mathbb{N}$ be subsets of $F$ such that $E_m \supset E_{m+1}$ and $\bigcap_{m\in\mathbb{N}} E_m = \emptyset$. Following 5.18, the function $f$ is strongly integrable over the set $E = E_1$. Given $w \in \mathcal{W}$, we choose a step function $h = \sum_{i=1}^{n} \chi_{G_i \otimes} a_i \in \mathcal{S}_{\mathfrak{R}}(X,\mathcal{P})$ as in the first part of 5.18, that is $\int_G h\,d\theta \in \mathcal{B}_w\big(\int_G f\,d\theta\big)$, and $\int_G f\,d\theta \le \int_G h\,d\theta + w$ holds for all $\theta \in \Theta$ and every subset $G \in \mathfrak{R}$ of $E$, in particular

$$\int_{E_m} f\,d\theta_n \le \int_{E_m} h\,d\theta_n + w$$

holds for all $m, n \in \mathbb{N}$. Thus

$$\varlimsup_{n\to\infty} \int_{E_m} f\,d\theta_n \le \varlimsup_{n\to\infty} \int_{E_m} h\,d\theta_n + w \le \int_{E_m} h\,d\theta + w$$

for every $m \in \mathbb{N}$, and consequently

$$\varliminf_{m\to\infty} \Big( \varlimsup_{n\to\infty} \int_{E_m} f\,d\theta_n \Big) \le \varliminf_{m\to\infty} \int_{E_m} h\,d\theta + w \le \mathfrak{O}\Big( \int_E h\,d\theta \Big) + w$$

by Proposition 4.18(b). Because $\int_E h\,d\theta$ is $w$-bounded relative to $\int_E f\,d\theta$, and because $\int_E f\,d\theta$ is bounded relative to $\int_F f\,d\theta$ by Proposition 4.15(c), we have $\mathfrak{O}\big(\int_E h\,d\theta\big) \preccurlyeq_w \mathfrak{O}\big(\int_F f\,d\theta\big)$ by Proposition I.5.13(a). The latter implies

$$\mathfrak{O}\Big(\int_E h\,d\theta\Big) \le \mathfrak{O}\Big(\int_F f\,d\theta\Big) + w.$$

Thus, summarizing,

$$\varliminf_{m\to\infty} \Big( \varlimsup_{n\to\infty} \int_{E_m} f\,d\theta_n \Big) \le \mathfrak{O}\Big( \int_F f\,d\theta \Big) + 2w.$$

This holds for all $w \in \mathcal{W}$ and all sequences of sets $E_k \in \mathfrak{R}$ such that $E_k \subset F$ and $\bigcap_{k\in\mathbb{N}} E_k$, and therefore demonstrates

$$\mathfrak{Rs}(\theta_n, F, f) \le \mathfrak{O}\Big( \int_F f\,d\theta \Big),$$

our claim. □

**Lemma 5.20.** *Let $\Theta = \{\theta, \theta_n\}_{n\in\mathbb{N}}$ be equibounded $\mathfrak{L}(\mathcal{P},\mathcal{Q})$-valued measures on $\mathfrak{R}$ such that $\theta_n \searrow \theta$. Let $E \in \mathfrak{R}$.*

*(a) Let $f_n, f, f^* \in \mathcal{F}_{(E,\Theta)}(X,\mathcal{P})$ and $v \in \mathcal{V}$. If both $f_n(x) \preccurlyeq_v f(x)$ and $f_n(x) \preccurlyeq_v f^*(x)$ holds $\Theta$-almost everywhere on $E$ for all $n \in \mathbb{N}$, then $\overline{\lim}_{n\to\infty} \int_E f_n\, d\theta_n \le \int_E f\, d\theta + \mathfrak{Rs}(\theta_n, E, f^*) + \mathfrak{D}\big(\sup\{\theta_E(v) \mid \theta \in \Theta\}\big)$.*

*(b) Let $f \in \mathcal{F}_{(E,\Theta)}(X,\mathcal{P})$. If $(\theta_n) \underset{\{f\}}{\overset{E}{\prec}} \theta$, then $\overline{\lim}_{n\to\infty} \int_E f\, d\theta_n \le \int_E f\, d\theta$.*

*Proof.* Let $\Theta = \{\theta, \theta_n\}_{n\in\mathbb{N}}$ be equibounded measures such that $\theta_n \searrow \theta$. As seen before, this implies

$$\overline{\lim_{n\to\infty}} \int_G h\, d\theta_n \le \int_G h\, d\theta$$

for every step function $h \in \mathcal{S}_{\mathfrak{R}}(X,\mathcal{P})$ and every $G \in \mathfrak{A}_{\mathfrak{R}}$. Let $E \in \mathfrak{R}$, $v \in \mathcal{V}$, and let $f_n, f, f^* \in \mathcal{F}_{(E,\Theta)}(X,\mathcal{P})$ such that both $f_n \preccurlyeq_v f$ and $f_n \preccurlyeq_v f^*$ holds $\Theta$-almost everywhere on $E$ for all $n \in \mathbb{N}$. Let us abbreviate

$$d = \mathfrak{D}\big(\sup\{\theta_E(v) \mid \theta \in \Theta\}\big) \in \mathcal{Q}.$$

Recall from Proposition I.5.11(b) that $\alpha d = d$ for all $\alpha > 0$. Given $w \in \mathcal{W}$ and $\varepsilon > 0$, we shall use Corollary 5.8 for the function $f$, with $\mathcal{P}_0 = \mathcal{P}$ and $\mathcal{V}_0 = \mathcal{V}$, in order to obtain a sequence $(h_k)_{k\in\mathbb{N}}$ of step functions satisfying 5.8(i), (ii) and (iii). We set

$$G_m = \{x \in E \mid f(x) \preccurlyeq_v h_k(x) \quad \text{for all} \quad k \ge m\}$$

for $m \in \mathbb{N}$. Following Theorem 1.6, the sets $G_m$ are contained in $\mathfrak{R}$, and we have $G_m \subset G_{m+1}$. If we set $G = \bigcup_{m\in\mathbb{N}} G_m$, then 5.8(ii) implies that $E \setminus G \in \mathfrak{Z}(\Theta)$, that is $\int_E g\, d\vartheta = \int_G g\, d\vartheta$ for the functions $g = f_n, f, f^*$ and all $\vartheta \in \Theta$. Because $f_n(x) \preccurlyeq_v h_m(x)$ holds $\Theta$-almost everywhere on $G_m$, and $f_n(x) \preccurlyeq_v f^*(x)$ holds $\Theta$-almost everywhere on $E$ for all $n \in \mathbb{N}$, and because $\mathfrak{D}(\vartheta_E(v)) \le d$ for all $\theta \in \Theta$, Proposition 4.16 yields

$$\int_{G_m} f_n\, d\vartheta \le \int_{G_m} h_m\, d\vartheta + d \qquad \text{and} \qquad \int_{(G\setminus G_m)} f_n\, d\vartheta \le \int_{G_m} f^*\, d\vartheta + d.$$

for all $\vartheta \in \Theta$ and $n \in \mathbb{N}$. Thus

$$\begin{aligned}\int_E f_n\, d\theta_n &= \int_{G_m} f_n\, d\theta_n + \int_{(G\setminus G_m)} f_n\, d\theta_n \\ &\le \int_{G_m} h_m\, d\theta_n + \int_{(G\setminus G_m)} f^*\, d\theta_n + d\end{aligned}$$

holds for all $m, n \in \mathbb{N}$. Let $E_m = G \setminus G_m$. Then $E_m \supset E_{m+1}$ and $\bigcap_{m\in\mathbb{N}} E_m = \emptyset$. Using this, we proceed with our argument. For a fixed $m \in \mathbb{N}$, we let $n$ tend to infinity in the preceding inequality, and obtain

$$\begin{aligned}\overline{\lim_{n\to\infty}}\int_E f\,d\theta_n &\le \overline{\lim_{n\to\infty}}\int_{G_m} h_m\,d\theta_n + \overline{\lim_{n\to\infty}}\int_{E_m} f^*\,d\theta_n + d\\ &\le \int_{G_m} h_m\,d\theta + \overline{\lim_{n\to\infty}}\int_{E_m} f^*\,d\theta_n + d\\ &\le \gamma\int_{G_m} f\,d\theta + \overline{\lim_{n\to\infty}}\int_{E_m} f^*\,d\theta_n + d + w.\end{aligned}$$

with some $1 \le \gamma \le 1+\varepsilon$. Finally, we let $m$ tend to infinity as well and use Proposition 4.18(a) for

$$\lim_{m\to\infty}\int_{G_m} f\,d\theta = \int_G f\,d\theta = \int_E f\,d\theta.$$

Thus

$$\begin{aligned}\overline{\lim_{n\to\infty}}\int_E f\,d\theta_n &\le \gamma\int_E f\,d\theta + \lim_{m\to\infty}\Big(\overline{\lim_{n\to\infty}}\int_{E_m} f^*\,d\theta_n\Big) + d + w\\ &\le \gamma\int_E f\,d\theta + \mathfrak{Rs}(\theta_n, E, f^*) + d + w\\ &\le \gamma\Big(\int_E f\,d\theta + \mathfrak{Rs}(\theta_n, E, f^*) + d\Big) + w.\end{aligned}$$

The last inequality holds for all $w \in \mathcal{W}$ and $\varepsilon > 0$, hence

$$\overline{\lim_{n\to\infty}}\int_E f\,d\theta_n \le \int_E f\,d\theta + \mathfrak{Rs}(\theta_n, E, f^*) + d,$$

since $\mathcal{Q}$ is endowed with the weak preorder.

For Part (b), we set $f_n = f = f^*$ in Part (a). If $(\theta_n) \overset{E}{\underset{\{f\}}{\prec}} \theta$, that is $\mathfrak{Rs}(\theta_n, E, f) \le \mathfrak{O}\big(\int_E f\,d\theta\big)$ holds in addition, then

$$\int_E f\,d\theta + \mathfrak{Rs}(\theta_n, E, f) \le \int_E f\,d\theta + \mathfrak{O}\Big(\int_E f\,d\theta\Big) = \int_E f\,d\theta$$

follows from Proposition I.5.14. Given $w \in \mathcal{W}$, we choose $v \in \mathcal{V}$ such that $\theta_E(v) \le w$ for all $\theta \in \Theta$. Then obviously $\mathfrak{O}\big(\sup\{\theta_E(v) \mid \theta \in \Theta\}\big) \le w$ holds as well. Part (a) therefore yields

$$\overline{\lim_{n\to\infty}}\int_E f\,d\theta_n \le \int_E f\,d\theta + \mathfrak{Rs}(\theta_n, E, f) + w \le \int_E f\,d\theta + w$$

for all $w \in \mathcal{W}$. Thus indeed

$$\overline{\lim_{n\to\infty}}\int_E f\,d\theta_n \le \int_E f\,d\theta,$$

since $\mathcal{Q}$ carries the weak preorder. □

Our upcoming convergence theorems will imply that the statements of Lemmas 5.14 and 5.20 do indeed extend to integrals over sets $F \in \mathfrak{A}_{\mathfrak{R}}$, if the concerned functions are contained in $\mathcal{F}_{(|F|,\{\theta_n,\theta\})}(X,\mathcal{P})$.

**Lemma 5.21.** *Let* $\Theta = \{\theta_n\}_{n\in\mathbb{N}}$ *be equibounded* $\mathfrak{L}(\mathcal{P},\mathcal{Q})$*-valued measures on* $\mathfrak{R}$ *such that* $\theta_n \longrightarrow \theta$ *for a measure* $\theta$*. Let* $F \in \mathfrak{A}_{\mathfrak{R}}$*. If* $f \in \mathcal{F}_{(F,\Theta)}(X,\mathcal{P})$*, then* $f \in \mathcal{F}_{(F,\Theta\cup\{\theta\})}(X,\mathcal{P})$*.*

*Proof.* We may assume that $F = X$, since for a function $f \in \mathcal{F}(X,\mathcal{P})$ integrability over $F$ means equivalently that the function $\chi_F \otimes f$ is integrable over $X$. Let $f \in \mathcal{F}_{(X,\Theta)}(X,\mathcal{P})$. Then $f \in \mathcal{F}_{(E,\Theta)}(X,\mathcal{P})$ for every $E \in \mathfrak{R}$, hence $f \in \mathcal{F}_{(E,\Theta\cup\{\theta\})}(X,\mathcal{P})$ by Lemma 5.14(a). Now let $E_0 \in \mathfrak{R}$ and $v \in \mathcal{V}$ be as in Definition 5.13(ii), that is $\theta|_{(X\setminus E_0)} = \theta_n|_{(X\setminus E_0)}$ holds for all $n \in \mathbb{N}$. Let $E \in \mathfrak{R}$ such that $E_0 \subset E$ and fix $n_0 \in \mathbb{N}$. Then

$$\int_E f\,d\theta = \int_{E_0} f\,d\theta + \int_{E\setminus E_0} f\,d\theta = \int_{E_0} f\,d\theta + \int_{E\setminus E_0} f\,d\theta_{n_0}.$$

Hence

$$\lim_{E\in\mathfrak{R}} \int_E f\,d\theta = \int_{E_0} f\,d\theta + \lim_{E\in\mathfrak{R}} \int_{E\setminus E_0} f\,d\theta_{n_0} = \int_{E_0} f\,d\theta + \int_{X\setminus E_0} f\,d\theta_{n_0}.$$

The function $f$ is therefore indeed integrable over $X$ with respect to $\theta$, and we infer that $f \in \mathcal{F}_{(E,\Theta\cup\{\theta\})}(X,\mathcal{P})$. □

**5.22 Convergence of Sequences in $\mathcal{F}(X,\mathcal{P})$.** In Section 3 we introduced several notions of pointwise convergence for sequences of $\mathcal{P}$-valued functions. They refer to the lower and upper relative topologies of $\mathcal{P}$, that is for a subset $F$ of $X$, a sequence $(f_n)_{n\in\mathbb{N}}$ and a function $f$ in $\mathcal{F}(X,\mathcal{P})$ we denote $f_n \nearrow_F f$ or $f_n \searrow_F f$ if for every $x \in F$, $v \in \mathcal{V}$ and $\varepsilon > 0$ there is $n_0$ such that

$$f(x) \in v_\varepsilon\big(f_n(x)\big) \qquad \text{or} \qquad f_n(x) \in v_\varepsilon\big(f(x)\big)$$

for all $n \geq n_0$, respectively. $f_n \underset{F}{\rightarrow} f$ means that both $f_n \nearrow_F f$ and $f_n \searrow_F f$. Correspondingly, if $\Theta$ is a family of measures on $\mathfrak{R}$, then we shall denote $f_n \nearrow_{a.e.F} f$, $f_n \searrow_{a.e.F} f$ or $f_n \underset{a.e.F}{\rightarrow} f$ if this convergence holds $\Theta$-almost everywhere on $F$, that is on a subset $F \setminus Z$ with some $Z \in \mathfrak{Z}(\Theta)$.

The following version of Fatou's lemma is the first of our main convergence theorems. It refers to lower convergence for both functions and measures.

**Theorem 5.23.** *Let* $\Theta = \{\theta, \theta_n\}_{n\in\mathbb{N}}$ *be equibounded* $\mathfrak{L}(\mathcal{P},\mathcal{Q})$*-valued measures on* $\mathfrak{R}$ *such that* $\theta_n \nearrow \theta$*. Let* $F \in \mathfrak{A}_{\mathfrak{R}}$*, and let* $f_n, f, f_*, f_{**} \in \mathcal{F}_{(|F|,\Theta)}(X,\mathcal{P})$ *such that* $(\theta_n) \underset{\{f_*\}}{\overset{F}{\prec}} \theta$*. Suppose that* $f_{**} \underset{a.e.F}{\leq} f_n + f_*$ *for all* $n \in \mathbb{N}$*, and that* $f_n \nearrow_{a.e.F} f$*. Then*

$$\int_F f\,d\theta \leq \varliminf_{n\to\infty} \int_F f_n\,d\theta_n + \mathfrak{O}\left(\int_F f_*\,d\theta\right).$$

*Proof.* Without loss of generality, we may assume that $F = X$. Indeed, the respective integrals over $F \in \mathfrak{A}_{\mathfrak{R}}$ equal the integrals over $X$ for the products of the concerned functions with $\chi_F$, and these products satisfy the conditions of the theorem with $X$ in place of $F$. Also, we may assume that the required convergence and boundedness properties hold everywhere on $X$ instead of $\Theta$-everywhere. Indeed, let $\mathfrak{Z}(\Theta)$ be the family of zero subsets of $X$. Then $f_n \nearrow_{a.e.X} f$ means that $f_n \nearrow_{(X \setminus Y)} f$ for some $Y \in \mathfrak{Z}(\theta)$. Using the fact that $\mathfrak{Z}(\Theta)$ contains countable unions of its members, we can find $Y' \in \mathfrak{Z}(\Theta)$ such that $\chi_{(X \setminus Y')} \otimes f_{**} \leq \chi_{(X \setminus Y')} \otimes (f_n + f_*)$ holds for all $n \in \mathbb{N}$ everywhere on $X$. Let $Z = Y \cup Y' \in \mathfrak{Z}(\Theta)$. The functions $f_n' = \chi_{(X \setminus Z)} \otimes f_n$, $f' = \chi_{(X \setminus Z)} \otimes f$ and $f_*' = \chi_{(X \setminus Z)} \otimes f_*$, then fulfill everywhere all the assumptions of the theorem and their respective integrals coincide with those of the given functions.

We shall proceed using these simplified assumptions of the theorem for the measures $\theta_n$ and $\theta$ and the functions $f_n, f, f_{**}$ and $f_*$. In a first step of this proof we shall discuss the respective integrals of the functions involved over a set $E \in \mathfrak{R}$. Let $w \in \mathcal{W}$ be fixed. Following Proposition 5.4, there is $s \in \mathcal{F}_{\mathfrak{R}}(X, \mathcal{V})$ and $\lambda > 0$ such that $\int_E s \, d\vartheta \leq \lambda w$ for all $\vartheta \in \Theta$ and both $0 \lesssim_{a.e.E} f + s$. Using a similar argument as above, that is the replacement of the functions $f_n$ and $f$ by suitable functions $f_n'$ and $f'$ which agree with the former ones $\Theta$-almost everywhere, we may also assume that the last relation holds indeed everywhere on $E$. Next we choose $0 < \varepsilon < \min\{1, \frac{1}{3\lambda}\}$.

According to Definition 5.3 (see also 4.12), we may assume that $s(x) \in \mathcal{V}$ for all $x \in X$. Thus, under the (now simplified) assumptions of the theorem, for every $x \in E$ there is $n_0 \in \mathbb{N}$ such that $f(x) \in \big(s(x)\big)_\varepsilon \big(f_n(x)\big)$ that is $f(x) \leq \gamma f_n(x) + \varepsilon s(x)$ for all $n \geq n_0$ with some $1 \leq \gamma \leq 1$. According to Lemma I.4.1(c), the latter implies that

$$f(x) \leq (1+\varepsilon) f_n(x) + \varepsilon(1+1+\varepsilon) s(x) \leq (1+\varepsilon) f_n(x) + 3\varepsilon s(x)$$

for all $n \geq n_0$. We choose a neighborhood $v \in \mathcal{V}$ such that $\vartheta_E(v) \leq w$ for all $\vartheta \in \Theta$. Following Theorem 1.6, all the sets

$$E_m = \{x \in E \mid f(x) \preccurlyeq_v (1+\varepsilon) f_n(x) + 3\varepsilon s(x) \quad \text{for all} \quad n \geq m\}$$

are in $\mathfrak{R}$, we have $E_m \subset E_{m+1}$, and $\bigcup_{m \in \mathbb{N}} E_m = E$ by the above. Thus

$$f(x) \preccurlyeq_v (1+\varepsilon) f_n(x) + 3\varepsilon s(x)$$

for all $x \in E_m$ and $n \geq m$. Now Proposition 4.16 yields that

$$\begin{aligned} \int_{E_m} f \, d\vartheta &\leq (1+\varepsilon) \int_{E_m} f_n \, d\vartheta + 3\varepsilon \int_{E_m} s \, d\vartheta + \mathfrak{O}\big(\vartheta(E_m, v)\big) \\ &\leq (1+\varepsilon) \int_{E_m} f_n \, d\vartheta + w \end{aligned}$$

holds for all $\vartheta \in \Theta$. The last part of the inequality follows, since $\int_{E_m} s\,d\vartheta \leq \int_E s\,d\vartheta \leq \lambda w$, $3\varepsilon\lambda < 1$ and $\mathfrak{O}(\vartheta(E_m, v)) \leq \mathfrak{O}(\vartheta(E, v)) \leq \varepsilon' w$ for all $\varepsilon' > 0$. Next we use $f_{**} \underset{a.e. X}{\leq} f_n + f_*$ for

$$\int_{E \setminus E_m} f_{**}\,d\vartheta \leq \int_{E \setminus E_m} f_n\,d\vartheta + \int_{E \setminus E_m} f_*\,d\vartheta,$$

multiply the latter by $(1+\varepsilon)$ and add it to the preceding inequality for

$$\int_{E_m} f\,d\vartheta + (1+\varepsilon) \int_{E \setminus E_m} f_{**}\,d\vartheta \leq (1+\varepsilon)\left(\int_E f_n\,d\vartheta + \int_{E \setminus E_m} f_*\,d\vartheta\right) + w.$$

The latter holds true for all $m \in \mathbb{N}$, $n \geq m$ and $\vartheta \in \Theta$. For fixed $m \in \mathbb{N}$, Lemma 5.14(b) yields together with I.5.19

$$\begin{aligned}
\int_{E_m} f\,d\theta + (1+\varepsilon) \int_{E \setminus E_m} f_{**}\,d\theta &\leq \underline{\lim}_{n\to\infty} \int_{E_m} f\,d\theta_n + (1+\varepsilon)\, \underline{\lim}_{n\to\infty} \int_{E \setminus E_m} f_{**}\,d\theta_n \\
&\leq \underline{\lim}_{n\to\infty} \left(\int_{E_m} f\,d\theta_n + (1+\varepsilon) \int_{E \setminus E_m} f_{**}\,d\theta_n\right) \\
&\leq (1+\varepsilon)\left(\underline{\lim}_{n\to\infty} \int_E f_n\,d\theta_n + \overline{\lim}_{n\to\infty} \int_{E \setminus E_m} f_*\,d\theta_n\right) \\
&\quad + w.
\end{aligned}$$

Now we let $m$ tend to infinity and apply Proposition 4.18(a) for

$$\lim_{m\to\infty} \int_{E_m} f\,d\theta = \int_E f\,d\theta$$

and 4.18(b) for

$$0 \leq \underline{\lim}_{m\to\infty} \int_{(E \setminus E_m)} f_*\,d\theta.$$

Moreover, the definition of the residual component in 5.16 together with our assumption $(\theta_n) \underset{\{f_*\}}{\overset{F}{\prec}} \theta$ yields

$$\underline{\lim}_{n\to\infty} \left(\overline{\lim}_{n\to\infty} \int_{E \setminus E_m} f_*\,d\theta_n\right) \leq \mathfrak{Rs}(\theta_n, X, f_*) \leq \mathfrak{O}\left(\int_X f_*\,d\theta\right).$$

The preceding inequality therefore leads to

$$\begin{aligned}
\int_E f\,d\theta &\leq \underline{\lim}_{m\to\infty} \int_{E_m} f\,d\theta + (1+\varepsilon)\, \underline{\lim}_{m\to\infty} \int_{(E \setminus E_m)} f_*\,d\theta \\
&\leq (1+\varepsilon)\, \underline{\lim}_{n\to\infty} \int_E f_n\,d\theta_n + \mathfrak{O}\left(\int_X f_*\,d\theta\right) + w.
\end{aligned}$$

Because this last inequality holds for all $w \in \mathcal{W}$ and $\varepsilon > 0$, and as $\mathcal{Q}$ carries the weak preorder, we infer that

$$\int_E f\,d\theta \leq \underline{\lim}_{n\to\infty} \int_E f_n\,d\theta_n + \mathfrak{O}\left(\int_X f_*\,d\theta\right).$$

Now in the second step of our proof, we shall extend the preceding inequality from integrals over sets $E \in \mathfrak{R}$ to the corresponding integrals over $X$. By our assumption there is $E_0 \in \mathfrak{R}$ such that $\theta|_{(X\setminus E_0)} \leq_{\mathcal{P}} \theta_n|_{(X\setminus E_0)}$ holds for all $n \in \mathbb{N}$. Following Proposition 5.12(c), the latter implies that $\int_F g\,d\theta \leq \int_F g\,d\theta_n$ for every $F \in \mathfrak{A}_{\mathfrak{R}}$ such that $F \subset X \setminus E_0$ and $g \in \mathcal{F}_{(F,\Theta)}(X,\mathcal{P})$. Recall that all functions involved in the theorem are in $\mathcal{F}_{(|X|,\Theta)}(X,\mathcal{P}_0)$, hence are integrable over complements of all sets in $\mathfrak{R}$. Using this, for every $E \in \mathfrak{R}$ such that $E_0 \subset E$ we infer that

$$\int_{(X\setminus E)} f_{**}\,d\theta \leq \int_{(X\setminus E)} f_n\,d\theta + \int_{(X\setminus E)} f_*\,d\theta \leq \int_{(X\setminus E)} f_n\,d\theta_n + \int_{(X\setminus E)} f_*\,d\theta,$$

hence

$$\int_E f_n\,d\theta_n + \int_{(X\setminus E)} f_{**}\,d\theta \leq \int_X f_n\,d\theta_n + \int_{(X\setminus E)} f_*\,d\theta$$

for all $n \in \mathbb{N}$. Thus using the above and the result of our first step we obtain

$$\begin{aligned}\int_E f\,d\theta + \int_{(X\setminus E)} f_{**}\,d\theta &\leq \underline{\lim}_{n\to\infty} \left(\int_E f_n\,d\theta_n + \int_{(X\setminus E)} f_{**}\,d\theta\right) + \mathfrak{O}\left(\int_X f_*\,d\theta\right)\\ &\leq \underline{\lim}_{n\to\infty} \int_X f_n\,d\theta_n + \int_{(X\setminus E)} f_*\,d\theta + \mathfrak{O}\left(\int_X f_*\,d\theta\right).\end{aligned}$$

Now we use the definition of the integral for

$$\int_X f\,d\theta = \lim_{E\in\mathfrak{R}} \int_E f\,d\theta$$

and Proposition 4.19 for

$$0 \leq \underline{\lim}_{E\in\mathfrak{R}} \int_{(X\setminus E)} f_{**}\,d\theta \qquad \text{and} \qquad \overline{\lim}_{E\in\mathfrak{R}} \int_{(X\setminus E)} f_*\,d\theta \leq \mathfrak{O}\left(\int_X f_*\,d\theta\right).$$

Finally, taking the limit over $E \in \mathfrak{R}$, and combining all of the above yields

$$\begin{aligned}\int_X f\,d\theta &\le \underline{\lim}_{E\in\mathfrak{R}} \int_E f\,d\theta + \underline{\lim}_{E\in\mathfrak{R}} \int_{(X\setminus E)} f_{**}\,d\theta \\ &\le \overline{\lim}_{E\in\mathfrak{R}} \left( \int_E f\,d\theta + \int_{(X\setminus E)} f_{**}\,d\theta \right) \\ &\le \underline{\lim}_{n\to\infty} \int_X f_n\,d\theta_n + \mathfrak{O}\left( \int_X f_*\,d\theta \right),\end{aligned}$$

since $\mathfrak{O}\left(\int_X f_*\,d\theta\right) + \mathfrak{O}\left(\int_X f_*\,d\theta\right) = \mathfrak{O}\left(\int_X f_*\,d\theta\right)$ by Proposition I.5.11. This completes our proof. □

Because cone-valued functions do in general not have additive inverses, we require a result corresponding to Theorem 5.23 with respect to upper convergence for both measures and functions.

**Theorem 5.24.** *Let* $\Theta = \{\theta, \theta_n\}_{n\in\mathbb{N}}$ *be equibounded* $\mathfrak{L}(\mathcal{P},\mathcal{Q})$*-valued measures on* $\mathfrak{R}$ *such that* $\theta_n \searrow \theta$. *Let* $F \in \mathfrak{A}_{\mathfrak{R}}$, *and let* $f_n, f, f^* \in \mathcal{F}_{(|F|,\Theta)}(X,\mathcal{P})$ *such that* $(\theta_n) \underset{\{f^*\}}{\overset{F}{\prec}} \theta$. *Suppose that* $f_n \underset{a.e.F}{\lesssim} f^*$ *for all* $n \in \mathbb{N}$, *and that* $f_n \underset{a.e.F}{\searrow} f$. *Then*

$$\overline{\lim}_{n\to\infty} \int_F f_n\,d\theta_n \le \int_F f\,d\theta + \mathfrak{O}\left( \int_F f^*\,d\theta \right).$$

*Proof.* Our argument will follow the lines of the proof of Theorem 5.23, though some substantial adaptations will be required. For the reasons given in 5.23, without loss of generality, we may assume that $F = X$, and that the stated convergence and boundedness properties hold everywhere on $X$ instead of $\Theta$-everywhere.

Suppose that the functions $f, f_n, f^*$ and the measures $\Theta = \{\theta, \theta_n\}$ fulfill these simplified assumptions of the theorem. Again, in a first step we shall discuss the respective integrals of the functions involved over a set $E \in \mathfrak{R}$. For this, let $w \in \mathcal{W}$ be fixed. Following Proposition 5.4, there is $s \in \mathcal{S}_{\mathfrak{R}}(X,\mathcal{V})$ and $\lambda > 0$ such that both $0 \underset{a.e.E}{\lesssim} f + s$ and $0 \underset{a.e.E}{\lesssim} f^* + s$ and $\int_E s\,d\vartheta \le \lambda w$ for all $\vartheta \in \Theta$. Moreover, we have $f_n \underset{a.e.E}{\lesssim} f^*$ for all $n \in \mathbb{N}$ by or assumption. Using a similar argument as before, we may assume that all these relations hold indeed everywhere on $E$. Next we choose $0 < \varepsilon < \min\{1, \frac{1}{2\lambda}\}$.

We may assume that $s(x) \in \mathcal{V}$ for all $x \in X$ (see 5.3 and 4.12). Thus, under the (now simplified) assumptions of the theorem, for every $x \in E$ there is $n_0 \in \mathbb{N}$ such that $f_n(x) \in \big(s(x)\big)_\varepsilon\big(f(x)\big)$ that is $f_n(x) \le \gamma f(x) + \varepsilon s(x)$ for all $n \ge n_0$ with some $1 \le \gamma \le 1$. According to Lemma I.4.1(b), the latter implies that

$$f_n(x) \le (1+\varepsilon) f(x) + 2\varepsilon s(x)$$

for all $n \ge n_0$. We choose a neighborhood $v \in \mathcal{V}$ such that $\vartheta_E(v) \le w$ for all $\vartheta \in \Theta$. Following Theorem 1.6, all the sets

$$E_m = \{x \in E \mid f_n(x) \preccurlyeq_v (1+\varepsilon) f(x) + 2\varepsilon s(x) \quad \text{for all} \quad n \ge m\}$$

are in $\mathfrak{R}$, we have $E_m \subset E_{m+1}$, and $\bigcup_{m\in\mathbb{N}} E_m = E$ by the above. Thus

$$f_n(x) \preccurlyeq_v (1+\varepsilon)f(x) + 2\varepsilon s(x) \qquad \text{and} \qquad f_n(x) \le f^*(x)$$

holds $\Theta$-almost everywhere on $E_m$ for all $n \ge m$. We fix $m \in \mathbb{N}$, recall that $\delta(1-2\varepsilon\lambda) > 0$, and that $\mathfrak{O}\big(\sup\{\theta_E(v) \mid \theta \in \Theta\}\big) \le \delta w$ for all $\delta > 0$, and use Lemma 5.20(a) for

$$\begin{aligned}\overline{\lim_{n\to\infty}} \int_{E_m} f_n \, d\theta_n &\le \int_{E_m} \Big((1+\varepsilon)f + 2\varepsilon s\Big)\, d\theta + \mathfrak{Rs}\big(\theta_n, E_m, f^*\big) + (1-2\varepsilon\lambda)w \\ &\le (1+\varepsilon)\int_{E_m} f\, d\theta + \mathfrak{O}\left(\int_X f^*\, d\theta\right) + w.\end{aligned}$$

The last part of this inequality follows, since $\int_{E_m} s\, d\theta \le \int_E s\, d\theta \le \lambda w$, and since

$$\mathfrak{Rs}\big(\theta_n, E_m, f^*\big) \le \mathfrak{Rs}\big(\theta_n, X, f^*\big) \le \mathfrak{O}\left(\int_X f^*\, d\theta\right)$$

by our assumption on the function $f^*$. Next we use $f_n \underset{a.e.E}{\lesssim} f^*$ for $\int_{(E\setminus E_m)} f_n\, d\vartheta \le \int_{(E\setminus E_m)} f^*\, d\vartheta$, and Lemma 5.20(b) for

$$\overline{\lim_{n\to\infty}} \int_{(E\setminus E_m)} f_n\, d\theta_n \le \overline{\lim_{n\to\infty}} \int_{(E\setminus E_m)} f^*\, d\theta_n \le \int_{(E\setminus E_m)} f^*\, d\theta.$$

Thus, using the limit rules from Lemma I.5.19, we obtain

$$\begin{aligned}\overline{\lim_{n\to\infty}} \int_E f_n\, d\theta_n &\le \overline{\lim_{n\to\infty}} \int_{E_m} f_n\, d\theta_n + \overline{\lim_{n\to\infty}} \int_{(E\setminus E_m)} f_n\, d\theta_n \\ &\le (1+\varepsilon)\int_{E_m} f\, d\theta + \int_{(E\setminus E_m)} f^*\, d\theta + \mathfrak{O}\left(\int_X f^*\, d\theta\right) + w.\end{aligned}$$

This holds true for all $m \in \mathbb{N}$. Now we let $m$ tend to infinity and apply Proposition 4.18(a) for

$$\lim_{m\to\infty} \int_{E_m} f\, d\theta = \int_E f\, d\theta$$

and 4.18(b) and Proposition I.5.11 for

$$\begin{aligned}\overline{\lim_{m\to\infty}} \int_{(E\setminus E_m)} f^*\, d\theta &\le \mathfrak{O}\left(\int_E f^*\, d\theta\right) \\ &\le \mathfrak{O}\left(\int_E f^*\, d\theta\right) + \mathfrak{O}\left(\int_{(X\setminus E)} f^*\, d\theta\right) = \mathfrak{O}\left(\int_X f^*\, d\theta\right).\end{aligned}$$

Combining all of the above, we obtain

$$\overline{\lim_{n\to\infty}} \int_E f_n \, d\theta_n \leq (1+\varepsilon) \int_E f \, d\theta + \mathfrak{O}\left( \int_X f^* \, d\theta \right) + w.$$

Because this inequality holds for all $w \in \mathcal{W}$ and $\varepsilon > 0$, and as $\mathcal{Q}$ carries the weak preorder, we infer that

$$\overline{\lim_{n\to\infty}} \int_E f_n \, d\theta_n \leq \int_E f \, d\theta + \mathfrak{O}\left( \int_X f^* \, d\theta \right).$$

Now in a second step, we shall extend the preceding inequality from integrals over sets $E \in \mathfrak{R}$ to the corresponding integrals over $X$. Following our definition of the convergence of measures in 5.13, there is $E_0 \in \mathfrak{R}$ such that $\theta_n|_{(X \setminus E_0)} \leq_{\mathcal{P}} \theta|_{(X \setminus E_0)}$ holds for all $n \in \mathbb{N}$. Following Proposition 5.12(c), the latter implies that $\int_F g \, d\theta_n \leq \int_F g \, d\theta$ for every $F \in \mathfrak{A}_{\mathfrak{R}}$ such that $F \subset X \setminus E_0$ and $g \in \mathcal{F}_{(F,\Theta)}(X, \mathcal{P})$. Using this, for every $E \in \mathfrak{R}$ such that $E_0 \subset E$ we infer that

$$\int_X f_n \, d\theta_n = \int_E f_n \, d\theta_n + \int_{(X \setminus E)} f_n \, d\theta_n \leq \int_E f_n \, d\theta_n + \int_{(X \setminus E)} f^* \, d\theta$$

for all $n \in \mathbb{N}$. Thus using the above and the result of our first step we obtain

$$\begin{aligned} \overline{\lim_{m\to\infty}} \int_X f_n \, d\theta_n &\leq \overline{\lim_{m\to\infty}} \int_E f_n \, d\theta_n + \int_{(X \setminus E)} f^* \, d\theta \\ &\leq \int_E f \, d\theta + \int_{(X \setminus E)} f^* \, d\theta + \mathfrak{O}\left( \int_X f^* \, d\theta \right). \end{aligned}$$

Now we use the definition of the integral for

$$\lim_{E \in \mathfrak{R}} \int_E f \, d\theta = \int_X f \, d\theta$$

and Proposition 4.19 for

$$\underline{\lim}_{E \in \mathfrak{R}} \int_{(X \setminus E)} f^* \, d\theta \leq \mathfrak{O}\left( \int_X f^* \, d\theta \right).$$

Combining all of these observations and taking the limit over $E \in \mathfrak{R}$ in the above inequality yields

$$\overline{\lim_{m\to\infty}} \int_X f_n \, d\theta_n \leq \int_X f \, d\theta + \mathfrak{O}\left( \int_X f^* \, d\theta \right),$$

since

$$\mathfrak{O}\left( \int_X f^* \, d\theta \right) + \mathfrak{O}\left( \int_X f^* \, d\theta \right) = \mathfrak{O}\left( \int_X f^* \, d\theta \right)$$

by Proposition I.5.1. This completes our proof. □

Note that for measures $\theta_n \searrow \theta$ and a stationary sequence of functions, that is $f_n = f^* = f \in \mathcal{F}_{(|F|,\Theta)}(X,\mathcal{P})$, such that $(\theta_n) \underset{\{f\}}{\overset{F}{\prec}} \theta$, Theorem 5.24 yields

$$\overline{\lim_{m\to\infty}} \int_F f\,d\theta_n \le \int_F f\,d\theta,$$

since $\int_F f\,d\theta + \mathfrak{O}\left(\int_F f\,d\theta\right) = \int_F f\,d\theta$ by Proposition I.5.14.

The combination of Theorems 5.23 and 5.24 leads to a version of Lebesgue's theorem on dominated convergence (see Proposition 18 in Chapter 11 of [178]). It refers to symmetric convergence for both measures and functions.

**Theorem 5.25.** *Let $\Theta = \{\theta_n\}_{n\in\mathbb{N}}$ be equibounded $\mathfrak{L}(\mathcal{P},\mathcal{Q})$-valued measures on $\mathfrak{R}$ such that $\theta_n \longrightarrow \theta$ for a measure $\theta$. Let $F \in \mathfrak{A}_{\mathfrak{R}}$, and let $f_n, f, f_{**}, f_*, f^{**}, f^* \in \mathcal{F}_{(|F|,\Theta)}(X,\mathcal{P})$ such that $(\theta_n) \underset{\{f_*,f^*\}}{\overset{F}{\prec}} \theta$. Suppose that $f_{**} \underset{a.e.F}{\lesssim} f_n + f_*$ and $f_n + f^{**} \underset{a.e.F}{\lesssim} f^*$ for all $n \in \mathbb{N}$, and that $f_n \underset{a.e.F}{\rightleftarrows} f$. Then*

$$\int_F f\,d\theta \le \underline{\lim_{n\to\infty}} \int_F f_n\,d\theta_n + \mathfrak{O}\left(\int_X f_*\,d\theta\right)$$

*and*

$$\overline{\lim_{n\to\infty}} \int_F f_n\,d\theta_n \le \int_F f\,d\theta + \mathfrak{O}\left(\int_F f^*\,d\theta\right).$$

*Proof.* Let the functions $f_n, f, f_{**}, f_*, f_{**} f^*$ and the measures $\Theta = \{\theta_n\}_{n\in\mathbb{N}}$ and $\theta$ be as in the assumptions of the theorem. Following Lemma 5.21, integrability with respect to $\Theta$ implies integrability with respect to $\bar{\Theta} = \Theta \cup \{\theta\}$. Our assumptions therefore imply those of Theorem 5.23, and we conclude that

$$\int_F f\,d\theta \le \underline{\lim_{n\to\infty}} \int_F f_n\,d\theta_n + \mathfrak{O}\left(\int_F f_*\,d\theta\right).$$

In order to apply Theorem 5.24 we set $g_n = f_n + f^{**}$ and $g = f + f^{**}$ Then $g_n, g \in \mathcal{F}_{(|F|,\bar{\Theta})}(X,\mathcal{P})$ and $g_n \underset{a.e.F}{\lesssim} f^*$ for all $n \in \mathbb{N}$. Moreover, $f_n \underset{a.e.}{\searrow} f$ implies that $g_n \underset{a.e.}{\searrow} g$, since the relative topologies were seen to be compatible with the algebraic operations in $\mathcal{P}$ (see Section I.4). The functions $g_n, g$ therefore fulfill the assumptions of Theorem 5.24, and we infer that

$$\begin{aligned}\overline{\lim_{n\to\infty}} \int_F f_n\,d\theta_n + \underline{\lim_{n\to\infty}} \int_F f^{**}\,d\theta_n &\le \overline{\lim_{n\to\infty}} \int_F (f_n + f^{**})\,d\theta_n \\ &\le \int_F f\,d\theta + \int_F f^{**} d\theta + \mathfrak{O}\left(\int_F f^*\,d\theta\right).\end{aligned}$$

Lemma 5.14(b) yields

$$\int_F f^{**}\,d\theta \le \underline{\lim_{n\to\infty}} \int_F f^{**}\,d\theta_n.$$

Thus using the cancellation law in I.5.10(a), we obtain

$$\overline{\lim_{n\to\infty}} \int_F f_n \, d\theta_n \leq \int_F f \, d\theta + \mathfrak{O}\left(\int_F f^* \, d\theta\right) + \mathfrak{O}\left(\int_F f^{**} \, d\theta\right).$$

Finally, the relations $f_n + f^{**} \underset{a.e.F}{\lesssim} f^*$ and $f_n \underset{a.e.F}{\nearrow} f$ imply that $f(x) + f^{**}(x) \preccurlyeq f^*(x)$ holds $\theta$-almost everywhere on $F$, and therefore

$$\int_F f \, d\theta + \int_F f^{**} \, d\theta \leq \int_F f^* \, d\theta$$

by Proposition 4.17. The element $\int_F f^{**} \, d\theta$ of $\mathcal{Q}$ is therefore bounded relative to the element $\int_F f^* d\theta$ (see Proposition I.4.11(b)), and Proposition I.5.14 yields that

$$\mathfrak{O}\left(\int_F f^{**} \, d\theta\right) + \mathfrak{O}\left(\int_F f^* \, d\theta\right) = \mathfrak{O}\left(\int_F f^* \, d\theta\right),$$

thus completing our argument. □

We may use the notions of boundedness from Chapter I.4.24 to formulate a special case of Theorem 2.25 that allows a stronger conclusion. Corresponding to I.4.24(iv) we shall say that a subset $\mathcal{A}$ of $\mathcal{F}(X, \mathcal{P})$ is *bounded above relative to a function* $f \in \mathcal{F}(X, \mathcal{P})$ if for every inductive limit neighborhood $\mathfrak{v}$ there are $\lambda, \rho \geq 0$ such that $g \leq \rho f + \lambda \mathfrak{v}$ holds for all $g \in \mathcal{A}$. Similarly we define boundedness below and (relative) boundedness almost everywhere on a set $F \in \mathfrak{A}_{\mathfrak{R}}$, as well as boundedness for nets and sequences in $\mathcal{F}(X, \mathcal{P})$. Recall the notations from I.4.24 and I.4.25.

**Corollary 5.26.** *Let $\theta$ be a bounded $\mathfrak{L}(\mathcal{P}, \mathcal{Q})$-valued measure on $\mathfrak{R}$. Let $F \in \mathfrak{A}_{\mathfrak{R}}$, and let $f_n, f^* \in \mathcal{F}_{(|F|,\theta)}(X, \mathcal{P})$ such that $\int_F f^* \, d\theta \in \mathcal{B}\big(\int_F f \, d\theta\big)$. Let $(f_n)_{n\in\mathbb{N}}$ be a sequence in $\mathcal{F}_{(|F|,\theta)}(X, \mathcal{P})$ that is $\theta$-almost everywhere on $F$ bounded below and bounded above relative to $f^*$. If $f_n \underset{a.e.F}{\longrightarrow} f$, then*

$$\lim_{n\to\infty} \int_F f_n \, d\theta = \int_F f \, d\theta.$$

*Proof.* This is an immediately consequence of Theorem 5.25: We set $\Theta = \{\theta\}$ and $f_{**} = f^{**} = 0$. Given $w \in \mathcal{W}$ there are $\lambda, \rho \geq 0$ and $n_0 \in \mathbb{N}$ such that $0 \underset{a.e.F}{\lesssim} f_n + \lambda \mathfrak{v}_w$ and $f_n \underset{a.e.F}{\lesssim} \rho f^* + \lambda \mathfrak{v}_w$ holds for all $n \geq n_0$. This means $0 \underset{a.e.F}{\lesssim} f_n + \lambda s$ and $f_n \underset{a.e.F}{\lesssim} \rho f^* + \lambda t$ for functions $s, t \in \mathcal{S}_{\mathfrak{R}}(X, \mathcal{V})$ such that both $\int_X s \, d\theta \leq w$ and $\int_X t \, d\theta \leq w$. Now we apply Theorem 5.25 with $\lambda s$ in place of $f_*$ and $\rho f^* + \lambda t$ in place of $f^*$ from 5.25. Then $\mathfrak{O}\big(\int_F \lambda s \, d\theta\big) \leq w$ and $\mathfrak{O}\big(\int_F (\rho f^* + \lambda t) \, d\theta\big) \leq \mathfrak{O}(f^*) + w$ by Proposition I.5.11. Because $\int_F f d\theta + \mathfrak{O}(f^*) = \int_F f \, d\theta$ by Proposition I.5.14 and our assumption on the function $f^*$, and because the neighborhood $w \in \mathcal{W}$ was arbitrarily chosen, our claim follows. □

An *elementary function* is a function $f = \varphi{\otimes}a \in \mathcal{F}(X,\mathcal{P})$, where $\varphi$ is a bounded measurable non-negative real-valued function supported by a set $E \in \mathfrak{R}$, and $a$ is an element of $\mathcal{P}$. Note that elementary functions are contained in $\mathcal{F}_{\mathfrak{R}}(X,\mathcal{P})$. Indeed, $\chi_{E}{\otimes}a \in \mathcal{F}_{\mathfrak{R}}(X,\mathcal{P})$ implies $\varphi{\otimes}a = \varphi{\otimes}(\chi_{E}{\otimes}a) \in \mathcal{F}_{\mathfrak{R}}(X,\mathcal{P})$ by Lemma 2.6. We make the following observations:

**Lemma 5.27.** *Let $\varphi$ be a bounded measurable non-negative real-valued function supported by a set in $\mathfrak{R}$. There is a sequence $(\varphi_n)_{n\in\mathbb{N}}$ of real-valued step functions converging uniformly on $X$ to $\varphi$ and such that $0 \le \varphi_n \le \varphi$ for all $n \in \mathbb{N}$.*

*Proof.* Let the function $\varphi$ be as stated, supported by the set $E \in \mathfrak{R}$. Without loss of generality, we may assume that $0 \le \varphi \le 1$. For $n \in \mathbb{N}$ and $i = 1, \ldots n$ we set

$$E_n^i = \left\{ x \in E \;\middle|\; \frac{i-1}{n} < \varphi(x) \le \frac{i}{n} \right\} \in \mathfrak{R}$$

and $\varphi_n = \sum_{i=1}^{n} \frac{i-1}{n}\, \chi_{E_n^i}$. Then $0 \le \varphi_n(x) \le \varphi(x) \le \varphi_n(x) + 1/n$ holds for all $x \in X$. □

**Corollary 5.28.** *Let $\theta$ be a bounded $\mathfrak{L}(\mathcal{P},\mathcal{Q})$-valued measure. Let $\varphi$ be a bounded non-negative real-valued function supported by a set in $\mathfrak{R}$, and let $(\varphi_n)_{n\in\mathbb{N}}$ be a sequence of measurable real-valued functions such that $0 \le \varphi_n \underset{a.e.X}{\le} \varphi$ for all $n \in \mathbb{N}$, converging $\theta$-almost everywhere to $\varphi$. Then for every $a \in \mathcal{P}$ the sequence $(\varphi_n{\otimes}a)_{n\in\mathbb{N}}$ in $\mathcal{F}(X,\mathcal{P})$ is bounded below and $\theta$-almost everywhere bounded above relative to $\varphi{\otimes}a$, the sequence $\left(\int_X \varphi_n{\otimes}a\, d\theta\right)_{n\in\mathbb{N}}$ in $\mathcal{Q}$ is bounded below and bounded above relative to the element $\int_X \varphi{\otimes}a\, d\theta$, and*

$$\varphi_n{\otimes}a \underset{a.e.X}{\longrightarrow} \varphi{\otimes}a \qquad \textit{and} \qquad \lim_{n\to\infty} \int_X \varphi_n{\otimes}a\, d\theta = \int_X \varphi{\otimes}a\, d\theta.$$

*Proof.* Let the function $\varphi$ be as stated, supported by the set $E \in \mathfrak{R}$. We may assume that $0 \le \varphi_n \underset{a.e.X}{\le} \varphi \le 1$ holds for all $n \in \mathbb{N}$. For $a \in \mathcal{P}$ we have $\varphi_n{\otimes}a \in \mathcal{F}_{\mathfrak{R}}(X,\mathcal{P})$ for all $n \in \mathbb{N}$ by Lemma 2.6. There is a set $Z \in \mathfrak{A}_{\mathfrak{R}}$ of measure $0$ such that the functions $\tilde{\varphi}_n = \chi_{(X\setminus Z)}{\otimes}\varphi_n$ converge pointwise everywhere to $\tilde{\varphi} = \chi_{(X\setminus Z)}{\otimes}\varphi$ and that $\tilde{\varphi}_n \le \tilde{\varphi} \le 1$ holds for all $n \in \mathbb{N}$. (For this, recall that a countable union of zero sets is again a zero set.) Theorem 1.7 guarantees that $\tilde{\varphi}$ is measurable, hence $\tilde{\varphi}{\otimes}a \in \mathcal{F}_{\mathfrak{R}}(X,\mathcal{P})$, and the function $\varphi{\otimes}a$ is integrable by 4.12. Let $x \in X \setminus Z$. If $\varphi(x) = 0$, then that $\varphi_n(x) = 0$ for all $n \in \mathbb{N}$ as well. If $\varphi(x) > 0$, then, given $v \in \mathcal{V}$ and $\varepsilon > 0$, there is $\lambda \ge 1$ such that $0 \le a + \lambda v$ and $n_0 \in \mathbb{N}$ such that $\varphi_n(x) \le \varphi(x) \le (1+\varepsilon)\varphi_n(x)$ for all $n \ge n_0$. Thus

$$\varphi_n(x)(a+\lambda v) \le \varphi(x)(a+\lambda v) \le \varphi(x)(a) + (1+\varepsilon)\lambda\varphi_n(x)v$$

and

$$\varphi(x)(a+\lambda v) \le (1+\varepsilon)\varphi_n(x)(a+\lambda v) \le (1+\varepsilon)\varphi_n(x)(a) + (1+\varepsilon)\lambda\varphi(x)v.$$

Now the cancellation law for positive elements (Lemma I.4.2 in [100]) yields

$$\varphi_n(x)a \le \varphi(x)a + 2\varepsilon\lambda\varphi_n(x)v \le \varphi(x)a + 2\varepsilon\lambda v$$

and

$$\varphi(x)a \le (1+\varepsilon)\varphi_n(x)a + 2\varepsilon\lambda\varphi(x)v \le \varphi_n(x)a + 2\varepsilon\lambda v.$$

This shows $\varphi_n(x)a \in v^s_{2\varepsilon\lambda}(\varphi(a))$ for all $n \ge n_0$ and demonstrates

$$\varphi_n(x)a \longrightarrow \varphi(x)a$$

in the symmetric relative topology of $\mathcal{P}$. Thus $\varphi_{n\otimes}a \underset{a.e.X}{\longrightarrow} \varphi_{\otimes}a$ holds as claimed.

Furthermore, given an inductive limit neighborhood $\mathfrak{v}$ there is $v \in \mathcal{V}$ such that $\chi_{E\otimes}v \le \mathfrak{v}$ and $\lambda \ge 0$ such that $0 \le a + \lambda v$. Then

$$0 \le \varphi_{n\otimes}(a+\lambda v) \le \varphi_{n\otimes}a + \lambda\mathfrak{v}$$

and

$$\varphi_{n\otimes}a \le \varphi_{n\otimes}(a+\lambda v) \underset{a.e.X}{\le} \varphi_{\otimes}(a+\lambda v) \le \varphi_{\otimes}a + \lambda\mathfrak{v}$$

for all $n \in \mathbb{N}$. The sequence $(\varphi_{n\otimes}a)_{n\in\mathbb{N}}$ in $\mathcal{F}(X,\mathcal{P})$ is therefore bounded below and $\theta$-almost everywhere bounded above relative to the function $\varphi_{\otimes}a$. Furthermore, for any $w \in \mathcal{W}$ we may choose the inductive limit neighborhood $\mathfrak{v}_w$. Then the above yields $0 \le \int_X \varphi_{n\otimes}a\, d\theta + \lambda w$ as well as $\int_X \varphi_{n\otimes}a\, d\theta \le \int_X \varphi_{\otimes}a\, d\theta + \lambda w$ for all $n \in \mathbb{N}$. Hence the sequence $\left(\int_X \varphi_{n\otimes}a\, d\theta\right)_{n\in\mathbb{N}}$ in $\mathcal{Q}$ is seen to be bounded below and bounded above relative to the element $\int_X \varphi_{\otimes}a\, d\theta$. The convergence statement for the sequence of integrals follows from Corollary 5.26. □

Corollary 5.28 in combination with Lemma 5.27 yields a strengthening of the result of Corollary 5.9, that is the approximation of integrable functions by a net of step functions, for elementary functions $f = \varphi_{\otimes}a \in \mathcal{F}(X,\mathcal{P})$ : There is a sequence $(h_n)_{n\in\mathbb{N}}$ of step functions that is bounded below and bounded above relative to $f$ such that

$$h_n \longrightarrow f \text{ and } \lim_{n\to\infty} \int_X h_n\, d\theta = \int_X f\, d\theta.$$

**5.29 Remarks.** (a) If $(\mathcal{Q},\mathcal{W})$ is the (simplified) standard lattice completion (see I.57) of some subcone $(\mathcal{Q}_0,\mathcal{W}_0)$, that is if $(\mathcal{Q},\mathcal{W})$ is a cone of $\overline{\mathbb{R}}$-valued functions on $\mathcal{P}^*$, then for elements $l,m,n \in \mathcal{Q}$ the statement $l \le m + \mathfrak{O}(n)$ means that $l(\mu) \le m(\mu)$ holds for all $\mu \in \mathcal{P}^*$ such that $n(\mu) < +\infty$. The convergence statements of Theorems 5.23 to 5.25 can then be read in this light. The conclusion of Theorem 5.25 means for example that

$$\left( \int_F f \, d\theta \right) (\mu) \leq \underline{\lim}_{n\to\infty} \left( \int_F f_n \, d\theta_n \right) (\mu)$$

holds for all $\mu \in \mathcal{P}^*$ such that $\left( \int_F f_* \, d\theta \right) (\mu) < +\infty$, and

$$\overline{\lim}_{n\to\infty} \left( \int_F f_n \, d\theta_n \right) (\mu) \leq \left( \int_F f \, d\theta \right) (\mu)$$

for all $\mu \in \mathcal{P}*$ such that $\left( \int_F f^* \, d\theta \right) (\mu) < +\infty$.

(b) The convergence statements in the preceding Theorems 5.23 to 5.25 refer to order convergence in $\mathcal{Q}$ for the concerned sequences of integrals. Stronger claims than those might state convergence in the lower, upper and symmetric topologies of $\mathcal{Q}$, respectively. In the context of our approach, such claims are however not valid in general, even for stationary sequences of measures, as the following simple example can show: Let $\mathfrak{R}$ be the $\sigma$-algebra of all Borel sets in $X = [0,1]$, let $\mathcal{P} = \overline{\mathbb{R}}$ with its usual order and locally convex cone topology. Let $\mathcal{Q}$ be the cone of all bounded below $\overline{\mathbb{R}}$-valued functions on $X$, endowed with the pointwise operations and order and the strictly positive constant functions $w$ as neighborhoods. Clearly $(\mathcal{Q}, \mathcal{W})$ is a locally convex complete lattice cone, and order convergence in $\mathcal{Q}$ means pointwise convergence for the concerned functions. We define an $\mathfrak{L}(\mathcal{P}, \mathcal{Q})$-valued measure $\theta$ on $\mathfrak{R}$, setting $\theta_E(\alpha) = \alpha \chi_E \in \mathcal{Q}$ for $E \in \mathfrak{R}$ and $\alpha \in \mathcal{P}$. It is then straightforward to check that $\int_X h \, d\theta = h$ holds for every $\mathcal{P}$-valued step function $h$ on $X$, that is the integral over $\theta$ yields the identity operator from $\mathcal{F}_{(|X|,\theta)}(X, \mathcal{P})$ into $\mathcal{Q}$. Now, if we consider the stationary sequences $\vartheta_n = \theta_n = \theta$ in Theorems 5.23 to 5.25, a review of the assumptions there reveals that only pointwise convergence is required for the sequences of functions $(f_n)_{n\in\mathbb{N}}$ and $(g_n)_{n\in\mathbb{N}}$ in $\mathcal{F}_{(|X|,\theta)}(X, \mathcal{P})$. Thus only pointwise, that is order convergence will result for their integrals in general. Note that in this example the measure $\theta$ is countably additive only with respect to order convergence in $\mathcal{Q}$, not with respect to the weak (see Section I.4.6) or indeed the symmetric relative topology of $\mathcal{Q}$. We shall demonstrate below (Theorem 5.36) that countable additivity for a measure with respect to the symmetric relative topology of $\mathcal{Q}$ in this situation would indeed imply the above stronger statement of convergence for the corresponding sequence of integrals. This shows in particular that no such measure can represent the identity operator from $\mathcal{F}_{(|X|,\theta)}(X, \mathcal{P})$ into $\mathcal{Q}$.

We shall in the following discuss some special cases where convergence with respect to the symmetric topology does indeed result from Theorems 5.23 to 5.25. For the sake of simplicity we shall restrict ourselves to stationary sequences of measures $\theta_n = \theta$ in this context. The preceding Remark 5.29(b) suggests that we shall need to impose further conditions for this purpose. One of these conditions will refer to the countable additivity of the measure $\theta$, another one will require the availability of sufficiently many order continuous linear functionals on $\mathcal{Q}$.

**5.30 Strong Additivity.** Countable additivity of an $\mathfrak{L}(\mathcal{P},\mathcal{Q})$-valued measure $\theta$ as introduced in Section 3 is meant with respect to order convergence in the locally convex complete lattice cone $(\mathcal{Q},\mathcal{W})$. In Theorem 3.11 we verified that in special cases this implies convergence in a stronger sense. In this context, we shall say that an $\mathfrak{L}(\mathcal{P},\mathcal{Q})$-valued measure $\theta$ is *strongly additive* if for every decreasing sequence $(E_n)_{n\in\mathbb{N}}$ of sets in $\mathfrak{R}$ such that $\bigcap_{n\in\mathbb{N}} E_n = \emptyset$, for $a \in \mathcal{P}$ and $w \in \mathcal{W}$ there is $n_0 \in \mathbb{N}$ such that

$$\theta_{E_n}(a) \leq \mathfrak{O}\big(\theta_{E_1}(a)\big) + w$$

holds for all $n \geq n_0$. Similarly, we shall say that a family $\Theta$ of $\mathfrak{L}(\mathcal{P},\mathcal{Q})$-valued measures is *uniformly strongly additive* if it is equibounded and if the above property holds with the same $n_0$ for all $\theta \in \Theta$.

Note that for strong additivity we do not require that a measure is countably additive with respect to the symmetric topology of $\mathcal{Q}$, since this would be overly restrictive. For $\mathcal{Q} = \overline{\mathbb{R}}$, for example, the element $+\infty$ is isolated, that is both open and closed in the symmetric topology of $\overline{\mathbb{R}}$. Thus, for a disjoint union $E = \bigcup_{i\in\mathbb{N}} E_i$ of sets in $\mathfrak{R}$ such that $\theta_E(a) = +\infty$ for $a \in \mathcal{P}$, countable additivity with respect to the symmetric topology would require that $\theta_{(\bigcup_{i=1}^n E_i)}(a) = +\infty$ for all $n$ greater than some $n_0 \in \mathbb{N}$. Lemma 5.31(b) will however imply that for a uniformly strongly additive family $\Theta$ of $\mathfrak{L}(\mathcal{P},\mathcal{Q})$-valued measures a requirement corresponding to 5.30 holds indeed with respect to the symmetric topology of $\mathcal{Q}$; more precisely: Given a decreasing sequence $(E_n)_{n\in\mathbb{N}}$ of sets in $\mathfrak{R}$ such that $\bigcap_{n\in\mathbb{N}} E_n = \emptyset$, $a \in \mathcal{P}$ and $w \in \mathcal{W}$, there is $n_0 \in \mathbb{N}$ such that

$$0 \leq \theta_{E_n}(a) + w \qquad \text{and} \qquad \theta_{E_n}(a) \leq \mathfrak{O}\big(\theta_{E_1}(a)\big) + w$$

holds for all $\theta \in \Theta$ and $n \geq n_0$.

There are several well-known results about strong additivity. Our version of Pettis' theorem, that is Theorem 3.11, (see Theorem IV.10.1 in [55]), states that in case that $(\mathcal{P},\mathcal{V})$ is a locally convex topological vector space and $(\mathcal{Q},\mathcal{W})$ is the standard lattice completion of some subcone $(\mathcal{Q}_0,\mathcal{W})$, every $\mathfrak{L}(\mathcal{P},\mathcal{Q}_0)$-valued measure is also strongly additive. The Vitali-Hahn-Saks theorem see Theorem III.7.2 in [55]) implies a theorem by Nikodým which states that every setwise convergent sequence of real- or Banach space-valued measures is in fact uniformly strongly additive (see Corollary III.7.4 and Theorem IV.10.6 in [55]). We shall investigate a few implications of strong additivity. Lemmas 5.31 and 5.32 will strengthen the corresponding statements from Proposition 4.18.

**Lemma 5.31.** *Suppose that the family $\Theta$ of $\mathfrak{L}(\mathcal{P},\mathcal{Q})$-valued measures is uniformly strongly additive. Let $E \in \mathfrak{R}$ and $f \in \mathcal{F}_{(E,\Theta)}(X,\mathcal{P})$.*

*(a) If $E_n \in \mathfrak{R}$ such that $E_n \subset E_{n+1}$ for all $n \in \mathbb{N}$, and $E = \bigcup_{n\in\mathbb{N}} E_n$, then for every $w \in \mathcal{W}$ there is $n_0 \in \mathbb{N}$ such that $\int_{E_n} f\,d\theta \leq \int_E f\,d\theta + w$ for all $\theta \in \Theta$ and $n \geq n_0$.*

*(b) If $E_n \in \mathfrak{R}$ such that $E \supset E_n \supset E_{n+1}$ for all $n \in \mathbb{N}$, and $\bigcap_{n\in\mathbb{N}} E_n = \emptyset$, then for every $w \in \mathcal{W}$ there is $n_0 \in \mathbb{N}$ such that $0 \le \int_{E_n} f\,d\theta + w$ for all $\theta \in \Theta$ and $n \ge n_0$.*

*Proof.* We shall first prove Part (b) of the lemma. Let $E_n \in \mathfrak{R}$ for $n \in \mathbb{N}$ be subsets of $E \in \mathfrak{R}$ such that $E_n \supset E_{n+1}$ and $\bigcap_{n\in\mathbb{N}} E_n = \emptyset$. In a first step, we shall consider a function $f \in \mathcal{F}_{\mathfrak{R}}(X,\mathcal{P})$. Let $w \in \mathcal{W}$. Because the family $\Theta$ is supposed to be equibounded, there is $v \in \mathcal{V}$ such that $\theta_E(v) \le w$ for all $\theta \in \Theta$. This implies $\mathfrak{O}\big(\theta_E(v)\big) \le \varepsilon w$ for all $\theta \in \Theta$ and $\varepsilon > 0$. Following Lemma 2.4(b), there is $\lambda \ge 0$ such that $0 \le \chi_{E^{\otimes}} f + \lambda \chi_{E^{\otimes}} v$. By 5.30 there is $n_0 \in \mathbb{N}$ such that

$$\theta_{E_n}(v) \le \mathfrak{O}\big(\theta_E(v)\big) + \frac{1}{2\lambda} w \le \frac{1}{\lambda} w$$

for all $\theta \in \Theta$ and $n \ge n_0$. This yields

$$0 \le \int_{E_n} (\chi_{E^{\otimes}} f + \lambda \chi_{E^{\otimes}} v)\,d\theta = \int_{E_n} f\,d\theta + \lambda \theta_{E_n}(v) \le \int_{E_n} f\,d\theta + w$$

for all $\theta \in \Theta$ and $n \ge n_0$. Now in the second and general step, let $f \in \mathcal{F}_{(E,\Theta)}(X,\mathcal{P})$. Given $w \in \mathcal{W}$ and $0 < \varepsilon \le 1/2$, let the functions $f_{(w,\varepsilon)} \in \mathcal{F}_{\mathfrak{R}}(X,\mathcal{P})$ and $s_{(w,\varepsilon)} \in \mathcal{F}_{\mathfrak{R}}(X,\mathcal{V})$ be as in the definition of integrability in 5.3, that is

$$f \underset{a.e.E}{\le} f_{(w,\varepsilon)} \underset{a.e.E}{\le} \gamma f + s_{(w,\varepsilon)} \qquad \text{and} \qquad \int_E s_{(w,\varepsilon)}\,d\theta \le \varepsilon w$$

for some $1 \le \gamma \le 1+\varepsilon$ and all $\theta \in \Theta$. Following our first step, there is $n_0 \in \mathbb{N}$ such that $0 \le \int_{E_n} f_{(w,\varepsilon)}\,d\theta + w/2$, hence $0 \le \gamma \int_{E_n} f\,d\theta + \left(\frac{1}{2}+\varepsilon\right) w$ for all $\theta \in \Theta$ and $n \ge n_0$. Because $\gamma \ge 1$ and $\varepsilon \le 1/2$ this yields

$$0 \le \int_{E_n} f\,d\theta + w$$

for all $\theta \in \Theta$ and $n \ge n_0$, our claim in Part (b). For Part (a), let $E_n \in \mathfrak{R}$ such that $E_n \subset E_{n+1}$ for all $n \in \mathbb{N}$, and $E = \bigcup_{n\in\mathbb{N}} E_n$. We set $F_n = E \setminus E_n \in \mathfrak{R}$ and have $F_n \supset F_{n+1}$ and $\bigcap_{n\in\mathbb{N}} F_n = \emptyset$. For a function $f \in \mathcal{F}_{(E,\Theta)}(X,\mathcal{P})$ we may now use Part (b) of the lemma: Given $w \in \mathcal{W}$, there is $n_0 \in \mathbb{N}$ such that $0 \le \int_{F_n} f\,d\theta + w$ holds for all $\theta \in \Theta$ and $n \ge n_0$. This yields

$$\int_{E_n} f\,d\theta \le \int_{E_n} f\,d\theta + \left(\int_{F_n} f\,d\theta + w\right) = \int_E f\,d\theta + w$$

for all $\theta \in \Theta$ and $n \ge n_0$, our claim in Part (a). □

**Lemma 5.32.** *Suppose that the family $\Theta$ of $\mathfrak{L}(\mathcal{P},\mathcal{Q})$-valued measures is uniformly strongly additive and that the function $f \in \mathcal{F}(X,\mathcal{P})$ is strongly integrable over $E \in \mathfrak{R}$ with respect to $\Theta$.*

*(a) If $E_n \in \mathfrak{R}$ are such that $E_n \subset E_{n+1}$ for all $n \in \mathbb{N}$, and $E = \bigcup_{n\in\mathbb{N}} E_n$, then for every $w \in \mathcal{W}$ there is $n_0 \in \mathbb{N}$ such that $\int_E f\,d\theta \le \int_{E_n} f\,d\theta + \mathfrak{O}\left(\int_E f\,d\theta\right) + w$ for all $\theta \in \Theta$ and $n \ge n_0$.*

*(b) If $E_n \in \mathfrak{R}$ are such that $E \supset E_n \supset E_{n+1}$ for all $n \in \mathbb{N}$, and $\bigcap_{n\in\mathbb{N}} E_n = \emptyset$, then for every $w \in \mathcal{W}$ there is $n_0 \in \mathbb{N}$ such that $\int_{E_n} f\,d\theta \le \mathfrak{O}\left(\int_E f\,d\theta\right) + w$ for all $\theta \in \Theta$ and $n \ge n_0$.*

*Proof.* Again, we shall first prove Part (b) of the Lemma. Let $f \in \mathcal{F}(X,\mathcal{P})$ be strongly integrable over $E \in \mathfrak{R}$ with respect to $\Theta$. For $w \in \mathcal{W}$, according to 5.18 there is a step function $h = \sum_{i=1}^m \chi_{F_i} \otimes a_i \in \mathcal{S}_{\mathfrak{R}}(X,\mathcal{P})$ such that

$$\int_G f\,d\theta \le \int_G h\,d\theta + w/3$$

and such that $\int_G h\,d\theta$ is $w$-bounded relative to $\int_G f\,d\theta$ for all $\theta \in \Theta$ and every subset $G \in \mathfrak{R}$ of $E$. Let $E_n \in \mathfrak{R}$ for $n \in \mathbb{N}$ be subsets of $E$ such that $E_n \supset E_{n+1}$ and $\bigcap_{n\in\mathbb{N}} E_n = \emptyset$. For every $n \in \mathbb{N}$ and $\theta \in \Theta$, we calculate

$$\int_{E_n} h\,d\theta = \sum_{i=1}^m \theta_{(E_n \cap F_i)}(a_i).$$

The measures in $\Theta$ are supposed to be uniformly strongly additive. Thus there is $n_0 \in \mathbb{N}$ such that

$$\theta_{(E_n \cap F_i)}(a_i) \le \mathfrak{O}\left(\theta_{(E \cap F_i)}(a_i)\right) + \frac{1}{3m}\,w$$

for all $n \ge n_0$, $\theta \in \Theta$ and $i = 1,\ldots,m$. Thus, using Proposition I.5.11

$$\int_{E_n} h\,d\theta \le \sum_{i=1}^m \mathfrak{O}\left(\theta_{(E\cap F_i)}(a_i)\right) + \frac{1}{3}w = \mathfrak{O}\left(\int_E h\,d\theta\right) + \frac{1}{3}\,w$$

and

$$\int_{E_n} h\,d\theta \le \mathfrak{O}\left(\int_E f\,d\theta\right) + \frac{2}{3}\,w,$$

since $\int_E h\,d\theta \in \mathcal{B}_w\left(\int_E f\,d\theta\right)$, which by Proposition I.5.13(a) implies that $\mathfrak{O}\left(\int_E h\,d\theta\right) \le \mathfrak{O}\left(\int_E h\,d\theta\right) + \varepsilon w$ for all $\varepsilon > 0$. Thus for all $n \ge n_0$ and $\theta \in \Theta$ we infer that

$$\int_{E_n} f\,d\theta \le \int_{E_n} h\,d\theta + \frac{1}{3}w \le \mathfrak{O}\left(\int_E f\,d\theta\right) + w,$$

that is Part (b) of our claim. For Part (a), let $E_n \in \mathfrak{R}$ such that $E_n \subset E_{n+1}$ for all $n \in \mathbb{N}$, and $E = \bigcup_{n\in\mathbb{N}} E_n$. We set $F_n = E \setminus E_n \in \mathfrak{R}$ and have $F_n \supset F_{n+1}$ and $\bigcap_{n\in\mathbb{N}} F_n = \emptyset$. For a function $f \in \mathcal{F}(X,\mathcal{P})$ that is strongly integrable over $E \in \mathfrak{R}$ with respect to $\Theta$ we may now use Part (b) of the lemma: Given $w \in \mathcal{W}$, there is $n_0 \in \mathbb{N}$ such that $\int_{F_n} f\,d\theta \le \mathfrak{O}\left(\int_E f\,d\theta\right) + w$

for all $\theta \in \Theta$ and $n \geq n_0$. This yields

$$\int_E f\,d\theta = \int_{E_n} f\,d\theta + \int_{F_n} f\,d\theta \leq \int_{E_n} f\,d\theta + \mathfrak{O}\left(\int_E f\,d\theta\right) + w$$

for all $\theta \in \Theta$ and $n \geq n_0$, our claim in Part (a). □

**5.33 Weakly Sequentially Compact Sets of Measures.** A family $\Theta$ of $\mathfrak{L}(\mathcal{P}, \mathcal{Q})$-valued measures is said to be *weakly sequentially compact* if every sequence $(\theta_n)_{n\in\mathbb{N}}$ in $\Theta$ contains a setwise convergent subsequence $(\theta_{n_k})_{k\in\mathbb{N}}$, that is $\theta_{n_k} \longrightarrow \theta$ for some measure $\theta$ on $\mathfrak{R}$ (see Definition II.3.18 in [55]). Note that we do not require that $\theta \in \Theta$. As a consequence, every subset of a sequentially compact set is again sequentially compact.

Theorem IV.9.1 in [55] provides a well-known criterion for weak sequential compactness of a family of finite real-valued measures defined on a $\sigma$-algebra $\mathfrak{R}$: Such a family $\Theta$ is weakly sequentially compact if and only if (i) $\Theta$ is equibounded, that is the total variation of its elements is bounded on $X$, and (ii) $\Theta$ is uniformly (strongly) additive. We shall use this to establish a criterion for sequential compactness of a family of functional-valued measures, that is for the case $\mathcal{Q} = \overline{\mathbb{R}}$.

**Lemma 5.34.** *Suppose that all elements of $\mathcal{P}$ are bounded and that $\mathcal{P}$ is separable in the symmetric relative $v$-topology for every $v \in \mathcal{V}$. Suppose that $X \in \mathfrak{R}$. Then every uniformly strongly additive family of $\mathcal{P}^*$-valued measures on $\mathfrak{R}$ is weakly sequentially compact.*

*Proof.* Let $\Theta$ be a uniformly strongly additive family of $\mathcal{P}^*$-valued measures on $\mathfrak{R}$. Because we assume that $X \in \mathfrak{R}$, and because uniform strong additivity includes equiboundedness (see 5.30), there is $v \in \mathcal{V}$ such that $\theta_X(v) \leq 1$ for all $\theta \in \Theta$. In a first step of our argument we fix an element $a \in \mathcal{P}$ and choose $\lambda \geq 0$ such that both $0 \leq a + \lambda v$ and $a \leq \lambda v$. The latter is possible because all elements of $\mathcal{P}$ are supposed to be bounded, which implies in particular that $\vartheta_E(a) < +\infty$ for every $\mathcal{P}^*$-valued measure $\vartheta$ and $E \in \mathfrak{R}$. For every $\theta \in \Theta$ we may therefore define a real-valued measure $\theta_a$ on $\mathfrak{R}$, setting $\theta_a(E) = \theta_E(a)$ for every $E \in \mathfrak{R}$. The above implies that $\theta_a(E) \leq \lambda\theta_E(v)$ and $0 \leq \theta_a(E) + \lambda\theta_E(v)$, hence $|\theta_a(E)| \leq \lambda\theta_E(v) < \lambda\theta_X(v) \leq \lambda$. Using this, we can estimate the usual (total) variation $\mathfrak{var}(\theta_a, X)$ of this measure (see Definition III.1.4 in [55]) as follows: For disjoint sets $E_1, \ldots, E_n \in \mathfrak{R}$ we have

$$\sum_{i=1}^{n} |\theta_{E_i}(a)| \leq \lambda \sum_{i=1}^{n} \theta_{E_i}(v) = \lambda\theta_{(\cup_{i=1}^{n} E_i)}(v) \leq \lambda,$$

hence

$$\mathfrak{var}(\theta_a, X) = \sup\left\{\sum_{i=1}^{n} |\theta_a(E_i)| \mid E_1, \ldots, E_n \in \mathfrak{R},\ \text{disjoint}\right\} \leq \lambda.$$

Thus for the family $\Theta_a = \{\theta_a \mid \theta \in \Theta\}$ of real-valued measures, firstly the total variation of its elements is bounded by $\lambda$, and secondly, the countable additivity on $\mathfrak{R}$ is uniform with respect to all measures in $\Theta_a$. The latter follows from our requirement that the family $\Theta$ is uniformly strongly countably additive. Indeed, let $E_n \in \mathfrak{R}$ such that $E_n \supset E_{n+1}$ and $\bigcap_{n\in\mathbb{N}} E_n = \emptyset$. Following 5.30, given $\varepsilon > 0$, there is $n_0 \in \mathbb{N}$ such that $\theta_{E_n}(a) \le \mathfrak{O}(\theta_E(a)) + \varepsilon$ for all $n \ge n_0$ and $\theta \in \Theta$. Because $\theta_E(a)$ is finite, we have $\mathfrak{O}(\theta_E(a)) = 0$. Thus

$$\theta_a(E_n) = \theta_{E_n}(a) \le \varepsilon$$

holds for all $\theta_a \in \Theta_a$ and $n \ge n_0$. Now the criterion from Theorem IV.9.1 in [55] (see the remark following 5.33) for weak sequential compactness of finite real-valued measures yields that the set $\Theta_a$ is indeed weakly sequentially compact. Now in the second step of our argument, following our assumption of the separability of $\mathcal{P}$, we choose a countable subset $\{a_n\}_{n\in\mathbb{N}}$ of $\mathcal{P}$ that is dense with respect to the symmetric relative $v$-topology. Let $(\theta_n)_{n\in\mathbb{N}}$ be a sequence in $\Theta$. We shall apply a diagonal procedure in order to construct a weakly convergent subsequence of $(\theta_n)_{n\in\mathbb{N}}$. For each $n \in \mathbb{N}$, the set $\Theta_{a_n}$ of real-valued measures was seen to be weakly sequentially compact. Thus there is a subsequence $(\theta_n^1)_{n\in\mathbb{N}}$ of $(\theta_n)_{n\in\mathbb{N}}$ and a real-valued measure $\vartheta^1$ such that ${\theta_n^1}_E(a_1) \to \vartheta^1(E)$ for all $E \in \mathfrak{R}$. Likewise, there is a subsequence $(\theta_n^2)_{n\in\mathbb{N}}$ of $(\theta_n^1)_{n\in\mathbb{N}}$ and a real-valued measure $\vartheta^2$ such that ${\theta_n^2}_E(a_2) \to \vartheta^2(E)$ for all $E \in \mathfrak{R}$. And so on... We choose the subsequence $(\theta_n^n)_{n\in\mathbb{N}}$ of $(\theta_n)_{n\in\mathbb{N}}$ and claim that this subsequence converges setwise towards some $\mathcal{P}^*$-valued measure $\vartheta$. Indeed, for every $i \in \mathbb{N}$ we have by our construction ${\theta_n^n}_E(a_i) \to \vartheta^i(E)$ for all $E \in \mathfrak{R}$. Let $\mathcal{P}_0$ be the subcone of $\mathcal{P}$ spanned by the elements $\{a_n\}_{n\in\mathbb{N}}$. By our assumption $\mathcal{P}_0$ is dense in $\mathcal{P}$ with respect to the symmetric relative $v$-topology. For a fixed $E \in \mathfrak{R}$ and $a = \sum_{i=1}^n \lambda_i a_i \in \mathcal{P}_0$ for $\lambda_i \ge 0$, set

$$\vartheta_E(a) = \lim_{n\to\infty} {\theta_n^n}_E(a) = \sum_{i=1}^n \lambda_i \vartheta^i(E) \in \mathbb{R}.$$

Clearly, $\vartheta_E$ is a linear functional on $\mathcal{P}_0$, and $a \le b + v$ for $a, b \in \mathcal{P}_0$ implies that

$$\theta_E(a) \le \theta_E(b) + \theta_E(v) \le \theta_E(b) + 1,$$

for all $\theta \in \Theta$. Using the limit rules, this shows in turn that $\vartheta_E(a) \le \vartheta_E(b) + 1$ holds as well. The linear functional $\vartheta_E : \mathcal{P}_0 \to \overline{\mathbb{R}}$ is therefore continuous with respect to the locally convex topology on $\mathcal{P}$ generated by the single neighborhood $v \in \mathcal{V}$, that is the neighborhood system $\mathcal{V}_v = \{\alpha v \mid \alpha > 0\}$, and can therefore be uniquely extended to a continuous linear functional on the whole cone $\mathcal{P}$ (see Theorem I.5.56). Moreover, for every $a \in \mathcal{P}$ and $0 < \varepsilon \le 1$ there is some $b \in \mathcal{P}_0$ such that both $a \in v_\varepsilon(b)$ and $b \in v_\varepsilon(a)$. This implies by the above that ${\theta_n^n}_E(a) \in v_\varepsilon({\theta_n^n}_E(b))$ and ${\theta_n^n}_E(b) \in v_\varepsilon({\theta_n^n}_E(a))$ for all $n \in \mathbb{N}$. There is $n_0 \in \mathbb{N}$ such that for all $n \ge n_0$ we have ${\theta_n^n}_E(b) \in v_\varepsilon(\vartheta_E(b))$ and $\vartheta_E(b) \in v_\varepsilon({\theta_n^n}_E(a))$. Now combining all of the above yields with Lemma I.4.1(a)

$$\theta^n_{n\,E}(a) \in v_\varepsilon\big(\theta^n_{n\,E}(b)\big) \subset v_{3\varepsilon}\big(\vartheta_E(b)\big) \subset v_{7\varepsilon}\big(\vartheta_E(a)\big)$$

and likewise

$$\vartheta_E(a) \in v_\varepsilon\big(\vartheta_E(b)\big) \subset v_{3\varepsilon}\big(\theta^n_{n\,E}(b)\big) \subset v_{7\varepsilon}\big(\theta^n_{n\,E}(a)\big)$$

for all $n \geq n_0$. This demonstrates that $\theta^n_{n\,E}(a) \to \vartheta_E(a)$ for all $a \in \mathcal{P}$. All left to show is that the mapping $E \mapsto \vartheta_E : \mathfrak{R} \to \mathcal{P}^*$ is countably additive, that is $\vartheta$ is indeed a $\mathcal{P}^*$-valued measure on $\mathfrak{R}$. For this, let $a \in \mathcal{P}$, and let $E_i \in \mathfrak{R}$, for $i \in \mathbb{N}$, be disjoint sets. Using the additivity of the measures $\theta^n_n$ and the limit rules, we have

$$\vartheta_{(\bigcup_{i=1}^{i_0} E_i)}(a) = \lim_{n\to\infty} \Big(\theta^n_{n\,(\bigcup_{i=1}^{i_0} E_i)}(a)\Big) = \sum_{i=1}^{i_0} \Big(\lim_{n\to\infty} \theta^n_{n\,E_i}(a)\Big) = \sum_{i=1}^{i_0} \vartheta_{E_i}(a).$$

for every $i_0 \in \mathbb{N}$. This shows finite additivity in particular. Given $\varepsilon > 0$, it follows from the uniform strong additivity of the measures in $\Theta$ together with Lemma 5.31(b) that there is $i_0 \in \mathbb{N}$ such that $|\theta_{(\bigcup_{i=i_0+1}^{\infty} E_i)}(a)| \leq \varepsilon$ holds for all $\theta \in \Theta$, hence also $|\vartheta_{(\bigcup_{i=i_0+1}^{\infty} E_i)}(a)| \leq \varepsilon$. This yields with the above

$$\begin{aligned} \Big|\vartheta_{(\bigcup_{i=1}^{\infty} E_i)}(a) - \sum_{i=1}^{i_0} \vartheta_{E_i}(a)\Big| &= \Big|\vartheta_{(\bigcup_{i=1}^{\infty} E_i)}(a) - \vartheta_{(\bigcup_{i=1}^{i_0} E_i)}(a)\Big| \\ &= \Big|\vartheta_{(\bigcup_{i=i_0+1}^{\infty} E_i)}(a)\Big| \leq \varepsilon. \end{aligned}$$

Because $\varepsilon > 0$ was arbitrarily chosen, this yields

$$\vartheta_{(\bigcup_{i=1}^{\infty} E_i)}(a) = \sum_{i=1}^{\infty} \vartheta_{E_i}(a).$$

Summarizing, we have verified that the subsequence $\big(\theta^n_n\big)_{n\in\mathbb{N}}$ of $\big(\theta_n\big)_{n\in\mathbb{N}}$ converges setwise towards the $\mathcal{P}^*$-valued measure $\vartheta$. □

In Section 3.9 we introduced the composition of an operator-valued measure $\theta$ with two linear operators. We shall now investigate integrals with respect to this type of measures. Let us recall our notations: Let $(\mathcal{P}, \mathcal{V})$ and $(\widetilde{\mathcal{P}}, \widetilde{\mathcal{V}})$ be full locally convex cones, and let $(\mathcal{Q}, \mathcal{W})$ and $(\widetilde{\mathcal{Q}}, \widetilde{\mathcal{W}})$ be locally convex complete lattice cones. For an $\mathfrak{L}(\mathcal{P}, \mathcal{Q})$-valued measure $\theta$, a continuous linear operator $S \in \mathfrak{L}(\widetilde{\mathcal{P}}, \mathcal{P})$ and an order continuous linear operator $U \in \mathfrak{L}(\mathcal{Q}, \widetilde{\mathcal{Q}})$, the $\mathfrak{L}(\widetilde{\mathcal{P}}, \widetilde{\mathcal{Q}})$-valued measure $(U \circ \theta \circ S)$ was defined as the set function

$$E \mapsto (U \circ \theta_E \circ S) : \mathfrak{R} \to \mathfrak{L}(\widetilde{\mathcal{P}}, \widetilde{\mathcal{Q}}).$$

For a $\widetilde{\mathcal{P}}$-valued function $f \in \mathcal{F}(X, \widetilde{\mathcal{P}})$ and a linear operator $S \in \mathfrak{L}(\widetilde{\mathcal{P}}, \mathcal{P})$ we denote by $S \circ f \in \mathcal{F}(X, \mathcal{P})$ the $\mathcal{P}$-valued function

$$x \mapsto S\big(f(x)\big) : X \to \mathcal{P}.$$

**Theorem 5.35.** *Let $(\mathcal{P},\mathcal{V})$ and $(\widetilde{\mathcal{P}},\widetilde{\mathcal{V}})$ be full locally convex cones, and let $(\mathcal{Q},\mathcal{W})$ and $(\widetilde{\mathcal{Q}},\widetilde{\mathcal{W}})$ be locally convex complete lattice cones. Let $\Theta$ be an equibounded family of $\mathfrak{L}(\mathcal{P},\mathcal{Q})$-valued measures, and let $\Upsilon \subset \mathfrak{L}(\mathcal{Q},\widetilde{\mathcal{Q}})$ be an equicontinuous family of continuous and order continuous linear operators. Let $S \in \mathfrak{L}(\widetilde{\mathcal{P}},\mathcal{P})$ such that $S$ is onto and $S(\widetilde{\mathcal{V}}) \subset \mathcal{V}$. Let*

$$\widetilde{\Theta} = \{(U \circ \theta) \mid U \in \Upsilon,\ \theta \in \Theta\} \qquad \textit{and} \qquad \widehat{\Theta} = \{(\theta \circ S) \mid \theta \in \Theta\}$$

*be the corresponding families of $\mathfrak{L}(\mathcal{P},\widetilde{\mathcal{Q}})$- and $\mathfrak{L}(\widetilde{\mathcal{P}},\mathcal{Q})$-valued composition measures on $\mathfrak{R}$. Let $F \in \mathfrak{A}_{\mathfrak{R}}$. If the function $f \in \mathcal{F}(X,\widetilde{\mathcal{P}})$ is integrable over $F$ with respect to $\widehat{\Theta}$, then the function $S \circ f \in \mathcal{F}(X,\mathcal{P})$ is integrable over $F$ with respect to $\widetilde{\Theta}$, and*

$$\int_F (S \circ f)\, d(U \circ \theta) = U\left(\int_F f\, d(\theta \circ S)\right)$$

*holds for all $\theta \in \Theta$ and $U \in \Upsilon$.*

*Proof.* We may assume that $F = X$. Let $\Theta, \Upsilon, S$ and $\widetilde{\Theta}, \widehat{\Theta}$ be as stated, and let $\theta \in \Theta$ and $U \in \Upsilon$. First, for a step function $h = \sum_{i=1}^{n} \chi_{E_i} \otimes \tilde{a}_i \in \mathcal{S}_{\mathfrak{R}}(X,\widetilde{\mathcal{P}})$ we have

$$\begin{aligned}
\int_X (S \circ h)\, d(U \circ \theta) &= \sum_{i=1}^{n} (U \circ \theta)_{E_i}(S(\tilde{a}_i)) \\
&= \sum_{i=1}^{n} U\big(\theta_{E_i}(S(\tilde{a}_i))\big) \\
&= U\left(\sum_{i=1}^{n} (\theta \circ S)_{E_i}(\tilde{a}_i)\right) \\
&= U\left(\int_X h\, d(\theta \circ S)\right).
\end{aligned}$$

Next we consider a function $f \in \mathcal{F}_{\mathfrak{R}}(X,\widetilde{\mathcal{P}})$. According to Theorem 1.8(c) the function $S \circ f \in \mathcal{F}_{\mathfrak{R}}(X,\mathcal{P})$ is also measurable. Let $\mathfrak{v}$ be an inductive limit neighborhood for $\mathcal{F}(X,\mathcal{P})$. Then for every $E \in \mathfrak{R}$ there is $v_E \in \mathcal{V}$ such that $\chi_E \otimes v_E \le \mathfrak{v}$. Correspondingly, there is $\tilde{v}_E \in \widetilde{\mathcal{V}}$ such that $S(\tilde{v}_E) \le v_E$ (see 2.2). Hence $S \circ (\chi_E \otimes \tilde{v}_E) \le \chi_E \otimes v_E \le \mathfrak{v}$. This shows that the convex set $\tilde{\mathfrak{v}}$ of all measurable $\widetilde{\mathcal{V}}$-valued functions $\tilde{s}$ such that $S \circ \tilde{s} \le \mathfrak{v}$ is a corresponding inductive limit neighborhood for $\mathcal{F}_{\mathfrak{R}}(X,\widetilde{\mathcal{P}})$. By 2.3 there is a step function $h \in \mathcal{S}_{\mathfrak{R}}(X,\widetilde{\mathcal{P}})$ such that $h \le f + \tilde{\mathfrak{v}}$. Then $S \circ h \in \mathcal{S}_{\mathfrak{R}}(X,\mathcal{P})$ and $S \circ h \le S \circ f + \mathfrak{v}$. This shows $S \circ f \in \mathcal{F}_{\mathfrak{R}}(X,\mathcal{P})$. Now let $E \in \mathfrak{R}$ and $(U \circ \theta) \in \widetilde{\Theta}$. Given $\tilde{w} \in \widetilde{\mathcal{W}}$ and $\varepsilon > 0$ we choose $w \in \mathcal{W}$ such that $U(s) \le U(t) + \tilde{w}$ whenever $s \le t + w$ for $s, t \in \mathcal{Q}$. Correspondingly, there is $v \in \mathcal{V}$ such that $\theta_E(v) \le w$, and $\tilde{v} \in \widetilde{\mathcal{V}}$ such that $S(\tilde{v}) \le v$. According to Corollary 2.8, given the inductive limit neighborhood $\tilde{\mathfrak{v}} = \{\chi_X \otimes \tilde{v}\}$ there is $1 \le \gamma \le 1 + \varepsilon$ and a bounded below sequence $(h_n)_{n \in \mathbb{N}}$ of step functions in $\mathcal{S}_{\mathfrak{R}}(X,\widetilde{\mathcal{P}})$ such that: (i) $h_n \le \gamma f + \chi_{X}\otimes\tilde{v}$ for all $n \in \mathbb{N}$ and (ii) for every $x \in E$ there is $n_0 \in \mathbb{N}$ such that $f(x) \le h_n(x)$ for all $n \ge n_0$. Thus (i˜) $S\circ h_n \le \gamma(S\circ f)+\chi_{X}\otimes v$ for all $n \in \mathbb{N}$ and (ii˜) for every $x \in E$ there is $n_0 \in \mathbb{N}$ such that $(S\circ f)(x) \le (S\circ h_n)(x)$ for all $n \ge n_0$. As $(\theta\circ S)_E(\tilde{v}) = \theta_E\big(S(\tilde{v})\big) \le \theta_E(v) \le w$ and $(U\circ\theta)_E(v) = U\big(\theta_E(v)\big) \le U(w) \le \tilde{w}$, this yields

$$\int_E h_n \, d(\theta\circ S) \le \gamma \int_E f \, d(\theta\circ S) + w$$

and

$$\int_E (S\circ h_n)\; d(U\circ\theta) \le \gamma \int_E (S\circ f)\; d(U\circ\theta) + \tilde{w}$$

for all $n \in \mathbb{N}$, as well as

$$\int_E f \, d(\theta\circ S) \le \varliminf_{n\to\infty} \int_E h_n \, d(\theta\circ S)$$

and

$$\int_E (S\circ f)\; d(U\circ\theta) \le \varliminf_{n\to\infty} \int_E (S\circ h_n)\; d(U\circ\theta)$$

with Theorem 5.23. Using our observation for order continuous linear operators from I.5.29 and the latter we infer that

$$\begin{aligned} U\left(\int_E f \, d(\theta\circ S)\right) &\le U\left(\varliminf_{n\to\infty} \int_E h_n \, d(\theta\circ S)\right) \\ &\le \varliminf_{n\to\infty} U\left(\int_E h_n \, d(\theta\circ S)\right) \\ &= \varliminf_{n\to\infty} \int_E (S\circ h_n)\; d(U\circ\theta) \\ &\le \gamma \int_E (S\circ f)\; d(U\circ\theta) + \tilde{w} \end{aligned}$$

and

$$\begin{aligned} \int_E (S\circ f)\; d(U\circ\theta) &\le \varliminf_{n\to\infty} \int_E (S\circ h_n)\; d(U\circ\theta) \\ &= \varliminf_{n\to\infty} U\left(\int_E h_n \, d(\theta\circ S)\right) \\ &\le \gamma\, U\left(\int_E f \, d(\theta\circ s)\right) + \tilde{w}. \end{aligned}$$

This holds true for all $\tilde{w} \in \widetilde{\mathcal{W}}$ and $\varepsilon > 0$ and therefore demonstrates

$$\int_E (S\circ f)\; d(U\circ\theta) = U\left(\int_E f \, d(\theta\circ S)\right)$$

for all $f \in \mathcal{F}_{\mathfrak{R}}(X,\widetilde{\mathcal{P}})$ and $\theta \in \Theta$ and $U \in \Upsilon$.

Next we observe that any set in $\mathfrak{A}_{\mathfrak{R}}$ of measure zero with respect to $\widehat{\Theta}$ is also of measure zero with respect to $\widetilde{\Theta}$. Indeed, if $(\theta \circ S)_E = 0$ for a set $E \in \mathfrak{R}$ and all $\theta \in \Theta$, then $\theta_E\big(S(\tilde{a})\big) = 0$ for all $\tilde{a} \in \widetilde{\mathcal{P}}$. As the operator $S$ is supposed to be surjective, this yields $\theta_E(a) = 0$ for all $a \in \mathcal{P}$, hence $\theta_E = 0$ and $(U \circ \theta)_E = 0$ for all $U \in \Upsilon$. Now let $f \in \mathcal{F}_{(X,\widehat{\Theta})}(X, \widetilde{\mathcal{P}})$. Let $E \in \mathfrak{R}$, let $\tilde{w} \in \widetilde{\mathcal{W}}$ and $\varepsilon > 0$. Because the family $\Upsilon$ was supposed to be equicontinuous, there is $w \in \mathcal{W}$ such that $U(s) \leq U(t) + \tilde{w}$ holds for all $U \in \Upsilon$ whenever $s \leq t + w$ for $s, t \in \mathcal{Q}$. Our definition in 5.3 of integrability with respect to the family $\widehat{\Theta}$ over the set $E \in \mathfrak{R}$ requires that there are functions $f_{(w,\varepsilon)} \in \mathcal{F}_{\mathfrak{R}}(X, \widetilde{\mathcal{P}})$ and $s_{(w,\varepsilon)} \in \mathcal{F}_R(X, \widetilde{\mathcal{V}})$ such that

$$f \underset{a.e.E}{\leq} f_{(w,\varepsilon)} \underset{a.e.E}{\leq} \gamma f + s_{(w,\varepsilon)} \qquad \text{and} \qquad \int_E s_{(w,\varepsilon)}\, d(\theta \circ S) \leq \varepsilon w$$

for some $1 \leq \gamma \leq 1 + \varepsilon$ and all $\theta \in \Theta$. Then $S \circ f_{(w,\varepsilon)} \in \mathcal{F}_{\mathfrak{R}}(X, \mathcal{P})$ and $S \circ s_{(w,\varepsilon)} \in \mathcal{F}_{\mathfrak{R}}(X, \mathcal{V})$ by our assumption that $S(\widetilde{\mathcal{V}}) \subset \mathcal{V}$. By the above we have

$$S \circ f \underset{a.e.E}{\leq} S \circ f_{(w,\varepsilon)} \underset{a.e.E}{\leq} \gamma(S \circ f) + (S \circ s_{(w,\varepsilon)})$$

and

$$\int_E (S \circ s_{(w,\varepsilon)})\, d(U \circ \theta) = U\left(\int_E s_{(w,\varepsilon)}\, d(\theta \circ S)\right) \leq \varepsilon \tilde{w}$$

for all $f \in \mathcal{F}_{\mathfrak{R}}(X, \widetilde{\mathcal{P}})$ and $\theta \in \Theta$ and $U \in \Upsilon$. By Definition 5.3, the function $S \circ f$ is therefore also integrable over $E$ with respect to the family $\widetilde{\Theta}$, and we have

$$\begin{aligned}
\int_E (S \circ f)\, d(U \circ \theta) &= \lim_{\substack{\varepsilon>0 \\ w \in \mathcal{W}}} \int_E (S \circ f_{(w,\varepsilon)})\, d(U \circ \theta) \\
&= \lim_{\substack{\varepsilon>0 \\ w \in \mathcal{W}}} U\left(\int_E f_{(w,\varepsilon)}\, d(\theta \circ S)\right) \\
&= U\left(\lim_{\substack{\varepsilon>0 \\ w \in \mathcal{W}}} \int_E f_{(w,\varepsilon)}\, d(\theta \circ S)\right) \\
&= U\left(\int_E f\, d(\theta \circ S)\right)
\end{aligned}$$

for every $\theta \in \Theta$ and $U \in \Upsilon$. Finally, we verify the second part of Definition 5.3, that is integrability over $F = X$. We have

$$\begin{aligned}\int_X (S \circ f)\, d(U \circ \theta) &= \lim_{E \in \mathfrak{R}} \int_E (S \circ f)\, d(U \circ \theta)\\ &= \lim_{E \in \mathfrak{R}} U\left(\int_E f\, d(\theta \circ S)\right)\\ &= U\left(\lim_{E \in \mathfrak{R}} \int_E f\, d(\theta \circ S)\right)\\ &= U\left(\int_X f\, d(\theta \circ S)\right)\end{aligned}$$

for all $\theta \in \Theta$ and $U \in \Upsilon$. Thus $S \circ f \in \mathcal{F}_{(X,\widetilde{\Theta})}(X, \mathcal{P})$, hence our claim. □

We shall in the following mainly use this result for the special case $\widetilde{\mathcal{P}} = \mathcal{P}$ and the identity operator for $S$, for $\widetilde{\mathcal{Q}} = \overline{\mathbb{R}}$ and an equicontinuous set $\Upsilon$ of order continuous linear functionals in $\mathcal{P}^*$.

Recall from Section I.5.32 that the order continuous linear functionals are said to support the separation property for a locally convex complete lattice cone $(\mathcal{Q}, \mathcal{W})$ if for every neighborhood $w \in \mathcal{W}$ we have $l \leq m + w$ for $l, m \in \mathcal{Q}$ whenever $\mu(l) \leq \mu(m) + 1$ holds for all order continuous lattice homomorphisms $\mu \in w^\circ$.

We are now prepared to formulate and prove a combined version of the Convergence Theorems 5.23, 5.24 and 5.25, that under additional assumptions yields convergence with respect to the upper, lower and symmetric topologies of $\mathcal{Q}$, respectively, for the concerned sequence of integrals. Because we shall deal only with bounded elements of $\mathcal{Q}$, we do not need to consider the relative topologies, since they coincide locally with the given topologies in this case (see Section I.4). Recall that for a sequence $(a_n)_{n \in \mathbb{N}}$ in $\mathcal{Q}$ convergence towards $a \in \mathcal{Q}$ in the upper, or lower topology of $\mathcal{Q}$ means that for every $w \in \mathcal{W}$ there is $n_0 \in \mathbb{N}$ such that

$$a_n \leq a + w, \qquad \text{or} \qquad a \leq a_n + w$$

holds for all $n \geq n_0$, respectively. Because these topologies are generally far from Hausdorff, limits need not be unique. Convergence in the symmetric topology combines convergence in both the upper and lower topologies.

**Theorem 5.36.** *Suppose that the order continuous linear functionals support the separation property for $\mathcal{Q}$. Let $\theta$ be a strongly additive $\mathfrak{L}(\mathcal{P}, \mathcal{Q})$-valued measure on $\mathfrak{R}$, and let $E \in \mathfrak{R}$. Let $f_n, f, f_{**}, f_*, f^{**}, f^* \in \mathcal{F}(X, \mathcal{P})$ be bounded-valued measurable functions, and suppose that for every $w \in \mathcal{W}$ there is $v \in \mathcal{V}$ such that $\theta_E(v) \leq w$ and such that these functions are $(\mathcal{P}, \mathcal{V}_0)$-based integrable over $E$ with respect to $\theta$ for the subsystem $\mathcal{V}_0 = \{\rho v \mid \rho > 0\}$ of $\mathcal{V}$. Suppose that the functions $f_*$ and $f^*$ are strongly integrable over $E$ with respect to $\theta$ and that their respective integrals are bounded in $\mathcal{Q}$.*

*(a) If* $f_{**}\underset{a.e.E}{\leq} f_n + f_*$ *for all* $n \in \mathbb{N}$, *and* $f_n \underset{a.e.E}{\nearrow} f$, *then*

$$\int_E f\,d\theta = \lim_{n\to\infty} \int_E f_n\,d\theta_n$$

*with respect to the lower topology of* $\mathcal{Q}$.

*(b) If* $f_n \underset{a.e.E}{\leq} f^*$ *for all* $n \in \mathbb{N}$, *and* $f_n \underset{a.e.E}{\searrow} f$, *then*

$$\int_E f\,d\theta = \lim_{n\to\infty} \int_E f_n\,d\theta_n$$

*with respect to the upper topology of* $\mathcal{Q}$.

*(c) If* $f_{**}\underset{a.e.E}{\leq} f_n + f_*$ *and* $f_n + f^{**}\underset{a.e.E}{\leq} f^*$ *for all* $n \in \mathbb{N}$, *and* $f_n \underset{a.e.E}{\longrightarrow} f$, *then*

$$\int_E f\,d\theta = \lim_{n\to\infty} \int_E f_n\,d\theta_n$$

*with respect to the symmetric topology of* $\mathcal{Q}$.

*Proof.* We shall deal with Parts (a), (b) and (c) simultaneously. By restricting the measure $\theta$ and all the functions involved to the set $E$, we may assume that $X = E \in \mathfrak{R}$. Let $\mathcal{G} = \{f_n, f, f_{**}, f_*, f^{**} f^*\}$ be the family of the functions used in our statement. This family is countable.

Suppose that contrary to our claim, the sequence $\left(\int_E f_n\,d\theta\right)_{n\in\mathbb{N}}$ does not converge towards $\int_E f\,d\theta$ in the (a) lower, (b) upper or (c) symmetric topology of $\mathcal{Q}$. Then there is $w \in \mathcal{W}$ and a subsequence $(f_{n_k})_{k\in\mathbb{N}}$ of $(f_n)_{n\in\mathbb{N}}$ such that either

$$\text{(a)} \int_E f\,d\theta \not\leq \int_E f_{n_k}\,d\theta + w \qquad \text{or} \qquad \text{(b)} \int_E f_{n_k}\,d\theta \not\leq \int_E f\,d\theta + w,$$

respectively, holds for all $k \in \mathbb{N}$. In case (c), we can find a subsequence $(f_{n_k})_{k\in\mathbb{N}}$ of $(f_n)_{n\in\mathbb{N}}$ either as in (a) or in (b). We have $\mu\left(\int_E f\,d\theta\right) < +\infty$ for all $\mu \in \mathcal{Q}^*$ since the integral of $f$ is supposed to be bounded in $\mathcal{Q}$. Let $\Upsilon$ be the family of all order continuous linear functionals in $w^\circ$ and $\Omega$ be the corresponding set $\{\mu \circ \theta \mid \mu \in \Upsilon,\}$ of $\mathcal{P}^*$-valued measures on $\mathfrak{R}$. Theorem 5.35 yields that the functions in $\mathcal{G}$ are integrable over $E$ with respect to the family $\Omega$ and that $\mu\left(\int_E g\,d\theta\right) = \int_E g\,d(\mu\circ\theta)$ holds for all $g \in \mathcal{G}$ and $\mu \in \Upsilon$. By our assumption the order continuous linear functionals support the separation property for $\mathcal{Q}$, thus there are functionals $\mu_k \in \Upsilon \subset w^\circ$ such that either

$$\text{(a)} \int_X f\,d(\mu_k\circ\theta) = \mu_k\left(\int_X f\,d\theta\right) > \mu_k\left(\int_X f_{n_k}\,d\theta\right) + 1 = \int_X f_{n_k}\,d(\mu_k\circ\theta) + 1$$

or

$$\text{(b)} \int_X f_{n_k}\,d(\mu_k\circ\theta) = \mu_k\left(\int_X f_{n_k}\,d\theta\right) > \mu_k\left(\int_X f\,d\theta\right) + 1 = \int_X f\,d(\mu_k\circ\theta) + 1$$

holds for all $k \in \mathbb{N}$, respectively. We shall proceed as follows:

There is $v \in \mathcal{V}$ such that $\theta_E(v) \leq w$ and such that the functions in $\mathcal{G}$ are $(\mathcal{P}, \mathcal{V}_0)$-based integrable over $E$ with respect to $\theta$ for the subsystem $\mathcal{V}_0 = \{\rho v \mid \rho > 0\}$ of $\mathcal{V}$. All functions in $\mathcal{G}$ are supposed to be measurable, thus their ranges are separable with respect to the symmetric relative $v$-topology by (M2) in Section 1.2. For every $g \in \mathcal{G}$, let $\mathcal{A}(g)$ be a countable dense subset in the range of $g$. Recall that by our assumption all elements of $\mathcal{A}(g)$ are bounded in $\mathcal{P}$. Following Definition 5.6, that is the $(\mathcal{P}, \mathcal{V}_0)$-based integrability of the functions in $\mathcal{G}$, for every $g \in \mathcal{G}$ and $n \in \mathbb{N}$ there is a function $g_n \in \mathcal{F}_{\mathfrak{R}}(X, \mathcal{P})$ and $s_n \in \mathcal{F}_{\mathfrak{R}}(X, \mathcal{V}_0)$ such that

$$g \underset{a.e.E}{\leq} g_n + s_n, \qquad g_n \underset{a.e.E}{\leq} \gamma_n g + s_n \quad \text{and} \quad \int_E s_n \, d\theta \leq \frac{1}{n} w$$

for some $1 \leq \gamma_n \leq 1 + 1/n$. The latter implies that $\int_E s_n \, d\omega \leq 1/n$ for all $\omega \in \Omega$. Again, measurability guarantees that there are countable dense subsets $\mathcal{A}(g_n)$ in the respective ranges of the functions $g_n$. Now, recalling the definition of the cone $\mathcal{F}_{\mathfrak{R}}(X, \mathcal{P})$ in Section 2.3, for every $n \in \mathbb{N}$ and $m \in \mathbb{N}$ there is a step function $h^m_{g_n} \in \mathcal{S}_{\mathfrak{R}}(X, \mathcal{P})$ such that $h^m_{g_n}(x) \leq g_n(x) + (1/m)v$ for all $x \in E$. Obviously, the range $\mathcal{A}(h^m_{g_n})$ of $h^m_{g_n}$ is finite. We denote by $\mathcal{B}$ the union of all the sets $\mathcal{A}(g)$, $\mathcal{A}(g_n)$ and $\mathcal{A}(h^m_{g_n})$, for $g \in \mathcal{G}$ and $n, m \in \mathbb{N}$, and by

$$\mathcal{C} = \left\{ \sum_{i=1}^{n} \rho_i b_i + \delta v \mid b_i \in \mathcal{B}, \ 0 \leq \rho_i \in \mathbb{Q}, \ 0 < \delta \in \mathbb{Q} \right\}.$$

This set is also countable, and all its elements are $v$-bounded in $\mathcal{P}$ by our assumption on the functions $g \in \mathcal{G}$. Finally, let $\mathcal{P}_0$ be the closure of $\mathcal{C}$ in $\mathcal{P}$ with respect to the symmetric relative $v$-topology. Then $\mathcal{P}_0$ is a subcone of $\mathcal{P}$, separable, and all of its elements are $v$-bounded, that is bounded with respect to the neighborhood subsystem $\mathcal{V}_0$, which itself is contained in $\mathcal{P}_0$. Moreover, the above shows that all functions in $\mathcal{G}$ are indeed $(\mathcal{P}_0, \mathcal{V}_0)$-based integrable (see 5.6) over $E$, hence over all subsets $G \in \mathfrak{R}$ of $E$, with respect to the family $\Omega$ of $\mathcal{P}^*$-valued measures. Proposition 5.10 now yields that all these functions are contained in $\mathcal{F}_{(E,\Omega_0)}(X, \mathcal{P}_0)$, where $\Omega_0$ denotes the family of the restrictions to $\mathcal{P}_0$ of the measures in $\Omega$, and that

$$\int_G g \, d(\mu \circ \theta)_0 = \int_G g \, d(\mu \circ \theta) = \mu \left( \int_G g \, d\theta \right)$$

holds for all $g \in \mathcal{G}$, $\mu \in \Upsilon$ and subsets $G \in \mathfrak{R}$ of $E$.

We proceed to apply Lemma 5.34 to the cone $(\mathcal{P}_0, \mathcal{V}_0)$ in order to show that the family $\Omega_0$ of $\mathcal{P}_0^*$-valued measures is weakly sequentially compact. As we mentioned before, the elements of $\mathcal{P}_0$ are bounded, and $\mathcal{P}_0$ is separable in the symmetric relative $v$-topology. For equiboundedness of the family $\Omega_0$, let $\varepsilon > 0$ be a neighborhood for $\mathbb{R}$. Correspondingly, we choose the neighborhood $\varepsilon v \in \mathcal{V}_0$ and conclude that

$$(\mu \circ \theta)_{0\,E}(v) = \mu\big(\theta_E(v)\big) \leq \mu(\varepsilon w) \leq \varepsilon$$

for all $(\mu \circ \theta)_0 \in \Omega_0$, since all functionals $\mu \in \Upsilon$ involved are contained in $w^\circ$. This shows that $\Omega_0$ is indeed equibounded. Likewise, $\Omega_0$ is seen to be uniformly strongly additive. Indeed, let $E_n \in \mathfrak{R}$ such that $E_n \supset E_{n+1}$ and $\bigcap_{n\in\mathbb{N}} E_n = \emptyset$. Following 5.30, given $\varepsilon > 0$, there is $n_0 \in \mathbb{N}$ such that $\theta_{E_n}(v) \leq \varepsilon w$ for all $n \geq n_0$ and $\theta \in \Theta$. Thus

$$(\mu \circ \theta)_{0\,E_n}(v) = (\mu \circ \theta)_{E_n}(v) = \mu\big(\theta_{E_n}(v)\big) \leq \mu\big(\theta_{E_n}(v)\big) \leq \varepsilon$$

for all $(\mu \circ \theta)_0 \in \Omega_0$ and $n \geq n_0$. Thus, following Lemma 5.34, the family $\Omega_0$ of $\mathcal{P}_0^*$-valued measures is weakly sequentially compact.

We may therefore assume that the sequence $\big((\mu_k \circ \theta)_0\big)_{k\in\mathbb{N}}$ from the first part of this proof converges setwise to some bounded $\mathcal{P}_0^*$-valued measure $\omega$. We abbreviate $\omega_k$ for $(\mu_k \circ \theta)_0$ and recall that either

$$\text{(a)} \int_X f\, d\omega_k > \int_X f_{n_k}\, d\omega_k + 1 \qquad \text{or} \qquad \text{(b)} \int_X f_{n_k}\, d\omega_k > \int_X f\, d\omega_k + 1$$

holds for all $k \in \mathbb{N}$, respectively.

Next we shall argue that $(\omega_k) \underset{\{f_*, f^*\}}{\overset{E}{\prec}} \omega$. In fact, we shall demonstrate that $\mathfrak{Rs}\big(\omega_k, E, g\big) = 0$ for every $g \in \{f_*, f^*\}$. For this, let $E_m \in \mathfrak{R}$ for $m \in \mathbb{N}$ be subsets of $E$ such that $E_m \supset E_{m+1}$ and $\bigcap_{n\in\mathbb{N}} E_m = \emptyset$. Let $\varepsilon > 0$. Because the function $g$ is supposed to be strongly integrable over $E$ with respect to $\theta$, Lemma 5.32(b) yields that for $\varepsilon > 0$ there is $m_0 \in \mathbb{N}$ such that $\int_{E_m} g\, d\theta \leq \mathfrak{O}\left(\int_E g\, d\theta\right) + \varepsilon w$ for all $m \geq m_0$. Because the element $\int_E g\, d\theta$ is supposed to be bounded in $\mathcal{Q}$, we infer that $\mathfrak{O}\left(\int_E g\, d\theta\right) = 0$ (see Proposition I.5.10(c)). We have

$$\int_{E_m} g\, d\omega_k = \mu_k\left(\int_{E_m} g\, d\theta\right) \leq \varepsilon$$

for all $m \geq m_0$ and $k \in \mathbb{N}$, since $\mu_k \in w^\circ$. Thus

$$\underline{\lim}_{m\to\infty}\left(\overline{\lim}_{k\to\infty} \int_{E_m} g\, d\omega_k\right) \leq \varepsilon$$

for all $\varepsilon > 0$, hence

$$\underline{\lim}_{m\to\infty}\left(\overline{\lim}_{k\to\infty} \int_{E_m} g\, d\omega_k\right) \leq 0.$$

This shows

$$\mathfrak{Rs}\big(\omega_k, E, g\big) = \sup_{(E_m)\in\mathfrak{F}} \left\{\underline{\lim}_{m\to\infty}\left(\overline{\lim}_{k\to\infty} \int_{E_m} g\, d\omega_k\right)\right\} = 0.$$

Now, finally, our preceding convergence theorems will yield a contradiction. We shall apply them to the cones $\mathcal{P}_0$ and $\overline{\mathbb{R}}$, the sequence of measures $(\omega_k)_{k\in\mathbb{N}}$ and $\omega$, and the given functions $f_n, f, f_{**}, f_*, f^{**} f^*$. First, Lemma 5.14(a) states that all functions involved are in $\mathcal{F}_{(E,\Omega_0\cup\{\omega\})}(X,\mathcal{P}_0)$. Moreover, Lemmas 5.14(b) and 5.20(b) demonstrate that

$$\int_E f\,d\omega = \lim_{k\to\infty}\int_E f\,d\omega_k.$$

In case (a), Theorem 5.23 yields

$$\int_F f\,d\omega \le \underline{\lim}_{k\to\infty}\int_F f_{n_k}\,d\omega_k$$

since $\mathfrak{O}\left(\int_F f_*\,d\theta\right) = 0$, contradicting our assumption at the start of this argument. Similarly, in case (b), Theorem 5.24 leads to

$$\overline{\lim_{k\to\infty}}\int_F f_{n_k}\,d\omega_k \le \int_F f\,d\omega,$$

contradicting the corresponding assumption for this case. In case (c), finally, Theorem 5.25 yields

$$\int_F f\,d\omega = \lim_{k\to\infty}\int_F f_{n_k}\,d\omega_k,$$

contradicting the assumptions of both cases (a) and (b). This completes our argument. □

As we established in I.5.57, every locally convex cone can be canonically embedded into a larger locally convex complete lattice cone whose order continuous lattice homomorphisms support the separation property. The corresponding requirement in Theorem 5.36 can therefore be met if we use this standard lattice completion for $\mathcal{Q}$. In Section 6 below we shall identify several special cases where Theorem 5.36 can be applied.

## 6. Examples and Special Cases

The generality of our approach to measures and integrals allows a wide range of settings, depending on the choices for the locally convex cones $(\mathcal{P},\mathcal{V})$ and $(\mathcal{Q},\mathcal{W})$. We shall present a selection of these special cases in this section. Throughout the following, we shall assume that $(\mathcal{P},\mathcal{V})$ is a quasi-full locally convex cone and that $(\mathcal{Q},\mathcal{V})$ is a locally convex complete lattice cone. $(\mathcal{P}_{\mathcal{V}},\mathcal{V})$ shall denote the standard full extension of $(\mathcal{P},\mathcal{V})$ into a full cone, as elaborated in Section 6 of Chapter I. $(\mathcal{Q}_0,\mathcal{W}_0)$, on the other hand, will stand for a locally convex cone whose standard lattice completion in the sense of I.5.57 is $(\mathcal{Q},\mathcal{W})$. We shall generally use the notations of the preceding sections. In particular, $\mathfrak{R}$ stands for a weak $\sigma$-ring of subsets of a set $X$, and $\theta$ is a bounded measure on $\mathfrak{R}$. The concepts of the preceding Sections 4 and 5, in particular our notions of integrability, will be applied to the full cone $(\mathcal{P}_{\mathcal{V}},\mathcal{V})$ instead of $(\mathcal{P},\mathcal{V})$.

Some of our general notions are considerably simplified in special cases. In the first set of examples we shall discuss the specific insertions for $\mathcal{P}$ and $\mathcal{Q}$ that lead to classical integration theory.

**6.1 The case $\mathcal{Q}=\overline{\mathbb{R}}$.** If we choose $\mathcal{Q}=\overline{\mathbb{R}}$ with the canonical order and the neighborhoods $\mathcal{V}=\{\varepsilon\in\mathbb{R}\mid\varepsilon>0\}$, then the values of the measure $\theta$ are linear functionals in the dual cone $\mathcal{P}^*$ of $\mathcal{P}$, and for each $a\in\mathcal{P}$ the mapping

$$E\mapsto\theta_E(a):\mathfrak{R}\to\overline{\mathbb{R}}$$

is an extended real-valued measure on $\mathfrak{R}$. The modulus of the measure $\theta$ is given by

$$|\theta|(E,v)=\sup\left\{\sum_{i=1}^{n}\theta_{E_i}(s_i)\,\middle|\, s_i\in\mathcal{P},\ s_i\le v,\ E_i\in\mathfrak{R}\text{ disjoint subsets of }E\right\},$$

which is an element of $\overline{\mathbb{R}}$, for $E\in\mathfrak{R}$ and $v\in\mathcal{V}$. Boundedness therefore means that for every $E\in\mathfrak{R}$ there is $v\in\mathcal{V}$ such that $|\theta|(E,v)<+\infty$. This coincides with Prolla's notion of *finite p-semivariation* in [155] (Ch. 5.5). A bounded measure can be extended to the full cone $(\mathcal{P}_{\mathcal{V}},\mathcal{V})$ as elaborated in Section 3.8. Integrals of $\mathcal{P}$-valued functions with respect to an $\mathfrak{L}(\mathcal{P},\overline{\mathbb{R}})$-, that is $\mathcal{P}^*$-valued measure are also in $\overline{\mathbb{R}}$. For a meaningful statement in our Convergence Theorems 5.22, 5.24 and 5.25 we need to enforce that $\int_F f_*\,d\theta<+\infty$ and $\int_F f^*\,d\theta<+\infty$ in this case.

**6.2 Extended Positive-Valued Functions and Measures.** We obtain classical integration theory for extended positive-valued functions with respect to extended positive-valued measures if we choose $\mathcal{P}=\overline{\mathbb{R}}_+$, endowed with the singleton neighborhood system $\mathcal{V}=\{0\}$ (see Example 1.2(b) in Chapter I), and $\mathcal{Q}=\overline{\mathbb{R}}$. The dual $\overline{\mathbb{R}}_+^*$ of $\overline{\mathbb{R}}_+$ consists of all elements of $\overline{\mathbb{R}}_+$ (via the usual multiplication) and the singular functional $\bar{0}$ such that $\bar{0}(\alpha)=0$ for all $\alpha\in\mathbb{R}_+$ and $\bar{0}(+\infty)=+\infty$. Every $\overline{\mathbb{R}}_+^*$-valued measure $\theta$ is therefore $\mathfrak{R}$-bounded and can be expressed as the sum of an $\overline{\mathbb{R}}_+$-valued measure $\theta^1$ in the usual sense and a measure $\theta^0$ that takes only the values $0$ and $\bar{0}$.

Because the symmetric relative topology renders the Euclidean topology on the interval $(0,+\infty)$, and the elements $0$ and $\infty$ as isolated points (see Example 4.18(a) in Chapter I), $\overline{\mathbb{R}}_+$ is separable in this topology. Our notion of measurability from Section 1 for $\overline{\mathbb{R}}_+$-valued functions therefore coincides with the usual one in this case. Continuity for an $\overline{\mathbb{R}}_+$-valued function defined on a topological space $X$ does however require that this function takes the values $0$ and $+\infty$ only on respective subsets of $X$ that are both open and closed.

Because $v = 0$ is the only neighborhood for $\overline{\mathbb{R}}_+$, according to Section 4, the integral of a measurable function $f$ over a set $F \in \mathfrak{A}_R$ with respect to a measure $\theta$ is defined as

$$\int_F f\, d\theta = \sup\left\{ \int_F h\, d\theta \mid h \in \mathcal{S}_{\mathfrak{R}}(X, \mathcal{P}),\ h \leq f \right\},$$

that is the classical definition of the integral.

**6.3 Extended Real-Valued Functions and Positive-Valued Measures.** We obtain classical integration theory for $\overline{\mathbb{R}}$-valued functions with respect to positive-valued measures if we choose $\mathcal{P} = \mathcal{Q} = \overline{\mathbb{R}}$. The dual $\overline{\mathbb{R}}^*$ of $\overline{\mathbb{R}}$ consists of all positive reals (via the usual multiplication) and the singular functional $\bar{0}$ such that $\bar{0}(\alpha) = 0$ for all $\alpha \in \mathbb{R}$ and $\bar{0}(+\infty) = +\infty$. Every $\overline{\mathbb{R}}^*$-valued measure $\theta$ is therefore $\mathfrak{R}$-bounded and can be expressed as the sum of a positive real-valued measure $\theta^1$ in the usual sense and a measure $\theta^0$ that takes only the values $0$ and $\bar{0}$. The notion of measurability from Section 1 for $\overline{\mathbb{R}}$-valued functions coincides with the usual one.

Let $f$ be a measurable and bounded below $\overline{\mathbb{R}}$-valued function, and let $F \in \mathfrak{A}_{\mathfrak{R}}$. For a neighborhood $w = \varepsilon \in \mathcal{W}$ the step functions $s \in \mathfrak{v}_\varepsilon$ are invertible, and for a step function $h \in \mathcal{S}_{\mathfrak{R}}(X, \overline{\mathbb{R}})$ such that $h \leq f + \mathfrak{v}_\varepsilon$ we have $h' \leq f$ with $h' = h - s \in \mathcal{S}_{\mathfrak{R}}(X, \overline{\mathbb{R}})$ and $\int_F h\, d\theta \leq \int_F h\, d\theta + \varepsilon$. This shows

$$\int_F^{(\varepsilon)} f\, d\theta \leq \sup\left\{ \int_F h\, d\theta \mid h \in \mathcal{S}_{\mathfrak{R}}(X, \mathcal{P}),\ h \leq f \right\} + \varepsilon$$

and consequently

$$\int_F f\, d\theta = \sup\left\{ \int_F h\, d\theta \mid h \in \mathcal{S}_{\mathfrak{R}}(X, \mathcal{P}),\ h \leq f \right\},$$

the usual definition.

**6.4 Real- or Complex-Valued Functions and Measures.** In the preceding example we integrated $\overline{\mathbb{R}}$-valued functions with respect to positive real-valued measures. Alternatively, we may consider real- or complex-valued functions, that is $\mathcal{P} = \mathbb{K}$ for $\mathbb{K} = \mathbb{R}$ or $\mathbb{K} = \mathbb{C}$ with the usual Euclidean topology and the equality as order. The vector space dual $\mathcal{P}^*_{\mathbb{K}}$, of $\mathbb{K}$ is of course $\mathbb{K}$ itself, whereas its dual $\mathcal{P}^*$ as a locally convex cone consists of the real parts of these evaluations (see Example I.2.1(c)). For $\mathcal{Q}$ we choose the simplified standard lattice completion $\widehat{\mathbb{K}}$ of $\mathbb{K}$ which consists of all bounded below $\overline{\mathbb{R}}$-valued functions on $\Gamma$, the unit circle of $\mathbb{K}$, endowed with the (strictly) positive constants as neighborhoods (see Example I.5.62(f)).

We consider $\mathbb{K}$-valued measures $E \mapsto \theta_{\mathbb{K}\,E} : \mathfrak{R} \to \mathbb{K}$ in this case, yielding continuous linear operators $\theta_E$ from $\mathbb{K}$ to $\widehat{\mathbb{K}}$ via the convention

$$\theta_E(a)(\gamma) = \mathfrak{Re}(\gamma a\, \theta_{\mathbb{K}\,E})$$

for $E \in \mathfrak{R}$, $a \in \mathbb{K}$ and $\gamma \in \Gamma$. According to 3.2 we calculate the modulus of such a measure for every $E \in \mathfrak{R}$ as

$$\begin{aligned}|\theta|(E,\mathbb{B})(\gamma) &= \sup\left\{\sum_{i=1}^{n} \mathfrak{Re}(\gamma a\, \theta_{\mathbb{K}\,E}) \;\middle|\; |a_i| \le 1,\ E_i \in \mathfrak{R} \text{ disjoint subsets of } E\right\}\\ &= \sup\left\{\sum_{i=1}^{n} |\theta_{E_i}| \;\middle|\; E_i \in \mathfrak{R} \text{ disjoint subsets of } E\right\}\end{aligned}$$

for all $\gamma \in \Gamma$, where $\mathbb{B} \in \mathcal{V}$ stands for the unit ball in $\mathbb{K}$. This is of course the usual notation for the *total variation* $\mathfrak{var}(\theta, E)$ of a real- or complex-valued measure on a set $E$ (see III.1.4 in [55] or Section 6.1 in [179]). A simple argument (see Lemmas III.1.5 and III.4.5 in [55]) shows that

$$|\theta|(E,\mathbb{B}) \le 4 \sup\big\{|\theta_G| \mid G \in \mathfrak{R},\ G \subset E\big\} < +\infty$$

for every $E \in \mathfrak{R}$ in this case. Hence any $\mathbb{K}$-valued measure is $\mathfrak{R}$-bounded in the sense of Section 3.6 and may therefore be extended to the standard full extension

$$\mathcal{P}_{\mathcal{V}} = \{a + \alpha\mathbb{B} \mid a \in \mathbb{K},\ \alpha \ge 0\}$$

of $\mathcal{P} = \mathbb{K}$, setting

$$\theta_E(a + \alpha\mathbb{B})(\gamma) = \theta_E(a)(\gamma) + \alpha|\theta|(E,\mathbb{B}) = \mathfrak{Re}(\gamma a\, \theta_{\mathbb{K}\,E}) + \alpha|\theta|(E,\mathbb{B})$$

for all $\gamma \in \Gamma$. The notion of measurability from Section 1 for $\mathbb{K}$-valued functions coincides with the usual one. A measurable function $f$ is contained in $\mathcal{F}_{\mathfrak{R}}(X,\mathbb{K})$ if on every set $E \in \mathfrak{R}$ it can be uniformly approximated by a sequence of step functions. It follows from our convergence theorems that the integral of $f$ over $E$ is the limit of the integrals of this sequence of step functions. Integrability in the sense of 4.12 and 4.13, however reaches beyond this requirement. Integrals of $\mathbb{K}$-valued functions are evaluated in $\widehat{\mathbb{K}}$, that is as $\overline{\mathbb{R}}$-valued functions on $\Gamma$. However, since according to Corollary 5.9 these integrals are elements of the order closure of the embedding of $\mathbb{K}$ into $\widehat{\mathbb{K}}$, hence are $\mathbb{K}$-linear by I.5.60(b). We may therefore identify the integral in the usual way with a number in $\mathbb{K}$, setting

$$\left\langle \int_F f\, d\theta \right\rangle_{\mathbb{R}} = \left(\int_F f\, d\theta\right)(1)$$

in the real, and

$$\left\langle \int_F f\,d\theta \right\rangle_{\mathbb{C}} = \left( \int_F f\,d\theta \right)(1) - i \left( \int_F f\,d\theta \right)(i)$$

in the complex case, respectively. Moreover, given $\gamma \in \Gamma$ we have

$$\left\langle \int_F \gamma h\,d\theta \right\rangle_{\mathbb{K}} = \gamma \left\langle \int_F f\,d\theta \right\rangle_{\mathbb{K}}$$

for every step function $h \in \mathcal{S}_{\mathfrak{R}}(X,\mathbb{K})$. Because $h \le f + \mathfrak{v}_w$ holds if and only if $\gamma h \le \gamma f + \mathfrak{v}_w$ for $h \in \mathcal{S}_{\mathfrak{R}}(X,\mathbb{K})$ and $f \in \mathcal{F}_{\mathfrak{R}}(X,\mathbb{K})$, we have

$$\left\langle \int_F^{(w)} \gamma f\,d\theta \right\rangle_{\mathbb{K}} = \gamma \left\langle \int_F^{(w)} f\,d\theta \right\rangle_{\mathbb{K}}.$$

Consequently, $\mathcal{F}_{(F,\theta)}(X,\mathbb{K})$ is a vector space over $\mathbb{K}$, and the mapping

$$f \mapsto \left\langle \int_F f\,d\theta \right\rangle_{\mathbb{K}} \;:\; \mathcal{F}_{(F,\theta)}(X,\mathbb{K}) \to \mathbb{K}$$

is linear over $\mathbb{K}$.

**6.5 The Case that $\mathcal{Q}$ Is the Standard Lattice Completion of Some Subcone $Q_0$.** Suppose that $(\mathcal{Q},\mathcal{W})$ is the standard lattice completion of a locally convex cone $(Q_0,\mathcal{W}_0)$ (see I.5.57), and suppose that the measure $\theta$ is indeed $\mathfrak{L}(\mathcal{P},\mathcal{Q}_0)$-valued. The closure of $\mathcal{Q}_0$ in $\mathcal{Q}$ with respect to the order topology was seen to be a subcone of the second dual $\mathcal{Q}_0^{**}$ (see Sections I.5.60 and I.7.3) in this case, and following Corollary 5.9, integrals of $(\mathcal{P},\mathcal{V})$-based integrable functions in $\mathcal{F}(X,\mathcal{P})$ are therefore elements of $\mathcal{Q}_0^{**}$. Stronger statements can be obtained for certain types of integrable functions. We shall develop these in the following remarks:

*Remarks 6.6.* Let $A$ be a relatively bounded subset of $\mathcal{P}$, that is, $A$ is bounded below, and bounded above relative to some element $a_0 \in \mathcal{P}$. Let $E \in \mathfrak{R}$. We observe the following:

(a) The convex hull of $A \cup \{0\}$, that is the set

$$\tilde{A} = \left\{ \sum_{i=1}^n \alpha_i a_i \;\middle|\; a_i \in A,\; 0 \le \alpha_i \in \mathbb{R},\; \sum_{i=1}^n \alpha_i \le 1 \right\}$$

is also bounded below and bounded above relative to $a_0$. Indeed, given $v \in \mathcal{V}$ let $\lambda, \rho \ge 0$ such that $0 \le a_0 + \lambda v$, $0 \le a + \lambda v$ and $a \le \rho a_0 + \lambda v$ for all $a \in A$. Then for any choice of $a_i \in A$ and $0 \le \alpha_i \in \mathbb{R}$ such that $\sum_{i=1}^n \alpha_i \le 1$ we have

$$0 \le \sum_{i=1}^n \alpha_i (a_i + \lambda v) \le \sum_{i=1}^n \alpha_i a_i + \lambda v$$

and

$$\sum_{i=1}^{n} \alpha_i a_i \leq \sum_{i=1}^{n} \alpha_i(\rho a_0 + \lambda v) + \rho\Big(1 - \sum_{i=1}^{n} \alpha_i\Big)(a_0 + \lambda v) \leq \rho a_0 + \lambda(1+\rho)v.$$

This yields our claim.

(b) For every $E \in \mathfrak{R}$ the set

$$\mathcal{Z}(A, E) = \left\{ \sum_{i=1}^{n} \theta_{E_i}(a_i) \mid a_i \in A,\ E_i \in \mathfrak{R} \text{ disjoint subsets of } E \right\}$$

is bounded below, and bounded above relative to the element $\theta_E(a_0)$, hence $\mathcal{Z}(A, E)$ is a relatively bounded subset of $\mathcal{Q}_0$. Indeed, given $w \in \mathcal{W}$ there is $v \in \mathcal{V}$ such that $|\theta|(E, v) = \theta_E(v) \leq w$. In turn, there are $\lambda, \rho \geq 0$ such that $0 \leq a + \lambda v$ and $a \leq \rho a_0 + \lambda v$ for all $a \in A$. We may also assume that $0 \leq \rho a_0 + \lambda v$. Now let $a_1, \ldots, a_n \in A$ and let $E_1, \ldots, E_n \in \mathfrak{R}$ be disjoint subsets of $E$. Then

$$0 \leq \sum_{i=1}^{n} \theta_{E_i}(a_i + \lambda v) = \sum_{i=1}^{n} \theta_{E_i}(a_i) + \lambda \sum_{i=1}^{n} \theta_{E_i}(v) \leq \sum_{i=1}^{n} \theta_{E_i}(a_i) + \lambda w$$

and

$$\sum_{i=1}^{n} \theta_{E_i}(a_i) \leq \sum_{i=1}^{n} \theta_{E_i}(\rho a_0 + \lambda v) \leq \theta_E(\rho a_0 + \lambda v) \leq \rho \theta_E(a_0) + \lambda w.$$

The set $\mathcal{Z}(A, E)$ is therefore bounded below and bounded above relative to the element $\theta_E(a_0)$, thus relatively bounded in $\mathcal{Q}_0$.

(c) Now recall from Section I.5.57 that the order topology of the standard lattice completion $\mathcal{Q}$ of $\mathcal{Q}_0$ coincides with the topology of pointwise convergence on the elements of $\mathcal{Q}_0^*$. Thus, according to I.7.3 the limit in $\mathcal{Q}$ with respect to order convergence of any net in the relatively bounded set $\mathcal{Z}(A, E) \subset \mathcal{Q}_0 \subset \mathcal{Q}$ from (b) is contained in the relative strong second dual $(\mathcal{Q}_0)_{sr}^{**}$ of $\mathcal{Q}_0$.

(d) Let $E \in \mathfrak{R}$, let $\varphi_1, \ldots, \varphi_n$ be non-negative measurable real-valued functions such that $\sum_{i=1}^{n} \varphi_i \leq \chi_E$ and let $a_1, \ldots, a_n \in A$. For each $i = 1, \ldots, n$ let $(\psi_k^i \otimes a_i)_{k \in \mathbb{N}}$ be a sequence of step functions approximating $\varphi_i \otimes a_i$ as in 5.27 and 5.28, that is

$$0 \leq \psi_k^i \leq \varphi_i \qquad \text{and} \qquad \lim_{k \to \infty} \int_X \psi_k^i \otimes a_i \, d\theta = \int_x \varphi_i \otimes a_i \, d\theta.$$

According to 5.27, these step functions $\psi_k^i \otimes a_i$ are of the type

$$\sum_{j=1}^{k} \chi_{E_j^{(i,k)}} \otimes (\alpha_j a_i),$$

with disjoint sets $E_j^{(i,k)} \in \mathfrak{R}$ whose union is $E$, with $0 \le \alpha_j \le 1$ and such that $\sum_{j=1}^{k} \alpha_j \chi_{E_j^{(i,k)}} \le \varphi_i$. For every $k \in \mathbb{N}$ let

$$h_k = \sum_{i=1}^{n} \psi_i^k \otimes a_i = \sum_{i=1}^{n} \sum_{j=1}^{k} \chi_{E_j^{(i,k)}} \otimes (\alpha_j a_i).$$

As

$$\sum_{i=1}^{n} \sum_{j=1}^{k} \alpha_j \chi_{E_j^{(i,k)}} \le \sum_{i=1}^{n} \varphi_i \le \chi_E,$$

the step function $h_k$ can be expressed as

$$h_k = \sum_{l=1}^{p} \chi_{F_l} \otimes b_l,$$

where $F_1, \ldots, F_p \in \mathfrak{R}$ are disjoint subsets of $E$ and $b_1, \ldots, b_p$ are suitable convex combinations of the elements of the relatively bounded set $\tilde{A} = A \cup \{0\}$ (see 6.6(a)); more precisely

$$b_l = \sum_{i=1}^{n} \beta_i a_i,$$

where $\beta_i$ is the sum of all those $\alpha_j$, for $j = 1, \ldots, k$, such that $F_l \subset E_j^{(i,k)}$. Thus the integral

$$\int_X h_k \, d\theta = \sum_{l=1}^{p} \theta_{F_l}(b_l)$$

is contained in the relatively bounded subset

$$\mathcal{Z}(\tilde{A}, E) = \left\{ \sum_{i=1}^{n} \theta_{E_i}(a_i) \;\middle|\; a_i \in \tilde{A},\ E_i \in \mathfrak{R} \text{ disjoint subsets of } E \right\}$$

of $\mathcal{Q}_0$. We have

$$\int_X \left( \sum_{i=1}^{n} \varphi_i \otimes a_i \right) d\theta = \lim_{k \to \infty} \int_X h_k \, d\theta,$$

hence according to (c), the integral of the function $\sum_{i=1}^{n} \varphi_i \otimes a_i$ is contained in the relative strong second dual $(\mathcal{Q}_0)^{**}_{sr}$ of $\mathcal{Q}_0$. The same applies to integrals of this function over sets $F \in \mathfrak{A}_R$, since the functions $\varphi_i$ may be replaced by the functions $\chi_F \varphi_i$ in the preceding argument.

(e) If for a function $f \in \mathcal{F}_{(F,\theta)}(X,\mathcal{P})$ there is a net $(f_j)_{j\in\mathcal{J}}$ consisting of functions $\sum_{i=1}^{n} \varphi_i \otimes a_i$ as in (d) such that $\int_F f\,d\theta = \lim_{j\in\mathcal{J}} \int_F f_j\,d\theta$, then according to (c), $\int_F f\,d\theta$ is also contained in $(\mathcal{Q}_0)^{**}_{sr}$.

We summarize:

**Proposition 6.7.** *Let $(\mathcal{P},\mathcal{V})$ and $(\mathcal{Q}_0,\mathcal{W}_0)$ be locally convex cones such that $(\mathcal{P},\mathcal{V})$ is quasi-full, and let $\theta$ be an $\mathfrak{L}(\mathcal{P},\mathcal{Q}_0)$-valued measure. Let $F \in \mathfrak{A}_{\mathfrak{R}}$.*

*(a) For every $(\mathcal{P},\mathcal{V})$-based integrable function in $f \in \mathcal{F}_{(F,\theta)}(X,\mathcal{P})$ the integral $\int_F f\,d\theta$ is contained in $\mathcal{Q}_0^{**}$, the second dual of $\mathcal{Q}_0$.*

*(b) Let $E \in \mathfrak{R}$ and let $A$ be a relatively bounded subset of $\mathcal{P}$. If for $f \in \mathcal{F}_{(F,\theta)}(X,\mathcal{P})$ there is a net $(f_j)_{j\in\mathcal{J}}$ consisting of functions $\sum_{i=1}^{n} \varphi_i \otimes a_i$, where $\varphi_i$ are non-negative measurable real-valued functions such that $\sum_{i=1}^{n} \varphi_i \le \chi_E$ and $a_i \in A$, and such that $\int_F f\,d\theta = \lim_{j\in\mathcal{J}} \int_F f_j\,d\theta$, then $\int_F f\,d\theta$ is contained in $(\mathcal{Q}_0)^{**}_{sr}$.*

We shall obtain a further strengthening of these observations in some special cases.

**6.8 Compact and Weakly Compact Measures.** Let $\theta$ be an $\mathfrak{L}(\mathcal{P},\mathcal{Q}_0)$-valued measure, where $(\mathcal{P},\mathcal{V})$ is a quasi-full and $(\mathcal{Q}_0,\mathcal{W}_0)$ is a locally convex cone such that $(\mathcal{Q},\mathcal{W})$ is its standard lattice completion. Such a measure $\theta$ is called *compact* (or *weakly compact*) if for every $E \in \mathfrak{R}$ and every relatively bounded subset $A$ of $\mathcal{P}$ the subset

$$\mathcal{Z}(A,E) = \left\{ \sum_{i=1}^{n} \theta_{E_i}(a_i) \;\middle|\; a_i \in A,\ E_i \in \mathfrak{R} \text{ disjoint subsets of } E \right\}$$

of $\mathcal{Q}_0$ is relatively compact in the symmetric relative topology (or in the weak topology $\sigma(\mathcal{Q}_0,\mathcal{Q}_0^*)$) of $\mathcal{Q}_0$ (see I.4.6).

Recall from Lemma I.4.7 that the symmetric relative topology is finer than $\sigma(\mathcal{Q}_0,\mathcal{Q}_0^*)$, and from I.5.57 that $\sigma(\mathcal{Q}_0,\mathcal{Q}_0^*)$ is finer than the induced order topology on $\mathcal{Q}_0$ which is however still Hausdorff. The latter two topologies coincide, if all elements of $\mathcal{Q}_0$ are bounded (see I.5.57). Moreover, $\sigma(\mathcal{Q}_0,\mathcal{Q}_0^*)$ coincides with its own relative topology (see I.4.6). We observe that every subset $\mathcal{Z}$ of $\mathcal{Q}_0$ which is relatively compact in the symmetric relative topology is also relatively weakly compact. Indeed, the closure $\overline{\mathcal{Z}}$ of $\mathcal{Z}$ with respect to the symmetric relative topology is contained in its closure $\overline{\mathcal{Z}}^w$ with respect to the weak topology. $\overline{\mathcal{Z}}$ is compact in the former, hence also in the latter topology, thus weakly closed since $\sigma(\mathcal{Q}_0,\mathcal{Q}_0^*)$ is Hausdorff. We infer that $\overline{\mathcal{Z}} = \overline{\mathcal{Z}}^w$, and our claim follows. Every compact measure $\theta$ is therefore also weakly compact.

For a set $E \in \mathfrak{R}$ and a relatively bounded subset $A \in \mathcal{P}$ we denote by

$$\mathcal{I}(A,E) = \left\{ \int_X \left( \sum_{i=1}^n \varphi_i \otimes a_i \right) d\theta \;\middle|\; a_i \in A,\ 0 \le \varphi_i \text{ measurable, } \sum_{i=1}^n \varphi_i \le \chi_E \right\}.$$

Clearly $\mathcal{Z}(A,E) \subset \mathcal{I}(A,E) \subset \mathcal{Q}$. Conversely, we observed in Remark 6.6(d) that $\mathcal{I}(A,E)$ is contained in the closure of $\mathcal{Z}(A,E)$ with respect to the order topology of $\mathcal{Q}$. If the measure $\theta$ is compact (or a weakly compact), then the (weak) closure $\overline{\mathcal{Z}(A,E)}^w$ of $\mathcal{Z}(A,E)$ is weakly compact and therefore also compact in the coarser induced order topology, and indeed closed in $\mathcal{Q}$ as the order topology is Hausdorff in this case. This demonstrates that

$$\mathcal{I}(A,E) \subset \overline{\mathcal{Z}(A,E)}^w \subset \mathcal{Q}_0$$

in this case. Consequently the set $\mathcal{I}(A,E)$ is also (weakly) compact in $\mathcal{Q}_0$.

We summarize:

**Proposition 6.9.** *Let $(\mathcal{P},\mathcal{V})$ and $(\mathcal{Q}_0,\mathcal{W}_0)$ be locally convex cones such that $(\mathcal{P},\mathcal{V})$ is quasi-full. An $\mathfrak{L}(\mathcal{P},\mathcal{Q}_0)$-valued measure $\theta$ is compact (or weakly compact), if and only if for every $E \in \mathfrak{R}$ and for every relatively bounded subset $A$ of $\mathcal{P}$,*

$$\left\{ \int_X \left( \sum_{i=1}^n \varphi_i \otimes a_i \right) d\theta \;\middle|\; a_i \in A,\ 0 \le \varphi_i \textit{ measurable, } \sum_{i=1}^n \varphi_i \le \chi_E \right\}$$

*is a relatively compact (or relatively weakly compact) subset of $\mathcal{Q}_0$.*

**Corollary 6.10.** *Let $(\mathcal{P},\mathcal{V})$ and $(\mathcal{Q}_0,\mathcal{W}_0)$ be locally convex cones such that $(\mathcal{P},\mathcal{V})$ is quasi-full and let $\theta$ be an $\mathfrak{L}(\mathcal{P},\mathcal{Q}_0)$-valued relatively compact measure. Let $E \in \mathfrak{R}$, $F \in \mathfrak{A}_R$ and let $A$ be a relatively bounded subset of $\mathcal{P}$. If for $f \in \mathcal{F}_{(F,\theta)}(X,\mathcal{P})$ there is a net $(f_j)_{j\in\mathcal{J}}$ consisting of functions $\sum_{i=1}^n \varphi_i \otimes a_i$, where $\varphi_i$ are non-negative measurable real-valued functions such that $\sum_{i=1}^n \varphi_i \le \chi_E$ and $a_i \in A$, and such that $\int_F f\,d\theta = \lim_{j\in\mathcal{J}} \int_F f_j\,d\theta$, then $\int_F f\,d\theta$ is contained in $\mathcal{Q}_0$.*

The following consequence of Theorem 3.15 yields that in certain special circumstances every bounded measure is weakly compact.

**Proposition 6.11.** *Suppose that $(\mathcal{P},\|\ \|)$ is a finite dimensional normed space and that $(\mathcal{Q},\mathcal{W})$ is the standard lattice completion of a Banach space $(\mathcal{Q}_0,\|\ \|)$. Then every bounded $\mathfrak{L}(\mathcal{P},\mathcal{Q}_0)$-valued measure is weakly compact.*

*Proof.* Let $(\mathcal{P},\|\ \|)$ and $(\mathcal{Q}_0,\|\ \|)$ be as stated and let $\theta$ be a bounded $\mathfrak{L}(\mathcal{P},\mathcal{Q}_0)$-valued measure on $\mathfrak{R}$. We consider both $\mathcal{P}$ and $\mathcal{Q}_0$ as normed spaces over $\mathbb{R}$. Given a basis $\{b_1,\ldots,b_m\}$ of $\mathcal{P}$, there is a constant $\rho > 0$ such that

$$\left\| \sum_{k=1}^{m} \beta_k b_k \right\| \geq \rho \left( \max_{k=1,\ldots,m} |\beta_k| \right)$$

for every choice of scalars $\beta_1, \ldots, \beta_m \in \mathbb{R}$ (see for example Lemma 2.4.1 in [107]). Now let $E \in \mathfrak{R}$ and let $A$ be a bounded subset of $\mathcal{P}$. According to the above then there exists $\lambda > 0$ such that

$$A \subset \left\{ \sum_{k=1}^{m} \beta_k b_k \;\middle|\; \beta_k \in \mathbb{R},\;\; |\beta_k| \leq \lambda \right\}.$$

We fix $1 \leq k \leq m$. Theorem 3.15 yields that the set

$$\mathcal{Z}_k = \{\theta_G(b_k) \mid G \in \mathfrak{R},\;\; G \subset E\}$$

is relatively compact in $\mathcal{Q}_0$ with respect to the weak topology $\sigma(\mathcal{Q}_0, \mathcal{Q}_0^*)$. Now let $E_i \in \mathfrak{R}$, for $i = 1, \ldots, n$, be disjoint subsets of $E$ and in a first step let $0 \leq \beta_k^1 \leq \beta_k^2 \ldots \leq \beta_k^n \leq 1$. Set $F_1 = \bigcup_{i=1}^n E_i$, $F_2 = \bigcup_{i=2}^n E_i$, and so on, and $F_n = E_n$. Then

$$\sum_{i=1}^{n} \beta_k^i \theta_{E_i}(b_k) = \beta_k^1 \theta_{F_1}(b_k) + \sum_{i=2}^{n} (\beta_k^i - \beta_k^{i-1}) \theta_{F_i}(b_k).$$

The element $\sum_{i=1}^n \beta_k^i \theta_{E_i}(b_k)$ is therefore contained in the convex hull $\widetilde{\mathcal{Z}_k}$ of the set $\mathcal{Z}_k$. Following a well-known theorem due to Krein (see Theorem IV.11.4 in [185]) this convex hull is again relatively weakly compact in $\mathcal{Q}_0$. So, obviously is the set $-\widetilde{\mathcal{Z}_k}$. Using this, we infer that indeed for every choice of $\beta_k^i \in \mathbb{R}$ such that $|\beta_k^i| \leq 1$ for all $i = 1, \ldots, n$ the element $\sum_{i=1}^n \beta_k^i \theta_{E_i}(b_k)$ is contained in relatively weakly compact set $\mathcal{Y}_k = \widetilde{\mathcal{Z}_k} + (-\widetilde{\mathcal{Z}_k})$.

Thus for every choice of elements $a_i = \sum_{k=1}^m \beta_k^i b_k \in A$ and disjoint subsets $E_i \in \mathfrak{R}$ of $E$ we have $|\beta_k^i| \leq \lambda$ for all $i = 1, \ldots, n$ and $k = 1, \ldots, m$, hence

$$\sum_{i=1}^{n} \theta_{E_i}(a_i) = \sum_{k=1}^{m} \sum_{i=1}^{n} \beta_k^i \theta_{E_i}(b_k) \in \lambda \left( \sum_{k=1}^{m} \mathcal{Y}_k \right).$$

As a finite sum of relatively weakly compact sets (see I.V.2 in [185]), the set on the right-hand side is also relatively weakly compact in $\mathcal{Q}_0$, and our claim follows. □

**6.12 The Case that $\mathcal{P}$ Is a Locally Convex Vector Space.** Let $(\mathcal{P}, \mathcal{V})$ be a locally convex topological vector space over $\mathbb{K} = \mathbb{R}$ or $\mathbb{K} = \mathbb{C}$, endowed with a basis $\mathcal{V}$ of balanced convex neighborhoods, that is subsets of $\mathcal{P}$. Equality is the order on $\mathcal{P}$, and involving the neighborhoods we have $a \leq b + v$ if $a - b \in v$ for $a, b \in \mathcal{P}$ and $v \in \mathcal{V}$. As a locally convex cone $(\mathcal{P}, \mathcal{V})$ is of course quasi-full (see I.6.1). The modulus of an $\mathfrak{L}(\mathcal{P}, \mathcal{Q})$-valued measure $\theta$ is given by

$$|\theta|(E,v) = \sup\left\{ \sum_{i=1}^{n} \theta_{E_i}(s_i) \;\middle|\; s_i \in v,\ E_i \in \mathfrak{R} \text{ disjoint subsets of } E \right\} \in \mathcal{Q}.$$

According to Lemma 2.5, the cone $\mathcal{F}_{\mathfrak{R}}(X,\mathcal{P})$ as introduced in 2.3 consists of those $\mathcal{P}$-valued functions that vanish outside some set $E \in \mathfrak{R}$ and may be uniformly approximated on $X$ by step functions; more precisely: for $f \in \mathcal{F}_{\mathfrak{R}}(X,\mathcal{P})$ there is $E \in \mathfrak{R}$ such that $f(x) = 0$ for all $x \in X \setminus E$ and for every $v \in \mathcal{V}$ there exists a step function $h \in \mathcal{S}_{\mathfrak{R}}(X,\mathcal{P})$ such that $h(x) - f(x) \in v$ for all $x \in X$. Any such function $f$ is measurable by Theorem 1.7. Consequently, the functions in $\mathcal{F}_{\mathfrak{R}}(X,\mathcal{P})$ are uniformly bounded on all sets in $\mathfrak{R}$. We have $\alpha f \in \mathcal{F}_{\mathfrak{R}}(X,\mathcal{P})$ whenever $f \in \mathcal{F}_{\mathfrak{R}}(X,\mathcal{P})$ and $\alpha \in \mathbb{K}$. Every measurable neighborhood-valued function $s \in \mathcal{F}(X,\mathcal{P}_{\mathcal{V}})$ is however contained $\mathcal{F}_{\mathfrak{R}}(X,\mathcal{P}_{\mathcal{V}})$, since its values are positive. For a positive real-valued measurable function $\varphi$ and a neighborhood $v \in \mathcal{V}$, for example, the function $\varphi \otimes v$ is measurable, hence in $\mathcal{F}(X,\mathcal{P}_{\mathcal{V}})$. Recall that $\mathcal{V}$-valued measurable functions are integrated using the canonical extension of the measure $\theta$ to the full cone $(\mathcal{P}_{\mathcal{V}},\mathcal{V})$ as elaborated in Section 3.8.

According to 4.12, a $\mathcal{P}$-valued function $f$ is integrable over a set $E \in \mathfrak{R}$ if for every $w \in \mathcal{W}$ and $\varepsilon > 0$ there are functions $f_{(w,\varepsilon)} \in \mathcal{F}_{\mathfrak{R}}(X,\mathcal{P}_{\mathcal{V}})$ and $s_{(w,\varepsilon)} \in \mathcal{F}_{\mathfrak{R}}(X,\mathcal{V})$ such that

$$f \underset{a.e.E}{\leq} f_{(w,\varepsilon)} \underset{a.e.E}{\leq} \gamma f + s_{(w,\varepsilon)}$$

and $\int_E s_{(w,\varepsilon)}\,d\theta \leq \varepsilon w$ for some $1 \leq \gamma \leq 1+\varepsilon$. A straightforward argument involving the uniform boundedness of the functions in $\mathcal{F}_{\mathfrak{R}}(X,\mathcal{P})$ leads to a slight simplification in this case, avoiding the relative topologies: A function $f \in \mathcal{F}(X,\mathcal{P}_{\mathcal{V}})$ is integrable over a set $E \in \mathfrak{R}$ if for every $w \in \mathcal{W}$ there are functions $f_w \in \mathcal{F}_{\mathfrak{R}}(X,\mathcal{P}_{\mathcal{V}})$ and $s_w \in \mathcal{F}_{\mathfrak{R}}(X,\mathcal{V})$ such that

$$\text{(I)} \qquad f \underset{a.e.E}{\leq} f_w \underset{a.e.E}{\leq} f + s_w \qquad \text{and} \qquad \int_E s_w\,d\theta \leq w.$$

The function $\alpha f$ is integrable over $E$ for any $\alpha \in \mathbb{K}$, whenever $f$ is. A function $f \in \mathcal{F}(X,\mathcal{P})$ is $(\mathcal{P},\mathcal{V})$-based integrable over $E \in \mathfrak{R}$ (see 5.6) if there are $f_w \in \mathcal{F}_{\mathfrak{R}}(X,\mathcal{P})$ and $s_w \in \mathcal{F}_{\mathfrak{R}}(X,\mathcal{V})$ such that

$$f \underset{a.e.E}{\leq} f_w + s_w, \qquad f_w \underset{a.e.E}{\leq} f + s_w \quad \text{and} \quad \int_E s_w\,d\theta \leq w.$$

Considering that the functions in $\mathcal{F}_{\mathfrak{R}}(X,\mathcal{P})$ can be approximated by step functions, this is equivalent to the following condition for integrability which is only slightly stronger than (I):

For every $w \in \mathcal{W}$ there is a step function $h_w \in \mathcal{S}_{\mathfrak{R}}(X,\mathcal{P})$ and $s_w \in \mathcal{F}_{\mathfrak{R}}(X,\mathcal{V})$ such that

$$\text{(BI 1)} \qquad f(x) - h_w(x) \in s_w(x) \;\; \text{a.e. on } E \quad \text{and} \quad \int_E s_w\,d\theta \leq w.$$

(Set $f_w = h_w + s_w \in \mathcal{F}_{\mathfrak{R}}(X, \mathcal{P}_\nu)$ in order to satisfy (I).) Condition (BI 1) yields indeed strong integrability in the meaning of Section 5.18, since it obviously implies that $f \leq h_w + s_w$, hence

$$\int_G f\, d\theta \leq \int_G h_w\, d\theta + \int_G s_w\, d\theta \leq \int_G h_w\, d\theta + w$$

for all subsets $G \in \mathfrak{R}$ of $E$. Somewhat stronger than (BI 1) is the following sufficient integrability condition: For $v \in \mathcal{V}$ let $\| \, \|_v$ denote the corresponding seminorm on $\mathcal{P}$, that is $\|a\|_v = \inf\{\lambda \geq 0 \mid a \in \lambda v\}$. We require that for every $v \in \mathcal{V}$ and $w \in \mathcal{W}$ there is a step function $h_{(v,w)} \in \mathcal{S}_{\mathfrak{R}}(X, \mathcal{P})$ such that the positive real-valued function $x \mapsto \|f(x) - h_{(v,w)}(x)\|_v$ is measurable and

(BI 2) $$\int_E \|f - h_{(v,w)}\|_v \otimes v \, d\theta \leq w.$$

Condition (BI 2) obviously implies (BI 1) since, given $w \in \mathcal{W}$ we choose any $v \in \mathcal{V}$ and set $s_w(x) = \|f(x) - h_{(v,w)}(x)\|\, v$. Then obviously $f(x) - h_{(v,w)}(x) \in s_w(x)$ holds for all $x \in E$, hence (BI 1). Moreover, a function $f \in \mathcal{F}(X, \mathcal{P})$ satisfying (BI 2) is $(\mathcal{P}, \mathcal{V}_0)$-based integrable over $E$ with respect to $\theta$ for every one-dimensional neighborhood subsystem $\mathcal{V}_0 = \{\rho v_0 \mid \rho > 0\}$, for $v_0 \in \mathcal{V}$. This is one of the requirements in Theorem 5.36. In the special case that $(\mathcal{P}, \mathcal{V})$ is a normed space, that is $\mathcal{V} = \{\rho \mathbb{B} \mid \rho > 0\}$, where $\mathbb{B}$ is the unit ball in $\mathcal{P}$, condition (BI 2) leads to the well-known notion of *Bochner (or Dunford and Schwartz) integrability* (see for example III.2.17 in [55] or II.2 in [43]). This will be further elaborated in Section 6.18 below.

In all of the above cases, integrability is then extended to sets $F \in \mathfrak{A}_{\mathfrak{R}}$ as in 4.13. Convergence for sequences of $\mathcal{P}$-valued functions as required in Theorems 5.23 to 5.25 and 5.34 refers to pointwise convergence with respect to the vector space topology of $\mathcal{P}$. If the measure $\theta$ is strongly additive and if the order continuous linear functionals on the locally convex complete lattice cone $(\mathcal{Q}, \mathcal{W})$ support the separation property (see I.5.32), then the strong convergence statements of Theorem 5.36 apply to functions satisfying (BI 2).

We already observed that the functions which are integrable over a set $E \in \mathfrak{R}$ with respect to any of the above criteria form also a vector space over $\mathbb{K}$ in this case.

Now suppose in addition to the above that $(\mathcal{Q}, \mathcal{W})$ is the standard lattice completion of some subcone $(\mathcal{Q}_0, \mathcal{W}_0)$ and the measure $\theta$ is $\mathfrak{L}(\mathcal{P}, \mathcal{Q}_0)$-valued (see 6.5). Then, according to Theorem 3.11 countable additivity for $\theta$ refers to the strong operator topology of $\mathfrak{L}(\mathcal{P}, \mathcal{Q}_0)$. Moreover, following Proposition 6.7(a), integrals of $(\mathcal{P}, \mathcal{V})$-based integrable functions in $\mathcal{F}(X, \mathcal{P})$ are elements of the second dual $\mathcal{Q}_0^{**}$ of $\mathcal{Q}_0$. We shall make a few supplementary observations for the case that $(\mathcal{Q}_0, \mathcal{W}_0)$ is indeed a vector space over $\mathbb{K} = \mathbb{R}$ or $\mathbb{K} = \mathbb{C}$:

(i) If $(\mathcal{Q}_0, \mathcal{W}_0)$ is a locally convex topological vector space over $\mathbb{K}$, then the $(\mathcal{P}, \mathcal{V})$-based integrable functions in $\mathcal{F}(X, \mathcal{P})$ form a vector space

$\mathcal{F}_{(F,\theta,BI1)}(X,\mathcal{P})$ over $\mathbb{K}$. The integrals of functions in $\mathcal{F}_{(F,\theta,BI1)}(X,\mathcal{P})$ for $F \in \mathfrak{A}_{\mathfrak{R}}$ are contained in the order closure of $\mathcal{Q}_0$ in $\mathcal{Q}$, hence are $\mathbb{K}$-linear (see I.5.60(b)) and therefore elements of the second vector space dual $\mathcal{Q}_{0\mathbb{K}}^{**}$ of $\mathcal{Q}_0$.

(ii) If the locally convex space $\mathcal{Q}_0$ is indeed topologically complete, and if a function $f \in \mathcal{F}(X,\mathcal{P})$ fulfills the integrability criterion (BI1), then for every $E \in \mathfrak{R}$ its integral $\int_E f\,d\theta$ in $\mathcal{Q}$ may be approximated in the symmetric (modular) topology of $\mathcal{Q}$ by a net $\left(\int_E h_i\,d\theta\right)_{i\in\mathcal{I}}$ of integrals over step functions. Integrals over step functions are however contained in the complete subspace $\mathcal{Q}_0$ of $\mathcal{Q}$. The Cauchy sequence $\left(\int_E h_n\,d\theta\right)_{n\in\mathbb{N}}$ is therefore convergent in $\mathcal{Q}_0$ and its limit, that is $\int_E f\,d\theta$ is also contained in $\mathcal{Q}_0$.

(iii) If the locally convex space $\mathcal{Q}_0$ is reflexive, then every bounded $\mathfrak{L}(\mathcal{P},\mathcal{Q}_0)$-valued measure $\theta$ is seen to be weakly compact. Indeed, for every $E \in \mathfrak{R}$ and every bounded subset $A$ of $\mathcal{P}$ the set

$$\mathcal{Z}(A,E) = \left\{ \sum_{i=1}^{n} \theta_{E_i}(a_i) \;\middle|\; a_i \in A,\ E_i \in \mathfrak{R} \text{ disjoint subsets of } E \right\}$$

from 6.8 is bounded in $\mathcal{Q}_0$ (see Remark 6.8(a)), hence relatively weakly compact, since this holds for all bounded subsets in reflexive spaces.

(iv) If both $\mathcal{P}$ and $\mathcal{Q}_0$ are locally convex topological vector spaces over $\mathbb{K}$, then we denote by $\mathfrak{L}_{\mathbb{K}}(\mathcal{P},\mathcal{Q}_0)$ the space of all continuous $\mathbb{K}$-linear operators from $\mathcal{P}$ into $\mathcal{Q}_0$. If the measure $\theta$ is indeed $\mathfrak{L}_{\mathbb{K}}(\mathcal{P},\mathcal{Q}_0)$-valued, then for every $F \in \mathfrak{A}_{\mathfrak{R}}$ the operator

$$f \mapsto \int_F f\,d\theta \;:\; \mathcal{F}_{(F,\theta,BI1)}(X,\mathcal{P}) \to \mathcal{Q}_{0\mathbb{K}}^{**}$$

is also linear over $\mathbb{K}$. According to I.5.60(d) we need to verify two conditions for this. The first one is obvious, because the additivity of the operator is given. Likewise, the second condition in I.5.60(d) is evident for all $\alpha \geq 0$. Thus all left to verify is that

$$\left(\int_F \gamma f\,d\theta\right)(\mu) = \left(\int_F f\,d\theta\right)(\gamma\mu).$$

holds for all $f \in \mathcal{F}_{(F,\theta,BI1)}(X,\mathcal{P})$, $\mu \in \mathcal{Q}_0^*$ and $\gamma \in \Gamma$, the unit circle in $\mathbb{K}$. Indeed, this obviously holds true for every step function $h \in \mathcal{S}_{\mathfrak{R}}(X,\mathcal{P})$. Because the neighborhoods in $\mathcal{V}$ and in $\mathcal{W}$ are supposed to be balanced, $h \leq f + \mathfrak{v}_w$ holds for $h \in \mathcal{S}_{\mathfrak{R}}(X,\mathcal{P})$ and $f \in \mathcal{F}_{\mathfrak{R}}(X,\mathcal{P})$ if and only if $\gamma h \leq \gamma f + \mathfrak{v}_w$. Therefore and because the lattice operations are taken pointwise in $\mathcal{Q}$, we infer that

$$\left(\int_F^{(w)} \gamma f\,d\theta\right)(\mu) = \left(\int_F^{(w)} f\,d\theta\right)(\gamma\mu).$$

Now Definition 4.13 yields our claim. We shall formulate this special case as a separate Proposition:

**Proposition 6.13.** *Let $(\mathcal{P},\mathcal{V})$ and $(\mathcal{Q}_0,\mathcal{W}_0)$ be locally convex topological vector spaces over $\mathbb{K}=\mathbb{R}$ or $\mathbb{K}=\mathbb{C}$ and let $\theta$ be a bounded $\mathfrak{L}_{\mathbb{K}}(\mathcal{P},\mathcal{Q}_0)$-valued measure. Then the functions in $\mathcal{F}(X,\mathcal{P})$ satisfying (BI1) form a vector space $\mathcal{F}_{(F,\theta,BI1)}(X,\mathcal{P})$ over $\mathbb{K}$, their integrals are contained in the second vector space dual $\mathcal{Q}_{0\mathbb{K}}^{**}$ of $\mathcal{Q}_0$, and the operator*

$$f\mapsto\int_F f\,d\theta\;:\;\mathcal{F}_{(F,\theta,BI1)}(X,\mathcal{P})\to\mathcal{Q}_{0\mathbb{K}}^{**}$$

*is linear over $\mathbb{K}$.*

**6.14 Algebra Homomorphisms.** Let us consider a special case of 6.13. Suppose that the locally convex vector spaces $(\mathcal{P},\mathcal{V})$ and $(\mathcal{Q}_0,\mathcal{W})$ are indeed topological algebras over $\mathbb{K}$, and that $(\mathcal{Q},\mathcal{W})$ is the standard lattice completion of $\mathcal{Q}_0$. A *topological algebra* $\mathcal{P}$ is an algebra and a locally convex topological vector space such that for a fixed element $a\in\mathcal{P}$ (or $b\in\mathcal{P}$) the linear operator $c\mapsto ac$ (or $c\mapsto cb$) from $\mathcal{P}$ into $\mathcal{P}$ is continuous (see for example 8.1 in [137]). Recall that for a linear operator continuity implies weak continuity. Thus $\mathcal{P}$ is also a topological algebra in its weak topology. Indeed, for a fixed $a\in\mathcal{P}$ and $\mu\in\mathcal{P}^*$, the mapping $c\mapsto\mu(ac):\mathcal{P}\to\mathbb{R}$ is a continuous linear functional. Thus, if the net $(c_i)_{i\in\mathcal{I}}$ in $\mathcal{P}$ converges weakly to $c\in\mathcal{P}$, then $\mu(ac_i)_{i\in\mathcal{I}}$ converges to $\mu(ac)$ in $\mathbb{R}$. The net $(ac_i)_{i\in\mathcal{I}}$ therefore converges weakly to $ac\in\mathcal{P}$.

Now suppose that $\theta$ is an $\mathfrak{R}$-bounded measure such that its values $\theta_E$ for all $E\in\mathfrak{R}$ of are continuous $\mathbb{K}$-linear operators from $\mathcal{P}$ to $\mathcal{Q}_0$ satisfying the following condition:

(A) $\theta_E(a)\,\theta_E(b)=\theta_E(ab)$ *and* $\theta_E(a)\,\theta_G(b)=0$ *for all* $a,b\in\mathcal{P}$ *and disjoint sets* $E,G\in\mathfrak{R}$.

Both requirements in Condition (A) may be reformulated and combined as

(A') $\theta_E(a)\,\theta_G(b)=\theta_{(E\cap G)}(ab)$ *for all* $E,G\in\mathfrak{R}$ *and* $a,b\in\mathcal{P}$.

Indeed, (A') implies (A), and if (A) holds, then for $a,b\in\mathcal{P}$ and $E,G\in\mathfrak{R}$ we have

$$\theta_E(a)\,\theta_G(b)=\big(\theta_{(E\setminus G)}(a)+\theta_{(E\cap G)}(a)\big)\big(\theta_{(G\setminus E)}(b)+\theta_{(E\cap G)}(b)\big)=\theta_{(E\cap G)}(ab),$$

hence (A'). Endowed with the canonical, that is pointwise multiplication, the $\mathcal{P}$-valued step functions form an algebra, and we obtain

$$\int_X(hl)\,d\theta=\left(\int_X h\,d\theta\right)\left(\int_X l\,d\theta\right)$$

for all $h,l\in\mathcal{S}_{\mathfrak{R}}(X,\mathcal{P})$ as an immediate consequence of (A). Indeed, the functions $h$ and $l$ can be expressed as $h=\sum_{i=1}^n\chi_{E_i\circledast}a_i$ and $l=\sum_{i=1}^n\chi_{E_i\circledast}b_i$ with disjoint sets $E_i\in\mathfrak{R}$ and elements $a_i,b_i\in\mathcal{P}$. Then $hl=\sum_{i=1}^n\chi_{E_i\circledast}a_ib_i$ and

$$\sum_{i=1}^{n} \theta_{E_i}(a_i b_i) = \left( \sum_{i=1}^{n} \theta_{E_i}(a_i) \right) \left( \sum_{i=1}^{n} \theta_{E_i}(b_i) \right),$$

that is our claim.

Now let us denote by $\mathcal{E}_{\mathfrak{R}}(X,\mathcal{P})$ the vector subspace of $\mathcal{F}(X,\mathcal{P})$ generated by all elementary functions. Recall that elementary functions are of the type $\varphi\otimes a$, where $\varphi$ is a bounded non-negative measurable real-valued function supported by a set in $\mathfrak{R}$, and $a$ is an element of $\mathcal{P}$. Obviously, $\mathcal{E}_{\mathfrak{R}}(X,\mathcal{P})$ forms also an algebra, as the product of two elementary functions $\varphi\otimes a$ and $\psi\otimes b$ is the elementary function $(\varphi\psi)\otimes(ab)$. We would like to establish that the integral defines a multiplicative operator on $\mathcal{E}_{\mathfrak{R}}(X,\mathcal{P})$ as well. However, because integrals of these functions are generally contained in the strong second dual $\mathcal{Q}_0^{**}$ of $\mathcal{Q}_0$ rather than in $\mathcal{Q}_0$ itself, we shall a introduce a continuation of the multiplication to $\mathcal{Q}_0^{**} \subset \mathcal{Q}$ in the following way: For elements $l, m \in \mathcal{Q}$ we denote by $l \bullet m$ the set of all elements $q \in \mathcal{Q}$ for which we can find nets $(l_i)_{i\in\mathcal{I}}$ and $(m_j)_{j\in\mathcal{J}}$ in $\mathcal{Q}_0 \subset \mathcal{Q}$ such that $\lim_{i\in\mathcal{I}} l_i = l$, $\lim_{j\in\mathcal{J}} m_j = m$ and

$$\underline{\lim}_{i\in\mathcal{I}}\,\underline{\lim}_{j\in\mathcal{J}}\, l_i m_j = \overline{\lim}_{i\in\mathcal{I}}\,\overline{\lim}_{j\in\mathcal{J}}\, l_i m_j = \underline{\lim}_{j\in\mathcal{J}}\,\underline{\lim}_{i\in\mathcal{I}}\, l_i m_j = \overline{\lim}_{j\in\mathcal{J}}\,\overline{\lim}_{i\in\mathcal{I}}\, l_i m_j = q.$$

Our introductory remark shows that for elements $l, m \in \mathcal{Q}_0$ we have $l \bullet m = \{lm\}$, since on $\mathcal{Q}_0 \subset \mathcal{Q}$ weak and order convergence coincide (see I.5.57). In general, the set $l \bullet m$ may be empty or contain more than one element of $\mathcal{Q}$. However, if $q \in l \bullet m$ and if $\mu \in \mathcal{Q}^*$ is a multiplicative linear functional, then

$$q(\mu) = \lim_{i\in\mathcal{I}} \lim_{j\in\mathcal{J}} \big(l_i m_j\big)(\mu) = \lim_{i\in\mathcal{I}} \lim_{j\in\mathcal{J}} l_i(\mu)\, m_j(\mu) = l(\mu)\, m(\mu).$$

Now let $f = \varphi\otimes a$ and $g = \psi\otimes b$ be two elementary functions. Their product $fg$ is the elementary function $(\varphi\psi)\otimes(ab)$. Let $(\varphi_n)_{n\in\mathbb{N}}$ and $(\psi_n)_{n\in\mathbb{N}}$ be the sequences of real-valued step functions converging to $\varphi$ and $\psi$ as in 5.27 and 5.28. Thus

$$\lim_{n\to\infty} \int_X \varphi_n\otimes a\, d\theta = \int_X f\, d\theta \qquad \text{and} \qquad \lim_{n\to\infty} \int_X \psi_n\otimes b\, d\theta = \int_X g\, d\theta$$

by 5.28. For every fixed $m \in \mathbb{N}$ the sequence $(\varphi_m \psi_n)_{n\in\mathbb{N}}$ converges pointwise to the function $\varphi_m\psi$, and we have $0 \le \varphi_m\psi_n \le \varphi_m\psi$ for all $n \in \mathbb{N}$. This shows

$$\lim_{n\to\infty} \left\{ \left( \int_X \varphi_m\otimes a \right) \left( \int_X \psi_n\otimes b\, d\theta \right) \right\} = \lim_{n\to\infty} \int_X (\varphi_m\psi_n)\otimes(ab)\, d\theta = \int_X (\varphi_m\psi)\otimes(ab)\, d\theta$$

by Corollary 5.28 and the above. Furthermore, the sequence $(\varphi_m\psi)_{m\in\mathbb{N}}$ converges pointwise to the function $\varphi\psi$, and we have $0 \le \varphi_m\psi \le \varphi\psi$ for all

$m \in \mathbb{N}$. Again using 5.28, this yields

$$\lim_{m\to\infty} \int_X (\varphi_m \psi)_{\otimes}(ab)\, d\theta = \int_X (\varphi\psi)_{\otimes}(ab)\, d\theta = \int_X (fg)\, d\theta,$$

hence

$$\lim_{m\to\infty} \lim_{n\to\infty} \left\{ \left( \int_X \varphi_{m\otimes} a \right) \left( \int_X \psi_{n\otimes} b\, d\theta \right) \right\} = \int_X (fg)\, d\theta.$$

Similarly, one verifies

$$\lim_{n\to\infty} \lim_{m\to\infty} \left\{ \left( \int_X \varphi_{m\otimes} a \right) \left( \int_X \psi_{n\otimes} b\, d\theta \right) \right\} = \int_X (fg)\, d\theta.$$

Thus indeed

$$\int_X (fg)\, d\theta \in \left( \int_X f\, d\theta \right) \cdot \left( \int_X g\, d\theta \right).$$

Finally, let $f, g \in \mathcal{E}_{\mathfrak{R}}(X, \mathcal{P})$, that is $f = \sum_{i=1}^{i_0} f_i$ and $g = \sum_{k=1}^{k_0} f_k$ with elementary functions $f_i, g_k$. For each of these functions there are approximating sequences $(h_n^i)_{n\in\mathbb{N}}$ and $(e_n^k)_{n\in\mathbb{N}}$ of step functions as in the preceding step of our argument. We set $h_n = \sum_{i=1}^{i_0} f_n^i$ and $e_n = \sum_{k=1}^{k_0} e_n^k$. The sequences

$$\left( \int_X h_n\, d\theta \right)_{n\in\mathbb{N}} \qquad \text{and} \qquad \left( \int_X e_n\, d\theta \right)_{n\in\mathbb{N}}$$

in $\mathcal{Q}_0$ then converge to $\int_X f\, d\theta$ and $\int_X g\, d\theta$, respectively. For all $n, m \in \mathbb{N}$ we have

$$\left( \int_X h_m\, d\theta \right) \left( \int_X e_n\, d\theta \right) = \sum_{i=1}^{i_o} \sum_{k=1}^{k_0} \int_X (h_m^i e_n^k)\, d\theta,$$

and for fixed $i$ and $k$

$$\lim_{n\to\infty} \lim_{n\to\infty} \int_X h_m^i e_n^k\, d\theta = \int_X (f_i g_k)\, d\theta$$

by the above. This yields

$$\lim_{m\to\infty} \lim_{n\to\infty} \left\{ \left( \int_X h_m\, d\theta \right) \left( \int_X e_n\, d\theta \right) \right\} = \sum_{i=1}^{i_o} \sum_{k=1}^{k_0} \int_X (f_i g_k)\, d\theta = \int_X (fg)\, d\theta.$$

Reversing the parts of $n$ and $m$ leads to the same result. Thus indeed

$$\int_X (fg)\, d\theta \in \left( \int_X f\, d\theta \right) \cdot \left( \int_X g\, d\theta \right)$$

holds for all $f, g \in \mathcal{E}_{\mathfrak{R}}(X, \mathcal{P})$, provided that the measure $\theta$ satisfies (A).

If both $(\mathcal{P},\mathcal{V})$ and $(\mathcal{Q}_0,\mathcal{W})$ are topological algebras with an involution, that is a continuous operator $a\mapsto a^*$ such that $(a+b)^*=a^*+b^*$, $(\alpha a)^*=\bar{\alpha}a^*, (a^*)^*=a$ and $(ab)^*=b^*a^*$ for $a,b$ in $\mathcal{P}$ or in $\mathcal{Q}_0$, respectively, and if the $\mathfrak{L}(\mathcal{P},\mathcal{Q}_0)$-valued measure $\theta$ satisfies

(A*) $\theta_E(a^*)=\big(\theta_E(a)\big)^*$ *for all* $E\in\mathfrak{R}$ *and* $a\in\mathcal{P}$

in addition to (A), then a similar property can be derived for the integrals of functions in $\mathcal{E}_{\mathfrak{R}}(X,\mathcal{P})$. Analogously to the above extension of the multiplication in $\mathcal{Q}_0$, for an elements $l\in\mathcal{Q}$ we denote by $l^\star$ the set of all elements $q\in\mathcal{Q}$ for which we can find a net $(l_i)_{i\in\mathcal{I}}$ in $\mathcal{Q}_0\subset\mathcal{Q}$ such that $\lim_{i\in\mathcal{I}}l_i=l$ and

$$\lim_{i\in\mathcal{I}} l_i^*=q.$$

The continuity of the involution in $\mathcal{Q}_0$ shows that for $l\in\mathcal{Q}_0$ we have $l^\star=\{l^*\}$. Otherwise, the set $l^\star$ may be empty or contain more than one element of $\mathcal{Q}$. Canonically, for a function $f\in\mathcal{F}(X,\mathcal{P})$ we denote by $f^*\in\mathcal{F}(X,\mathcal{P})$ the function $x\mapsto\big(f(x)\big)^*$. Then an argument similar to that for the multiplication yields

$$\int_X f^*\,d\theta\in\left(\int_X f\,d\theta\right)^\star$$

for all $f\in\mathcal{E}_{\mathfrak{R}}(X,\mathcal{P})$ and every $\mathfrak{L}(\mathcal{P},\mathcal{Q}_0)$-valued measure $\theta$ which satisfies (A) and (A*).

Because both $\chi_{F\circledast}f,\chi_{F\circledast}g\in\mathcal{E}_{\mathfrak{R}}(X,\mathcal{P})$ whenever $f,g\in\mathcal{E}_{\mathfrak{R}}(X,\mathcal{P})$ and $F\in\mathfrak{A}_{\mathfrak{R}}$, and because $\int_F f\,d\theta=\int_X\chi_{F\circledast}f\,d\theta$, the above properties apply also to integrals over measurable subsets $F$ of $X$.

We formulate this as a further proposition:

**Proposition 6.15.** *Let* $(\mathcal{P},\mathcal{V})$ *and* $(\mathcal{Q}_0,\mathcal{W}_0)$ *be topological algebras over* $\mathbb{K}=\mathbb{R}$ *or* $\mathbb{K}=\mathbb{C}$ *and let* $\theta$ *be a bounded* $\mathfrak{L}_{\mathbb{K}}(\mathcal{P},\mathcal{Q}_0)$*-valued measure such that* $\theta_E(a)\,\theta_G(b)=\theta_{(E\cap G)}(ab)$ *holds for all* $E,G\in\mathfrak{R}$ *and* $a,b\in\mathcal{P}$*. Then*

$$\int_X (fg)\,d\theta\in\left(\int_X f\,d\theta\right)\cdot\left(\int_X g\,d\theta\right)$$

*holds for all* $f,g\in\mathcal{E}_{\mathfrak{R}}(X,\mathcal{P})$*. If both* $(\mathcal{P},\mathcal{V})$ *and* $(\mathcal{Q}_0,\mathcal{W})$ *are topological algebras with an involution* $a\mapsto a^*$ *and if* $\theta$ *satisfies* $\theta_E(a^*)=\big(\theta_E(a)\big)^*$ *for all* $E\in\mathfrak{R}$ *and* $a\in\mathcal{P}$*, then*

$$\int_X f^*\,d\theta\in\left(\int_X f\,d\theta\right)^\star$$

*holds for all* $f\in\mathcal{E}_{\mathfrak{R}}(X,\mathcal{P})$.

*The case that* $\mathcal{Q}_0 = \mathbb{K}$. If $\mathcal{Q}_0 = \mathbb{K}$, that is if the values $\theta_E$ of the measure $\theta$ are $\mathbb{K}$-linear functionals on the algebra $\mathcal{P}$, then Condition (A) means that all functionals $\theta_E$ are multiplicative and that for disjoint sets $E, G \in \mathfrak{R}$ we have either $\theta_E = 0$ or $\theta_G = 0$. Thus $\theta$ takes at most one non-zero value, that is some multiplicative $\mathbb{K}$-linear functional in $\mathcal{P}^*$. In special cases (see also Section 4.7 in Chapter III below) we infer that $\theta$ is indeed some point evaluation measure. Condition (A*) for an algebra with involution means that $\theta_E(a^*) = \overline{\theta_E(a)}$ holds for all $E \in \mathfrak{R}$ and $a \in \mathcal{P}$.

**6.16 Lattice Homomorphisms.** In Section 5.1 of Chapter I we defined a locally convex $\vee$-semilattice cone to be a locally convex cone $(\mathcal{P}, \mathcal{V})$ with the following properties: The order in $\mathcal{P}$ is antisymmetric, for any two elements $a, b \in \mathcal{P}$ their supremum $a \vee b$ exists in $\mathcal{P}$ and

($\vee$1) $(a+c) \vee (b+c) = a \vee b + c$ *holds for all* $a, b, c \in \mathcal{P}$.
($\vee$2) $a \le c+v$ *and* $b \le c+w$ *for* $a, b, c \in \mathcal{P}$ *and* $v, w \in \mathcal{V}$ *implies that* $a \vee b \le c + (v+w)$.

In case that the locally convex cone $(\mathcal{P}, \mathcal{V})$ is quasi-full, ($\vee$2) may be replaced by the somewhat simpler condition

($\vee 2'$) $a \le v$ *for* $a \in \mathcal{P}$ *and* $v \in \mathcal{V}$ *implies that* $a \vee 0 \le v$.

Indeed, suppose that ($\vee$1) and ($\vee 2'$) hold in a quasi-full cone $(\mathcal{P}, \mathcal{V})$, and that $a \le c+v$ and $b \le c+w$ for $a, b, c \in \mathcal{P}$ and $v, w \in \mathcal{V}$. Then $a \le c+s$ and $b \le c+t$ for some elements $s \le v$ and $t \le w$ by (QF1) in I.6.1. By ($\vee 2'$) we have $s \vee 0 \le v$ and $t \vee 0 \le w$ as well. Now $a \le c + s \vee 0 + t \vee 0$ and $b \le c + s \vee 0 + t \vee 0$ implies

$$a \vee b \le c + s \vee 0 + t \vee 0 \le c + (v+w)$$

as required in ($\vee$2). Recall from Proposition I.5.2 that in a locally convex $\vee$-semilattice cone the lattice operation, that is the mapping $(a, b) \mapsto a \vee b : \mathcal{P} \times \mathcal{P} \to \mathcal{P}$ is continuous with respect to the symmetric relative topology.

Topological vector lattices and locally convex complete lattice cones in the sense of I.5 are locally convex $\vee$-semilattice cones. Further specific examples include $\overline{\mathbb{R}}$ and $\overline{\mathbb{R}}_+$ (Examples I.1.4(a) and (b)) and cones of non-empty convex subsets of a topological vector space with the set-inclusion as order (Example I.1.4(c)). The supremum of two convex sets is their convex hull in this case while infima do not always exist.

In the following let us suppose that $(\mathcal{P}, \mathcal{V})$ is a quasi-full locally convex $\vee$-semilattice cone and that $\theta$ is an $\mathfrak{R}$-bounded $\mathfrak{L}(\mathcal{P}, \mathcal{Q})$-valued measure whose values $\theta_E$ for all $E \in \mathfrak{R}$ of are continuous linear operators from $\mathcal{P}$ to $\mathcal{Q}$ satisfying the following condition:

(L) $\theta_E(a) \vee \theta_E(b) = \theta_E(a \vee b)$ *and* $\theta_E(a) \vee \theta_G(b) = \theta_E(a) + \theta_G(b)$ *for all* $a, b \ge 0$ *in* $\mathcal{P}$ *and disjoint sets* $E, G \in \mathfrak{R}$.

We shall verify below that (L) implies that its first requirement, that is to say

$$\theta_E(a) \vee \theta_E(b) = \theta_E(a \vee b),$$

holds indeed for all, not only the positive elements of $\mathcal{P}$. First we observe that Condition (L) implies

$$\text{(i)} \qquad \theta_E(a) \wedge \theta_G(b) \leq \mathfrak{O}\,(\theta_E(a) \vee \theta_G(b))$$

for disjoint sets $E, G \in \mathfrak{R}$ and $0 \leq a, b \in \mathcal{P}$, as well as

$$\text{(ii)} \qquad \sup_{i=1,\ldots,n} \theta_{E_i}(a_i) = \sum_{i=1}^{n} \theta_{E_i}(a_i)$$

for disjoint sets $E_i \in \mathfrak{R}$ and $a_i \geq 0$ in $\mathcal{P}$. For (i), let $E, G \in \mathfrak{R}$ be disjoint and $0 \leq a, b \in \mathcal{P}$. Then

$$\theta_E(a) \vee \theta_G(b) = \theta_E(a) + \theta_G(b) = \theta_E(a) \vee \theta_G(b) + \theta_E(a) \wedge \theta_G(b)$$

by (L) and Proposition I.5.3. This yields our claim via the cancellation rule in I.5.10(a). We shall prove (ii) by induction: For $n = 1$ there is nothing to prove. Suppose our claim holds for $n \in \mathbb{N}$, and let $E_1, \ldots, E_{n+1} \in \mathfrak{R}$ be disjoint sets, and $0 \leq a_1, \ldots, a_{n+1} \in \mathcal{P}$. The inequality

$$\sup_{i=1,\ldots,n+1} \theta_{E_i}(a_i) \leq \sum_{i=1}^{n+1} \theta_{E_i}(a_i)$$

is obvious. For the converse, using Proposition I.5.3 we infer

$$\begin{aligned} \sum_{i=1}^{n+1} \theta_{E_i}(a_i) &= \sup_{i=1,\ldots,n} \theta_{E_i}(a_i) + \theta_{E_{n+1}}(a_{n+1}) \\ &= \sup_{i=1,\ldots,n+1} \theta_{E_i}(a_i) + \sup_{i=1,\ldots,n} \theta_{E_i}(a_i) \wedge \theta_{E_{n+1}}(a_{n+1}). \end{aligned}$$

We have

$$\begin{aligned} &\sup_{i=1,\ldots,n} \theta_{E_i}(a_i) \wedge \theta_{E_{n+1}}(a_{n+1}) \\ &\qquad \leq \sup_{i=1,\ldots,n} \left(\theta_{E_i}(a_i) \wedge \theta_{E_{n+1}}(a_{n+1})\right) + \mathfrak{O}\left(\sup_{i=1,\ldots,n+1} \theta_{E_i}(a_i)\right) \end{aligned}$$

by Proposition I.5.15(b), and for each $i = 1, \ldots, n$

$$\theta_{E_i}(a_i) \wedge \theta_{E_{n+1}}(a_{n+1}) \leq \mathfrak{O}\,(\theta_{E_i}(a_i) \vee \theta_{E_{n+1}}(a_{n+1})) \leq \mathfrak{O}\left(\sup_{i=1,\ldots,n+1} \theta_{E_i}(a_i)\right)$$

by (i). Thus Propositions I.5.10(c) and I.5.11 yield

$$\sum_{i=1}^{n+1} \theta_{E_i}(a_i) \leq \sup_{i=1,\ldots,n+1} \theta_{E_i}(a_i)$$

as claimed. Next we shall verify that

$$\text{(iii)}\qquad \left(\sum_{i=1}^{n}\theta_{E_i}(a_i)\right)\vee\left(\sum_{i=1}^{n}\theta_{E_i}(b_i)\right)=\sum_{i=1}^{n}\theta_{E_i}(a_i\vee b_i)$$

holds for disjoint sets $E_i\in\mathfrak{R}$ and $a_i,b_i\in\mathcal{P}$. Indeed, let $E=\bigcup_{i=1}^{n}E_i$. Given $w\in\mathcal{W}$ there is $v\in\mathcal{V}$ such that $\theta_E(v)\le w$ and $\lambda\ge 0$ such that $0\le a_i+\lambda v$ and $0\le b_i+\lambda v$ for all $i=1,\dots,n$. Because the locally convex cone $\mathcal{P}$ is supposed to be quasi-full, there are $s_i,t_i\in\mathcal{P}$ such that $s_i,t_i\le\lambda v$ and $0\le a_i+s_i$ and $0\le b_i+t_i$. Then $s_i\vee 0,\ t_i\vee 0\le\lambda v$ by our assumptions for a semi lattice cone. We set $s=\sum_{i=1}^{n}(s_i\vee 0)+(t_i\vee 0)$ and conclude that $0\le s\le n\lambda v$ as well as $0\le a_i+s$ and $0\le b_i+s$ for all $i=1,\dots,n$. Using this and (ii) from above, we conclude that

$$\begin{aligned}
&\left(\sum_{i=1}^{n}\theta_{E_i}(a_i)\right)\vee\left(\sum_{i=1}^{n}\theta_{E_i}(b_i)\right)+\theta_E(s)\\
&\qquad=\left(\sum_{i=1}^{n}\theta_{E_i}(a_i+s)\right)\vee\left(\sum_{i=1}^{n}\theta_{E_i}(b_i+s)\right)\\
&\qquad=\sup_{i=1,\dots,n}\theta_{E_i}(a_i+s)\vee\sup_{i=1,\dots,n}\theta_{E_i}(b_i+s)\\
&\qquad=\sup_{i=1,\dots,n}\theta_{E_i}(a_i+s)\vee\theta_{E_i}(b_i+s)\\
&\qquad=\sup_{i=1,\dots,n}\theta_{E_i}\big((a_i+s)\vee(b_i+s)\big)\\
&\qquad=\sum_{i=1}^{n}\theta_{E_i}\big((a_i+s)\vee(b_i+s)\big)\\
&\qquad=\sum_{i=1}^{n}\theta_{E_i}(a_i\vee b_i)\,+\,\theta_E(s).
\end{aligned}$$

Considering that $\mathfrak{O}\big(\theta_E(s)\big)\le w$ and that $w\in\mathcal{W}$ was arbitrarily chosen, now the cancellation law from Proposition I.5.10(a) yields (iii). Note that (iii) implies a strengthening of the first requirement in (L): $\theta_E(a)\vee\theta_E(b)=\theta_E(a\vee b)$ holds for all $E\in\mathfrak{R}$ and all (not necessarily positive) elements $a,b\in\mathcal{P}$.

The supremum $f\vee g\in\mathcal{F}(X,\mathcal{P})$ of two functions $f,g\in\mathcal{F}(X,\mathcal{P})$ is canonically defined as the mapping $x\mapsto f(x)\vee g(x)$. If we take into account the continuity of the lattice operation in $\mathcal{P}$, then Theorem 1.4 yields immediately that the supremum of two measurable functions is again measurable, and consequently, a brief review of 2.3 confirms that the subcone $\mathcal{F}_{\mathfrak{R}}(X,\mathcal{P})$ of $\mathcal{F}(X,\mathcal{P})$ is closed for suprema. As an immediate consequence of (iii) we infer that

$$\int_X (h\vee l)\,d\theta=\left(\int_X h\,d\theta\right)\vee\left(\int_X l\,d\theta\right)$$

holds for all step functions $h,l\in\mathcal{S}_{\mathfrak{R}}(X,\mathcal{P})$. Now let us denote by $\mathcal{S}_{\mathfrak{R}}^{\sigma}(X,\mathcal{P})$ the subcone of all functions $f\in\mathcal{F}_{\mathfrak{R}}(X,\mathcal{P})$ for which there exists a sequence

$(h_n)_{n\in\mathbb{N}}$ of step functions that is bounded below and bounded above relative to $f$ and such that $h_n \longrightarrow f$. According to Corollary 5.26, this implies $\lim\limits_{n\to\infty} \int_X h_n\,d\theta = \int_X f\,d\theta$. Lemma 5.27 and Corollary 5.28 yield in particular that $\mathcal{E}_{\mathfrak{R}}(X,\mathcal{P})$, the subcone generated by all elementary functions, is contained in $\mathcal{S}^{\sigma}_{\mathfrak{R}}(X,\mathcal{P})$. We proceed to establish that the integral with respect to a measure satisfying (L) defines a $\vee$-semilattice homomorphism (see I.5.30) from $\mathcal{S}^{\sigma}_{\mathfrak{R}}(X,\mathcal{P})$ into $\mathcal{Q}$:

Let $f,g \in \mathcal{S}^{\sigma}_{\mathfrak{R}}(X,\mathcal{P})$, and let $(h_n)_{n\in\mathbb{B}}$ and $(l_n)_{n\in\mathbb{B}}$ be the corresponding sequences of step functions approaching $f$ and $g$ as required above. Because of the continuity (with respect to the symmetric relative topology) of the lattice operation in $\mathcal{P}$, this implies $h_n \vee l_n \longrightarrow f \vee g$, that is the sequence $(h_n \vee l_n)_{n\in\mathbb{N}}$ of step functions converges pointwise to the function $f \vee g \in \mathcal{F}(X,\mathcal{P})$. We shall proceed to verify that this sequence is bounded below and bounded above relative to $f \vee g$, hence the function $f \vee g$ is also contained in $\mathcal{S}^{\sigma}_{\mathfrak{R}}(X,\mathcal{P})$. Indeed, let $\mathfrak{v}$ be an inductive limit neighborhood for $\mathcal{F}(X,\mathcal{P})$. There are $\lambda,\rho,\sigma \geq 0$ such that all of the following hold true: $0 \leq f + \lambda\mathfrak{v}$, $0 \leq g + \lambda\mathfrak{v}$ (see Lemma 2.4(a)), as well as $0 \leq h_n + \lambda\mathfrak{v}$, $0 \leq l_n + \lambda\mathfrak{v}$, $h_n \leq \rho f + \lambda\mathfrak{v}$ and $l_n \leq \sigma g + \lambda\mathfrak{v}$ for all $n \in \mathbb{N}$. We may indeed assume that $\sigma = \rho$, since otherwise, for example if $\sigma < \rho$, we can suitably adjust

$$l_n \leq (\sigma g + \lambda\mathfrak{v}) + (\rho - \sigma)(g + \lambda\mathfrak{v}) = \rho g + \lambda' v.$$

Using this, we argue as follows: Firstly, the preceding conditions imply that $0 \leq h_n \vee l_n + \lambda\mathfrak{v}$ holds for all $n \in \mathbb{N}$. Secondly, there are $\overline{\mathcal{V}}$-valued functions $s_n, t_n \in \mathfrak{v}$ such that $h_n \leq \rho f + \lambda s_n$ and $l_n \leq \rho g + \lambda t_n$. Let $x \in X$. Because $\mathcal{P}$ is quasi-full, there are $0 \leq u_n, v_n \in \mathcal{P}$ such that $u_n \leq s_n(x)$, $v_n \leq t_n(x)$ and $h_n(x) \leq \rho f(x) + \lambda u_n$ and $l_n(x) \leq \rho g(x) + \lambda v_n$. Thus both $h_n(x), l_n(x) \leq \rho(f \vee g)(x) + \lambda(u_n + v_n)$ and therefore

$$(h_n \vee l_n)(x) \leq \rho(f \vee g)(x) + \lambda(u_n + v_n) \leq \rho(f \vee g)(x) + \lambda(s_n + t_n)(x).$$

This shows $h_n \vee l_n \leq \rho(f \vee g) + 2\lambda\mathfrak{v}$ for all $n \in \mathbb{N}$ and verifies our claim. We therefore have

$$\lim_{n\to\infty} \int_X h_n\,d\theta = \int_X f\,d\theta, \qquad \lim_{n\to\infty} \int_X l_n\,d\theta = \int_X g\,d\theta$$

and

$$\lim_{n\to\infty} \int_X h_n \vee l_n\,d\theta = \int_X f \vee g\,d\theta$$

by Corollary 5.26. As

$$\begin{aligned}\lim_{n\to\infty} \int_X h_n \vee l_n\,d\theta &= \lim_{n\to\infty} \left( \left( \int_X h_n\,d\theta \right) \vee \left( \int_X l_n\,d\theta \right) \right) \\ &= \lim_{n\to\infty} \left( \int_X h_n\,d\theta \right) \vee \lim_{n\to\infty} \left( \int_X l_n\,d\theta \right)\end{aligned}$$

by the above and by Proposition I.5.25(a), we conclude that

$$\int_X (f \vee g)\, d\theta = \Big(\int_X f\, d\theta\Big) \vee \Big(\int_X g\, d\theta\Big)$$

holds for all functions $f, g \in \mathcal{S}^{\sigma}_{\mathfrak{R}}(X, \mathcal{P})$, provided that the measure $\theta$ satisfies (L). In other words, the integral with respect to $\theta$ defines a $\vee$-semilattice homomorphism from $\mathcal{S}^{\sigma}_{\mathfrak{R}}(X, \mathcal{P})$ to $\mathcal{Q}$ in the sense of I.5.30. Because both functions $\chi_{F} \otimes f$ and $\chi_{F} \otimes g$ are elements of $\mathcal{S}^{\sigma}_{\mathfrak{R}}(X, \mathcal{P})$ whenever $f, g \in \mathcal{E}_{\mathfrak{R}}(X, \mathcal{P})$ and $F \in \mathfrak{A}_{\mathfrak{R}}$, and because $\int_F f\, d\theta = \int_X \chi_F \otimes f\, d\theta$, this applies also to integrals over measurable subsets $F$ of $X$.

We summarize:

**Proposition 6.17.** *Let $(\mathcal{P}, \mathcal{V})$ be a quasi-full locally convex $\vee$-semilattice cone and let $\theta$ be a bounded $\mathfrak{L}_{\mathbb{K}}(\mathcal{P}, \mathcal{Q}_0)$-valued measure such that $\theta_E(a) \vee \theta_E(b) = \theta_E(a \vee b)$ and $\theta_E(a) \vee \theta_G(b) = \theta_E(a) + \theta_G(b)$ for all $a, b \geq 0$ in $\mathcal{P}$ and disjoint sets $E, G \in \mathfrak{R}$ Then*

$$\int_X (f \vee g)\, d\theta = \Big(\int_X f\, d\theta\Big) \vee \Big(\int_X g\, d\theta\Big)$$

*holds for all functions $f, g \in \mathcal{S}^{\sigma}_{\mathfrak{R}}(X, \mathcal{P})$.*

*The case that $\mathcal{Q} = \overline{\mathbb{R}}$.* If $\mathcal{Q} = \overline{\mathbb{R}}$, that is if the values $\theta_E$ of the measure $\theta$ are elements of $\mathcal{P}^*$, then Condition (L) means that (i) all functionals $\theta_E$ are lattice homomorphisms and (ii) for disjoint sets $E, G \in \mathfrak{R}$ we have either $\theta_E = 0$ or $\theta_G = 0$.

Similar concepts and results could obviously developed for locally convex $\wedge$-semilattice cones as defined in Section I.5.1 and $\wedge$-semilattice homomorphisms (see I.5.30).

**6.18 Cone-Valued Functions and Positive Real-Valued Measures.** If $\mathcal{P}$ is a subcone of $\mathcal{Q}$, and if the topology induced onto $\mathcal{P}$ by the neighborhood system $\mathcal{W}$ of $\mathcal{Q}$ is equivalent to the topology induced by its given neighborhood system $\mathcal{V}$, then for every $\rho \in \mathbb{R}_+$ the mapping

$$a \mapsto \rho a : \mathcal{P} \to \mathcal{Q},$$

defines a continuous linear operator. Thus every $\mathbb{R}_+$-valued measure $\theta$ on $\mathfrak{R}$, that is

$$E \mapsto \theta_E : \mathfrak{R} \to \mathbb{R}_+$$

is an operator-valued measure in the sense of Section 3. In particular, $\sigma$-additivity in our sense follows from $\sigma$-additivity for the $\mathbb{R}_+$-valued measure $\theta$ in the usual sense using Proposition I.5.22. Indeed, let $E_i \in \mathfrak{R}$ be disjoint sets, $E = \bigcup_{i=1}^{\infty} E_i$ and set $F_n = \bigcup_{i=1}^{n} E_i$. Then $\theta_E = \lim_{n \to \infty} \theta_{F_n} \in \mathbb{R}_+$. For $\sigma$-additivity of $\theta$ as an $\mathfrak{L}(\mathcal{P}, \mathcal{Q})$-valued measure, we shall first consider the case that $\theta_E = 0$. Then $\theta_{F_n} = 0$ for all $n \in \mathbb{N}$, as this sequence is increasing. For any $a \in \mathcal{P}$ this means

$$\sum_{i=1}^{\infty} \theta_{E_i}(a) = \lim_{n\to\infty} \theta_{F_n}(a) = \theta_E(a) = 0.$$

Otherwise, Proposition I.5.22 yields

$$\sum_{i=1}^{\infty} \theta_{E_i}(a) = \lim_{n\to\infty} \theta_{F_n}(a) = \lim_{n\to\infty} \left(\theta_{F_n} a\right) = \left(\lim_{n\to\infty} \theta_{F_n}\right) a = \theta_E\, a = \theta_E(a)$$

as well. For $E \in \mathfrak{R}$ and $w \in \mathcal{W}$ there is $v \in \mathcal{V}$ such that $a \le b + v$ for $a, b \in \mathcal{P}$ implies $a \le b + w$. Then the modulus of the measure $\theta$ is given by

$$\begin{aligned} |\theta|(E,v) &= \sup\left\{ \sum_{i=1}^{n} \theta_{E_i}\, s_i \;\middle|\; s_i \le v,\ E_i \in \mathfrak{R} \text{ disjoint subsets of } E \right\} \\ &\le \sup\left\{ \sum_{i=1}^{n} \theta_{E_i} \;\middle|\; E_i \in \mathfrak{R} \text{ disjoint subsets of } E \right\} w \;\le\; \theta_E\, w. \end{aligned}$$

The $\mathfrak{L}(\mathcal{P},\mathcal{Q})$-valued measure $\theta$ is therefore bounded in the sense of Section 3.6 and can be extended to the full cone $(\mathcal{P}_{\mathcal{V}}, \mathcal{V})$ (see Section 3.8). In case that $(\mathcal{Q},\mathcal{W})$ is indeed the standard lattice completion of $(\mathcal{P},\mathcal{V})$ as introduced in I.5.57, then Corollary 5.9 (see also 6.5) yields that the integrals of integrable functions in $\mathcal{F}(X,\mathcal{P})$ are indeed elements of the second dual $\mathcal{P}^{**}$ of $\mathcal{P}$.

**6.19 Vector-Valued Functions and Real- or Complex-Valued Measures.** Let $(\mathcal{P},\mathcal{V})$ be a locally convex topological vector space over $\mathbb{K} = \mathbb{R}$ or $\mathbb{K} = \mathbb{C}$, endowed with a basis $\mathcal{V}$ of balanced convex neighborhoods, and let $(\mathcal{Q},\mathcal{W})$ be the standard lattice completion of $(\mathcal{P},\mathcal{V})$, as defined in Section 5.57 of Chapter I. Then the topology induced by $\mathcal{W}$ onto the embedding of $\mathcal{P}$ into $\mathcal{Q}$ is equivalent to the topology induced by its given neighborhood system $\mathcal{V}$ (see I.5.57). For each $\rho \in \mathbb{K}$ the mapping

$$a \mapsto \rho a : \mathcal{P} \to \mathcal{Q},$$

is therefore a continuous linear operator. Thus every $\mathbb{K}$-valued measure $\theta$ on $\mathfrak{R}$, that is

$$E \mapsto \theta_E : \mathfrak{R} \to \mathbb{K}$$

is an operator-valued measure in the sense of Section 3. For $\sigma$-additivity, let $E_i \in \mathfrak{R}$ be disjoint sets, $E = \bigcup_{i=1}^{\infty} E_i$ and set $F_n = \bigcup_{i=1}^{n} E_i$. Then $\theta_E = \lim_{n\to\infty} \theta_{F_n} \in \mathbb{K}$, and $\lim_{n\to\infty} \left(\theta_{F_n} a\right) = \left(\lim_{n\to\infty} \theta_{F_n}\right) a$ holds for all $a \in \mathcal{P}$, since $(\mathcal{P},\mathcal{V})$ is a topological vector space, hence the scalar multiplication is continuous. The $\mathfrak{L}(\mathcal{P},\mathcal{Q})$-valued measure $\theta$ is indeed strongly additive in the sense of 5.32 since for every decreasing sequence $(E_n)_{n\in\mathbb{N}}$ of sets in $\mathfrak{R}$

such that $\bigcap_{n\in\mathbb{N}} E_n = \emptyset$ and $\varepsilon > 0$ there is $n_0 \in \mathbb{N}$ such that $|\theta_{E_n}| \leq \varepsilon$ for all $n \geq n_0$. Because for $a \in \mathcal{P}$ and $w \in \mathcal{W}$ there is $v \in \mathcal{V}$ and $\lambda \geq 0$ such that $a \leq \lambda v \leq \lambda w$, hence

$$\theta_{E_n}(a) = \theta_{E_n} a \leq \lambda w,$$

holds for all $n \geq n_0$. The latter follows since the neighborhoods in $\mathcal{V}$ are balanced and convex for $\mathcal{P}$. Recall from 6.4 that the total variation $\mathfrak{var}(\theta, E)$ of a real- or complex-valued measure $\theta$ on is always finite. For $E \in \mathfrak{R}$ and $w \in \mathcal{W}$ there is $v \in \mathcal{V}$ such that $a \leq b + v$, that is $a - b \in v$ for $a, b \in \mathcal{P}$ implies $a \leq b + w$. We have $\gamma s \in |\gamma| v$ for all $\gamma \in \mathbb{K}$ whenever $s \in v$ for $s \in \mathcal{P}$ and $v \in \mathcal{V}$. According to 6.12 the modulus of the $\mathfrak{L}(\mathcal{P}, \mathcal{Q})$-valued measure $\theta$ is therefore given by

$$\begin{aligned} |\theta|(E, v) &= \sup\left\{ \sum_{i=1}^{n} \theta_{E_i} s_i \;\middle|\; s_i \in v,\ E_i \in \mathfrak{R} \text{ disjoint subsets of } E \right\} \\ &\leq \sup\left\{ \sum_{i=1}^{n} |\theta_{E_i}| \;\middle|\; E_i \in \mathfrak{R} \text{ disjoint subsets of } E \right\} w \\ &= \mathfrak{var}(\theta, E)\, w. \end{aligned}$$

The $\mathfrak{L}(\mathcal{P}, \mathcal{Q})$-valued measure $\theta$ is therefore bounded in the sense of Section 3.6.

Integrability for $\mathcal{P}$-valued functions had been characterized in 6.12. Integrals of functions that satisfy Condition (BI 1) are elements of the second vector space dual $\mathcal{P}_{\mathbb{K}}^{**}$ of $\mathcal{P}$ (see 6.12(i)). According to 6.12(iv), the operator

$$f \mapsto \int_F f \, d\theta \; : \; \mathcal{F}_{(F,\theta,BI1)}(X, \mathcal{P}) \to \mathcal{P}_{\mathbb{K}}^{**}$$

is linear over $\mathbb{K}$. Integrals of functions that satisfy Condition (BI 2) from 6.12 are indeed elements of the closure with respect to the symmetric topology of $\mathcal{P}$ in $\mathcal{P}_{\mathbb{K}}^{**}$. In case of a topologically complete locally convex vector space $\mathcal{P}$, this closure coincides with $\mathcal{P}$.

Neighborhood-valued measurable functions are integrated using the canonical extension of the measure $\theta$ to the full cone $(\mathcal{P}_{\mathcal{V}}, \mathcal{V})$ as elaborated in Section 3.8. For a positive real-valued measurable function $\varphi$ and a neighborhood $v \in \mathcal{V}$, for example, the function $\varphi{\otimes}v$ is measurable, hence in $\mathcal{F}(X, \mathcal{P}_{\mathcal{V}})$. According to the above for every $F \in \mathfrak{R}$ its integral may be estimated as $\int_F \varphi{\otimes}v \, d\theta \leq \left(\int_F \varphi \, d\,\mathfrak{var}(\theta)\right) w$, where $\mathfrak{var}(\theta)$ is the positive real-valued measure $E \mapsto \mathfrak{var}(\theta, E) : \mathfrak{R} \to \mathbb{R}$ and $w \in \mathcal{W}$ is a neighborhood such that $a \leq b + v$, that is $a - b \in v$ for $a, b \in \mathcal{P}$ implies $a \leq b + w$.

Because the locally convex complete lattice cone $(\mathcal{Q}, \mathcal{W})$ allows sufficiently many order continuous linear functionals, that is the order continuous lattice homomorphisms on $\mathcal{Q}$ support the separation property (see I.5.32 and I.5.57), the strong convergence statements of Theorem 5.36 apply to functions satisfying Condition (BI 2) from 6.12.

Let us consider the special case that $(\mathcal{P},\mathcal{V})$ is a normed space, that is $\mathcal{V}=\{\rho\mathbb{B}\mid\rho>0\}$, where $\mathbb{B}$ is the unit ball in $\mathcal{P}$. A vector-valued function $f\in\mathcal{F}(X,\mathcal{P})$ is called *Bochner (or Dunford and Schwartz) integrable* over a set $E\in\mathfrak{R}$ with respect to a scalar-valued measure $\theta$ (see for example III.2.17 in [55] or II.2 in [43]) if for every $\varepsilon>0$ there is a step function $h_\varepsilon\in\mathcal{S}_{\mathfrak{R}}(X,\mathcal{P})$ such that the mapping $x\mapsto\|f(x)-h_\varepsilon(x)\|$ is measurable and

$$\int_E\|f-h_\varepsilon\|\,d\,\mathfrak{var}(\theta)\le\varepsilon.$$

Indeed, if the $\mathcal{P}$-valued function $f$ is Bochner integrable, then given $w\in\mathcal{W}$ there is $\varepsilon>0$ such that $a\le b+\varepsilon\mathbb{B}$, that is $\|a-b\|\le\varepsilon\mathbb{B}$ for $a,b\in\mathcal{P}$ implies $a\le b+w$. We set $h_{(\mathbb{B},w)}=h_\varepsilon\in\mathcal{S}_{\mathfrak{R}}(X,\mathcal{P})$ and compute

$$\int_E\|f-h_\varepsilon\|\circledast\mathbb{B}\,d\theta\le\int_E\|f-h_\varepsilon\|\,d\,\mathfrak{var}(\theta)\le w$$

by our preceding considerations, hence (BI 2) from 6.12 holds for $f$.

**6.20 Operator-Valued Functions and Operator-Valued Measures.** Let $\mathcal{N}$ and $\mathcal{H}$ be cones, and let $\mathfrak{Z}$ and $\mathfrak{Y}$ be families of subsets of $\mathcal{N}$ and of $\mathcal{H}$, directed upward by set inclusion. Furthermore, let $(\mathcal{M},\mathcal{U})$ and $(\mathcal{L},\mathcal{R})$ be two locally convex cones, and for the respective cones $L(\mathcal{N},\mathcal{M})$ and $L(\mathcal{H},\mathcal{L})$ of linear operators consider the neighborhoods $V_{(Z,u)}$ for $Z\in\mathfrak{Z}$ and $u\in\mathcal{U}$, and $W_{(Y,r)}$ for $Y\in\mathfrak{Y}$ and $r\in\mathcal{R}$ (see Section I.7); that is $S\le U+V_{(Z,u)}$ or $R\le T+W_{(Y,r)}$ for operators $S,U\in L(\mathcal{N},\mathcal{M})$ or $R,T\in L(\mathcal{H},\mathcal{L})$, respectively, if

$$S(z)\le U(z)+u\quad\text{for all } z\in Z,\qquad\text{or}\qquad R(y)\le T(y)+r\quad\text{for all } y\in Y.$$

Let $\mathfrak{H}(\mathcal{N},\mathcal{M})$ be a subcone of $L(\mathcal{N},\mathcal{M})$ such that all its elements are bounded below with respect to the neighborhoods $V_{(Z,u)}$ and such that the resulting locally convex cone $\big(\mathfrak{H}(\mathcal{N},\mathcal{M}),\mathfrak{V}\big)$ is quasi-full. Similarly, let $\mathfrak{H}(\mathcal{H},\mathcal{L})$ be a subcone of $L(\mathcal{H},\mathcal{L})$ whose elements are bounded below with respect to the neighborhoods $W_{(Y,r)}$ and denote the resulting locally convex cone by $\big(\mathfrak{H}(\mathcal{H},\mathcal{L}),\mathfrak{W}\big)$. Let $\big(\widehat{\mathfrak{H}}(\mathcal{H},\mathcal{L}),\widehat{\mathfrak{W}}\big)$ be a locally convex complete lattice cone containing the latter, for example its (simplified) standard lattice completion (see Sections I.5.57 and I.7). Now in the context of our general theory we may consider integrals for $\mathfrak{H}(\mathcal{N},\mathcal{M})$-valued functions with respect to bounded $\mathfrak{L}\big(\mathfrak{H}(\mathcal{N},\mathcal{M}),\,\widehat{\mathfrak{H}}(\mathcal{H},\mathcal{L})\big)$-valued measures.

This is indeed a rather unwieldy setting. It does however facilitate a considerably wider choice of applications for our theory, as we shall see in Sections 6.22 to 6.23 below. Moreover, note that this point of view generalizes our original one since the given cones $(\mathcal{P},\mathcal{V})$ and $(\mathcal{Q},\mathcal{W})$ may be considered as cones of linear operators from $\mathbb{R}_+$ to $\mathcal{P}$ or to $\mathcal{Q}$, respectively (see Example I.7.1(c)).

We shall study two useful special cases in further detail:

(i) *The case* $\mathcal{N} = \mathcal{H}$ *and* $\mathfrak{Z} = \mathfrak{Y}$. In this case every linear operator $T \in L(\mathcal{M}, \mathcal{L})$ may be reinterpreted as a linear operator $\overline{T}$ from $\mathfrak{H}(\mathcal{N}, \mathcal{M})$ into $L(\mathcal{N}, \mathcal{L})$ mapping the operator $U \in \mathfrak{H}(\mathcal{N}, \mathcal{M})$ into the operator $T \circ U \in L(\mathcal{N}, \mathcal{L})$; that is

$$(T \circ U)(z) = T\big(U(z)\big) \in \mathcal{L} \qquad \text{for all} \qquad z \in \mathcal{N}.$$

In order to guarantee that the operator $T \circ U$ is bounded below with respect to the neighborhoods $W_{(Y,r)} \in \mathfrak{W}$, and that the operator

$$\overline{T} : \mathfrak{H}(\mathcal{N}, \mathcal{M}) \to L(\mathcal{N}, \mathcal{L})$$

is continuous with regard to the respective neighborhood systems for these cones, we shall require that $T$ itself is continuous from $(\mathcal{M}, \mathcal{U})$ into $(\mathcal{L}, \mathcal{R})$, that is $T \in \mathfrak{L}(\mathcal{M}, \mathcal{L})$. Indeed, for $Z \in \mathfrak{Z}$ and $r \in \mathcal{R}$ there is $u \in \mathcal{U}$ such that $T(a) \le T(b) + r$ whenever $a \le b + u$ for $a, b \in \mathcal{M}$. Then for operators $S, U \in \mathfrak{H}(\mathcal{N}, \mathcal{M})$ such that $S \le U + V_{(\mathfrak{Z},u)}$ we have $S(z) \le U(z) + r$, hence $(T \circ S)(z) \le (T \circ U)(z) + r$ for all $z \in Z$. This shows $\overline{T}(S) \le \overline{T}(U) + W_{(Z,r)}$. Moreover, as for every $S \in \mathfrak{H}(\mathcal{N}, \mathcal{M})$ we have $0 \le S + \lambda V_{(Z,u)}$ for some $\lambda \ge 0$, the above implies that $0 \le \overline{T}(S) + \lambda W_{(Z,r)}$.

In this way, an $\mathfrak{L}(\mathcal{M}, \mathcal{L})$-valued measure $\theta$ on $\mathfrak{R}$ may be reinterpreted as an $\mathfrak{L}\big(\mathfrak{H}(\mathcal{N}, \mathcal{M}), \widehat{\mathfrak{H}}(\mathcal{N}, \mathcal{L})\big)$-valued measure, where $(\widehat{\mathfrak{H}}(\mathcal{N}, \mathcal{L}), \mathcal{W})$ is a locally convex complete lattice cone containing all the operators $\theta_E \circ U$ for $E \in \mathfrak{R}$ and $U \in \mathfrak{H}(\mathcal{N}, \mathcal{M})$. We are using the above identification of a continuous linear operator $T \in \mathfrak{L}(\mathcal{M}, \mathcal{L})$ with a continuous linear operator $\overline{T} \in \mathfrak{L}\big(\mathfrak{H}(\mathcal{N}, \mathcal{M}), \widehat{\mathfrak{H}}(\mathcal{N}, \mathcal{L})\big)$. We proceed to calculate the modulus of such a measure: For $E \in \mathfrak{R}$ and $V_{(Z,u)} \in \mathfrak{V}$ we have

$$\begin{aligned}&|\theta|\big(E, V_{(Z,u)}\big)\\ &\qquad = \sup\left\{\sum_{i=1}^{n} \theta_{E_i} \circ S_i \;\middle|\; S_i \le V_{(Z,u)},\ E_i \in \mathfrak{R} \text{ disjoint subsets of } E\right\}.\end{aligned}$$

The supremum on the right-hand side is taken in the locally convex complete lattice cone $\widehat{\mathfrak{H}}(\mathcal{N}, \mathcal{L})$. For $\mathfrak{R}$-boundedness of this measure we require that for every $r \in \mathcal{R}$ and $Z \in \mathfrak{Z}$ there is $u \in \mathcal{U}$ such that $|\theta|\big(E, V_{(Z,u)}\big) \le W_{(Z,r)}$. Note that for $\mathcal{N} = \mathbb{R}_+$ and $\mathfrak{Z} = \{\{1\}\}$, that is for $\big(\mathfrak{H}(\mathcal{N}, \mathcal{M}), \mathfrak{V}\big)$ and $\big(\mathfrak{H}(\mathcal{N}, \mathcal{L}), \mathfrak{W}\big)$ being isomorphic to the given cones $(\mathcal{M}, \mathcal{U})$ and $(\mathcal{L}, \mathcal{R})$ we have $|\theta|\big(E, V_{(\{1\},u)}\big) = |\theta|(E, u)|$. Countable additivity for the $\mathfrak{L}\big(\mathfrak{H}(\mathcal{N}, \mathcal{M}), \widehat{\mathfrak{H}}(\mathcal{N}, \mathcal{L})\big)$-valued measure $\theta$ requires that for disjoint sets $E_i \in \mathfrak{R}$ for every $U \in \mathfrak{H}(\mathcal{N}, \mathcal{M})$ the series

$$\theta_{(\bigcup_{i \in \mathbb{N}} E_i)} \circ U = \sum_{i=1}^{\infty} \big(\theta_{E_i} \circ U\big)$$

is order convergent in $\widehat{\mathfrak{H}}(\mathcal{N},\mathcal{L})$. In case that $\widehat{\mathfrak{H}}(\mathcal{N},\mathcal{L})$ is the simplified standard lattice completion of $\mathfrak{H}(\mathcal{N},\mathcal{L})$ as constructed in I.7.1, this means that for disjoint sets $E_i \in \mathfrak{R}$

$$\mu\Big(\theta_{(\bigcup\limits_{i\in\mathbb{N}} E_i)}(U(a))\Big) = \sum_{i=1}^{\infty} \mu\Big(\theta_{E_i}(U(a))\Big)$$

holds for all $U \in \mathfrak{H}(\mathcal{N},\mathcal{M})$, $a \in \bigcup_{Z\in\mathfrak{Z}} Z$ and $\mu \in \mathcal{L}^*$. Also in this case, Corollary 5.9 yields together with Remark I.7.1 that integrals of $(\mathfrak{H}(\mathcal{N},\mathcal{M}),\mathfrak{V})$-based integrable functions are indeed linear operators from $\mathcal{N}$ into $\mathcal{L}^{**}$, the second dual of $\mathcal{L}$.

(ii) *The case* $\mathcal{M} = \mathcal{L}$ *and* $\mathcal{U} = \mathcal{R}$. In this case every linear operator $T \in L(\mathcal{H},\mathcal{N})$ may be reinterpreted as a linear operator $\widetilde{T}$ from $\mathfrak{H}(\mathcal{N},\mathcal{M})$ into $L(\mathcal{H},\mathcal{M})$, mapping the operator $U \in \mathfrak{H}(\mathcal{N},\mathcal{M})$ into the operator $U \circ T \in L(\mathcal{H},\mathcal{M})$; that is

$$(U \circ T)(z) = U\big(T(z)\big) \in \mathcal{M} \qquad \text{for all} \qquad z \in \mathcal{H}.$$

In order to guarantee that the operator $U \circ T$ is bounded below with respect to the neighborhoods $W_{(Y,r)} \in \mathfrak{W}$, and that the operator

$$\widetilde{T} : \mathfrak{H}(\mathcal{N},\mathcal{M}) \to L(\mathcal{H},\mathcal{M})$$

is continuous with regard to the respective neighborhood systems, we shall require that for every $Y \in \mathfrak{Y}$ there is some $Z \in \mathfrak{Z}$ such that $f(Y) \subset Z$. Indeed, for $Y \in \mathfrak{Y}$ and $u \in \mathcal{U}$ let $Z \in \mathfrak{Z}$ such that $f(Y) \subset Z$. Then for operators $S, U \in \mathfrak{H}(\mathcal{N},\mathcal{M})$ such that $S \le U + V_{(\mathfrak{Z},u)}$ we have $S(z) \le U(z) + u$ for all $z \in Z$, hence $\big(S \circ T\big)(y) \le \big(U \circ T\big)(z) + u$ for all $y \in Y$. This shows $\widetilde{T}(S) \le \widetilde{T}(U) + W_{(Y,u)}$. Moreover, as for every $S \in \mathfrak{H}(\mathcal{N},\mathcal{M})$ we have $0 \le S + \lambda V_{(Z,u)}$ for some $\lambda \ge 0$, the above implies that $0 \le \overline{T}(S) + \lambda W_{(Y,u)}$.

In this way, an $L(\mathcal{H},\mathcal{N})$-valued measure $\theta$ satisfying the above requirement may be reinterpreted as an $\mathfrak{L}(\mathfrak{H}(\mathcal{N},\mathcal{M}), \widehat{\mathfrak{H}}(\mathcal{H},\mathcal{M}))$-valued measure, where $(\widehat{\mathfrak{H}}(\mathcal{H},\mathcal{M}), \widehat{\mathcal{W}})$ is a locally convex complete lattice cone containing all the operators $U \circ \theta_E$ for $E \in \mathfrak{R}$ and $U \in \mathfrak{H}(\mathcal{N},\mathcal{M})$, and using the above identification. The modulus of such a measure is calculated for $E \in \mathfrak{R}$ and $V_{(Z,u)} \in \mathfrak{V}$ as

$$|\theta|\big(E, V_{(Z,u)}\big) = \sup\left\{ \sum_{i=1}^{n} S_i \circ \theta_{E_i} \;\middle|\; S_i \le V_{(Z,u)},\ E_i \in \mathfrak{R} \text{ disjoint subsets of } E \right\}.$$

The supremum on the right-hand side is taken in the locally convex complete lattice cone $\widehat{\mathfrak{H}}(\mathcal{H},\mathcal{M})$. For $\mathfrak{R}$-boundedness of this measure we require that for every $u \in \mathcal{U}$ and $Y \in \mathfrak{Y}$ there is $Z \in \mathfrak{Z}$ such that $|\theta|\big(E, V_{(Z,u)}\big) \le W_{(Y,u)}$. Countable additivity for the measure $\theta$ requires that for disjoint sets $E_i \in \mathfrak{R}$ for every $U \in \mathfrak{H}(\mathcal{N},\mathcal{M})$ the series

$$U \circ \theta_{(\bigcup_{i\in\mathbb{N}} E_i)} = \sum_{i=1}^{\infty} (U \circ \theta_{E_i})$$

is order convergent in $\mathfrak{H}(\mathcal{H},\mathcal{M})$. In case that $\widehat{\mathfrak{H}}(\mathcal{H},\mathcal{M})$ is the simplified standard lattice completion of $\mathfrak{H}(\mathcal{H},\mathcal{M})$ as constructed in I.7.1, Corollary 5.9 and Remark I.7.1 yield that integrals of $(\mathfrak{H}(\mathcal{N},\mathcal{M}),\mathfrak{V})$-based integrable functions are linear operators from $\mathcal{H}$ into $\mathcal{M}^{**}$, the second dual of $\mathcal{M}$. If both $\mathcal{H}$ and $\mathcal{M}$ are vector spaces over $\mathbb{K}=\mathbb{R}$ or $\mathbb{K}=\mathbb{C}$, then these integrals are indeed $\mathbb{K}$-linear operators from $\mathcal{M}$ into the second vector space dual $\mathcal{M}^{**}_{\mathbb{K}}$ of $\mathcal{M}$ (see I.7.1).

**6.21 Positive, Real or Complex-Valued Functions and Operator-Valued Measures.** This is a special case for the preceding section. Let $(\mathcal{P},\mathcal{V})$ and $(\mathcal{Q},\mathcal{W})$ be locally convex cones, and let $\mathbb{K}=\mathbb{R}_+$, or $\mathbb{K}=\mathbb{R}$ or $\mathbb{K}=\mathbb{C}$ if $\mathcal{P}$ and $\mathcal{Q}$ are indeed locally convex topological vector spaces over $\mathbb{R}$ or $\mathbb{C}$, respectively, endowed with their symmetric topologies. We choose $\mathcal{N}=\mathcal{M}=\mathcal{P}$ and $\mathfrak{H}(\mathcal{N},\mathcal{M})=\mathbb{K}$ in the setting of Section 6.20. Depending on the suitable choice for the family $\mathfrak{Z}$ of bounded below subsets of $\mathcal{P}$, the following upper neighborhoods for an element $\alpha\in\mathbb{K}$ will render $\mathbb{K}$ into a quasi-full locally convex (see Example I.7.2(c)): For $\mathbb{K}=\mathbb{R}_+$ the family of all $\mathbb{B}^u_\varepsilon(\alpha)=[0,\alpha+\varepsilon]$ for $\varepsilon>0$, or the single neighborhood and $\mathbb{B}^u_0(\alpha)=[0,\alpha]$, both yielding the natural order; for $\mathbb{K}=\mathbb{R}$ or $\mathbb{K}=\mathbb{C}$ the Euclidean neighborhoods $\mathbb{B}_\varepsilon(\alpha)=\{\beta\in\mathbb{K}\mid |\beta-\alpha|\le\varepsilon\}$ with equality as the order on $\mathbb{K}$. In order to deal with these cases simultaneously, let us denote by $\mathbb{B}$ either $\mathbb{B}^u_1(0),\mathbb{B}^u_0(0)$ or $\mathbb{B}_1(\alpha)$, that is the respective unit neighborhoods of $0\in\mathbb{K}$, and let $\Gamma=\{0\},\Gamma=\{0,1\}$ or $\Gamma=\{\gamma\in\mathbb{K}\mid |\gamma|=1\}\}$ be the corresponding units spheres.

We set $\mathcal{L}=\mathcal{Q}$ and use the special case (i) in Section 6.20 in order to integrate $\mathbb{K}$-valued functions with respect to an $\mathfrak{L}(\mathcal{P},\mathcal{Q})$-valued measure. For $\widehat{\mathfrak{H}}(\mathcal{P},\mathcal{Q})$ we choose the simplified standard lattice completion of $\mathfrak{L}(\mathcal{P},\mathcal{Q})$. For $E\in\mathfrak{R}$ and the above neighborhoods we calculate the modulus of an $\mathfrak{L}(\mathcal{P},\mathcal{Q})$-valued measure $\theta$ as follows:

$$|\theta|(E,\mathbb{B}) = \sup\left\{ \sum_{i=1}^{n} \gamma_i\theta_{E_i} \;\middle|\; \gamma_i\in\Gamma, \quad E_i\in\mathfrak{R} \text{ disjoint subsets of } E \right\}.$$

The supremum on the right-hand side of these expressions is taken in the locally convex complete lattice cone $\widehat{\mathfrak{H}}(\mathcal{P},\mathcal{Q})$, that is a cone of $\overline{\mathbb{R}}$-valued functions with the pointwise algebraic and lattice operations. For $\mathbb{K}=\mathbb{R}_+$ and $\mathbb{B}=\mathbb{B}^u_0$ we have of course $|\theta|(E,\mathbb{B})=0$. For the remaining cases boundedness of the $\mathfrak{L}(\mathbb{K},\mathfrak{L}(\mathcal{P},\mathcal{Q}))$-valued measure $\theta$ requires that for every $E\in\mathfrak{R}$, the modulus $|\theta|(E,\mathbb{B})$ is bounded in $\widehat{\mathfrak{H}}(\mathcal{P},\mathcal{Q})$ with respect to all neighborhoods $W_{(Z,w)}$ for $Z\in\mathfrak{Z}$ and $w\in\mathcal{W}$. Let us recall the construction in I.7 of the standard lattice completion $\widehat{\mathfrak{H}}(\mathcal{P},\mathcal{Q})$ of $\mathfrak{L}(\mathcal{P},\mathcal{Q})$ to understand this further: The elements of $\widehat{\mathfrak{H}}(\mathcal{P},\mathcal{Q})$ are $\overline{\mathbb{R}}$-valued functions on the set

$\Upsilon = \big(\bigcup_{Z\in\mathfrak{Z}} Z\big) \times \mathcal{Q}^*$. An element $\varphi \in \widehat{\mathfrak{H}}(\mathcal{P},\mathcal{Q})$, that is an $\overline{\mathbb{R}}$-valued function on $\Upsilon$ is bounded relative to a neighborhood $W_{(Z,w)}$ if there is $\lambda \geq 0$ such that $\varphi(a,\mu) \leq \lambda$ holds for all $a \in Z$ and $\mu \in w^\circ$. Thus for boundedness of the measure $\theta$ we require that for every choice of disjoint subsets $E_i \in \mathfrak{R}$ of $E$ and $\gamma_i \subset \Gamma$ we have

$$\sum_{i=1}^{n} \mu\big(\gamma_i E_i(a)\big) = \sum_{i=1}^{n} \mathfrak{Re}(\gamma_i)\mu\big(E_i(a)\big) \leq \lambda$$

for all $a \in Z$ and $\mu \in w^\circ$; or equivalently, that for every $Z \in \mathfrak{Z}$ the subset

$$\left\{ \sum_{i=1}^{n} \theta_{E_i}(a) \;\middle|\; E_i \in \mathfrak{R} \text{ disjoint subsets of } E,\;\; a \in Z \right\}$$

is bounded above in $\mathcal{Q}$. Indeed, in case that $\mathbb{K} = \mathbb{R}$ or $\mathbb{K} = \mathbb{C}$, both $(\mathcal{P},\mathcal{V})$ and $(\mathcal{Q},\mathcal{W})$ are locally convex vector spaces, and we have $\gamma\mu \in w^\circ$ for all $\gamma \in \Gamma$ whenever $\mu \in w^\circ$ for $w \in \mathcal{W}$.

Recall that all sets $Z \in \mathfrak{Z}$ are required to be bounded below in $\mathcal{P}$. The choice of all these sets for $\mathfrak{Z}$ results in the uniform operator topology for $\mathfrak{L}(\mathcal{P},\mathcal{Q})$ $\big($see I,7.1(i)$\big)$. If the sets in $\mathfrak{Z}$ are also bounded above (as is indeed implied in the case that $\mathcal{P}$ is a locally convex vector space in its symmetric topology), then boundedness of $\theta$ as an $\mathfrak{L}\big(\mathbb{K},\mathfrak{L}(\mathcal{P},\mathcal{Q})\big)$-valued measure is already implied by its boundedness as an $\mathfrak{L}(\mathcal{P},\mathcal{Q})$-valued measure. Indeed, given $E \in \mathfrak{R}$ and $w \in \mathcal{W}$ there is $v \in \mathcal{V}$ such that $|\theta|(E,v) \leq w$ (see Sections 3.2 to 3.6). Then for every $Z \in \mathfrak{Z}$ there is $\lambda \geq 0$ such that $z \leq \lambda v$ for all $z \in \mathfrak{Z}$. This implies the above condition for the boundedness of $\theta$.

If $\mathfrak{Z}$ consists of all finite subsets of $\mathcal{P}$, that is if we consider the strong operator topology for $\mathfrak{L}(\mathcal{P},\mathcal{Q})$ $\big($see I,7.1(ii)$\big)$, then boundedness is a much weaker condition for $\theta$: For every $a \in \mathcal{P}$ the subset

$$\left\{ \sum_{i=1}^{n} \theta_{E_i}(a) \;\middle|\; E_i \in \mathfrak{R} \text{ disjoint subsets of } E \right\}$$

is required to be bounded above in $\mathcal{Q}$.

Countable additivity for the $\mathfrak{L}\big(\mathbb{K},\mathfrak{L}(\mathcal{P},\mathcal{Q})\big)$-valued measure $\theta$ demands that for disjoint sets $E_i \in \mathfrak{R}$

$$\mu\big(\theta_{(\bigcup_{i\in\mathbb{N}} E_i)}(a)\big) = \sum_{i=1}^{\infty} \mu\big(\theta_{E_i}(a)\big)$$

holds for all $a \in \bigcup_{Z\in\mathfrak{Z}} Z$ and $\mu \in \mathcal{Q}^*$.

Our notion of measurability for $\mathbb{K}$-valued functions coincides with the usual one (see also Examples 6.3 and 6.4). For $\mathbb{K} = \mathbb{R}_+$ all measurable $\mathbb{K}$-valued functions are in $\mathcal{F}(X,\mathbb{K})$, hence integrable. For $\mathbb{K} = \mathbb{R}$ or $\mathbb{K} = \mathbb{C}$ with the Euclidean topology, a measurable $\mathbb{K}$-valued function is in $\mathcal{F}(X,\mathbb{K})$ if on every set $E \in \mathfrak{R}$ it can be uniformly approximated by step functions. This implies of course strong integrability in the sense of 5.18. Because $\widehat{\mathfrak{H}}(\mathcal{P},\mathcal{Q})$ was supposed to be the simplified standard lattice completion of $\mathfrak{L}(\mathcal{P},\mathcal{Q})$, the integral to a function $\varphi \in \mathcal{F}(X,\mathbb{K})$ with respect to an $\mathfrak{L}(\mathcal{P},\mathcal{Q})$-valued measure over a set $E \in \mathfrak{R}$ is a linear operator from $\mathcal{P}$ into $Q^{**}$, contained in the closure of $\mathfrak{L}(\mathcal{P},\mathcal{Q})$ in $\widehat{\mathfrak{H}}(\mathcal{P},\mathcal{Q})$ with respect to the symmetric relative topology. Thus, if the cone $\mathfrak{L}(\mathcal{P},\mathcal{Q})$ is topologically complete with respect to this topology, then this integral is indeed an element of $\mathfrak{L}(\mathcal{P},\mathcal{Q})$.

Let us proceed to discuss the convergence theorems from Section 5: For the sake of simplicity we shall restrict ourselves to the case of a single measure, that is $\theta_n = \theta$ for all $n \in \mathbb{N}$ in Theorems 5.23 to 5.25: Let $(\varphi_n)_{n\in\mathbb{N}}$ be a sequence of integrable $\mathbb{K}$-valued functions that converges pointwise $\theta$-almost everywhere on a set $F \in \mathfrak{A}_{\mathfrak{R}}$ to a function $\varphi$ in the symmetric relative topology of $\mathbb{K}$. This is of course the usual (Euclidean) notion of convergence, except for the case of $\mathbb{K} = \mathbb{R}_+$ endowed with the neighborhood $\mathbb{B}_0^u$ which renders $0 \in \mathbb{R}_+$ into an isolated point (see Example I.4.37(b)). The boundedness conditions from Theorem 5.25 for the sequence $(\varphi_n)_{n\in\mathbb{N}}$ read somewhat differently for the different choices for $\mathbb{K}$: We set $\varphi_{**} = \varphi_* = 0$ in all cases. For $\mathbb{K} = \mathbb{R}_+$ we require that $\varphi_n \underset{a.e.F}{\leq} \varphi^*$ holds for all $n \in \mathbb{N}$ with some integrable function $\varphi^*$. For $\mathbb{K} = \mathbb{R}$ or $\mathbb{K} = \mathbb{C}$ with the Euclidean topology and the order we use an integrable positive-valued function $\varphi^*$ and the function $f^* = \varphi^* {\otimes} \mathbb{B}$ whose values are in the full cone $\mathbb{K}_{\mathcal{V}} = \{\alpha + \rho\mathbb{B} \mid \alpha \in \mathbb{K},\ \rho \geq 0\}$ to which Theorem 5.25 applies in this case. We therefore require that $|\varphi_n| \underset{a.e.F}{\leq} \varphi^*$ holds for all $n \in \mathbb{N}$ in this case. The assumptions of Theorem 5.25 are now satisfied. Let $T_n = \int_F \varphi_n \, d\theta$, $T = \int_F \varphi \, d\theta$ and $T^* = \int_F \varphi^* \, d\theta$, or $T^* = \int_F (\varphi^* {\otimes} \mathbb{B}) \, d\theta$ in case $\mathbb{K} = \mathbb{R}$ or $\mathbb{K} = \mathbb{C}$. These integrals are in general elements of $\widehat{\mathfrak{H}}(\mathcal{P},\mathcal{Q})$. The conclusion of Theorem 5.25 now states that

$$T \leq \underline{\lim}_{n\to\infty} T_n \qquad \text{and} \qquad \overline{\lim}_{n\to\infty} T_n \leq T + \mathfrak{D}(T^*)$$

in $\widehat{\mathfrak{H}}(\mathcal{P},\mathcal{Q})$, that is

$$T(a,\mu) \leq \underline{\lim}_{n\to\infty} T_n(a,\mu)$$

for all $a \in \bigcup_{Z\in\mathfrak{Z}} Z$ and $\mu \in \mathcal{P}^*$, and indeed

$$T(a,\mu) = \lim_{n\to\infty} T_n(a,\mu)$$

whenever $T^*(a,\mu) < +\infty$. Note that for linear operators $T \in \mathfrak{L}(\mathcal{P},\mathcal{Q})$ as elements of $\widehat{\mathfrak{H}}(\mathcal{P},\mathcal{Q})$ we have $T(a,\mu) = \mu\big(T(a)\big)$.

Now let us investigate the additional assumptions of Theorem 5.36 which will lead to convergence of $(T_n)_{n\in\mathbb{N}}$ towards $T$ in the symmetric topology of $\widehat{\mathfrak{H}}(\mathcal{P},\mathcal{Q})$ : We require that $F = E$ is in $\mathfrak{R}$. Strong additivity of the $\mathfrak{L}\big(\mathbb{K},\mathfrak{L}(\mathcal{P},\mathcal{Q})\big)$-valued measure $\theta$ in the sense of 5.30 means that for every decreasing sequence $(E_n)_{n\in\mathbb{N}}$ of sets in $\mathfrak{R}$ such that $\bigcap_{n\in\mathbb{N}} E_n = \emptyset$, for

$Z \in \mathfrak{Z}$ and $w \in \mathcal{W}$ there is $n_0 \in \mathbb{N}$ such that

$$\theta_{E_n}(a) \leq \mathfrak{D}\big(\theta_{E_1}(a)\big) + w$$

holds for all $a \in Z$ and $n \geq n_0$. Recall that in case $\mathbb{K} = \mathbb{R}$ or $\mathbb{K} = \mathbb{C}$ we assume that both $\mathcal{P}$ and $\mathcal{Q}$ are locally convex vector spaces in their respective symmetric topologies, thus $\mathfrak{D}\big(\theta_{E_1}(a)\big) = 0$, and $\theta_{E_n}(a) \leq w$ implies that $\theta_{E_n}(\gamma a) \leq w$ for all $\gamma \in \Gamma$. The above therefore means that the sequence $(\theta_{E_n})_{n \in \mathbb{N}}$ of linear operators converges to $0$ in the symmetric topology of $\big(\mathfrak{L}(\mathcal{P},\mathcal{Q}), \mathfrak{W}\big)$. We also need to require that the functions $\varphi_n, \varphi$ and $\varphi^*$ or $\varphi^*_{\otimes}\mathbb{B}$ are $(\mathbb{K},\mathfrak{V})$-based integrable in the sense of 5.6. Measurability in the classical sense and boundedness below almost everywhere on the set $E$ is sufficient for this. This condition also yields strong integrability for the functions $\varphi^*$ or $\varphi^*_{\otimes}\mathbb{B}$. Finally, according to 5.36 we require that the element $T^* = \int_E \varphi^*\, d\theta$ or $T^* = \int_E (\varphi^*_{\otimes}\mathbb{K})\, d\theta$ is bounded in $\widehat{\mathfrak{H}}(\mathcal{P},\mathcal{Q})$. Under these additional assumptions then Theorem 5.36 yields

$$T = \lim_{n \to \infty} T_n$$

in the symmetric topology of $\widehat{\mathfrak{H}}(\mathcal{P},\mathcal{Q})$. If as in most cases of interest the integrals $T_n$ and $T$ are actually elements of $\mathfrak{L}(\mathcal{P},\mathcal{Q})$, then we infer convergence in the symmetric operator topology of $(\mathfrak{L}(\mathcal{P},\mathcal{Q}), \mathfrak{W})$.

*Operator algebras.* If $\mathcal{H} = \mathcal{P} = \mathcal{Q}$ is a locally convex topological vector space, then the space of continuous linear operator $\mathfrak{L}(\mathcal{P},\mathcal{P})$ forms a topological algebra, endowed with the composition of operators as its multiplication (see 6.4). We integrate $\mathbb{K}$-valued functions with respect to an $\mathfrak{L}(\mathcal{P},\mathcal{P})$-valued measure $\theta$ in this case. The values of the integrals are contained in the simplified standard completion $\widehat{\mathfrak{H}}(\mathcal{P},\mathcal{P})$ of $\mathfrak{L}(\mathcal{P},\mathcal{P})$. For the integral to determine a multiplicative linear operator from $\mathcal{E}_{\mathfrak{R}}(X,\mathbb{K}) = \mathcal{F}_{\mathfrak{R}}(X,\mathbb{K})$ into $\widehat{\mathfrak{H}}(\mathcal{P},\mathcal{P})$ in the sense of Example 6.14 we need to require that the measure $\theta$ satisfies Condition (A), that is $\theta_E(a)\,\theta_E(b) = \theta_E(ab)$ and $\theta_E(a)\,\theta_G(b) = 0$ holds for all $a, b \in \mathbb{K}$ and disjoint sets $E, G \in \mathfrak{R}$. As $\theta_E(a) = a\theta_E$ in this case, Condition (A) reads as follows:

(A) $(\theta_E)^2 = \theta_E$ *and* $\theta_E\,\theta_G = 0$ *for disjoint sets* $E, G \in \mathfrak{R}$,

that is the operators $\theta_E \in \mathfrak{L}(\mathcal{P},\mathcal{P})$ are required to be idempotent and pairwise orthogonal for disjoint sets $E, G \in \mathfrak{R}$.

*Spectral Measures.* For a concrete example, let $\mathcal{H} = \mathcal{P} = \mathcal{Q}$ be a complex Hilbert space with unit ball $\mathbb{U}$ and the neighborhood system $\mathcal{V}=\{\rho\mathbb{U} \mid \rho>0\}$. Let $\mathfrak{R}$ be a weak $\sigma$-ring, and as in spectral theory, let $\theta$ be a projection-valued measure on $\mathfrak{R}$. We consider $\theta$ as an $\mathfrak{L}\big(\mathbb{C}, \mathfrak{L}(\mathcal{H},\mathcal{H})\big)$-valued measure in the above sense. Such a measure is seen to be $\mathfrak{R}$-bounded, even if we choose the uniform operator topology for $\mathfrak{L}(\mathcal{H},\mathcal{H})$ (see I.7.2(i)), that is the family of all bounded subsets of $\mathcal{H}$ for $\mathfrak{Z}$. Indeed, let $a \in \mathcal{H}$ such that

$\|a\| \le 1$ and let $E_i \in \mathfrak{R}$, for $i = 1, \ldots, n$ be disjoint sets. For a spectral measure the $\theta_{E_i}$ are projections onto mutually orthogonal subspaces of $\mathcal{P}$. Thus the elements $a_i = \theta_{E_i}(a)$ are orthogonal and $\sum_{i=1}^{n} \|a_i\|^2 \le \|a\|^2 = 1$ by the Bessel inequality (see Theorem 1 in I.5 of [82]). Thus

$$\Big\|\sum_{i=1}^{n} \theta_{E_i}(a)\Big\|^2 = \Big\|\sum_{i=1}^{n} a_i\Big\|^2 = \sum_{i=1}^{n} \|a_i\|^2 \le 1.$$

The set

$$\left\{ \sum_{i=1}^{n} \theta_{E_i}(a) \;\middle|\; E_i \in \mathfrak{R} \text{ disjoint subsets of } E, \quad a \in \mathcal{H}, \ \|a\| \le 1 \right\}$$

is therefore indeed bounded above in $\mathcal{H}$. Countable additivity for a spectral measure is however required only with respect to the strong operator topology for $\mathfrak{L}(\mathcal{H}, \mathcal{H})$, which arises if we choose the family of all finite subsets of $\mathcal{H}$ for $\mathfrak{Z}$. (Because projection operators in $\mathfrak{L}(\mathcal{H}, \mathcal{H})$ are of norm 1, countable additivity with respect to the uniform operator topology can of course only apply to finite sums of such operators.) Theorem 5.36 therefore yields convergence in the strong but not in the uniform operator topology of $\mathfrak{L}(\mathcal{H}, \mathcal{H})$ for spectral measures.

Spectral measures satisfy Condition (A) from above and also Condition (A*) from 6.14, that is $\theta_E(a^*) = \big(\theta_E(a)\big)^*$ for all $E \in \mathfrak{R}$ and $a \in \mathcal{P}$. As $\theta_E(a^*) = \bar{a}\theta_E$ and $\big(\theta_E(a)\big)^* = \big(a\theta_E\big)^* = \bar{a}\big(\theta_E\big)^*$ in this case, this is equivalent to

(A*) $\theta_E = \big(\theta_E\big)^*$ *for all* $E \in \mathfrak{R}$.

This condition holds because the projection operators $\theta_E \in \mathfrak{L}(\mathcal{H}, \mathcal{H})$ are self-adjoint. The linear operator $f \mapsto \int_X f\, d\theta$ is therefore multiplicative on the space $\mathcal{F}_{\mathfrak{R}}(X, \mathbb{C})$ of bounded measurable $\mathbb{K}$-valued functions and preserves the involution, that is $\int_X f^*\, d\theta = \big(\int_X f\, d\theta\big)^*$.

**6.22 Operator-Valued Functions and Cone-Valued Measures.** This is again a special case of 6.20. Let $\mathcal{P}$ be a cone, $(\mathcal{Q}, \mathcal{W})$ a locally convex complete lattice cone. We choose $\mathcal{N} = \mathcal{P}$, $\mathcal{M} = \mathcal{L} = \mathcal{Q}$ and $\mathcal{H} = \mathbb{R}_+$ in the setting of 6.20 and use the special case (ii). For $\mathfrak{Z}$ we choose a family of subsets of $\mathcal{P}$, directed upward by set inclusion such that $\bigcup_{Z \in \mathfrak{Z}} Z = \mathcal{P}$, and suppose that the locally convex cone $(\mathfrak{H}(\mathcal{P}, \mathcal{Q}), \mathfrak{V})$ of linear operators from $\mathcal{P}$ into $\mathcal{Q}$ is quasi-full. Let $\mathfrak{Y}$ consist of the singleton subset $\{1\}$ of $\mathbb{R}_+$. Then the locally convex cone $\big(L(\mathbb{R}_+, \mathcal{Q}), \mathfrak{W}\big)$ is isomorphic to $(\mathcal{Q}, \mathcal{W})$ (see Example I.7.2(d)), hence a locally convex complete lattice cone. Similarly, because the cone $\mathcal{P}$ can be identified with the cone $L(\mathbb{R}_+, \mathcal{P})$, we may consider the elements of $\mathcal{P}$ to be linear operators from some quasi-full cone $\mathfrak{H}(\mathcal{P}, \mathcal{Q})$ into $L(\mathbb{R}_+, \mathcal{Q})$, that is into $\mathcal{Q}$. Our choice for the families $\mathfrak{Z}$ and $\mathfrak{Y}$ guarantees that these operators are continuous (see 6.20 (ii)). Using these settings, case (ii) from 6.20 therefore permits us to consider $\mathfrak{H}(\mathcal{P},\mathcal{Q})$-valued functions together with $\mathcal{P}$-valued measures. Countable additivity requires for a $\mathcal{P}$-valued measure $\theta$ that for disjoint sets $E_i \in \mathfrak{R}$ and for every linear operator $T \in \mathfrak{H}(\mathcal{P},\mathcal{Q})$ the series

$$T\Big(\theta_{(\bigcup_{i\in\mathbb{N}} E_i)}\Big) = \sum_{i=1}^{\infty} T(\theta_{E_i})$$

is order convergent in $\mathcal{Q}$. The modulus of $\theta$ is calculated for $E \in \mathfrak{R}$ and $V_{(Z,w)} \in \mathfrak{V}$ as

$$|\theta|\big(E, V_{(Z,w)}\big)$$
$$= \sup\left\{ \sum_{i=1}^{n} T_i\big(\theta_{E_i}\big) \;\middle|\; T_i \le V_{(Z,w)},\ E_i \in \mathfrak{R} \text{ disjoint subsets of } E \right\}.$$

$\mathfrak{R}$-boundedness in the sense of Section 3.6 requires that for every $E \in \mathfrak{R}$ and $u \in \mathcal{W}$ there is $V_{(Z,w)} \in \mathfrak{V}$ such that $|\theta|\big(E, V_{(Z,w)}\big) \le u$. A bounded $\mathcal{P}$-valued measure then integrates $\mathfrak{H}(\mathcal{P},\mathcal{Q})$-valued functions, and the values of these integrals are elements of $L(\mathbb{R}_+,\mathcal{Q})$, that is $\mathcal{Q}$ itself. If $(\mathcal{Q},\mathcal{W})$ is indeed the standard lattice completion of some locally convex cone $\mathcal{Q}_0$ and if the concerned function is $\big(\mathfrak{H}(\mathcal{P},\mathcal{Q}_0)\big)$-based integrable, then its integral is an element of the subcone $\mathcal{Q}_0^{**}$ of $\mathcal{Q}$.

Let us further consider the special case that $(\mathcal{P},\mathcal{V})$ is a locally convex vector space, that $(\mathcal{Q},\mathcal{W})$ is the standard lattice completion of a locally convex vector space $(\mathcal{Q}_0,\mathcal{W}_0)$ and that $\mathfrak{H}(\mathcal{P},\mathcal{Q}) \subset \mathfrak{L}(\mathcal{P},\mathcal{Q}_0)$. Then countable additivity of an $\mathcal{P}$-valued measure $\theta$ is guaranteed by weak convergence of the concerned series $\sum_{i=1}^{n} \theta_{E_i}$ in $\mathcal{P}$ in this case. Indeed, weak convergence in $\mathcal{P}$ implies weak convergence in $\mathcal{Q}$ for the series $\sum_{i=1}^{n} \theta_{E_i}(T) = \sum_{i=1}^{n} T(\theta_{E_i})$ for every operator $T \in \mathfrak{L}(\mathcal{P},\mathcal{Q}_0)$. (see IV.2.1 in [185]). Weak convergence in $\mathcal{Q}_0$, however, coincides with order convergence in $\mathcal{Q}$ in this case (see I.5.57) as required for countable additivity. Moreover, Theorem 3.11 (or Corollary 3.13), that is our version of Pettis' theorem yields that for a vector-valued measure, countable additivity with respect to weak convergence implies countable additivity with respect to strong convergence, that is convergence in the symmetric topology of $\mathcal{P}$. Every such measure is therefore strongly additive in the sense of 5.30.

**6.23 Positive, Real or Complex-Valued Functions and Cone- or Vector-Valued Measures.** This is a special case for the preceding section. Let $(\mathcal{P},\mathcal{V})$ be a locally convex cone and let and $(\mathcal{Q},\mathcal{W})$ be its standard lattice completion, Let $\mathbb{K} = \mathbb{R}_+$, or $\mathbb{K} = \mathbb{R}$ or $\mathbb{K} = \mathbb{C}$ if $\mathcal{P}$ is indeed a locally convex vector space over $\mathbb{R}$ or $\mathbb{C}$, respectively, endowed with its symmetric topology. We choose $\mathfrak{H}(\mathcal{P},\mathcal{Q}) = \mathbb{K}$ endowed with one of the suitable topologies arising from the choice for the family $\mathfrak{Z}$ of bounded below subsets of $\mathcal{P}$ (see Example I.7.2(c) and 6.21 above), that is topologies generated by the neighborhoods $\mathbb{B}$ as discussed for the respective cases in 6.21. We shall also use the notation for the unit sphere $\Gamma$ from 6.21. Using this, the modulus of a $\mathcal{P}$-valued measure $\theta$ is given by

$$|\theta|(E,\mathbb{B}) = \sup\left\{\sum_{i=1}^{n} \gamma_i\theta_{E_i} \;\middle|\; \gamma_i \in \Gamma,\;\; E_i \in \mathfrak{R} \text{ disjoint subsets of } E\right\}.$$

The supremum on the right-hand side of this expression is taken in the locally convex complete lattice cone $\mathcal{Q}$, that is a cone of $\overline{\mathbb{R}}$-valued functions with the pointwise algebraic and lattice operations. Boundedness is of course guaranteed in the case of $\mathbb{B} = \mathbb{B}_0^u$. In the remaining cases it requires (see the corresponding detailed argument in 6.21) that the set

$$\left\{\sum_{i=1}^{n} \theta_{E_i} \;\middle|\; E_i \in \mathfrak{R} \text{ disjoint subsets of } E\right\}$$

is bounded in $\mathcal{P}$. We will be able to verify that every $\mathcal{P}$-valued measure $\theta$ is $\mathfrak{R}$-bounded in this instance. For this call to mind that the elements $\theta_E \in \mathcal{P}$, for all $E \in \mathfrak{R}$, are considered to be continuous linear operators from $\mathbb{K}$ into $\mathcal{N}$, thus are required to be bounded elements of $\mathcal{P}$. Furthermore, recall from I.5.57 that the neighborhood system $\mathcal{V}$ for $\mathcal{P}$ is a generating subset of the neighborhood system $\mathcal{W}$ for the standard lattice completion $\mathcal{Q}$ of $\mathcal{P}$. For $E \in \mathfrak{R}$ let us consider the subset

$$A = \{\theta_{E'} \mid E' \in \mathfrak{R},\; E' \subset E\}$$

of $\mathcal{P}$. We shall use Proposition I.4.25 (which is derived from the Uniform Boundedness Theorem 3.4 in [172]) in order to verify that $A$ is bounded above in $\mathcal{P}$. For this, let $\mu \in \mathcal{P}^*$. Because the elements of $\mathcal{P}^*$ are also order continuous linear functionals on the standard lattice completion $\mathcal{Q}$ of $\mathcal{P}$, we know from 3.9 that $\mu \circ \theta$ is an $\mathfrak{L}(\mathbb{K}, \overline{\mathbb{R}})$-valued, that is an $\overline{\mathbb{R}}$-valued measure on $\mathfrak{R}$. This measure is indeed real-valued, since the elements $\theta_{E'}$, for all $E' \in \mathfrak{R}$, were seen to be bounded elements of $\mathcal{P}$. A countably additive real-valued measure is however known to be bounded, that is

$$\big\{\mu(a) \mid a \in A\big\} = \big\{(\mu \circ \theta)_{E'} \mid E' \in \mathfrak{R},\; E' \subset E\big\}$$

is a bounded subset of $\mathbb{R}$. Because this holds true for all linear functionals $\mu \in \mathcal{P}^*$, Proposition I.4.25 yields that the set $A$ is bounded above relative to $0 \in \mathcal{P}$, that is bounded above, as claimed. Thus, given $v \in \mathcal{V}$, there is indeed $\lambda \geq 0$ such that $\theta_{E'} \leq \lambda v$ holds for all subsets $E' \in \mathfrak{R}$ of $E$. We claim that this implies $|\theta|(E, V) \leq 4\lambda v$. For this let us recall the construction of the standard lattice completion $(\mathcal{Q}, \mathcal{W})$ of $(\mathcal{P}, \mathcal{V})$. Its elements are $\overline{\mathbb{R}}$-valued functions $\varphi$ on the dual $\mathcal{P}^*$ of $\mathcal{P}$, and we have $\varphi \leq v$ if $\varphi(\mu) \leq 1$ for all $\mu \in v^\circ$. For any such $\mu \in v^\circ$, $\mu \circ \theta$ was seen to be a real-valued countably additive measure on $\mathfrak{R}$. As $(\mu \circ \theta)_{E'} \leq \lambda$ for all subsets $E' \in \mathfrak{R}$ of $E$, we know that is total variation on $E$, that is $\mathfrak{var}(\mu \circ \theta, E)$ is bounded by the constant $4\lambda$ (see 6.4). Thus

$$|\theta|(E, V)(\mu) = \mathfrak{var}(\mu \circ \theta, E) \leq 4\lambda$$

for all $\mu \in v^\circ$. This demonstrates $|\theta|(E, V) \leq 4\lambda v$, as claimed.

Integrals of $\mathbb{K}$-valued functions with respect to a $\mathcal{P}$-valued measure $\theta$ were seen to be elements of $\mathcal{P}^{**}$. If $(\mathcal{P}, \mathcal{V})$ is indeed a locally convex vector space that is complete in its symmetric topology and if as required in some integrability conditions in the literature (see for example IV.10.7 in [55]) the $\mathbb{K}$-valued function $\varphi$ can be approximated by a sequence of step functions converging pointwise almost everywhere towards $\varphi$ and such that the sequence of integrals over these step functions is convergent in $\mathcal{P}$, then this additional requirement guarantees that the value of the integral of $\varphi$ is also contained in $\mathcal{P}$ rather than in $\mathcal{P}^{**}$.

Let us discuss the convergence theorems from Section 5: Let $(\varphi_n)_{n \in \mathbb{N}}$ be a sequence of integrable $\mathbb{K}$-valued functions that converges pointwise $\theta$-almost everywhere on a set $F \in \mathfrak{A}_{\mathfrak{R}}$ to a function $\varphi$ in the symmetric relative topology of $\mathbb{K}$. This is the usual notion of convergence, except for the case of $\mathbb{K} = \mathbb{R}_+$ endowed with the neighborhood $\mathbb{B}_0^u$ which renders $0 \in \mathbb{R}_+$ as an isolated point. The boundedness conditions from Theorem 5.25 are as follows: We set $\varphi_{**} = \varphi_* = 0$ in all cases. For $\mathbb{K} = \mathbb{R}_+$ we require that $\varphi_n \underset{a.e.F}{\leq} \varphi^*$ holds for all $n \in \mathbb{N}$ with some integrable function $\varphi^*$. For $\mathbb{K} = \mathbb{R}$ or $\mathbb{K} = \mathbb{C}$ with the Euclidean topology and the order we use an integrable positive-valued function $\varphi^*$ and the function $f^* = \varphi^* \otimes \mathbb{B}$ whose values are in the full cone $\mathbb{K}_{\mathcal{V}} = \{\alpha + \rho\mathbb{B} \mid \alpha \in \mathbb{K},\ \rho \geq 0\}$ to which Theorem 5.25 applies. We require that $|\varphi_n| \underset{a.e.F}{\leq} \varphi^*$ holds for all $n \in \mathbb{N}$ in this case. The assumptions of Theorem 5.25 are now satisfied. Let $a_n = \int_F \varphi_n \, d\theta$, $a = \int_F \varphi \, d\theta$ and $a^* = \int_F \varphi^* \, d\theta$, or $a^* = \int_F (\varphi^* \otimes \mathbb{K}) \, d\theta$ in case $\mathbb{K} = \mathbb{R}$ or $\mathbb{K} = \mathbb{C}$. These integrals are in general elements of $\mathcal{P}^{**}$. The conclusion of Theorem 5.25 now states that

$$a \leq \underline{\lim}_{n \to \infty} a_n \qquad \text{and} \qquad \overline{\lim}_{n \to \infty} a_n \leq a + \mathfrak{O}(a^*)$$

in $\mathcal{P}^{**}$, that is

$$a(\mu) \leq \underline{\lim}_{n \to \infty} a_n(\mu)$$

for all $\mu \in \mathcal{P}^*$, and indeed

$$a(\mu) = \lim_{n \to \infty} a_n(\mu)$$

whenever $a^*(\mu) < +\infty$. For elements $a \in \mathcal{P} \subset \mathcal{P}^{**}$ we have $a(\mu) = \mu(a)$. If $(\mathcal{P}, \mathcal{V})$ is indeed a locally convex topological vector space and if $F = E \in \mathfrak{R}$, then the assumptions of Theorem 5.36 apply: The measure $\theta$ is strongly additive by Theorem 3.11. Measurability in the classical sense and boundedness below almost everywhere on the set $E$ is sufficient for the functions $\varphi_n, \varphi$ and $\varphi^*{\otimes}\mathbb{B}$ to be $(\mathbb{K},\mathfrak{V})$-based integrable in the sense of 5.6. The latter is indeed strongly integrable in the sense of 5.18. All integrals involved are elements of $\mathcal{P}$ and Theorem 5.36 yields

$$a = \lim_{n\to\infty} a_n$$

in the symmetric, that is the given topology of $\mathcal{P}$.

*Algebra-valued measures.* If $\mathcal{P}$ is a topological algebra, that is a locally convex topological vector space over $\mathbb{K} = \mathbb{R}$ or $\mathbb{K} = \mathbb{C}$ with a compatible multiplication, then Conditions (A) and (A*) from 6.14 read as follows:

(A) $(\theta_E)^2 = \theta_E$ *and* $\theta_E\,\theta_G = 0$ *for disjoint sets* $E, G \in \mathfrak{R}$.
(A*) $(\theta_E)^* = \theta_E$ *for all* $E \in \mathfrak{R}$.

According to 6.14, Condition (A) guarantees the multiplicativity of the integral as an operator from $\mathcal{E}_{\mathfrak{R}}(X,\mathbb{K}) = \mathcal{F}_{\mathfrak{R}}(X,\mathbb{K})$ into $\mathcal{P}^{**}$, that is $\int_X (fg)\,d\theta \in \left(\int_X f\,d\theta\right)\bullet\left(\int_X g\,d\theta\right)$, Condition (A*) the compatibility with an involution, that is $\int_X f^*\,d\theta = \left(\int_X f\,d\theta\right)^*$ for $\mathbb{K}$-valued functions $f, g \in \mathcal{F}_{\mathfrak{R}}(X,\mathbb{K})$.

*Lattice-valued measures.* Now suppose that $\mathcal{P}$ is a lattice cone over $\mathbb{K} = \mathbb{R}$ or $\mathbb{K} = \mathbb{R}_+$ in the sense of 6.16, that is a quasi-full locally convex cone containing suprema for any two of its elements and satisfying the properties specified in 6.16. For the integral to determine a $\vee$-semilattice homomorphism from $\mathcal{E}_{\mathfrak{R}}(X,\mathbb{K}) = \mathcal{F}_{\mathfrak{R}}(X,\mathbb{K})$ into $\mathcal{P}^{**}$ in the sense of 6.16 we need to require that the measure $\theta$ satisfies Condition (L), that is $\theta_E(a)\vee\theta_E(b) = \theta_E(a\vee b)$ and $\theta_E(a)\vee\theta_G(b) = \theta_E(a)+\theta_G(b)$ holds for all $0 \le a, b \in \mathbb{K}$ and disjoint sets $E, G \in \mathfrak{R}$. As $\theta_E(a) = a\theta_E$ in this case, Condition (L) reads as follows:

(L) $\theta_E \ge 0$ *and* $\theta_E \vee \theta_G = \theta_E + \theta_G$ *for disjoint sets* $E, G \in \mathfrak{R}$,

that is the elements $\theta_E \in \mathcal{P}$ are positive and mutually disjoint for disjoint sets $E, G \in \mathfrak{R}$. According to 6.16, Condition (L) guarantees that the integral as an operator from $\mathcal{S}^{o}_{\mathfrak{R}}(X,\mathbb{K}) = \mathcal{F}_{\mathfrak{R}}(X,\mathbb{K})$ into $\mathcal{P}^{**}$ is a $\vee$-semilattice homomorphism, that is $\int_X (f\vee g)\,d\theta = \left(\int_X f\,d\theta\right)\vee\left(\int_X g\,d\theta\right)$ for functions $f, g \in \mathcal{F}_{\mathfrak{R}}(X,\mathbb{K})$.

**6.24 Positive Linear Operators on Cones of $\overline{\mathbb{R}}$-Valued Functions.** Let $\mathcal{P} = \overline{\mathbb{R}}$, let $X$ and $\mathfrak{R}$ be as before, and let $\mathcal{W}$ be a neighborhood system for $\mathcal{F}(X,\overline{\mathbb{R}})$, consisting of non-negative functions $w \in \mathcal{F}(X,\overline{\mathbb{R}})$. Let $\mathcal{Q} = \mathcal{F}_{\mathcal{W}}(X,\overline{\mathbb{R}})$ be the subcone of functions in $\mathcal{F}(X,\overline{\mathbb{R}})$ that are bounded below with respect to $\mathcal{W}$. Then $\left(\mathcal{F}_{\mathcal{W}}(X,\overline{\mathbb{R}}),\mathcal{W}\right)$ is a full locally convex complete lattice cone, provided that for every $x \in X$ there is $w \in \mathcal{W}$ such that $w(x) < +\infty$ (see Example I.5.7(c)). There are two distinct types of continuous linear operators from $\overline{\mathbb{R}}$ into $\mathcal{F}_{\mathcal{W}}(X,\overline{\mathbb{R}})$. Firstly, for a non-negative real-valued function $\varphi$ such that both $\varphi, -\varphi \in \mathcal{F}_{\mathcal{W}}(X,\overline{\mathbb{R}})$, let $T_\varphi(a) = a\varphi$

for $a \in \overline{\mathbb{R}}$. (In particular, this means $T_\varphi(+\infty)(x) = +\infty$ for all $x \in X$ such that $\varphi(x) > 0$ and $T_\varphi(+\infty)(x) = 0$ else.) Secondly, for a function $\psi \in \mathcal{F}_\mathcal{W}(X, \overline{\mathbb{R}})$ that takes only the values $0$ and $+\infty$, set $T_\psi^0(a) = 0$ for $a \in \mathbb{R}$ and $T_\psi^0(+\infty) = \psi$. Then every linear operator $T \in \mathfrak{L}(\overline{\mathbb{R}}, \mathcal{F}_\mathcal{W}(X, \overline{\mathbb{R}}))$ can be expressed as $T = T_\varphi + T_\psi^0$ with some $\varphi, \psi \in \mathcal{F}_\mathcal{W}(X, \overline{\mathbb{R}})$ as above. Consequently, an $\mathfrak{L}(\overline{\mathbb{R}}, \mathcal{F}_\mathcal{W}(X, \overline{\mathbb{R}}))$-valued measure $\theta$ on $\mathfrak{R}$ can be expressed as a sum of two $\mathcal{F}_\mathcal{W}(X, \overline{\mathbb{R}})$-valued measures $\theta^1$ and $\theta^0$, both yielding functions in $\mathcal{F}_\mathcal{W}(X, \overline{\mathbb{R}})$, and such that for each $E \in \mathfrak{R}$ the function $\theta_E^1$ is positive and both $\theta_E^1, -\theta_E^1 \in \mathcal{F}_\mathcal{W}(X, \overline{\mathbb{R}})$, and the function $\theta_E^0$ takes only the values $0$ and $+\infty$. For a step function

$$h = \sum_{i=1}^{n} \chi_{E_i} a_i \in \mathcal{S}_\mathfrak{R}(X, \overline{\mathbb{R}}),$$

where $a_1, \ldots, a_n \in \overline{\mathbb{R}}$, we have in particular

$$\int_X h \, d\theta = \sum_{i=1}^{n} a_i \theta_{E_i}^1 + \sum_{\substack{i=1,\ldots,n \\ \text{s.th.} a_i = +\infty}} \theta_{E_i}^0 \in \mathcal{F}_\mathcal{W}(X, \overline{\mathbb{R}}).$$

On $\mathcal{F}_\mathfrak{R}(X, \overline{\mathbb{R}})$, the mapping

$$f \mapsto \int_X f \, d\theta \; : \; \mathcal{F}_\mathfrak{R}(X, \overline{\mathbb{R}}) \to \mathcal{F}_\mathcal{W}(X, \overline{\mathbb{R}})$$

defines a linear operator, continuous with respect to the locally convex cone topologies induced by the neighborhood system $\mathcal{W}$, that is

$$\int_X f \, d\theta \le \int_X g \, d\theta + w \qquad \text{whenever} \qquad f \le g + \mathfrak{v}_w,$$

for $f, g \in \mathcal{F}_\mathfrak{R}(X, \overline{\mathbb{R}})$. Recall from Section 4 that $\mathfrak{v}_w$ consists of all step functions $s = \sum_{i=1}^n \chi_{E_i} \circledast a_i$ for $0 < a_i \in \mathbb{R}$ such that $\int_X s \, d\theta = \sum_{i=1}^n a_i \theta_{E_i}^1 \le w$.

According to 6.17, the linear operator determined by the integral is indeed a $\vee$-semilattice homomorphism, if Condition (L) holds, that is if $\theta_E(a) \vee \theta_E(b) = \theta_E(a \vee b)$ and $\theta_E(a) \vee \theta_G(b) = \theta_E(a) + \theta_G(b)$ for all $a, b \ge 0$ in $\overline{\mathbb{R}}$ and disjoint sets $E, G \in \mathfrak{R}$. The first part of this condition holds always true for an $\mathfrak{L}(\overline{\mathbb{R}}, \mathcal{F}_\mathcal{W}(X, \overline{\mathbb{R}}))$-valued measure $\theta$ as introduced above, since the operators involved, $T_\varphi$ and $T_\psi^0$, are defined using non-negative functions $\varphi$ and $\psi$. Let us investigate the second part of the condition in (L): For disjoint sets $E, G \in \mathfrak{R}$ let $\theta_E = T_{\varphi_E} + T_{\psi_E}^0$ and $\theta_G = T_{\varphi_G} + T_{\psi_G}^0$. Then (L) requires that the functions $\varphi_E$ and $\varphi_G$ are orthogonal, that is $\varphi_E(x)\varphi_G(x) = 0$ for all $x \in X$. (There are no additional conditions for the functions $\psi_E$ and $\psi_G$.) If this condition is satisfied, then we have

$$\int_X (f \vee g) \, d\theta = \left( \int_X f \, d\theta \right) \vee \left( \int_X g \, d\theta \right)$$

for all $f, g \in \mathcal{F}_\mathfrak{R}(X, \overline{\mathbb{R}})$.

**6.25 Bounded Linear Operators on Spaces of Real- or Complex-Valued Functions.** Now let $\mathcal{P} = \mathbb{K}$ for $\mathbb{K} = \mathbb{R}$ or $\mathbb{K} = \mathbb{C}$, endowed with the equality as order and the usual topology, that is $\mathcal{V} = \{\rho\mathbb{B} \mid \rho > 0\}$, and $a \leq b + \rho\mathbb{B}$ if $|a-b| \leq \rho$ for $a, b \in \mathbb{K}$. Let $X$ and $\mathfrak{R}$ be as before. Let $\mathcal{W}$ be a system of nonnegative $\overline{\mathbb{R}}$-valued functions on $X$, closed for addition and multiplication by (strictly) positive scalars and directed downward. Suppose that for every $x \in X$ there is $v \in \mathcal{V}$ such that $v(x) < +\infty$. Let $\mathcal{Q}_0 = \mathcal{F}_{\mathcal{W}}(X, \mathbb{K})$ be the vector space over $\mathbb{K}$ of all functions $f \in \mathcal{F}(X, \mathbb{K})$ that are bounded with respect to the functions in $\mathcal{W}$, that is for every $w \in \mathcal{W}$ there is $\lambda \geq 0$ such that $|f(x)| \leq \lambda w(x)$ for all $x \in X$. The above condition on $\mathcal{W}$ guarantees that for every $x \in X$ the point evaluation $\varepsilon_x$ is contained in the vector space dual $\mathcal{Q}^*_{0\mathbb{K}}$ of $\mathcal{Q}_0$. Let $\mathcal{Q}$ be the standard lattice completion of $Q_0$. We shall consider an $\mathfrak{L}(\mathbb{K}, \mathcal{Q}_0)$-valued measure $\theta$ such that for all $E \in \mathfrak{R}$ the operators $\theta_E \in \mathfrak{L}(\mathbb{K}, \mathcal{Q}_0)$ are linear over $\mathbb{K}$. According to 6.12(iii) then the operator

$$f \mapsto \int_X f\, d\theta \,:\, \mathcal{F}_{\mathfrak{R}}(X, \mathbb{K}) \to \mathcal{Q}$$

is linear over $\mathbb{K}$ in the sense that

$$\left(\int_X af\, d\theta\right)(\mu) = \left(\int_X f\, d\theta\right)(a\mu)$$

for every $f \in \mathcal{F}_{(F,\theta)}(X, \mathcal{P})$, $\mu \in \mathcal{Q}^*_0$ and $a \in \mathbb{K}$. If we set

$$\left(\int_F f\, d\theta\right)(x) \equiv \left(\int_F f\, d\theta\right)(\varepsilon_x) - i\left(\int_F f\, d\theta\right)(i\varepsilon_x)$$

for $x \in X$, then these integrals may be reinterpreted as $\mathbb{K}$-valued functions on $X$ and the integral is a $\mathbb{K}$-linear operator from $\mathcal{F}_{\mathfrak{R}}(X, \mathbb{K})$ into $\mathcal{F}_{\mathcal{W}}(X, \mathbb{K})$. Moreover,

$$\left|\int_F f\, d\theta\right| \leq w \qquad \text{holds whenever} \qquad f \leq \mathfrak{v}_w$$

for $f \in \mathcal{F}_{\mathfrak{R}}(X, \mathbb{K})$ and $w \in \mathcal{W}$. (Recall that $\mathfrak{v}_w$ consists of all step functions $s = \sum_{i=1}^n \alpha_i \chi_{E_i} \otimes \mathbb{B}$ for $0 < \alpha_i \in \mathbb{R}$ such that $\int_X s\, d\theta = \sum_{i=1}^n \alpha_i |\theta|(E_i, \mathbb{B}) \leq w$.)

Obviously, $\mathbb{K}$-linear operators in $\mathfrak{L}\big(\mathbb{K}, \mathcal{F}_{\mathcal{W}}(X, \mathbb{K})\big)$ correspond to functions $\varphi \in \mathcal{F}_{\mathcal{W}}(X, \mathbb{K})$. They operate as

$$T_\varphi(a) = a\varphi \qquad \text{for} \quad a \in \mathbb{K}.$$

An $\mathfrak{L}\big(\mathbb{K}, \mathcal{F}_{\mathcal{W}}(X, \mathbb{K})\big)$-valued measure $\theta$ on $\mathfrak{R}$ may therefore be considered as an $\mathcal{F}_{\mathcal{W}}(X, \mathbb{K})$-valued set function on $\mathfrak{R}$. Boundedness means that for every $E \in \mathfrak{R}$ and $w \in \mathcal{W}$ there is $\rho \geq 0$ such that

$$|\theta|(E,\mathbb{B}) = \sup\left\{\sum_{i=1}^{n} |\theta_{E_i}| \;\middle|\; E_i \in \mathfrak{R} \text{ disjoint subsets of } E\right\} \le \rho w.$$

Measurability for a function in $\mathcal{F}(X,\mathbb{K})$ in the sense of Section 1 coincides with measurability in the usual sense.

Both $\mathcal{P} = \mathbb{K}$ and $\mathcal{Q}_0 = \mathcal{F}_{\mathcal{W}}(X,\mathbb{K})$ are indeed topological algebras. Thus according to 6.14, the integral is an algebra homomorphism if Condition (A) holds, that is if $\theta_E(a)\theta_E(b) = \theta_E(ab)$ and $\theta_E(a)\theta_G(b) = 0$ holds for all $a,b \in \mathbb{K}$ and disjoint sets $E,G \in \mathfrak{R}$. The first part of this condition means that the function $\theta_E \in \mathcal{F}_{\mathcal{W}}(X,\mathbb{K})$ takes only the values $0$ or $1$, i.e. is the characteristic function of some subset $\Phi(E)$ of $X$. The second part of (A) requires that for disjoint sets $E,G \in \mathfrak{R}$ the functions $\theta_E$ and $\theta_G$ are orthogonal, that is $\theta_E(x)\theta_G(x) = 0$ for all $x \in X$, that is the sets $\Phi(E)$ and $\Phi(G)$ are disjoint. If this condition is satisfied, then we have

$$\int_X (fg)\, d\theta = \left(\int_X f\, d\theta\right)\left(\int_X g\, d\theta\right)$$

for all $f,g \in \mathcal{F}_{\mathfrak{R}}(X,\mathbb{K})$. The extension of the multiplication from $\mathcal{Q}_0$ to $\mathcal{Q}$ that was introduced in 6.14 implies pointwise multiplication for the corresponding $\mathbb{K}$-valued functions. Thus, under Condition (A) the integral defines a $\mathbb{K}$-linear bounded and multiplicative operator from $\mathcal{F}_{\mathfrak{R}}(X,\mathbb{K})$ into $\mathcal{F}_{\mathcal{W}}(X,\mathbb{K})$. It also preserves the involution since Condition (A*) is obviously implied by (A).

## 7. Extended Integrability

We can further extend integrability to a wider class of functions $f \in \mathcal{F}(X,\mathcal{P})$. Obviously, if there is $g \in \mathcal{F}_{(F,\theta)}(X,\mathcal{P})$ such that both $f+g$ and $g$ are contained in $\mathcal{F}_{(F,\theta)}(X,\mathcal{P})$ and if the element $\int_F g\, d\theta$ is invertible in $\mathcal{Q}$, then we may set

$$\int_F f\, d\theta = \int_F (f+g)\, d\theta - \int_F g\, d\theta.$$

The class of these functions $f \in \mathcal{F}(X,\mathcal{P})$ will be denoted by $\mathfrak{F}_{(F,\theta)}(X,\mathcal{P})$.

The following is straightforward to verify:

**Theorem 7.1.** *$\mathfrak{F}_{(F,\theta)}$ is a subcone of $\mathcal{F}(X,\mathcal{P})$ containing $\mathcal{F}_{(F,\theta)}(X,\mathcal{P})$. If $f,g \in \mathfrak{F}_{(F,\theta)}$ and $0 \le \alpha \in \mathbb{R}$, then*

*(a)* $\int_F (\alpha f)\, d\theta = \alpha \int_F f\, d\theta$
*(b)* $\int_F (f+g)\, d\theta = \int_F f\, d\theta + \int_F g\, d\theta$
*(c)* $\int_F f\, d\theta \le \int_F g\, d\theta + w$ *whenever* $f \le g + \mathfrak{v}_w$ *for* $w \in \mathcal{W}$.

## 8. Notes and Remarks

The beginnings of modern measure theory date back to the late 19th century, some of the foundations being laid by Riemann, Harnack, Peano, Jordan, Borel, Baire, Lebesgue, Carathéodory and Radon, to name just a few of the mathematicians involved. Excellent expositions about the early history of measure theory can be found in the works of Lebesgue [114], [115] and [116], Carathéodory [30], Hahn and Rosenthal [80], Halmos [83] and Saks [182]. Vector-valued measure theory originated in the first half of the twentieth century in treatises by Clarkson, Bochner, Dunford, Morse, Pettis and Gelfand among others. Since its appearance in 1977 the book by Diestel and Uhl [43] about vector measures has become a standard reference on the subject and is also often cited in this text. It contains various sections with detailed surveys of the history of the field. There is also an extensive literature on finitely additive measures. The books by Dunford and Schwartz [55], [56], [57] and Diestel and Uhl [43] contain some sections about these. However, finitely additive measures appear to be less suitable for analytic purposes, and we therefore do not address them in this text.

The *(total) variation* of a Banach space-valued measure $\theta$ on a $\sigma$-field $\mathfrak{R}$ is usually defined as the positive $\overline{\mathbb{R}}$-valued set-function $|\theta|$ by

$$|\theta|(E) = \sup\left\{ \sum_{i=1}^{n} \|\theta(E_i)\| \;\middle|\; E_i \imath \mathfrak{R} \text{ disjoint subsets of } E \right\}$$

for $E \in \mathfrak{R}$ (See III.1.4 in [55] or I.1.4 in [43]). The *semivariation* of a vector-valued measure was introduced by Gowurin [74] and is given by

$$\|\theta\|(E) = \sup\left\{ \left\| \sum_{i=1}^{n} \gamma_i \theta(E_i) \right\| \;\middle|\; |\gamma_i| \leq 1,\;\; E_i \in \mathfrak{R} \text{ disjoint subsets of } E \right\},$$

Clearly $\|\theta\|(E) \leq |\theta|(E)$, and every countably additive vector measure is known to be bounded, that is $|\theta|(X) < +\infty$ (see IV.10.2 in [55]). The set-function $|\theta|$ is seen to be $\sigma$-additive, whereas $\|\theta\|$ is generally only subadditive. On the other hand, the definition of the modulus of $\theta$ from Section 3.2, if applied to this situation, reads as

$$|\theta|(E, \mathbb{B}) = \sup\left\{ \sum_{i=1}^{n} \gamma_i \theta_{E_i} \;\middle|\; |\gamma_i| \leq 1,\; E_i \in \mathfrak{R} \text{ disjoint subsets of } E \right\},$$

where $\mathbb{B}$ denotes the unit ball of $\mathcal{P} = \mathbb{R}$ or $\mathcal{P} = \mathbb{C}$. Recall that $|\theta|(E, \mathbb{B})$ is an element of the standard lattice completion of the given Banach space, that is an $\overline{\mathbb{R}}$-valued function on its dual unit ball $\mathbb{B}^*$. Since this function is non-negative, it cannot be considered as an element of the second dual of this Banach space. However, its supremum norm $\| |\theta|(E,\mathbb{B}) \|$ as a function on $\mathbb{B}^*$ is the semivariation of the measure. Indeed,

$$\begin{aligned}
&\| |\theta|(E,\mathbb{B}) \| \\
&= \sup_{\mu\in\mathbb{B}^*} \left( \sup \left\{ \sum_{i=1}^{n} \gamma_i \theta_{E_i} \;\middle|\; |\gamma_i| \le 1,\ E_i \in \mathfrak{R} \text{ disjoint subsets of } E \right\} \right)(\mu) \\
&= \sup_{\mu\in\mathbb{B}^*} \sup \left\{ \left( \sum_{i=1}^{n} \gamma_i \theta_{E_i} \right)(\mu) \;\middle|\; |\gamma_i| \le 1,\ E_i \in \mathfrak{R} \text{ disjoint subsets of } E \right\} \\
&= \sup \left\{ \sup_{\mu\in\mathbb{B}^*} \left( \sum_{i=1}^{n} \gamma_i \theta_{E_i} \right)(\mu) \;\middle|\; |\gamma_i| \le 1,\ E_i \in \mathfrak{R} \text{ disjoint subsets of } E \right\} \\
&= \sup \left\{ \left\| \sum_{i=1}^{n} \gamma_i \theta_{E_i} \right\| \;\middle|\; |\gamma_i| \le 1,\ E_i \in \mathfrak{R} \text{ disjoint subsets of } E \right\} \\
&= \|\theta\|(E).
\end{aligned}$$

This observation establishes the relationship between the modulus of a vector-valued measure according to Section 3.6 and its classical semivariation. However, while the modulus is a countably additive set-function, the semivariation, as its norm is only subadditive. Boundedness in the sense of Section 3.6 means that $\sup_{\mu\in\mathbb{B}^*} |\theta|(X,\mathbb{B})(\mu) < +\infty$, hence that the semivariation $\|\theta\|(X)$ is finite. Boundedness guarantees that the linear operators from $\mathcal{P}$ into $\mathcal{Q}$ which are the values of the measure can be extended to linear operators from the standard full extension $\mathcal{P}_\nu$ into $\mathcal{Q}$.

In the literature there is no shortage of different concepts of integrability for scalar-valued functions with respect to vector-valued functions or measures and variations in the resulting definitions of the integral. The best known are perhaps those by Bochner [19], Pettis [144], Bartle [8], [9] and Dunford and Schwartz [55]. There are also some corresponding differences in the definition of measurability. Again, a comprehensive treatment of the relevant definitions and their implications can be found in Chapter II of Diestel and Uhl [43]. Due to its well-understood properties, the Bochner integral is probably most used in applications. Not surprisingly, our very general approach in this chapter covers many of the above-mentioned notions. This is because we are using locally convex cones in our settings and order convergence for most of our definitions and results, and order convergence is generally weaker than the originally given topological convergence. Stronger results are pointed out when possible. Since $\mathcal{Q}$, the range of the integrals, is required to be a locally convex complete lattice cone, if applied to the case of a vector space, the results of this chapter often refer to the second dual of this vector space (Section 6.5). This situation is well-understood for vector-valued measures.

It would probably be worthwhile, though demanding, to explore the Radon-Nikodým property in the settings of this chapter. This refers to a special case of the Application 6.20 from above. Let $\mathcal{P}$ and $\mathcal{Q}$ be locally convex cones satisfying our standard assumptions, and let $\mu$ be a scalar-valued (positive, real or complex-valued) measure on $\mathfrak{R}$. These scalars can be interpreted as linear operators from $\mathfrak{L}(\mathcal{P},\mathcal{Q})$ into itself. One can therefore integrate certain $\mathfrak{L}(\mathcal{P},\mathcal{Q})$-valued functions with respect to $\mu$, and the integral is evaluated in the standard completion $\widehat{\mathfrak{L}}(\mathcal{P},\mathcal{Q})$ of $\mathfrak{L}(\mathcal{P},\mathcal{Q})$. Given a suitable function $\varphi$ of this type, this can be used to define an $\mathfrak{L}(\mathcal{P},\mathcal{Q})$-valued set function by

$$\theta_E = \int_E \varphi \, d\mu \qquad \text{for} \quad E \in \mathfrak{R}.$$

The convergence theorems then will guarantee that $\theta$ is countably additive and indeed an $\mathfrak{L}(\mathcal{P},\mathcal{Q})$-valued measure. Now investigations would have to be carried out, under which conditions a given $\mathfrak{L}(\mathcal{P},\mathcal{Q})$-valued measure $\theta$ can be expressed in this way using a given scalar-valued measure $\mu$ and some $\mathfrak{L}(\mathcal{P},\mathcal{Q})$-valued density function $\varphi$.

# Chapter III
# Measures on Locally Compact Spaces

This final chapter of the book is concerned with topological measure theory and integral representation. The domain $X$ of the operator-valued measure $\theta$ will be a locally compact topological space and the $\sigma$-ring $\mathfrak{R}$ will consist of all relatively compact Borel subsets of $X$. As usual, measures on topological spaces are supposed to fulfill certain regularity requirements. Sections 1 and 2 of this chapter will probe different notions of continuity for locally convex cone-valued functions on a topological space. Inductive limit-type topologies lead to the identification of certain cones of continuous functions. This is motivated by the concept of weighted spaces of continuous real-valued functions which is due to Nachbin [136] and Prolla [155]. Continuous linear operators on such cones of cone-valued functions will be investigated in Section 4. The main result is a generalized Riesz-type integral representation theorem for this type of operators in Section 5. Section 6 contains a long list of special cases and examples, including a generalization of the classical Spectral representation theorem for normal linear operators on a complex Hilbert space.

Generally, the notations introduced in Chapters I and II will be used. As usual, for a subset $Y$ of a topological space $X$, the sets $\overline{Y}$, $Y^{\circ}$ and $\partial Y = \overline{Y} \setminus Y^{\circ}$ denote its topological closure, interior and boundary in $X$, respectively. The *core support* of a cone-valued function $f$ on $X$ is the set $\{x \in X \mid f(x) \neq 0\}$ and denoted by $\operatorname{supp}^*(f)$. Its closure, $\operatorname{supp}(f)$ is the usual support of $f$.

Throughout the following, let $(\mathcal{P}, \mathcal{V})$ be a locally convex cone.

## 1. Relatively Continuous Cone-Valued Functions

Due to the possible presence of unbounded elements in a locally convex cone $(\mathcal{P}, \mathcal{V})$, continuity for $\mathcal{P}$-valued functions on a topological space $X$ with respect to any of the given topologies of $\mathcal{P}$ is a rather restrictive requirement. For example, even if $\varphi$ is a continuous real-valued positive function on $X$, the mapping $\varphi_{\otimes}a \in \mathcal{F}(X,\mathcal{P})$ need not be continuous if the element $a$ is not bounded (see Proposition I.4.2(iii) and Remark I.4.38(b)). We shall provide an example for this below. As with our definition of measurability in Section II.1 we shall therefore use the slightly coarser relative topologies on $\mathcal{P}$ which will better suit our purposes. Continuity with respect to the symmetric relative topology, *r-continuity* for short, turns out to be a sufficiently generous concept. We shall denote the set of all r-continuous functions in $\mathcal{F}(X,\mathcal{P})$ by $\mathcal{C}^r(X,\mathcal{P})$. Clearly, every $\mathcal{P}$-valued function $f \in \mathcal{F}(X,\mathcal{P})$ which is continuous with respect to the given symmetric topology of $\mathcal{P}$ is also r-continuous, but the reverse does not hold true if the values of $f$ are not bounded in $\mathcal{P}$. However, if $f(x)$ is bounded for $x \in X$, then continuity at $x$ coincides for both topologies, as their neighborhood systems for $f(x) \in \mathcal{P}$ are equivalent (see Proposition I.4.2(iv)). We observe that a function $f \in \mathcal{F}(X,\mathcal{P})$ is r-continuous at a point $x \in X$ if and only if for every $v \in \mathcal{V}$ and $\varepsilon > 0$ there is a neighborhood $U$ of $x$ such that $f(y) \in v_\varepsilon\big(f(z)\big)$ for all $y,z \in U$. Similarly, we shall say that a function $f \in \mathcal{F}(X,\mathcal{P})$ is *r-lower* respectively *r-upper continuous* at a point $x \in X$ if it is continuous with respect to the lower or upper relative topologies. Obviously, r-continuity is the combination of both r-lower and r-upper continuity.

**Lemma 1.1.** *The r-lower, the r-upper and the r-continuous functions form subcones of* $\mathcal{F}(X,\mathcal{P})$.

*Proof.* Clearly, $\alpha f$ is r-lower (r-upper) continuous whenever $f$ is and $\alpha \geq 0$. It is however less obvious that these properties are preserved by addition. For this, let $f,g \in \mathcal{F}(X,\mathcal{P})$ and $x \in X$. For $v \in \mathcal{V}$ and $\varepsilon > 0$ there is $\lambda \geq 0$ such that both $0 \leq f(x) + \lambda v$ and $0 \leq g(x) + \lambda v$.

First suppose that both functions $f,g \in \mathcal{F}(X,\mathcal{P})$ are r-upper continuous at $x$. We set $\delta = 1/(2+2\lambda)$ and choose a neighborhood $U$ of $x$ such that both $f(y) \in v_\delta\big(f(x)\big)$ and $g(y) \in v_\delta\big(g(x)\big)$ for all $y \in U$. Following Lemma I.4.1(d), we conclude that

$$f(y) + g(y) \in v_\varepsilon\big(f(x) + g(x)\big)$$

for all $y \in U$, hence the function $f+g$ is indeed r-upper continuous at $x$.

Now for the second case, suppose that both $f,g \in \mathcal{F}(X,\mathcal{P})$ are r-lower continuous at $x$. We set $\delta = \min\{1, 1/(4+2\lambda)\}$ and choose a neighborhood $U$ of $x$ such that both $f(y) \in \big(f(x)\big)v_\delta$, that is $f(x) \in v_\delta\big(f(y)\big)$, and $g(y) \in \big(g(x)\big)v_\delta$, that is $g(x) \in v_\delta\big(g(y)\big)$ for all $y \in U$. Following Lemma I.4.1(c), this implies $0 \leq f(y) + (\lambda+\delta)v$ and $0 \leq g(y) + (\lambda+\delta)v$. Now using Lemma I.4.1(d), as $2\delta(1+\lambda+\delta) \leq \varepsilon$, we conclude that

$$f(x) + g(x) \in v_\varepsilon\big(f(y) + g(y)\big)$$

for all $y \in U$, hence the function $f+g$ is r-lower continuous at $x$. □

**Lemma 1.2.** *An invertible function* $f \in \mathcal{F}(X,\mathcal{P})$ *is r-lower (or r-upper) continuous if and only if its inverse* $-f \in \mathcal{F}(X,\mathcal{P})$ *is r-upper (or r-lower) continuous.*

*Proof.* Let $f \in \mathcal{F}(X,\mathcal{P})$ be invertible, and let $x \in X$. As the element $f(x) \in \mathcal{P}$ is invertible, hence bounded, the notions of continuity and r-continuity coincide for $f$ (see Proposition I.4.2(iv)). We have $f(y) \in v\big(f(x)\big)$, that is $f(y) \le f(x)+v$ if and only if $-f(x) \le -f(y)+v$ that is $-f(x) \in v\big(-f(y)\big)$. This demonstrates that the function $f$ is upper continuous at $x$ if and only if $-f$ is lower continuous at $x$, and vice versa. □

If $(\mathcal{P},\mathcal{V})$ is a locally convex topological vector space over $\mathbb{K}=\mathbb{R}$ or $\mathbb{K}=\mathbb{C}$, then Lemma 1.2 applies to all functions in $\mathcal{F}(X,\mathcal{P})$. Moreover, then $f(y) \in v\big(f(x)\big)$, that is $f(y)-f(x) \in v(0)$ for a function $f \in \mathcal{F}(X,\mathcal{P})$ and $x,y \in X$ implies that $\gamma\big((f(y)-f(x)\big) \in v(0)$, that is $\gamma f(y) \in v\big(\gamma f(x)\big)$ for all $\gamma \in \Gamma$, the unit circle in $\mathbb{K}$. Thus $f$ is upper continuous at $x$ if and only if $\gamma f$ is upper continuous of $x$ for all $\gamma \in \Gamma$ (see I.1.4(d)). This yields:

**Lemma 1.3.** *If* $(\mathcal{P},\mathcal{V})$ *is a locally convex topological vector space over* $\mathbb{K}=\mathbb{R}$ *or* $\mathbb{K}=\mathbb{C}$, *then the notions of r-lower, r-upper and r-continuity coincide, and the r-continuous functions form a* $\mathbb{K}$*-linear subspace of* $\mathcal{F}(X,\mathcal{P})$.

We return to the general case:

**Lemma 1.4.** *For an r-lower, r-upper or r-continuous function* $f \in \mathcal{F}(X,\mathcal{P})$ *and for* $v \in \mathcal{V}$, *there is an upper semicontinuous, lower semicontinuous or continuous positive real-valued function* $\varphi$ *on* $X$, *respectively, such that* $f+\varphi \otimes v \ge 0$. *In particular, if* $f$ *is r-lower continuous, then for every compact subset* $K$ *of* $X$ *and* $v \in \mathcal{V}$ *there is* $\lambda \ge 0$ *such that* $f(x)+\lambda v \ge 0$ *for all* $x \in K$.

*Proof.* For a $f \in \mathcal{F}(X,\mathcal{P})$ and $v \in \mathcal{V}$ set $\psi(x)=\inf\{\lambda \ge 0 \mid f(x)+\lambda v \ge 0\}$. For fixed $x,y \in X$ and $\varepsilon>0$ such that $f(x) \in v_\varepsilon\big(f(y)\big)$ we have $\gamma f(y)+\varepsilon v \ge f(x)$, hence $\gamma f(y)+(2\varepsilon+\psi(x))v \ge 0$ for some $1 \le \gamma \le 1+\varepsilon$. Thus

$$\psi(y) \le (1/\gamma)(\psi(x)+2\varepsilon) \le \psi(x)+2\varepsilon.$$

If $f$ is r-lower continuous at $x$ then there is a neighborhood $U$ of $x$ such that $f(y) \in \big(f(x)\big)v_\varepsilon$, that is $f(x) \in v_\varepsilon\big(f(y)\big)$ for all $y \in U$. The above shows that the function $\psi$ is upper semicontinuous in this case. A similar argument shows that $\psi$ is lower semicontinuous if $f$ is r-upper continuous. The combination of both yield continuity. Our claim holds with the function $\varphi=\psi+1$. □

Given a neighborhood $v \in \mathcal{V}$, in Section I.4 we defined a disjoint partition of $\mathcal{P}$ into its $v$-boundedness components, subsets $\mathcal{B}_v^s(a)$ that are both open and closed with respect to the symmetric relative topology of $\mathcal{P}$ and whose elements are $v$-bounded relative to each other (Proposition I.4.19). The inverse images of these boundedness components under an r-continuous mapping $f$ therefore yield a corresponding disjoint partition of $X$ into both open and closed segments.

**Proposition 1.5.** *Let $f \in \mathcal{C}^r(X, \mathcal{P})$. For every $v \in \mathcal{V}$ there is a disjoint partition of $X$ into segments which are both open and closed and such that the values of $f$ at points in the same segment are $v$-bounded relative to each other.*

If the function $f \in \mathcal{C}^r(X, \mathcal{P})$ has a compact support, then a simple compactness argument shows that there can be only finitely many segments in the above partition of $X$. Moreover, for every such segment $X_i \subset X$ the function $f_i = \chi_{X_i \otimes} f$ is also r-continuous, and $f$ may be expressed as the sum of these functions. The values of each $f_i$ are $v$-bounded relative to each other at points in $X_i$.

Similarly, the global boundedness components $\mathcal{B}^s(a)$ provide a disjoint partition of $\mathcal{P}$ into closed (but not necessarily also open) subsets. However, if the range of an r-continuous function $f$ is covered by only finitely many of these boundedness components, then their inverse images furnish a corresponding finite disjoint partition of $X$ into closed subsets. Finiteness implies that these sets are also open, hence the statement of Proposition 1.5 also holds for the global boundedness components in this case. The same holds true, if the locally convex cone $(\mathcal{P}, \mathcal{V})$ is locally connected in the symmetric relative topology (see Section I.4.20), as its boundedness components are open and closed in this case.

**Proposition 1.6.** *Let $f \in \mathcal{C}^r(X, \mathcal{P})$. There is a disjoint partition of $X$ into closed segments such that the values of $f$ at points in the same segment are bounded relative to each other. If either $\mathcal{P}$ is locally connected in the symmetric relative topology or if the range of $f$ is covered by finitely many global boundedness components of $\mathcal{P}$, then these segments are also open.*

Let us recall that the *quasi-component of a point* $x$ in a topological space $X$ is the intersection of all closed and open subsets of $X$ which contain the point $x$. The quasi-components contain the *components*, that is the maximal connected subsets of $X$, and constitute a decomposition of $X$ into pairwise disjoint and closed subsets (see VIII.26 in [198] or VI.1 in [59]). In compact spaces, the quasi-components and components coincide (see Theorem 6.1.22 in [59]). In locally connected spaces they coincide and are both open and closed (see Corollary 27.10 in [198]). Hence in locally connected compact spaces they form a finite partition into disjoint open, closed and connected subsets.

**Corollary 1.7.** *Two points $x, y \in X$ are contained in the same quasi-component of $X$ if and only if $f(x)$ and $f(y)$ are bounded relative to each other for every r-continuous function $f$ from $X$ into a locally convex cone $(\mathcal{P}, \mathcal{V})$.*

*Proof.* If the points $x, y \in X$ are not contained in the same quasi-component of $X$, then there is an open and closed subset $U$ of $X$ such that $x \in U$ but $y \notin U$. For $\mathcal{P} = \overline{\mathbb{R}}$ the function $f$ defined as $f(x) = 0$ for $x \in U$ and $f(x) = +\infty$ else, is in $\mathcal{C}^r(X, \overline{\mathbb{R}})$, and the values of $f$ in $x$ and in $y$ are obviously not bounded relative to each other. For the converse, suppose that for $x, y \in X$ and $f \in \mathcal{C}^r(X, \mathcal{P})$ for some convex cone $(\mathcal{P}, \mathcal{V})$ the function values $f(x)$ and $f(y)$ are not bounded relative to each other, that is $f(x)$ and $f(y)$ are not $v$-bounded relative to each other for some neighborhood $v \in \mathcal{V}$. Following Proposition 1.5 there is an open and closed subset of $X$ containing $x$ but not $y$. Then $x$ and $y$ are not contained in the same quasi-component of $X$. □

**Proposition 1.8.** *Let $f = \varphi \otimes g \in \mathcal{F}(X, \mathcal{P})$ for a positive real-valued continuous function $\varphi$ and $g \in \mathcal{F}(X, \mathcal{P})$.*

(a) *The function $f$ is r-lower continuous at all points $x \in X$ where $g$ is r-lower continuous.*
(b) *The function $f$ is r-upper continuous at all points $x \in X$ where $g$ is r-upper continuous and where either $\varphi(x) > 0$ or $g(x)$ is bounded in $\mathcal{P}$.*

*Proof.* For Part (a), suppose that the function $g$ is r-lower continuous at $x \in X$. Let $v \in \mathcal{V}$ and $\varepsilon > 0$. There is $\lambda \geq 0$ such that $0 \leq g(x) + \lambda v$. We choose any $\delta > 0$. Because $g$ is r-lower continuous and $\varphi$ is r-continuous at $x$, there is a neighborhood $U$ of $x$ such that $\varphi(x) \leq (1+\delta)\varphi(y)$, $\varphi(y) \leq \varphi(x) + \delta$ and $g(y) \in \big(g(x)\big)v_\delta$, that is $g(x) \in v_\delta\big(g(y)\big)$ for all $y \in U$. Then for all $z \in U$ this yields with some $1 \leq \gamma \leq 1+\delta$

$$\begin{aligned}\varphi(x)\big(g(x) + \lambda v\big) &\leq (1+\delta)\varphi(y)\big(\gamma g(y) + (\lambda+\delta)v\big) \\ &\leq \gamma(1+\delta)\varphi(y)g(y) + \big(\varphi(x)+\delta\big)(1+\delta)(\lambda+\delta)v.\end{aligned}$$

We may choose $\delta > 0$ sufficiently small such that both $\gamma(1+\delta) \leq (1+\delta)^2 \leq 1+\varepsilon$ and $\big(\varphi(x)+\delta\big)(1+\delta)(\lambda+\delta) < \lambda\varphi(x) + \varepsilon$. Then, using the cancellation property for the positive element $v$ (see Lemma I.4.2 in [100]), we conclude that

$$\varphi(x)g(x) \leq \gamma(1+\delta)\varphi(z)g(y) + \varepsilon v,$$

that is $f(x) \in v_\varepsilon\big(f(y)\big)$, that is $f(y) \in \big(f(x)\big)v_\varepsilon$ holds for all $y \in U$. Hence $f$ is indeed r-lower continuous at $x$.

Now for Part (b), suppose that $g$ is r-upper continuous at $x \in X$, and for the first part of our claim, that $\varphi(x) > 0$. Let $v \in \mathcal{V}$ and $\varepsilon > 0$. There is $\lambda \geq 0$ such that $0 \leq g(x) + \lambda v$. We choose any $\delta > 0$. Because $g$ is r-upper continuous and $\varphi$ is continuous at $x$, and as $\varphi(x) > 0$, there is a neighborhood $U$ of $x$ such that $\varphi(y) \leq (1+\delta)\varphi(x)$, $\varphi(x) \leq \varphi(y) + \delta$ and $g(y) \in v_\delta\big(g(x)\big)$ for all $y \in U$. Following Lemma I.4.1(c), the latter implies $0 \leq g(y) + (\lambda+\delta)v$ for all $y \in U$. Then for all $y \in U$ this yields with some $1 \leq \gamma \leq 1+\delta$

$$\begin{aligned}
\varphi(y)g(y) + \varphi(x)(\lambda+\delta)v \\
&\le \varphi(y)g(y) + \big(\varphi(y)+\delta\big)(\lambda+\delta)v \\
&\le \varphi(y)\big(g(y) + (\lambda+\delta)v\big) + \delta(\lambda+\delta)v \\
&\le (1+\delta)\varphi(x)\big(\gamma g(x) + (\lambda+2\delta)v\big) + \delta(\lambda+\delta)v \\
&\le \gamma(1+\delta)\varphi(x)g(x) + \big((1+\delta)(\lambda+2\delta)\varphi(x) + \delta(\lambda+\delta)\big)v.
\end{aligned}$$

Again, we may choose $\delta$ sufficiently small such that both $\gamma(1+\delta) \le (1+\delta)^2 \le 1+\varepsilon$ and $(1+\delta)(\lambda+2\delta)\varphi(x)+\delta(\lambda+\delta) < (\lambda+\delta)\varphi(x)+\varepsilon$. Then, using the cancellation property for the positive element $v$ we conclude that

$$\varphi(y)g(y) \le \gamma(1+\delta)\varphi(z)g(z) + \varepsilon v,$$

that is $f(y) \in v_\varepsilon\big(f(x)\big)$ holds for all $y \in U$. Hence $f$ is indeed r-upper continuous at $x$.

For the second part of our claim in (b), suppose that $g$ is r-upper continuous at $x \in X$, that $\varphi(x) = 0$, hence $f(x) = 0$, and that $g(x)$ is bounded in $\mathcal{P}$, and let $v \in \mathcal{V}$. There is $\lambda \ge 0$ such that $g(x) \le \lambda v$. We choose a neighborhood $U$ of $x$ such that $0 \le \varphi(y) \le 1/(2\lambda+1)$ and $g(y) \in v_1\big(g(x)\big)$, for all $y \in U$. With some $1 \le \gamma \le 2$, the latter means

$$g(y) \le \gamma g(x) + v \le (\gamma\lambda+1)v \le (2\lambda+1)v.$$

Then

$$\varphi(y)g(y) \le \varphi(y)(2\lambda+1)v \le v,$$

hence $f(y) \in v\big(f(x)\big) \subset v_\varepsilon\big(f(x)\big)$ for all $\varepsilon > 0$ and $y \in U$. Thus $f = \varphi_\otimes g$ is indeed r-upper continuous at $x$. □

**Proposition 1.9.** *Let* $f = \sum_{i=1}^n \varphi_{i\otimes} g_i$ *for positive real-valued continuous functions* $\varphi_i$ *and* $g_i \in \mathcal{C}^r(X,\mathcal{P})$. *The function* $f$ *is r-continuous if and only if for every* $v \in \mathcal{V}$ *there is a disjoint partition of* $X$ *into segments which are both open and closed and such that the values of* $f$ *at points in the same segment are* $v$*-bounded relative to each other.*

*Proof.* Let $f = \sum_{i=1}^n \varphi_{i\otimes} g_i$ be as stated, and for each $i \in \mathcal{I} = \{1,\ldots,n\}$ let $O_i = \{x \in X \mid \varphi(x) > 0\}$ be the (open) core support of the positive real-valued function $\varphi_i$. If the function $f$ is r-continuous, then Proposition 1.5 yields the existence of a partition of $X$ with the claimed properties for each $v \in \mathcal{V}$.

For the converse, Proposition 1.8(a) shows that each of the functions $\varphi_{i\otimes} g_i$ is r-lower continuous, hence following Lemma 1.1, the function $f$ is also r-lower continuous on $X$. Now let us assume that for every $v \in \mathcal{V}$ there exists a disjoint partition of $X$ into open and closed segments such that the values of $f$ at points in the same segment are $v$-bounded relative to each other. We shall proceed to verify that $f$ is r-upper continuous in this case. For this, let $x \in X$, let $\mathcal{I}_1 = \{i \in \mathcal{I} \mid x \in \partial O_i\}$ and $\mathcal{I}_2 = \{i \in \mathcal{I} \mid x \notin \partial O_i\}$, where the open sets $O_i \subset X$, $i = 1,\ldots,n$, are defined as above and $\partial O_i$ is the topological boundary of $O_i$. Set

$$h_1 = \sum_{i \in \mathcal{I}_1} \varphi_{i \circledcirc} g_i \qquad \text{and} \qquad h_2 = \sum_{i \in \mathcal{I}_2} \varphi_{i \circledcirc} g_i.$$

Then $f = h_1 + h_2$ and $h_1(x) = 0$. For every $i \in \mathcal{I}_2$ the function $\varphi_i$ either vanishes on a neighborhood of $x$, or $\varphi_i(x) > 0$. In both cases, Proposition 1.8(b) shows that the function $\varphi_{i \circledcirc} g_i$ is r-upper continuous at $x$. In turn, Lemma 1.1 yields that the function $h_2$ is also r-upper continuous at $x$. For the function $h_1$ we continue our argument as follows: Let $v \in \mathcal{V}$ and $\varepsilon > 0$. By our assumption on $f$ there is a neighborhood $O$ of $x$ such that all values of $f$ on $O$ are $v$-bounded relative to $f(x)$. Moreover, because the functions $g_i$ are r-continuous, following Proposition 1.5 we may assume in addition that the values of $g_i$ on $O$ are $v$-bounded relative to $g_i(x)$ for all $i \in \mathcal{I}_1$. For every $i \in \mathcal{I}_1$, on the other hand, there is an element $y_i \in O$ such that $\varphi_i(y_i) > 0$. Thus

$$f(y_i) = \sum_{i=1}^{n} \varphi_i(y_i) g_i(y_i) \in \mathcal{B}_v\big(f(x)\big)$$

implies that $g_i(y_i) \in \mathcal{B}_v\big(f(x)\big)$, as the boundedness components were seen to be faces in $\mathcal{P}$ (see Proposition I.4.11(b)). There is $\lambda > 0$ such that $g_i(x) \in v_\lambda\big(f(x)\big)$ for all $i \in \mathcal{I}_1$. We may also assume that $0 \leq f(x) + \lambda v$. Now we choose $\delta = \min\{\frac{1}{1+\lambda}, \frac{\varepsilon}{(1+\lambda)(3+\lambda)}\}$ and a neighborhood $U$ of $x$ such that $\varphi_i(y) \leq \delta/n$, $g_i(y) \in v_\delta\big(g_i(x)\big)$ for all $i \in \mathcal{I}_1$, and $h_2(y) \in v_\delta\big(h_2(x)\big)$ holds for all $y \in U$. Recall that $f(x) = h_2(x)$ as $h_1(x) = 0$. Thus the latter implies with Lemma I.4.1(b)

$$h_2(y) \leq (1+\delta) f(x) + \delta(1+\lambda) v.$$

Using I.4.1(a) and our assumption that $\delta(\lambda+1) \leq 1$, the former yields

$$g_i(y) \in v_{(\delta+\lambda+\delta\lambda)}\big(f(x)\big) \subset v_{(1+\lambda)}\big(f(x)\big)$$

for all $y \in U$ and $i \in \mathcal{I}_1$. Hence, again using I.4.1(b)

$$g_i(y) \leq (2+\lambda) f(x) + (1+\lambda)^2 v \leq (2+\lambda)\big(f(x) + (1+\lambda)\, v\big)$$

and therefore, as $f(x) + (1+\lambda)\, v \geq 0$ and $\varphi_i(y) \leq \delta/n$

$$\begin{aligned} h_1(y) &= \sum_{i \in \mathcal{I}_1} \varphi_i(y) g_i(y) \\ &\leq \sum_{i \in \mathcal{I}_1} \varphi_i(y)(2+\lambda)\big(f(x) + (1+\lambda)\, v\big) \\ &\leq \delta(2+\lambda)\big(f(x) + (1+\lambda)\, v\big) \end{aligned}$$

for all $y \in U$. Thus, combining

$$\begin{aligned} f(y) &= h_1(y) + h_2(y) \\ &\leq \big(\delta(2+\lambda) + (1+\delta)\big) f(x) + \big(\delta(2+\lambda)(1+\lambda) + \delta(1+\lambda)\big) v \\ &\leq \big(1 + \delta(3+\lambda)\big) f(x) + \delta(1+\lambda)(3+\lambda) v. \end{aligned}$$

Because

$$1 \leq 1 + \delta(3+\lambda) \leq 1+\varepsilon \qquad \text{and} \qquad \delta(1+\lambda)(3+\lambda) \leq \varepsilon,$$

this demonstrates $f(y) \in v_\varepsilon\big(f(x)\big)$ for all $y \in U$. The function $f$ is therefore r-upper continuous at $x$, as claimed. □

**1.10 Elementary Functions.** We are particularly interested in *elementary functions* of the type $\varphi{\otimes}a$ for a positive real-valued continuous function $\varphi$ on $X$ and an element $a \in \mathcal{P}$, and in sums of functions of this type. These functions are r-lower continuous by Proposition 1.8(a). If the element $a$ is unbounded, then the inverse image under $\varphi{\otimes}a$ of the boundedness component $\mathcal{B}^s(a)$ is the set

$$\operatorname{supp}^*(\varphi) = \{x \in X \mid \varphi(x) > 0\},$$

that is the core support of the function $\varphi$. This set is open since $\varphi$ is continuous and indeed also closed whenever $\varphi{\otimes}a$ is r-continuous, since $\mathcal{B}^s(a)$ is closed in the symmetric relative topology of $X$.

Because the range of a function $f = \sum_{i=1}^n \varphi_i{\otimes}a_i$ is covered by the finitely many global boundedness components $\mathcal{B}^s(a_\mathcal{I},)$ where $a_\mathcal{I} = \sum_{i\in\mathcal{I}} a_i$ and $\mathcal{I} \subset \{1,\ldots,n\}$, the second part of Proposition 1.6 applies. In combination with Proposition 1.9, we obtain the following characterization of r-continuous elementary functions:

**Proposition 1.11.** *Let $\varphi, \varphi_i$ be positive real-valued continuous functions on $X$ and let $a, a_i \in \mathcal{P}$.*

*(a) The elementary function $\varphi{\otimes}a$ is relatively continuous if and only if either the element $a \in \mathcal{P}$ is bounded or the set $\{x \in X \mid \varphi(x) > 0\}$ is both open and closed in $X$.*
*(b) The function $f = \sum_{i=1}^n \varphi_i{\otimes}a_i$ is r-continuous if and only if there is a disjoint partition of $X$ into segments which are both open and closed and such that the values of $f$ at points in the same segment are bounded relative to each other.*

*Examples 1.12.* Let $\mathcal{P}$ be the cone of all real-valued continuous functions on $\mathbb{R}$ which are uniformly bounded below, endowed with the pointwise algebraic operations and order. With the neighborhood system $\mathcal{V}$ consisting of all strictly positive constant functions in $\mathcal{P}$, then $(\mathcal{P},\mathcal{V})$ becomes a full locally convex cone. The function $f \in \mathcal{P}$ such that $f(t) = t^2$ for $t \in \mathbb{R}$ is obviously not bounded in $\mathcal{P}$. If we choose $X = [0,1)$ and the real-valued function

$\varphi(x) = x$ on $X$, then the $\mathcal{P}$-valued function $\varphi \otimes f$ is r-continuous on $(0,1)$ but not at $x = 0$. Continuity with respect to any of the given locally convex topologies on $\mathcal{P}$ fails at all points of $X$.

## 2. Cone-Valued Functions on Locally Compact Spaces

We shall further specialize the assumptions of the previous section. Throughout the following we shall assume that $X$ is a locally compact Hausdorff space. Let $\mathfrak{K}$ be the family of all compact subsets of $X$, and by $\mathfrak{K}_0$ the subfamily of $\mathfrak{K}$ consisting of those subsets of $X$ that are both open and compact. We shall denote the cone of all continuous positive real-valued functions on $X$ with compact support by $\mathcal{K}(X)$, and by $\mathcal{K}_0(X)$ the subcone of those functions $\varphi \in \mathcal{K}(X)$ such that $\mathrm{supp}^*(\varphi) \in \mathfrak{K}_0$. The latter notion is motivated by the observation in Proposition 1.11. Note that every $0 \neq \varphi \in \mathcal{K}_0(X)$ attains a minimal non-zero value. The functions in $\mathcal{K}_0(X)$ may alternatively be characterized as continuous $\mathbb{R}_+$-valued functions with compact support, where continuity is required with respect to the topology of $\overline{\mathbb{R}}_+$ as introduced in Example I.1.4(b). Throughout the following, let $\mathfrak{R}$ be the weak $\sigma$-ring of all relatively compact Borel subsets of $X$. Then the corresponding $\sigma$-algebra $\mathfrak{A}_{\mathfrak{R}}$, consisting of all sets $A \subset X$ such that $A \cap E \in \mathfrak{R}$ for all $E \in \mathfrak{R}$, is just the $\sigma$-algebra of all Borel subsets of $X$ (see Lemma 13.9 in [178]).

As before, $(\mathcal{P}, \mathcal{V})$ is a locally convex cone without any further requirements.

**2.1 Inductive Limit Topologies.** Recall from Section II.2.2 that an $\mathfrak{R}$-compatible inductive limit neighborhood is a convex subset $\mathfrak{v}$ of measurable functions in $\mathcal{F}(X, \overline{\mathcal{V}})$ such that for every $E \in \mathfrak{R}$ there is $v_E \in \mathcal{V}$ such that $\chi_E \otimes v_E \leq \mathfrak{v}$, that is $\chi_K \otimes v \leq s$ for some $s \in \mathfrak{v}$. An *inductive limit topology* on $\mathcal{F}(X, \mathcal{P})$ is generated by a system $\mathfrak{V}$ of $\mathfrak{R}$-compatible inductive limit neighborhoods, closed for addition and multiplication by strictly positive scalars and directed downward with respect to the order relation $\leq$ as defined in II.2.2. In this case $\mathfrak{R}$-compatibility means that for every compact set $K$ we have $\chi_K \otimes v \leq \mathfrak{v}$ for some $v \in \mathcal{V}$.

An $\mathfrak{R}$-compatible inductive limit neighborhood $\mathfrak{v}$ is called *r-lower, r-upper* or *r-continuous* if all its elements $s \in \mathfrak{v}$ are r-lower, r-upper or r-continuous $\overline{\mathcal{V}}$-valued functions, respectively. (Recall that $(\overline{\mathcal{V}}, \mathcal{V})$ itself is also a locally convex cone.) Correspondingly, an inductive limit topology $\mathfrak{V}$ on $\mathcal{F}(X, \mathcal{P})$ is called *r-lower, r-upper* or *r-continuous* if it contains a basis of *r-lower, r-upper* or *r-continuous* neighborhoods, respectively. For our further investigations, r-lower continuous topologies will be of particular interest. Considering this, we observe that for a non-negative $\overline{\mathbb{R}}$-valued lower semicontinuous function $\varphi$ and some $v \in \overline{\mathcal{V}}$, the neighborhood-valued function $\varphi \otimes v$ is r-lower continuous (see 1.8(a)). In this context, $0 \cdot v = 0$ and $+\infty \cdot v = \infty$. (Recall from II.2.1 that $\overline{\mathcal{V}}$ contains $0 \in \mathcal{P}$ as well as the maximal element $\infty$).

We record a wide variety of r-lower continuous inductive limit topologies, the finest of which is the standard inductive limit topology for functions on a locally compact space, consisting of all $\mathfrak{R}$-compatible inductive limit neighborhoods. This topology is r-lower continuous, as for every inductive limit neighborhood $\mathfrak{v}$ there is an r-lower continuous neighborhood $\mathfrak{u}$ such that $\mathfrak{u} \leq \mathfrak{v}$. Indeed, for every relatively compact open set $O \subset X$ choose $v_O \in \mathcal{V}$ such that $\chi_O \otimes v_O \leq s$ for some $s \in \mathfrak{v}$, and let $\mathfrak{u}$ consist of all convex combinations of the functions $\chi_O \otimes v_O$. If, for another example, the elements of $\mathfrak{V}$ are just the singleton sets containing the constant mappings $x \mapsto v$ for $v \in \mathcal{V}$, then $\mathfrak{V}$ generates the topology of uniform convergence. If, on the other hand, these singleton sets consist of mappings $x \mapsto v$ for $x \in K$ and $x \mapsto \infty$ else, for some $K \in \mathfrak{K}$, then we obtain the topology of compact convergence. If we use finite sets instead of compact ones in the last example, the topology of pointwise convergence emerges. All these topologies are obviously r-lower continuous. (See also Examples 2.11 below.)

**2.2 Functions that Vanish at Infinity.** Given an inductive limit topology, we shall say that a function $f \in \mathcal{F}(X, \mathcal{P})$ *vanishes at infinity (relative to a system $\mathfrak{V}$ of inductive limit neighborhoods)* if for every $\mathfrak{v} \in \mathfrak{V}$ there is $K \in \mathfrak{K}$ such that

$$\chi_{(X \setminus K)} \otimes f \leq \mathfrak{v} \qquad \text{and} \qquad 0 \leq \chi_{(X \setminus K)} \otimes f + \mathfrak{v}.$$

Lemma 1.4 shows that every r-continuous function $f$ that vanishes at infinity is bounded below relative to the neighborhoods in $\mathfrak{V}$. Indeed, for $\mathfrak{v} \in \mathfrak{V}$, choose the compact set $K \subset X$ from above. There is $v \in \mathcal{V}$ such that $\chi_K \otimes v \leq \mathfrak{v}$. Following 1.4 there is $\lambda \geq 0$ such that $0 \leq f(x) + \lambda v$ for all $x \in K$, that is $0 \leq \chi_K \otimes f + \lambda \mathfrak{v}$. This demonstrates

$$0 \leq \big(\chi_{(X \setminus K)} \otimes f + \mathfrak{v}\big) + \big(\chi_K \otimes f + \lambda \mathfrak{v}\big) = f + (\lambda + 1)\mathfrak{v}.$$

**2.3 The Cones $\mathcal{E}(X, \mathcal{P})$ and $\mathcal{E}_0(X, \mathcal{P})$.** We shall use $\mathcal{E}(X, \mathcal{P})$ to denote the subcone of $\mathcal{F}(X, \mathcal{P})$ consisting of all finite sums of elementary functions $\varphi \otimes a$, defined by an element $a \in \mathcal{P}$ and a continuous positive real-valued function $\varphi \in \mathcal{K}(X)$. Recall that for a function $\varphi \otimes a \in \mathcal{E}(X, \mathcal{P})$ to be r-continuous, it is necessary and sufficient that either the element $a \in \mathcal{P}$ is bounded or that $\varphi \in \mathcal{K}_0(X)$. We shall denote the subcone of $\mathcal{E}(X, \mathcal{P})$ generated by all r-continuous elementary functions by $\mathcal{E}_0(X, \mathcal{P})$. Note that $\mathcal{E}_0(X, \mathcal{P})$ does not necessarily contain all r-continuous functions in $\mathcal{E}(X, \mathcal{P})$, which had been characterized in Proposition 1.11(b).

**2.4 The Cones $\mathcal{F}_{\mathfrak{V}}(X, \mathcal{P})$ and $\mathcal{F}_{\mathfrak{V}_0}(X, \mathcal{P})$.** We denote the closure of $\mathcal{E}(X, \mathcal{P})$ in $\mathcal{F}(X, \mathcal{P})$ with respect to the symmetric relative topology generated by the inductive limit neighborhoods in $\mathfrak{V}$, by $\mathcal{F}_{\mathfrak{V}}(X, \mathcal{P})$. It consists of those functions $f \in \mathcal{F}(X, \mathcal{P})$ such that for every $\mathfrak{v} \in \mathfrak{V}$ and $\varepsilon > 0$ there is some $g \in \mathcal{E}(X,\mathcal{P})$ such that

$$f \leq \gamma g + \varepsilon\mathfrak{v} \qquad \text{and} \qquad g \leq \gamma' f + \varepsilon\mathfrak{v}$$

for some $1 \leq \gamma, \gamma' \leq 1+\varepsilon$; that is $f \in \mathfrak{v}^s_\varepsilon(g)$. Recall from Section 2 that the relative topologies of a locally convex cone are in general no longer locally convex cone topologies. These topologies are however still compatible with the addition and multiplication by positive scalars as shown in Lemma I.4.1(d). The closure $\mathcal{F}_{\mathfrak{V}}(X,\mathcal{P})$ of the subcone $\mathcal{E}(X,\mathcal{P})$ is therefore again a subcone of $\mathcal{F}(X,\mathcal{P})$. Moreover, since all functions in $\mathcal{F}_{\mathfrak{V}}(X,\mathcal{P})$ obviously vanish at infinity, hence are bounded below relative to the neighborhoods in $\mathfrak{V}$, we infer that $\big(\mathcal{F}_{\mathfrak{V}}(X,\mathcal{P}),\mathfrak{V}\big)$ forms a locally convex cone. Similarly, we shall denote the closure (with respect to the same topology) of $\mathcal{E}_0(X,\mathcal{P})$ in $\mathcal{F}(X,\mathcal{P})$ by $\mathcal{F}_{\mathfrak{V}_0}(X,\mathcal{P})$.

**2.5 The Cones $\mathcal{C}^r_{\mathfrak{V}}(X,\mathcal{P})$ and $\mathcal{C}^r_{\mathfrak{V}_0}(X,\mathcal{P})$.** We are particularly interested in the respective subcones of r-continuous functions in $\mathcal{F}_{\mathfrak{V}}(X,\mathcal{P})$ and $\mathcal{F}_{\mathfrak{V}_0}(X,\mathcal{P})$ which we shall denote by $\mathcal{C}^r_{\mathfrak{V}}(X,\mathcal{P})$ and $\mathcal{C}^r_{\mathfrak{V}_0}(X,\mathcal{P})$, respectively. A complete characterization of their elements appears to be difficult to achieve, due to the generality of our approach and the wide variety of choices for inductive limit topologies. The following Theorem 2.6, however, provides a rather general and powerful criterion for an r-continuous function to belong to either of these cones. Recall that $\mathcal{B}$ denotes the subcone of all bounded elements of $\mathcal{P}$.

**Theorem 2.6.** *Let $f \in \mathcal{C}^r(X,\mathcal{P})$ and let $S = \{x \in X \mid f(x) \in \mathcal{B}\}$.*

*(a) If for every $\mathfrak{v} \in \mathfrak{V}$ there is $K \in \mathfrak{K}$ such that $\partial K \subset S^\circ$ and such that $\chi_{(X\setminus K)} \circledast f \leq \mathfrak{v}$ and $0 \leq \chi_{(X\setminus K)} \circledast f + \mathfrak{v}$, then $f \in \mathcal{C}^r_{\mathfrak{V}}(X,\mathcal{P})$.*

*(b) If for every $\mathfrak{v} \in \mathfrak{V}$ there are $K \in \mathfrak{K}$ and a subset $K_0 \in \mathfrak{K}_0$ of $K$ such that $K \setminus K_0 \subset S^\circ$, and such that $\chi_{(X\setminus K)} \circledast f \leq \mathfrak{v}$ and $0 \leq \chi_{(X\setminus K)} \circledast f + \mathfrak{v}$, then $f \in \mathcal{C}^r_{\mathfrak{V}_0}(X,\mathcal{P})$.*

*Proof.* We shall proceed in several steps. First we observe that the condition of Part (b) implies the condition of Part (a). Indeed, $K_0 \subset K$ and $K_0 \in \mathfrak{K}_0$ implies that $K_0 \subset K^\circ$, hence

$$\partial K = K \setminus K^\circ \subset K \setminus K_0 \subset S^\circ,$$

as required in (a). Next we consider a few special cases for functions in $\mathcal{C}^r(X,\mathcal{P})$.

(i) In a first step we assume that the function $f \in \mathcal{C}^r(X,\mathcal{P})$ has a compact support $K \in \mathfrak{K}$. Let $\mathfrak{v} \in \mathcal{V}$ and $\varepsilon > 0$. Let $U$ be a relatively compact open set containing $K$ and let $v \in \mathcal{V}$ such that $\chi_U \circledast v \leq \mathfrak{v}$. Following Lemma 1.4 there is $\lambda \geq 0$ such that $0 \leq f(x) + \lambda v$ for all $x \in X$. We proceed to employ a partition of the unit for the compact set $K$. Let $\varepsilon' = \frac{\varepsilon}{1+\lambda}$ and $\gamma = 1+\varepsilon'$. As $f$ is r-continuous, there are open subsets $O_1,\ldots,O_n \in \mathfrak{R}$

of $U$ whose union covers $K$, and such that $f(y) \in (v_{\varepsilon'})(f(x))$, that is (using Lemma I.4.1(b))

$$f(y) \leq \gamma f(x) + \varepsilon v$$

whenever $x, y \in O_i$ for some $i = 1, \ldots, n$. There is a corresponding set of functions $\varphi_1, \ldots, \varphi_n \in \mathcal{K}(X)$ such that $\operatorname{supp}(\varphi_i) \subset O_i$, and for $\varphi = \sum_{i=1}^n \varphi_i$ we have $0 \leq \varphi(x) \leq 1$ for all $x \in X$, and $\varphi(x) = 1$ for all $x \in K$. We choose $a_i = f(x_i)$ for some $x_i \in O_i$. All the functions $\varphi_{i \otimes} a_i$ are in $\mathcal{E}(X, \mathcal{P})$, and if all the values of $f$ are bounded in $\mathcal{P}$, then these functions are even contained in $\mathcal{E}_0(X, \mathcal{P})$. We observe that

$$\varphi_i(x) a_i \leq \varphi_i(x) \gamma f(x) + \varphi_i(x) \varepsilon v \qquad \text{and} \qquad \varphi_i(x) f(x) \leq \varphi_i(x) \gamma a_i + \varphi_i(x) \varepsilon v$$

holds for all $x \in X$ and set $g = \sum_{i=1}^n \varphi_{i \otimes} a_i \in \mathcal{E}(X, \mathcal{P})$. If in addition all the values of $f$ are bounded, then $g$ is contained in $\mathcal{E}_0(X, \mathcal{P})$. The above yields

$$g(x) \leq \varphi(x) \gamma f(x) + \varphi(x) \varepsilon v \qquad \text{and} \qquad \varphi(x) f(x) \leq \gamma g(x) + \varphi(x) \varepsilon v$$

for all $x \in X$, hence

$$g(x) \leq \gamma f(x) + \varepsilon v \qquad \text{and} \qquad f(x) \leq \gamma g(x) + \varepsilon v$$

for all $x \in K$. For $x \in U \setminus K$ we have $f(x) = 0$ and $0 \leq \varphi(x) \leq 1$, hence by the above

$$g(x) \leq \varepsilon v = \gamma f(x) + \varepsilon v \qquad \text{and} \qquad f(x) = 0 \leq \gamma g(x) + \varepsilon v.$$

As $f(x) = g(x) = 0$ for all $x \notin U$, this yields

$$g \leq \gamma f + \varepsilon\, \chi_{U \otimes} v \leq \gamma f + \varepsilon \mathfrak{v} \qquad \text{and} \qquad f \leq \gamma g + \varepsilon\, \chi_{U \otimes} v \leq \gamma g + \varepsilon \mathfrak{v},$$

thus $f \in \mathfrak{v}_\varepsilon(g)$ and $g \in \mathfrak{v}_\varepsilon(f)$. As $g \in \mathcal{E}(X, \mathcal{P})$ and as $\mathfrak{v} \in \mathfrak{V}$ and $\varepsilon > 0$ were arbitrarily chosen, this yields $f \in \mathcal{C}^r_{\mathfrak{V}}(X, \mathcal{P})$.

(ii) If the $f \in \mathcal{C}^r(X, \mathcal{P})$ has a compact support and if in addition all of its values are bounded, then the function $g$ from step (i) was indeed seen to be an element of $\mathcal{E}_0(X, \mathcal{P})$. This yields $f \in \mathcal{C}^r_{\mathfrak{V}_0}(X, \mathcal{P})$.

(iii) In the next case, let $K \in \mathfrak{K}_0$ and $v \in \mathcal{V}$, and let $f \in \mathcal{C}^r(X, \mathcal{P})$ be a function whose support is contained in $K$ and whose values at points in $K$ are $v$-bounded relative to each other. Let $0 < \varepsilon \leq 1$. A simple compactness argument shows that we can find $\lambda \geq 0$ such that $f(x) \in v_\lambda(f(y))$ for all $x, y \in K$. Following Lemma 1.4 we may also assume that $0 \leq f(x) + \lambda v$ holds for all $x \in K$. By Lemma I.4.1(b) this implies

$$f(x) \leq (1 + \lambda)(f(y) + \lambda v)$$

for all $x, y \in K$. We set $\varepsilon' = \frac{\varepsilon}{2(\lambda+1)}$, $\gamma = 1 + \varepsilon'$ and as in (i) choose an open cover by subsets $O_1, \ldots, O_n$ of $K$ such that $f(y) \in v_{\varepsilon'}(f(x))$, that is following Lemma I.4.1(b)

$$f(y) \leq (1+\varepsilon')f(x) + \varepsilon'(1+\lambda)v = \gamma f(x) + \frac{\varepsilon}{2}v.$$

whenever $x, y \in O_i$ for some $i = 1, \dots, n$. Let $\varphi_1, \dots, \varphi_n \in \mathcal{K}(X)$ be the corresponding partition of the unit, that is $\operatorname{supp}(\varphi_i) \subset O_i$, and for $\varphi = \sum_{i=1}^n \varphi_i$ we have $0 \leq \varphi(x) \leq 1$ for all $x \in X$, and $\varphi(x) = 1$ for all $x \in K$. We choose $a_i = f(x_i)$ for some $x_i \in O_i$. But as the elements $a_i$ are not necessarily bounded in $\mathcal{P}$, the functions $\varphi_{i} \otimes a_i$ need not be r-continuous at points $x$ where $\varphi_i(x) = 0$. We therefore use the functions $\psi_i = \varphi_i + \varepsilon'' \chi_K$ instead, where $\varepsilon'' = \frac{\varepsilon'}{2n(1+\lambda)}$. Because $\operatorname{supp}^*(\psi_i) = K \in \mathfrak{K}_0$, we have $\psi_i \in \mathcal{K}_0(X)$, hence the functions $\psi_{i} \otimes a_i$ are r-continuous and contained in $\mathcal{E}_0(X, \mathcal{P})$. For each $i = 1, \dots, n$ we observe that

$$a_i \leq \gamma f(x) + \frac{\varepsilon}{2}v \qquad \text{and} \qquad f(x) \leq \gamma a_i + \frac{\varepsilon}{2}v$$

for all $x \in O_i$ and

$$a_i \leq (1+\lambda)(f(x) + \lambda v) \qquad \text{and} \qquad 0 \leq a_i + \lambda v$$

for all $x \in K$. Therefore, as $\operatorname{supp}(\varphi_i) \subset O_i$, we have

$$\psi_i(x)a_i = \varphi_i(x)a_i + \varepsilon'' a_i \leq \varphi_i(x)\left(\gamma f(x) + \frac{\varepsilon}{2}v\right) + \frac{\varepsilon'}{2n}\left(f(x) + \lambda v\right)$$

and

$$\varphi_i(x)f(x) \leq \varphi_i(x)\left(\gamma a_i + \frac{\varepsilon}{2}v\right) + \gamma\varepsilon''(a_i + \lambda v) = \gamma\psi_i(x)a + \left(\varphi_i(x)\frac{\varepsilon}{2} + \gamma\lambda\varepsilon''\right)v$$

for all $x \in K$. Next we consider the function $g = \sum_{i=1}^n \psi_{i} \otimes a_i \in \mathcal{E}_0(X, \mathcal{P})$. As $\sum_{i=1}^n \varphi_i(x) = 1$ for all $x \in K$, we obtain from the above

$$g(x) \leq (\gamma+\varepsilon')f(x) + \left(\frac{\varepsilon}{2} + \frac{\varepsilon'\lambda}{2}\right)v \qquad \text{and} \qquad f(x) \leq \gamma g(x) + \left(\frac{\varepsilon}{2} + n\gamma\lambda\varepsilon''\right)v$$

for all $x \in K$, and even for all $x \in X$, as $f(x) = g(x) = 0$ for all $x \in X \setminus K$. Next we observe that both

$$\frac{\varepsilon'\lambda}{2} \leq \frac{\varepsilon}{2} \qquad \text{and} \qquad n\gamma\lambda\varepsilon'' = \frac{\gamma\lambda\varepsilon'}{2(1+\lambda)} \leq \frac{\gamma\varepsilon'}{2} \leq \varepsilon' \leq \frac{\varepsilon}{2}.$$

With $1 \leq \gamma' = \gamma + \varepsilon' = 1 + 2\varepsilon' \leq 1 + \varepsilon$, this shows

$$g \leq \gamma' f + \varepsilon \chi_{K} \otimes v \qquad \text{and} \qquad f \leq \gamma g + \varepsilon \chi_{K} \otimes v.$$

(iv) Now let $K \in \mathfrak{K}_0$ and let $f \in \mathcal{C}^r(X, \mathcal{P})$ be a function whose support is contained in $K$. Let $\mathfrak{v} \in \mathfrak{V}$ and let $0 < \varepsilon \leq 1$. There is $v \in \mathcal{V}$ such that $\chi_{K\otimes}v \le \mathfrak{v}$, and following the remark after Proposition 1.5, $f$ can be expressed as the sum of finitely many functions $f_i \in \mathcal{C}^r(X,\mathcal{P})$, for $i = 1,\ldots,n$, whose support is contained in subsets $K_i \in \mathfrak{K}_0$ of $K$ respectively, and such that the values of $f_i$ at points in $K_i$ are $v$-bounded relative to each other. Let $\lambda \ge 0$ such that $0 \le f_i + \lambda\mathfrak{v}$ for all $i = 1,\ldots,n$. Now we apply step (iii) to each of the functions $f_i$ with the neighborhood $v$ and $\varepsilon' = \frac{\varepsilon}{n(1+\lambda)}$ instead of $\varepsilon$ : We find $g_i \in \mathcal{E}_0(X,\mathcal{P})$ such that

$$g_i \le \gamma_i' f_i + \varepsilon' \chi_{K_i\otimes}v \qquad \text{and} \qquad f_i \le \gamma_i g_i + \varepsilon' \chi_{K_i\otimes}v$$

with $1 \le \gamma_i, \gamma_i' \le 1+\varepsilon'$, that is $g_i \in \mathfrak{v}_{\varepsilon'}(f_i)$ and $f_i \in \mathfrak{v}_{\varepsilon'}(g_i)$ for all $i = 1,\ldots,n$. We set $g = g_1 + \ldots + g_n \in \mathcal{E}_0(X,\mathcal{P})$. Then Lemma I.4.1(d) yields indeed $g \in \mathfrak{v}_\varepsilon(f)$ and $f \in \mathfrak{v}_\varepsilon(g)$. Because this may be obtained for any choice of $\mathfrak{v} \in \mathcal{V}$ and $\varepsilon \ge 0$, we conclude that $f \in \mathcal{C}^r_{\mathfrak{V}_0}(X,\mathcal{P})$.

(v) Now we consider the case of a function $f \in \mathcal{C}^r(X,\mathcal{P})$ that fulfills the assumptions of Part (a). Let $\mathfrak{v} \in \mathfrak{V}$ and $\varepsilon > 0$. There is $\lambda \ge 0$ such that $0 \le f + \lambda\mathfrak{v}$. We set $\varepsilon' = \frac{\varepsilon}{2(1+\lambda)}$ and corresponding to the neighborhood $\varepsilon'\mathfrak{v}$ choose the set $K \in \mathfrak{K}$ with the assumed properties. There is a relatively compact open set $O \subset S$ containing $\partial K$. Then $K$ is contained in the relatively compact open set $U = O \cup K^\circ$. Thus we find a continuous real-valued function $\varphi$ on $X$ such that $0 \le \varphi(x) \le 1$ for all $x \in X$, such that the support of $\varphi$ is contained in $U$ and $\varphi(x) = 1$ for all $x \in K$. We set $f_1 = \varphi_\otimes f$ and $f_2 = (1-\varphi)_\otimes f$ and use Proposition 1.8 to verify that both functions are r-continuous. Indeed, $f_1$ is r-continuous at all points $x \in X$ where $\varphi(x) > 0$ or $f(x)$ is bounded. If $\varphi(x) = 0$ and $f(x)$ is unbounded, then we have neither $x \in K$ nor $x \in O$. Thus $x \notin U$, hence $x \notin \operatorname{supp}(\varphi)$. Thus $\varphi(y) = 0$ throughout a neighborhood of $x$, and $f_1$ is continuous at $x$ as well. Similarly, the function $f_2$ is r-continuous at all points $x \in X$ where $\varphi(x) < 1$ or $f(x)$ is bounded. But $\varphi(x) = 1$ implies that $x \in U$, and if $f(x)$ is unbounded, even $x \in K^\circ$. Therefore $\varphi(y) = 1$ throughout a neighborhood of $x$, and $f_2$ is continuous at $x$ as well. Using this, we observe that the function $f_1$ is r-continuous and has a compact support. It is therefore contained in $\mathcal{C}^r_{\mathfrak{V}}(X,\mathcal{P})$, as shown in step (i). Thus there is $g \in \mathcal{E}(X,\mathcal{P})$ such that $f_1 \in \mathfrak{v}^s_{\varepsilon'}(g)$. The function $f_2$, on the other hand, vanishes on $K$, hence $f_2 \le \varepsilon'\mathfrak{v}$ and $0 \le f_2 + \varepsilon'\mathfrak{v}$, that is $f_2 \in \mathfrak{v}^s_{\varepsilon'}(0)$ by our assumption. Finally, as $f = f_1 + f_2$, Lemma I.4.1(d) yields

$$f \in \mathfrak{v}^s_{2\varepsilon'(1+\lambda)}(g) = \mathfrak{v}^s_\varepsilon(g).$$

Because $v \in \mathcal{V}$ and $\varepsilon > 0$ were arbitrary, this shows $f \in \mathcal{C}^r_{\mathfrak{V}}(X,\mathcal{P})$.

(vi) Finally, we consider the case of a function $f \in \mathcal{C}^r(X,\mathcal{P})$ that fulfills the assumptions of Part (b). We shall adapt some of the arguments from step (v). Let $\mathfrak{v} \in \mathfrak{V}$ and $\varepsilon > 0$. There is $\lambda \ge 0$ such that $0 \le f + \lambda\mathfrak{v}$. We set $\varepsilon' = \frac{\varepsilon}{3(1+\lambda)}$ and corresponding to the neighborhood $\varepsilon'\mathfrak{v}$ choose the set $K \in \mathfrak{K}$ and the subset $K_0 \in \mathfrak{K}_0$ of $K$ with the assumed properties. Because $K_0$ is both open and closed, both functions $f_1 = \chi_{(X\setminus K_0)}\otimes f$ and $f_2 = \chi_{K_0}\otimes f$ are r-continuous. The support of $f_2$ is contained in $K_0 \in \mathfrak{K}_0$, thus $f_2 \in \mathcal{C}^r_{\mathfrak{V}_0}(X,\mathcal{P})$ as was shown in step (iv). For the function $f_1$ we shall argue as follows: There is a relatively compact open set $U$ in $S$ containing the compact set $K' = K\setminus K_0$. Thus we find $\varphi \in \mathcal{K}(X)$ such that $0 \le \varphi(x) \le 1$ for all $x \in X$, the support of $\varphi$ is contained in $U$ and $\varphi(x) = 1$ for all $x \in K'$. We set $f_1' = \varphi\otimes f_1$ and $f_1'' = (1-\varphi)\otimes f_1$ and as in step (v) use Proposition 1.8 to verify that both functions are r-continuous. The function $f_1'$ is r-continuous, has bounded values and compact support and following step (ii) is therefore contained in $\mathcal{C}^r_{\mathfrak{V}_0}(X,\mathcal{P})$. The function $f_1''$ vanishes on $K'$ and on $K_0$, hence on $K$, and $f_1'' \le \varepsilon'\mathfrak{v}$ and $0 \le f_1'' + \varepsilon'\mathfrak{v}$, that is $f_2'' \in \mathfrak{v}^s_{\varepsilon'}(0)$ holds by our assumption. As $f_1', f_2 \in \mathcal{C}^r_{\mathfrak{V}_0}(X,\mathcal{P})$, there are $g_1, g_2 \in \mathcal{E}_0(X,\mathcal{P})$ such that

$$f_1' \in \mathfrak{v}^s_{\varepsilon'}(g_1) \qquad \text{and} \qquad f_2 \in \mathfrak{v}^s_{\varepsilon'}(g_2).$$

Finally, as $f = f_1' + f_1'' + f_2$, Lemma I.4.1(d) yields with $g = g_1 + g_2 \in \mathcal{E}_0(X,\mathcal{P})$

$$f \in \mathfrak{v}^s_{\varepsilon' 3(1+\lambda)}(g) = \mathfrak{v}^s_{\varepsilon}(g).$$

Because $v \in \mathcal{V}$ and $\varepsilon > 0$ were arbitrary, this shows $f \in \mathcal{C}^r_{\mathfrak{V}_0}(X,\mathcal{P})$. □

**Corollary 2.7.** *Suppose that the function* $f \in \mathcal{C}^r(X,\mathcal{P})$ *vanishes at infinity.*

(a) *If there is* $K \in \mathfrak{K}$ *such that* $f$ *takes only bounded values on* $X \setminus K$, *then* $f \in \mathcal{C}^r_{\mathfrak{V}}(X,\mathcal{P})$.

(b) *If there is* $K_0 \in \mathfrak{K}_0$ *such that* $f$ *takes only bounded values on* $X \setminus K_0$, *then* $f \in \mathcal{C}^r_{\mathfrak{V}_0}(X,\mathcal{P})$.

*Proof.* (a) Let $f \in \mathcal{C}^r(X,\mathcal{P})$ and $K \in \mathfrak{K}$ be as stated, that is $X\setminus K \subset S = \{x \in X \mid f(x) \in \mathcal{B}\}$. For the criterion in Theorem 2.6(a), given $\mathfrak{v} \in \mathfrak{V}$, as the function $f$ vanishes at infinity, there is $K_1 \in \mathfrak{K}$ such that $\chi_{(X\setminus K_1)}\otimes f \le \mathfrak{v}$ and $0 \le \chi_{(X\setminus K_1)}\otimes f + \mathfrak{v}$, Indeed, we may choose $K_1$ such that $K$ is contained in its interior $K_1^\circ$. As $X \setminus K \subset S^\circ$ in this case, we have $\partial K_1 = K_1 \setminus K_1^\circ \subset X \setminus K \subset S^\circ$ as required for the set $K_1 \in \mathfrak{K}$ in 2.6(a).

(b) For the criterion of Theorem 2.6(b) we have $X \setminus K_0 \subset S^\circ$ and may choose the set $K \in \mathfrak{K}$ from the criterion such that $K_0 \subset K$. Then $K\setminus K_0 \subset S^\circ$ as required. □

In some special cases the criteria of Theorem 2.6 and Corollary 2.7 can be simplified or are even necessary for a function $f \in \mathcal{C}^r(X,\mathcal{P})$ to belong to $\mathcal{C}^r_{\mathfrak{V}_0}(X,\mathcal{P})$ or $\mathcal{C}^r_{\mathfrak{V}}(X,\mathcal{P})$, respectively.

**Corollary 2.8.** *If either* $(\mathcal{P},\mathcal{V})$ *is locally connected in the symmetric relative topology or if the range of the function* $f \in \mathcal{C}^r(X,\mathcal{P})$ *is covered by finitely many global boundedness components of* $\mathcal{P}$, *then the conditions on* $f$ *in* 2.7(a) *and* 2.7(b) *are equivalent.*

*Proof.* In this case, Proposition 1.6 provides a partition of $X$ into open and closed components such that the values of $f$ at points in the same component are globally bounded relative to each other. In particular, the set $S = \{x \in X \mid f(x) \in \mathcal{B}\}$ is both open and closed in $X$. Condition 2.7(a) states that its complement $K_0 = X \setminus S$ is relatively compact, hence in $\mathfrak{K}_0$, since it is both open and closed. Thus the criterion in 2.7(b) holds as well. □

We shall say that an inductive limit neighborhood $\mathfrak{v}$ *yields boundedness at infinity* if $f \leq g + \mathfrak{v}$ for functions $f, g \in \mathcal{F}(X, \mathcal{P})$ implies that there is $K \in \mathfrak{K}$ such that $f(x) \in \mathcal{B}\big(g(x)\big)$ holds for all $x \in X \setminus K$.

**Corollary 2.9.** *If* $\mathfrak{V}$ *contains a neighborhood which yields boundedness at infinity, then the condition in* 2.7(b) *is also necessary for a function to belong to* $\mathcal{C}^r_{\mathfrak{V}}(X, \mathcal{P})$. *If in addition,* $(\mathcal{P}, \mathcal{V})$ *is locally connected in the symmetric relative topology, then* $\mathcal{C}^r_{\mathfrak{V}_0}(X, \mathcal{P})$ *and* $\mathcal{C}^r_{\mathfrak{V}}(X, \mathcal{P})$ *coincide.*

*Proof.* Suppose that $\mathfrak{V}$ contains a neighborhood $\mathfrak{v}_0$ which yields boundedness at infinity. We shall verify that every $f \in \mathcal{C}^r_{\mathfrak{V}}(X, \mathcal{P})$ satisfies 2.7(b). Obviously, $f$ is continuous and vanishes at infinity. For the neighborhood $\mathfrak{v}_0 \in \mathfrak{V}$ and $\varepsilon = 1$ there is a function $g \in \mathcal{E}(X, \mathcal{P})$ such that $f \in (\mathfrak{v}_0)^s_\varepsilon(g)$, that is

$$f \leq \gamma g + \mathfrak{v}_0 \qquad \text{and} \qquad g \leq \gamma' f + \mathfrak{v}_0$$

for some $1 \leq \gamma, \gamma' \leq 2$. Thus there is $K \in \mathfrak{K}$ such that $f(x) \in \mathcal{B}\big(g(x)\big)$ and $g(x) \in \mathcal{B}\big(f(x)\big)$ for all $x \in X \setminus K$. The function $g \in \mathcal{E}(X, \mathcal{P})$ has a compact support $K_g \in \mathfrak{K}$. From this we infer that $f$ takes only bounded values on $X \setminus (K \cup K_g)$ and therefore satisfies 2.7(a).

If in addition, $(\mathcal{P}, \mathcal{V})$ is locally connected in the symmetric relative topology, then Conditions 2.7(a) and 2.7(b) are equivalent by 2.8 and imply that a function $f \in \mathcal{C}^r(X, \mathcal{P})$ that vanishes at infinity and fulfills these criteria is contained in $\mathcal{C}^r_{\mathfrak{V}_0}(X, \mathcal{P}) \subset \mathcal{C}^r_{\mathfrak{V}}(X, \mathcal{P})$. Every $f \in \mathcal{C}^r_{\mathfrak{V}}(X, \mathcal{P})$, on the other hand satisfies 2.7(a) by our preceding argument. □

We shall consider some further special cases in the following remarks and examples.

*Remarks 2.10.* (a) If all elements of the locally convex cone $(\mathcal{P}, \mathcal{V})$ are bounded, then we infer from Corollary 2.7 that the cones $\mathcal{C}^r_{\mathfrak{V}_0}(X, \mathcal{P})$ and $\mathcal{C}^r_{\mathfrak{V}}(X, \mathcal{P})$ coincide and consist of all r-continuous (hence continuous) functions that vanish at infinity.

(b) Proposition 1.9 and Corollary 2.7(b) imply in particular that all r-continuous functions in $\mathcal{E}(X, \mathcal{P})$ are contained in $\mathcal{C}^r_{\mathfrak{V}_0}(X, \mathcal{P})$. Indeed, for an r-continuous function $f \in \mathcal{E}(X, \mathcal{P})$, the set $K_0 = \{x \in X \mid f(x) \notin \mathcal{B}\}$ is both open and closed by 1.9. As $f$ has a compact support, that is $K_0 \in \mathfrak{K}_0$, Corollary 2.7(b) yields $f \in \mathcal{C}^r_{\mathfrak{V}_0}(X, \mathcal{P})$.

(c) If the space $X$ is indeed compact, then $\mathcal{C}^r_{\mathfrak{V}_0}(X, \mathcal{P}) = \mathcal{C}^r_{\mathfrak{V}}(X, \mathcal{P}) = \mathcal{C}^r(X, \mathcal{P})$. Indeed, in Theorem 2.6(b) we may choose $K = K_0 = X$, and our criterion holds trivially for every $f \in \mathcal{C}^r(X, \mathcal{P})$.

(d) If the locally compact space $X$ is not compact but connected, then the criterion in Theorem 2.6(b) requires that $K_0 = \emptyset$, that is $K \subset S$. Following Corollary 1.7 a function $f \in \mathcal{C}^r(X, \mathcal{P})$ therefore fulfills this criterion if it vanishes at infinity and all of its values are bounded elements of $\mathcal{P}$. Depending on the choice for the inductive limit topology $\mathfrak{V}$, this requirement may however not be a necessary for $f$ to belong to $\mathcal{C}^r_{\mathfrak{V}}(X, \mathcal{P})$.

(e) If $X$ is a discrete space (for example $X = \mathbb{N}$), then we may choose $K_0 = K$ in the criterion of Theorem 2.6(b). $\mathcal{C}^r_{\mathfrak{V}_0}(X, \mathcal{P})$ and $\mathcal{C}^r_{\mathfrak{V}}(X, \mathcal{P})$ therefore coincide and consist of all functions in $\mathcal{C}^r(X, \mathcal{P})$ that vanish at infinity.

*Examples 2.11.* (a) We obtain the finest inductive limit topology if $\mathfrak{V}$ consists of all $\mathfrak{R}$-compatible inductive limit neighborhoods. This topology was seen to be r-lower continuous (see 2.1). Theorem 2.6 provides sufficient criteria for a function to belong to $\mathcal{C}^r_{\mathfrak{V}_0}(X, \mathcal{P})$ or $\mathcal{C}^r_{\mathfrak{V}}(X, \mathcal{P})$. If the locally compact space $X$ is $\sigma$-compact, that is the union of a sequence of compact sets, then we can obtain a (nearly) complete description of the cone $\mathcal{C}^r_{\mathfrak{V}}(X, \mathcal{P})$ in this case: It follows from 2.7(a) that every r-continuous function with compact support is contained in $\mathcal{C}^r_{\mathfrak{V}}(X, \mathcal{P})$. On the other hand we shall verify that for every $f \in \mathcal{C}^r_{\mathfrak{V}}(X, \mathcal{P})$ there is $K \in \mathfrak{K}$ such that $f(x) \approx 0$, that is $f(x) \preccurlyeq 0$ and $0 \preccurlyeq f(x)$, holds for all $x \in X \backslash K$. Indeed, a simple argument (see Proposition 1, Section 8.5 in [178]) shows that there is an increasing sequence of relatively compact open subsets $Y_n$ of $X$ such that $X = \bigcup_{i=1}^{\infty} Y_n$. Let us assume to the contrary of our claim that there is a sequence $x_n \in Y_{n+1} \setminus Y_n$ such that $f(x_n) \not\approx 0$. Then there is a corresponding sequence $(v_n)_{n \in \mathbb{N}}$ in $\mathcal{V}$ such that either $f(x_n) \not\leq v_n$ or $0 \not\leq f(x_n) + v_n$. Now we consider the inductive limit neighborhood $\mathfrak{v} = \{s\}$ containing the single function $s$ such that $s(x) = v_n$ for $x \in Y_{n+1} \setminus Y_n$. Because every compact subset $K$ of $X$ is contained in the union of finitely many of the open sets $Y_i$, hence $\chi_{(X \backslash K)} \circledast f \leq \mathfrak{v}$ and $0 \leq \chi_{(X \backslash K)} \circledast f + \mathfrak{v}$ can not hold for any $K \in \mathfrak{K}$. Thus $f$ does not vanish at infinity and is therefore not contained in $\mathcal{C}^r_{\mathfrak{V}}(X, \mathcal{P})$.

(b) The topology of uniform convergence is generated by the neighborhood system $\mathfrak{V}$ whose elements are just the singleton sets $\mathfrak{v}_{(v)}$, corresponding to a neighborhood $v \in \mathcal{V}$, and containing the constant mapping $x \mapsto v$ for $v \in \mathcal{V}$.

(c) The topology of compact convergence is generated by the neighborhood system $\mathfrak{V}$ whose elements are all singleton sets $\mathfrak{v}_{(v,K)}$, corresponding to a neighborhood $v \in \mathcal{V}$ and a compact subset $K \subset X$, and containing the mapping $x \mapsto v$ for $x \in K$ and $x \mapsto \infty$ else. This topology was also seen to be r-lower continuous.

(d) The topology of pointwise convergence is generated by a neighborhood system $\mathfrak{V}$ whose elements are all singleton sets $\mathfrak{v}_{(v,Y)}$, corresponding to a neighborhood $v \in \mathcal{V}$ and a finite subset $Y \subset X$, and containing the mapping $x \mapsto v$ for $x \in Y$ and $x \mapsto \infty$ else. It is r-lower continuous. In this case $\mathcal{C}^r_{\mathfrak{V}}(X, \mathcal{P}) = \mathcal{C}^r(X, \mathcal{P})$, as for a given neighborhood $\mathfrak{v} \in \mathfrak{V}$ corresponding to a finite subset $Y \subset X$ we may choose $K = K_0 = Y \in \mathfrak{K}_0$ in the criterion of Theorem 2.6(b).

(e) A family $\Omega$ of non-negative real-valued upper semicontinuous functions on $X$ is called a *family of weights* (see [136] and [155]) if for all $\omega_1, \omega_2 \in \Omega$ there are $\omega_3 \in \Omega$ and $\rho > 0$ such that $\omega_1 \leq \rho\,\omega_3$ and $\omega_2 \leq \rho\,\omega_3$. We obtain an r-lower continuous inductive limit topology in the sense of 2.1 in the following way: For $v \in \mathcal{V}$ and $\omega \in \Omega$ we set $s_{\omega,v}(x) = \infty$ if $\omega(x) = 0$ and $s_{\omega,v}(x) = \big(1/\omega(x)\big)\, v$ else. Thus $\mathfrak{V}_\Omega = \big\{\mathfrak{v}_{\omega,v} = \{s_{\omega,v}\} \mid v \in \mathcal{V},\ \omega \in \Omega\big\}$ forms a basis for an inductive limit topology. For functions $f, g \in \mathcal{F}(X, \mathcal{P})$ we have

$$f \leq g + \mathfrak{v}_{\omega,v} \qquad \text{if} \qquad \omega(x) f(x) \leq \omega(x) g(x) + v \quad \text{for all } x \in X.$$

The family of weight functions may be chosen in a way which yields the preceding Examples 2.11(b), (c) and (d). Further examples can be found in [155].

(f) Let $X = \mathbb{N}$ with the discrete topology and $\mathcal{P} = \overline{\mathbb{R}}$. For $1 \leq p \leq \infty$, let the inductive limit neighborhood $\mathfrak{v}_p$ consist of all positive real-valued sequences $(\alpha_n)_{n\in\mathbb{N}}$ in the unit ball of the sequence space $l^p$. Let $\mathcal{V} = \{\varepsilon\mathfrak{v}_p \mid \varepsilon > 0\}$. Then $\mathcal{C}^r_{\mathfrak{V}_0}(\mathbb{N}, \overline{\mathbb{R}}) = \mathcal{C}^r_{\mathfrak{V}}(\mathbb{N}, \overline{\mathbb{R}})$ consists of all sequences $(x_n)_{n\in\mathbb{N}}$ in $\overline{\mathbb{R}}$ such that for every $\varepsilon > 0$ there is $n_0 \in \mathbb{N}$ such that $|x_n| \leq \alpha_n$ for all $n \geq n_0$ with some sequence $(\alpha_n)_{n\in\mathbb{N}} \in \varepsilon\mathfrak{v}_p$. The latter means that $\big(\sum_{n_0}^{\infty} |x_n|^p\big)^{(1/p)} \leq \varepsilon$ for $p < \infty$ and $\sup_{n \geq n_0} |x_n| \leq \varepsilon$ for $p = \infty$. Apart from taking the value $+\infty$ finitely many times, $(x_n)_{n\in\mathbb{N}}$ itself is therefore a sequence in $l^p$.

## 3. Continuous Linear Operators on Cones of Functions

As before, let $X$ be a locally compact Hausdorff space, $(\mathcal{P}, \mathcal{V})$ a locally convex cone, and let $\mathfrak{V}$ be a system of inductive limit neighborhoods for $\mathcal{F}(X, \mathcal{P})$ as introduced in the preceding section. Let $(\mathcal{Q}, \mathcal{W})$ be a locally convex complete lattice cone. We shall proceed to investigate continuous linear operators defined on $(\mathcal{F}_{\mathfrak{V}}(X, \mathcal{P}), \mathfrak{V})$ or any of the subcones introduced in Section 2, into $(\mathcal{Q}, \mathcal{W})$. We are particularly interested in extensions of these operators. If in addition $(\mathcal{Q}, \mathcal{W})$ is a full cone, if all elements of $\mathcal{Q}$ other than $+\infty$ are invertible, and $\mathcal{W} =$ consists of the (strictly) positive multiples of a single neighborhood $w \in \mathcal{W}$, then a Hahn-Banach type argument (see Theorem I.5.55) yields that every continuous linear operator on a subcone of $\mathcal{F}_{\mathfrak{V}}(X, \mathcal{P})$ can be extended to a continuous linear operator on $\mathcal{F}_{\mathfrak{V}}(X, \mathcal{P})$. For more general ranges $(\mathcal{Q}, \mathcal{W})$, however, this need no longer be the case. We proceed to investigate a few further cases were such extensions are possible.

First, as an immediate consequence of Theorem I.5.56 we obtain:

**Theorem 3.1.** *Let $\mathfrak{V}$ be a system of inductive limit neighborhoods for $\mathcal{F}(X, \mathcal{P})$.*

(a) *Every continuous linear operator* $T : \mathcal{E}_0(X,\mathcal{P}) \to \mathcal{Q}$ *can be uniquely extended to* $\mathcal{F}_{\mathfrak{V}_0}(X,\mathcal{P})$.

(b) *Every continuous linear operator* $T : \mathcal{E}(X,\mathcal{P}) \to \mathcal{Q}$ *can be uniquely extended to* $\mathcal{F}_{\mathfrak{V}}(X,\mathcal{P})$.

Our following extension results are however less obvious. We shall have to impose some additional requirements on the locally convex cone $(\mathcal{P},\mathcal{V})$ and the inductive limit neighborhood system $\mathfrak{V}$.

For our first result we need to require that the locally convex cone $(\mathcal{P},\mathcal{V})$ is both quasi-full and has uniform boundedness components (see Section I.4.23). Both requirements can be combined into the following conditions:

(QU1) $a \le b+v$ *for* $a,b \in \mathcal{P}$ *and* $v \in \mathcal{V}$ *if and only if* $a \le b+s$ *for some bounded element* $s \in \mathcal{P}$ *such that* $s \le v$.

(QU2) $a \le u+v$ *for* $a \in \mathcal{P}$ *and* $u,v \in \mathcal{V}$ *if and only if* $a \le s+t$ *for some bounded elements* $s,t \in \mathcal{P}$ *such that* $s \le u$ *and* $t \le v$.

Obviously, every (ordered) locally convex topological vector space satisfies (QU1) and (QU2); likewise, every full locally convex cone whose neighborhood system consists only of the positive multiples of a single neighborhood.

**Theorem 3.2.** *Suppose that* $(\mathcal{P},\mathcal{V})$ *is quasi-full and has uniform boundedness components. Let* $\mathfrak{V}$ *be a system of inductive limit neighborhoods such that for all* $\mathfrak{v} \in \mathfrak{V}$ *and every* $s \in \mathfrak{v}$ *the set* $K_s = \{x \in X \mid s(x) = \infty\}$ *is contained in* $\mathfrak{K}_0$. *Then every continuous linear operator* $T : \mathcal{E}_0(X,\mathcal{P}) \to \mathcal{Q}$ *can be extended to* $\mathcal{E}(X,\mathcal{P})$.

*Proof.* Following Theorem 3.1, the operator $T$ can be extended to a continuous linear operator on $\mathcal{C}^r_{\mathfrak{V}_0}(X,\mathcal{P})$, which by Corollaries 2.7 to 2.9 together with Proposition I.4.22 coincides with $\mathcal{C}^r_{\mathfrak{V}}(X,\mathcal{P})$ in this case, and consists of all r-continuous functions in $\mathcal{F}(X,\mathcal{P})$ that vanish at infinity and take unbounded values only on a compact subset of $X$. For any neighborhood $w \in \mathcal{W}$ there is some $\mathfrak{u}_w \in \mathfrak{V}$ such that $T(f) \le T(g)+w$ holds whenever $f \le g+\mathfrak{u}_w$ for $f,g \in \mathcal{C}^r_{\mathfrak{V}_0}(X,\mathcal{P})$. In a first step, for a function $f \in \mathcal{E}(X,\mathcal{P})$ we set

$$\text{(i)}\quad T(f) = \inf\left\{T(f+g) \mid 0 \le g \in \mathcal{E}(X,\mathcal{P}) \text{ such that } f+g \in \mathcal{C}^r_{\mathfrak{V}_0}(X,\mathcal{P})\right\}.$$

For $f \in \mathcal{E}(X,\mathcal{P}) \cap \mathcal{C}^r_{\mathfrak{V}_0}(X,\mathcal{P})$, this definition is of course consistent with the given value of $T$. As usual, the infimum over the empty set, is set to be $\infty \in \mathcal{Q}$. On the other hand, if for a function $f \in \mathcal{E}(X,\mathcal{P})$ there is $0 \le g \in \mathcal{E}(X,\mathcal{P})$ such that $f+g \in \mathcal{C}^r_{\mathfrak{V}_0}(X,\mathcal{P})$, then the above infimum exists as the concerned subset of $\mathcal{Q}$ is bounded below. Indeed, for $f \in \mathcal{E}(X,\mathcal{P})$ and every $w \in \mathcal{W}$ we find $\lambda \ge 0$ such that $0 \le f+\lambda\mathfrak{u}_w$, hence $0 \le (f+g)+\lambda\mathfrak{u}_w$ for every $0 \le g \in \mathcal{E}(X,\mathcal{P})$ such that $f+g \in \mathcal{C}^r_{\mathfrak{V}_0}(X,\mathcal{P})$. This implies $0 \le T(f+g)+w$. All left to show is that this extension of the operator $T$ to the cone $\mathcal{E}(X,\mathcal{P})$ is again linear and continuous. For this, $T(\alpha f) = \alpha T(f)$ and $T(f+g) \le T(f)+T(g)$ for $f,g \in \mathcal{E}(X,\mathcal{P})$ and $\alpha \ge 0$ are evident from our definition. The reverse of the latter inequality is however less obvious. We shall proceed as follows:

(ii) Proposition 1.9 states that a function $g = \sum_{i=1}^n \varphi_{i \otimes} a_i \in \mathcal{E}(X, \mathcal{P})$ is r-continuous, hence in $\mathcal{C}^r_{\mathfrak{V}_0}(X, \mathcal{P})$ if and only if the inverse images under $g$ of all boundedness components of $\mathcal{P}$ are both open and closed subsets of $X$. Because these inverse images are determined by the core supports of the functions $\varphi_i$, it is evident that the above criterion remains valid if we replace the functions $\varphi_i \in \mathcal{K}(X)$ by strictly positive multiples; precisely: If the function $g = \sum_{i=1}^n \varphi_{i \otimes} a_i \in \mathcal{E}(X, \mathcal{P})$ is r-continuous, and if $\lambda_1, \ldots, \lambda_n > 0$, then the function $g = \sum_{i=1}^n \lambda_i \varphi_{i \otimes} a_i \in \mathcal{E}(X, \mathcal{P})$ is also r-continuous. We shall make use of this observation in the following.

(iii) Now let $f = \sum_{i=1}^n \varphi_{i \otimes} a_i \in \mathcal{E}(X, \mathcal{P})$. Given $w \in \mathcal{W}$ and a compact set $K$ which contains the support of $f$, there is $v \in \mathcal{V}$ such that $\chi_{K \otimes} v \leq \mathfrak{u}_w$ and $\lambda \geq 0$ such that $0 \leq a_i + \lambda^{\frac{1}{2}} v$ for all $i = 1, \ldots, n$. We may also assume that $0 \leq \varphi_i(x) \leq \lambda^{\frac{1}{2}}/n$ holds for all $x \in X$ and $i = 1, \ldots, n$. By our assumption on $\mathcal{P}$ then we can find elements $s_i \in \mathcal{P}$ such that $s_i \leq \lambda^{\frac{1}{2}} v$ and $a_i + s_i \geq 0$ for all $i = 1, \ldots, n$. As $\mathcal{P}$ has uniform boundedness components, the elements $s_i$ are even (globally) bounded, hence the function $f' = \sum_{i=1}^n \varphi_{i \otimes} s_i$ is in $\mathcal{E}_0(X, \mathcal{P})$. Furthermore, $f + f' \geq 0$, and as

$$\varphi_i(x) s_i \leq \varphi_i(x) \lambda^{\frac{1}{2}} v \leq \frac{\lambda}{n} v$$

holds for all $x \in X$ and $i = 1, \ldots, n$, we have $f' \leq \lambda v$, hence $T(f') \leq \lambda w$.

(iv) Now let $f, g \in \mathcal{E}(X, \mathcal{P})$ and $0 \leq h \in \mathcal{E}(X, \mathcal{P})$ such that $(f+g)+h \in \mathcal{C}^r_{\mathfrak{V}_0}(X, \mathcal{P})$. Given $w \in \mathcal{W}$ choose functions $f', g' \in \mathcal{E}_0(X, \mathcal{P})$ as in (iii), that is $(f + f'), (g + g') \geq 0$ and $T(f'), T(g') \leq \lambda v$ for some $\lambda \geq 0$. Then $f + (g + g' + h) \in \mathcal{C}^r_{\mathfrak{V}_0}(X, \mathcal{P})$ as well, and likewise $f + \varepsilon(g + g' + h) \in \mathcal{C}^r_{\mathfrak{V}_0}(X, \mathcal{P})$ for every $0 < \varepsilon \leq 1/2$, as we had observed in step (ii). Because $0 \leq \varepsilon(g + g' + h)$, this yields

$$T(f) \leq T\big(f + \varepsilon(g + g' + h)\big)$$

by (i). Similarly, we obtain

$$T(g) \leq T\big(g + \varepsilon(f + f' + h)\big).$$

As the operator $T$ is linear on the cone $\mathcal{C}^r_{\mathfrak{V}_0}(X, \mathcal{P})$, and as all the functions $f + \varepsilon(g + g' + h)$, $g + \varepsilon(f + f' + h)$, $(1 + \varepsilon)(f + g) + 2\varepsilon h$ and $f' + g'$ are in $\mathcal{C}^r_{\mathfrak{V}_0}(X, \mathcal{P})$, adding the preceding inequalities yields

$$\begin{aligned} T(f) + T(g) &\leq T\big(f + \varepsilon(g + g' + h)\big) + T\big(g + \varepsilon(f + f' + h)\big) \\ &= T\big((1 + \varepsilon)(f + g) + \varepsilon(f' + g') + 2\varepsilon h\big) \\ &\leq T\big((1 + \varepsilon)(f + g) + 2\varepsilon h\big) + \varepsilon T(f' + g') \\ &\leq (1 + \varepsilon) T\big((f + g) + h\big) + \varepsilon \lambda w. \end{aligned}$$

Because both $w \in \mathcal{W}$ and $0 < \varepsilon \le 1$ were arbitrarily chosen, we infer that $T(f) + T(g) \le T\big((f+g)+h\big)$. Now the definition of $T(f+g)$ in (i) yields $T(f)+T(g) \le T\big(f+g\big)$, hence together with the above, the linearity of the operator $T$ on $\mathcal{E}(X,\mathcal{P})$.

(v) All left to show is the continuity of $T$. For this, let $w \in \mathcal{W}$ and $f, g \in \mathcal{E}(X,\mathcal{P})$ such that $f \le g + \mathfrak{u}_w$, that is $f \le g + s$ for some $s \in \mathfrak{v}$. By our assumption on the neighborhood system $\mathfrak{V}$, the set $K_s = \{x \in X \mid s(x) = \infty\}$ is contained in $\mathfrak{K}_0$.

In a first case, suppose that $f(x) = g(x) = 0$ for all $x \in K_s$. Let $\lambda \ge 0$ such that $0 \le T(g) + \lambda v$ and let $0 \le h \in \mathcal{E}(X,\mathcal{P})$ such that $g + h \in \mathcal{C}^r_{\mathfrak{V}_0}(X,\mathcal{P})$ and let $\varepsilon > 0$. Let $g' = (1+\varepsilon)g + h$ and $f' = f + \varepsilon g + h$. Then $g'$ is again r-continuous, and $f' \le g' + s$. As $s(x) \neq \infty$ for all $x \in X \backslash K_s$ and as $\mathcal{P}$ has uniform boundedness components, we conclude that $f'(x) \in \mathcal{B}(g_1(x))$ for all $x \in X$. On the other hand $g'(x) = (1+\varepsilon)g(x) + h(x)$ is evidently bounded relative to $g(x)+h(x)$, hence relative to $f'(x) = f(x)+\varepsilon g(x)+h(x)$ as well. This demonstrates that both $f'$ and $g'$ create the same inverses of boundedness components in $X$, and following Proposition 8.8 therefore $f'$ is r-continuous as $g'$ is. Then by (i)

$$\begin{aligned} T(f) &\le T(f) + \varepsilon T(g) + \varepsilon\lambda w \le T(f') + \varepsilon\lambda w \\ &\le T(g') + (1+\varepsilon\lambda)w \le (1+\varepsilon)T(g+h) + (1+\varepsilon\lambda)w. \end{aligned}$$

This holds for all $\varepsilon > 0$, and we conclude that $T(f) \le T(g+h) + w$. Taking the infimum over all such functions $0 \le h \in \mathcal{E}(X,\mathcal{P})$ yields

$$T(f) \le T(g) + w.$$

To prepare our second case, suppose that a function $h \in \mathcal{E}(X,\mathcal{P})$ is supported by $K_s$. Because the range of $h$ is spanned by finitely many elements $a_1, \ldots, a_n \in \mathcal{P}$, we have $h(x) \in \mathcal{B}(a)$, where $a = a_1 + \ldots + a_n$ for all $x \in K_s$. Note that $\chi_{K_s} \in \mathcal{K}_0(X)$. Let $\lambda \ge 0$ such that $0 \le T(\chi_{K_s} \otimes a) + \lambda w$ and choose $\varepsilon > 0$. We set $h' = h + \chi_{K_s} \otimes a \in \mathcal{E}(X,\mathcal{P})$. Then all the values of $h$ on $K_s$ are bounded relative to each other, hence $h' \in \mathcal{C}^r_{\mathfrak{V}_0}(X,\mathcal{P})$ by Proposition 1.9. As $h' \le \varepsilon s$, we have $T(h') \le \varepsilon w$, hence

$$T(h) \le T(h) + \varepsilon T(\chi_{K_s} \otimes a) + \varepsilon\lambda w = T(h') + \varepsilon\lambda w \le \varepsilon(1+\lambda)w$$

for all $\varepsilon > 0$. Likewise

$$0 \le T(h) + T(\chi_{K_s} \otimes a) \le T(h) + \varepsilon w$$

as $\chi_{K_s} \otimes a \le \varepsilon s$ for all $\varepsilon > 0$. Now in our second case, suppose that both functions $f, g \in \mathcal{E}(X,\mathcal{P})$ are supported by $K_s$. Then our preparing remarks yield

$$T(f) \le T(g) + \varepsilon w$$

for all $\varepsilon > 0$.

Now, finally let us consider the general case. Both functions $\chi_1 = \chi_{(X\setminus K_s)}$ and $\chi_2 = \chi_{K_s}$ are r-continuous. The functions $f_1 = \chi_1 \otimes f$, $f_2 = \chi_2 \otimes f$, $g_1 = \chi_1 \otimes g$ and $g_2 = \chi_2 \otimes g$ are therefore all contained in $\mathcal{E}(X, \mathcal{P})$. Our first step shows that $T(f_1) \leq T(g_1) + w$, the second step $T(f_2) \leq T(g_2) + \varepsilon w$ for all $\varepsilon > 0$. Combining, this yields

$$T(f) = T(f_1) + T(f_2) \leq T(g_1) + T(g_2) + (1+\varepsilon)w = T(g) + (1+\varepsilon)w,$$

for all $\varepsilon > 0$, hence $T(f) \leq T(g) + w$, as claimed. □

The following extension result applies in particular to locally convex topological vector spaces $(\mathcal{P}, \mathcal{V})$. Recall the construction of the canonical embedding of a quasi-full locally convex $(\mathcal{P}, \mathcal{V})$ cone into the full cone $(\mathcal{P}_{\mathcal{V}}, \mathcal{V})$, that is its standard full extension, as elaborated in Section 6.2 of Chapter I.

**Theorem 3.3.** *Suppose that $(\mathcal{P}, \mathcal{V})$ is quasi-full and that all elements of $\mathcal{P}$ are bounded. Let $\mathfrak{V}$ be a system of r-lower continuous inductive limit neighborhoods for $\mathcal{F}(X, \mathcal{P})$. Then every continuous linear operator $T : \mathcal{E}(X, \mathcal{P}) \to \mathcal{Q}$ can be extended to a continuous linear operator from $\mathcal{E}(X, \mathcal{P}_{\mathcal{V}})$ to $\mathcal{Q}$.*

*Proof.* Note that under the conditions of the theorem the cones $\mathcal{E}_0(X, \mathcal{P})$ and $\mathcal{E}(X, \mathcal{P})$ coincide and that all functions in $\mathcal{E}(X, \mathcal{P})$ are r-continuous (see Proposition 1.11(a)) and indeed continuous with respect to the given symmetric topology of $\mathcal{P}$. The $\overline{\mathcal{V}}$-valued functions comprising the neighborhood system $\mathfrak{V}$ are r-lower continuous. The functions in $\mathcal{E}(X, \mathcal{V})$, on the other hand, satisfy the following: For a neighborhood-valued function $\varphi \otimes v \in \mathcal{E}(X, \mathcal{V})$, a point $x \in X$ and $\gamma > 1$ there is a neighborhood $U_x$ of $x$ such that $(1/\gamma)\varphi(x) \leq \varphi(y)$, hence $(1/\gamma)\varphi(x)v \leq \varphi(y)v$ for all $y \in U_x$. Consequently, for a function $s \in \mathcal{E}(X, \mathcal{V})$, $x \in X$ and $\gamma > 1$ one finds a neighborhood $U_x$ of $x$ such that $(1/\gamma)s(x) \leq s(y)$ for all $y \in U_x$.

Now let $T : \mathcal{E}(X, \mathcal{P}) \to \mathcal{Q}$ be a continuous linear operator. For a $\mathcal{P}_{\mathcal{V}}$-valued function $f \in \mathcal{E}(X, \mathcal{P}_{\mathcal{V}})$ we set

$$T(f) = \sup\{T(h) \mid h \in \mathcal{E}(X, \mathcal{P}),\ h \leq f\}.$$

For $\mathcal{V} \subset \mathcal{P}$, that is $f \in \mathcal{E}(X, \mathcal{P})$, this definition is clearly consistent, and immediately yields

$$T(\alpha f) = \alpha T(f) \qquad \text{and} \qquad T(f+g) \geq T(f) + T(g)$$

for $f, g \in \mathcal{E}(X, \mathcal{P}_{\mathcal{V}})$ and $\alpha \geq 0$. Subadditivity, on the other hand, is not as obvious. To prepare our argument for this, let us consider the following construction:

Let $f, g \in \mathcal{E}(X, \mathcal{P}_{\mathcal{V}})$, let $u$ be an r-lower continuous $\overline{\mathcal{V}}$-valued function, and let $h \in \mathcal{E}(X, \mathcal{P})$ such that $h \leq f+g+u$. Then $f = f'+s$ and $g = g'+t$ for functions $f', g' \in \mathcal{E}(X, \mathcal{P})$ and $s', g' \in \mathcal{E}(X, \mathcal{V})$. We choose a compact set $K \in \mathfrak{K}$ which supports all the functions involved, $f, g, h, s$ and $t$. Let $v \in \mathcal{V}$ and $\varepsilon > 0$. There is $\lambda \geq 1$ such that both $0 \leq f' + \lambda \chi_{K \otimes} v$ and $0 \leq g' + \lambda \chi_{k \otimes} v$. We set $\gamma = 1 + \varepsilon/\lambda \leq 1 + \varepsilon$. For every $x \in K$, following or assumption that $\mathcal{P}$ is quasi-full, there are elements $a_x, b_x$ and $c_x$ in $\mathcal{P}$ such that $a_x \leq s(x)$, $b_x \leq t(x)$, $c_x \leq u(x)$ and

$$h(x) \leq f'(x) + g'(x) + a_x + b_x + c_x.$$

The different types of continuity of the concerned functions guarantees that there is an open neighborhood $U_x$ of $x$ in $X$ such that

$$(1/\gamma) a_x \leq s(y), \qquad (1/\gamma) b_x \leq t(y) \qquad \text{and} \qquad c_x \leq \gamma u(y) + \varepsilon v,$$

as well as

$$h(y) \leq f'(y) + g'(y) + a_x + b_x + c_x + \varepsilon v$$

holds for all $y \in U_x$. In turn there is a finite cover $U_{x_1}, \ldots, U_{x_n}$ of these open sets for the compact set $K$ and a corresponding partition of the unit, consisting of continuous real-valued functions $\varphi_1, \ldots, \varphi_n$ such that $\operatorname{supp}(\varphi_i) \subset U_{x_i}$ for all $i = 1, \ldots, n$ and $\varphi_1(x) + \ldots + \varphi_n(x) = 1$ for all $x \in K$. We observe that

$$(1/\gamma)\varphi_i(x) a_{x_i} \leq \varphi_i(x) s(x), \qquad (1/\gamma)\varphi_i(x) b_{x_i} \leq \varphi_i(x) t(x)$$

and

$$\varphi_i c_{x_i} \leq \varphi_i(x)\big(\gamma u(x) + \varepsilon v\big),$$

as well as

$$\varphi_i(x) h(x) \leq \varphi_i(x)\big(f'(x) + g'(x) + a_x + b_x + c_x + \varepsilon v\big)$$

holds for all $x \in X$ and $i = 1, \ldots, n$. We set

$$f'' = f' + \frac{1}{\gamma} \sum_{i=1}^{n} \varphi_{i \otimes} a_{x_i} \qquad \text{and} \qquad g'' = g' + \frac{1}{\gamma} \sum_{i=1}^{n} \varphi_{i \otimes} b_{x_i}.$$

Then $f'', g'' \in \mathcal{E}(X, \mathcal{P})$ and

$$f'' \leq f' + s = f \qquad \text{and} \qquad g'' \leq g' + t = g$$

as well as

$$\begin{aligned} h &\leq f' + g' + \sum_{i=1}^{n} \varphi_{i \otimes}(a_{x_i} + b_{x_i} + c_{x_i}) + \varepsilon \chi_{K \otimes} v \\ &\leq \left( f' + \sum_{i=1}^{n} \varphi_{i \otimes} a_{x_i} \right) + \left( g' + \sum_{i=1}^{n} \varphi_{i \otimes} b_{x_i} \right) + \gamma u + 2\varepsilon \chi_{K \otimes} v. \end{aligned}$$

We observe that

$$f' \leq f' + (\varepsilon/\lambda)(f' + \lambda\chi_{K\circledast v}) = \gamma f' + \varepsilon\chi_{K\circledast v}$$

and, likewise

$$g' \leq \gamma g' + \varepsilon\chi_{K\circledast v}.$$

Combining with the above, this yields

$$h \leq \gamma(f'' + g'' + u) + 3\varepsilon\chi_{K\circledast v}.$$

We shall in the sequel use this construction of the functions $f'', g'' \in \mathcal{E}(X,\mathcal{P})$ in order to prove both subadditivity and continuity of the extended operator $T$:

For subadditivity, we choose $f, g \in \mathcal{E}(X, \mathcal{P}_{\mathcal{V}})$ and set $u = 0$ in the above. Let $h \in \mathcal{E}(X,\mathcal{P})$ such that $h \leq f + g$. Given $w \in \mathcal{W}$ for the set $K \in \mathfrak{K}$ from above there is an inductive limit neighborhood $\mathfrak{v} \in \mathcal{V}$ such that $T(j) \leq T(l) + w$ holds whenever $j \leq l + \mathfrak{v}$ for functions $j, l \in \mathcal{E}(X,\mathcal{P})$, and there is $v \in \mathcal{V}$ such that $\chi_{K\circledast v} \leq \mathfrak{v}$. With this insertion for $v \in \mathcal{V}$ and any $\varepsilon > 0$ we construct the functions $f'', g'' \in \mathcal{E}(X,\mathcal{P})$ as before. Then

$$T(f'') \leq T(f) \qquad \text{and} \qquad T(g'') \leq T(g),$$

since $f'' \leq f$ and $g'' \leq g$, and

$$T(h) \leq \gamma\big(T(f'') + T(g'')\big) + 3\varepsilon w \leq \gamma\big(T(f) + T(g)\big) + 3\varepsilon w,$$

since $h \leq \gamma f'' + \gamma g'' + 3\varepsilon\mathfrak{v}$, and consequently

$$T(h) \leq T(f) + T(g),$$

because $w \in \mathcal{W}$ and $\varepsilon > 0$ were arbitrarily chosen. Finally, taking the supremum over all $h \in \mathcal{E}(X,\mathcal{P})$ such that $h \leq f + g$ on the left hand side of the last inequality yields indeed

$$T(f + g) \leq T(f) + T(g)$$

by our definition of the extension of the operator $T$.

For continuity of this extension, let $w \in \mathcal{W}$ and $\mathfrak{v} \in \mathfrak{V}$ such that $T(j) \leq T(l) + w$ holds whenever $j \leq l + \mathfrak{v}$ for functions $j, l \in \mathcal{E}(X,\mathcal{P})$. We shall verify that this property is preserved by the extension of $T$ to the cone $\mathcal{E}(X, \mathcal{P}_{\mathcal{V}})$. Indeed, let $f, g \in \mathcal{E}(X, \mathcal{P}_{\mathcal{V}})$ such that $f \leq g + \mathfrak{v}$, that is $f \leq g + u$ for some $\overline{\mathcal{V}}$-valued r-lower continuous function $u \leq \mathfrak{v}$. Let $h \in \mathcal{E}(X,\mathcal{P})$ such that $h \leq f$. Then $h \leq g + u$ by the definition of the order in the extended cone $\mathcal{P}_{\mathcal{V}}$ as defined in Section I.6.2. We shall now use the above construction with the function $f = 0$, hence $f'' = 0$: Given $v \in \mathcal{V}$ such that $\chi_{K\circledast v} \leq \mathfrak{v}$ and $\varepsilon > 0$, there is $g'' \in \mathcal{E}(X,\mathcal{P})$, satisfying $g'' \leq g$ and

$$h \leq \gamma(g'' + u) + 3\varepsilon\chi_{K^{\otimes}}v \leq \gamma g'' + (\gamma + 3\varepsilon)\mathfrak{v}.$$

This yields

$$T(h) \leq \gamma T(g'') + (\gamma + 3\varepsilon)w \leq \gamma T(g) + (\gamma + 3\varepsilon)w,$$

and consequently

$$T(h) \leq T(g) + w,$$

because $\varepsilon > 0$ was arbitrarily chosen. Finally, taking the supremum over all $h \in \mathcal{E}(X, \mathcal{P})$ such that $h \leq f$ on the left hand side of the last inequality yields indeed

$$T(f) \leq T(g) + w,$$

thus completing our argument. □

**Proposition 3.4.** *Suppose that* $(\mathcal{P}, \mathcal{V})$ *is quasi-full and that* $X$ *carries the discrete topology. Let* $\mathfrak{V}$ *be a system of r-lower continuous inductive limit neighborhoods for* $\mathcal{F}(X, \mathcal{P})$. *Then every continuous linear operator* $T : \mathcal{E}(X, \mathcal{P}) \to \mathcal{Q}$ *can be extended to a continuous linear operator from* $\mathcal{E}(X, \mathcal{P}_{\mathcal{V}})$ *to* $\mathcal{Q}$.

*Proof.* This is a straightforward consequence of Theorem I.6.3. $\mathfrak{R}$ consists of all finite subsets of $X$ in this case. For every $x \in X$ the operator $T_x : \mathcal{P} \to \mathcal{Q}$, that is $a \mapsto T(\chi_{\{x\}} \otimes a)$ is linear and continuous. According to I.6.3 there is a continuous linear extension to $\mathcal{P}_{\mathcal{V}}$, that is $T_x : \mathcal{P}_{\mathcal{V}} \to \mathcal{Q}$. Since all compact subsets of $X$ are finite, the functions $f \in \mathcal{E}(X, \mathcal{P}_{\mathcal{V}})$ are of the type $f = \sum_{i=1}^{n} \chi_{\{x_i\}} \otimes a_i$ for $x_i \in X$ and $a_i \in \mathcal{P}_{\mathcal{V}}$. Then the formula

$$T\left(\sum_{i=1}^{n} \chi_{\{x_i\}} \otimes a_i\right) = \sum_{i=1}^{n} T_{x_i}(a_i)$$

provides the stated extension of $T$ to $\mathcal{E}(X, \mathcal{P}_{\mathcal{V}})$. □

## 4. Measures on Locally Compact Spaces

We shall now return to the concepts of integration theory from Chapter II. As in the preceding sections, let $X$ be a locally compact Hausdorff space. Let $\mathfrak{R}$ be the weak $\sigma$-ring of all relatively compact Borel subsets of $X$ and correspondingly, $\mathfrak{A}_{\mathfrak{R}}$ the $\sigma$-algebra of all Borel subsets of $X$. As before, $(\mathcal{P}, \mathcal{V})$ is a locally convex cone.

**Proposition 4.1.** *Let* $f \in \mathcal{F}(X, \mathcal{P})$.

*(a) If* $f$ *is r-continuous, then* $f$ *is measurable.*

*(b) If* $f$ *is r-lower continuous and measurable, then* $\chi_{E}\otimes f \in \mathcal{F}_{\mathfrak{R}}(X,\mathcal{P})$ *for every* $E \in \mathfrak{R}$.

*Proof.* (a) The symmetric relative topology of $\mathcal{P}$ is the common refinement of all symmetric relative $v$-topologies, and therefore an r-continuous function $f \in \mathcal{F}(X,\mathcal{P})$ is continuous with respect to any of these topologies. It is therefore clear that $f^{-1}(O) \in \mathfrak{A}_{\mathfrak{R}}$ for any set $O$ in $\mathcal{P}$ which is open in any of the relative $v$-topologies. This is Condition (M1) from II.1.2. Furthermore, since all sets $E \in \mathfrak{R}$ are relatively compact in $X$, the sets $f(E)$ are relatively compact in every symmetric relative $v$-topology. But these topologies are generated by pseudo-metrics (see Section I.4), and relatively compact sets are therefore separable, hence (M2). For Part (b), let $f \in \mathcal{F}(X,\mathcal{P})$ be r-lower continuous and measurable, and let $E \in \mathfrak{R}$. First we recall from Theorem II.1.8(a) that the function $\chi_{E}\otimes f$ is also measurable. The closure $\bar{E}$ of $E$ is compact and also contained in $\mathfrak{R}$. We shall have to verify that $f$ can be reached from below by step functions in the sense of II.2.3: Given an inductive limit neighborhood $\mathfrak{v}$ there is $v \in \mathcal{V}$ such that $\chi_{\bar{E}}\otimes v \leq \mathfrak{v}$. By Lemma 1.4 then there is $\lambda \geq 0$ such that $0 \leq f(x)+\lambda v$ for all $x \in \bar{E}$. Given $0 \leq \varepsilon \leq \frac{1}{1+\lambda}$, as the function $f$ is r-lower continuous, a compactness argument yields that there are $x_1,\ldots,x_n \in \bar{E}$ and corresponding neighborhoods $E_1\ldots,E_n \in \mathfrak{R}$ of these points, covering $\bar{E}$ and such that $f(x) \in \big(f(x_i)\big)v_\varepsilon$, that is $f(x_i) \in v_\varepsilon\big(f(x)\big)$ whenever $x \in E_i$. We set $a_i = f(x_i)$, and for any $x \in E_i$ we have $a_i \leq \gamma f(x) + \varepsilon v$ for some $1 \leq \gamma \leq 1+\varepsilon$, hence $a_i \leq (1+\varepsilon)f(x)+\varepsilon(1+\lambda)v$ by Lemma I.4.1(b), and $(1+\varepsilon)^{-1}a_i \leq f(x)+v$. We set $F_1 = E \cap E_1$ and $F_i = (E\cap E_i)\setminus \cup_{k=1}^{i-1}F_k$ for $i=2,\ldots,n$. Then

$$h = (1+\varepsilon)^{-1}\sum_{i=1}^{n} \chi_{F_i}\otimes a_i \in \mathcal{S}_{\mathfrak{R}}(X,\mathcal{P}),$$

and

$$h \leq \chi_{E}\otimes f + \chi_{E}\otimes v \leq \chi_{E}\otimes f + \mathfrak{v}.$$

Our claim follows. □

Part (b) of Proposition 4.1 applies in particular to the functions in $\mathcal{E}(X,\mathcal{P})$ since every function of the type $\varphi\otimes a$ for a continuous positive real-valued function $\varphi$ and $a \in \mathcal{P}$ is measurable by Theorem II.1.8(a) and r-lower continuous by Proposition 1.8(b).

In addition to the general assumptions of the preceding we shall for the remainder of this section require that $(\mathcal{P},\mathcal{V})$ is a quasi-full locally convex cone, and that $(\mathcal{Q},\mathcal{W})$ is a locally convex complete lattice cone.

**4.2 Regularity of Measures.** Following the usual terminology, we shall say that an operator-valued measure $\theta : \mathfrak{R} \to \mathfrak{L}(\mathcal{P},\mathcal{Q})$ is *inner regular* or *outer regular* on a set $E \in \mathfrak{R}$ if, with respect to the order convergence in $\mathcal{Q}$,

$$\theta_E(a) = \lim_{K\subset E}\theta_K(a) \qquad \text{or} \qquad \theta_E(a) = \lim_{O\supset E}\theta_O(a)$$

holds for all $a \in \mathcal{P}$, respectively. The limits are taken over the upward directed family of all compact sets of $K \subset E$ in the first case, and the downward directed family of all relatively compact open sets $O \supset E$ in the second case. A measure is *outer* or *inner regular* if it is outer or inner regular for all $E \in \mathfrak{R}$, respectively. A measure which is both outer regular for all $E \in \mathfrak{R}$ and inner regular for all open sets $E \in \mathfrak{R}$ is called *quasi regular.*

**Lemma 4.3.** *Let $\theta$ be a bounded measure on $\mathfrak{R}$. Let $E, F \in \mathfrak{R}$ be disjoint sets and let $a \in \mathcal{P}$.*

*(a) If $\theta$ is outer regular for $F$ and inner regular for $E \cup F$, then*

$$\overline{\lim_{K \subset E}}\, \theta_K(a) \leq \theta_E(a) \leq \underline{\lim}_{K \subset E}\, \theta_K(a) + \mathfrak{O}\big(\theta_F(a)\big).$$

*(b) If $\theta$ is inner regular for $F$ and outer regular for $E \cup F$, then*

$$\theta_E(a) \leq \underline{\lim}_{O \supset E}\, \theta_O(a) \leq \overline{\lim_{O \supset E}}\, \theta_O(a) \leq \theta_E(a) + \mathfrak{O}\big(\theta_F(a)\big).$$

*These limits are taken over the upward directed family of all compact sets $K \subset E$ or over the downward directed family of all relatively compact open sets $O \supset E$, respectively.*

*Proof.* Because the $\mathfrak{R}$-bounded measure $\theta$ can be extended to $\mathcal{P}_{\mathcal{V}}$, we may assume that the locally convex cone $(\mathcal{P}, \mathcal{V})$ is full. For Part (a), let $E, F \in \mathfrak{R}$ be disjoint sets and suppose that $\theta$ is outer regular for $F$ and inner regular for $E \cup F$. In a first step let us consider an element $0 \leq a \in \mathcal{P}$. Let $S$ be any compact subset of $E \cup F$ and let $O \in \mathfrak{R}$ be any open set containing $F$. Let $K = S \setminus O$. Then $K$ is compact, $K \subset E$, and as $S = K \cup \big(S \cap O\big)$ we have

$$\theta_S(a) = \theta_K(a) + \theta_{(S \cap O)}(a) \leq \theta_K(a) + \theta_O(a),$$

hence

$$\theta_S(a) \leq \sup\{\theta_K(a) \mid K \subset E,\ K \in \mathfrak{K}\} + \theta_O(a) = \lim_{K \subset E} \theta_K(a) + \theta_O(a).$$

First taking the supremum over all compact subsets $S$ of $E \cup F$ and using the inner regularity of $\theta$ for the set $E \cup F$ on the left-hand side of this inequality, and then taking the infimum over all open sets $O \in \mathfrak{R}$ containing $F$ and using the outer regularity of $\theta$ for the set $F$ on the right-hand side leads to

$$\theta_{(E \cup F)}(a) \leq \lim_{K \subset E} \theta_K(a) + \theta_F(a).$$

As $\theta_{(E \cup F)}(a) = \theta_E(a) + \theta_F(a)$, the cancellation law from Proposition I.5.10(a) yields

$$\theta_E(a) \leq \lim_{K \subset E} \theta_K(a) + \mathfrak{O}\big(\theta_F(a)\big).$$

For the general case, let $a \in \mathcal{P}$, and for a neighborhood $w \in \mathcal{W}$ we choose $v \in \mathcal{V}$ such that $\theta_{(E \cup F)}(v) \leq w$ and $\lambda \geq 0$ such that $0 \leq a + \lambda v$. Then using the above we infer that

$$\begin{aligned} \theta_E(a) + \theta_E(\lambda v) &\leq \underline{\lim}_{K \subset E} \theta_K(a + \lambda v) + \mathfrak{D}\big(\theta_F(a + \lambda v)\big) \\ &\leq \underline{\lim}_{K \subset E} \theta_K(a) + \underline{\lim}_{K \subset E} \theta_K(\lambda v) + \mathfrak{D}\big(\theta_F(a + \lambda v)\big) \\ &\leq \underline{\lim}_{K \subset E} \theta_K(a) + \theta_E(\lambda v) + \mathfrak{D}\big(\theta_F(a)\big) + \mathfrak{D}\big(\theta_F(\lambda v)\big). \end{aligned}$$

For the latter we used Proposition I.5.11. Now because both $\mathfrak{D}\big(\theta_E(\lambda v)\big) \leq w$ and $\mathfrak{D}\big(\theta_F(\lambda v)\big) \leq w$, the cancellation law in Proposition I.5.10(a) yields

$$\theta_E(a) \leq \underline{\lim}_{K \subset E} \theta_K(a) + \mathfrak{D}\big(\theta_F(a)\big) + 2w.$$

This holds for all $w \in \mathcal{W}$ and therefore demonstrates

$$\theta_E(a) \leq \underline{\lim}_{K \subset E} \theta_K(a) + \mathfrak{D}\big(\theta_F(a)\big),$$

hence the right-hand part of the inequality in Lemma 4.3(a). For the left-hand part, using Lemma I.5.19 and the above we observe that

$$\begin{aligned} \overline{\lim}_{K \subset E} \theta_K(a) + \theta_E(\lambda v) &\leq \overline{\lim}_{K \subset E} \theta_K(a) + \underline{\lim}_{K \subset E} \theta_K(\lambda v) + \mathfrak{D}\big(\theta_F(\lambda v)\big) \\ &\leq \overline{\lim}_{K \subset E} \theta_K(a + \lambda v) + w \\ &\leq \theta_E(a + \lambda v) + w \\ &= \theta_E(a) + \theta_E(\lambda v) + w. \end{aligned}$$

Thus

$$\overline{\lim}_{K \subset E} \theta_K(a) \leq \theta_E(a) + 2w$$

using the cancellation law in I.5.10(a). This holds for all $w \in \mathcal{W}$, yielding

$$\overline{\lim}_{K \subset E} \theta_K(a) \leq \theta_E(a),$$

our claim in Part (a). The argument for Part (b) of Lemma 4.3 follows along similar lines: Let $E, F \in \mathfrak{R}$ be disjoint and suppose that $\theta$ is inner regular for $F$ and outer regular for $E \cup F$, and let $0 \leq a \in \mathcal{P}$. Let $G \in \mathfrak{R}$ be an open set containing $E \cup F$ and let $K$ be a compact subset of $F$. The set $O = G \setminus K \in \mathfrak{R}$ is open and contains $E$. Hence

$$\theta_K(a) + \underline{\lim}_{O \supset E} \theta_O(a) \leq \theta_K(a) + \theta_{(G \setminus K)}(a) = \theta_G(a),$$

and therefore by the inner regularity of $\theta$ for $F$ and the outer regularity for $E \cup F$,

$$\theta_F(a) + \lim_{O \supset E} \theta_O(a) \leq \theta_{(E \cup F)}(a) = \theta_F(a) + \theta_E(a).$$

The cancellation law I.5.10(a) yields

$$\lim_{O \supset E} \theta_O(a) \leq \theta_E(a) + \mathfrak{O}\big(\theta_F(a)\big).$$

For the general case, let $a \in \mathcal{P}$, and for $w \in \mathcal{W}$ choose $v \in \mathcal{V}$ such that $\theta_G(v) \leq w$ and $\lambda \geq 0$ such that $0 \leq a + \lambda v$. Using the above we infer that

$$\begin{aligned} \overline{\lim_{O \supset E}}\, \theta_O(a) + \theta_E(\lambda v) &\leq \overline{\lim_{O \supset E}}\, \theta_O(a) + \lim_{O \supset E} \theta_O(\lambda v) \\ &\leq \lim_{O \supset E} \theta_O(a + \lambda v) \\ &\leq \theta_E(a + \lambda v) + \mathfrak{O}\big(\theta_F(a + \lambda v)\big) \\ &= \theta_E(a) + \theta_E(\lambda v) + \mathfrak{O}\big(\theta_F(a)\big) + \mathfrak{O}\big(\theta_F(\lambda v)\big). \end{aligned}$$

Now the cancellation law I.5.10(a) yields

$$\overline{\lim_{O \supset E}}\, \theta_O(a) \leq \theta_E(a) + \mathfrak{O}\big(\theta_F(a)\big) + 2w$$

for all $w \in \mathcal{W}$, hence

$$\overline{\lim_{O \supset E}}\, \theta_O(a) \leq \theta_E(a) + \mathfrak{O}\big(\theta_F(a)\big).$$

For the left-hand part of the inequality in 4.3(b), using I.5.19 and the above we observe that

$$\begin{aligned} \theta_E(a) + \theta_E(\lambda v) &\leq \underline{\lim_{O \supset E}}\, \theta_O(a + \lambda v) \\ &\leq \underline{\lim_{O \supset E}}\, \theta_O(a) + \overline{\lim_{O \supset E}}\, \theta_O(\lambda v) \\ &\leq \underline{\lim_{O \supset E}}\, \theta_O(a) + \theta_E(\lambda v) + \mathfrak{O}\big(\theta_F(\lambda v)\big). \end{aligned}$$

Thus

$$\theta_E(a) \leq \underline{\lim_{O \supset E}}\, \theta_O(a) + w$$

for all $w \in \mathcal{W}$, yielding

$$\theta_E(a) \leq \underline{\lim_{O \supset E}}\, \theta_O(a),$$

our claim in Part (b). □

**Proposition 4.4.** *Let $\theta$ be a bounded measure on $\mathfrak{R}$. Let $E \in \mathfrak{R}$ and $a \in \mathcal{P}$.*

*(a) If $\theta$ is outer regular, then*

$$\overline{\lim_{K \subset E}}\, \theta_K(a) \le \theta_E(a) \le \underline{\lim}_{K \subset E}\, \theta_K(a) + \mathfrak{O}\big(\theta_{(G \setminus E)}(a)\big)$$

*for any set $G \in \mathfrak{R}$ containing $E$ such that $\theta$ is inner regular for $G$.*

*(b) If $\theta$ is inner regular, then*

$$\theta_E(a) \le \underline{\lim}_{O \supset E}\, \theta_O(a) \le \overline{\lim_{O \supset E}}\, \theta_O(a) \le \theta_E(a) + \mathfrak{O}\big(\theta_{(G \setminus E)}(a)\big)$$

*for any set $G \in \mathfrak{R}$ containing $E$ such that $\theta$ is outer regular for $G$.*

*(c) If $\theta$ is quasi regular, then*

$$\overline{\lim_{K \subset E}}\, \theta_K(a) \le \theta_E(a) \le \underline{\lim}_{K \subset E}\, \theta_K(a) + \mathfrak{O}\big(\theta_{(E \setminus G)}(a)\big)$$

*for any subset $G \in \mathfrak{R}$ of $E$ such that $\theta$ is inner regular for $G$.*

*Proof.* Let $\theta$ be a bounded measure, let $E \in \mathfrak{R}$ and $a \in \mathcal{P}$. We may again assume that the cone $(\mathcal{P}, \mathcal{V})$ is full. Parts (a) and (b) of the Proposition follow directly from Parts (a) and (b) of Lemma 4.3, respectively, if we set $F = G \setminus E$. For Part (c), suppose that $\theta$ is quasi regular and let $H \in \mathfrak{R}$ be an open set containing $E$. With $F = H \setminus E \in \mathfrak{R}$ the assumptions of Lemma 4.3(a) are satisfied, and we have

$$\begin{aligned}\overline{\lim_{K \subset E}}\, \theta_K(a) \le \theta_E(a) &\le \underline{\lim}_{K \subset E}\, \theta_K(a) + \mathfrak{O}\big(\theta_{(H \setminus E)}(a)\big)\\ &\le \underline{\lim}_{K \subset E}\, \theta_K(a) + \mathfrak{O}\big(\theta_H(a)\big).\end{aligned}$$

The latter follows from Proposition II.4.15(c) and holds for all open sets $H \in \mathfrak{R}$ containing $E$. We have

$$\underline{\lim}_{H \supset E}\, \mathfrak{O}\big(\theta_H(a)\big) \le \mathfrak{O}\Big(\underline{\lim}_{H \supset E}\, \theta_H(a)\Big) = \mathfrak{O}\big(\theta_E(a)\big)$$

by Proposition I.5.24 and by the outer regularity of $\theta$. This shows

$$\theta_E(a) \le \underline{\lim}_{K \subset E}\, \theta_K(a) + \mathfrak{O}\big(\theta_E(a)\big).$$

Next, for $w \in \mathcal{W}$ let $v \in \mathcal{V}$ such that $\theta_E(v) \le w$ and $\lambda \ge 0$ such that $0 \le a + \lambda v$. Let $G \in \mathfrak{R}$ be a subset of $E$ such that $\theta$ is inner regular for $G$. Then

$$\theta_G(a) \le \theta_G(a + \lambda v) = \underline{\lim}_{K \subset G}\, \theta_K(a + \lambda v) \le \underline{\lim}_{K \subset E}\, \theta_K(a + \lambda v) \le \underline{\lim}_{K \subset E}\, \theta_K(a) + \lambda w.$$

This shows $\theta_G(a) \in \mathcal{B}(\lim_{K\subset E} \theta_K(a))$ and implies that

$$\lim_{K\subset E} \theta_K(a) + \mathfrak{O}(\theta_G(a)) = \lim_{K\subset E} \theta_K(a)$$

using Proposition I.5.14. Because $\mathfrak{O}(\theta_E(a)) = \mathfrak{O}(\theta_G(a)) + \mathfrak{O}(\theta_{(E\setminus G)}(a))$ by I.5.11(a), we have indeed demonstrated that

$$\theta_E(a) \leq \lim_{K\subset E} \theta_K(a) + \mathfrak{O}(\theta_{(E\setminus G)}(a))$$

holds as claimed. □

Note that the set $G \in \mathfrak{R}$ in Proposition 4.4 may, for example, be chosen as $\overline{E}$ in Part (a), as any open set containing $E$ in Part (b) and as $E^\circ$ or as any closed subset of $E$ in Part (c).

For a set $E \in \mathfrak{R}$ and a function $\varphi \in \mathcal{K}(X)$ we abbreviate $E \prec \varphi$ if $\chi_E \leq \varphi$, and $\varphi \prec E$ if $\varphi \leq \chi_E$ and $\operatorname{supp}(\varphi) \subset E$. Note that the families $\{\varphi \in \mathcal{K}(X) \mid \varphi \prec E\}$ and $\{\varphi \in \mathcal{K}(X) \mid \varphi \prec E\}$ are directed downward and upward, respectively. Both are therefore suitable index sets for nets.

The order topology for a locally convex complete lattice cone was defined in Section I.5.43.

**Proposition 4.5.** *Let $\theta$ be a bounded measure on $\mathfrak{R}$.*

*(a) If $\theta$ is inner regular for an open set $0 \in \mathfrak{R}$, then*

$$\theta_O(a) = \lim_{\varphi\prec O} \int_X \varphi_{\otimes} a \, d\theta \qquad \textit{for all} \quad a \in \mathcal{P}.$$

*(b) If $\theta$ is outer regular for a compact set $K \in \mathfrak{R}$, then*

$$\theta_K(a) = \lim_{\varphi\succ K} \int_X \varphi_{\otimes} a \, d\theta \qquad \textit{for all} \quad a \in \mathcal{P}.$$

*(c) If $\theta$ is quasi regular, then for every $E \in \mathfrak{R}$ and every open set $O \in \mathfrak{R}$ containing $E$ there exists a net $(\varphi_i)_{i\in\mathcal{I}}$ in $\mathcal{K}(X)$ such that $\varphi_i \prec O$ for all $i \in \mathcal{I}$, and*

$$\theta_E(a) = \lim_{i\in\mathcal{I}} \int_X \varphi_{i\otimes} a \, d\theta \qquad \textit{for all} \quad a \in \mathcal{P}$$

*in the order topology of $\mathcal{Q}$.*

*Proof.* We may assume that the cone $(\mathcal{P}, \mathcal{V})$ is full. Let $\theta$ be a bounded measure on $\mathfrak{R}$. Propositions 1.8(a) and 4.1(b) imply that all functions in $\mathcal{E}(X, \mathcal{P})$ are integrable over any set $F \in \mathfrak{A}_R$ with respect to $\theta$. For Part (a) of our proposition, let $O \in \mathfrak{R}$ be an open set such that $\theta$ is inner regular for $O$. Let us first consider the case of a positive element $0 \leq a \in \mathcal{P}$. Let $K$ be a compact subset of $O$. Following Urysohn's lemma (see 2.12 in [179]) there is $\varphi \in \mathcal{K}(X)$ such that $K \prec \varphi \prec O$. Thus

$$\theta_K(a) \leq \int_X \varphi \circledast a \, d\theta \leq \theta_E(a).$$

This shows $\theta_K(a) \leq \lim_{\varphi \prec O} \int_X \varphi \circledast a \, d\theta$, hence

$$\theta_O(a) = \lim_{K \subset O} \theta_K(a) \leq \lim_{\varphi \prec O} \int_X \varphi \circledast a \, d\theta \leq \theta_O(a),$$

and therefore $\theta_O(a) = \lim_{\varphi \prec O} \int_X \varphi \circledast a \, d\theta$. Now for the general case, let $a \in \mathcal{P}$. For $w \in \mathcal{W}$ there is $v \in \mathcal{V}$ such that $\theta_O(v) \leq w$, and there is $\lambda \geq 0$ such that $a + \lambda v \geq 0$. As argued before, we have

$$\theta_O(a + \lambda v) = \lim_{\varphi \prec O} \int_X \varphi \circledast (a + \lambda v) \, d\theta \qquad \text{and} \qquad \theta_O(v) = \lim_{\varphi \prec O} \int_X \varphi \circledast v \, d\theta.$$

Thus using Lemma I.5.19

$$\begin{aligned} \overline{\lim_{\varphi \prec O}} \int_X \varphi \circledast a \, d\theta + \lambda \theta_O(v) &= \overline{\lim_{\varphi \prec O}} \int_X \varphi \circledast a \, d\theta + \underline{\lim_{\varphi \prec O}} \int_X \varphi \circledast v \, d\theta \\ &\leq \lim_{\varphi \prec O} \int_X \varphi \circledast (a + \lambda v) \, d\theta \\ &= \theta_O(a) + \lambda \theta_O(v). \end{aligned}$$

The cancellation law in Proposition I.5.10(a) now yields

$$\overline{\lim_{\varphi \prec O}} \int_X \varphi \circledast a \, d\theta \leq \theta_O(a) + \varepsilon w$$

for all $\varepsilon > 0$, since $\theta_O(v) \leq w$. The latter holds true for all $w \in \mathcal{W}$ and therefore demonstrates

$$\overline{\lim_{\varphi \prec O}} \int_X \varphi \circledast a \, d\theta \leq \theta_O(a).$$

Similarly, one argues that

$$\begin{aligned}\theta_O(a)+\lambda\theta_O(v) &= \lim_{\varphi\prec O}\int_X \varphi_{\otimes}(a+\lambda v)\,d\theta \\ &\le \underline{\lim}_{\varphi\prec O}\int_X \varphi_{\otimes}a\,d\theta + \lambda\,\overline{\lim}_{\varphi\prec O}\int_X \varphi_{\otimes}v\,d\theta \\ &= \underline{\lim}_{\varphi\prec O}\int_X \varphi_{\otimes}a\,d\theta + \lambda\theta_O(v)\end{aligned}$$

implies

$$\theta_O(a) \le \underline{\lim}_{\varphi\prec O}\int_X \varphi_{\otimes}a\,d\theta.$$

This completes the proof of Part (a).

For Part (b), suppose that $\theta$ is outer regular for the compact set $K \in \mathfrak{R}$. First, let $0 \le a \in \mathcal{P}$. The net $\left(\int_X \varphi_{\otimes}a\,d\theta\right)_{\varphi\succ K}$ is decreasing and bounded below, hence convergent in $\mathcal{Q}$. For every open set $O \in \mathfrak{R}$ containing $K$ there is $\varphi \in \mathcal{K}(X)$ such that $K \prec \varphi \prec O$. This shows $\int_X \varphi_{\otimes}a\,d\theta \le \theta_O(a)$, hence $\lim_{\varphi\succ K}\int_X \varphi_{\otimes}a\,d\theta \le \theta_O(a)$ and therefore

$$\lim_{\varphi\succ K}\int_X \varphi_{\otimes}a\,d\theta \le \theta_K(a)$$

by the outer regularity of $\theta$ for the set $K$. For the converse inequality, let $\varphi \succ K$ and $\gamma > 1$. There is an open set $O \supset K$ such that $O \prec \gamma\varphi$, hence $\theta_K(a) \le \theta_O(a) \le \gamma\int_X \varphi_{\otimes}a\,d\theta$. As this holds for all $\varphi \succ K$ we infer that

$$\theta_K(a) \le \gamma \lim_{\varphi\succ K}\int_X \varphi_{\otimes}a\,d\theta.$$

Finally, as $\gamma > 1$ was arbitrarily chosen, and as $\mathcal{Q}$ is a locally convex complete lattice cone, this yields together with the above $\theta_K(a) = \lim_{\varphi\succ K}\int_X \varphi_{\otimes}a\,d\theta$. Now for the general case, let $a \in \mathcal{P}$. Given any $w \in \mathcal{W}$ there is $v \in \mathcal{V}$ such that $\theta_K(v) \le w$ and $\lambda \ge 0$ such that $0 \le a+\lambda v$. As we observed before, the latter yields

$$\theta_K(a+\lambda v) = \lim_{\varphi\succ K}\int_X \varphi_{\otimes}(a+\lambda v)\,d\theta \qquad\text{and}\qquad \theta_K(v) = \lim_{\varphi\succ K}\int_X \varphi_{\otimes}v\,d\theta.$$

Thus

$$\begin{aligned}\overline{\lim}_{\varphi\succ K}\int_X \varphi_{\otimes}a\,d\theta + \lambda\theta_K(v) &= \overline{\lim}_{\varphi\succ K}\int_X \varphi_{\otimes}a\,d\theta + \underline{\lim}_{\varphi\succ K}\int_X \varphi_{\otimes}v\,d\theta \\ &\le \lim_{\varphi\succ K}\int_X \varphi_{\otimes}(a+\lambda v)\,d\theta \\ &= \theta_K(a)+\lambda\theta_K(v).\end{aligned}$$

As $\theta_K(v) \leq \lambda w$, the cancellation law in Proposition I.5.10 yields

$$\overline{\lim_{\varphi \succ K}} \int_X \varphi \otimes a \, d\theta \leq \theta_K(a) + \varepsilon w \qquad \text{and} \qquad \theta_K(a) \leq \underline{\lim_{\varphi \succ K}} \int_X \varphi \otimes a \, d\theta + \varepsilon w$$

for all $\varepsilon > 0$. As $w \in \mathcal{W}$ was arbitrary, this yields $\theta_K(a) = \lim_{\varphi \succ K} T(\varphi \otimes a)$.

For Part (c), suppose that $\theta$ is quasi regular and let $E \in \mathfrak{R}$. Let $O_0 \in \mathfrak{R}$ be an open set containing $E$. Then for every open set $O \in \mathfrak{R}$ such that $E \subset O \subset O_0$, following Part (a), the element $\theta_O(a) \in \mathcal{Q}$ is the limit of the net $\left(\int_X \varphi \otimes a \, d\theta\right)_{\varphi \prec O}$. The outer regularity of $\theta$ yields

$$\theta_E(a) = \lim_{O_0 \supset O \supset E} \theta_O(a).$$

Now a well-known diagonal principle for convergent nets (see for example I.6.A in [59] or 11D in [198]) yields that there is a diagonal net $\left(\int_X \varphi_i \otimes a \, d\theta\right)_{i \in \mathcal{I}}$ converging to $\theta_E(a)$ in the order topology of $\mathcal{Q}$. More precisely: Let $\mathcal{O}$ be the family of all open sets $O$ such that $O_0 \supset O \supset E$. Then the index set $\mathcal{I}$ of the diagonal net consists of all ordered pairs $(O, f)$, where $O \in \mathcal{O}$ and $f : \mathcal{O} \to \mathcal{K}(X)$ is a mapping such that $f(O) \prec O$ for all $O \in \mathcal{O}$. This index set is ordered by $(O_1, f_1) \leq (O_2, f_2)$ if $O_1 \supset O_2$ and $f_1(O) \leq f_2(O)$ for all $O \in \mathcal{O}$. If we set $\varphi_i = f(O) \in \mathcal{K}(X)$ for $i = (O, f) \in \mathcal{I}$, then it can be easily verified that $\theta_E(a) = \lim_{i \in \mathcal{I}} \int_X \varphi_i \otimes a \, d\theta$. □

Note that the limit in 4.5(c) refers to the order topology of $\mathcal{Q}$, not necessarily to order convergence as defined in I.5.18.

**Corollary 4.6.** *Let $\theta$ and $\vartheta$ be bounded quasi regular measures on $\mathfrak{R}$ and let $a \in \mathcal{P}$. If $\int_X \varphi \otimes a \, d\theta = \int_X \varphi \otimes a \, d\vartheta$ for all $\varphi \in \mathcal{K}(X)$, then $\theta_E(a) = \vartheta_E(a)$ for all $E \in \mathfrak{R}$. In particular, the measures $\theta$ and $\vartheta$ coincide, provided that their integrals for all functions in $\mathcal{E}(X, \mathcal{P})$ coincide.*

*Proof.* We have $\theta_O(a) = \vartheta_O(a)$ for all open sets $O \in \mathfrak{R}$ by Proposition 4.5(a), hence $\theta_E(a) = \vartheta_E(a)$ for all sets $E \in \mathfrak{R}$ due to outer regularity. □

**4.7 Measures as Continuous Linear Operators.** We shall say that a bounded operator-valued measure $\theta : \mathfrak{R} \to \mathfrak{L}(\mathcal{P}, \mathcal{Q})$ is *continuous relative to a system $\mathfrak{V}$ of inductive limit neighborhoods* if for every $w \in \mathcal{W}$ there is $\mathfrak{v} \in \mathfrak{V}$ such that

$$\int_X f \, d\theta \leq \int_X g \, d\theta + w \qquad \text{whenever} \qquad f \leq g + \mathfrak{v}$$

for all functions $f, g \in \mathcal{F}_{(X,\theta)}(X, \mathcal{P})$, that is the subcone of all functions in $\mathcal{F}(X, \mathcal{P})$ that are integrable over $X$. Note that every $\mathfrak{R}$-bounded measure is continuous relative to some system of inductive limit neighborhoods, that is the system $\mathfrak{V} = \{\mathfrak{v}_w \mid w \in \mathcal{W}\}$ (see Section II.4).

**Proposition 4.8.** *Let $\theta$ be continuous relative to the system $\mathfrak{V}$ of inductive limit neighborhoods.*

(a) *If for $E \in \mathfrak{R}$ and every $\mathfrak{v} \in \mathfrak{V}$ there is $s \in \mathfrak{v}$ such that $s(x) = \infty$ for all $x \in E$, then $\theta_E = 0 \in \mathfrak{L}(\mathcal{P}, \mathcal{Q})$.*

(b) *Every $f \in \mathcal{F}_{\mathfrak{V}}(X, \mathcal{P})$ is integrable over every Borel set $F \in \mathfrak{A}_{\mathfrak{R}}$ with respect to $\theta$.*

(c) *Every $f \in \mathcal{F}_{\mathfrak{V}_0}(X, \mathcal{P})$ is strongly integrable over every Borel set $F \in \mathfrak{A}_{\mathfrak{R}}$ with respect to $\theta$.*

*Proof.* Suppose that $\theta$ is continuous relative to $\mathfrak{V}$. Let $E \in \mathfrak{R}$ as in Part (a). For $w \in \mathcal{W}$ choose the inductive limit neighborhood $\mathfrak{v} \in \mathfrak{V}$ as in 4.7. Following our assumption then there is $s \in \mathfrak{v}$, such that $s(x) = \infty$ for all $x \in E$. Then for every $a \in \mathcal{P}$ we have $\chi_E \otimes a \leq s$ and $0 \leq \chi_E \otimes a + s$. Thus $\int_X \chi_E \otimes a \, d\theta \leq w$ and $0 \leq \int_X \chi_E \otimes a \, d\theta + w$. As $w \in \mathcal{W}$ was arbitrarily chosen, this shows $\theta_E(a) = \int_X \chi_E \otimes a \, d\theta = 0$, as claimed in (a).

For (b), let $f \in \mathcal{F}_{\mathfrak{V}}(X, \mathcal{P})$. According to Definitions II.4.12 and II.4.13 we shall first verify integrability with respect to $\theta$ over a set $E \in \mathfrak{R}$. For this, let $w \in \mathcal{W}$ and $0 < \varepsilon \leq 1$. There is an inductive limit neighborhood $\mathfrak{v} \in \mathfrak{V}$ such that $\int_X g \, d\theta \leq \int_X h \, d\theta + (\varepsilon/6)w$ holds whenever $g \leq h + \mathfrak{v}$ for integrable functions $g, h \in \mathcal{F}(X, \mathcal{P})$. By our definition of the cone $\mathcal{F}_{\mathfrak{V}}(X, \mathcal{P})$ as the closure of $\mathcal{E}(X, \mathcal{P})$ with respect to the symmetric relative topology generated by $\mathfrak{V}$ there is $g \in \mathcal{E}(X, \mathcal{P})$ such that both $f \in \mathfrak{v}_\varepsilon^s(g)$, that is

$$f \leq \gamma g + s \qquad \text{and} \qquad g \leq \gamma' f + s'$$

for some $1 \leq \gamma, \gamma' \leq 1 + \varepsilon$ and $s, s' \in \mathfrak{v}$. The functions in $\mathfrak{v}$ are supposed to be measurable, but can take the values $0$ and $\infty \in \overline{\mathcal{V}}$. To remedy this, we choose $v \in \mathcal{V}$ such that $(\varepsilon/2)\chi_E \otimes v \leq \mathfrak{v}$, and for any $s \in \mathfrak{v}$ we set $\tilde{s} = \chi_{(X \setminus F)} \otimes s + \chi_X \otimes v$, where $F \in \mathfrak{A}_{\mathfrak{R}}$ is the set of all points in $X$ where the function $s$ takes the value $\infty$. Then the function $\tilde{s}$ is $\mathcal{V}$-valued and measurable, hence contained in $\mathcal{F}_{\mathfrak{R}}(X, \mathcal{V})$. Moreover, according to Lemma II.4.4 we have

$$\int_E \tilde{s} \, d\theta = \sup \left\{ \int_E h \, d\theta \;\middle|\; h \in \mathcal{S}_{\mathfrak{R}}(X, \mathcal{P}), \quad h \leq \tilde{s} \right\} \leq \frac{\varepsilon}{3} w,$$

since $h \leq \tilde{s}$ implies that $\chi_E \otimes h \leq s + \chi_E \otimes v \leq 2\mathfrak{v}$, hence $\int_E h \, d\theta \leq (\varepsilon/3)w$. For Definition II.4.12 now we choose

$$f_{(w,\varepsilon)} = \gamma g + \tilde{s} \qquad \text{and} \qquad s_{(w,\varepsilon)} = (\gamma + 1)\tilde{s} \leq 3\tilde{s}.$$

Then $f_{(w,\varepsilon)} \in \mathcal{F}_{\mathfrak{R}}(X, \mathcal{P}_{\mathcal{V}})$ by Proposition 4.1(b) and $s_{(w,\varepsilon)} \in \mathcal{F}_{\mathfrak{R}}(X, \mathcal{V})$. Moreover,

$$f(x) \leq f_{(w,\varepsilon)}(x) \leq \gamma f(x) + s_{(w,\varepsilon)}(x)$$

holds for all $x \in X \setminus F$, hence

$$f \underset{a.e.E}{\leq} f_{(w,\varepsilon)} \underset{a.e.E}{\leq} \gamma f + s_{(w,\varepsilon)} \qquad \text{and} \qquad \int_E s_{(w,\varepsilon)}\, d\theta \leq \varepsilon w,$$

since $F$ was seen to be a set of measure $0$ in Part (a). This shows integrability of $f$ over the set $E \in \mathfrak{R}$ by II.4.12. For integrability over a set $F \in \mathfrak{A}_{\mathfrak{R}}$ we have to verify that the limit

$$\int_F f\, d\theta = \lim_{E \in \mathfrak{R}} \int_{(E \cap F)} f\, d\theta$$

exists in $\mathcal{Q}$. According to Proposition I.5.41 $(\mathcal{Q}, \mathcal{W})$ is complete with respect to the symmetric relative topology of $\mathcal{Q}$, and according to I.5.42, convergence in this topology implies order convergence as required in the definition of the integral. It suffices therefore to verify that $\big(\int_{(E\cap F)} f\, d\theta\big)_{E \in \mathfrak{R}}$ forms a Cauchy net in $\mathcal{Q}$. For this, let $w \in \mathcal{W}$ and $0 < \varepsilon \leq 1$. There is $\mathfrak{v} \in \mathfrak{V}$ such that $\int_X j\, d\theta \leq \int_X l\, d\theta + w$ whenever $j \leq l + \mathfrak{v}$ for integrable functions $j, l \in \mathcal{F}(X, \mathcal{P})$, and in turn there is $g \in \mathcal{E}(X, \mathcal{P})$ such that

$$f \leq \gamma g + \frac{\varepsilon}{3}\mathfrak{v} \qquad \text{and} \qquad g \leq \gamma' f + \frac{\varepsilon}{3}\mathfrak{v}$$

for some $1 \leq \gamma, \gamma' \leq 1 + \varepsilon/3$. Let $E_0 \in \mathfrak{R}$ be a set that contains the support of the function $g$. Then for $E_1, E_2 \in \mathfrak{R}$ such that both $E_0 \subset E_1$ and $E_0 \subset E_2$ we have $\int_{E_1} g\, d\theta = \int_{E_2} g\, d\theta$, hence

$$\int_{E_1} f\, d\theta \leq \gamma \int_{E_1} g\, d\theta + \frac{\varepsilon}{3} w = \gamma \int_{E_2} g\, d\theta + \frac{\varepsilon}{3} w \leq \gamma\gamma' \int_{E_2} f\, d\theta + \varepsilon w,$$

and likewise

$$\int_{E_2} f\, d\theta \leq \gamma\gamma' \int_{E_1} f\, d\theta + \varepsilon w.$$

As $1 \leq \gamma\gamma' \leq 1 + \varepsilon$, this shows $\int_{E_1} f\, d\theta \in \mathfrak{v}^s_\varepsilon\big(\int_{E_2} f\, d\theta\big)$. The net $\big(\int_{(E\cap F)} f\, d\theta\big)_{E \in \mathfrak{R}}$ is therefore indeed a Cauchy net, hence the function $f$ is integrable over $F$ as claimed.

For Part (c), all left to show is that, according to the definition of strong integrability in II.5.18, for every $f \in \mathcal{F}_{\mathfrak{V}_0}(X, \mathcal{P})$, and $w \in \mathcal{W}$ there is a step function $h \in \mathcal{S}_{\mathfrak{R}}(X, \mathcal{P})$ such that $\int_E f\, d\theta \leq \int_E h\, d\theta + w$ and $\int_E h\, d\theta$ is $w$-bounded relative to $\int_E f\, d\theta$ for all $E \in \mathfrak{R}$. Let us first consider the case that $f \in \mathcal{E}_0(X, \mathcal{P})$. Let $K$ be the compact support of $f$ and let $v \in \mathcal{V}$ such that $|\theta|(K, v) \leq w$. There is $\lambda \geq 0$ such that $0 \leq f(x) + \lambda v$ for all $x \in K$. Because the function $f$ is r-continuous, given $0 < \varepsilon \leq 1/(1+\lambda)$, there is a partition of $K$ into finitely many disjoint sets $E_1, \ldots, E_n \in \mathfrak{R}$ such that $f(x) \in v_\varepsilon\big(f(y)\big)$, that is (see Lemma I.4.1(b))

$$f(x) \le (1+\varepsilon)f(y) + \varepsilon(1+\lambda)v \le (1+\varepsilon)f(y) + v$$

whenever $x, y$ are contained in the same component $E_i$. We choose $x_i \in E_i$, $a_i = (1+\varepsilon)f(x_i)$ and $h = \sum_{i=1}^{n} \chi_{E_i} \otimes a_i \in \mathcal{S}_{\mathfrak{R}}(X, \mathcal{P})$. Then

$$f \le h + \chi_K \otimes v \qquad \text{and} \qquad h \le (1+\varepsilon)^2 f + 2\chi_K \otimes v.$$

Thus indeed $\int_E f\, d\theta \le \int_E h\, d\theta + w$ and $\int_E h\, d\theta \le (1+\varepsilon)^2 \int_E f\, d\theta + 2w$, hence $\int_E h\, d\theta + w$ is $w$-bounded relative to $\int_E f\, d\theta$, for all $E \in \mathfrak{R}$.

Now for the general case, that is $f \in \mathcal{F}_{\mathfrak{V}_0}(X, \mathcal{P})$, first choose $\mathfrak{v} \in \mathcal{V}$ such that $\int_X f\, d\theta \le \int_X g\, d\theta + w/3$ whenever $f \le g + \mathfrak{v}$ for $f, g \in \mathcal{F}_{\mathfrak{R}}(X, \mathcal{P})$. For a given function $f \in \mathcal{F}_{\mathfrak{V}_0}(X, \mathcal{P})$ and $\mathfrak{v} \in \mathfrak{V}$ there is $g \in \mathcal{E}_0(X, \mathcal{P})$ such that both

$$f \le \gamma g + \mathfrak{v} \qquad \text{and} \qquad g \le \gamma' f + \mathfrak{v}$$

for some $1 \le \gamma, \gamma' \le 2$. For the function $g \in \mathcal{E}_0(X, \mathcal{P})$, however, we did verify the existence of a step function $h \in \mathcal{S}_{\mathfrak{R}}(X, \mathcal{P})$ with the required properties, that is $\int_E g\, d\theta \le \int_E h\, d\theta + w/3$ and $\int_E h\, d\theta$ is $w$-bounded relative to $\int_E g\, d\theta$ for all $E \in \mathfrak{R}$. Set $h' = \gamma h \in \mathcal{S}_{\mathfrak{R}}(X, \mathcal{P})$. Then

$$\int_E f\, d\theta \le \int_E \gamma g\, d\theta + w/3 \le \int_E h'\, d\theta + w$$

for all $E \in \mathfrak{R}$. Moreover, $\int_E h'\, d\theta$ is $w$-bounded relative to $\int_E g\, d\theta$ which in turn is $w$-bounded relative to $\int_E f\, d\theta$. Our claim follows. □

In this way, a $\mathfrak{V}$-continuous $\mathfrak{L}(\mathcal{P}, \mathcal{Q})$-valued Borel measure $\theta$ defines a continuous linear operator $T$ from $\mathcal{F}_{\mathfrak{V}}(X, \mathcal{P})$ into $\mathcal{Q}$, that is

$$f \mapsto \int_F f\, d\theta \; : \; \mathcal{F}_{\mathfrak{V}}(X, \mathcal{P}) \to \mathcal{Q}.$$

Recall from Sections 6.12, 6.14 and 6.16 of Chapter II that in special cases certain additional properties transfer from the measure $\theta$ to this operator:

If $\mathcal{P}$ is a locally convex topological vector space over $\mathbb{K} = \mathbb{R}$ or $\mathbb{K} = \mathbb{C}$ and if the operators $\theta_E \in \mathfrak{L}(\mathcal{P}, \mathcal{Q})$ are linear over $\mathbb{K}$, then $T$ is also linear over $\mathbb{K}$ (see II.6.12).

If $\mathcal{P}$ is a topological algebra and if $\mathcal{Q}$ is the standard lattice completion of another topological algebra, then Condition (A) from II.6.14 guarantees that $T$ is an algebra homomorphism on some algebra of integrable functions. Condition (A*) yields compatibility with an involution. For the special case $\mathcal{Q} = \mathbb{K}$, that is for a functional-valued measure $\theta$ we observed in a remark after II.6.15 that (A) means that the functionals $\theta_E$ are multiplicative and either $\theta_E = 0$ or $\theta_G = 0$ holds for disjoint sets $E, G \in \mathfrak{R}$. The latter holds, because otherwise we could find elements $a, b \in \mathcal{P}$ such that $\theta_E(a) = \theta_G(b) = 1$, contradicting the requirement $\theta_E(a)\theta_G(b) = 0$. Consequently, $\theta$ takes at most one non-zero value, that is some multiplicative $\mathbb{K}$-linear functional in $\mu_0 \in \mathcal{P}^*$. If $\theta \ne 0$ is inner regular, this leads to the following conclusion: First we observe that $\theta_{(E\cap G)} = \mu_0$ whenever $\theta_E = \theta_G = \mu_0$ for $E, G \in \mathfrak{R}$. Let $K_0$ be the intersection of all compact sets $K \in \mathfrak{R}$ such that $\theta_K = \mu_0$. A compactness argument together with the above shows that $K_0 \neq \emptyset$. Inner regularity then implies that $\theta_E = 0$ whenever $E \in \mathfrak{R}$ is disjoint to $K_0$. $\theta_E = \mu_0$ on the other hand, implies that $K_0 \subset E$. Indeed, every compact subset $K$ of $E$ is disjoint from $K_0$, hence $\theta_K = 0$. Thus $\theta_E = \theta_{(E\cap K_0)}$ for every $E \in \mathfrak{R}$ and therefore $\theta_{K_0} = \mu_0$. Now the assumption that $K_0$ might contain more than one point is easily contradicted. We infer that $K_0 = \{x_0\}$ for some $x_0 \in X$, and for $E \in \mathfrak{R}$ we have $\theta_E = \mu_0$ if $x_0 \in E$, and $\theta_E = 0$ else. $\theta$ is therefore a point-evaluation measure in this case.

If $\mathcal{P}$ is a $\vee$-semilattice cone, then Condition (L) from II.6.16 implies that $T$ is a $\vee$-semilattice homomorphism on some $\vee$-semilattice of integrable functions. Moreover, if $\mathcal{Q} = \overline{\mathbb{R}}$ and $\theta$ is inner regular, then an argument similar to the above shows that under Condition (L) $\theta$ is a point evaluation, its only non-zero value being a $\vee$-semilattice homomorphism in $\mathcal{P}^*$.

Conversely, we shall demonstrate in our main result of the following section that for an r-lower continuous inductive limit topology, every continuous linear operator from $\mathcal{F}_{\mathfrak{V}}(X, \mathcal{P})$ into $\mathcal{Q}$ can be represented by a unique quasi regular measure. In Section 6 we shall show that in special cases like the above certain additional properties of the operator transfer to this measure.

## 5. Integral Representation

For the following integral representation theorem we shall consider continuous linear operators $T : \mathcal{F}_{\mathfrak{V}}(X, \mathcal{P}) \to \mathcal{Q}$, where $(\mathcal{P}, \mathcal{V})$ is a full and $(\mathcal{Q}, \mathcal{W})$ is a locally convex complete lattice cone such that the order continuous linear functionals support the separation property for $\mathcal{Q}$. $\mathfrak{V}$ is a system of r-lower continuous inductive limit neighborhoods for $\mathcal{F}(X, \mathcal{P})$. The locally convex cones $(\mathcal{F}_{\mathfrak{V}}(X, \mathcal{P}), \mathfrak{V})$ and $(\mathcal{F}_{\mathfrak{V}_0}(X, \mathcal{P}), \mathfrak{V})$ were introduced and investigated in Section 2. We took particular effort to characterize their respective subcones of r-continuous functions. In Section 3 we investigated various situations where continuous linear operators defined only on a subcone of $\mathcal{F}_{\mathfrak{V}}(X, \mathcal{P})$ can be extended. These results apply in particular if $\mathcal{Q} = \overline{\mathbb{R}}$ or if $(\mathcal{P}, \mathcal{V})$ is the standard full extension of some locally convex vector subspace $(\mathcal{P}_0, \mathcal{V})$ and if the operator is defined on a suitable cone of $\mathcal{P}_0$-valued functions (see Theorems 3.2 and 3.3).

**Representation Theorem 5.1.** *Let $(\mathcal{P}, \mathcal{V})$ be a full locally convex cone and let $(\mathcal{Q}, \mathcal{W})$ be a locally convex complete lattice cone such that the order continuous linear functionals support the separation property for $\mathcal{Q}$. Let $X$ be a locally compact Hausdorff space and let $\mathfrak{V}$ be a basis for an r-lower continuous inductive limit topology on $\mathcal{F}(X, \mathcal{P})$. Then every continuous linear operator $T : \mathcal{F}_{\mathfrak{V}}(X, \mathcal{P}) \to \mathcal{Q}$ can be represented as an integral on $X$. More precisely: There exists a unique bounded quasi regular $\mathfrak{L}(\mathcal{P}, \mathcal{Q})$-valued* *measure* $\theta$ *on the weak* $\sigma$*-ring* $\mathfrak{R}$ *of all relatively compact Borel subsets of* $X$ *such that* $\theta$ *is continuous relative to* $\mathfrak{V}$*, all functions in* $\mathcal{F}_{\mathfrak{V}}(X,\mathcal{P})$ *are integrable, all functions in* $\mathcal{F}_{\mathfrak{V}_0}(X,\mathcal{P})$ *are strongly integrable with respect to* $\theta$*, and*

$$\int_X \varphi{\otimes}a\, d\theta \le T(\varphi{\otimes}a) \le \int_X \varphi{\otimes}a\, d\theta + \mathfrak{O}\big(\theta_A(a)\big)$$

*holds for all* $\varphi \in \mathcal{K}(X)$ *and* $a \in \mathcal{P}$*, where* $A$ *denotes the compact support of the function* $\varphi$*. Thus*

$$\int_X f\, d\theta \le T(f) \qquad \text{and indeed} \qquad \int_X g\, d\theta = T(g)$$

*holds for all* $f \in \mathcal{F}_{\mathfrak{V}}(X,\mathcal{P})$ *and all* $g \in \mathcal{F}_{\mathfrak{V}_0}(X,\mathcal{P})$*, respectively.*

*Proof.* Under the assumptions of this theorem, let $T : \mathcal{F}_{\mathfrak{V}}(X,\mathcal{P}) \to \mathcal{Q}$ be a continuous linear operator, that is for every $w \in \mathcal{W}$ there is an r-lower continuous neighborhood $\mathfrak{u}_w \in \mathfrak{V}$ such that

$$T(f) \le T(g) + w \qquad \text{whenever} \qquad f \le g + \mathfrak{u}_w$$

for $f,g \in \mathcal{F}_{\mathfrak{V}}(X,\mathcal{P})$. We proceed to construct a bounded inner regular $\mathfrak{L}(\mathcal{P},\mathcal{Q})$-valued measure $\theta$ on $\mathfrak{R}$, the weak $\sigma$-ring of all relatively compact Borel subsets of $X$. We shall follow some of the main lines of the standard proof for the Riesz representation theorem (see for example [178]), though the presence of unbounded elements in $\mathcal{P}$, and the non-availability of negatives will complicate matters considerably. We shall first list a few notations and abbreviations: For a set $E \in \mathfrak{R}$ and a function $\varphi \in \mathcal{K}(X)$ we write $E \prec \varphi$ if $\chi_E \le \varphi$, and $\varphi \prec E$ if $\varphi \le \chi_E$ and $\operatorname{supp}(\varphi) \subset E$.

We are ready to begin with our step-by-step construction of the measure $\theta$. In a first step, we shall define the measure $\theta$ on all relatively compact open subsets of $X$. For a relatively compact open set $O \in \mathfrak{R}$ and an element $a \in \mathcal{P}$ we set

(i) $$\theta_O(a) = \lim_{\varphi \prec O} T(\varphi{\otimes}a).$$

This limit is taken over the net whose index set is the upward directed family of all functions $\varphi \in \mathcal{K}(X)$ such that $\varphi \prec O$. For a positive element $0 \le a \in \mathcal{P}$ this limit obviously exists, as the net $\big(T(\varphi{\otimes}a)\big)_{\varphi \prec O}$ is increasing. We shall use the criterion in Proposition I.5.23 to verify that this limit exists in general for an element $a \in \mathcal{P}$. To prepare this, let us observe the following:

(ii) If $\chi_O{\otimes}a \le \chi_O{\otimes}b + \mathfrak{u}_w$ for $0 \le a,b \in \mathcal{P}$, $w \in \mathcal{W}$ and an open set $O \in \mathfrak{R}$, then $\theta_O(a) \le \theta_O(b) + w$. In particular, if $\chi_O{\otimes}v \le \mathfrak{u}_w$ for some $v \in \mathcal{V}$, then $\theta_O(v) \le w$.

Indeed, let $0 \le a,b \in \mathcal{P}$, let $O \in \mathfrak{R}$ be an open set and let $\chi_O{\otimes}a \le \chi_O{\otimes}b + \mathfrak{u}_w$, that is $\chi_O{\otimes}a \le \chi_O{\otimes}b + s$ for some $s \in \mathfrak{u}_w$. Let $\varphi \prec O$. Then

$$\varphi_{\otimes}a \leq \varphi_{\otimes}b + \varphi_{\otimes}s \leq \varphi_{\otimes}b + \mathfrak{u}_w.$$

By the continuity of the operator $T$, this implies

$$T(\varphi_{\otimes}a) \leq T(\varphi_{\otimes}b) + w.$$

Taking the limit over all functions $\varphi \prec O$ in this last inequality yields (ii) (see Lemma I.5.20(c)).

Now let $a \in \mathcal{P}$. Given any neighborhood $w \in \mathcal{W}$ there is $v \in \mathcal{V}$ such that $\chi_{O}{\otimes}v \leq \mathfrak{u}_w$, and there is $\lambda \geq 0$ such that $0 \leq a + \lambda v$. As argued before, the nets $\big(T(\varphi_{\otimes}\lambda v)\big)_{\varphi\prec O}$ and $\big(T(\varphi_{\otimes}a)+T(\varphi_{\otimes}\lambda v)\big)_{\varphi\prec O}$ are convergent in $\mathcal{Q}$, and $\theta_O(v) \leq w$ by (ii). Thus

$$\lim_{\varphi\prec O} T(\varphi_{\otimes}\lambda v) = \theta_O(\lambda v) = \lambda\theta_O(v) \leq \lambda w.$$

Now Proposition I.5.23 yields the convergence of the net $\big(T(\varphi_{\otimes}a)\big)_{\varphi\prec O}$.

Let us observe that $\theta_{\emptyset} = 0$ since $\varphi = 0$ is the only function in $\mathcal{K}(X)$ such that $\operatorname{supp}(\varphi) \subset \emptyset$. Because $T$ is linear on $\mathcal{E}(X,\mathcal{P})$, we have

$$T\big(\varphi_{\otimes}(\alpha a)\big) = \alpha T\big(\varphi_{\otimes}a\big) \qquad \text{and} \qquad T\big(\varphi_{\otimes}(a+b)\big) = T\big(\varphi_{\otimes}a\big) + T\big(\varphi_{\otimes}b\big)$$

for an open set $O \in \mathfrak{R}$, for all $a,b \in \mathcal{P}$, $\alpha \geq 0$ and $\varphi \prec O$. Following Lemmas I.5.19 and I.5.21 this implies

$$\theta_O(\alpha a) = \lim_{\varphi\prec O} T\big(\varphi_{\otimes}(\alpha a)\big) = \alpha \lim_{\varphi\prec O} T(\varphi_{\otimes}a) = \alpha\theta_O(a)$$

and

$$\theta_O(a+b) = \lim_{\varphi\prec O} T\big(\varphi_{\otimes}(a+b)\big) = \lim_{\varphi\prec O} T(\varphi_{\otimes}a) + \lim_{\varphi\prec O} T(\varphi_{\otimes}b) = \theta_O(a) + \theta_O(a).$$

Moreover, if $a \leq b$ for $a,b \in \mathcal{P}$, then $\varphi_{\otimes}a \leq \varphi_{\otimes}b$, hence $T(\varphi_{\otimes}a) \leq T(\varphi_{\otimes}b)$, and therefore $\theta_O(a) \leq \theta_O(b)$. In this way $\theta_O$ defines a monotone linear operator from $\mathcal{P}$ to $\mathcal{Q}$. Obviously, this operator is also continuous: Given $w \in W$ we choose $v \in \mathcal{V}$ such that $\chi_{O}{\otimes}v \leq \mathfrak{u}_w$. Then $\theta_O(v) \leq w$ by (ii), and $a \leq b+v$ for elements $a,b \in \mathcal{P}$ therefore implies

$$\theta_O(a) \leq \theta_O(b+v) \leq \theta_O(b) + w.$$

This yields

(iii) $\theta_O \in \mathfrak{L}(\mathcal{P},\mathcal{Q})$ for all open sets $O \in \mathfrak{R}$.

Next we observe

(iv) If $0 \leq a$ for $a \in \mathcal{P}$, and if $O \subset U$ for open sets $O,U \in \mathfrak{R}$, then $0 \leq \theta_O(a) \leq \theta_U(a)$.

For this, let $a, O, U$ be as in (iv). Then

$$\theta_O(a) = \lim_{\varphi \prec O} T(\varphi \otimes a) = \sup_{\varphi \prec O} \{T(\varphi \otimes a)\} \leq \sup_{\varphi \prec U} \{T(\varphi \otimes a)\} = \theta_U(a).$$

(v) If $0 \leq a$ for $a \in \mathcal{P}$, and if $O_i \in \mathfrak{R}$ are open sets such that $O = \bigcup_{i \in \mathbb{N}} O_i \in \mathfrak{R}$, then $\theta_O(a) \leq \sum_{i=1}^{\infty} \theta_{O_i}(a)$.

Indeed, let $\varphi \prec O$. Because the support of $\varphi$ is a compact subset of $O$, it is covered by finitely many of the open sets $O_i$. Thus there are functions $\varphi_1, \ldots, \varphi_n \in \mathcal{K}(X)$ such that $\varphi_i \prec O_i$ and $\sum_{i=1}^{n} \varphi_i(x) = 1$ for all $x \in \operatorname{supp}(\varphi)$. Then $\varphi_i \varphi \prec O_i$ and $\varphi = \sum_{i=1}^{n} \varphi_i \varphi$. Using (iv) we infer

$$T(\varphi \otimes a) = \sum_{i=1}^{n} T\big((\varphi_i \varphi) \otimes a\big) \leq \sum_{i=1}^{n} \theta_{O_i}(a) \leq \sum_{i=1}^{\infty} \theta_{O_i}(a).$$

Taking the supremum, that is the limit over all $\varphi \prec O$, yields our claim.

(vi) If the open sets $O_i \in \mathfrak{R}$ are pairwise disjoint and if $O = \bigcup_{i \in \mathbb{N}} O_i \in \mathfrak{R}$, then $\theta_O = \sum_{i=1}^{\infty} \theta_{O_i}$, that is $\theta_O(a) = \sum_{i=1}^{\infty} \theta_{O_i}(a)$ for all $a \in \mathcal{P}$.

Let $O, O_i \in \mathfrak{R}$ be as in (vi). First suppose that $0 \leq a$ for $a \in \mathcal{P}$. We fix $n \in \mathbb{N}$ and choose functions $\varphi_i \prec O_i$ for $i = 1, \ldots, n$. Then $\varphi = \sum_{i=1}^{n} \varphi_i \prec O$, hence

$$\sum_{i=1}^{n} T(\varphi_i \otimes a) = T(\varphi \otimes a) \leq \theta_O(a).$$

Taking the suprema overall such choices of the functions $\varphi_i \prec O_i$ yields

$$\sum_{i=1}^{n} \theta_{O_i}(a) \leq \theta_O(a).$$

This holds for all $n \in \mathbb{N}$, and as

$$\sum_{i=1}^{\infty} \theta_{O_i}(a) = \sup_{n \in \mathbb{N}} \left\{ \sum_{i=1}^{n} \theta_{O_i}(a) \right\},$$

we infer that $\sum_{i=1}^{\infty} \theta_{O_i}(a) \leq \theta_O(a)$. Together with (v), this yields

$$\sum_{i=1}^{\infty} \theta_{O_i}(a) = \theta_O(a).$$

Now for the general case, let $a \in \mathcal{P}$ and $w \in \mathcal{W}$. There is $v \in \mathcal{V}$ such that $\chi_O \otimes v \leq \mathfrak{u}_w$ and $\lambda \geq 0$ such that $0 \leq a + \lambda v$. Together with the limit rules in Lemma I.5.19, the above shows

$$\begin{aligned}\overline{\lim_{n\to\infty}}\Big(\sum_{i=1}^{n}\theta_{O_i}(a)\Big)+\theta_O(\lambda v)&=\overline{\lim_{n\to\infty}}\Big(\sum_{i=1}^{n}\theta_{O_i}(a)\Big)+\lim_{n\to\infty}\Big(\sum_{i=1}^{n}\theta_{O_i}(v)\Big)\\&\le\overline{\lim_{n\to\infty}}\Big(\sum_{i=1}^{n}\theta_{O_i}(a+\lambda v)\Big)\\&=\sum_{i=1}^{\infty}\Big(\theta_{O_i}(a+\lambda v)\Big)\\&=\theta_O(a)+\theta_O(\lambda v)\end{aligned}$$

and

$$\begin{aligned}\theta_O(a)+\theta_O(\lambda v)&=\underline{\lim}_{n\to\infty}\Big(\sum_{i=1}^{n}\theta_{O_i}(a+\lambda v)\Big)\\&\le\underline{\lim}_{n\to\infty}\Big(\sum_{i=1}^{n}\theta_{O_i}(a)\Big)+\lim_{n\to\infty}\Big(\sum_{i=1}^{n}\theta_{O_i}(v)\Big)\\&=\underline{\lim}_{n\to\infty}\Big(\sum_{i=1}^{n}\theta_{O_i}(a)\Big)+\theta_O(\lambda v).\end{aligned}$$

Because $\theta_O(\lambda v)\le\lambda w$ by (ii), the cancellation law in Proposition I.5.9 yields that

$$\overline{\lim_{n\to\infty}}\Big(\sum_{i=1}^{n}\theta_{O_i}(a)\Big)\le\theta_O(a)+\varepsilon w\qquad\text{and}\qquad\theta_O(a)\le\underline{\lim}_{n\to\infty}\Big(\sum_{i=1}^{n}\theta_{O_i}(a)\Big)+\varepsilon w$$

for all $\varepsilon>0$. As $w\in\mathcal{W}$ was arbitrarily chosen, we infer that

$$\sum_{i=1}^{\infty}\theta_{O_i}(a)=\theta_O(a),$$

hence our claim follows.

Now, in the second step of the construction of the measure $\theta$ we proceed to extend the definition of $\theta$ to all relatively compact subsets of $X$. For a relatively compact set $E\subset X$ and an element $a\in\mathcal{P}$ we set

(vii) $$\theta_E(a)=\lim_{O\supset E}\theta_O(a).$$

This limit is taken over the net whose index set is the downward directed family of all open sets $O\in\mathfrak{R}$ containing $E$. For a positive element $0\le a\in\mathcal{P}$ this limit obviously exists, since the net $\big(\theta_O(a)\big)_{O\supset E}$ is decreasing in $\mathcal{Q}$ as we established in (iv). For the general case we shall again use the criterion in Proposition I.5.23 to verify convergence. First we observe the following:

(viii) If $\chi_{E\otimes}a \leq \chi_{E\otimes}b + \mathfrak{u}_w$ for $0 \leq a, b \in \mathcal{P}$, $w \in \mathcal{W}$ and a set $E \in \mathfrak{R}$, then $\theta_E(a) \leq \theta_E(b) + w$. In particular, if $\chi_{E\otimes}v \leq \mathfrak{u}_w$ for some $v \in \mathcal{V}$, then $\theta_E(v) \leq w$.

To verify this, let $0 \leq a, b \in \mathcal{P}$, $w \in \mathcal{W}$ and $E \in \mathfrak{R}$, and suppose that $\chi_{E\otimes}a \leq \chi_{E\otimes}b + \mathfrak{u}_w$, that is $\chi_{E\otimes}a \leq \chi_{E\otimes}b + s$ for some r-lower continuous $\overline{\mathcal{V}}$-valued function $s \in \mathfrak{u}_w$. Let $O \in \mathfrak{R}$ be an open set such that $E \subset O$. There is $v \in \mathcal{V}$ such that $\chi_{O\otimes}v \leq \mathfrak{u}_w$. Let $\varepsilon > 0$. By the r-lower continuity of the function $s + \chi_{O\otimes}b$ (see Lemma 1.1), for every $x \in E$ there is an open neighborhood $U_x \subset O$ of $x$ such that $a \leq (1+\varepsilon)\big(s(y) + b\big) + \varepsilon v$ for all $y \in U_x$. If we set $U = \bigcup_{x \in E} U_x$, then $U$ is an open subset of $O$, hence $U \in \mathfrak{R}$, and we have

$$\begin{aligned} \chi_{U\otimes}a &\leq (1+\varepsilon)(s + \chi_{U\otimes}b) + \varepsilon\, \chi_{U\otimes}v \\ &\leq (1+\varepsilon)\chi_{U\otimes}b + (1+2\varepsilon)\mathfrak{u}_w. \end{aligned}$$

This shows

$$\theta_U(a) \leq (1+\varepsilon)\theta_U(b) + (1+2\varepsilon)w \leq (1+2\varepsilon)\big(\theta_U(b) + w\big)$$

by (ii). Because $\varepsilon > 0$ was arbitrarily chosen and because $(\mathcal{Q}, \mathcal{W})$ is endowed with its weak preorder, this yields $\theta_U(a) \leq \theta_U(b) + w$. Taking the limit (infimum) over all open sets $E \subset U \subset O$ in this last inequality demonstrates (viii).

Now in order to show convergence of the net in (vii) in the general case, let $E \subset X$ be relatively compact, and let $a \in \mathcal{P}$. Given any neighborhood $w \in \mathcal{W}$ there is $v \in \mathcal{V}$ such that $\chi_{E\otimes}v \leq \mathfrak{u}_w$, and there is $\lambda \geq 0$ such that $0 \leq a + \lambda v$. As we remarked before, the net $\big(\theta_O(\lambda v)\big)_{O \supset E}$ is convergent, and $\lim_{O \supset E} \theta_O(\lambda v) = \theta_E(\lambda v) \leq \lambda w$ by (viii). Likewise, the net $\big(\theta_O(a) + \theta_O(\lambda v)\big)_{O \supset E}$ is decreasing, hence convergent in $\mathcal{Q}$. Thus Proposition I.5.23 yields indeed the convergence of the net $\big(\theta_O(a)\big)_{O \supset E}$.

The linearity of $\theta_E$ as an operator from $\mathcal{P}$ into $\mathcal{Q}$ follows immediately from the corresponding properties of the operators $\theta_O$, for $O \supset E$, as stated in (iii), together with the limit rules in Lemma I.5.19. If $a \leq b$ for $a, b \in \mathcal{P}$, then $\theta_O(a) \leq \theta_O(b)$ for all $O \supset E$, thus $\theta_E(a) \leq \theta_E(b)$. The operator $\theta_E$ is therefore monotone and also continuous: Given $w \in W$ we choose $v \in \mathcal{V}$ such that $\chi_{E\otimes}v \leq \mathfrak{u}_w$. Then $\theta_E(v) \leq w$ by (viii), and $a \leq b + v$ for elements $a, b \in \mathcal{P}$ implies

$$\theta_E(a) \leq \theta_E(b + v) \leq \theta_E(b) + w.$$

This yields

(ix) $\theta_E \in \mathfrak{L}(\mathcal{P}, \mathcal{Q})$ for all relatively compact sets $E \in \mathfrak{R}$.

We have

(x) If $0 \leq a$ for $a \in \mathcal{P}$, and if $E \subset F$ for relatively compact sets $E, F \subset X$, then $0 \leq \theta_E(a) \leq \theta_F(a)$.

Indeed, let $a, E, F$ be as in (x). Then

$$\theta_E(a) = \lim_{O \supset E} \theta_O(a) = \inf_{O \supset E} \{\theta_O(a)\} \leq \inf_{O \supset F} \{\theta_O(a)\} = \theta_F(a).$$

Next we shall verify that $\theta$ is inner regular on all open sets in $\mathfrak{R}$, that is

(xi) $\theta_O(a) = \lim_{K \subset O} \theta_K(a)$ for every open set $O \in \mathfrak{R}$ and $a \in \mathcal{P}$.

The limit in (x) is taken over the upward directed family of all compact sets $K \subset O$. To verify this, let $O \in \mathfrak{R}$ be open. First, let us consider an element $0 \leq a \in \mathcal{P}$. In this case the net $(\theta_K(a))_{K \subset O}$ is increasing, hence convergent in $\mathcal{Q}$, and obviously $\lim_{K \subset O} (\theta_K(a)) \leq \theta_O(a)$. For the reverse inequality, let $\varphi \prec O$ and $K = \operatorname{supp}(\varphi) \subset O$. Then $T(\varphi \otimes a) \leq \theta_U(a)$ for every open set $K \subset U \in \mathfrak{R}$, hence $T(\varphi \otimes a) \leq \theta_K(a)$, and therefore $T(\varphi \otimes a) \leq \lim_{K \subset O} \theta_K(a)$. Since this holds for all functions $\varphi \prec O$, we conclude that $\theta_O(a) \leq \lim_{K \subset O} \theta_K(a)$ holds as well.

Now for the general case let $a \in \mathcal{P}$. Given any $w \in \mathcal{W}$ there is $v \in \mathcal{V}$ such that $\chi_O \otimes v \leq \mathfrak{u}_w$ and $\lambda \geq 0$ such that $0 \leq a + \lambda v$. As we observed before, this yields

$$\lim_{K \subset O} \big(\theta_K(a) + \theta_K(\lambda v)\big) = \theta_O(a) + \theta_O(\lambda v)$$

as well as

$$\lim_{K \subset O} \theta_K(v) = \theta_O(v).$$

Using the preceding and the limit rules in Lemma I.5.19 we infer that

$$\begin{aligned} \overline{\lim_{K \subset O}}\, \theta_K(a) + \theta_O(\lambda v) &= \overline{\lim_{K \subset O}}\, \theta_K(a) + \underline{\lim_{K \subset O}}\, \theta_K(\lambda v) \\ &\leq \overline{\lim_{K \subset O}} \big(\theta_K(a) + \theta_K(\lambda v)\big) \\ &= \theta_O(a) + \theta_O(\lambda v) \end{aligned}$$

and

$$\begin{aligned} \theta_O(a) + \theta_O(\lambda v) &= \underline{\lim_{K \subset O}} \big(\theta_K(a) + \theta_K(\lambda v)\big) \\ &\leq \underline{\lim_{K \subset O}}\, \theta_K(a) + \overline{\lim_{K \subset O}}\, \theta_K(\lambda v) \\ &= \underline{\lim_{K \subset O}}\, \theta_K(a) + \theta_O(\lambda v). \end{aligned}$$

As $\theta_K(v) \leq \lambda w$ by (xi), we conclude using the cancellation law in Proposition I.5.10 that

$$\overline{\lim_{K \subset O}}\, \theta_K(a) \leq \theta_O(a) + \varepsilon w \qquad \text{and} \qquad \theta_O(a) \leq \underline{\lim_{K \subset O}}\, \theta_K(a) + \varepsilon w$$

for all $\varepsilon > 0$. As $w \in \mathcal{W}$ was arbitrary, this shows indeed

$$\theta_O(a) = \lim_{K \subset O} \theta_O(a).$$

Moreover, for compact subsets of $X$ we establish:

(xii) $\theta_K(a) = \lim_{\varphi \succ K} T(\varphi \otimes a)$ for every compact set $K \in \mathfrak{K}$ and $a \in \mathcal{P}$.

This limit is taken over the downward directed family of all functions $\varphi \in \mathcal{K}(X)$ such that $K \prec \varphi$, that is $\chi_K \leq \varphi$. To verify this, let $K \in \mathfrak{K}$. First, let $0 \leq a \in \mathcal{P}$. The net $\big(T(\varphi \otimes a)\big)_{\varphi \succ K}$ is decreasing and bounded below, hence convergent in $\mathcal{Q}$. For every open set $O \in \mathfrak{O}$ containing $K$ there is $\varphi \in \mathcal{K}(X)$ such that $K \prec \varphi \prec O$. This shows $T(\varphi \otimes a) \leq \theta_O(a)$, hence $\lim_{\varphi \succ K} T(\varphi \otimes a) \leq \theta_O(a)$ and therefore

$$\lim_{\varphi \succ K} T(\varphi \otimes a) \leq \theta_K(a)$$

by (vii). For the converse inequality, let $\varphi \succ K$ and $\gamma > 1$. There is an open set $O \supset K$ such that $O \prec \gamma\varphi$, hence $T(\psi \otimes a) \leq \gamma T(\varphi \otimes a)$ for every $\psi \prec O$. Therefore $\theta_O(a) \leq \gamma T(\varphi \otimes a)$ and $\theta_K(a) \leq \gamma T(\varphi \otimes a)$ by (vii) and (x). As this holds for all $\varphi \succ K$ we infer that

$$\theta_K(a) \leq \gamma \lim_{\varphi \succ K} T(\varphi \otimes a).$$

Finally, as $\gamma > 1$ was arbitrarily chosen, and as $\mathcal{Q}$ is a locally convex complete lattice cone, this yields together with the above that

$$\theta_K(a) = \lim_{\varphi \succ K} T(\varphi \otimes a).$$

Now for the general case, let $a \in \mathcal{P}$. Given any $w \in \mathcal{W}$ there is $v \in \mathcal{V}$ such that $\chi_K \otimes v \leq \mathfrak{u}_w$ and $\lambda \geq 0$ such that $0 \leq a + \lambda v$. As we observed before, the latter yields

$$\lim_{\varphi \succ K} \big(T(\varphi \otimes a) + T(\varphi \otimes \lambda v)\big) = \theta_K(a) + \theta_K(\lambda v).$$

Using the preceding and the limit rules in Lemma I.5.19 we infer that

$$\overline{\lim_{\varphi \succ K}}\, T(\varphi \otimes a) + \theta_K(\lambda v) \leq \theta_K(a) + \theta_K(\lambda v) \leq \underline{\lim_{\varphi \succ K}}\, T(\varphi \otimes a)) + \theta_K(\lambda v).$$

As $\theta_K(v) \leq \lambda w$ by (viii) we conclude using the cancellation law in Proposition I.5.10 that

$$\overline{\lim_{\varphi \succ K}}\, T(\varphi \otimes a) \leq \theta_K(a) + \varepsilon w \qquad \text{and} \qquad \theta_K(a) \leq \underline{\lim_{\varphi \succ K}}\, T(\varphi \otimes a) + \varepsilon w$$

for all $\varepsilon > 0$. As $w \in \mathcal{W}$ was arbitrary, this yields $\theta_K(a) = \lim_{\varphi \succ K} T(\varphi \otimes a)$.

(xiii) If $0 \leq a$ for $a \in \mathcal{P}$, and if $E_i \subset X$ such that $E = \bigcup_{i \in \mathbb{N}} E_i$ is relatively compact, then $\theta_E(a) \leq \sum_{i=1}^{\infty} \theta_{E_i}(a)$.

To verify this, we shall make use of our assumption that the order continuous linear functionals support the separation property for $\mathcal{Q}$. We shall use Proposition I.5.35: Let $E, E_i$ be as in (xiii), and $0 \le a \in \mathcal{P}$. Let $U \in \mathfrak{R}$ be a relatively compact open set containing $E$. For every $i \in \mathbb{N}$ let

$$A_i = \{\theta_{O_i}(a) \mid O_i \in \mathfrak{R},\ O_i \text{ open},\ E_i \subset O_i \subset U\} \subset \mathcal{Q}.$$

These sets $A_i$ are directed downward, and because all of their elements are positive, the series $\sum_{i=1}^{\infty} \inf A_i$ is obviously convergent since its partial sums form an increasing sequence. The assumptions of I.5.35 are therefore satisfied, and we have

$$\inf\left\{\sum_{i=1}^{\infty} A_i\right\} = \sum_{i=1}^{\infty} \inf A_i = \sum_{i=1}^{\infty} \theta_{E_i}(a)$$

by I.5.35 and (vi). Now let $\sum_{i=1}^{\infty} \theta_{O_i}(a) \in \sum_{i=1}^{\infty} A_i$, that is $E_i \subset O_i \subset U$ for open sets $O_i \in \mathfrak{R}$, and let $O = \bigcup_{i\in\mathbb{N}} O_i$. Then $E \subset O \subset U$, hence $O$ is open and relatively compact. Thus

$$\theta_E(a) \le \theta_O(a) \le \sum_{i=1}^{\infty} \theta_{O_i}(a),$$

by (x) and by (v), and therefore

$$\theta_E(a) \le \inf\left\{\sum_{i=1}^{\infty} A_i\right\} = \sum_{i=1}^{\infty} \theta_{E_i}(a).$$

as claimed in (xiii).

(xiv) If $E_i \subset X$ such that $E = \bigcup_{i\in\mathbb{N}} E_i$ is relatively compact, and if $\theta_E(a) \ge \sum_{i=1}^{\infty} \theta_{E_i}(a)$ holds for all $0 \le a \in \mathcal{P}$, then $\theta_E = \sum_{i=1}^{\infty} \theta_{E_i}$, that is $\theta_E(a) = \sum_{i=1}^{\infty} \theta_{E_i}(a)$ for all $a \in \mathcal{P}$.

Indeed, the assumptions of (xiv) imply together with (xiii) that $\theta_E(a) = \sum_{i=1}^{\infty} \theta_{E_i}(a)$ holds for all $0 \le a \in \mathcal{P}$. Then a similar argument as in the second part of (vi) verifies our claim.

We proceed to define and investigate measurability with respect to $\theta$ for relatively compact subsets of $X$.

(xv) A relatively compact subset $E$ of $X$ is said to be *measurable* if $\theta_F = \theta_{(F\cap E)} + \theta_{(F\setminus E)}$ holds for all relatively compact sets $F \subset X$.

In the light of (xii), for measurability of a set $E$ we shall only have to check that $\theta_F(a) \ge \theta_{(F\cap E)}(a) + \theta_{(F\setminus E)}(a)$ holds for all relatively compact subsets $F \subset X$ and all $0 \le a \in \mathcal{P}$. Moreover:

(xvi) A relatively compact subset $E$ of $X$ is measurable if and only if $\theta_O(a) \ge \theta_{(O\cap E)}(a) + \theta_{(O\setminus E)}(a)$ holds for all open sets $O \in \mathfrak{R}$ and all $0 \le a \in \mathcal{P}$.

The necessity of this condition is obvious. For its sufficiency, let $0 \le a \in \mathcal{P}$ and let $E, F \subset X$ be relatively compact. Let $O \in \mathfrak{R}$ be open such that $O \supset F$. Then

$$\theta_O(a) \ge \theta_{(O\cap E)}(a) + \theta_{(O\setminus E)}(a) \ge \theta_{(F\cap E)}(a) + \theta_{(F\setminus E)}(a).$$

Taking the infimum on the left-hand side over all such sets $O \supset F$ yields $\theta_F(a) \ge \theta_{(F\cap E)}(a) + \theta_{(F\setminus E)}(a)$ by (vii), hence our claim by the preceding remark.

(xvii) Every relatively compact open set is measurable.

Let $E \in \mathfrak{R}$ be open. We shall verify the criterion in (xvi): For this, let $O \in \mathfrak{R}$ be open and let $0 \le a \in \mathcal{P}$. Then $E \cap O$ is open. Let $\varphi \prec E \cap O$ for $\varphi \in \mathcal{K}(X)$ and let $K = \operatorname{supp}(\varphi) \in \mathfrak{K}$. The set $O \setminus K \supset O \setminus E$ is also open, and for every $\psi \prec O \setminus K$ we have $\psi + \varphi \prec O$, hence

$$\theta_O(a) \ge T(\psi \otimes a) + T(\varphi \otimes a)$$

by (i). Taking the supremum over all such functions $\psi \prec O \setminus K$ on the right-hand side of the last inequality yields by (i)

$$\theta_O(a) \ge \theta_{(O\setminus K)}(a) + T(\varphi \otimes a) \ge \theta_{(O\setminus E)}(a) + T(\varphi \otimes a).$$

Now taking the supremum over all functions $\varphi \prec E \cap O$ yields

$$\theta_E(a) \ge \theta_{(E\setminus O)}(a) + \theta_{(E\cap O)}(a),$$

hence our claim.

(xviii) The measurable relatively compact subsets of $X$ form a weak $\sigma$-ring $\widetilde{\mathfrak{R}}$.

We shall verify properties (R1), (R2) and (R3) from II.1.1 for the collection $\widetilde{\mathfrak{R}}$ of all measurable relatively compact subsets of $X$. Obviously, $\emptyset \in \widetilde{\mathfrak{R}}$, that is (R1). If $E_1, E_2 \in \widetilde{\mathfrak{R}}$, then for every relatively compact subset $F$ of $X$ we have

$$\theta_F = \theta_{(F\cap E_1)} + \theta_{(F\setminus E_1)} \qquad \text{and} \qquad \theta_{(F\setminus E_1)} = \theta_{((F\setminus E_1)\cap E_2)} + \theta_{((F\setminus E_1)\setminus E_2)},$$

by the measurability of $E_1$ and $E_2$, respectively. Because $(F \setminus E_1) \setminus E_2 = F \setminus (E_1 \cup E_2)$, this yields

$$\begin{aligned} \theta_F &= \theta_{(F\cap E_1)} + \theta_{((F\setminus E_1)\cap E_2)} + \theta_{((F\setminus(E_1\cup E_2))} \\ &= \theta_{(F\cap E_1)} + \theta_{((F\cap(E_2\setminus E_1))} + \theta_{((F\setminus(E_1\cup E_2))}. \end{aligned}$$

First we observe that $\big(F\cap(E_1\cup E_2)\big)\cap E_1 = F\cap E_1$ and $\big(F\cap(E_1\cup E_2)\big)\setminus E_1 = (F \setminus E_1) \cap E_2$. We have

$$\theta_{(F\cap(E_1\cup E_2))} = \theta_{(F\cap E_1)} + \theta_{((F\setminus E_1)\cap E_2)},$$

by the measurability of $E_1$, hence

$$\theta_F = \theta_{(F\cap(E_1\cup E_2))} + \theta_{((F\setminus(E_1\cup E_2))}$$

by the above. This shows the measurability of the set $E_1 \cup E_2$. Secondly, we observe that $\bigl(F \setminus (E_2 \setminus E_1)\bigr) \cap E_1 = F \cap E_1$ and $\bigl(F \setminus (E_2 \setminus E_1)\bigr) \setminus E_1 = F \setminus (E_1 \cup E_2)$. This shows

$$\theta_{(F\setminus(E_2\setminus E_1))} = \theta_{(F\cap E_1)} + \theta_{(F\setminus(E_1\cup E_2)},$$

by the measurability of $E_1$, hence

$$\theta_F = \theta_{(F\cap(E_2\setminus E_1))} + \theta_{((F\setminus(E_2\setminus E_2))}$$

again using the above. This shows the measurability of the set $E_2 \setminus E_1$, hence (R2). All left to show is (R3). By induction the union of finitely many measurable subsets of $X$ is again measurable. Let $(E_i)_{i\in\mathbb{N}}$ be a sequence of disjoint sets in $\widetilde{\mathfrak{R}}$ such that $E = \bigcup_{i=1}^{\infty} E_i$ is again relatively compact. Let $G_n = \bigcup_{i=1}^{n} E_i$. Then $G_n \in \widetilde{\mathfrak{R}}$ by the preceding. For $0 \leq a \in \mathcal{P}$ and any relatively compact set $F \subset X$ we have for all $n \in \mathbb{N}$

$$\theta_F(a) = \theta_{F\cap G_n}(a) + \theta_{(F\setminus G_n)}(a) \geq \theta_{F\cap G_n}(a) + \theta_{(F\setminus E)}(a).$$

Now $G_n \cap E_n = E_n$ and $G_n \setminus E_n = G_{n-1}$. As $E_n$ is measurable, this shows

$$\theta_{F\cap G_n}(a) = \theta_{F\cap E_n}(a) + \theta_{F\cap G_{n-1}}(a).$$

Induction then leads to

$$\theta_{F\cap G_n}(a) = \sum_{i=1}^{n} \theta_{F\cap E_i}(a),$$

hence

$$\theta_F(a) \geq \sum_{i=1}^{n} \theta_{F\cap E_i}(a) + \theta_{(F\setminus E)}(a)$$

for all $n \in \mathbb{N}$, and therefore

$$\begin{aligned}\theta_F(a) &\geq \sum_{i=1}^{\infty} \theta_{F\cap E_i}(a) + \theta_{(F\setminus E)}(a)\\ &\geq \theta_{F\cap E}(a) + \theta_{(F\setminus E)}(a)\end{aligned}$$

by (xiii). This demonstrates $E \in \widetilde{\mathfrak{R}}$ by (xvi).

(xix) $\mathfrak{R} \subset \widetilde{\mathfrak{R}}$.

Let $\mathfrak{A}_{\widetilde{\mathfrak{R}}} = \{A \subset X \mid A \cap E \in \widetilde{\mathfrak{R}} \quad \text{for all} \quad E \in \widetilde{\mathfrak{R}}\}$ be the $\sigma$-algebra generated by $\widetilde{\mathfrak{R}}$. In order to demonstrate that $\mathfrak{A}_{\widetilde{\mathfrak{R}}}$ contains all Borel subsets of $X$, it suffices to sow that it contains all open sets. For this, let $U \subset X$ be an open set. We shall use (xvi) to demonstrate that $U \cap E \in \widetilde{\mathfrak{R}}$ for all $E \in \widetilde{\mathfrak{R}}$. Let $O \subset X$ be relatively compact and open, and let $0 \leq a \in \mathcal{P}$. Then the set $O \cap U$ is open and relatively compact, hence in $\widetilde{\mathfrak{R}}$ by (xvii). In turn, the set $O \cap U \cap E$ in also in $\widetilde{\mathfrak{R}}$, since every weak $\sigma$-ring is closed for finite intersections. As $O \setminus (O \cap U \cap E) = O \setminus (U \cap E)$, the latter yields

$$\theta_O = \theta_{(O \cap U \cap E)} + \theta_{(O \setminus (U \cap E))}.$$

This show that the set $U \cap E$ is indeed measurable. Now, finally, let $E$ be a relatively compact Borel subset of $X$, that is $E \in \mathfrak{R}$, and let $O \subset X$ be relatively compact such that $E \subset O$. Then $O \in \widetilde{\mathfrak{R}}$ by (xvii) and $E \in \mathfrak{A}_{\widetilde{\mathfrak{R}}}$ by the above. Thus $E = E \cap O \in \widetilde{\mathfrak{R}}$. This yields $\mathfrak{R} \subset \widetilde{\mathfrak{R}}$ as claimed.

(xx) If the sets $E_i \in \widetilde{\mathfrak{R}}$ are pairwise disjoint and if $E = \bigcup_{i \in \mathbb{N}} E_i \in \widetilde{\mathfrak{R}}$, then $\theta_E = \sum_{i=1}^{\infty} \theta_{E_i}$, that is $\theta_E(a) = \sum_{i=1}^{\infty} \theta_{E_i}(a)$ for all $a \in \mathcal{P}$.

First let $E_1$ and $E_2$ be disjoint sets in $\widetilde{\mathfrak{R}}$. Then the measurability of $E_2$ implies that

$$\theta_{(E_1 \cup E_2)} = \theta_{((E_1 \cup E_2) \cap E_2)} + \theta_{((E_1 \cup E_2) \setminus E_2)} = \theta_{E_1} + \theta_{E_2}.$$

Finite additivity of $\theta$ on $\widetilde{\mathfrak{R}}$ follows by induction. If $E \in \widetilde{\mathfrak{R}}$ is the disjoint union of the sets $E_i \in \widetilde{\mathfrak{R}}$ for $i \in \mathbb{N}$, then for every $0 \leq a \in \mathcal{P}$ and every $n \in \mathbb{N}$ we have

$$\theta_E(a) \geq \theta_{(\cup_{i=1}^n E_i)}(a) = \sum_{i=1}^{n} \theta_{E_i}(a),$$

and therefore

$$\theta_E(a) \geq \sup_{n \in \mathbb{N}} \sum_{i=1}^{n} \theta_{E_i}(a) = \sum_{i=1}^{\infty} \theta_{E_i}(a).$$

Our claim follows from (xiv).

Summarizing our observations from (xviii), (xix), (xx), (ix), (vii) and (xi) we realize that

(xxi) $\theta$ is a quasi regular $\mathfrak{R}$-bounded $\mathfrak{L}(\mathcal{P}, \mathfrak{Q})$-valued measure on $\mathfrak{R}$.

Furthermore,

(xxii) $\theta$ is continuous relative to the given r-lower continuous system $\mathfrak{V}$ of inductive limit neighborhoods; more precisely: for $w \in \mathcal{W}$ and integrable functions $f, g \in \mathcal{F}_{(X,\theta)}(X, \mathcal{P})$ such that $f \leq g + \mathfrak{u}_w$ we have $\int_X f \, d\theta \leq \int_X g \, d\theta + w$.

In order to verify (xxii), given $w \in \mathcal{W}$, recall that $\mathfrak{u}_w \in \mathfrak{V}$ is a corresponding r-lower continuous neighborhood such that

$$T(f) \leq T(g) + w \qquad \text{whenever} \qquad f \leq g + \mathfrak{u}_w$$

for $f, g \in \mathcal{F}_{\mathfrak{V}}(X, \mathcal{P})$. In a first step we shall generalize the statement of (viii) to (not necessarily positive) elements $a, b \in \mathcal{P}$. For this, let $a, b \in \mathfrak{R}$, $E \in \mathfrak{R}$ and suppose that $\chi_{E} \otimes a \leq \chi_{E} \otimes b + \mathfrak{u}_w$. For $u \in \mathcal{W}$ choose $v \in \mathcal{V}$ such that $\theta_E(v) \leq u$ and $\lambda \geq 0$ such that both $0 \leq a + \lambda v$ and $0 \leq b + \lambda v$. Then $\chi_{E} \otimes (a + \lambda v) \leq \chi_{E} \otimes (b + \lambda v) + \mathfrak{u}_w$, hence $\theta_E(a + \lambda v) \leq \theta_E(a + \lambda v) + w$ by (viii). Using the cancellation law from Proposition I.5.10(a) for the element $\theta_E(\lambda v) \leq \lambda u$ yields $\theta_E(a) \leq \theta_E(b) + w + \varepsilon u$ for all $\varepsilon > 0$. Because this holds for all $u \in \mathcal{W}$, we have indeed $\theta_E(a) \leq \theta_E(b) + w$.

Now in the second step of our argument for (xxii), let $s$ be any of the $\overline{\mathcal{V}}$-valued functions in $\mathfrak{u}_w$. By the r-lower continuity of $s$ the set

$$O_s = \{x \in X \mid s(x) = +\infty\}$$

is open in $X$. Moreover, for any subset $E \in \mathfrak{R}$ of $O_s$ and any $a \in \mathcal{P}$ we have $\chi_{E} \otimes a \leq \varepsilon s$ as well as $0 \leq \chi_{E} \otimes a + \varepsilon s$, and therefore by the argument from our first step $\theta_E(a) \leq \varepsilon w$ as well as $0 \leq \theta_E(a) + \varepsilon w$ for all $\varepsilon > 0$. This shows

$$\int_{O_s} f \, d\theta \leq \varepsilon w \qquad \text{as well as} \qquad 0 \leq \int_{O_s} f \, d\theta + \varepsilon w$$

for every integrable function $f \in \mathcal{F}(X, \mathcal{P})$ and all $\varepsilon > 0$.

Next we shall verify that $\int_E s \, d\theta \leq w$ holds for every subset $E \in \mathfrak{R}$ of $X \setminus O_s$. First we recall from Proposition 4.1(b) that $\chi_{E} \otimes s \in \mathcal{F}_{\mathfrak{R}}(X, \mathcal{P})$, since the function $s$ is r-lower continuous and $\chi_{E} \otimes s$ is $\mathcal{P}$-valued. We shall calculate the integral of $s$ over the set $E$ according to Definition II.4.9. For this, let $u \in \mathcal{W}$ and as in II.4 let $\mathfrak{v}_u = \left\{ r \in \mathcal{S}_{\mathfrak{R}}(X, \mathcal{V}) \mid \int_X r \, d\theta \leq u \right\}$, and correspondingly

$$\int_E^{(u)} s \, d\theta = \sup \left\{ \int_E h \, d\theta \;\middle|\; h \in \mathcal{S}_{\mathfrak{R}}(X, \mathcal{P}), \;\; h \leq s + \mathfrak{v}_u \right\}.$$

Let $h \in \mathcal{S}_{\mathfrak{R}}(X, \mathcal{P})$ such that $h \leq s + \mathfrak{v}_u$, that is $h \leq s + r$ for some $r \in \mathfrak{v}_u$. We may express both step functions $h$ and $r$ as

$$h = \sum_{i=1}^{n} \chi_{E_i} \otimes a_i \qquad \text{and} \qquad r = \sum_{i=1}^{n} \chi_{E_i} \otimes v_i$$

with $a_i \in \mathcal{P}$, $v_i \in \mathcal{V}$ and disjoint sets $E_i \in \mathfrak{R}$. Let $\{K_i\}_{i=1}^{n}$ be a family of compact sets such that $K_i \subset (E_i \cap E)$. We have $a_i \leq s(x) + v_i$ for all $x \in K_i$. Let $O \in \mathfrak{R}$ be an open set containing $E$ and let $v \in \mathcal{V}$ such that $\chi_{O} \otimes v \leq \mathfrak{u}_w$ and let $\varepsilon > 0$. Recall from Lemma 1.1 that the functions $s + \chi_{X} \otimes v_i$ for $i = 1, \ldots, n$ are also r-lower continuous. Thus there is a family $\{O_i\}_{i=1}^{n}$ of disjoint open sets $O_i \in \mathfrak{R}$ such that $K_i \subset O_i \subset O$ and

$a_i \leq (1+\varepsilon)(s(x)+v_i)+\varepsilon v$ holds for all $x \in O_i$. There exist functions $\varphi_i \in \mathcal{K}(X)$ such that $K_i \prec \varphi_i \prec O_i$ for $i = 1, \ldots, n$. Then

$$\varphi_{i\otimes}a_i \leq (1+\varepsilon)\varphi_{i\otimes}s + (1+\varepsilon)\varphi_{i\otimes}v_i + \varepsilon\varphi_{i\otimes}v,$$

and we set

$$f = \sum_{i=1}^{n} \varphi_{i\otimes}a_i \qquad \text{and} \qquad g = (1+\varepsilon)\sum_{i=1}^{n} \varphi_{i\otimes}v_i.$$

Because $\sum_{i=1}^{n} \varphi_{i\otimes}s \leq s \leq \mathfrak{u}_w$ and $\sum_{i=1}^{n} \varphi_{i\otimes}s \leq \chi_{0\otimes}v \leq \mathfrak{u}_w$, we have $f \leq g + (1+2\varepsilon)\mathfrak{u}_w$, and as $f, g \in \mathcal{E}(X,\mathcal{P}) \subset \mathcal{F}_{\mathfrak{V}}(X,\mathcal{P})$, the latter implies

$$\sum_{i=1}^{n} T(\varphi_{i\otimes}a_i) = T(f) \leq T(g) + (1+2\varepsilon)w = (1+\varepsilon)\sum_{i=1}^{n} T(\varphi_{i\otimes}v_i) + (1+2\varepsilon)w.$$

Next, while keeping the sets $K_i$ and $O_i$ fixed, for each $i = 1, \ldots, n$ we consider the downward directed net of all functions $K_i \prec \varphi_i \prec O_i$ and recall from (xii) that

$$\theta_{K_i}(a_i) = \lim_{\varphi_i \succ K_i} T(\varphi_{\otimes}a_i) \qquad \text{and} \qquad \theta_{K_i}(v_i) = \lim_{\varphi_i \succ K_i} T(\varphi_{\otimes}v_i).$$

Carrying out this limit process in the preceding inequality then leads to

$$\sum_{i=1}^{n} \theta_{K_i}(a_i) \leq (1+\varepsilon)\sum_{i=1}^{n} \theta_{K_i}(v_i) + (1+2\varepsilon)w.$$

The latter holds for all $\varepsilon > 0$, hence

$$\sum_{i=1}^{n} \theta_{K_i}(a_i) \leq \sum_{i=1}^{n} \theta_{K_i}(v_i) + w.$$

Moreover, as $K_i \subset E_i$, we have

$$\sum_{i=1}^{n} \theta_{K_i}(v_i) \leq \sum_{i=1}^{n} \theta_{E_i}(v_i) = \int_X r\, d\theta \leq u,$$

hence

$$\sum_{i=1}^{n} \theta_{K_i}(a_i) \leq u + w.$$

Now, for each $i = 1, \ldots, n$ we consider the upward directed net of all $K_i \subset (E_i \cap E)$ and recall from Proposition 4.4(c) that

$$\theta_{(E_i \cap E)}(a_i) \leq \lim_{K_i \subset (E_i \cap E)} \theta_{K_i}(a_i) + \mathfrak{O}\big(\theta_{(E_i \cap E)}(a_i)\big).$$

From the first step of our argument we recall that $\chi_{E_i \otimes a_i} \le \chi_{E_i \otimes v_i} + \mathfrak{u}_w$ implies that $\theta_{E_i}(a_i) \le \theta_{E_i}(v_i) + w \le u + w$, hence

$$\mathfrak{O}\big(\theta_{(E_i \cap E)}(a_i)\big) \le \mathfrak{O}\big(\theta_{E_i}(a_i)\big) \le \varepsilon(u+w)$$

for all $\varepsilon > 0$. Taking the limit over all such nets of sets $K_i \subset E_i$ leads to

$$\int_E h\, d\theta = \sum_{i=1}^{n} \theta_{E_i}(a_i) \le (u+v) + \varepsilon(u+w) = (1+\varepsilon)(u+w)$$

for all $\varepsilon > 0$, and therefore

$$\int_E h\, d\theta = \sum_{i=1}^{n} \theta_{E_i}(a_i) \le u + v.$$

This demonstrates

$$\int_E s\, d\theta \le \int_E^{(u)} s\, d\theta \le u + w,$$

for all $u \in \mathcal{W}$, hence

$$\int_E s\, d\theta \le w.$$

Now in the third and final step of this argument, let $f, g \in \mathcal{F}_{(X,\theta)}(X, \mathcal{P})$ and suppose that $f \le g + \mathfrak{u}_w$, that is $f \le g + s$ for some $s \in \mathfrak{u}_w$. Let $O_s = \{x \in X \mid s(x) = +\infty\}$. For every $E \in \mathfrak{R}$ both functions $f$ and $g$ are integrable over the sets $E \cap O_s \in \mathfrak{R}$ and $E \setminus O_s \in \mathfrak{R}$. We have

$$\int_{(E \cap O_s)} f\, d\theta \le \varepsilon w \le \int_{(E \cap O_s)} g\, d\theta + 2\varepsilon w$$

for all $\varepsilon > 0$ by the above, and

$$\begin{aligned} \int_{(E \setminus O_s)} f\, d\theta &\le \int_{(E \setminus O_s)} (g+s)\, d\theta \\ &\le \int_{(E \setminus O_s)} g\, d\theta + \int_{(E \setminus O_s)} s\, d\theta \le \int_{(E \setminus O_s)} g\, d\theta + w. \end{aligned}$$

Thus

$$\begin{aligned} \int_E f\, d\theta &= \int_{(E \cap O_s)} f\, d\theta + \int_{(E \setminus O_s)} f\, d\theta \\ &\le \int_{(E \cap O_s)} g\, d\theta + \int_{(E \setminus O_s)} g\, d\theta + (1+2\varepsilon) w \\ &= \int_E g\, d\theta + (1+2\varepsilon) w \end{aligned}$$

for all $\varepsilon > 0$, hence

$$\int_E f\, d\theta \leq \int_E g\, d\theta + w.$$

This holds for all $E \in \mathfrak{R}$ and yields

$$\int_X f\, d\theta \leq \int_X g\, d\theta + w$$

by our definition of the integral over $X$ in II.4.13. This completes our proof for (xxii).

Note that Parts (b) and (c) of Proposition 4.8 now yield that all functions in $\mathcal{F}_{\mathfrak{V}}(X,\mathcal{P})$ or in $\mathcal{F}_{\mathfrak{V}_0}(X,\mathcal{P})$ are integrable or indeed strongly integrable over every Borel set $F \subset X$ with respect to the measure $\theta$. All left to show is that $\theta$ represents the given operator $T$ as stated in our Theorem.

(xxiii) Let $\varphi{\otimes}a \in \mathcal{E}(X,\mathcal{P})$ for $\varphi \in \mathcal{K}(X)$ and $a \in \mathcal{P}$. Then

$$\int_X \varphi{\otimes}a\, d\theta \leq T(\varphi{\otimes}a) \leq \int_X \varphi{\otimes}a\, d\theta + \mathfrak{O}\big(\theta_A(a)\big)$$

where $A \in \mathfrak{R}$ denotes the support of the function $\varphi$.

We may assume that $0 \leq \varphi \leq 1$ an denote by $A \subset X$ the compact support of $\varphi$. Let us first consider the case that $a \geq 0$. We fix $n \geq 2$ in $\mathbb{N}$, and for $i \geq 1$ define compact sets

$$A_i^n = \left\{ x \in X \mid \varphi(x) \geq \frac{i}{n} \right\}.$$

Then $A = A_0^n \supset A_1^n \supset \ldots \supset A_{n+1} = \emptyset$. For $i \geq 0$ we define $\psi_i^n \in \mathcal{K}(X)$ by

$$\psi_i^n(x) = \begin{cases} \frac{1}{n}, & \text{if } x \in A_{i+1}^n \\ \varphi(x) - \frac{i}{n}, & \text{if } x \in A_i^n \setminus A_{i+1}^n \\ 0, & \text{if } x \notin A_i^n \end{cases}$$

Then $A_{i+1}^n \prec n\psi_i^n \prec A_i^n$. All the functions $\psi_i^n{\otimes}a$ are contained in $\mathcal{E}(X,\mathcal{P})$, hence

$$\frac{1}{n}\,\theta_{A_{i+1}^n}(a) \leq T(\psi_i^n{\otimes}a) \leq \frac{1}{n}\,\theta_{A_i^n}(a).$$

The first part of the last inequality follows from (xii), the second part from (vi) and the fact that $n\psi_i^n \prec O$ for every open set $O \in \mathfrak{R}$ containing $A_i^n$.

Let $x \in A_0^n$ such that $\frac{k}{n} \leq \varphi(x) < \frac{k+1}{n}$ for some $k = 0, \ldots, (n+1)$, that is $x \in A_k^n \setminus A_{k+1}^n$. Then $\psi_i^n(x) = \frac{1}{n}$ for all $i = 0, \ldots, (k-1)$, $\psi_k^n(x) = \varphi(x) - \frac{k}{n}$ and $\psi_i^n(x) = 0$ for all $i = k+1, \ldots, (n+1)$. This shows

$$\sum_{i=0}^{n} \psi_i^n(x) = \frac{k}{n} + \left(\varphi(x) - \frac{k}{n}\right) = \varphi(x).$$

and

$$\sum_{i=1}^{n} \psi_i^n(x) = \begin{cases} 0 & \text{if } \varphi(x) < \frac{1}{n} \\ \varphi(x) - \frac{1}{n} & \text{if } \varphi(x) \geq \frac{1}{n} \end{cases}$$

We set $\varphi_n = \sum_{i=1}^n \psi_i$ and conclude from the above that $\varphi_n \otimes a \leq \varphi \otimes a$ and $\varphi_n \nearrow_X \varphi$ as $n$ tends to infinity, in the sense of Section II.5.22. Indeed, given $x \in X$, $v \in \mathcal{V}$ and $\varepsilon > 0$, we have either $\varphi(x) = 0$, hence $\varphi(x)a \in v_\varepsilon\big(\varphi_n(x)a\big)$ for all $n \in \mathbb{N}$, or there is $n_0 \in \mathbb{N}$ such that $\varphi(x) \leq (1+\varepsilon)\varphi_n(x)$, hence $\varphi(x)a \leq (1+\varepsilon)\varphi_n(x)a$ and therefore $\varphi(x)a \in v_\varepsilon\big(\varphi_n(x)a\big)$ for all $n \geq n_0$. We may therefore apply or Convergence Theorem II.5.25 with the measures $\theta_n = \theta$ and the functions $f_n = \varphi_n \otimes a$, $f = \varphi \otimes a$, and $f_* = f_{**} = 0$. This yields

$$\int_X \varphi \otimes a \, d\theta \leq \underline{\lim}_{n \to \infty} \int_X \varphi_n \otimes a \, d\theta,$$

hence

$$\int_X \varphi \otimes a \, d\theta = \lim_{n \to \infty} \int_X \varphi_n \otimes a \, d\theta,$$

as $\varphi_n \otimes a \leq \varphi \otimes a$ for all $n \in \mathbb{N}$. Next we choose the step functions

$$h_n = \frac{1}{n} \sum_{i=1}^{n} \chi_{A_i^n} \otimes a \in \mathcal{S}_{\mathfrak{R}}(X, \mathcal{P}).$$

As $\psi_i^n \otimes a \leq \frac{1}{n} \chi_{A_i^n} \otimes a$ for all $i = 0, \ldots, n$, we have $\varphi_n \otimes a \leq h_n$, and consequently

$$\begin{aligned} \int_X \varphi_n \otimes a \, d\theta \leq \int_X h_n \, d\theta = \frac{1}{n} \sum_{i=1}^{n} \theta_{A_i^n}(a) = \frac{1}{n} \sum_{i=0}^{n} \theta_{A_{i+1}^n}(a) \\ \leq \sum_{i=0}^{n} T(\psi_i^n \otimes a) = T\left( \sum_{i=0}^{n} \psi_i^n \otimes a \right) = T(\varphi \otimes a). \end{aligned}$$

Thus the above demonstrates that

$$\int_X \varphi \otimes a \, d\theta = \lim_{n \to \infty} \int_X \varphi_n \otimes a \, d\theta \leq T(\varphi \otimes a).$$

For the reverse inequality, we observe that

$$\begin{aligned} T(\varphi \otimes a) &= \sum_{i=0}^{n} T(\psi_i^n \otimes a) \\ &\leq \frac{1}{n} \sum_{i=0}^{n} \theta_{A_i^n}(a) = \frac{1}{n} \sum_{i=1}^{n} \theta_{A_i^n}(a) + \frac{1}{n} \theta_A(a). \end{aligned}$$

As

$$h_n = \frac{1}{n}\sum_{i=1}^{n} \chi_{A_i}{\otimes} a \le \varphi_n{\otimes} a \le \varphi{\otimes} a$$

implies

$$\int_X h_n\, d\theta = \frac{1}{n}\sum_{i=1}^{n} \theta_{A_i}(a) \le \int_X \varphi{\otimes} a\, d\theta,$$

we realize that

$$T(\varphi{\otimes} a) \le \int_X \varphi{\otimes} a\, d\theta + \frac{1}{n}\,\theta_A(a)$$

holds for all $n \in \mathbb{N}$. This shows

$$T(\varphi{\otimes} a) \le \int_X \varphi{\otimes} a\, d\theta + \mathfrak{O}\big(\theta_A(a)\big).$$

Now suppose that the element $a \in \mathcal{P}$ is not necessarily positive. Given $w \in \mathcal{W}$ there is $\mathfrak{u}_w \in \mathfrak{V}$ such that $T(f) \le T(g) + w$ whenever $f \le g + \mathfrak{u}_w$ for $f, g \in \mathcal{F}_{\mathfrak{V}}(X, \mathcal{P})$. In turn, there is $v \in \mathcal{V}$ such that $\chi_A{\otimes} v \le \mathfrak{u}_w$ and $\lambda \ge 0$ such that $0 \le a + \lambda v$. Then

$$\begin{aligned}\int_X \varphi{\otimes}(a+\lambda v)\, d\theta &\le T\big(\varphi{\otimes}(a+\lambda v)\big)\\ &\le \int_X \varphi{\otimes}(a+\lambda v)\, d\theta + \mathfrak{O}\big(\theta_A(a+\lambda v)\big)\end{aligned}$$

and

$$\int_X \varphi{\otimes} v\, d\theta \le T(\varphi{\otimes} v) \le \int_X \varphi{\otimes} v\, d\theta + \mathfrak{O}\big(\theta_A(v)\big)$$

holds by the preceding argument for positive elements. Thus, firstly,

$$\begin{aligned}\int_X \varphi{\otimes} a\, d\theta + \lambda \int_X \varphi{\otimes} v\, d\theta &\le T(\varphi{\otimes} a) + \lambda T(\varphi{\otimes} v)\\ &\le T(\varphi{\otimes} a) + \lambda \int_X \varphi{\otimes} v\, d\theta + \mathfrak{O}\big(\theta_A(v)\big),\end{aligned}$$

hence

$$\int_X \varphi{\otimes} a\, d\theta \le T(\varphi{\otimes} a) + \mathfrak{O}\Big(\int_X \varphi{\otimes} v\, d\theta\Big) + \mathfrak{O}\big(\theta_A(v)\big),$$

by the cancellation law from Proposition I.5.10(a), and because $\varphi{\otimes} v \le \chi_A{\otimes} v \le \mathfrak{u}_w$ we have $\int_X \varphi{\otimes} v\, d\theta \le \theta_A(v) \le w$, hence

$$\int_X \varphi{\otimes} a\, d\theta \le T(\varphi{\otimes} a) + \varepsilon w$$

for all $\varepsilon > 0$. This holds for all $w \in \mathcal{W}$ and therefore yields

$$\int_X \varphi\otimes a \, d\theta \leq T(\varphi\otimes a).$$

Secondly, we have by the above

$$\begin{aligned} T(\varphi\otimes a) + \lambda \int_X \varphi\otimes v \, d\theta &\leq T(\varphi\otimes a) + \lambda T(\varphi\otimes v) \\ &\leq \int_X \varphi\otimes a \, d\theta + \lambda \int_X \varphi\otimes v \, d\theta + \mathfrak{O}\big(\theta_A(a)\big) + \mathfrak{O}\big(\theta_A(v)\big). \end{aligned}$$

Again using the cancellation law from Proposition I.5.10(a), this yields

$$T(\varphi\otimes a) \leq \int_X \varphi\otimes a \, d\theta + \mathfrak{O}\left(\int_X \varphi\otimes v \, d\theta\right) + \mathfrak{O}\big(\theta_A(a)\big) + \mathfrak{O}\big(\theta_A(v)\big),$$

that is

$$T(\varphi\otimes a) \leq \int_X \varphi\otimes a \, d\theta + \mathfrak{O}\big(\theta_A(a)\big) + \varepsilon w$$

for all $\varepsilon > 0$. This holds for all $w \in \mathcal{W}$ and therefore yields

$$T(\varphi\otimes a) \leq \int_X \varphi\otimes a \, d\theta + \mathfrak{O}\big(\theta_A(a)\big)$$

as claimed.

(xxiv) Let $\varphi\otimes a \in \mathcal{E}_0(X, \mathcal{P})$ for $\varphi \in \mathcal{K}(X)$ and $a \in \mathcal{P}$. Then

$$\int_X \varphi\otimes a \, d\theta = T(\varphi\otimes a).$$

For this, let $\varphi\otimes a \in \mathcal{E}_0(X, \mathcal{P})$ such that $0 \leq \varphi \leq 1$ and let $A \in \mathfrak{R}$ be the compact support of $\varphi$. Following Proposition 1.11(a) we have to distinguish two cases: In the first case, let us assume that the element $a \in \mathcal{P}$ is bounded in $\mathcal{P}$. Then $\chi_A\otimes a$ is bounded in $\big(\mathcal{F}_{\mathfrak{V}}(X, \mathcal{P}), \mathfrak{V}\big)$, hence $\theta_A(a)$ is bounded in $\mathcal{Q}$. This shows $\mathfrak{O}\big(\theta_A(a)\big) = 0$ and therefore

$$\int_X \varphi\otimes a \, d\theta = T(\varphi\otimes a)$$

by (xxiii). If on the other hand, the element $a \in \mathcal{P}$ is unbounded in $\mathcal{P}$, then the set $\{x \in X \mid \varphi(x) > 0\}$ is both open and closed in $X$, hence coincides with the support $A$ of $\varphi$. Moreover, we have $\varphi(x) \geq \rho$ for all $x \in A$ with some $\rho > 0$. Thus $\chi_A \leq (1/\rho)\varphi$. Given $w \in \mathcal{W}$ let $v \in \mathcal{V}$ such that $\chi_A\otimes v \leq \mathfrak{u}_w$ and $\lambda \geq 0$ such that $0 \leq a + \lambda v$. Then

$$\chi_A(a) \leq \chi_A\otimes(a + \lambda v) \leq \frac{1}{\rho}\,\varphi\otimes(a + \lambda v),$$

hence

$$\theta_A(a) \le \frac{1}{\rho}\int_X \varphi_{\otimes}(a+\lambda v)\,d\theta \le \frac{1}{\rho}\int_X \varphi_{\otimes}a\,d\theta + \frac{\lambda}{\rho}w.$$

This shows $\theta_A(a) \in \mathcal{B}_w\big(\int_X \varphi_{\otimes}a\,d\theta\big)$ for all $w \in \mathcal{W}$, hence $\theta_A(a) \in \mathcal{B}\big(\int_X \varphi_{\otimes}a\,d\theta\big)$ and therefore

$$\int_X \varphi_{\otimes}a\,d\theta + \mathfrak{O}\big(\theta_A(a)\big) = \int_X \varphi_{\otimes}a\,d\theta$$

by Proposition I.5.14. Thus (xxiii) yields our claim in this case as well.

Thus finally, both $\theta$ and $T$ represent continuous linear operators from the locally convex cone $\big(\mathcal{F}_{\mathfrak{V}}(X,\mathcal{P}),\mathfrak{V}\big)$ into the locally convex complete lattice cone $(\mathcal{Q},\mathcal{W})$. Both operators coincide on the subcone $\mathcal{E}_0(X,\mathcal{P})$, hence on its closure $\mathcal{F}_{\mathfrak{V}_0}(X,\mathcal{P})$. Moreover, as $\int_X f\,d\theta \le T(f)$ holds for all functions $f$ in the dense subcone $\mathcal{E}(X,\mathcal{P})$ of $\mathcal{F}(X,\mathcal{P})$, this inequality holds for all $f \in \mathcal{F}(X,\mathcal{P})$.

Now all left to demonstrate is the uniqueness of the representing measure. This will follow with Proposition 4.5. Let $\vartheta$ be any representing measure for the operator $T : \mathcal{F}_{\mathfrak{V}}(X,\mathcal{P}) \to \mathcal{Q}$ satisfying the stated properties. Then

(xxv) $\theta_E(a) = \lim_{i\in\mathcal{I}} \int_X \varphi_{i\otimes}a\,d\vartheta$ for every $E \in \mathfrak{R}$ and $a \in \mathcal{P}$,

where $\mathcal{I}$ consists of all ordered pairs $(O,f)$, where $O \in \mathcal{O}$, the collection of all open supersets $O \in \mathfrak{R}$ of $E$, and $f : \mathcal{O} \to \mathcal{K}(X)$ is a mapping such that $f(O) \prec O$ for all $O \in \mathcal{O}$. This index set is ordered by $(O_1,f_1) \le (O_2,f_2)$ if $O_1 \supset O_2$ and $f_1(O) \le f_2(O)$ for all $O \in \mathcal{O}$. We set $\varphi_i = f(O) \in \mathcal{K}(X)$ for $i = (O,f) \in \mathcal{I}$. Let $A_i$ denote the support of the function $\varphi_i \in \mathcal{K}(X)$. Then $A_i \subset O$, hence $\mathfrak{O}\big(\theta_{A_i}(a)\big) \le \mathfrak{O}\big(\theta_O(a)\big)$ by I.5.11. Thus following Proposition 4.5(c) and the properties stated in Theorem 5.1

$$\overline{\lim_{i\in\mathcal{I}}}\, \mathfrak{O}\big(\theta_{A_i}(a)\big) \le \overline{\lim_{O\in\mathcal{O}}}\, \mathfrak{O}\big(\theta_O(a)\big) \le \mathfrak{O}\Big(\lim_{O\in\mathcal{O}} \theta_O(a)\Big) = \mathfrak{O}\big(\theta_E(a)\big)$$

by Proposition I.5.24. This yields

$$\begin{aligned}
\overline{\lim_{i\in\mathcal{I}}}\, T(\varphi_{i\otimes}a) &\le \overline{\lim_{i\in\mathcal{I}}}\left(\int_X \varphi_{i\otimes}a\,d\vartheta + \mathfrak{O}\big(\theta_{A_i}(a)\big)\right)\\
&\le \overline{\lim_{i\in\mathcal{I}}} \int_X \varphi_{i\otimes}a\,d\vartheta + \overline{\lim_{i\in\mathcal{I}}}\, \mathfrak{O}\big(\theta_{A_i}(a)\big)\\
&\le \theta_E(a) + \mathfrak{O}\big(\theta_E(a)\big)\\
&= \theta_E(a)\\
&= \lim_{i\in\mathcal{I}} \int_X \varphi_{i\otimes}a\,d\vartheta\\
&\le \underline{\lim_{i\in\mathcal{I}}}\, T(\varphi_{i\otimes}a)
\end{aligned}$$

in the order topology of $\mathcal{Q}$. The value $\vartheta_E(a) \in \mathcal{Q}$, for all $a \in \mathcal{P}$ and $E \in \mathfrak{R}$, is therefore uniquely determined by the operator $T$. This completes our argument. □

*Remarks 5.2.* (a) Let us recollect and summarize the main steps in the construction of the quasi regular representing measure $\theta$ for the operator $T : \mathcal{F}_{\mathfrak{V}}(X, \mathcal{P}) \to \mathcal{Q}$ in Theorem 5.1. Using Proposition 4.5 we obtain:

(i) For every open set $O \in \mathfrak{R}$ and $a \in \mathcal{P}$ we have

$$\theta_O(a) = \lim_{\varphi \prec O} \int_X \varphi \otimes a \, d\theta = \lim_{\varphi \prec O} T(\varphi \otimes a).$$

(ii) For every compact set $K \in \mathfrak{R}$ and $a \in \mathcal{P}$ we have

$$\theta_K(a) = \lim_{\varphi \succ K} \int_X \varphi \otimes a \, d\theta = \lim_{\varphi \succ K} T(\varphi \otimes a).$$

(iii) For every $E \in \mathfrak{R}$ and every open set $O \in \mathfrak{R}$ containing $E$ there is a net $(\varphi_i)_{i \in \mathcal{I}}$ in $\mathcal{K}(X)$ such that $\varphi_i \prec 0$ for all $i \in \mathcal{I}$, and

$$\theta_E(a) = \lim_{i \in \mathcal{I}} \int_X \varphi_i \otimes a \, d\theta = \lim_{i \in \mathcal{I}} T(\varphi_i \otimes a)$$

for all $a \in \mathcal{P}$ in the order topology of $\mathcal{Q}$.

(b) Part (a) yields the following implication: Let $\theta$ be the representing measure for the operator $T : \mathcal{F}_{\mathfrak{V}}(X, \mathcal{P}) \to \mathcal{Q}$. Let $O \in \mathfrak{R}$ be an open set. Part(iii) of 5.2(a) then implies that for every subset $E \in \mathfrak{R}$ of $O$ and every $a \in \mathcal{P}$ the element $\theta_E(a) \in \mathcal{Q}$ is contained in the order closure in $\mathcal{Q}$ of the image under $T$ of the subset

$$\mathcal{A} = \{\varphi \otimes a \mid \varphi \in \mathcal{K}(X), \ \varphi \prec O\}$$

of $\mathcal{E}(X, \mathcal{P})$. We observe that this set is relatively bounded in $\mathcal{F}_{\mathfrak{V}}(X, \mathcal{P})$. Indeed,let $\psi \in \mathcal{K}(X)$ such that $O \prec \psi$. Let $\mathfrak{v} \in \mathfrak{V}$ and choose $v \in \mathcal{V}$ such that $\chi_F \otimes v \le \mathfrak{v}$, where $F \in \mathfrak{R}$ denotes the support of $\psi$. Let $\lambda \geq 0$ such that $0 \leq a + \lambda v$. Then for every function $\varphi \in \mathcal{K}(X)$ such that $\varphi \prec O$ we have $\varphi \leq \psi$ and

$$0 \leq \varphi \otimes (a + \lambda v) \leq \psi \otimes (a + \lambda v) \leq \psi \otimes a + \lambda \mathfrak{v},$$

and therefore

$$0 \leq \varphi \otimes a + \lambda \mathfrak{v} \qquad \text{and} \qquad \varphi \otimes a \leq \psi \otimes a + \lambda \mathfrak{v}.$$

The set $\mathcal{A}$ is therefore bounded below and bounded above relative to the function $\psi \otimes a$, thus relatively bounded in $\mathcal{F}_{\mathfrak{V}}(X, \mathcal{P})$ according to I.4.24.

Recall from a remark in I.4.24 that every continuous linear operator maps relatively bounded sets into relatively bounded sets.

## 6. Special Cases and Applications

We shall discuss a range of special cases and applications of Theorem 5.1 in this final section. Most of these settings had been earlier dealt with in Section 6 of Chapter II. Throughout the following, we shall assume that $(\mathcal{P},\mathcal{V})$ is a quasi-full locally convex cone and that $(\mathcal{Q},\mathcal{V})$ is a locally convex complete lattice cone whose order continuous linear functionals support the separation property. We shall apply Theorem 5.1 with the standard full extension $(\mathcal{P}_{\mathcal{V}},\mathcal{V})$ in place of $\mathcal{P}$ and assume that the linear operator $T:\mathcal{F}_{\mathfrak{V}}(X,\mathcal{P})\to\mathcal{Q}$ can be extended to $\mathcal{F}_{\mathfrak{V}}(X,\mathcal{P}_{\mathcal{V}})$. In Section 3 we investigated various situations where these extensions are guaranteed, in particular if $\mathcal{Q}=\overline{\mathbb{R}}$ or if $(\mathcal{P},\mathcal{V})$ is a locally convex vector space. Throughout the following $(\mathcal{Q}_0,\mathcal{W}_0)$ will stand for a locally convex cone whose standard lattice completion in the sense of I.5.57 is $(\mathcal{Q},\mathcal{W})$. We shall use the notations of Chapters I, II and the preceding sections of Chapter III.

**6.1 The Case that $\mathcal{Q}$ Is the Standard Lattice Completion of Some Subcone $\mathcal{Q}_0$.** Suppose that $(\mathcal{Q},\mathcal{W})$ is the standard lattice completion of a locally convex cone $(Q_0,\mathcal{W}_0)$ (see I.5.57) and that $T\big(\mathcal{F}_{\mathfrak{V}}(X,\mathcal{P})\big)\subset\mathcal{Q}_0$. Then $T$ maps relatively bounded subsets of $\mathcal{F}_{\mathfrak{V}}(X,\mathcal{P})$ into relatively bounded subsets of $\mathcal{Q}_0$. Hence the bounded quasi regular $\mathfrak{L}(\mathcal{P},\mathcal{Q})$-valued measure $\theta$ that represents $T$ as in 5.1 takes its values $\theta_E(a)$ for $E\in\mathfrak{R}$ and $a\in\mathcal{P}$, in the order closure in $\mathcal{Q}$ of relatively bounded subsets of $\mathcal{Q}_0$ (see Remark 5.2(b)). According to I.5.57 these values are therefore elements of $\mathcal{Q}_0^{**}$, the second dual of $\mathcal{Q}_0$, and according to I.7.3 indeed contained in the dual of $\mathcal{Q}_0^*$, if $\mathcal{Q}_0^*$ is endowed with the topology generated by the family $\mathfrak{Z}$ of all relatively bounded subsets of $\mathcal{Q}_0$. This cone was introduced as the (relative) strong second dual $\mathcal{Q}_{0\,sr}^{**}$ of $\mathcal{Q}_0$ in Section I.7.3.

**Corollary 6.2.** *Suppose that $(\mathcal{Q},\mathcal{W})$ is the standard lattice completion of some locally convex cone $(Q_0,\mathcal{W}_0)$. Then, under the assumptions of Theorem 5.1, the representing measure for a linear operator $T:\mathcal{F}_{\mathfrak{V}}(X,\mathcal{P})\to\mathcal{Q}_0$, is $\mathfrak{L}(\mathcal{P},\mathcal{Q}_{0\,sr}^{**})$-valued.*

We may obtain a further strengthening of this observation in special cases:

**6.3 Compact and Weakly Compact Operators.** Suppose that as in 6.1 $(\mathcal{Q},\mathcal{W})$ is the standard lattice completion of a locally convex cone $(Q_0,\mathcal{W}_0)$. A linear operator $T:\mathcal{E}(X,\mathcal{P})\to\mathcal{Q}$ is called *compact* (or *weakly compact*) if for every $E\in\mathfrak{R}$ and every relatively bounded subset $A$ of $\mathcal{P}$ the image under $T$ of the set

$$\left\{ \sum_{i=1}^{n} \varphi_{i\otimes} a_i \;\middle|\; a_i \in A, \;\; \varphi_i \in \mathcal{K}(X), \;\; \sum_{i=1}^{n} \varphi_i \leq \chi_E \right\}$$

is relatively compact in the symmetric relative topology (or in the weak topology $\sigma(\mathcal{Q}_0, \mathcal{Q}_0^*)$) of $\mathcal{Q}_0$. (For the definition of the weak topology see I.4.6.) Recall from Lemma I.4.7 that the symmetric relative topology is finer than $\sigma(\mathcal{Q}_0, \mathcal{Q}_0^*)$, and from I.5.57 that $\sigma(\mathcal{Q}_0, \mathcal{Q}_0^*)$ is finer than the induced order topology on $\mathcal{Q}_0$ which is however still Hausdorff. The latter two topologies coincide, if all elements of $\mathcal{Q}_0$ are bounded (see I.5.57). Moreover, $\sigma(\mathcal{Q}_0, \mathcal{Q}_0^*)$ coincides with its own relative topology (see I.4.6). Every subset $\mathcal{M}$ of $\mathcal{Q}_0$ which is relatively compact in the symmetric relative topology is also relatively weakly compact. Indeed, the closure $\overline{\mathcal{M}}$ of $\mathcal{M}$ with respect to the symmetric relative topology is contained in its closure $\overline{\mathcal{M}}^w$ with respect to the weak topology. $\overline{\mathcal{M}}$ is compact in the former, hence also in the latter topology, thus weakly closed since $\sigma(\mathcal{Q}_0, \mathcal{Q}_0^*)$ is Hausdorff. We infer that $\overline{\mathcal{M}} = \overline{\mathcal{M}}^w$, and our claim follows. Every compact operator is therefore also weakly compact.

*Remarks 6.4.* (a) For any $E \in \mathfrak{R}$ and every relatively bounded subset $A$ of $\mathcal{P}$ the set

$$\mathcal{A} = \left\{ \sum_{i=1}^{n} \varphi_{i\otimes} a_i \mid a_i \in A, \;\; \varphi_i \in \mathcal{K}(X), \;\; \sum_{i=1}^{n} \varphi_i \leq \chi_E \right\}$$

from 6.2 is relatively bounded in $\mathcal{F}_{\mathfrak{V}}(X, \mathcal{P})$. Indeed, let $\psi \in \mathcal{K}(X)$ such that $E \prec \psi$. There is $a_0 \in \mathcal{P}$ such that $A$ is bounded above relative to $a_0$. Now let $\mathfrak{v} \in \mathfrak{V}$ and choose $v \in \mathcal{V}$ such that $\chi_{F\otimes} v \leq \mathfrak{v}$, where $F \in \mathfrak{R}$ is the support of the function $\psi$. Let $\lambda, \rho \geq 0$ such that $0 \leq a_0 + \lambda v$, $0 \leq a + \lambda v$ and $a \leq \rho a_0 + \lambda v$ for all $a \in A$. Now let $\varphi_1, \ldots, \varphi_n \in \mathcal{K}(X)$ such that $\sum_{i=1}^{n} \varphi_i \leq \chi_E$ and let $a_1, \ldots a_n \in A$. Then

$$0 \leq \sum_{i=1}^{n} \varphi_{i\otimes}(a_i + \lambda v) \leq \sum_{i=1}^{n} \varphi_{i\otimes} a_i + \lambda \chi_{E\otimes} v \leq \sum_{i=1}^{n} \varphi_{i\otimes} a_i + \lambda \mathfrak{v}$$

and

$$\sum_{i=1}^{n} \varphi_{i\otimes} a_i \leq \sum_{i=1}^{n} \varphi_{i\otimes}(\rho a_0 + \lambda v) \leq \psi_{\otimes}(\rho a_0 + \lambda v) \leq \rho \psi_{\otimes} a_0 + \lambda \chi_{F\otimes} v \leq \rho \psi_{\otimes} a_0 + \lambda \mathfrak{v}.$$

The set $\mathcal{A}$ is therefore bounded below and bounded above relative to the function $\psi_{\otimes} a_0$, thus relatively bounded in $\mathcal{F}_{\mathfrak{V}}(X, \mathcal{P})$.

(b) As a consequence of (a), every continuous linear operator $T : \mathcal{F}_{\mathfrak{V}}(X, \mathcal{P}) \to \mathcal{Q}_0$ that maps relatively bounded subsets of $\mathcal{F}_{\mathfrak{V}}(X, \mathcal{P})$ into relatively (weakly) compact subsets of $\mathcal{Q}_0$, that is every continuous linear operator which is (weakly) compact in the usual sense, is also (weakly) compact in the sense of 6.3. The converse of this statement holds true in special cases: Suppose that $X$ is compact, that is $X \in \mathfrak{K}$, and that $\mathcal{F}_{\mathfrak{V}}(X,\mathcal{P})$ carries the topology of uniform convergence generated by a single neighborhood; more precisely: The inductive limit neighborhood system $\mathfrak{V}$ consists of multiples of a single neighborhood $\mathfrak{v}$ which in turn contains the single constant function $x \mapsto v$ for a fixed neighborhood $v \in \mathcal{V}$, that is $\mathfrak{V} = \{\varepsilon\mathfrak{v} \mid \varepsilon > 0\}$, where $\mathfrak{v} = \{\chi_{X \otimes} v\}$. Then $\mathcal{F}_{\mathfrak{V}}(X,\mathcal{P}) = \mathcal{C}^r(X,\mathcal{P})$ (see Remark 2.10(c)). In this situation, suppose that the operator $T : \mathcal{F}_{\mathfrak{V}}(X,\mathcal{P}) \to \mathcal{Q}_0$ is (weakly) compact in the sense of 6.2 and let $\mathcal{B}$ be a relatively bounded subset of $\mathcal{F}_{\mathfrak{V}}(X,\mathcal{P})$, that is there is $g \in \mathcal{F}_{\mathfrak{V}}(X,\mathcal{P})$ and $\lambda \geq 0$ such that $0 \leq f + \lambda\chi_{X \otimes} v$ and $f \leq g + \lambda\chi_{X \otimes} v$ for all $f \in \mathcal{B}$. We may also assume that $0 \leq g + \lambda\chi_{X \otimes} v$. We proceed to construct a corresponding subset $\mathcal{A}$ of $\mathcal{E}(X,\mathcal{P})$ in the following way: Because $\mathcal{F}_{\mathfrak{V}}(X,\mathcal{P})$ is the closure of the subcone $\mathcal{E}(X,\mathcal{P})$, for every $f \in \mathcal{B}$ and every $0 < \varepsilon \leq 1$ we can find an element $h_{(f,\varepsilon)} \in \mathcal{E}(X,\mathcal{P})$ such that $h_{(f,\varepsilon)} \in \mathfrak{v}^s_\varepsilon(f)$. According to Lemma I.4.1(b) and (c) this implies

$$h_{(f,\varepsilon)} \leq (1+\varepsilon)f + \varepsilon(1+\lambda)\chi_{X \otimes} v \qquad \text{and} \qquad f \leq (1+\varepsilon)h_{(f,\varepsilon)} + \varepsilon(2+\lambda)\chi_{X \otimes} v.$$

Using an argument involving a partition of the unit, we may indeed assume that the functions $h_{(f,\varepsilon)}$ are all of the type

$$\sum_{i=1}^{n} \varphi_{i \otimes} a_i, \qquad \text{where} \qquad a_i \in \mathcal{P} \quad \text{and} \quad \sum_{i=1}^{n} \varphi_i = 1,$$

and for each $i = 1, \ldots, n$ there is a point $x_i \in X$ such that $\varphi_i(x_i) = 1$. Similarly, there is a function $l = \sum_{k=1}^{m} \psi_{k \otimes} b_k \in \mathcal{E}(X,\mathcal{P})$ such that $l \in \mathfrak{v}^s_1(g)$, which implies by I.4.1(b) and (c)

$$l \leq 2g + (1+\lambda)\chi_{X \otimes} v \qquad \text{and} \qquad g \leq 2l + (2+\lambda)\chi_{X \otimes} v.$$

Now let

$$\mathcal{A} = \{h_{(f,\varepsilon)} \mid f \in \mathcal{B},\ \varepsilon > 0\}.$$

Following our construction, $\mathcal{B}$ is contained in the closure $\overline{\mathcal{A}}$ of $\mathcal{A}$, taken in $\mathcal{F}_{\mathfrak{V}}(X,\mathcal{P})$ with respect to its symmetric relative inductive limit topology. Moreover, for any function $h_{(f,\varepsilon)} = \sum_{i=1}^{n} \varphi_{i \otimes} a_i \in \mathcal{A}$, for any $i = 1, \ldots, n$, we have $h(x_i) = a_i$ for some $x_i \in X$, hence

$$0 \leq f(x_i) + \lambda v \leq 2h(x_i) + (2+\lambda v) + \lambda v = 2a_i + 2(1+\lambda)v,$$

and

$$0 \leq a_i + (1+\lambda)v.$$

Furthermore,

$$a_i = h(x_i) \leq 2f(x_i) + (1+\lambda)v \leq 2g(x_i) + (1+3\lambda)v \leq 2l(x_i) + (5+5\lambda)v,$$

where $l(x_i) = \sum_{k=1}^m \psi_k(x_i)b_k$. There is $\rho \geq 0$ such that $0 \leq b_k + \rho v$ for $k = 1, \ldots, m$. We set $b = b_1 + \ldots + b_m$ and realize that

$$\sum_{k=1}^m \psi_k(x_i)b_k \leq \sum_{k=1}^m \Big(\psi_k(x_i)b_k + (1 - \psi_k(x_i))(b_k + \rho v)\Big) \leq b + m\rho v.$$

Thus

$$a_i \leq 2b + (5 + 5\lambda + m\rho)v$$

for the elements $a_i \in \mathcal{P}$ from above. We conclude that the set $A \subset \mathcal{P}$ of all values of the functions in $\mathcal{A}$ is bounded below and bounded above relative to the element $b \in \mathcal{P}$. Therefore $T(\mathcal{A})$ is contained in the relatively (weakly) compact image under $T$ of the set

$$\left\{ \sum_{i=1}^n \varphi_i \otimes a_i \;\middle|\; a_i \in A,\;\; \varphi_i \in \mathcal{K}(X),\;\; \sum_{i=1}^n \varphi_i \leq 1 \right\}.$$

Because the operator $T : \mathcal{F}_{\mathfrak{V}}(X, \mathcal{P}) \to \mathcal{Q}_0$ is also continuous if we consider the respective symmetric relative topologies on $\mathcal{F}_{\mathfrak{V}}(X, \mathcal{P})$ and $\mathcal{Q}_0$ (see Proposition I.4.5), we have

$$T(\mathcal{B}) \subset T(\overline{\mathcal{A}}) \subset \overline{T(\mathcal{A})},$$

where $\overline{T(\mathcal{A})}$ denotes the closure of $T(\mathcal{A})$ in $\mathcal{Q}_0$ with respect to the symmetric relative topology. But this topology is finer than the weak topology $\sigma(\mathcal{Q}_0, \mathcal{Q}_0^*)$ (see Lemma I.4.7), hence $\overline{T(\mathcal{A})}$ is contained in the weak closure of $T(\mathcal{A})$ which was seen to be (weakly) compact. Thus $T(\mathcal{B})$ is also relatively (weakly) compact in $\mathcal{Q}_0$, and the operator $T : \mathcal{F}_{\mathfrak{V}}(X, \mathcal{P}) \to \mathcal{Q}_0$ is (weakly) compact in the usual sense, as claimed.

For the next corollary, recall the introduction of various operator topologies from Example I.7.2(a). In particular, the strong operator topology on $\mathfrak{L}(\mathcal{P}, \mathcal{Q})$ is generated by the family $\mathfrak{Z}$ of all finite subsets of $\mathcal{P}$ and the symmetric topology of $\mathcal{Q}$ (see I.7.3(a)(ii)); the weak operator topology by the family $\mathfrak{Z}$ of all finite subsets of $\mathcal{P}$ and the weak topology $\sigma(\mathcal{Q}, \mathcal{Q}^*)$ of $\mathcal{Q}$ (see I.7.3(a)(iii)). We shall use these definitions with the subcone $\mathcal{Q}_0$ of $\mathcal{Q}$ in place of $\mathcal{Q}$.

**Corollary 6.5.** *Suppose that $(\mathcal{Q}, \mathcal{W})$ is the standard lattice completion of some locally convex cone $(\mathcal{Q}_0, \mathcal{W}_0)$ and that, under the assumptions of Theorem 5.1, the operator $T : \mathcal{F}_{\mathfrak{V}}(X, \mathcal{P}) \to \mathcal{Q}_0$ is compact (or weakly compact) on $\mathcal{E}(X, \mathcal{P})$. Then the following holds:*

*(a) The representing measure $\theta$ for $T$ is $\mathfrak{L}(\mathcal{P}, \mathcal{Q}_0)$-valued.*
*(b) For every $E \in \mathfrak{R}$ and $a \in \mathcal{P}$ the set $\{\theta_G(a) \mid G \in \mathfrak{R},\; G \subset E\}$ is compact (or weakly compact) in $\mathcal{Q}_0$.*

(c) *For every* $E \in \mathfrak{R}$ *the set* $\{\theta_G \mid G \in \mathfrak{R},\ G \subset E\}$ *of linear operators from* $\mathcal{P}$ *into* $\mathcal{Q}_0$ *is relatively compact in* $\mathfrak{L}(\mathcal{P}, \mathcal{Q}_0)$*, if endowed with the symmetric relative strong (or with the symmetric weak) operator topology.*

*Proof.* Suppose that the operator $T : \mathcal{F}_{\mathfrak{V}}(X, \mathcal{P}) \to \mathcal{Q}_0$ is compact (or weakly compact) in the sense of 6.3 and recall that compactness implies weak compactness. Let $E \in \mathfrak{R}$ and let $O \in \mathfrak{R}$ be any open set containing $E$.

For Parts (a) and (b), let $a \in \mathcal{P}$ and let $G \in \mathfrak{R}$ be a subset of $E$. According to 5.2(b) the element $\theta_G(a) \in \mathcal{Q}$ is contained in the order closure in $\mathcal{Q}$ of the image under $T$ of the subset $\mathcal{A} = \{\varphi{\otimes}a \mid \varphi \in \mathcal{K}(X),\ \psi \prec O\}$ of $\mathcal{E}(X, \mathcal{P})$. This image is relatively compact (or relatively weakly compact) in $\mathcal{Q}_0$ by 6.3. Let $\overline{T(\mathcal{A})}$ and $\overline{T(\mathcal{A})}^w$ denote its closure in $\mathcal{Q}_0$ with respect to the symmetric relative topology and with respect to the weak topology $\sigma(\mathcal{Q}_0, \mathcal{Q}_0^*)$, respectively. Let us first consider the weakly compact case: Then $\overline{T(\mathcal{A})}^w$ is weakly compact. Recall from I.5.57 that the weak topology $\sigma(\mathcal{Q}_0, \mathcal{Q}_0^*)$ is generally finer than the induced order topology on $\mathcal{Q}_0$ which is however still Hausdorff. The subset $\overline{T(\mathcal{A})}^w$ of $\mathcal{Q}_0$ is therefore also compact in the coarser induced order topology, and indeed closed in $\mathcal{Q}$ as the order topology is Hausdorff in this case. Similarly, in the compact case, because the symmetric relative topology on $\mathcal{Q}_0$ is also finer than the induced order topology on $\mathcal{Q}_0$, one argues that the set $\overline{T(\mathcal{A})}$ is closed in $\mathcal{Q}$ in the order topology. This demonstrates in particular that $\theta_G(a) \in \overline{T(\mathcal{A})} \subset \mathcal{Q}_0$ in the compact case, or in $\theta_G(a) \in \overline{T(\mathcal{A})}^w \subset \mathcal{Q}_0$ in the weakly compact case. The set $\{\theta_G \mid G \in \mathfrak{R},\ G \subset E\}$ is therefore compact (or weakly compact) and contained in $\mathcal{Q}_0$ as claimed.

For Part (c), let

$$\mathcal{O} = \{\theta_G \mid G \in \mathfrak{R},\ G \subset E\} \subset \mathfrak{L}(\mathcal{P}, \mathcal{Q}_0),$$

and for every $a \in \mathcal{P}$ let

$$\mathcal{O}(a) = \{\theta_G(a) \mid G \in \mathfrak{R},\ G \subset E\}.$$

According to Part (b), this set is relatively (weakly) compact in $\mathcal{Q}_0$. Its closure $\overline{\mathcal{O}(a)}$ in $\mathcal{Q}_0$ (taken either in the symmetric relative or in the weak topology) is therefore (weakly) compact. Following Tychonoff's theorem the set $\mathcal{C} = \prod_{a \in \mathcal{P}} \overline{\mathcal{O}(a)}$ is compact in the product of the symmetric relative (or weak) topology. Now let $(S_i)_{i \in \mathcal{I}}$ be a net in $\mathcal{O}$. Then the mapping

$$i \mapsto \big(S_i(a)\big)_{a \in \mathcal{P}} : \mathcal{I} \to \mathcal{C}$$

is a net in $\mathcal{C}$ which because of compactness permits a convergent subnet

$$j \mapsto \big(S_j(a)\big)_{a \in \mathcal{P}} : \mathcal{J} \to \mathcal{C}.$$

Let $\big(S(a)\big)_{a\in\mathcal{P}} \in \mathcal{C}$ be the limit of this net. We claim that the mapping $a \mapsto S(a) : \mathcal{P} \to \mathcal{Q}_0$ is a continuous linear operator. Indeed, linearity follows directly from the linearity of the operators $S_j$ in the approximating net. For continuity, given $w \in \mathcal{W}_0$ there is $v \in \mathcal{V}$ such that $|\theta|(E,v) \le w$. Then $a \le b+v$ for $a,b \in \mathcal{P}_0$ implies that $\theta_G(a) \le \theta_G(b)+w$ for all $\theta_G \in \mathcal{O}$. Because all the operators $S_j$ are contained in $\mathcal{O}$, this shows that $S(a) \le S(b)+w$ holds as well. Thus $S \in \mathfrak{L}(\mathcal{P},\mathcal{Q}_0)$, and the net $(S_j)_{j\in\mathcal{J}}$ converges to $S$ in the relative strong (or weak) operator topology of $\mathfrak{L}(\mathcal{P},\mathcal{Q}_0)$. This demonstrates, as claimed, that the set $\mathcal{O}$ is relatively compact in this topology. □

For the next corollary recall the definition of a compact and a weakly compact measure from Chapter II.6.8.

**Corollary 6.6.** *Suppose that $(\mathcal{Q},\mathcal{W})$ is the standard lattice completion of some locally convex cone $(\mathcal{Q}_0,\mathcal{W}_0)$ and that all elements of $\mathcal{Q}_0$ are bounded. Under the assumptions of Theorem 5.1, the linear operator $T : \mathcal{F}_{\mathfrak{V}}(X,\mathcal{P}) \to \mathcal{Q}_0$ is compact (or weakly compact) on $\mathcal{E}(X,\mathcal{P})$ if and only if its representing measure $\theta$ is $\mathfrak{L}(\mathcal{P},\mathcal{Q}_0)$-valued and compact (or weakly compact). In this case, and if all elements of $\mathcal{P}$ are bounded as well, the measure $\theta$ is countably additive with respect to the strong operator topology.*

*Proof.* Let us first assume that the operator $T : \mathcal{F}_{\mathfrak{V}}(X,\mathcal{P}) \to \mathcal{Q}_0$ is compact (or weakly compact), and let $\theta$ be its $\mathfrak{L}(\mathcal{P},\mathcal{Q}_0)$-valued representing measure. Given $E \in \mathfrak{R}$ and a relatively bounded subset $A$ of $\mathcal{P}$, let $O \in \mathfrak{R}$ be an open set containing $E$. Then the image under $T$ of the set

$$\mathcal{A} = \left\{ \sum_{i=1}^{n} \varphi_i \otimes a_i \;\middle|\; a_i \in A,\ \ \varphi_i \in \mathcal{K}(X),\ \ \sum_{i=1}^{n} \varphi_i \le \chi_O \right\}$$

is relatively compact (or relatively weakly compact) in $\mathcal{Q}_0$. Let us recall our earlier remarks: In the second, that is the relatively weakly compact case, the weak closure $\overline{T(\mathcal{A})}^w$ of $T(\mathcal{A})$ in $\mathcal{Q}_0$ is weakly compact. In the first, that is the relatively compact case, the closure $\overline{T(\mathcal{A})}$ of $T(\mathcal{A})$ in the symmetric topology coincides with $\overline{T(\mathcal{A})}^w$ and is compact in both topologies. Now consider disjoint compact subsets $K_1,\ldots,K_n$ of $E$, and let $O_1,\ldots,O_n$ be disjoint open subsets of $O$ such that $K_j \subset O_j$ for $j=1,\ldots,n$. Let $a_1,\ldots,a_n \in A$. For each $j=1,\ldots,n$ let $(\varphi_{i_j})_{i_j\in\mathcal{I}_j}$ be a net in $\mathcal{K}(X)$ as in 5.2(a)(iii), with $K_j$ in place of $E$ and $O_j$ in place of $O$, that is $\varphi_{i_j} \prec 0_j$ for all $i_j \in \mathcal{I}_j$ and $j=1,\ldots,n$, as well as $\theta_{K_j}(a) = \lim_{i_j\in\mathcal{I}_j} T(\varphi_{i_j}\otimes a)$ for all $a \in \mathcal{P}$. Let $\mathcal{I} = \mathcal{I}_1 \times \ldots \times \mathcal{I}_n$, endowed with the componentwise order, and for $i=(i_1,\ldots,i_n)$ set

$$f_i = \sum_{j=1}^{n} \varphi_{i_j} \otimes a_j.$$

Then $f_i \in \mathcal{A}$ for all $i \in \mathcal{I}$ and

$$\lim_{i\in\mathcal{I}} T(f_i) = \sum_{j=1}^{n} \lim_{i\in\mathcal{I}} T(\varphi_i \otimes a_j) = \sum_{j=1}^{n} \lim_{i_j\in\mathcal{I}_j} T(\varphi_{i_j} \otimes a_j) = \sum_{j=1}^{n} \theta_{K_j}(a_j)$$

by Proposition I.5.22 and Lemma I.5.20(d), since for every $j = 1, \ldots, n$ the net $\big(T(\varphi_i \otimes a_j)\big)_{i\in\mathcal{I}}$ can be considered to be a subnet of $\big(T(\varphi_{i_j} \otimes a_j)\big)_{i_j\in\mathcal{I}_j}$. This yields that the element $\sum_{j=1}^{n} \theta_{K_j}(a_j)$ is contained in the (compact or weakly compact) closure $\overline{T(\mathcal{A})}^{w}$ of $T(\mathcal{A})$. Now let $E_1, \ldots, E_n \in \mathfrak{R}$ be disjoint subsets of $E$ and let $a_1, \ldots, a_n \in A$. Because all elements of $\mathcal{Q}_0$ are supposed to be bounded, Proposition 4.4 yields that $\theta_{E_j}(a_j) = \lim_{K_j \subset E_j} \theta_{K_j}(a_j)$ for every $j = 1, \ldots, n$. Now an argument similar to the above yields

$$\lim_{\substack{K_1 \subset E_1, \\ \cdots \\ K_n \subset E_n}} \sum_{j=1}^{n} \theta_{K_j}(a_j) = \sum_{j=1}^{n} \lim_{K_j \subset E_j} \theta_{K_j}(a_j) = \sum_{j=1}^{n} \theta_{E_j}(a_j).$$

From this we infer that the element $\sum_{j=1}^{n} \theta_{E_j}(a_j)$ is also contained in $\overline{T(\mathcal{A})}^{w}$. As a subset of $\overline{T(\mathcal{A})}^{w}$ the set

$$\left\{ \sum_{i=1}^{n} \theta_{E_i}(a_i) \;\middle|\; a_i \in A,\ E_i \in \mathfrak{R} \text{ disjoint subsets of } E \right\}$$

is therefore also relatively compact (or weakly compact), hence the measure $\theta$ is compact (or weakly compact) in the sense of II.6.8. If the elements of $\mathcal{P}$ are also bounded, then our version of Pettis' theorem, that is Theorem II.3.11 applies: The representing $\mathfrak{L}(\mathcal{P}, \mathcal{Q}_0)$-valued measure $\theta$ is countably additive with respect to the strong operator topology of $\mathfrak{L}(\mathcal{P}, \mathcal{Q}_0)$ in this case.

Now suppose that the measure $\theta$ is compact (or weakly compact), let $E \in \mathfrak{R}$ and let $A$ be a relatively bounded subset of $\mathcal{P}$. According to Proposition II.6.9 then

$$\left\{ \sum_{i=1}^{n} \int_X \varphi_i \otimes a_i \;\middle|\; a_i \in A,\ 0 \le \psi_i \text{ measurable},\ \sum_{i=1}^{n} \varphi_i \le \chi_E \right\}$$

is a relatively compact (or relatively weakly compact) subset of $\mathcal{Q}_0$. As all elements of $\mathcal{Q}_0$ are bounded, Theorem 5.1 yields that $T(\varphi \otimes a) = \int_X \varphi \otimes a \, d\theta$ holds for all $\varphi \otimes a \in \mathcal{E}(X, \mathcal{P})$. Thus the subset

$$\left\{ \sum_{i=1}^{n} T(\varphi_i \otimes a_i) \;\middle|\; a_i \in A,\ \ \varphi_i \in \mathcal{K}(X),\ \ \sum_{i=1}^{n} \varphi_i \le \chi_E \right\}$$

of the above is also relatively compact (or relatively weakly compact), hence our claim. □

Proposition II.6.11 states that in special circumstances every bounded $\mathfrak{L}(\mathcal{P},\mathcal{Q}_0)$-valued measure is weakly compact. In combination with Corollary 6.6 this yields:

**Corollary 6.7.** *Suppose that* $(\mathcal{P},\|\ \|)$ *is a finite dimensional normed space and that* $(\mathcal{Q},\mathcal{W})$ *is the standard lattice completion of a Banach space* $(\mathcal{Q}_0,\|\ \|)$*. For a linear operator* $T:\mathcal{F}_{\mathfrak{V}}(X,\mathcal{P})\to\mathcal{Q}_0$ *and its representing measure* $\theta$ *the following are equivalent:*

*(a)* $T$ *is weakly compact on* $\mathcal{E}(X,\mathcal{P})$*.*
*(b)* $\theta$ *is weakly compact.*
*(c)* $\theta$ *is* $\mathfrak{L}(\mathcal{P},\mathcal{Q}_0)$*-valued.*

*Each of these properties implies that the measure* $\theta$ *is countably additive with respect to the strong operator topology.*

*Remark 6.8.* If $X$ carries the discrete topology, then all functions $\chi_{E\otimes}a$ for $E\in\mathfrak{R}$ and $a\in\mathcal{P}$ are contained in $\mathcal{E}_0(X,\mathcal{P})\subset\mathcal{F}_{\mathfrak{V}}(X,\mathcal{P})$. Hence, for a continuous linear operator $T:\mathcal{F}_{\mathfrak{V}}(X,\mathcal{P})\to\mathcal{Q}_0$ its representing measure $\theta$ is given by $\theta_E(a)=T(\chi_{E\otimes}a)$ for all $E\in\mathfrak{R}$ and $a\in\mathcal{P}$. The measure $\theta$ is therefore $\mathfrak{L}(\mathcal{P},\mathcal{Q}_0)$-valued for any choice of the operator $T$. This example shows that the implication (c) $\Rightarrow$ (a) from Corollary 6.7 does not hold in more general circumstances.

**6.9 Locally Convex Topological Vector Spaces.** The case that both $(\mathcal{P},\mathcal{V})$ and $(\mathcal{Q}_0,\mathcal{W}_0)$ are locally convex topological vector spaces over $\mathbb{K}=\mathbb{R}$ or $\mathbb{K}=\mathbb{C}$ is of particular interest. We shall assume that $(\mathcal{Q},\mathcal{W})$ is the standard lattice completion of $(\mathcal{Q}_0,\mathcal{W}_0)$ and that the neighborhoods in $\mathcal{V}$ and in $\mathcal{W}_0$ are balanced and convex. The cones $\mathcal{F}_{\mathfrak{V}}(X,\mathcal{P})$ and $\mathcal{F}_{\mathfrak{V}_0}(X,\mathcal{P})$ then coincide, and Theorems 3.3 and 3.1 yield that a continuous linear operator $T:\mathcal{F}_{\mathfrak{V}}(X,\mathcal{P})\to\mathcal{Q}_0$ can be extended into a continuous linear operator $T:\mathcal{F}_{\mathfrak{V}}(X,\mathcal{P}_{\mathcal{V}})\to\mathcal{Q}$. We shall denote the strong second dual of a locally convex topological vector space $(\mathcal{N},\mathcal{U})$ by $\mathcal{N}_s^{**}$, that is the dual of $\mathcal{N}^*$ if the latter is endowed with the topology that is generated by the family $\mathfrak{Z}$ of all bounded subsets of $\mathcal{N}$ (see Section I.7.3). Recall from Example I.2.1(d) that there is a canonical correspondence between the dual cone $\mathcal{N}^*$ of a vector space $\mathcal{N}$ (considered as a locally convex cone) and its usual vector space dual, that is the space of all continuous $\mathbb{K}$-linear functionals on $\mathcal{N}$. The same holds for the second dual cone. Thus, in our context, $\mathcal{Q}_{0s}^{**}$ may be considered to be either the strong second dual cone of $(\mathcal{Q}_0,\mathcal{W}_0)$ as a locally convex cone, hence a subcone of $\mathcal{Q}$ (recall the remarks in I.7.3), or the strong second vector space dual of $(\mathcal{Q}_0,\mathcal{W}_0)$. The statements of Theorem 5.1 (with $\mathcal{P}_{\mathcal{V}}$ in place of $\mathcal{P}$) can therefore be reformulated using only the underlying vector spaces $\mathcal{P}$ and $\mathcal{Q}_0$ and their vector space duals.

If the operator $T:\mathcal{F}_{\mathfrak{V}}(X,\mathcal{P})\to\mathcal{Q}_0$ is indeed linear over $\mathbb{K}$, then Remark 5.2(a) yields that the operators $\theta_E\in\mathfrak{L}(\mathcal{P},\mathcal{Q}_{0s}^{**})$ are also linear over $\mathbb{K}$. In order to prove this claim, let $E\in\mathfrak{R}$. According to 5.2(a)(iii) there is a net $(\varphi_i)_{i\in\mathcal{I}}$ in $\mathcal{K}(X)$ such that $\theta_E(a) = \lim_{i\in\mathcal{I}} T(\varphi_{i\otimes}a)$ for all $a \in \mathcal{P}$ in the order topology of $\mathcal{Q}$, that is

$$\theta_E(a)(\mu) = \lim_{i\in\mathcal{I}} T(\varphi_{i\otimes}a)(\mu)$$

for all $a \in \mathcal{P}$ and $\mu \in \mathcal{Q}_0^*$. Thus for $a \in \mathcal{P}$ and $\alpha \in \mathbb{K}$, as $T(\varphi_{i\otimes}a) \in \mathcal{Q}_0$, we have

$$\theta_E(\alpha a)(\mu) = \lim_{i\in\mathcal{I}} \mu\big(T(\varphi_{i\otimes}(\alpha a))\big) = \alpha \lim_{i\in\mathcal{I}} \mu\big(T(\varphi_{i\otimes}a)\big) = \alpha\theta_E(a)(\mu)$$

for all $\mu \in \mathcal{Q}_0^*$; that is $\theta_E(\alpha a) = \alpha\theta_E(a)$ as an element of $\mathcal{Q}_{0s}^{**} \subset \mathcal{Q}$. In the light of this and the preceding observation the measure $\theta$ can be interpreted as an element of $\mathfrak{L}_{\mathbb{K}}(\mathcal{P}, \mathcal{Q}_{0s}^{**})$, that is the space of all continuous $\mathbb{K}$-linear operators from the vector space $\mathcal{P}$ into the vector space $\mathcal{Q}_{0s}^{**}$.

We shall formulate this important special case as a corollary. For our notions of various choices for operator topologies we refer to Section I.7.2(a). Recall that in case that both $(\mathcal{P}, \mathcal{V})$ and $(\mathcal{Q}_0, \mathcal{W}_0)$ are locally convex topological vector spaces and $\mathcal{Q}_0$ is reflexive, then every continuous linear operator $T : \mathcal{F}_{\mathfrak{V}}(X, \mathcal{P}) \to \mathcal{Q}_0$ is weakly compact (see Corollary VI.4.3 in [55]).

**Corollary 6.10.** *Let $(\mathcal{P}, \mathcal{V})$ and $(\mathcal{Q}_0, \mathcal{W}_0)$ be locally convex topological vector spaces over $\mathbb{K} = \mathbb{R}$ or $\mathbb{K} = \mathbb{C}$. Let $X$ be a locally compact Hausdorff space and let $\mathfrak{V}$ be a basis for a symmetric r-lower continuous inductive limit topology on $\mathcal{F}(X, \mathcal{P})$. Then every continuous $\mathbb{K}$-linear operator $T : \mathcal{F}_{\mathfrak{V}}(X, \mathcal{P}) \to \mathcal{Q}_0$ can be represented as an integral on $X$. More precisely: There exists a unique bounded $\mathfrak{L}_{\mathbb{K}}(\mathcal{P}, \mathcal{Q}_{0s}^{**})$-valued measure $\theta$ on the weak $\sigma$-ring $\mathfrak{R}$ of all relatively compact Borel subsets of $X$ with the following properties: $\theta$ is countably additive and quasi regular with respect to the weak* operator topology of $\mathfrak{L}_{\mathbb{K}}(\mathcal{P}, \mathcal{Q}_{0s}^{**})$, continuous relative to $\mathfrak{V}$, all functions in $\mathcal{F}_{\mathfrak{V}}(X, \mathcal{P})$ are strongly integrable with respect to $\theta$, and*

$$\int_X f\,d\theta = T(f) \qquad \text{for all} \qquad f \in \mathcal{F}_{\mathfrak{V}}(X, \mathcal{P}).$$

*The operator $T$ is compact (or weakly compact) on $\mathcal{E}(X, \mathcal{P})$, if and only if the measure $\theta$ is $\mathfrak{L}_{\mathbb{K}}(\mathcal{P}, \mathcal{Q}_0)$-valued and compact (or weakly compact). In this case $\theta$ is countably additive with respect to the strong operator topology of $\mathfrak{L}_{\mathbb{K}}(\mathcal{P}, \mathcal{Q}_0)$.*

This corollary generalizes quite a few of the standard results that can be found in the literature. Theorem IV.7.2 in [55] (or Theorem 1 in Chapter VI.2 of [43]), for example, establishes an integral representation for a continuous linear operator from the space of continuous real-valued functions on a compact space into a Banach space. The representing integral takes its values in the second dual of this Banach space. The corresponding stronger results for weakly compact and compact operators can be found in Theorems IV.7.3

and IV.7.7 in [55] or in Theorems 5 and 18 in Section VI.2 of [43]. These are obviously special cases of our Corollary 6.4. Indeed, let $\mathcal{P} = \mathbb{K}$ for $\mathbb{K} = \mathbb{R}$ or $\mathbb{K} = \mathbb{C}$ with the Euclidean topology. The values $\theta_E$ for $E \in \mathfrak{R}$ of the representing measure $\theta$ from Corollary 6.4 then are $\mathbb{K}$-linear operators from $\mathbb{K}$ into $\mathcal{Q}_{0s}^{**}$, that is elements of $\mathcal{Q}_{0s}^{**}$ or indeed elements of $\mathcal{Q}_0$ if the represented operator $T$ is weakly compact. The strong operator topology of $\mathfrak{L}_{\mathbb{K}}(\mathbb{K}, \mathcal{Q}_0)$ corresponds to the given (symmetric) topology of $\mathcal{Q}_0$.

We proceed to discuss a two special cases of Corollary 6.10:

(i) *The case that* $(\mathcal{Q}_0, \mathcal{W}_0) = \big(\mathcal{F}_{\mathfrak{U}}(Y, \mathbb{K}), \mathfrak{U}\big)$. Let $(\mathcal{P}, \mathcal{V})$ be a vector space over $\mathbb{K}$ and let $(\mathcal{Q}_0, \mathcal{W}_0) = \big(\mathcal{F}_{\mathfrak{U}}(Y, \mathbb{K}), \mathfrak{U}\big)$ in its symmetric topology, where $Y$ is a second locally compact space and $\mathfrak{U}$ is a basis for a symmetric r-lower continuous inductive limit topology on $\mathcal{F}(Y, \mathbb{K})$. As before, we shall assume that $(\mathcal{Q}, \mathcal{W})$ is the standard lattice completion of $(\mathcal{Q}_0, \mathcal{W}_0)$. Recall that the neighborhoods for $\mathbb{K}$ are the strictly positive multiples of the unit ball $\mathbb{B}$ in $\mathbb{K}$. We shall assume in addition that for every $y \in Y$ there is a neighborhood $\mathfrak{u}_y \in \mathfrak{U}$ such that $s(y) \leq \mathbb{B}$ for all $s \in \mathfrak{u}$. This condition implies that all point evaluations $\varepsilon_y$, that is the mappings

$$g \mapsto g(y) : \mathcal{F}_{\mathfrak{U}}(Y, \mathbb{K}) \to \mathbb{K}$$

are continuous $\mathbb{K}$-linear functionals on $\mathcal{F}_{\mathfrak{U}}(Y, \mathbb{K})$, Hence elements of the vector space dual of $\mathcal{F}_{\mathfrak{U}}(Y, \mathbb{K})$. Consequently, considering $\big(\mathcal{F}_{\mathfrak{U}}(Y, \mathbb{K}), \mathfrak{U}\big)$ as a locally convex cone, for every $y \in Y$ and $\alpha \in \mathbb{K}$ the real-valued linear functional $\alpha\varepsilon_y$, that is the mapping

$$g \mapsto \mathfrak{Re}\big(\alpha g(y)\big) : \mathcal{F}_{\mathfrak{U}}(Y, \mathbb{K}) \to \mathbb{R}$$

is an element of its dual cone $\mathcal{F}_{\mathfrak{U}}(Y, \mathbb{K})^*$. Every element $l$ of the second (vector space) dual $\mathcal{Q}_{0s}^{**}$ of $\mathcal{Q}_0$ can be projected onto a $\mathbb{K}$-valued function $\varphi_l$ on $Y$ in a canonical way: We set $\varphi_l(y) = l(\varepsilon_y)$ for all $y \in Y$. If applied to this situation, Corollary 6.10 yields an integral representation for a continuous linear operator

$$T : \mathcal{F}_{\mathfrak{V}}(X, \mathcal{P}) \to \mathcal{F}_{\mathfrak{U}}(Y, \mathbb{K}).$$

Let $\theta$ be its representing measure. Following the preceding remark, the elements of $\mathfrak{L}_{\mathbb{K}}(\mathcal{P}, \mathcal{Q}_{0s}^{**})$, that is the values $\theta_E$ of $\theta$, can be identified with $\mathbb{K}$-linear operators from $\mathcal{P}$ into $\mathcal{F}(Y, \mathbb{K})$. The integral of a $\mathcal{P}$-valued step function $h = \sum_{i=1}^{n} \chi_{E_i \otimes} a_i \in \mathcal{S}_{\mathfrak{R}}(X, \mathcal{P})$ then is given by

$$\int_X h \, d\theta = \sum_{i=1}^{n} \theta_{E_i}(a_i) \in \mathcal{F}(Y, \mathbb{K}),$$

since the evaluations $\theta_{E_i}(a_i)$ of the measure $\theta$ are functions in $\mathcal{F}(Y, \mathbb{K})$.

For every $y \in Y$ and $\alpha \in \mathbb{K}$ the functional $\alpha\varepsilon_y$ is an order continuous lattice homomorphism from $\mathcal{Q}$ into $\overline{\mathbb{R}}$. Thus the $\mathcal{P}^*$-valued composition measure

$$E \mapsto \alpha\varepsilon_y \circ \theta_E \,:\, \mathfrak{R} \to \mathcal{P}^*$$

is well defined (see II.3.9). According to Theorem II.5.35 we have

$$\left(\int_X f\, d\theta\right)(\alpha\varepsilon_y) = \int_X f\, d(\alpha\varepsilon_y \circ \theta)$$

for every $f \in \mathcal{F}_{\mathfrak{U}}(Y, \mathbb{K})$. Consequently, for every $y \in Y$ the measure $\vartheta_y$ on $\mathfrak{R}$, that is

$$E \mapsto (\varepsilon_y \circ \theta) \qquad \text{or} \qquad E \mapsto (\varepsilon_y \circ \theta) - i(i\varepsilon_y \circ \theta)$$

in the real or the complex case, respectively, is $\mathfrak{L}_{\mathbb{K}}(\mathcal{P}, \mathbb{K})$-valued, that is its values are in the vector space dual of $\mathcal{P}$. If we interpret the integral which is an element of $\mathcal{Q}_{0s}^{**}$ as a $\mathbb{K}$-valued function on $Y$, this yields

$$T(f)(y) = \left(\int_X f\, d\theta\right)(y) = \int_X f\, d\vartheta_y$$

for all $f \in \mathcal{F}_{\mathfrak{V}}(X, \mathcal{P})$ and all $y \in Y$, where

$$\int_X f\, d\vartheta_y = \int_X f\, d(\varepsilon_y \circ \theta) = \left(\int_X f\, d\theta\right)(\varepsilon_y)$$

in the real, and

$$\begin{aligned} \int_X f\, d\vartheta_y &= \int_X f\, d(\varepsilon_y \circ \theta) - i \int_X f\, d(i\varepsilon_y \circ \theta) \\ &= \left(\int_X f\, d\theta\right)(y) - i\left(\int_X f\, d\theta\right)(i\varepsilon_y) \end{aligned}$$

in the complex case. Finally, let us consider the mapping $y \mapsto \vartheta_y$ from the locally compact space $Y$ into the set of $\mathfrak{L}_{\mathbb{K}}(\mathcal{P}, \mathbb{K})$-valued measures. For this let us assume in addition that the representation measure $\theta$ from above is indeed $\mathfrak{L}_{\mathbb{K}}\big(\mathcal{P}, \mathcal{C}_{\mathfrak{U}}(Y, \mathbb{K})\big)$-valued. This is guaranteed, for example, if the range of the operator $T$ is contained in $\mathcal{C}_{\mathfrak{U}}(Y, \mathbb{K})$ and if $T$ is weakly compact. Then the mapping $y \mapsto \vartheta_y$ is seen to be continuous with respect to the given topology of $Y$ and the topology of setwise convergence for $\mathfrak{L}_{\mathbb{K}}(\mathcal{P}, \mathbb{K})$-valued measures (see II.5.13). Indeed, let $(y_i)_{i \in \mathcal{I}}$ be a net in $Y$ converging towards $y \in Y$. Then for every $E \in \mathfrak{R}$ and $a \in \mathcal{P}$ we have

$$\lim_{i \in \mathcal{I}} \vartheta_{y_i\, E}(a) = \lim_{i \in \mathcal{I}} \theta_E(a)(y_i) = \theta_E(a)(y)$$

by the continuity of the function $\theta_E(a) \in \mathcal{C}_{\mathfrak{U}}(Y, \mathbb{K})$. This yields our claim.

(ii) *The case that* $\mathcal{P} = \mathbb{K}$ *and* $(\mathcal{Q}_0, \mathcal{W}_0) = \big(\mathcal{F}_{\mathfrak{U}}(Y, \mathbb{K}), \mathfrak{U}\big)$. This special case of the preceding one is of particular interest. Under the same conditions on $(\mathcal{F}_{\mathfrak{U}}(Y,\mathbb{K}),\mathfrak{U})$ we obtain an integral representation for a continuous linear operator

$$T : \mathcal{F}_{\mathfrak{V}}(X,\mathbb{K}) \to \mathcal{F}_{\mathfrak{U}}(Y,\mathbb{K}).$$

The values $\theta_E$ of the representing measure $\theta$, that is the elements of $\mathfrak{L}_{\mathbb{K}}(\mathcal{P},\mathcal{Q}_{0s}^{**})$ may now be identified with functions $\psi \in \mathcal{F}(Y,\mathbb{K})$, acting as

$$\alpha \mapsto \alpha\psi : \mathbb{K} \to \mathcal{F}(Y,\mathbb{K}).$$

The integral of a $\mathbb{K}$-valued step function $h = \sum_{i=1}^{n} \alpha_i \chi_{E_i} \in \mathcal{S}_{\mathfrak{R}}(X,\mathbb{K})$ then is given by

$$\int_X h\, d\theta = \sum_{i=1}^{n} \alpha_i\, \theta_{E_i},$$

where the evaluations $\theta_{E_i}$ of the measure $\theta$ are functions in $\mathcal{F}(Y,\mathbb{K})$. For $Y = X$ and $\mathfrak{U} = \mathfrak{V}$, for example, the identity operator on $\mathcal{F}_{\mathfrak{V}}(X,\mathbb{K})$ is represented by the measure

$$E \mapsto \chi_E \;:\; \mathfrak{R} \to \mathcal{F}(X,\mathbb{K}).$$

For comprehensiveness we shall formulate this special case as a further corollary.

**Corollary 6.11.** *Let $\mathbb{K} = \mathbb{R}$ or $\mathbb{K} = \mathbb{C}$. Let $X$ and $Y$ be locally compact Hausdorff spaces and let $\mathfrak{V}$ and $\mathfrak{U}$ be bases for symmetric r-lower continuous inductive limit topologies on $\mathcal{F}(X,\mathbb{K})$ and $\mathcal{F}(Y,\mathbb{K})$, respectively. Suppose that for every $y \in Y$ there is $\mathfrak{u}_y \in \mathfrak{U}$ such that $s(y) \le \mathbb{B}$ for all $s \in \mathfrak{u}$. Then every continuous $\mathbb{K}$-linear operator $T : \mathcal{F}_{\mathfrak{V}}(X,\mathbb{K}) \to \mathcal{F}_{\mathfrak{U}}(Y,\mathbb{K})$ can be represented as an integral on $X$. More precisely: There exists a unique bounded quasi regular $\mathcal{F}(Y,\mathbb{K})$-valued measure $\theta$ on the weak $\sigma$-ring $\mathfrak{R}$ of all relatively compact Borel subsets of $X$ with the following properties: $\theta$ is countably additive and quasi regular with respect to pointwise convergence on $X$ for the functions in $\mathcal{F}(Y,\mathbb{K})$, continuous relative to $\mathfrak{V}$, all functions in $\mathcal{F}_{\mathfrak{V}}(X,\mathcal{P})$ are strongly integrable with respect to $\theta$, and*

$$\int_X f\, d\theta = T(f) \qquad \textit{for all} \qquad f \in \mathcal{F}_{\mathfrak{V}}(X,\mathbb{K}).$$

*The operator $T$ is compact (or weakly compact) on $\mathcal{E}(X,\mathcal{P})$, if and only if the measure $\theta$ is $\mathcal{F}_{\mathfrak{U}}(Y,\mathbb{K})$-valued and compact (or weakly compact). In this case $\theta$ is countably additive with respect to the symmetric topology of $\mathcal{F}_{\mathfrak{U}}(Y,\mathbb{K})$.*

**6.12 Algebra Homomorphisms.** Now suppose that the locally convex vector spaces $(\mathcal{P},\mathcal{V})$ and $(\mathcal{Q}_0,\mathcal{W}_0)$ are indeed topological algebras over $\mathbb{K} = \mathbb{R}$ or $\mathbb{K} = \mathbb{C}$ and that the continuous linear operator $T : \mathcal{F}_{\mathfrak{V}}(X,\mathcal{P}) \to \mathcal{Q}_0$ is multiplicative on $\mathcal{E}(X,\mathcal{P})$. As before we assume that $(\mathcal{Q},\mathcal{W})$ is the standard lattice completion of $(\mathcal{Q}_0,\mathcal{W}_0)$. We shall verify that the representing measure

$\theta$ satisfies Condition (A) from II.6.14 in this case. First let us recall some facts and techniques that were introduced in II.6.14: A topological algebra $\mathcal{A}$ is an algebra and a locally convex topological vector space such that for a fixed element $a \in A$ (or $b \in A$) the linear operator $c \mapsto ac$ (or $c \mapsto cb$) from $A$ into $A$ is continuous (see for example 8.1 in [137]). Because for a linear operator continuity implies weak continuity, $A$ is also a topological algebra in its weak topology. Moreover, on $\mathcal{Q}_0 \subset \mathcal{Q}$ order convergence and week convergence coincide (see I.5.57). An extension of the multiplication from $\mathcal{Q}_0$ to $\mathcal{Q}$ was introduced in II.6.14: For $l, m \in \mathcal{Q}$ we denote by $l \bullet m$ the set of all elements $q \in \mathcal{Q}$ for which we can find nets $(l_i)_{i \in \mathcal{I}}$ and $(m_j)_{j \in \mathcal{J}}$ in $\mathcal{Q}_0 \subset \mathcal{Q}$ such that $\lim_{i \in \mathcal{I}} l_i = l$, $\lim_{j \in \mathcal{J}} m_j = m$ and

$$\underline{\lim}_{i \in \mathcal{I}} \underline{\lim}_{j \in \mathcal{J}} l_i m_j = \overline{\lim}_{i \in \mathcal{I}} \overline{\lim}_{j \in \mathcal{J}} l_i m_j = \underline{\lim}_{j \in \mathcal{J}} \underline{\lim}_{i \in \mathcal{I}} l_i m_j = \overline{\lim}_{j \in \mathcal{J}} \overline{\lim}_{i \in \mathcal{I}} l_i m_j = q.$$

The set $l \bullet m$ may be empty or contain more than one element of $\mathcal{Q}$. For $l, m \in \mathcal{Q}_0$ we have $l \bullet m = \{lm\}$. Similarly, if the algebra $\mathcal{Q}_0$ has an involution $a \mapsto a^*$, then for every $l \in \mathcal{Q}$ we denote by $l^\star$ the set of all elements $q \in \mathcal{Q}$ for which we can find a net $(l_i)_{i \in \mathcal{I}}$ in $\mathcal{Q}_0 \subset \mathcal{Q}$ such that $\lim_{i \in \mathcal{I}} l_i = l$ and

$$\lim_{i \in \mathcal{I}} l_i^* = q.$$

Obviously, if $\mathcal{P}$ is an algebra, so is $\mathcal{E}(X, \mathcal{P})$, endowed with the canonical, that is pointwise multiplication for $\mathcal{P}$-valued functions. Now suppose that the continuous linear operator $T : \mathcal{F}_{\mathfrak{V}}(X, \mathcal{P}) \to \mathcal{Q}_0$ is multiplicative on $\mathcal{E}(X, \mathcal{P})$ and let $\theta$ be its representing measure from Corollary 6.10. We shall establish that $\theta$ satisfies

(A) $\theta_E(ab) \in \theta_E(a) \bullet \theta_E(b)$ *and* $0 \in \theta_E(a) \bullet \theta_G(b)$ *for all* $a, b \in \mathcal{P}$ *and disjoint sets* $E, G \in \mathfrak{R}$.

For this, let $E \in \mathfrak{R}$ and $a, b \in \mathcal{P}$. Let $U \in \mathfrak{R}$ be an open set containing $\overline{E}$ and choose the net $(\varphi_i)_{i \in \mathcal{I}}$ in $\mathcal{K}(X)$ as in Proposition 4.5(c) with $U$ in place of $O$, that is $\varphi_i \prec U$ for all $i \in \mathcal{I}$ and

$$\theta_E(a) = \lim_{i \in \mathcal{I}} T(\varphi_{i \otimes} a), \quad \theta_E(b) = \lim_{i \in \mathcal{I}} T(\varphi_{i \otimes} b) \quad \text{and} \quad \theta_E(ab) = \lim_{i \in \mathcal{I}} T(\varphi_{i \otimes}(ab)).$$

Recall from 4.5(c) that the index set $\mathcal{I}$ consists of all ordered pairs $(O, f)$, where $O \in \mathcal{O}$, the family of all open sets $O \in \mathfrak{R}$ such that $E \subset O$ and $\overline{O} \subset U$, and $f : \mathcal{O} \to \mathcal{K}(X)$ is a mapping such that $f(O) \prec O$ for all $O \in \mathcal{O}$. $\mathcal{I}$ is ordered by $(O_1, f_1) \le (O_2, f_2)$ if $O_1 \supset O_2$ and $f_1(O) \le f_2(O)$ for all $O \in \mathcal{O}$. We have $\varphi_i = f(O) \in \mathcal{K}(X)$ for $i = (O, f) \in \mathcal{I}$.

Now let $w \in \mathcal{W}$ and $v \in \mathcal{V}$ such that $\theta_U(v) \le w$. There is $\lambda \ge 0$ such that $0 \le ab + \lambda v$. For a first step in our argument we fix a set $O_0 \in \mathcal{O}$ and a compact subset $K_0$ of $E$. Let $\varphi_0 \in \mathcal{K}(X)$ such that $K_0 \prec \varphi_0 \prec O_0$. Let $i_0 = (O_0, f_0) \in \mathcal{I}$, where $f_0 : \mathcal{O} \to \mathcal{K}(X)$ is a mapping such that $K_0 \prec f_0(O) \prec O$ for all $O \in \mathcal{O}$. Thus $\chi_{K_0} \le \varphi_i \varphi_0 \le \chi_{O_0}$ for all $i \ge i_0$.

This yields

$$\chi_{K_0 \otimes}(ab + \lambda v) \leq \varphi_i \varphi_{0 \otimes}(ab + \lambda v),$$

hence

$$\begin{aligned} \theta_{K_0}(ab) + \lambda \theta_{K_0}(v) &\leq T\big(\varphi_i \varphi_{0\otimes}(ab + \lambda v)\big) \\ &\leq T(\varphi_{i\otimes} a)\, T(\varphi_{0\otimes} b) + \lambda \theta_{O_0}(v). \end{aligned}$$

The latter demonstrates

$$\theta_{K_0}(ab) + \lambda \theta_{K_0}(v) \leq \underline{\lim}_{i \in \mathcal{I}} \Big( T(\varphi_{i\otimes} a)\, T(\varphi_{0\otimes} b) \Big) + \lambda \theta_{O_0}(v)$$

and, similarly,

$$\overline{\lim}_{i \in \mathcal{I}} \Big( T(\varphi_{i\otimes} a)\, T(\varphi_{0\otimes} b) \Big) + \lambda \theta_{K_0}(v) \leq \theta_{O_0}(ab) + \lambda \theta_{O_0}(v).$$

Moreover, since $K_0 \leq \varphi_j \prec O_0$ holds for all $i_0 \leq j \in \mathcal{I}$, the above yields indeed

$$\theta_{K_0}(ab) + \lambda \theta_{K_0}(v) \leq \underline{\lim}_{j \in \mathcal{I}}\, \underline{\lim}_{i \in \mathcal{I}} \Big( T(\varphi_{i\otimes} a)\, T(\varphi_{j\otimes} b) \Big) + \lambda \theta_{O_0}(v)$$

and

$$\overline{\lim}_{j \in \mathcal{I}}\, \overline{\lim}_{i \in \mathcal{I}} \Big( T(\varphi_{i\otimes} a)\, T(\varphi_{j\otimes} b) \Big) + \lambda \theta_{K_0}(v) \leq \theta_{O_0}(ab) + \lambda \theta_{O_0}(v).$$

Finally, we take the supremum and the infimum over all compact sets $K_0 \subset E$ and open sets $E \subset O_0$, respectively, in the preceding inequalities. We have

$$\inf_{O_0 \supset E} \theta_{O_0}(v) = \theta_E(v) \leq \sup_{K_0 \subset E} \theta_{K_0}(v) + \mathfrak{O}\big(\theta_U(v)$$

by Proposition 4.4. Using this, the cancellation law from Proposition I.5.10(a), the fact that $\theta_E(v) \leq \theta_U(v) \leq w$ and that the neighborhood $w \in \mathcal{W}$ was arbitrarily chosen, we obtain

$$\theta_E(ab) = \underline{\lim}_{j \in \mathcal{I}}\, \underline{\lim}_{i \in \mathcal{I}} \Big( T(\varphi_{i\otimes} a)\, T(\varphi_{j\otimes} b) \Big) = \overline{\lim}_{j \in \mathcal{I}}\, \overline{\lim}_{i \in \mathcal{I}} \Big( T(\varphi_{i\otimes} a)\, T(\varphi_{j\otimes} b) \Big).$$

Similarly, one verifies

$$\theta_E(ab) = \underline{\lim}_{i \in \mathcal{I}}\, \underline{\lim}_{j \in \mathcal{I}} \Big( T(\varphi_{i\otimes} a)\, T(\varphi_{j\otimes} b) \Big) = \overline{\lim}_{i \in \mathcal{I}}\, \overline{\lim}_{j \in \mathcal{I}} \Big( T(\varphi_{i\otimes} a)\, T(\varphi_{j\otimes} b) \Big).$$

The first part of (A) follows. For the second part, let $E, G \in \mathfrak{R}$ be disjoint sets and let $a, b \in \mathcal{P}$. Let $O \in \mathfrak{R}$ be an open set containing both $E$ and $G$, and choose the nets $(\varphi_i)_{i \in \mathcal{I}}$ and $(\psi_j)_{i j \in \mathcal{J}}$ in $\mathcal{K}(X)$ as in Proposition 4.5(c) for the sets $E$ and $G$, respectively, that is $\varphi_i, \psi_j \prec O$ for all $i \in \mathcal{I}$ and $j \in \mathcal{J}$ and

$$\theta_E(a) = \lim_{i \in \mathcal{I}} T(\varphi_{i \otimes} a) \qquad \text{and} \qquad \theta_G(b) = \lim_{j \in \mathcal{J}} T(\psi_{j \otimes} b).$$

Given $w \in \mathcal{W}$ we choose $v \in \mathcal{V}$ such that $\theta_O(v) \leq w$ and $\lambda \geq 0$ such that $0 \leq a + \lambda v$. For any choice of open sets $U, V \subset O$ such that $E \subset U$ and $G \subset V$ there are $i_0 \in \mathcal{I}$ and $j_0 \in \mathcal{J}$ such that $\varphi_i \prec U$ and $\psi_j \prec V$, hence $\varphi_i \psi_j \prec U \cap V$ for all $i \geq i_0$ and $j \geq j_0$. Thus

$$0 \leq T\big(\varphi_i \psi_{j \otimes}(ab + \lambda v)\big) \leq T(\varphi_{i \otimes} a)\, T(\psi_{j \otimes} a) + \lambda \theta_{(U \cap V)}(v)$$

and

$$T(\varphi_{i \otimes} a)\, T(\psi_{j \otimes} a) \leq T\big(\varphi_i \psi_{j \otimes}(ab + \lambda v)\big) \leq \theta_{(U \cap V)}(ab + \lambda v).$$

This shows

$$0 \leq \underline{\lim}_{j \in \mathcal{J}} \underline{\lim}_{i \in \mathcal{I}} \Big( T(\varphi_{i \otimes} a)\, T(\psi_{j \otimes} a) \Big) + \lambda \theta_{(U \cap V)}(v)$$

and

$$\overline{\lim}_{j \in \mathcal{J}} \overline{\lim}_{i \in \mathcal{I}} \Big( T(\varphi_{i \otimes} a)\, T(\psi_{j \otimes} a) \Big) \leq \theta_{(U \cap V)}(ab + \lambda v).$$

Next we observe that

$$\begin{aligned} \theta_{(E \cup G)}(c) + \theta_{(U \cap V)}(c) &\leq \theta_{(U \cup V)}(c) + \theta_{(U \cap V)}(c) \\ &= \theta_U(c) + \theta_V(c) \end{aligned}$$

holds for all elements $c \geq 0$ in the standard full extension $\mathcal{P}_{\mathcal{V}}$ of $\mathcal{P}$. Taking the infimum over all open sets $U$ and $V$ containing the given sets $E$ and $G$, respectively, yields

$$\theta_{(E \cup G)}(c) + \inf_{\substack{U \supset E, \\ V \supset G}} \theta_{(U \cap V)}(c) = \theta_E(c) + \theta_G(c) = \theta_{(E \cup G)}(c),$$

hence

$$0 \leq \inf_{\substack{U \supset E, \\ V \supset G}} \theta_{(U \cap V)}(c) \leq \mathfrak{D}\big(\theta_{(E \cup G)}(c)\big) \leq \mathfrak{D}\big(\theta_O(c)\big)$$

by Proposition I.5.10(a). As $\mathcal{P}$ is a vector space, its elements are bounded, and we have

$$\mathfrak{D}\big(\theta_O(ab + v)\big) = \mathfrak{D}\big(\theta_O(v)\big) \leq w.$$

Using this in the above inequalities together with the fact that the neighborhood $w \in \mathcal{W}$ was arbitrarily chosen then yields

$$\underline{\lim}_{j \in \mathcal{J}} \underline{\lim}_{i \in \mathcal{I}} \Big( T(\varphi_{i \otimes} a)\, T(\psi_{j \otimes} a) \Big) = \overline{\lim}_{j \in \mathcal{J}} \overline{\lim}_{i \in \mathcal{I}} \Big( T(\varphi_{i \otimes} a)\, T(\psi_{j \otimes} a) \Big) = 0.$$

Similarly, one verifies

$$\underline{\lim}_{i \in \mathcal{J}} \underline{\lim}_{j \in \mathcal{I}} \Big( T(\varphi_{i \otimes} a)\, T(\psi_{j \otimes} a) \Big) = \overline{\lim}_{i \in \mathcal{J}} \overline{\lim}_{j \in \mathcal{I}} \Big( T(\varphi_{i \otimes} a)\, T(\psi_{j \otimes} a) \Big) = 0,$$

hence the second part of (A).

If $(\mathcal{P},\mathcal{V})$ is a topological algebra with an involution $a \mapsto a^*$, then canonically, for a function $f \in \mathcal{F}(X,\mathcal{P})$ we denote the mapping $x \mapsto \big(f(x)\big)^*$ by $f^*$. If $(\mathcal{Q}_0,\mathcal{W})$ has also an involution and if the continuous linear operator $T : \mathcal{E}(X,\mathcal{P}) \to \mathcal{Q}_0$ is compatible with the respective involutions, that is if $T(f^*) = \big(T(f)\big)^*$ holds for all $f \in \mathcal{E}(X,\mathcal{P})$, then the representing measure $\theta$ satisfies

(A*) $\theta_E(a^*) \in \big(\theta_E(a)\big)^\star$ *for all* $E \in \mathfrak{R}$ *and* $a \in \mathcal{P}$.

(See Section II.6.14.) This claim is readily verified: For $E \in \mathfrak{R}$ and $a \in \mathcal{P}$ let the net $(\varphi_i)_{i\in\mathcal{I}}$ in $\mathcal{K}(X)$ be as in Proposition 4.5(c), that is

$$\theta_E(a) = \lim_{i\in\mathcal{I}} T\big(\varphi_{i\otimes} a\big)$$

and

$$\theta_E(a^*) = \lim_{i\in\mathcal{I}} T(\varphi_{i\otimes} a^*) = \lim_{i\in\mathcal{I}} \big(T(\varphi_{i\otimes} a)\big)^*.$$

According to our previously introduced notation, this means $\theta_E(a^*) \in \big(\theta_E(a)\big)^\star$.

We summarize:

**Corollary 6.13.** *Let $(\mathcal{P},\mathcal{V})$ and $(\mathcal{Q}_0,\mathcal{W}_0)$ be topological algebras over $\mathbb{K}=\mathbb{R}$ or $\mathbb{K}=\mathbb{C}$. Let $X$ be a locally compact Hausdorff space and let $\mathfrak{V}$ be a basis for a symmetric r-lower continuous inductive limit topology on $\mathcal{F}(X,\mathcal{P})$. Then every continuous $\mathbb{K}$-linear operator $T : \mathcal{F}_{\mathfrak{V}}(X,\mathcal{P}) \to \mathcal{Q}_0$ that is multiplicative on $\mathcal{E}(X,\mathcal{P})$ can be represented as an integral on $X$. More precisely: There exists a unique bounded $\mathfrak{L}_{\mathbb{K}}(\mathcal{P},\mathcal{Q}_{0s}^{**})$-valued measure $\theta$ on the weak $\sigma$-ring $\mathfrak{R}$ of all relatively compact Borel subsets of $X$ with the following properties: $\theta$ is countably additive and quasi regular with respect to the weak* operator topology of $\mathfrak{L}_{\mathbb{K}}(\mathcal{P},\mathcal{Q}_{0s}^{**})$. We have*

$$\theta_E(ab) \in \theta_E(a)\boldsymbol{\cdot}\theta_E(b) \qquad \text{and} \qquad 0 \in \theta_E(a)\boldsymbol{\cdot}\theta_G(b)$$

*for all $a,b \in \mathcal{P}$ and disjoint sets $E,G \in \mathfrak{R}$. If $T$ is compatible with respective involutions in $\mathcal{P}$ and $\mathcal{Q}_0$, then $\theta_E(a^*) \in \big(\theta_E(a)\big)^\star$ holds in addition. The measure $\theta$ is continuous relative to $\mathfrak{V}$, all functions in $\mathcal{F}_{\mathfrak{V}}(X,\mathcal{P})$ are strongly integrable with respect to $\theta$, and*

$$\int_X f\,d\theta = T(f) \qquad \text{for all} \qquad f \in \mathcal{F}_{\mathfrak{V}}(X,\mathcal{P}).$$

*The operator $T$ is compact (or weakly compact) on $\mathcal{E}(X,\mathcal{P})$, if and only if the measure $\theta$ is $\mathfrak{L}_{\mathbb{K}}(\mathcal{P},\mathcal{Q}_0)$-valued and compact (or weakly compact). In this case $\theta$ is countably additive with respect to the strong operator topology of $\mathfrak{L}_{\mathbb{K}}(\mathcal{P},\mathcal{Q}_0)$.*

We proceed to discuss three special cases of Corollary 6.13:

(i) *The case that* $\mathcal{Q}_0 = \mathbb{K}$. If $\mathcal{Q}_0 = \mathbb{K}$, that is if the values $\theta_E$ of the representation measure $\theta$ are $\mathbb{K}$-linear functionals on the algebra $\mathcal{P}$, then Condition (A) means that all functionals $\theta_E$ are multiplicative and that for disjoint sets $E, G \in \mathfrak{R}$ we have either $\theta_E = 0$ or $\theta_G = 0$. In the conclusion of Section 4 we inferred that $\theta$ is therefore indeed some point evaluation measure in this case. (Inner regularity of $\theta$ is guaranteed by the fact that all values in $\mathcal{Q}_0$ are bounded (see Proposition 4.4(a).)

(ii) *The case that* $(\mathcal{Q}_0, \mathcal{W}_0) = (\mathcal{F}_{\mathfrak{U}}(Y, \mathbb{K}), \mathfrak{U})$. This situation was considered earlier as a special case of Corollary 6.10. The conditions on $(\mathcal{F}_{\mathfrak{U}}(Y, \mathbb{K}), \mathfrak{U})$ from 6.10(i) guarantee that the point evaluations $\varepsilon_y$ are elements of the vector space dual of $\mathcal{F}_{\mathfrak{U}}(Y, \mathbb{K})$. We obtain additional information if $(\mathcal{P}, \mathcal{V})$ is topological algebra and if the operator

$$T : \mathcal{F}_{\mathfrak{V}}(X, \mathcal{P}) \to \mathcal{F}_{\mathfrak{U}}(Y, \mathbb{K})$$

is multiplicative on $\mathcal{E}(X, \mathcal{P})$. Let $\theta$ be its representing measure from Corollary 6.13. Recall that the values $\theta_E$ of $\theta$ may be identified with $\mathbb{K}$-linear operators from $\mathcal{P}$ into $\mathcal{F}(Y, \mathbb{K})$. The extended multiplication $\bullet$ in $\mathcal{Q}$ then corresponds to the pointwise multiplication in $\mathcal{F}(Y, \mathbb{K})$. Thus for every $y \in Y$ the mapping

$$f \mapsto T(f)(y) \; : \; \mathcal{F}_{\mathfrak{V}}(X, \mathcal{P}) \to \mathbb{K}$$

is a continuous linear functional on $\mathcal{F}_{\mathfrak{V}}(X, \mathcal{P})$ which is multiplicative on $\mathcal{E}(X, \mathcal{P})$. The representation measure $\delta_y$ (see the remark in 6.10(i)) for this functional therefore satisfies Conditions (A), and following case (i) is some point evaluation measure on $X$. More precisely: There exist mappings $y \mapsto x_y$ and $y \mapsto \mu_y$ from $Y$ into $X$ and into the set of continuous multiplicative linear functionals on $\mathcal{P}$, respectively, such that

$$T(f)(y) = \int_X f \, d\vartheta_y = \mu_y\big(f(x_y)\big)$$

for all $f \in \mathcal{C}^r_{\mathfrak{V}}(X, \mathcal{P})$ and all $y \in Y$. Now let us assume in addition that the range of the operator $T$ is contained in $\mathcal{C}_{\mathfrak{U}}(Y, \mathbb{K})$ and consider the mapping

$$y \mapsto (x_y, \mu_y) \; : \; Y \to X \times \mathcal{P}^*.$$

Let $(y_i)_{i \in \mathcal{I}}$ be a net in $Y$ converging towards $y \in Y$. Then

$$\lim_{i \in \mathcal{I}} \mu_{y_i}\big(f(x_{y_i})\big) = \lim_{i \in \mathcal{I}} T(f)(y_i) = T(f)(y) = \mu_y\big(f(x_y)\big)$$

for all $f \in \mathcal{F}_{\mathfrak{V}}(X, \mathcal{P})$. Thus the mapping $y \mapsto (x_y, \mu_y)$ is continuous with respect to the given topology of $Y$ and the weak topology induced on $X \times \mathcal{P}^*$ by the functions $f \in \mathcal{F}_{\mathfrak{V}}(X, \mathcal{P})$ acting on $X \times \mathcal{P}^*$ in the canonical way, that is $(x, \mu) \mapsto \mu\big(f(x)\big) \in \mathbb{K}$. (See the definition of the weak topology in 8.9

in [198] or 1.4.8 in [59].) Two special cases are of particular interest. For these let us assume in addition that for every $y \in Y$ there is $f \in \mathcal{F}_{\mathfrak{V}}(X, \mathcal{P})$ such that $T(f)(y) \neq 0$. (i) If $\mathcal{P} = \mathbb{K}$, then the identity is the only no-zero multiplicative linear functional on $\mathcal{P}$, that is $\mu_y = 1$ for all $y \in Y$. Then the mapping $y \mapsto x_y : Y \to X$ is continuous with respect to the given topology of $Y$ and the weak topology on $X$ generated by the functions in $\mathcal{F}_{\mathfrak{V}}(X, \mathcal{P})$. (ii) If $X = \{x\}$ is a singleton set, then $\mathcal{F}_{\mathfrak{V}}(X, \mathcal{P}) = \mathcal{P}$ and the mapping $y \mapsto x_y = x$ is constant. Given $a \in \mathcal{P}$ there is $f \in \mathcal{F}_{\mathfrak{V}}(X, \mathcal{P})$ such that $f(x) = a$, hence the above shows that the mapping $y \mapsto \mu_y : Y \to \mathcal{P}^*$ is continuous with respect to the given topology of $Y$ and the topology $w(\mathcal{P}^*, \mathcal{P})$ of $\mathcal{P}^*$.

We formulate this observation as a further corollary:

**Corollary 6.14.** *Let $(\mathcal{P}, \mathcal{V})$ be a topological algebra over $\mathbb{K} = \mathbb{R}$ or $\mathbb{K} = \mathbb{C}$. Let $X$ and $Y$ be locally compact Hausdorff spaces and let $\mathfrak{V}$ and $\mathfrak{U}$ be bases for symmetric r-lower continuous inductive limit topologies on $\mathcal{F}(X, \mathcal{P})$ and $\mathcal{F}(Y, \mathbb{K})$, respectively. Suppose that for every $y \in Y$ there is $\mathfrak{u}_y \in \mathfrak{U}$ such that $s(y) \leq \mathbb{B}$ for all $s \in \mathfrak{u}$. Then for every continuous $\mathbb{K}$-linear operator $T : \mathcal{F}_{\mathfrak{V}}(X, \mathcal{P}) \to \mathcal{F}(Y, \mathbb{K})$ that is multiplicative on $\mathcal{E}(X, \mathcal{P})$ there exist mappings $y \mapsto x_y$ and $y \mapsto \mu_y$ from $Y$ into $X$ and into the set of continuous multiplicative linear functionals on $\mathcal{P}$, respectively, such that*

$$T(f)(y) = \mu_y\big(f(x_y)\big)$$

*for all $f \in \mathcal{C}^r_{\mathfrak{V}}(X, \mathcal{P})$ and all $y \in Y$. If the range of the operator $T$ is contained in $\mathcal{C}_{\mathfrak{U}}(Y, \mathbb{K})$, then the mapping $y \mapsto (x_y, \mu_y)$ is continuous with respect to the given topology of $Y$ and the weak topology induced on $X \times \mathcal{P}^*$ by the functions $f \in \mathcal{F}_{\mathfrak{V}}(X, \mathcal{P})$.*

(iii) *$B^*$-algebras.* Let $\mathcal{P} = \mathbb{C}$ and let $(\mathcal{Q}_0, \|\ \|)$ be a commutative $B^*$-algebra, that is a commutative complex Banach algebra with identity and an involution $a \mapsto a^*$ satisfying $\|a^* a\| = \|a^*\| \|a\|$ (see IX.3.1 in [56]). Let $X$ be the spectrum of $\mathcal{Q}_0$, that is the $\sigma(Q_0^*, \mathcal{Q}_0)$-compact subset of all multiplicative linear functionals in $\mathcal{Q}_0^*$. According to the Gelfand-Naĭmark theorem (see Theorem IX.3.7 in [56] or Theorem III.16.1 in [137]) then $\mathcal{Q}_0$ is isometrically $*$-isomorphic to $\mathcal{C}(X, \mathbb{C})$, that is to say there is a continuous multiplicative linear operator $T : \mathcal{C}(X, \mathbb{C}) \to \mathcal{Q}_0$ such that $T(f^*) = T(f)^*$ for all $f \in \mathcal{C}(X, \mathbb{C})$. As $\mathcal{P} = \mathbb{C}$, the complex linear operators from $\mathcal{P}$ into $\mathcal{Q}_{0s}^{**}$ are indeed the elements $a$ of $\mathcal{Q}_{0s}^{**}$, acting as $\alpha \mapsto \alpha a : \mathbb{C} \to \mathcal{Q}_{0s}^{**}$. Thus according to Corollary 6.13, $T$ can be represented by a $\mathcal{Q}_{0s}^{**}$-valued measure $\theta$ with the stated properties. In addition, we have Condition (A*) from 6.12, that is $\overline{\alpha}\theta_E \in (\alpha\theta_E)^\star$ for all $E \in \mathfrak{R}$ and $\alpha \in \mathbb{K}$. As $(\alpha\theta_E)^\star = \overline{\alpha}\theta_E^\star$ (see II.6.14), this yields $\theta_E \in \theta_E^\star$. On the subset $X$ of $\mathcal{Q}_0^*$ of all multiplicative linear functionals, for each $E \in \mathfrak{R}$ the element $\theta_E$ is therefore a real-valued function taking only the values $0$ or $1$, that is the characteristic function of some subset $\Phi(E)$ of $X$. Moreover, for disjoint sets $E, G \in \mathfrak{R}$ the subsets $\Phi(E)$ and $\Phi(G)$ are disjoint.

Summarizing, there is an $\mathcal{Q}_{0s}^{**}$-valued measure $\theta$ on the spectrum $X$ of $\mathcal{Q}_0$ whose values $\theta_E$ yield characteristic functions of subsets of $X$, disjoint for disjoint sets $E, G \in \mathfrak{R}$, and such that the integrals of the functions in $\mathcal{C}(X, \mathbb{C})$ with respect to $\theta$ are the elements of $\mathcal{Q}_0$.

**6.15 Lattice Homomorphisms.** Now suppose that the quasi-full locally convex cone $(\mathcal{P}, \mathcal{V})$ is indeed a locally convex $\vee$-semilattice cone (see Section 5.1 of Chapter I and the additional remarks concerning quasi-full cones in Section II.6.16), that is the order in $\mathcal{P}$ is antisymmetric, for any two elements $a, b \in \mathcal{P}$ their supremum $a \vee b$ exists in $\mathcal{P}$ and

($\vee$1) $(a + c) \vee (b + c) = a \vee b + c$ *holds for all* $a, b, c \in \mathcal{P}$.
($\vee$2′) $a \leq v$ *for* $a \in \mathcal{P}$ *and* $v \in \mathcal{V}$ *implies that* $a \vee 0 \leq v$.

Topological vector lattices and locally convex complete lattice cones in the sense of I.5 are of course locally convex $\vee$-semilattice cones in this sense. Further examples include $\mathbb{R}$ and $\overline{\mathbb{R}}_+$ (Examples I.1.4(a) and (b)) and cones of non-empty convex subsets of a topological vector space with the set-inclusion as order (Example I.1.4(c)).

The supremum $f \vee g \in \mathcal{F}(X, \mathcal{P})$ of two functions $f, g \in \mathcal{F}(X, \mathcal{P})$ is canonically defined as the mapping $x \mapsto f(x) \vee g(x)$. A brief review of Definition 2.4 confirms that the subcone $\mathcal{F}_{\mathfrak{V}}(X, \mathcal{P})$ of $\mathcal{F}(X, \mathcal{P})$ is indeed closed for suprema. As in Section II.6.16 we denote by $\mathcal{S}_{\mathfrak{R}}^{\sigma}(X, \mathcal{P})$ the subcone of all functions $f \in \mathcal{F}_{\mathfrak{R}}(X, \mathcal{P})$ for which there exists a sequence $(h_n)_{n \in \mathbb{N}}$ of step functions that is bounded below and bounded above relative to $f$ and such that $h_n \longrightarrow f$. According to Corollary II.5.26, this implies that $\lim_{n \to \infty} \int_X h_n \, d\theta = \int_X f \, d\theta$ holds for every $\mathfrak{R}$-bounded $\mathfrak{L}(\mathcal{P}, \mathcal{Q})$-valued measure $\theta$. We shall denote the intersection of $\mathcal{S}_{\mathfrak{R}}^{\sigma}(X, \mathcal{P})$ and $\mathcal{F}_{\mathfrak{V}}(X, \mathcal{P})$ by $\mathcal{S}_{\mathfrak{V}}^{\sigma}(X, \mathcal{P})$. Lemma II.5.27 and Corollary II.5.28 yield that $\mathcal{E}(X, \mathcal{P})$ is contained in $\mathcal{S}_{\mathfrak{V}}^{\sigma}(X, \mathcal{P})$.

Now suppose that the continuous linear operator $T : \mathcal{F}_{\mathfrak{V}}(X, \mathcal{P}) \to \mathcal{Q}$ is a $\vee$-semilattice homomorphism on $\mathcal{S}_{\mathfrak{V}}^{\sigma}(X, \mathcal{P})$ in the sense of I.5.30, that is

$$T(f \vee g) = T(f) \vee T(g)$$

holds for all $f, g \in \mathcal{S}_{\mathfrak{V}}^{\sigma}(X, \mathcal{P})$, and that $T$ can be extended to $\mathcal{F}_{\mathfrak{V}}(X, \mathcal{P}_v)$. Let $\theta$ be the representing measure for $T$ from Theorem 5.1. We shall establish that $\theta$ satisfies the following condition:

(L) $\theta_E(a) \vee \theta_E(b) = \theta_E(a \vee b)$ *for all* $E \in \mathfrak{R}$ *and* $a, b \in \mathcal{P}$, *and*
$\theta_E(a) \vee \theta_G(b) \leq \theta_E(a) + \theta_G(b) \leq \theta_E(a) \vee \theta_G(b) + \mathfrak{O}\big(\theta_{(E \cup G)}(a \vee b)\big)$
*for all* $a, b \geq 0$ *in* $\mathcal{P}$ *and disjoint sets* $E, G \in \mathfrak{R}$.

(Recall the corresponding Condition (L) from II.6.16, which is equivalent to the above in case that all elements of $\mathcal{P}$ are bounded, but slightly stronger in its second part for the general case.)

In order to verify (L), let $E \in \mathfrak{R}$ and $a, b \in \mathcal{P}$. Let $O \in \mathfrak{R}$ be an open set containing $E$ and choose the net $(\varphi_i)_{i \in \mathcal{I}}$ in $\mathcal{K}(X)$ as in Proposition 4.5(c), that is $\varphi_i \prec O$ for all $i \in \mathcal{I}$ and

$$\theta_E(a) = \lim_{i \in \mathcal{I}} T(\varphi_{i\otimes}a), \quad \theta_E(b) = \lim_{i \in \mathcal{I}} T(\varphi_{i\otimes}b) \quad \text{and} \quad \theta_E(a \vee b) = \lim_{i \in \mathcal{I}} T(\varphi_{i\otimes}(a \vee b)).$$

We have $\varphi_{i\otimes}(a \vee b) = (\varphi_{i\otimes}a) \vee (\varphi_{i\otimes}b)$, hence $T(\varphi_{i\otimes}(a \vee b)) = T(\varphi_{i\otimes}a) \vee T(\varphi_{i\otimes}b)$, since $T$ is a $\vee$-semilattice homomorphism on $\mathcal{S}_{\mathfrak{V}}^{\sigma}(X, \mathcal{P})$. Now the continuity of the lattice operation in $\mathcal{Q}$ (see Proposition I.5.25(a)) yields

$$\lim_{i \in \mathcal{I}} \Big( T(\varphi_{i\otimes}a) \vee T(\varphi_{i\otimes}b) \Big) = \Big( \lim_{i \in \mathcal{I}} T(\varphi_{i\otimes}a) \Big) \vee \Big( \lim_{i \in \mathcal{I}} T(\varphi_{i\otimes}b) \Big),$$

hence $\theta_E(a) \vee \theta_E(b) = \theta_E(a \vee b)$ as claimed in the first part of (L).

For the second part of (L), let $E, G \in \mathfrak{R}$ be disjoint sets an let $0 \leq a, b \in \mathcal{P}$. Because

$$\theta_E(a) + \theta_G(b) = \theta_E(a) \vee \theta_G(b) + \theta_E(a) \wedge \theta_G(b)$$

by Proposition I.5.3, and because $\theta_E(a), \theta_G(b) \geq 0$, we have

$$\theta_E(a) \vee \theta_G(b) \leq \theta_E(a) + \theta_G(b).$$

For the right-hand side of the inequality in (L) let $O \in \mathfrak{R}$ be an open set containing both $E \cup G$, and choose the nets $(\varphi_i)_{i \in \mathcal{I}}$ and $(\psi_j)_{ij \in \mathcal{J}}$ in $\mathcal{K}(X)$ as in Proposition 4.5(c) for the sets $E$ an $G$, respectively, that is $\varphi_i, \psi_j \prec O$ for all $i \in \mathcal{I}$ and $j \in \mathcal{J}$ and

$$\theta_E(a) = \lim_{i \in \mathcal{I}} T(\varphi_{i\otimes}a) \qquad \text{and} \qquad \theta_G(b) = \lim_{j \in \mathcal{J}} T(\psi_{j\otimes}b).$$

Before we proceed, let us observe that in a semilattice cone $\mathcal{P}$ we always have

$$\alpha a + \beta b \leq \alpha a \vee \beta b + (\alpha \wedge \beta)(a \vee b)$$

for any choice of positive elements $a, b \in \mathcal{P}$ and $\alpha, \beta \geq 0$. Indeed, if $\beta \leq \alpha$, then $\alpha a \leq \alpha a \vee \beta b$ and $\beta b \leq (\alpha \wedge \beta)(a \vee b)$, hence our claim. Using this, we infer that

$$\varphi_{i\otimes}a + \psi_{j\otimes}b \leq (\varphi_{i\otimes}a) \vee (\psi_{j\otimes}b) + (\varphi_i \wedge \psi_j)_{\otimes}(a \vee b),$$

hence

$$T(\varphi_{i\otimes}a) + T(\psi_{j\otimes}b) = T(\varphi_{i\otimes}a) \vee T(\psi_{j\otimes}b) + T\big((\varphi_i \wedge \psi_j)_{\otimes}(a \vee b)\big).$$

For any choice of open sets $U, V \subset O$ such that $E \subset U$ and $G \subset V$ there are $i_0 \in \mathcal{I}$ and $j_0 \in \mathcal{J}$ such that $\varphi_i \prec U$ and $\psi_j \prec V$, hence $\varphi_i \wedge \psi_j \prec U \cap V$ and

$$T\big((\varphi_i \wedge \psi_j)_{\otimes}(a \vee b)\big) \leq \theta_{(U \cap V)}(a \vee b)$$

for all $i \geq i_0$ and $j \geq j_0$. Now passing to the limits over $i \in \mathcal{I}$ and $j \in \mathcal{J}$ in the preceding equation and using Proposition I.5.25(a) leads to

$$\theta_E(a) + \theta_G(b) \le \theta_E(a) \vee \theta_G(b) + \theta_{(U\cap V)}(a \vee b)$$

for any choice of such open sets $U$ and $V$. Thus

$$\theta_E(a) + \theta_G(b) \le \theta_E(a) \vee \theta_G(b) + \inf_{\substack{U\supset E,\\ V\supset G}} \theta_{(U\cap V)}(a \vee b).$$

We abbreviate $c = a \vee b \ge 0$ and observe that

$$\begin{aligned} \theta_{(E\cup G)}(c) + \theta_{(U\cap V)}(c) &\le \theta_{(U\cup V)}(c) + \theta_{(U\cap V)}(c) \\ &= \theta_U(c) + \theta_V(c). \end{aligned}$$

Taking the infimum over all open sets $U$ and $V$ containing the given sets $E$ and $G$, respectively, yields

$$\theta_{(E\cup G)}(c) + \inf_{\substack{U\supset E,\\ V\supset G}} \theta_{(U\cap V)}(c) \le \theta_E(c) + \theta_G(c) = \theta_{(E\cup G)}(c),$$

hence

$$\inf_{\substack{U\supset E,\\ V\supset G}} \theta_{(U\cap V)}(c) \le \mathfrak{O}\big(\theta_{(E\cup G)}(c)\big)$$

by Proposition I.5.10(a). Together with the above, this yields

$$\theta_E(a) + \theta_G(b) \le \theta_E(a) \vee \theta_G(b) + \mathfrak{O}\big(\theta_{(E\cup G)}(a \vee b)\big)$$

as claimed. We summarize:

**Corollary 6.16.** *Let $(\mathcal{P}, \mathcal{V})$ be a quasi-full semilattice cone and suppose that the order continuous linear functionals support the separation property for $\mathcal{Q}$. Let $X$ be a locally compact Hausdorff space and let $\mathfrak{V}$ be a basis for a symmetric r-lower continuous inductive limit topology on $\mathcal{F}(X, \mathcal{P})$. Let $T : \mathcal{S}^{\sigma}_{\mathfrak{V}}(X, \mathcal{P}) \to \mathcal{Q}$ be a continuous linear $\vee$-semilattice homomorphism that can be extended to a continuous linear operator on $\mathcal{F}_{\mathfrak{V}}(X, \mathcal{P}_{\mathsf{v}})$. Then there exists a unique bounded quasi regular $\mathfrak{L}(\mathcal{P}_{\mathsf{v}}, \mathcal{Q})$-valued measure $\theta$ on the weak $\sigma$-ring $\mathfrak{R}$ of all relatively compact Borel subsets of $X$ with the following properties: $\theta$ is continuous relative to $\mathfrak{V}$ and satisfies Condition* (L)*, all functions in $\mathcal{F}_{\mathfrak{V}}(X, \mathcal{P})$ are integrable, all functions in $\mathcal{F}_{\mathfrak{V}_0}(X, \mathcal{P})$ are strongly integrable with respect to $\theta$, and*

$$\int_X \varphi_{\otimes} a \, d\theta \le T(\varphi_{\otimes} a) \le \int_X \varphi_{\otimes} a \, d\theta + \mathfrak{O}\big(\theta_A(a)\big)$$

*holds for all $\varphi \in \mathcal{K}(X)$ and $a \in \mathcal{P}$, where $A$ denotes the compact support of the function $\varphi$. Thus*

$$\int_X f \, d\theta \le T(f) \qquad \textit{and indeed} \qquad \int_X g \, d\theta = T(g)$$

*holds for all $f \in \mathcal{F}_{\mathfrak{V}}(X, \mathcal{P})$ and all $g \in \mathcal{F}_{\mathfrak{V}_0}(X, \mathcal{P})$, respectively.*

Significant simplifications occur if $(\mathcal{P},\mathcal{V})$ is indeed a topological vector lattice over $\mathbb{R}$. We shall formulate these as an additional corollary.

**Corollary 6.17.** *Let $(\mathcal{P},\mathcal{V})$ be a topological vector lattice and suppose that the order continuous linear functionals support the separation property for $\mathcal{Q}$. Let $X$ be a locally compact Hausdorff space and let $\mathfrak{V}$ be a basis for a symmetric r-lower continuous inductive limit topology on $\mathcal{F}(X,\mathcal{P})$. Let $T: \mathcal{F}_{\mathfrak{V}}(X,\mathcal{P}) \to \mathcal{Q}$ be a continuous linear operator that is a $\vee$-semilattice homomorphism on $\mathcal{S}^{\sigma}_{\mathfrak{V}}(X,\mathcal{P})$. Then there exists a unique bounded $\mathfrak{L}(\mathcal{P},\mathcal{Q})$-valued measure $\theta$ on the weak $\sigma$-ring $\mathfrak{R}$ of all relatively compact Borel subsets of $X$ with the following properties: $\theta$ is countably additive and quasi regular. We have*

$$\theta_E(a \vee b) = \theta_E(a) \vee \theta_E(b) \qquad \text{and} \qquad \theta_E(a) \vee \theta_G(b) = \theta_E(a) + \theta_E(b)$$

*for all $0 \le a, b \in \mathcal{P}$ and disjoint sets $E, G \in \mathfrak{R}$. The measure $\theta$ is continuous relative to $\mathfrak{V}$, all functions in $\mathcal{F}_{\mathfrak{V}}(X,\mathcal{P})$ are strongly integrable with respect to $\theta$, and*

$$\int_X f\, d\theta = T(f) \qquad \text{for all} \qquad f \in \mathcal{F}_{\mathfrak{V}}(X,\mathcal{P}).$$

We shall discuss two special cases of Corollary 6.17:

(i) *The case that $(\mathcal{P},\mathcal{V})$ is a topological vector lattice and that $\mathcal{Q} = \overline{\mathbb{R}}$.* If $\mathcal{Q} = \overline{\mathbb{R}}$, that is if the values $\theta_E$ of the measure $\theta$ are elements of $\mathcal{P}^*$, then Condition (L) means that (i) all functionals $\theta_E$ are $\vee$-semilattice homomorphisms and (ii) for disjoint sets $E, G \in \mathfrak{R}$ we have either $\theta_E = 0$ or $\theta_G = 0$. In the conclusion of Section 4 we inferred that $\theta$ is therefore indeed some point evaluation measure in this case. (Inner regularity of $\theta$ is guaranteed by the fact that all values of $\theta$ are bounded (see Proposition 4.4(c).)

(ii) *The case that $(\mathcal{P},\mathcal{V})$ is a topological vector latticc and that $(\mathcal{Q},\mathcal{W}) = (\mathcal{F}_{\mathfrak{U}}(Y,\overline{\mathbb{R}}),\mathfrak{U})$.* Let $(\mathcal{Q},\mathcal{W}) = (\mathcal{F}_{\mathfrak{U}}(Y,\overline{\mathbb{R}}),\mathfrak{U})$, where $Y$ is a second locally compact space and $\mathfrak{U}$ is a basis for a symmetric r-lower continuous inductive limit topology on $\mathcal{F}(Y,\overline{\mathbb{R}})$. Then $(\mathcal{F}_{\mathfrak{U}}(Y,\overline{\mathbb{R}}),\mathfrak{U})$ is a locally convex complete lattice cone. We shall assume in addition that for every $y \in Y$ there is a neighborhood $\mathfrak{u}_y \in \mathfrak{U}$ such that $s(y) \le 1$ for all $s \in \mathfrak{u}$. This condition implies that all point evaluations $\varepsilon_y$, that is the mappings

$$g \mapsto g(y) : \mathcal{F}_{\mathfrak{U}}(Y,\overline{\mathbb{R}}) \to \overline{\mathbb{R}}$$

are continuous linear functionals on $\mathcal{F}_{\mathfrak{U}}(Y,\overline{\mathbb{R}})$. Let $T : \mathcal{F}_{\mathfrak{V}}(X,\mathcal{P}) {\to} \mathcal{F}_{\mathfrak{U}}(Y,\mathbb{R})$ be a continuous linear operator that is a $\vee$-semilattice homomorphism on $\mathcal{S}^{\sigma}_{\mathfrak{V}}(X,\mathcal{P})$ and let $\theta$ be its representing measure from Corollary 6.17. The values $\theta_E$ of $\theta$ are linear operators from $\mathcal{P}$ into $\mathcal{F}(Y,\mathbb{R})$. Then for every $y \in Y$ the mapping

$$f \mapsto T(f)(y) : \mathcal{F}_{\mathfrak{V}}(X, \mathcal{P}) \to \mathbb{R}$$

is a continuous linear functional on $\mathcal{F}_{\mathfrak{V}}(X, \mathcal{P})$ which is a $\vee$-semilattice homomorphism on $\mathcal{S}^{\sigma}_{\mathfrak{V}}(X, \mathcal{P})$. Following case (i), this functional is represented by some point evaluation measure on $X$. Consequently, there exist mappings $y \mapsto x_y$ and $y \mapsto \mu_y$ from $Y$ into $X$ and into the set of continuous linear $\vee$-semilattice homomorphisms on $\mathcal{P}$, respectively, such that

$$T(f)(y) = \mu_y\big(f(x_y)\big)$$

for all $f \in \mathcal{C}^r_{\mathfrak{V}}(X, \mathcal{P})$ and all $y \in Y$. Now let us assume in addition that the range of the operator $T$ is contained in $\mathcal{C}_{\mathfrak{U}}(Y, \mathbb{R})$. the mapping

$$y \mapsto (x_y, \mu_y) : Y \to X \times \mathcal{P}^*$$

is continuous with respect to the given topology of $Y$ and the weak topology which is generated on $X \times \mathcal{P}^*$ by the functions $f \in \mathcal{F}_{\mathfrak{V}}(X, \mathcal{P})$. Two special cases are of particular interest. (i) If $\mathcal{P} = \mathbb{R}$, then a continuous linear $\vee$-semilattice homomorphism on $\mathcal{P}$ is the multiplication with some non-negative real number, that is $\mu_y = \varphi(y)$ for some real-valued function $\varphi : Y \to [0, +\infty)$, and we have

$$T(f)(y) = \varphi(y)\big(f(x_y)\big)$$

for all $f \in \mathcal{C}^r_{\mathfrak{V}}(X, \mathcal{P})$ and $y \in Y$. Moreover, the mapping $y \mapsto \varphi(y)\varepsilon_{x_y} : Y \to \mathcal{C}^r_{\mathfrak{V}}(X, \mathcal{P})^*$ is continuous with respect to the given topology of $Y$ and the topology $\sigma\big(\mathcal{F}_{\mathfrak{V}}(X, \mathcal{P})^*, \mathcal{F}_{\mathfrak{V}}(X, \mathcal{P})\big)$ on $\mathcal{F}_{\mathfrak{V}}(X, \mathcal{P})^*$. (ii) If $X = \{x\}$ is a singleton set, then $\mathcal{F}_{\mathfrak{V}}(X, \mathcal{P}) = \mathcal{P}$ and the mapping $y \mapsto x_y = x$ is constant. Given $a \in \mathcal{P}$ there is $f \in \mathcal{F}_{\mathfrak{V}}(X, \mathcal{P})$ such that $f(x) = a$, hence the above shows that the mapping $y \mapsto \mu_y : Y \to \mathcal{P}^*$ is continuous with respect to the given topology of $Y$ and the topology $w(\mathcal{P}^*, \mathcal{P})$ of $\mathcal{P}^*$.

We formulate this observation as a further corollary:

**Corollary 6.18.** *Let $(\mathcal{P}, \mathcal{V})$ be a topological vector lattice. Let $X$ and $Y$ be locally compact Hausdorff spaces and let $\mathfrak{V}$ and $\mathfrak{U}$ be bases for symmetric r-lower continuous inductive limit topologies on $\mathcal{F}(X, \mathcal{P})$ and $\mathcal{F}(Y, \overline{\mathbb{R}})$, respectively. Suppose that for every $y \in Y$ there is $\mathfrak{u}_y \in \mathfrak{U}$ such that $s(y) \leq 1$ for all $s \in \mathfrak{u}$. Let $T : \mathcal{F}_{\mathfrak{V}}(X, \mathcal{P}) \to \mathcal{F}(Y, \mathbb{R})$ be a continuous linear operator that is a $\vee$-semilattice homomorphism on $\mathcal{S}^{\sigma}_{\mathfrak{V}}(X, \mathcal{P})$ and can be extended to $\mathcal{F}_{\mathfrak{V}}(X, \mathcal{P}_{\mathcal{V}})$. Then there exist mappings $y \mapsto x_y$ and $y \mapsto \mu_y$ from $Y$ into $X$ and into the set of continuous $\vee$-semilattice homomorphisms on $\mathcal{P}$, respectively, such that*

$$T(f)(y) = \mu_y\big(f(x_y)\big)$$

*for all $f \in \mathcal{C}^r_{\mathfrak{V}}(X, \mathcal{P})$ and all $y \in Y$. If the range of the operator $T$ is contained in $\mathcal{C}_{\mathfrak{U}}(Y, \mathbb{R})$, then the mapping $y \mapsto (x_y, \mu_y)$ is continuous with respect to the given topology of $Y$ and the weak topology induced on $X \times \mathcal{P}^*$ by the functions $f \in \mathcal{F}_{\mathfrak{V}}(X, \mathcal{P})$.*

**6.19 The Case $\mathcal{P} = \overline{\mathbb{R}}$.** An elementary function $\varphi_{\otimes}(+\infty)$ with $\varphi \in \mathcal{K}(X)$ is the characteristic function $\chi_{O^{\otimes}}(+\infty)$ for some relatively compact open subset $O$ of $X$ in this case. This function is relatively continuous if and only if the set $O \in \mathfrak{R}$ is both open and compact (see Proposition 1.11(b)). For real-valued functions, on the other hand, continuity with respect to the symmetric relative topology of $\mathfrak{R}$ coincides with the usual (Euclidean) notion. The cone $\mathcal{E}(X, \overline{\mathbb{R}})$ therefore consists of all sums of continuous real-valued functions and characteristic functions $\chi_{O^{\otimes}}(+\infty)$ for some open set $O \in \mathfrak{R}$. The subcone $\mathcal{E}_0(X, \overline{\mathbb{R}})$ of $\mathcal{E}(X, \overline{\mathbb{R}})$ consists of all sums of continuous real-valued functions and characteristic functions $\chi_{O^{\otimes}}(+\infty)$ for some open and compact set $O \in \mathfrak{R}$.

The values $\theta_E$ for $E \in \mathfrak{R}$ of a representing measure in the sense of 5.1 are continuous linear operators from $\overline{\mathbb{R}}$ into some locally convex complete lattice cone $\mathcal{Q}$ in this case. Such an operator $\theta_E$ maps real numbers into invertible (and therefore bounded) elements of $\mathcal{Q}$; more precisely: We have $\theta_E(\alpha) = \alpha l$ with some positive invertible element $0 \le l \in \mathcal{Q}$ for all $\alpha \in \mathbb{R}$. The image of $+\infty \in \overline{\mathbb{R}}$ on the other hand is some zero component $q_\infty \in \mathcal{Q}$, (that is $q_\infty \geq 0$ and $\alpha q_\infty = q_\infty$ for all $\alpha > 0$, see I.5.8) such that $\theta_E(\alpha) \le q_\infty$ for all $\alpha \in \mathbb{R}$. If $(\mathcal{Q}, \mathcal{W}) = \big(\mathcal{F}_{\mathfrak{U}}(Y, \overline{\mathbb{R}}), \mathfrak{U}\big)$, where $Y$ is a second locally compact space and $\mathfrak{U}$ is a basis for a symmetric r-lower continuous inductive limit topology on $\mathcal{F}(Y, \overline{\mathbb{R}})$, then its invertible elements are real-valued, whereas its zero components are functions on $Y$ that take only the values $0$ and $+\infty$.

**6.20 The Case $\mathcal{P} = \overline{\mathbb{R}}_+$.** Here we choose $\mathcal{P} = \overline{\mathbb{R}}_+$, endowed with the singleton neighborhood system $\mathcal{V} = \{0\}$ (see Example 1.4(b) in Chapter I). The symmetric relative topology on $\mathcal{P} = \overline{\mathbb{R}}_+$ coincides with the Euclidean topology on $(0, +\infty)$ and renders both $0$ and $+\infty$ as isolated points (see Example I.4.37(b)). The cone $\mathcal{E}(X, \overline{\mathbb{R}})$ therefore consists of all sums of continuous non-negative real-valued functions and characteristic functions $\chi_{O^{\otimes}}(+\infty)$ for some open set $O \in \mathfrak{R}$. The subcone $\mathcal{E}_0(X, \overline{\mathbb{R}})$ of $\mathcal{E}(X, \overline{\mathbb{R}})$ consists of all such functions that take each of the values $0$ and $+\infty$ on both open and compact subsets of $X$, respectively. An operator $\theta_E$, corresponding to an $\mathfrak{L}(\overline{\mathbb{R}}_+, \mathcal{Q})$-valued measure $\theta$, maps positive reals $\alpha \mapsto \alpha l$ with some positive (not necessarily invertible) element $0 \le l \in \mathcal{Q}$ The image of $+\infty \in \overline{\mathbb{R}}$ is again some zero component $q_\infty \in \mathcal{Q}$, such that $\theta_E(\alpha) \le q_\infty$ for all $\alpha > 0$.

**6.21 The Case $\mathcal{Q} = \overline{\mathbb{R}}$.** We choose $\mathcal{Q} = \overline{\mathbb{R}}$ with the canonical order and the neighborhoods $\mathcal{V} = \{\varepsilon \in \mathbb{R} \mid \varepsilon > 0\}$ and consider the representation of a continuous linear functional on the cone $\mathcal{F}_{\mathfrak{V}}(X, \mathcal{P})$ by a $\mathcal{P}^*$-valued measure (see Section II.6.1). We may use any locally convex cone $(\mathcal{P}, \mathcal{V})$ in this case, since a linear functional on the cone $\mathcal{F}_{\mathfrak{V}}(X, \mathcal{P})$ can be extended into a linear functional on $\mathcal{F}_{\mathfrak{V}}(X, \mathcal{P}_{\mathcal{V}})$ where $(\mathcal{P}_{\mathcal{V}}, \mathcal{V})$ denotes the standard full extension of $\mathcal{P}$. We may than use Theorem 5.1 for this full cone $\mathcal{P}_{\mathcal{V}}$ and obtain:

**Corollary 6.22.** *Let $(\mathcal{P},\mathcal{V})$ be a locally convex cone. Let $X$ be a locally compact Hausdorff space and let $\mathfrak{V}$ be a basis for an r-lower continuous inductive limit topology on $\mathcal{F}(X,\mathcal{P})$. Then every continuous linear functional $\mu \in \mathcal{F}_{\mathfrak{V}}(X,\mathcal{P})^*$ can be represented as an integral on $X$. More precisely: There exists a bounded quasi regular $\mathcal{P}^*$-valued measure $\theta$ on the weak $\sigma$-ring $\mathfrak{R}$ of all relatively compact Borel subsets of $X$ such that $\theta$ is continuous relative to $\mathfrak{V}$. All functions in $\mathcal{F}_{\mathfrak{V}}(X,\mathcal{P})$ are integrable with respect to $\theta$, and*

$$\int_X f\,d\theta \le \mu(f) \qquad \text{for all} \qquad f \in \mathcal{F}_{\mathfrak{V}}(X,\mathcal{P}).$$

*All functions in $\mathcal{F}_{\mathfrak{V}_0}(X,\mathcal{P})$ are strongly integrable with respect to $\theta$, and*

$$\int_X f\,d\theta = \mu(f) \qquad \text{for all} \qquad f \in \mathcal{F}_{\mathfrak{V}_0}(X,\mathcal{P}).$$

This corollary recovers and generalizes the result from Theorem 4.2 in [171].

**6.23 Sequence Cones.** In order to obtain sequence cones we choose $X = \mathbb{N}$ with the discrete topology (see also Example 2.11(f)). Let $(\mathcal{P},\mathcal{V})$ be a quasi-full locally convex cone. The functions in $\mathcal{E}(\mathbb{N},\mathcal{P}) = \mathcal{E}_0(\mathbb{N},\mathcal{P})$ then are finite sequences in $\mathcal{P}$. Depending on our choice for the inductive limit neighborhood system $\mathfrak{V}$ we obtain a variety of sequence cones for $\mathcal{F}_{\mathfrak{V}}(\mathbb{N},\mathcal{P})$, for example $l^p$-type cones as elaborated in 2.11(f). An $\mathfrak{L}(\mathcal{P},\mathcal{Q})$-valued measure $\theta$ on $\mathfrak{R}$, that is the collection of finite subsets of $\mathbb{N}$, corresponds to a sequence $(\theta_i)_{i\in\mathbb{N}}$ of operators in $\mathfrak{L}(\mathcal{P},\mathcal{Q})$. Every such measure is bounded, as for $E = \{x_1,\ldots,x_n\} \in \mathfrak{R}$ and each neighborhood $w \in \mathcal{W}$ there is $v \in \mathcal{V}$ such that $\theta_{x_i}(s) \le (1/n)w$ whenever $s \le v$ for $s \in \mathcal{P}$ and all $i = 1,\ldots,n$. Hence $|\theta|(E,v) \le w$. According to Proposition 3.4 every continuous linear operator $T : \mathcal{F}_{\mathfrak{V}}(\mathbb{N},\mathcal{P}) \to \mathcal{Q}$ can be extended to $\mathcal{F}_{\mathfrak{V}}(\mathbb{N},\mathcal{P}_{\mathcal{V}})$ in this case. Thus, as stated in Theorem 5.1 there exists a unique sequence $(\theta_i)_{i\in\mathbb{N}}$ of linear operators in $\mathfrak{L}(\mathcal{P},\mathcal{Q})$ such that

$$T\big((a_i)_{i\in\mathbb{N}}\big) = \sum_{i=1}^{\infty} \theta_i(a_i)$$

holds for every element (sequence) $(a_i)_{i\in\mathbb{N}} \in \mathcal{F}_{\mathfrak{V}}(\mathbb{N},\mathcal{P})$.

**6.24 The Convergence Theorems.** For a continuous linear operator $T : \mathcal{F}_{\mathfrak{V}}(X,\mathcal{P}) \to \mathcal{Q}$ and its $\mathfrak{L}(\mathcal{P},\mathcal{Q})$-valued representing measure $\theta$, the convergence theorems of Chapter II.5 if applied to the measure $\theta$ may be reinterpreted in terms of the operator $T$. Theorem II.5.25, for example, yields the following:

**Corollary 6.25.** *Let $(\mathcal{P},\mathcal{V})$ be a full locally convex cone and let $(\mathcal{Q},\mathcal{W})$ be a locally convex complete lattice cone such that the order continuous linear* *functionals support the separation property for* $\mathcal{Q}$. *Let* $X$ *be a locally compact Hausdorff space and let* $\mathfrak{V}$ *be a basis for an r-lower continuous inductive limit topology on* $\mathcal{F}(X,\mathcal{P})$. *Let* $T:\mathcal{F}_{\mathfrak{V}}(X,\mathcal{P})\to\mathcal{Q}$ *be a continuous linear operator than can be extended to* $\mathcal{F}_{\mathcal{V}}(X,\mathcal{P}_{\mathcal{V}})$. *Let* $f_n,f,f_{**},f_*,f^{**},f^*\in\mathcal{E}_0(X,\mathcal{P}_{\mathcal{V}})$ *such that* $f_{**}\le f_n+f_*$ *and* $f_n+f^{**}\le f^*$ *for all* $n\in\mathbb{N}$, *and that*

$$f_n\longrightarrow f.$$

*Then*

$$T(f)\le\varliminf_{n\to\infty}T(f_n)+\mathfrak{O}\left(T(f_*)\right)$$

*and*

$$\varlimsup_{n\to\infty}T(f_n)\le T(f)+\mathfrak{O}\left(T(f^*)\right).$$

*If* $(\mathcal{Q},\mathcal{W})$ *is the standard lattice completion of a locally convex cone* $(\mathcal{Q}_0,\mathcal{W}_0)$, *and the range of* $T$ *is contained in* $\mathcal{Q}_0$, *then the above convergence statements refer to weak convergence in* $\mathcal{Q}_0$. *If* $\mathcal{P}$ *and* $\mathcal{Q}_0$ *are indeed locally convex vector spaces and if the operator* $T$ *is weakly compact, then*

$$\lim_{n\to\infty}T(f_n)=T(f)$$

*holds with respect to the symmetric topology of* $\mathcal{Q}_0$.

The last part of this corollary follows from Theorem II.5.36. It assumptions are satisfied if $\mathcal{P}$ and $\mathcal{Q}_0$ are locally convex vector spaces and if the operator $T$ is weakly compact.

**6.26 The Case that $\mathcal{Q}$ Is the Standard Lattice Completion of Some Operator Cone.** Suppose that $\mathcal{Q}=\widehat{\mathfrak{H}}(\mathcal{N},\mathcal{M})$ is the simplified standard lattice completion of some locally convex cone $\mathfrak{H}(\mathcal{N},\mathcal{M})$ of linear operators from a cone $\mathcal{N}$ into a locally convex cone $(\mathcal{M},\mathcal{W})$ as introduced in Section I.7.1, endowed with a locally convex cone topology generated by a family $\mathfrak{Z}$ of subsets of $\mathcal{N}$. We suppose that the union of the sets $Z\in\mathfrak{Z}$ is all of $\mathcal{N}$. Let $(\mathcal{P},\mathcal{V})$ be a quasi-full locally convex cone and let $T:\mathcal{F}_{\mathfrak{V}}(X,\mathcal{P})\to\mathfrak{H}(\mathcal{N},\mathcal{M})$ be a continuous linear operator that permits an extension $T:\mathcal{F}_{\mathfrak{V}}(X,\mathcal{P}_{\mathcal{V}})\to\widehat{\mathfrak{H}}(\mathcal{N},\mathcal{M})$. Let $\theta$ be the bounded quasi regular $\mathfrak{L}(\mathcal{P}_{\mathcal{V}},\widehat{\mathfrak{H}}(\mathcal{N},\mathcal{M}))$-valued measure $\theta$ that represents the operator $T$ as in Theorem 5.1. Recall from I.7.1 that the elements $\varphi$ of $\widehat{\mathfrak{H}}(\mathcal{N},\mathcal{M})$ are $\overline{\mathbb{R}}$-valued functions on $\mathcal{N}\times\mathcal{M}^*$, and the neighborhood system $\widehat{\mathfrak{W}}$ for $\widehat{\mathfrak{H}}(\mathcal{N},\mathcal{M})$ is generated by the functions $\varphi_{(Z,w)}$ for $Z\in\mathfrak{Z}$ and $w\in\mathcal{W}$ such that $\varphi_{(Z,w)}(z,\mu)=1$ if $(z,\mu)\in Z\times w^\circ$ and $\varphi_{(Z,w)}(z,\mu)=+\infty$, else. According to Remark 5.2(b), for fixed $E\in\mathfrak{R}$ and $a\in\mathcal{P}$ the value $\theta_E(a)$ of the representing measure $\theta$ is contained in the order closure in $\widehat{\mathfrak{H}}(\mathcal{N},\mathcal{M})$ of the image $T(\mathcal{A})$ of some relatively bounded subset $\mathcal{A}$ of $\mathcal{F}_{\mathfrak{V}}(X,\mathcal{P})$. We shall proceed to verify that the elements of this order closure, that is in particular the element $\theta_E(a)$, may again be identified with linear operators from $\mathcal{N}$

into $\mathcal{M}^{**}_{sr}$, the (relative) strong second dual of $\mathcal{M}$. For this, suppose that the set $\mathcal{A}$ is bounded below and bounded above relative to the function $f_0 \in \mathcal{F}_{\mathfrak{V}}(X, \mathcal{P})$. Then, obviously, the set $T(\mathcal{A}) \subset \mathfrak{H}(\mathcal{N}, \mathcal{M})$ is bounded below and bounded above relative to the operator $L_0 = T(f_0)$ in $\mathfrak{H}(\mathcal{N}, \mathcal{M})$. Let $\varphi \in \widehat{\mathfrak{H}}(\mathcal{N}, \mathcal{M})$ be in the order closure of $T(\mathcal{A})$, and let $(\varphi_i)_{i\in\mathcal{I}}$ be a net in $T(\mathcal{A})$ converging to $\varphi$ with respect to the order topology. This implies pointwise convergence on $\mathcal{N} \times \mathcal{M}^*$ and yields the following:

(i) For every $z \in \mathcal{N}$ the mapping $\mu \mapsto \varphi_i(z, \mu) : \mathcal{M}^* \to \overline{\mathbb{R}}$ is linear for every $i \in \mathcal{I}$, hence the mapping $\mu \mapsto \varphi(z, \mu) : \mathcal{M}^* \to \overline{\mathbb{R}}$ is linear as well, hence may be interpreted as an element of $\mathcal{M}^{**}$, the second dual of $\mathcal{M}$.

(ii) More precisely, for a fixed element $z \in \mathcal{N}$ the set $T(\mathcal{A})(z) = \{L(z) \mid L \in T(\mathcal{A})\}$ is a relatively bounded subset of $\mathcal{M}$. Indeed, by the above $T(\mathcal{A})$ is bounded below and bounded above relative to the operator $L_0 = T(f_0) \in \mathfrak{H}(\mathcal{N}, \mathcal{M})$. This implies that $T(\mathcal{A})(z)$ is bounded below and bounded above relative to the element $L_0(z) \in \mathcal{M}$. To demonstrate this, let $w \in \mathcal{W}$. There is $Z \in \mathfrak{Z}$ such that $z \in Z$, and according to I.4.24(iv) there are $\lambda, \rho \geq 0$ such that $0 \leq L + \lambda V_{(Z,w)}$ and $L \leq \rho L_0 + \lambda V_{(Z,w)}$ for all $L \in \mathcal{L}$. As $z \in Z$, this implies in particular that $0 \leq L(z) + w$ and $L(z) \leq \rho L_0(z) + w$ for all $L \in T(\mathcal{A})$, hence our claim. Because the mapping $\mu \mapsto \varphi(z, \mu) : \mathcal{M}^* \to \overline{\mathbb{R}}$ is the pointwise limit of the mappings $\mu \mapsto \varphi_i(z, \mu) : \mathcal{M}^* \to \overline{\mathbb{R}}$, according to I.7.3, the former may therefore be identified with an element of $\mathcal{M}^{**}_{sr}$, the (relative) strong second dual of $\mathcal{M}$.

(iii) For every $\mu \in \mathcal{M}^*$, the mapping $z \mapsto \varphi(z, \mu) : \mathcal{N} \to \overline{\mathbb{R}}$ is clearly linear as it is the pointwise limit of linear mappings. Thus we realize that for every $E \in \mathfrak{R}$ and $a \in \mathcal{P}$ the value $\theta_E(a)$ of the representing measure $\theta$ is indeed a linear operator from $\mathcal{N}$ into $\mathcal{M}^{**}_{sr}$.

(iv) Now let us suppose in addition that all the operators in $\mathfrak{H}(\mathcal{N}, \mathcal{M})$ are bounded on $\mathfrak{Z}$, that is for every $L \in \mathfrak{H}(\mathcal{N}, \mathcal{M})$ and $Z \in \mathfrak{Z}$ the set $L(Z)$ is bounded in $\mathcal{M}$. Let $\mathfrak{Y}$ be the family of all neighborhoods $w^\circ \subset \mathcal{M}^*$ for $w \in \mathcal{W}$, and let us endow the cone $\mathcal{M}^{**}_{sr}$ of linear functionals on $\mathcal{M}^*$ with the neighborhood system $\mathcal{W}^{**}$ generated by this family. In this way, $(\mathcal{M}, \mathcal{W})$ is a subcone of $(\mathcal{M}^{**}_{sr}, \mathcal{W}^{**})$. We shall proceed to establish that, under these circumstances, the operator $\theta_E(a) \in L(\mathcal{N}, \mathcal{M}^{**}_{sr})$ is also bounded on $\mathfrak{Z}$. In order to demonstrate this claim, let $Z \in \mathfrak{Z}$ and $w^\circ \in \mathfrak{Y}$. There are $\lambda, \rho \geq 0$ such that $0 \leq L + \lambda V_{(Z,w)}$ and $L \leq \rho L_0 + \lambda V_{(Z,w)}$ for all $L \in T(\mathcal{A})$, and there is $\sigma \geq 0$ such that $L_0(z) \leq \sigma w$ for all $z \in Z$. This implies in particular that

$$0 \leq L(z) + \lambda w \qquad \text{and} \qquad L(z) \leq L_0(z) + \lambda w \leq (\lambda + \sigma) w$$

holds for all $z \in Z$ and all $L \in T(\mathcal{A})$. Because the operator $\theta_E(a)$ is the pointwise limit (as functions on $\mathcal{N} \times \mathcal{M}^*$) of operators in $T(\mathcal{A})$, the same relations hold true for the operator $\theta_E(a) \in L(\mathcal{N}, \mathcal{M}^{**}_{sr})$, that is our claim.

(v) Let $(\widehat{\mathcal{M}}, \widehat{\mathcal{W}})$ be the standard lattice completion of $(\mathcal{M}, \mathcal{W})$, that is a cone of $\overline{\mathbb{R}}$-valued functions as elaborated in I.5.57. For a fixed element $z \in \mathcal{N}$ let us denote by $T^z$ the operator from $\mathcal{F}_{\mathfrak{V}}(X, \mathcal{P}_{\!\vee})$ into $\widehat{\mathcal{M}}$ that maps an element $f \in \mathcal{F}_{\mathfrak{V}}(X, \mathcal{P}_{\!\vee})$ to the function

$$\mu \mapsto T(f)(z,\mu) \,:\, \mathcal{M}^* \to \overline{\mathbb{R}}.$$

This function is indeed an element of $\widehat{\mathcal{M}}$. (Recall that $T(f)$ is an element of $\widehat{\mathfrak{H}}(\mathcal{N},\mathcal{M})$, that is an $\overline{\mathbb{R}}$-valued function on $\mathcal{N} \times \mathcal{M}^*$.) The operator $T^z : \mathcal{F}_{\mathfrak{V}}(X,\mathcal{P}_\vee) \to \widehat{\mathcal{M}}$ is clearly linear and continuous with respect to the given neighborhood systems $\mathfrak{V}$ and $\widehat{\mathcal{W}}$ for $\mathcal{F}_{\mathfrak{V}}(X,\mathcal{P}_\vee)$ and $\widehat{\mathcal{M}}$, respectively. Indeed, let $Z \in \mathfrak{Z}$ such that $z \in Z$ and $w \in \mathcal{W}$. By the continuity of the operator

$$T : \mathcal{F}_{\mathfrak{V}}(X,\mathcal{P}_\vee) \to \widehat{\mathfrak{H}}(\mathcal{N},\mathcal{M})$$

there is $\mathfrak{v} \in \mathfrak{V}$ such that

$$T(f) \leq T(g) + V_{(Z,w)}$$

whenever $f \leq g + \mathfrak{v}$ for $f,g \in \mathcal{F}_{\mathfrak{V}}(X,\mathcal{P}_\vee)$. This implies in particular that

$$T(f)(z,\mu) \leq T(g)(z,\mu) + 1,$$

that is

$$T^z(f)(\mu) \leq T(g)(\mu) + 1$$

for all $\mu \in w^\circ$. Likewise, for a fixed element $z \in \mathcal{N}$ and the representing measure $\theta$ we may define an $\mathfrak{L}(\mathcal{P}_\vee,\widehat{\mathcal{M}})$-valued measure $\theta^z$ in the following way: For every fixed $E \in \mathfrak{R}$ let us denote by $\theta^z_E$ the operator from $\mathcal{P}_\vee$ into $\widehat{\mathcal{M}}$ that maps an element $a \in \mathcal{P}_\vee$ to the function

$$\mu \mapsto \theta_E(a)(z,\mu) \,:\, \mathcal{M}^* \to \overline{\mathbb{R}}.$$

As before, it is easy to verify that this operator is linear and continuous with respect to the given neighborhood systems of $\mathcal{P}_\vee$ and $\mathcal{M}$. Thus the set function

$$E \mapsto \theta^z_E \,:\, \mathfrak{R} \to \mathfrak{L}(\mathcal{P}_\vee,\widehat{\mathcal{M}})$$

is an $\mathfrak{L}(\mathcal{P}_\vee,\widehat{\mathcal{M}})$-valued measure on $\mathfrak{R}$ in the sense of Section II.3. Countable additivity and boundedness follow immediately from the corresponding properties of the $\mathfrak{L}(\mathcal{P}_\vee,\widehat{\mathfrak{H}}(\mathcal{N},\mathcal{M}))$-valued measure $\theta$.

It can now be shown using Theorem II.5.35 that the measure $\theta^z$ represents the operator $T^z$ in the sense of Theorem 5.1. We omit the details.

(vi) If in addition to the assumptions of (iv), for every $z \in \mathcal{N}$ the restriction of the operator $T^z : \mathcal{F}_{\mathfrak{V}}(X,\mathcal{P}_\vee) \to \widehat{\mathcal{M}}$ to the subcone $\mathcal{P}$, that is the operator $T^z : \mathcal{F}_{\mathfrak{V}}(X,\mathcal{P}) \to \mathcal{M}$ is compact (or weakly compact), then a similar argument as in Corollary 6.5 shows that the values $\theta_E(a)$ of the representing measure $\theta$, for $E \in \mathfrak{R}$ and $a \in \mathcal{P}$, are indeed linear operators in $L(\mathcal{N},\mathcal{M})$ that are bounded on $\mathfrak{Z}$. Moreover, for every $E \in \mathfrak{R}$ the set $\{\theta^z_G \mid G \in \mathfrak{R},\ G \subset E\}$ of linear operators from $\mathcal{P}$ into $\mathcal{M}$ is seen to be relatively compact in $\mathfrak{L}(\mathcal{P},\mathcal{M})$ if endowed with the symmetric strong (or with the symmetric weak) operator topology.

(vii) If the values $\theta_E(a)$ of the representing measure $\theta$ are linear operators in $L(\mathcal{N},\mathcal{M})$ that are bounded on $\mathfrak{Z}$ (see (vi) or the *reflexive* case that $\mathcal{M}_s^{**}=\mathcal{M}$), and if all elements of $\mathcal{P}$ are bounded, then our version of Pettis' theorem, that is Theorem II.3.11 applies: For every fixed element $z\in\mathcal{N}$ the $\mathfrak{L}(\mathcal{P},\mathcal{M})$-valued measure $\theta^z$ is countably additive with respect to the strong operator topology of $\mathfrak{L}(\mathcal{P},\mathcal{M})$ in this case.

We shall formulate the special case that the cones $\mathcal{P},\mathcal{N}$ and $\mathcal{M}$ are indeed locally convex vector spaces over $\mathbb{K}=\mathbb{R}$ or $\mathbb{K}=\mathbb{C}$ as another corollary of Theorem 5.1. We shall assume that all the operators involved are linear over $\mathbb{K}$ in this case and say that a linear operator $L:\mathcal{N}\to\mathcal{M}$ is *bounded* if it maps bounded subsets of $\mathcal{N}$ into bounded subsets of $\mathcal{M}$. Note that this notion of boundedness does not guarantee that the operator $L$ is continuous. We shall use the family of all bounded subsets of $\mathcal{N}$ for $\mathfrak{Z}$. This family $\mathfrak{Z}$ generates the uniform operator topology (see I.7.2(i)) on $\mathfrak{H}_{\mathbb{K}}(\mathcal{N},\mathcal{M})$, the space of all bounded $\mathbb{K}$-linear operators from $\mathcal{N}$ into $\mathcal{M}$. Theorems 3.3 and 3.1 yield that a continuous linear operator $T:\mathcal{F}_{\mathfrak{V}}(X,\mathcal{P})\to\widehat{\mathfrak{H}}(\mathcal{N},\mathcal{M})$ can be extended into a continuous linear operator $T:\mathcal{F}_{\mathfrak{V}}(X,\mathcal{P}_{\mathcal{V}})\to\widehat{\mathfrak{H}}(\mathcal{N},\mathcal{M})$ in this case. A similar argument as in 6.9 demonstrates that the values $\theta_E(a)$ for $E\in\mathfrak{R}$ and $a\in\mathcal{P}$ of the representing measure theta are indeed bounded $\mathbb{K}$-linear operators from $\mathcal{N}$ into $\mathcal{M}_s^{**}$ in this case.

**Corollary 6.27.** *Let $(\mathcal{P},\mathcal{V})$, $(\mathcal{N},\mathcal{U})$ and $(\mathcal{M},\mathcal{W})$ be locally convex topological vector spaces over $\mathbb{K}=\mathbb{R}$ or $\mathbb{K}=\mathbb{C}$. Let $\mathfrak{H}_{\mathbb{K}}(\mathcal{N},\mathcal{M})$ be the space of all bounded $\mathbb{K}$-linear operators from $\mathcal{N}$ into $\mathcal{M}$, endowed with the uniform operator topology. Let $X$ be a locally compact Hausdorff space and let $\mathfrak{V}$ be a basis for an r-lower continuous inductive limit topology on $\mathcal{F}(X,\mathcal{P})$. Then every continuous $\mathbb{K}$-linear operator $T:\mathcal{F}_{\mathfrak{V}}(X,\mathcal{P})\to\mathfrak{H}_{\mathbb{K}}(\mathcal{N},\mathcal{M})$ can be represented as an integral on $X$. More precisely: There exists a unique bounded measure $\theta$ on the weak $\sigma$-ring $\mathfrak{R}$ of all relatively compact Borel subsets of $X$ whose values $\theta_E$ for $E\in\mathfrak{R}$ are continuous $\mathbb{K}$-linear operators from $\mathcal{P}$ into $\mathfrak{H}_{\mathbb{K}}(\mathcal{N},\mathcal{M}_s^{**})$, the space of all bounded $\mathbb{K}$-linear operators from $\mathcal{N}$ into the strong second dual $\mathcal{M}_s^{**}$ of $\mathcal{M}$, with the following properties: For every $z\in\mathcal{N}$ the $\mathfrak{L}(\mathcal{P},\mathcal{M})$-valued measure $\theta^z$ is countably additive and quasi regular with respect to the weak* operator topology of $\mathfrak{L}_{\mathbb{K}}(\mathcal{P},\mathcal{M}_s^{**})$. $\theta$ is continuous relative to $\mathfrak{V}$, all functions in $\mathcal{F}_{\mathfrak{V}}(X,\mathcal{P})$ are strongly integrable with respect to $\theta$, and*

$$\int_X f\,d\theta=T(f)\qquad \textit{for all}\qquad f\in\mathcal{F}_{\mathfrak{V}}(X,\mathcal{P}).$$

*If for every $z\in\mathcal{N}$ the operator $T^z$ is compact (or weakly compact), then the values of $\theta$ are indeed continuous linear operators from $\mathcal{P}$ into $\mathfrak{H}_{\mathbb{K}}(\mathcal{N},\mathcal{M})$, and in this case for every $z\in\mathcal{N}$ the measure $\theta^z$ is countably additive with respect to the strong operator topology of $\mathfrak{L}_{\mathbb{K}}(\mathcal{P},\mathcal{M})$. Moreover, for every $E\in\mathfrak{R}$ the set $\{\theta_G^z\mid G\in\mathfrak{R},\ G\subset E\}$ of linear operators from $\mathcal{P}$ into*

$\mathcal{M}$ *is relatively compact in* $\mathfrak{L}(\mathcal{P},\mathcal{M})$ *endowed with the strong (or the weak) operator topology.*

The following special case of Corollary 6.27 is of particular interest:

*The case that* $(\mathcal{P},\mathcal{V})$ *is a topological algebra and that* $\mathcal{N}=\mathcal{M}$. Suppose that $(\mathcal{P},\mathcal{V})$ is a topological algebra over $\mathbb{K}$ and that $\mathcal{N}=\mathcal{M}$, that is a locally convex topological vector space over $\mathbb{K}$ and endowed with the neighborhood system $\mathcal{W}$. The vector space $\mathfrak{H}_{\mathbb{K}}(\mathcal{N})=\mathfrak{L}_{\mathbb{K}}(\mathcal{N})$ of all bounded $\mathbb{K}$-linear operators on $\mathcal{N}$ then forms a topological (non-commutative) algebra, where the canonical multiplication is the composition of the concerned operators. Let $T:\mathcal{F}_{\mathfrak{V}}(X,\mathcal{P})\to\mathfrak{H}_{\mathbb{K}}(\mathcal{N})$ be a continuous linear operator that is multiplicative on $\mathcal{E}(X,\mathcal{P})$. Its $\mathfrak{L}_{\mathbb{K}}\big(\mathcal{P},\mathfrak{H}_{\mathbb{K}}(\mathcal{N})\big)$-valued representation measure $\theta$ then satisfies condition (A) from 6.12. In order to understand this condition, let us investigate how the operator multiplication extends to the standard lattice completion $\widehat{\mathfrak{H}}_{\mathbb{K}}(\mathcal{N})$ of $\mathfrak{H}_{\mathbb{K}}(\mathcal{N})$ in the sense of 6.12. First let us recall that an operator $L\in\mathfrak{H}_{\mathbb{K}}(\mathcal{N})$ is represented as an element of $\widehat{\mathfrak{H}}_{\mathbb{K}}(\mathcal{N})$, that is a real-valued function on $\mathcal{N}\times\mathcal{N}^*$ as $(z,\mu)\mapsto\mathfrak{Re}\big(\mu(L(z))\big)$. Now let $L,M\in\mathfrak{H}_{\mathbb{K}}(\mathcal{N})$. $(z,\mu)\in\mathcal{N}\times\mathcal{N}^*$ we have

$$\big(L\circ M\big)(z,\mu)=\mathfrak{Re}\Big(\mu\big(L(M(z))\big)\Big)=L\big(M(z),\mu\big).$$

Now let $\widetilde{L}\in\mathfrak{L}_{\mathbb{K}}(\mathcal{N},\mathcal{N}_s^{**})$. Its adjoint operator $\widetilde{L}^*$ maps $\mathcal{N}_s^{***}$ into $\mathcal{N}^*$. For every $\mu\in\mathcal{N}_s^{***}$ we have $\widetilde{L}^*(\mu)(z)=\mu\big(\widetilde{L}(z)\big)$ for all $z\in\mathcal{N}$. Similarly, the second adjoint $L^{**}$ maps $\mathcal{N}^{**}$ into $\mathcal{N}_s^{****}$. This yields for the representation of $\widetilde{L}$ as an element of $\widehat{\mathfrak{H}}_{\mathbb{K}}(\mathcal{N})$

$$\widetilde{L}(z,\mu)=\mathfrak{Re}\Big(\widetilde{L}^*(\mu)(z)\Big)$$

for all $(z,\mu)\in\mathcal{N}\times\mathcal{N}^*$. If both $L,M\in\mathfrak{L}_{\mathbb{K}}(\mathcal{N},\mathcal{N}_s^{**})$, then the composition operator $L^{**}\circ M$ is in $\mathfrak{L}_{\mathbb{K}}(\mathcal{N},\mathcal{N}_s^{****})$ and can be represented as an element of the standard lattice completion $\widehat{\mathfrak{H}}_{\mathbb{K}}(\mathcal{N})$ as

$$\big(L^{**}\circ M\big)(z,\mu)=\mathfrak{Re}\Big(\mu\big((L^{**}\circ M)(z)\big)\Big).$$

for $(z,\mu)\in\mathcal{N}\times\mathcal{N}^*$. The latter operation is well defined, since $\mathcal{N}^*\subset\mathcal{N}_s^{***}$ and the elements of $\mathcal{N}_s^{****}$ are linear functionals on $\mathcal{N}_s^{***}$. Now let $\widetilde{L},\widetilde{M}\in\mathfrak{L}_{\mathbb{K}}(\mathcal{N},\mathcal{N}_s^{**})\subset\widehat{\mathfrak{H}}_{\mathbb{K}}(\mathcal{N})$ and let $(L_i)_{i\in\mathcal{I}}$ and $(M_j)_{j\in\mathcal{J}}$ be nets in $\mathfrak{H}_{\mathbb{K}}(\mathcal{N})$ such that $\lim_{i\in\mathcal{I}}L_i=\widetilde{L}$ and $\lim_{j\in\mathcal{J}}M_j=\widetilde{M}$ with respect to order convergence in $\widehat{\mathfrak{H}}_{\mathbb{K}}(\mathcal{N})$, that is pointwise convergence on $\mathcal{N}\times\mathcal{N}^*$. Then for all $(z,\mu)\in\mathcal{N}\times\mathcal{N}^*$ this implies that

$$\lim_{i\in\mathcal{I}}L_i\big(M_j(z),\mu\big)=\widetilde{L}\big(M_j(z),\mu\big)$$

for every $j \in \mathcal{J}$, and

$$\begin{aligned}\lim_{j\in\mathcal{J}} \widetilde{L}(M_j(z),\mu) &= \lim_{j\in\mathcal{J}} \mathfrak{Re}\Big(\widetilde{L}^*(\mu)\big(M_j(z)\big)\Big) = \lim_{j\in\mathcal{J}} M_j\big(z,\widetilde{L}^*(\mu)\big) = \widetilde{M}\big(z,\widetilde{L}^*(\mu)\big)\\ &= \mathfrak{Re}\Big(\widetilde{L}^*(\mu)\big(\widetilde{M}(z)\big)\Big) = \mathfrak{Re}\Big(\mu\big((\widetilde{L}^{**}\circ\widetilde{M})(z)\big)\Big) = (\widetilde{L}^{**}\circ\widetilde{M})(z,\mu).\end{aligned}$$

Thus

$$\lim_{j\in\mathcal{J}}\lim_{i\in\mathcal{I}}\big(L_i\circ M_j\big)(z,\mu) = (\widetilde{L}^{**}\circ\widetilde{M})(z,\mu).$$

Similarly, one verifies

$$\lim_{i\in\mathcal{I}}\lim_{j\in\mathcal{J}}\big(L_i\circ M_j\big)(z,\mu) = (\widetilde{L}^{**}\circ\widetilde{M})(z,\mu).$$

The extension of the multiplication to elements $L, M \in \widehat{\mathfrak{H}}_{\mathbb{K}}(\mathcal{N})$ is therefore given by

$$L\bullet M = L^{**}\circ M$$

and Condition (A) from 6.12 for the representing measure $\theta$ then reads as follows:

(A') $\theta_E(ab) = \theta_E(a)^{**}\circ\theta_E(b)$ and $\theta_E(a)^{**}\circ\theta_G(b) = 0$ *for all* $a,b\in\mathcal{P}$ *and disjoint sets* $E,G\in\mathfrak{R}$.

*The case that* $\mathcal{P}=\mathbb{K}$ *and that* $\mathcal{N}=\mathcal{M}$. This case is of particular interest as it will lead to the Spectral representation theorem. If $\mathcal{P}=\mathbb{K}$, then linear operators from $\mathcal{P}$ into $\mathfrak{H}_{\mathbb{K}}(\mathcal{N})$ are indeed elements $L$ of $\mathfrak{H}_{\mathbb{K}}(\mathcal{N})$ acting as $\alpha\mapsto\alpha L:\mathbb{K}\to\mathfrak{H}_{\mathbb{K}}(\mathcal{N})$. We shall assume in addition that the representation measure $\theta$ is indeed $\mathfrak{L}_{\mathbb{K}}\big(\mathbb{K},\mathfrak{H}_{\mathbb{K}}(\mathcal{N})\big)$-, that is $\mathfrak{H}_{\mathbb{K}}(\mathcal{N})$-valued. According to Corollary 6.27, this circumstance is guaranteed if the operator $T$ is weakly compact or if the locally convex vector space $\mathcal{N}$ is reflexive. Condition (A) then is further simplified and reads:

(A") $\theta_E^2 = \theta_E$ and $\theta_E\circ\theta_G = 0$ *for disjoint sets* $E,G\in\mathfrak{R}$.

We shall formulate this as an additional corollary:

**Corollary 6.28.** *Let* $(\mathcal{N},\mathcal{U})$ *be a locally convex topological vector space over* $\mathbb{K}=\mathbb{R}$ *or* $\mathbb{K}=\mathbb{C}$ *and let* $\mathfrak{H}_{\mathbb{K}}(\mathcal{N})$ *be the space of all bounded* $\mathbb{K}$*-linear operators on* $\mathcal{N}$, *endowed with the uniform operator topology. Let* $X$ *be a locally compact Hausdorff space and let* $\mathfrak{V}$ *be a basis for an r-lower continuous inductive limit topology on* $\mathcal{F}(X,\mathcal{P})$. *Let* $T:\mathcal{F}_{\mathfrak{V}}(X,\mathbb{K})\to\mathfrak{H}_{\mathbb{K}}(\mathcal{N})$ *be a continuous multiplicative* $\mathbb{K}$*-linear operator, and suppose that either* $T$ *is weakly compact or that* $\mathcal{N}$ *is reflexive. Then there exists a unique bounded* $\mathfrak{H}_{\mathbb{K}}(\mathcal{N})$*-valued measure* $\theta$ *on the weak* $\sigma$*-ring* $\mathfrak{R}$ *of all relatively compact Borel subsets with the following properties: For every* $z\in\mathcal{N}$ *the* $\mathcal{N}$*-valued measure* $\theta^z$ *is countably additive and quasi regular with respect to the given topology of* $\mathcal{N}$. *We have*

$$\theta_E^2 = \theta_E \qquad \textit{and} \qquad \theta_E \circ \theta_G = 0$$

*for disjoint sets* $E, G \in \mathfrak{R}$. *The measure* $\theta$ *is continuous relative to* $\mathfrak{V}$,*, all functions in* $\mathcal{F}_{\mathfrak{V}}(X, \mathcal{P})$ *are strongly integrable with respect to* $\theta$, *and*

$$\int_X f\, d\theta = T(f) \qquad \textit{for all} \qquad f \in \mathcal{F}_{\mathfrak{V}}(X, \mathcal{P}).$$

*If for every* $z \in \mathcal{N}$ *the operator* $T^z$ *is compact (or weakly compact), then for every* $E \in \mathfrak{R}$ *the subset* $\{\theta_G^z \mid G \in \mathfrak{R},\ G \subset E\}$ *of* $\mathcal{N}$ *relatively compact (or relatively weakly compact).*

This corollary leads directly to the spectral representation theorem, our final application.

**6.29 The Spectral Theorem.** We choose a complex Hilbert space $(\mathcal{H}, \langle , \rangle)$ for $\mathcal{N}$ in Corollary 6.28. Let $L$ be a normal continuous linear operator on $\mathcal{H}$, that is $L \circ L^* = L^* \circ L$, and let $X \subset \mathbb{C}$ be its compact spectrum. For the neighborhood system $\mathfrak{V}$ we choose the collection of singleton sets $\mathfrak{v}$, each containing a constant function $x \mapsto \alpha$ for some $\alpha > 0$. Thus $\mathcal{F}_{\mathfrak{V}}(X, \mathcal{P}) = \mathcal{C}(X, \mathbb{C})$, endowed with the topology of uniform convergence. The closed subalgebra $\Lambda$ of $\mathfrak{H}_{\mathbb{K}}(\mathcal{H}) = \mathfrak{L}_{\mathbb{K}}(\mathcal{H})$ generated by the identity operator $I$ and the two elements $L$ and $L^*$ is a commutative $B^*$-algebra with involution (see IX.3.1 in [56]), and its spectrum coincides with the spectrum of $L$ (see Corollaries IX.3.10 and IX.3.11 in [56]). According to the Gelfand-Naĭmark theorem (see Theorem IX.3.7 in [56] or Theorem III.16.1 in [137]) then $\Lambda$ is isometrically *-isomorphic to $\mathcal{C}(X, \mathbb{C})$, that is to say there is a continuous multiplicative linear operator $T : \mathcal{C}(X, \mathbb{C}) \to \mathfrak{L}_{\mathbb{K}}(\mathcal{H})$ that maps the constant function $x \mapsto 1$ into $I$ and the identity function $x \mapsto x$ into the operator $L$. Furthermore, $T(f^*) = T(f)^*$ holds for all $f \in \mathcal{C}(X, \mathbb{C})$. According to Corollary 6.27, $T$ can be represented by an $\mathfrak{L}_{\mathbb{K}}(\mathcal{H})$-valued measure $\theta$ with the stated properties. In addition, we have Condition (A*) from 6.12, that is $\overline{\alpha}\theta_E = (\alpha\theta_E)^*$ for all $E \in \mathfrak{R}$ and $\alpha \in \mathbb{K}$. As $(\alpha\theta_E)^* = \overline{\alpha}\theta_E^*$, this yields $\theta_E = \theta_E^*$. Summarizing, Conditions (A) and (A*) demonstrate that the operators $\theta_E$ are indeed projections and that $\theta_E$ and $\theta_G$ are orthogonal whenever $E, G \in \mathfrak{R}$ are disjoint sets. This is of course the classical Spectral representation theorem for normal operators on a Hilbert space (see Theorem X.2.1 in [56] or Theorem II.44.1 in [82]).

## 7. Notes and Remarks

The topology $\tau$ of a locally convex topological vector space $\mathcal{N}$ is called the *inductive limit* of the topologies $\tau_i$ of a directed (with respect to inclusion) family $\{\mathcal{N}_i\}_{i \in \mathcal{I}}$ of subspaces of $\mathcal{N}$, if $\mathcal{N} = \bigcup_{i \in \mathcal{I}} \mathcal{N}_i$, and if $\tau$ is the finest locally convex topology on $\mathcal{N}$ whose trace topology on each of the subspaces

$\mathcal{N}_i$ is coarser than $\tau_i$ (see II.6 in [185]). The inductive limit is called *strict*, if these trace topologies coincide with the $\tau_i$. Now let $\mathcal{N}$ be the space of all continuous real-valued functions with compact support on a locally compact Hausdorff space $X$, and for any compact subset $E$ of $X$ let $\mathcal{N}_E$ be the subspace consisting of all functions in $\mathcal{N}$ whose support is contained in $E$. If each subspace $\mathcal{N}_E$ is endowed with the supremum norm for the functions on $E$, then the corresponding inductive limit topology on $\mathcal{N}$ can be described by the following system of 0-neighborhoods: Let $\mathfrak{v}$ be a convex subset of non-negative lower semicontinuous $\overline{\mathbb{R}}$-valued functions on $X$ such that for every compact subset $E$ of $X$ there is $\varepsilon > 0$ and $s \in \mathfrak{v}$ such that $\varepsilon \chi_E \leq s$, where $\chi_E$ denotes the characteristic function of the set $E$. The corresponding 0 -neighborhood $\mathfrak{v}_{\mathcal{N}}$ in $\mathcal{N}$ then is given by

$$\mathfrak{v}_N = \big\{ f \in \mathcal{N} \bigm| |f| \leq s \quad \text{for some } s \in \mathfrak{v} \big\}.$$

It is now straightforward to verify that the family of all these sets $\mathfrak{v}_{\mathcal{N}}$ establishes a 0-neighborhood system for the inductive limit topology of $\mathcal{N}$. This is of course our approach to inductive limit topologies on cones of functions from Section 2.1. The neighborhood system $\mathcal{V}$ of $\mathcal{P} = \mathbb{R}$ consists of all positive constants in this case, and the system $\mathfrak{V}$ of all convex sets $\mathfrak{v}$ of $\overline{\mathcal{V}}$ -valued functions from above defines an inductive limit topology on $\mathcal{N}$ in the sense of 2.1. The functions in $\mathcal{N}$ of course vanish at infinity.

The concept of weighted spaces of continuous real-valued functions on a locally compact set is due to Nachbin [136] and Prolla [155]. In brief, it works as follows: A family $\mathcal{W}$ of non-negative real-valued upper semicontinuous functions on a locally compact Hausdorff space $X$ is called a *family of weights* if for all $w_1, w_2 \in \mathcal{W}$ there are $w_3 \in \mathcal{W}$ and $\rho > 0$ such that

$$w_1 \leq \rho w_3 \qquad \text{and} \qquad w_2 \leq \rho w_3.$$

The associated the subspace of $C(X)$

$$C_{\mathcal{W}}(X) \;=\; \{ f \in C(X) \mid wf \;\text{ vanishes at infinity for all } w \in \mathcal{W} \},$$

together with the locally convex topology generated by the seminorms

$$p_w(f) \;=\; \sup \big\{ |wf(x)| \bigm| x \in X \big\}$$

for $w \in \mathcal{W}$ and $f \in C_{\mathcal{W}}$, is called a *weighted space of functions*. The neighborhood system $\mathcal{V}$ of $\mathcal{P} = \mathbb{R}$ consists of all positive constants in this case. The $\overline{\mathcal{V}}$-valued functions $s_w(x) = 1/w(x)$, for $w \in \mathcal{W}$, are lower semicontinuous, hence bounded below by a positive constant on every relatively compact subset of $X$. Thus, as explained in Example 2.11(e), the inductive limit neighborhoods $\mathfrak{v}_w = \{s_w\}$, for all $w \in \mathcal{W}$, model this situation in the settings of Section 2.1 and lead to a representation of a weighted space of functions as a cone of real-valued functions endowed with a lower semicontinuous inductive limit topology.

In this way, the concept of inductive limit topologies from Section 2.1 combines the corresponding classical notion with the notion of weighted spaces of functions.

The classical Riesz representation theorem [164] is of course the prototype of integral representation theorems for linear operators on function spaces. It states that every positive linear functional on the space $\mathcal{N}$ of all continuous real-valued functions with compact support on a locally compact Hausdorff space $X$ can be represented by the integral with respect to some regular positive Borel measure on $X$ (Theorem 13.23 in [178]). Every such positive positive linear functional is of course continuous with respect to the inductive limit topology from above, and an immediate generalization of this theorem states that every real-valued linear functional on $\mathcal{N}$, which is continuous with respect to the inductive limit topology, can be represented by a regular Borel measure on $X$. There are a number of generalizations of this result, most of them are concerned with operators on spaces of continuous real-valued functions and representations by vector-valued measures. One of the better-known versions is the representation theorem by Bartle-Dunford-Schwartz ([7], Theorem iV.5 in [43]). It states that a weakly compact continuous linear operator from a space of continuous functions on a compact space (endowed with the supremum norm) into some Banach space $\mathcal{L}$ can be represented by the integral with respect to a regular $\mathcal{L}$-valued Borel measure (see [26]). In case that the operator $T$ is not weakly compact, the representing measure has to be allowed to take values in the second dual of $\mathcal{L}$. These results are recovered as special cases of Theorem 5.1 in Section 6 (see Corollary 6.5) of this chapter.

For further studies, it might be interesting and possibly rewarding, to investigate potential adaptations of Choquet's theorem to cone- or vector-valued functions. In this case, one looks for an integral representation of a linear operator which is defined only on a given subspace of continuous functions. There are usually many measures doing this, but some can be singled out for having a particularly small and identifiable support in $X$. Choquet's theorem (see [4] or [148]) states that, given a subspace $\mathcal{L}$ of the space $\mathcal{N}$ of continuous real-valued functions on a compact Hausdorff space $X$ and a continuous linear functional $\mu : \mathcal{L} \to \mathbb{R}$, there is a real-valued regular Borel measure $\theta$ on $X$ such that $\mu(f) = \int f \, d\theta$ holds for all $f \in \mathcal{L}$, and such that $\theta$ is supported (in a delicate way) by some subset of $X$, called the Choquet boundary of $\mathcal{L}$. Though classical by now, this is still considered to be a deep result, and the arguments involved are quite subtle, in particular if the compact space $X$ carries a non-metric topology. The order structure of the involved function and measure spaces is used extensively, since the sought after representing measures are maximal in some sense. This book uses order structures as its main means to treat integration theory and might therefore lead to a line of approach for the generalization of Choquet's theorem to the vector- or cone-valued case.

# List of Symbols

## Standard Symbols

| | |
|---|---|
| $\mathbb{N} = \{1, 2, 3, \ldots\}$ | The natural numbers |
| $\mathbb{Z} = \{\ldots -1, 0, 1, 2, \ldots\}$ | The integer numbers |
| $\mathbb{R}$ | The real numbers |
| $\overline{\mathbb{R}} = \mathbb{R} \cup \{+\infty\}$ | The extended real numbers |
| $\overline{\mathbb{R}}_+ = \{\alpha \in \overline{\mathbb{R}} \mid \alpha \geq 0\}$ | The non-negative extended real numbers |
| $\mathbb{C}$ | The complex numbers |
| $\Gamma = \{\gamma \in \mathbb{C} \mid \|\gamma\| = 1\}$ | The unit circle of $\mathbb{C}$ |
| $\mathbb{K}$ | Stands for either $\mathbb{R}$ or $\mathbb{C}$ |
| $\mathbb{B}$ | Unit ball of a normed space |
| $\mathbb{B}^*$ | Dual unit ball of a normed space |

## Special Symbols

| | |
|---|---|
| $\mathcal{P}, \mathcal{Q}, \mathcal{N}, \mathcal{M}$ | Cones, I.1 |
| $\mathcal{V}, \mathcal{W}, \mathcal{U}$ | Abstract neighborhood systems for cones, I.1 |
| $(\mathcal{P}, \mathcal{V}),\ (\mathcal{Q}, \mathcal{W}), \ldots$ | Locally convex cones, I.1 |
| $\overline{\mathcal{V}} = \mathcal{V} \cup \{0, \infty\}$ | augmented neighborhood system, I.1.4, II.2.2 |
| $\mathcal{F}(X, \mathcal{P})$ | Cone of all $\mathcal{P}$-valued functions on $X$, I.1.4 |
| $\left(\mathcal{F}_{\widehat{\mathcal{V}}_b}(X, \mathcal{P}), \widehat{\mathcal{V}}\right)$ | Locally convex cone of $\mathcal{P}$-valued functions on $X$, I.1.4 |
| $(\widehat{\mathcal{P}}, \widehat{\mathcal{V}})$ | Standard completion of a locally convex cone $(\mathcal{P}, \mathcal{V})$, I.5.57 |
| $(\mathcal{P}_{\mathcal{V}}, \mathcal{V})$ | Standard full extension of a quasi-full locally convex cone $(\mathcal{P}, \mathcal{V})$, I.6.2 |
| $\sigma(\mathcal{P}, \mathcal{P}^*)$ | Weak topology on $\mathcal{P}$, I.4.6 |
| $\mathcal{O}(\mathcal{P})$ | Order topology on $\mathcal{P}$, I.5.43 |

| | |
|---|---|
| $o(\mathcal{P}, \mathcal{P}^*)$ | Weak order topology on $\mathcal{P}$, I.5.49 |
| $L(\mathcal{N}, \mathcal{M})$ | Cone of linear operators form $\mathcal{N}$ to $\mathcal{M}$, I.7 |
| $\mathfrak{L}(\mathcal{P}, \mathcal{Q})$ | Cone of continuous linear operators form $\mathcal{P}$ to $\mathcal{Q}$, II.3 |
| $\mathfrak{V}_{(\mathfrak{Z},\mathcal{W})}$ | Neighborhood system for $L(\mathcal{N}, \mathcal{M})$, I.7 |
| $(\mathfrak{H}(\mathcal{N}, \mathcal{M}), \mathfrak{V})$ | Locally convex cone of linear operators from $\mathcal{N}$ to $\mathcal{M}$, I.7 |
| $\mathcal{P}_w^{**}, \mathcal{P}_s^{**}, \mathcal{P}_{sr}^{**}, \mathcal{P}_{sl}^{**}, \mathcal{P}^{**}$ | Second duals of a locally convex cone, I.7.3 |

## Integral-Related Special Symbols

| | |
|---|---|
| $\mathfrak{R}$ | Weak $\sigma$-ring of subsets, II.1.1 |
| $\mathfrak{A}_{\mathfrak{R}}$ | $\sigma$-algebra of measurable subsets, II.1.1 |
| $\chi_E$ | Characteristic function of a subset $E \subset X$, II.1.1 |
| $\mathcal{S}_{\mathfrak{R}}(X, \mathcal{P})$ | Cone of all $\mathcal{P}$-valued step functions supported by $\mathfrak{R}$, II.1.1 |
| $\mathcal{E}_{\mathfrak{R}}(X, \mathcal{P})$ | Subcone generated by all $\mathcal{P}$-valued elementary functions, II.6.16 |
| $\mathcal{F}_{\mathfrak{R}}(X, \mathcal{P})$ | A cone of measurable functions, II.2.3 |
| $\lvert\theta\rvert(E, v)$ | The modulus of the measure $\theta$, II.3.2 |
| $\int_F f\,d\theta$ | Integral of a function $f \in \mathcal{F}_{\mathfrak{R}}(X, \mathcal{P})$ over a set $F \in \mathfrak{A}_{\mathfrak{R}}$, II.4.9 |
| $\int_F f\,d\theta$ | Integral of a function $f \in \mathcal{F}(X, \mathcal{P})$ over a set $F \in \mathfrak{A}_{\mathfrak{R}}$, II.4.13 |
| $\mathcal{F}_{(F,\theta)}(X, \mathcal{P})$ | Functions in $\mathcal{F}(X, \mathcal{P})$ that are integrable over $F \in \mathfrak{A}_{\mathfrak{R}}$, II.4.13 |
| $\mathcal{F}_{(F,\Theta)}(X, \mathcal{P})$ | Functions integrable with respect to a family of measures, II.5.3 |
| $\mathcal{F}_{(\lvert F\rvert,\Theta)}(X, \mathcal{P})$ | Functions integrable with respect to a family of measures, II.5.3 |
| $(\mathcal{F}_{(F,\Theta)}(X, \mathcal{P}), \mathfrak{V}(F, \Theta))$ | Locally convex cone of integrable functions, II.5.5 |
| $\mathfrak{Rs}(\theta_n, F, f)$ | Residual component of a function, II.5.16 |
| $\mathfrak{var}\,(\theta_a, X)$ | Variation of a real-valued measure, II.5.34 |
| $\mathcal{Z}(A, E)$, $\mathcal{I}(A, E)$ | II.6.8 |

## Order Relations

| | |
|---|---|
| $\leq$ | Standard order relation (reflexive, transitive and compatible with algebraic operations) |
| $\preccurlyeq$ | (global) weak preorder, I.3 |
| $\preccurlyeq_v$ | local weak preorder (referring to a neighborhood $v \in \mathcal{V}$), I.3 |

| | |
|---|---|
| $\preceq$ | (global) preorder, I.8 |
| $\preceq_v$ | local preorder (referring to a neighborhood $v \in \mathcal{V}$), I.8 |
| $\underset{a.e.F}{\leq}$ | Almost everywhere order relation for functions, II.4.11 |
| $\leq_{\mathcal{P}}$ | Order relation for measures, II.5.11 |
| $\overset{F}{\underset{\mathfrak{F}}{\preccurlyeq}}$ | Order relation for measures, II.5.17 |

## Operations on Elements

| | |
|---|---|
| $\mathcal{B}(a)$, $(a)\mathcal{B}$, $\mathcal{B}^s(a)$ | Boundedness components of $a$, I.4.9 |
| $\mathcal{B}_v(a)$, $(a)\mathcal{B}_v$, $\mathcal{B}_v^s(a)$ | Local boundedness components of $a$ (referring to a neighborhood $v \in \mathcal{V}$), I.4 |
| $\mathfrak{O}(a)$ | Zero component of $a$, I.5.8 |
| $\mathfrak{O}(a \searrow b)$ | Zero component of $a$ relative to $b$, I.5.16 |

## Operations on Sets

| | |
|---|---|
| $\downarrow A = \{x \in E \mid x \leq a \text{ for some } a \in A\}$ | decreasing hull of a set $A \subset \mathcal{P}$, I.1.4 |
| $\uparrow A = \{x \in E \mid x \geq a \text{ for some } a \in A\}$ | increasing hull of a set $A \subset \mathcal{P}$, |
| $conv(A)$ | Convex hull of a set $A \subset \mathcal{P}$, I.5.7 |
| $Ex(A)$ | Set of extreme points of a convex set $A \subset \mathcal{P}$, I.5.33 |
| $\overline{A}$, $A^\circ$, $\partial A = \overline{A} \setminus A$ | topological closure, interior and boundary of a set $A$ |
| $\overline{A}^{(l)}$ | Closure of $A$ with respect to the lower relative topology, I.4.24 |
| $\overline{A}^{(u)}$ | Closure of $A$ with respect to the upper relative topology, I.4.24 |
| $v(A)$, $(A)v$ | upper and lower relative neighborhoods of a set $A \subset \mathcal{P}$, I.4.28 |

## Convergence

| | |
|---|---|
| $\underline{\lim}_{i \in \mathcal{I}}\, a_i$, $\overline{\lim}_{i \in \mathcal{I}}\, a_i$, | Order convergence for a net $(a_i)_{i \in \mathcal{I}}$, I.5.18 |
| $f_n \overset{v}{\underset{F}{\searrow}} f$, $f_n \overset{v}{\underset{F}{\nearrow}} f$, $f_n \overset{v}{\underset{F}{\rightarrow}} f$ | Upper, lower and symmetric pointwise convergence for functions on a set $F$, II.1.7 |
| $f_n \underset{a.e.F}{\searrow} f$, $f_n \underset{a.e.F}{\nearrow} f$, $f_n \underset{a.e.F}{\longrightarrow} f$ | Upper, lower and symmetric almost everywhere pointwise convergence for functions on a set $F$, II.5.22 |

| | |
|---|---|
| $\theta_n \nearrow \theta$, $\theta_n \searrow \theta$, $\theta_n \longrightarrow \theta$ | Upper, lower and symmetric convergence of sequences of measures, II.5.13 |

## Symbols Related to Continuous Cone-valued Functions

| | |
|---|---|
| $\mathrm{supp}(f) = \overline{\{x \in X \mid f(x) \neq 0\}}$ | The support of a function, III.1 |
| $\mathrm{supp}^*(f) = \{x \in X \mid f(x) \neq 0\}$ | The core support of a function, III.1 |
| $\mathcal{C}^r(X,\mathcal{P})$ | Cone of r-continuous $\mathcal{P}$-valued functions, III.1 |
| $\varphi_{\otimes} a$ | Elementary function, III.1.10 |
| $\mathfrak{K}$ | Family of all compact subsets of $X$, III.2 |
| $\mathfrak{K}_0$ | Family of all both open and compact subsets of $X$, III.2 |
| $\mathcal{K}(X)$ | Continuous positive real-valued functions with support in $\mathfrak{K}$, III.2 |
| $\mathcal{K}_0(X)$ | Continuous positive real-valued functions with core support in $\mathfrak{K}_0$, III.2 |
| $\mathcal{E}(X,\mathcal{P})$, $\mathcal{E}_0(X,\mathcal{P})$ | Cones generated by elementary functions, III.2.3 |
| $\mathcal{F}_{\mathfrak{V}}(X,\mathcal{P})$, $\mathcal{F}_{\mathfrak{V}_0}(X,\mathcal{P})$ | Closures of $\mathcal{E}(X,\mathcal{P})$ and $\mathcal{E}_0(X,\mathcal{P})$, III.2.4 |
| $\mathcal{C}^r_{\mathfrak{V}}(X,\mathcal{P})$, $\mathcal{C}^r_{\mathfrak{V}_0}(X,\mathcal{P})$ | The r-continuous functions in $\mathcal{F}_{\mathfrak{V}}(X,\mathcal{P})$, and $\mathcal{F}_{\mathfrak{V}_0}(X,\mathcal{P})$, III.2.5 |
| $E \prec \varphi$, $\varphi \prec E$ | for $E \in \mathfrak{R}$ and $\varphi \in \mathcal{K}(X)$, III.4.5 |

# References


1. L.V. Ahlfors, *Complex Analysis*, McGraw-Hill, New York, 1953.
2. A.D. Alexandroff, *Additive set functions in abstract spaces*, Mat. Sbornik N.S. **50** (1940), 307–348.
3. E.M. Alfsen, *On a general theory of integration based on order*, Math. Scand **6** (1958), 67–79.
4. E.M. Alfsen, *Compact convex sets and boundary integrals*, Ergebnisse der Mathematik und ihrer Grenzgebiete, vol. 57, Springer Verlag, Heidelberg-Berlin-New York, 1971.
5. B. Anger and C. Portenier, *Radon integrals*, Birkhäuser, Boston-Basel-Berlin, 1992.
6. R.F. Arens, *Representation of functionals by integrals*, Duke Math. Journal **17** (1950), 499–506.
7. R.G. Bartle, N. Dunford and J.T. Schwartz, *Weak compactness and vector measures*, Studia Mathematica **15** (1956), 337–352.
8. R.G. Bartle, *A general bilinear vector integral*, Canadian Journal of Mathematics **7** (1955), 298–305.
9. R.G. Bartle, *The elements of integration*, Wiley, New York, 1966.
10. J. Batt, *Integraldarstellungen linearer Transformationen und schwache Kompaktheit*, Math. Ann. **147** (1967), 291–304.
11. J. Batt and E.J. Berg, *On weak compactness in spaces of vector-valued measures and Bochner integrable functions in connection with the Radon-Nikodým property of Banach spaces*, Arch. Math. (Basel) **19** (1974), 285–304.
12. H. Bauer, *Über die Beziehungen einer abstrakten Theorie des Riemann-Integrals zur Theorie Radonscher Maße*, Mathematische Zeitschrift **65** (1956), 448–482.
13. H. Bauer, *Funktionenkegel und Integralungleichungen*, Sitz. Ber. math. naturw. Kl. Bayer. Akad. Wiss. München (1978), 53–61.
14. H. Bauer, *Maß- und Integrationstheorie*, De Gruyter, Berlin-New York, 1990.
15. G. Birkhoff, *Integration of functions with values in Banach spaces*, Transactions of the American Mathematical Society **38** (1935), 357–378.
16. E. Bishop and K. de Leeuw, *The representation of linear functionals by measures on sets of extreme points*, Ann. Inst. Fourier (Grenoble) **9** (1959), 305–331.
17. T.S. Blyth, *Lattices and ordered algebraic structures*, Springer Verlag, Heidelberg-Berlin-New York, 2005.
18. N. Boboc, G. Bucur and A. Cornea, *Order and convexity in Potential Theory: H-cones*, Lecture Notes in Mathematics, vol. 853, Springer Verlag, Heidelberg-Berlin-New York, 1981.
19. S. Bochner, *Integration von Funktionen deren Werte die Elemente eines Vektorraumes sind*, Fund. Math. **20** (1933), 262–276.
20. S. Bochner, *Absolut-additive abstrakte Mengenfunktionen*, Fund. Math. **21** (1933), 211–213.

21. S. Bochner and A.E. Taylor, *Linear functionals on certain spaces of abstractly-valued functions*, Annals of Mathematics (2) **39** (1938), 913–944.
22. S. Bochner, *Additive set functions on groups*, Annals of Mathematics (2) **40** (1939), 769–799.
23. W.M. Bogdanowicz, *A generalization of the Lebesgue-Bochner-Stieltjes integral and a new approach to the theory of integration.*, Proc. Nat. Acad. Sci. U.S.A. **53** (1965), 492–498.
24. W.M. Bogdanowicz, *Integral representation of linear continuous operators from the space of Lebesgue-Bochner-Stieltjes summable functions into any Banach space*, Proc. Nat. Acad. Sci. U.S.A. **54** (1965), 351–354.
25. N. Bourbaki, *Éléments de mathématique*, Intégration, Livre VI, Fascicule XIII, Hermann, Paris, 1965.
26. N. Bourbaki, *Éléments de mathématique*, Topologie générale, Chap. V–X, Hermann, Paris, 1974.
27. R.D. Bourgin, *Geometric aspects of convex sets with the RadonNikodým property*, Lecture Notes in Mathematics, vol. 993, Springer Verlag, Heidelberg-Berlin-New York, 1983.
28. J.K. Brooks, *Weak compactness in the space of vector measures*, Bulletin of the American Mathematical Society **78** (1972), 284–287.
29. J.K. Brooks and N. Dinculeanu, *Strong additivity, absolute continuity and compactness in spaces of measures*, Journal Math. Anal. Appl. **45** (1974), 156–175.
30. C. Carathéodory, *Vorlesungen über reelle Funktionen*, Teubner, Berlin and Leipzig, 1918.
31. C. Castaing and M. Valadier, *Convex analysis and measurable multifunctions*, Lecture Notes in Mathematics, vol. 580, Springer Verlag, Heidelberg-Berlin-New York, 1977.
32. G. Choquet, *Theory of capacities*, Ann. Inst Fourier **5** (1955), 131–295.
33. G. Choquet and J. Deny, *Ensembles semi-réticulés et ensembles réticulés de fonctions continues*, J. Math. Pures Appl. **36** (1957), 179–189.
34. G. Choquet, *Mesures coniques, affines et cylindriques*, Symposia Mathematica **II** (1969), 145–182.
35. D.L. Cohn, *Measure theory*, Birkhäuser, Boston-Basel-Berlin, 1980.
36. P.J. Daniell, *A general form of integral*, Ann. Math. **19** (1917/18), 279–294.
37. R.B. Darst, *A note on abstract integration*, Transactions of the American Mathematical Society **99** (1961), 292–297.
38. R.B. Darst, *A note on integration of vector-valued functions*, Proceedings of the American Mathematical Society **13** (1962), 858–863.
39. M.M. Day, *Normed linear spaces*, 3rd edition, Springer Verlag, Berlin-Heidelberg-New York, 1973.
40. M. De Wilde and M.T. De Wilde-Nibes, *Intégration par rapport des mesures à valeurs vectorielles*, Revue Roumaine Math. Pures et Appl. **13** (1968), 1529–1537.
41. J. Diestel, *Applications of weak compactness and bases to vector measures and vectorial integration*, Revue Roumaine Math. Pures et Appl. **18** (1973), 211–224.
42. J. Diestel and J.J. Uhl, *The Radon Nikodým theorem for Banach space-valued measures*, Rocky Mountain Journal of Mathematics **6** (1976), 1–46.
43. J. Diestel and J.J. Uhl, *Vector measures*, Mathematical surveys, vol. 15, American Mathematical Society, Providence, 1977.
44. J. Dieudonné, *Sur le théorème de Lebesgue-Nikodým*, Canadian Journal of Mathematics **3** (1951), 129–139.
45. J. Dieudonné, *Sur la convergence des suites des mesures de Radon*, Anais. Acad. Brasil. **23** (1951), 21–38.
46. N. Dinculeanu, *Sur la représentation intégrale des certaines opérations linéaires. I*, C.R. Acad. Sci. Paris **245** (1957), 1203–1205.
47. N. Dinculeanu, *Sur la représentation intégrale des certaines opérations linéaires. II*, Compositio Math. **14** (1959), 1–22.

48. N. Dinculeanu, *Sur la représentation intégrale des certaines opérations linéaires. III*, Proceedings of the American Mathematical Society **10** (1959), 59–68.
49. N. Dinculeanu and I. Klunánek, *On vector measures*, Proceedings of the London Mathematical Society **17** (1967), 505–512.
50. N. Dinculeanu and P.W. Lewis, *Regularity of Baire measures*, Proceedings of the American Mathematical Society **26** (1970), 92–94.
51. J.L. Doob, *Measure theory*, Graduate Texts in Mathematics, vol. 143, Springer Verlag, Heidelberg-Berlin-New York, 1994.
52. N. Dunford, *Integration in general analysis*, Transactions of the American Mathematical Society **37** (1935), 441–453.
53. N. Dunford, *Integration and linear operations*, Transactions of the American Mathematical Society **40** (1936), 474–494.
54. N. Dunford, *Integration of vector-valued functions*, Bulletin of the American Mathematical Society **43** (1937).
55. N. Dunford and J. Schwartz, *Linear operators, Part I, General Theory*, Wiley Classics Library Edition, John Wiley & Sons, New York-Chichester-Brisbane-Toronto-Singapore, 1988.
56. N. Dunford and J. Schwartz, *Linear operators, Part II, Spectral Theory*, Wiley Classics Library Edition, John Wiley & Sons, New York-Chichester-Brisbane-Toronto-Singapore, 1988.
57. N. Dunford and J. Schwartz, *Linear operators, Part III, Spectral Operators*, Wiley Classics Library Edition, John Wiley & Sons, New York-Chichester-Brisbane-Toronto-Singapore, 1988.
58. R.E. Edwards, *Vector-valued measures and bounded variation in Hilbert space*, Math. Scand. **3** (1955), 90–96.
59. R. Engelking, *General topology*, Monografie Matematyczne, Polish Scientific Publishers, Warszawa, 1977.
60. B. Faires, *On Vitali-Hahn-Saks type theorems*, Bulletin of the American Mathematical Society **80** (1974), 679–674.
61. H. Federer, *Geometric measure theory*, Die Grundlehren der mathematischen Wissenschaften, vol. 153, Springer, New York, 1969.
62. C. Foiaş and I. Singer, *Some remarks on the representation of linear operators in spaces of vector-valued continuous functions*, Révue Roumaine Math. Pures Appl. **5** (1960), 729–752.
63. B. Fuchssteiner and W. Lusky, *Convex cones*, vol. 56, North Holland Mathematical Studies, 1981.
64. A.D. Gadzhiev, *Positive linear operators in weighted spaces of functions in several variables*, (Russian) Izv. Akad Nauk. Azerbaidzan SSSR, ser. Fiz.-Tehn. Mat. Nauk **4** (1980), 32–37.
65. P. Gänssler, *Compactness and sequential compactness in spaces of measures*, Zeitschrift für Wahrscheinlichkeitstheorie und verwandte Gebiete **17** (1964), 381–410.
66. H.G. Garnier and J. Schmets, *Théorie de l'intégration par rapport à une mesure dans un espace linéaire à semi-normes*, Bull. Soc. Roy. Sci. Liège **33** (1971), 124–146.
67. G. Gierz and K. Keimel, *Halbstetige Funktionen und stetige Verbände*, Bremen, 1981, pp. 59–67.
68. G. Gierz, K.H. Hofmann, K. Keimel, J.D. Lawson, M.W. Mislove and D.S. Scott, *A compendium of continuous lattices*, Springer Verlag, Berlin-Heidelberg-New York, 1980.
69. G. Gierz, K.H. Hofmann, K. Keimel, J.D. Lawson, M.W. Mislove and D.S. Scott, *Continuous lattices and domains*, Encyclopedia of Mathematics and its Applications, vol. 93, Cambridge University Press, 2003.
70. H.H. Goldstine, *Linear functionals and integrals in abstract spaces*, Bulletin of the American Mathematical Society **47** (1941), 615–620.

71. R.K. Goodrich, *A Riesz representation theorem in the setting of locally convex spaces*, Transactions of the American Mathematical Society **131** (1968), 246–258.
72. R.K. Goodrich, *A Riesz representation theorem*, Proceedings of the American Mathematical Society **24** (1970), 629–636.
73. G.G. Gould, *Integration over vector-valued measures*, Proceedings of the London Mathematical Society (3) **15** (1965), 193–225.
74. M. Gowurin, *Über die Stieltjesche Integration abstrakter Funktionen*, Fund. Math. **27** (1936), 254–268.
75. W.H. Graves, *On the theory of vector measures*, American Mathematical Society, Providence, 1977.
76. W.H. Graves (Editor), *Proceedings on the conference on integration, topology, and geometry in linear spaces*, American Mathematical Society, Providence, 1980.
77. N.E. Gretsky, *Representation theorems on Banach function spaces*, vol. 84, Memoirs of the American Mathematical Society, 1968.
78. N.E. Gretsky and J.J. Uhl, *Bounded linear operators on Banach function spaces of vector-valued functions*, Transactions of the American Mathematical Society **167** (1972), 263–277.
79. A. Grothendieck, *Sur les applications linéaires faiblement compactes d'espaces du type $C(K)$*, Canadian Journal of Mathematics **5** (1953), 129–173.
80. H. Hahn and A. Rosenthal, *Set functions*, University of New Mexico Press, Albuquerque, 1948.
81. P.R. Halmos, *The range of a vector measure*, Bulletin of the American Mathematical Society **54** (1948), 416–421.
82. P.R. Halmos, *Introduction to Hilbert space and the theory of spectral multiplicity*, Second edition, Chelsea Publishing Company, New York, 1957.
83. P.R. Halmos, *Measure theory*, Springer Verlag, Heidelberg-Berlin-New York, 1974.
84. U. an der Heiden, *On the representation of linear functionals by finitely additive set functions*, Arch. Math. **30** (1978), 210–214.
85. E. Helley, *Über lineare Funktionaloperationen*, S.B.K. Akademie der Wissenschaften, Wien (IIa) **121** (1912), 265–297.
86. E. Hewitt and K.A. Ross, *Abstract harmonic analysis I*, Die Grundlehren der mathematischen Wissenschaften, vol. 115, Springer Verlag, Berlin-Göttingen-Heidelberg, 1963.
87. E. Hewitt and K. Stromberg, *Real and abstract analysis*, Springer, New York, 1965.
88. T.H. Hildebrandt, *On bounded functional operations*, Transactions of the American Mathematical Society **36** (1934), 868–875.
89. T.H. Hildebrandt, *Integration in abstract spaces*, Bulletin of the American Mathematical Society **59** (1953), 111–139.
90. J. Hoffman-Jørgensen, *Vector measures*, Math. Scand. **28** (1971), 5–32.
91. H. Ionescu Tulcea, *On measurability, pointwise convergence and compactness*, Bulletin of the American Mathematical Society **80** (1974), 231–236.
92. K. Jacobs, *Measure and integral*, Academic Press, New York, 1978.
93. S. Kakutani, *Concrete representation of abstract (L)-spaces and the mean ergodic theorem*, Annals of Mathematics (2) **42** (1941), 523–537.
94. S. Kakutani, *Concrete representation of abstract (M)-spaces. (A characterization of the space of continuous functions)*, Annals of Mathematics (2) **42** (1941), 994–1024.
95. N.J. Kalton, *Topologies on Riesz groups and applications to measure theory*, Proceedings of the London Mathematical Society (3) **28** (1974), 253–273.
96. L.V. Kantorovič, *Lineare halbgeordnete Räume*, Receuil. Math. **2** (1937), 121–168.
97. L.V. Kantorovič, *Linear operations in semi-ordered spaces*, Mat. Sb. (49) **7** (1940), 209–284.
98. L.V. Kantorovič, *Functional analysis in partially ordered spaces (Russian)*, Moscow-Leningrad, 1950.
99. S. Kaplan, *On weak compactness in the space of Radon measures*, Journal of Functional Analysis **5** (1970), 259–298.

100. K. Keimel and W. Roth, *Ordered cones and approximation*, Lecture Notes in Mathematics, vol. 1517, Springer Verlag, Heidelberg-Berlin-New York, 1992.
101. J.L. Kelley, *Banach spaces with the extension property*, Transactions of the American Mathematical Society **72** (1952), 323–326.
102. J. Kisynski, *Remark on strongly additive set functions*, Fund. Math. **63** (1968), 327–332.
103. I. Kluvánek, *On the theory of vector measures*, I, Mat. Fyz. Časopis. Sloven. Akad. Vied. **11** (1961), 173–191.
104. I. Kluvánek, *Intégrale vectorielle de Daniell*, I, Mat. Fyz. Časopis. Sloven. Akad. Vied. **15** (1965), 141–161.
105. I. Kluvánek, *Completion of vector measure spaces*, Revue Roumaine Math. Pures Appl. **12** (1967), 1483–1488.
106. H. König, *On the basic extension theorem in measure theory*, Math. Zeitschrift **190** (1985), 83–94.
107. E. Kreyszig, *Introductory functional analysis with applications*, John Wiley & Sons, New York-Chichester-Brisbane-Toronto-Singapore, 1978.
108. K. Kunisawa, *Some theorems on abstractly-valued functions in an abstract space*, Proc. Imp. Acad. Tokyo **16** (1940), 68–72.
109. T. Kuo, *Weak convergence of vector measures on F-spaces*, Mathematische Zeitschrift **143 (2)** (1975), 175–180.
110. I. Labuda, *Sur quelques généralisations des théorèmes de Nikodým et de Vitali-Hahn-Saks*, Bull. Acad. Polon. Sci. Sér. Sci. Math. Astronom. Phys. **20** (1972), 447–456.
111. I. Labuda, *Sur le théorème de Bartle-Dunford-Schwartz*, Bull. Acad. Polon. Sci. Sér. Sci. Math. Astronom. Phys. **20** (1972), 549–553.
112. D. Landers and L. Rogge, *Equicontinuity and convergence of measures*, Manuscripta Math. **5** (1971), 123–131.
113. D. Landers, *Connectedness properties of the range of vector and semi measures*, Manuscripta Math. **9** (1973), 105–112.
114. H. Lebesgue, *Intégrale Longueur, Aire*, Ann. di Mat. **7** (1902), 231–359.
115. H. Lebesgue, *Sur les intégrales singulières*, Ann. de Toulouse (3) **1** (1909), 25–117.
116. H. Lebesgue, *Leçons sur l'intégration et la recherche des fonctions primitives*, Gauthiers-Villars, Paris, 1904, Second Edition 1928.
117. J. Lembcke, *Reguläre Maße mit einer gegebenen Familie von Bildmaßen*, Sitz. Ber. Bayer. Akad. Wiss. Math.-Naturwiss. Kl. (1977), 61–115.
118. D.R. Lewis, *Integration with respect to vector measures*, Pacific Journal of Mathematics **33** (1970), 157–165.
119. P.W. Lewis, *Extension of operator-valued set functions with finite semivariation*, Proceedings of the American Mathematical Society **22** (1969), 563–569.
120. P.W. Lewis, *Some regularity conditions on measures with finite semivariation*, Revue Roumaine Math. Pures Appl. **15** (1970), 575–384.
121. P.W. Lewis, *Vector measures and topology*, Revue Roumaine Math. Pures Appl. **16** (1971), 1201–1209.
122. J. Lindenstrauss, *Extension of compact operators*, Memoirs of the American Mathematical Society **48** (1964).
123. J. Lindenstrauss and L. Tzafriri, *Classical Banach spaces*, Lecture Notes in Mathematics, vol. 338, Springer Verlag, Berlin-Heidelberg-New York, 1973.
124. G.G. Lorentz and D.G. Wertheim, *Representation of linear functionals on Köthe spaces*, Canadian Journal of Mathematics **5** (1953), 568–575.
125. W.A.J. Luxemburg and A.C. Zaanen, *Compactness of integral operators in Banach function spaces*, Mathematische Annalen **149** (1963), 150–180.
126. A. Lyapunov, *Sur les fonctions-vecteurs complètement additives*, (Russian) Izv. Akad. Nauk SSSR Ser. Mat. **4** (1940), 465–478.
127. A. Lyapunov, *Sur les fonctions-vecteurs complètement additives*, (Russian) Izv. Akad. Nauk SSSR Ser. Mat. **10** (1946), 277–279.

128. A. Markoff, *On mean values and exterior densities*, Math. Sb. **46** (1938), 165–191.
129. B. Maurey, *Intégration dans les espaces p-normés*, Ann. Scuola Norm. Sup. Pisa (4) **26** (1972), 911–931.
130. E.J. McShane, *Linear functionals on certain Banach spaces*, Proceedings of the American Mathematical Society **1** (1950), 402–408.
131. M. Métivier, *On strong measurability of Banach-valued functions*, Proceedings of the American Mathematical Society **21** (1969), 747–748.
132. P. Meyer-Nieberg, *Banach lattices*, Springer Verlag, Heidelberg-Berlin-New York, 1991.
133. K. Musial, *Absolute continuity of vector measures*, Colloq. Math. **27** (1973), 319–321.
134. L. Nachbin, *A theorem of the Hahn-Banach type for linear transformations*, Transactions of the American Mathematical Society **68** (1950), 28–46.
135. L. Nachbin, *Topology and order*, D. Van Nostrand Co., Princeton, 1965.
136. L. Nachbin, *Elements of approximation theory*, D. Van Nostrand Co., Princeton, 1967.
137. M. A. Naĭmark, *Normed algebras*; translated from the second Russian edition by Leo F. Born, Wolters-Nordhoff Publishing, Groningen, The Netherlands, 1972.
138. M. Nakamura, *Notes on Banach space. IX: Vitali-Hahn-Saks' theorem and K-spaces*, Tôhoku Math. Journal (2) **1** (1949), 106–108.
139. J. von Neumann, *Zur allgemeinen Theorie des Masses*, Fund. Math. **13** (1929), 73–116.
140. O.M. Nikodým, *Sur une généralisation des itégrales de M.J. Radon*, Fund. Math. **15** (1930), 131–179.
141. O.M. Nikodým, *Contibution à la théorie des fonctionelles linéaires en connection avec la théorie de la mesure des ensembles abstraits*, Mathematica (Cluj) **5** (1931), 130–141.
142. S. Ohba, *Extensions of vector measures*, Yokohama Math. Journal **21** (1973), 61–66.
143. W. Orlicz, *Absolute continuity of vector-valued finitely additive set functions. I*, Studia Math. **30** (1968), 121–133.
144. B.J. Pettis, *On integration in vector spaces*, Transactions of the American Mathematical Society **44** (1938), 277–304.
145. B.J. Pettis, *Linear functionals and completely additive set functions*, Duke Math. Journal **4** (1938), 552–565.
146. B.J. Pettis, *On the extension of measures*, Ann. Math. **54** (1951), 186–197.
147. J. Pfanzagl, *Convergent sequences of regular measures*, Manuscripta Math. **4** (1971), 91–98.
148. R.R. Phelps, *Lectures on Choquet's theorem*, Van Nostrand Math. Studies, vol. 7, Van Nostrand, Princeton, 1966.
149. R.S. Phillips, *Integration in a convex linear topological space*, Transactions of the American Mathematical Society **47** (1940), 114–145.
150. R.S. Phillips, *On linear transformations*, Transactions of the American Mathematical Society **48** (1940), 516–541.
151. A.G. Pinsker, *The space of convex sets of a locally convex space*, (Russian) Trudy Leningrad Inzh.-Ekon. In-ta **63** (1966), 13–17.
152. D. Pollard and F. Topsøe, *A unified approach to Riesz type representation theorems*, Studia Math. **54** (1975), 173–190.
153. C. Portenier, *Formes linéaires positives et mesures*, Sém. Choquet, 10e année (1970/71), Comm. no. 6.
154. G.B. Price, *The theory of integration*, Transactions of the American Mathematical Society **47** (1940), 1–50.
155. J.B. Prolla, *Approximation of vector-valued functions*, Mathematical Studies, vol. 25, North Holland, 1977.
156. D. Przeworska-Rolewicz and S. Rolewicz, *On integrals with values in a complete linear metric space*, Studia Math. **26** (1966), 121–131.
157. M.G. Rabinovich, *An embedding theorem for spaces of convex sets*, (Russian) Sibirsk. Mat. Zh. **8** (1967), 367–383.

158. J. Radon, *Theorie und Anwendungen der absolut additiven Mengenfunktionen*, Sitzungsber. Österr. Akad. Wiss. Math.-Naturwiss. Kl. **122** (1913), 1295–1438.
159. J. Radon, *Über lineare Funktionaltransformationen und Funktionalgleichungen*, Sitzungsber. Österr. Akad. Wiss. Math.-Naturwiss. Kl. **128** (1919), 1083–1121.
160. M.M. Rao, *Remarks on a Radon-Nikodým theorem for vector measures*, Vector and Operator Valued Measures and Applications (Proc. Sympos., Snowbird Resort, Alta, Utah, 1972) (1973), Academic Press, New York, 303–317.
161. C.E. Rickart, *Integration in a convex linear topological space*, Transactions of the American Mathematical Society **52** (1942), 498–521.
162. B. Riemann, *Über die Darstellbarkeit einer Funktion durch eine trigonometrische Reihe*, In: Gesammelte mathematische Werke, Teubner, Leipzig, 1876, pp. 213–250.
163. F. Riesz, *Sur une espèce de géométrie analytique des systèmes de fonctions sommables*, C. R. Acad. Sci. Paris **144** (1907), 1409–1411.
164. F. Riesz, *Sur les opérations fonctionelles linéaire*, C. R. Acad. Sci. Paris **149** (1909), 974–977.
165. F. Riesz, *Untersuchungen über Systeme integrierbarer Funktionen*, Mathematische Annalen **69** (1910), 449–497.
166. U. Rønnow, *On integral representation of vector-valued measures*, Math. Scand. **21** (1967), 45–53.
167. W. Roth, *Integral type linear functionals on ordered cones*, Transactions of the American Mathematical Society **343** (1996), 5065–5085.
168. W. Roth, *Real and complex linear extensions for locally convex cones*, Journal of Functional Analysis **151** (1997), no. 2, 437–454.
169. W. Roth, *Locally convex cones as generalizations of locally convex vector spaces*, Far East Journal of Mathematical Sciences, Special Volume, Part II (1998), 215–245.
170. W. Roth, *A uniform boundedness theorem for locally convex cones*, Proceedings of the American Mathematical Society **126** (1998), no. 7, 1973–1982.
171. W. Roth, *A Riesz representation theorem for cone-valued functions*, Abstract and Applied Analysis **4** (1999), no. 4, 209–229.
172. W. Roth, *Hahn-Banach type theorems for locally convex cones*, Journal of the Australian Math. Soc. (Series A) **68** (2000), no. 1, 104–125.
173. W. Roth, *Inner products on ordered cones*, New Zealand Journal of Mathematics **30** (2001), 157–175.
174. W. Roth, *Integral representations for continuous linear functionals in operator-initiated topologies*, Positivity **6** (2002), no. 2, 115–127.
175. W. Roth, *Separation properties for locally convex cones*, Journal of Convex Analysis **9** (2002), no. 1, 301–307.
176. W. Roth, *Boundedness and connectedness components for locally convex cones*, New Zealand Journal of Mathematics **34** (2005), 143–158.
177. W. Roth, *Locally convex lattice cones*, to appear in Journal of Convex Analysis.
178. H.L. Royden, *Real analysis*, Macmillan Publishing Company, New York/Collier Macmillan Publishing, London, 1980.
179. W. Rudin, *Real and complex analysis*, McGraw-Hill Inc., New York, 1974.
180. V.I. Rybakov, *the Radon-Nikodým theorem and integral representation for vector measures*, Dokl. Akad. Nauk SSSR **180** (1968), 282–285.
181. V.I. Rybakov, *Theorem of Bartle, Dunford and Schwartz on vector-valued measures*, Mat. Zametik **7** (1970), 147–151.
182. S. Saks, *Theory of the integral*, Second Edition, vol. 7, Monografje Matematycne, Warsaw, 1937.
183. S. Saks, *Integration in abstract metric spaces*, Duke Mathematical Journal **4** (1938), 408–411.
184. H.H. Schäfer, *Banach lattices and positive operators*, Springer Verlag, Heidelberg-Berlin-New York, 1974.
185. H.H. Schäfer, *Topological vector spaces*, Springer Verlag, Heidelberg-Berlin-New York, 1980.

186. L. Schwartz, *Mesures de Radon sur des espaces topologiques arbitraires*, Cours de 3ème cycle, Institut H. Poincaré, Paris (1964/65).
187. G. Schwarz, *Variations on vector measure*, Pacific Journal of Mathematics **23** (1967), 373–375.
188. D.S. Scott, *Continuous lattices*, In F.W. Lawvere, editor, Toposes, algebraic geometry and logic, Dalhouse University, Halifax, Nova Scotia, January 16–19, 1971, in Lecture Notes in Mathematics, vol. 274, Springer-Verlag, Berlin-Heidelberg-New York, 1972, pp. 97–136.
189. I.A. Segal and R.A. Kunze, *Integrals and operators*, Die Grundlehren der mathematischen Wissenschaften, vol. 228, Springer Verlag, Heidelberg-Berlin-New York, 1978.
190. S. Shuchat, *Integral representation theorems in topological vector spaces*, Transactions of the American Mathematical Society **172** (1972), 373–397.
191. I. Singer, *Sur les applications linéaires intégrales des espaces de fonctions continues à valeurs vectorielles*, Acta Math. Acad. Sci. Hungar. **11** (1960), 3–13.
192. M. Sion, *A theory of semi group-valued measures*, Lecture Notes in Mathematics, vol. 355, Springer Verlag, Heidelberg-Berlin-New York, 1973.
193. K. Swong, *A representation theory of continuous linear maps*, Math. Ann. **155** (1964), 270–291.
194. M. Talagrand, *Pettis integral and measure theory*, Memoirs of the American Mathematical Society **307** (1984).
195. F. Topsøe, *Topology and measure*, Lecture Notes in Mathematics, vol. 133, Springer Verlag, Heidelberg-Berlin-New York, 1970.
196. F. Topsøe, *Further results on integral representation*, Studia Math. **55** (1976), 239–245.
197. D. Vogt, *Integrationstheorie in p-normierten Räumen*, Math. Ann. **173** (1967), 219–232.
198. S. Willard, *General topology*, Addison-Wesley Publishing Company, Reading, Massachusetts, 1970.

# Index

# Lecture Notes in Mathematics

For information about earlier volumes
please contact your bookseller or Springer
LNM Online archive: springerlink.com

Vol. 1782: C.-H. Chu, A. T.-M. Lau, Harmonic Functions on Groups and Fourier Algebras (2002)
Vol. 1783: L. Grüne, Asymptotic Behavior of Dynamical and Control Systems under Perturbation and Discretization (2002)
Vol. 1784: L. H. Eliasson, S. B. Kuksin, S. Marmi, J.-C. Yoccoz, Dynamical Systems and Small Divisors. Cetraro, Italy 1998. Editors: S. Marmi, J.-C. Yoccoz (2002)
Vol. 1785: J. Arias de Reyna, Pointwise Convergence of Fourier Series (2002)
Vol. 1786: S. D. Cutkosky, Monomialization of Morphisms from 3-Folds to Surfaces (2002)
Vol. 1787: S. Caenepeel, G. Militaru, S. Zhu, Frobenius and Separable Functors for Generalized Module Categories and Nonlinear Equations (2002)
Vol. 1788: A. Vasil'ev, Moduli of Families of Curves for Conformal and Quasiconformal Mappings (2002)
Vol. 1789: Y. Sommerhäuser, Yetter-Drinfel'd Hopf algebras over groups of prime order (2002)
Vol. 1790: X. Zhan, Matrix Inequalities (2002)
Vol. 1791: M. Knebusch, D. Zhang, Manis Valuations and Prüfer Extensions I: A new Chapter in Commutative Algebra (2002)
Vol. 1792: D. D. Ang, R. Gorenflo, V. K. Le, D. D. Trong, Moment Theory and Some Inverse Problems in Potential Theory and Heat Conduction (2002)
Vol. 1793: J. Cortés Monforte, Geometric, Control and Numerical Aspects of Nonholonomic Systems (2002)
Vol. 1794: N. Pytheas Fogg, Substitution in Dynamics, Arithmetics and Combinatorics. Editors: V. Berthé, S. Ferenczi, C. Mauduit, A. Siegel (2002)
Vol. 1795: H. Li, Filtered-Graded Transfer in Using Noncommutative Gröbner Bases (2002)
Vol. 1796: J.M. Melenk, hp-Finite Element Methods for Singular Perturbations (2002)
Vol. 1797: B. Schmidt, Characters and Cyclotomic Fields in Finite Geometry (2002)
Vol. 1798: W.M. Oliva, Geometric Mechanics (2002)
Vol. 1799: H. Pajot, Analytic Capacity, Rectifiability, Menger Curvature and the Cauchy Integral (2002)
Vol. 1800: O. Gabber, L. Ramero, Almost Ring Theory (2003)
Vol. 1801: J. Azéma, M. Émery, M. Ledoux, M. Yor (Eds.), Séminaire de Probabilités XXXVI (2003)
Vol. 1802: V. Capasso, E. Merzbach, B. G. Ivanoff, M. Dozzi, R. Dalang, T. Mountford, Topics in Spatial Stochastic Processes. Martina Franca, Italy 2001. Editor: E. Merzbach (2003)
Vol. 1803: G. Dolzmann, Variational Methods for Crystalline Microstructure – Analysis and Computation (2003)
Vol. 1804: I. Cherednik, Ya. Markov, R. Howe, G. Lusztig, Iwahori-Hecke Algebras and their Representation Theory. Martina Franca, Italy 1999. Editors: V. Baldoni, D. Barbasch (2003)
Vol. 1805: F. Cao, Geometric Curve Evolution and Image Processing (2003)
Vol. 1806: H. Broer, I. Hoveijn. G. Lunther, G. Vegter, Bifurcations in Hamiltonian Systems. Computing Singularities by Gröbner Bases (2003)
Vol. 1807: V. D. Milman, G. Schechtman (Eds.), Geometric Aspects of Functional Analysis. Israel Seminar 2000-2002 (2003)
Vol. 1808: W. Schindler, Measures with Symmetry Properties (2003)
Vol. 1809: O. Steinbach, Stability Estimates for Hybrid Coupled Domain Decomposition Methods (2003)
Vol. 1810: J. Wengenroth, Derived Functors in Functional Analysis (2003)
Vol. 1811: J. Stevens, Deformations of Singularities (2003)
Vol. 1812: L. Ambrosio, K. Deckelnick, G. Dziuk, M. Mimura, V. A. Solonnikov, H. M. Soner, Mathematical Aspects of Evolving Interfaces. Madeira, Funchal, Portugal 2000. Editors: P. Colli, J. F. Rodrigues (2003)
Vol. 1813: L. Ambrosio, L. A. Caffarelli, Y. Brenier, G. Buttazzo, C. Villani, Optimal Transportation and its Applications. Martina Franca, Italy 2001. Editors: L. A. Caffarelli, S. Salsa (2003)
Vol. 1814: P. Bank, F. Baudoin, H. Föllmer, L.C.G. Rogers, M. Soner, N. Touzi, Paris-Princeton Lectures on Mathematical Finance 2002 (2003)
Vol. 1815: A. M. Vershik (Ed.), Asymptotic Combinatorics with Applications to Mathematical Physics. St. Petersburg, Russia 2001 (2003)
Vol. 1816: S. Albeverio, W. Schachermayer, M. Talagrand, Lectures on Probability Theory and Statistics. Ecole d'Eté de Probabilités de Saint-Flour XXX-2000. Editor: P. Bernard (2003)
Vol. 1817: E. Koelink, W. Van Assche (Eds.), Orthogonal Polynomials and Special Functions. Leuven 2002 (2003)
Vol. 1818: M. Bildhauer, Convex Variational Problems with Linear, nearly Linear and/or Anisotropic Growth Conditions (2003)
Vol. 1819: D. Masser, Yu. V. Nesterenko, H. P. Schlickewei, W. M. Schmidt, M. Waldschmidt, Diophantine Approximation. Cetraro, Italy 2000. Editors: F. Amoroso, U. Zannier (2003)
Vol. 1820: F. Hiai, H. Kosaki, Means of Hilbert Space Operators (2003)
Vol. 1821: S. Teufel, Adiabatic Perturbation Theory in Quantum Dynamics (2003)
Vol. 1822: S.-N. Chow, R. Conti, R. Johnson, J. Mallet-Paret, R. Nussbaum, Dynamical Systems. Cetraro, Italy 2000. Editors: J. W. Macki, P. Zecca (2003)
Vol. 1823: A. M. Anile, W. Allegretto, C. Ringhofer, Mathematical Problems in Semiconductor Physics. Cetraro, Italy 1998. Editor: A. M. Anile (2003)
Vol. 1824: J. A. Navarro González, J. B. Sancho de Salas, $\mathcal{C}^\infty$ – Differentiable Spaces (2003)

Vol. 1825: J. H. Bramble, A. Cohen, W. Dahmen, Multiscale Problems and Methods in Numerical Simulations, Martina Franca, Italy 2001. Editor: C. Canuto (2003)

Vol. 1826: K. Dohmen, Improved Bonferroni Inequalities via Abstract Tubes. Inequalities and Identities of Inclusion-Exclusion Type. VIII, 113 p, 2003.

Vol. 1827: K. M. Pilgrim, Combinations of Complex Dynamical Systems. IX, 118 p, 2003.

Vol. 1828: D. J. Green, Gröbner Bases and the Computation of Group Cohomology. XII, 138 p, 2003.

Vol. 1829: E. Altman, B. Gaujal, A. Hordijk, Discrete-Event Control of Stochastic Networks: Multimodularity and Regularity. XIV, 313 p, 2003.

Vol. 1830: M. I. Gil', Operator Functions and Localization of Spectra. XIV, 256 p, 2003.

Vol. 1831: A. Connes, J. Cuntz, E. Guentner, N. Higson, J. E. Kaminker, Noncommutative Geometry, Martina Franca, Italy 2002. Editors: S. Doplicher, L. Longo (2004)

Vol. 1832: J. Azéma, M. Émery, M. Ledoux, M. Yor (Eds.), Séminaire de Probabilités XXXVII (2003)

Vol. 1833: D.-Q. Jiang, M. Qian, M.-P. Qian, Mathematical Theory of Nonequilibrium Steady States. On the Frontier of Probability and Dynamical Systems. IX, 280 p, 2004.

Vol. 1834: Yo. Yomdin, G. Comte, Tame Geometry with Application in Smooth Analysis. VIII, 186 p, 2004.

Vol. 1835: O.T. Izhboldin, B. Kahn, N.A. Karpenko, A. Vishik, Geometric Methods in the Algebraic Theory of Quadratic Forms. Summer School, Lens, 2000. Editor: J.-P. Tignol (2004)

Vol. 1836: C. Năstăsescu, F. Van Oystaeyen, Methods of Graded Rings. XIII, 304 p, 2004.

Vol. 1837: S. Tavaré, O. Zeitouni, Lectures on Probability Theory and Statistics. Ecole d'Eté de Probabilités de Saint-Flour XXXI-2001. Editor: J. Picard (2004)

Vol. 1838: A.J. Ganesh, N.W. O'Connell, D.J. Wischik, Big Queues. XII, 254 p, 2004.

Vol. 1839: R. Gohm, Noncommutative Stationary Processes. VIII, 170 p, 2004.

Vol. 1840: B. Tsirelson, W. Werner, Lectures on Probability Theory and Statistics. Ecole d'Eté de Probabilités de Saint-Flour XXXII-2002. Editor: J. Picard (2004)

Vol. 1841: W. Reichel, Uniqueness Theorems for Variational Problems by the Method of Transformation Groups (2004)

Vol. 1842: T. Johnsen, A. L. Knutsen, $K_3$ Projective Models in Scrolls (2004)

Vol. 1843: B. Jefferies, Spectral Properties of Noncommuting Operators (2004)

Vol. 1844: K.F. Siburg, The Principle of Least Action in Geometry and Dynamics (2004)

Vol. 1845: Min Ho Lee, Mixed Automorphic Forms, Torus Bundles, and Jacobi Forms (2004)

Vol. 1846: H. Ammari, H. Kang, Reconstruction of Small Inhomogeneities from Boundary Measurements (2004)

Vol. 1847: T.R. Bielecki, T. Björk, M. Jeanblanc, M. Rutkowski, J.A. Scheinkman, W. Xiong, Paris-Princeton Lectures on Mathematical Finance 2003 (2004)

Vol. 1848: M. Abate, J. E. Fornaess, X. Huang, J. P. Rosay, A. Tumanov, Real Methods in Complex and CR Geometry, Martina Franca, Italy 2002. Editors: D. Zaitsev, G. Zampieri (2004)

Vol. 1849: Martin L. Brown, Heegner Modules and Elliptic Curves (2004)

Vol. 1850: V. D. Milman, G. Schechtman (Eds.), Geometric Aspects of Functional Analysis. Israel Seminar 2002-2003 (2004)

Vol. 1851: O. Catoni, Statistical Learning Theory and Stochastic Optimization (2004)

Vol. 1852: A.S. Kechris, B.D. Miller, Topics in Orbit Equivalence (2004)

Vol. 1853: Ch. Favre, M. Jonsson, The Valuative Tree (2004)

Vol. 1854: O. Saeki, Topology of Singular Fibers of Differential Maps (2004)

Vol. 1855: G. Da Prato, P.C. Kunstmann, I. Lasiecka, A. Lunardi, R. Schnaubelt, L. Weis, Functional Analytic Methods for Evolution Equations. Editors: M. Iannelli, R. Nagel, S. Piazzera (2004)

Vol. 1856: K. Back, T.R. Bielecki, C. Hipp, S. Peng, W. Schachermayer, Stochastic Methods in Finance, Bressanone/Brixen, Italy, 2003. Editors: M. Fritelli, W. Runggaldier (2004)

Vol. 1857: M. Émery, M. Ledoux, M. Yor (Eds.), Séminaire de Probabilités XXXVIII (2005)

Vol. 1858: A.S. Cherny, H.-J. Engelbert, Singular Stochastic Differential Equations (2005)

Vol. 1859: E. Letellier, Fourier Transforms of Invariant Functions on Finite Reductive Lie Algebras (2005)

Vol. 1860: A. Borisyuk, G.B. Ermentrout, A. Friedman, D. Terman, Tutorials in Mathematical Biosciences I. Mathematical Neurosciences (2005)

Vol. 1861: G. Benettin, J. Henrard, S. Kuksin, Hamiltonian Dynamics – Theory and Applications, Cetraro, Italy, 1999. Editor: A. Giorgilli (2005)

Vol. 1862: B. Helffer, F. Nier, Hypoelliptic Estimates and Spectral Theory for Fokker-Planck Operators and Witten Laplacians (2005)

Vol. 1863: H. Führ, Abstract Harmonic Analysis of Continuous Wavelet Transforms (2005)

Vol. 1864: K. Efstathiou, Metamorphoses of Hamiltonian Systems with Symmetries (2005)

Vol. 1865: D. Applebaum, B.V. R. Bhat, J. Kustermans, J. M. Lindsay, Quantum Independent Increment Processes I. From Classical Probability to Quantum Stochastic Calculus. Editors: M. Schürmann, U. Franz (2005)

Vol. 1866: O.E. Barndorff-Nielsen, U. Franz, R. Gohm, B. Kümmerer, S. Thorbjønsen, Quantum Independent Increment Processes II. Structure of Quantum Lévy Processes, Classical Probability, and Physics. Editors: M. Schürmann, U. Franz, (2005)

Vol. 1867: J. Sneyd (Ed.), Tutorials in Mathematical Biosciences II. Mathematical Modeling of Calcium Dynamics and Signal Transduction. (2005)

Vol. 1868: J. Jorgenson, S. Lang, Pos$_n$(R) and Eisenstein Series. (2005)

Vol. 1869: A. Dembo, T. Funaki, Lectures on Probability Theory and Statistics. Ecole d'Eté de Probabilités de Saint-Flour XXXIII-2003. Editor: J. Picard (2005)

Vol. 1870: V.I. Gurariy, W. Lusky, Geometry of Müntz Spaces and Related Questions. (2005)

Vol. 1871: P. Constantin, G. Gallavotti, A.V. Kazhikhov, Y. Meyer, S. Ukai, Mathematical Foundation of Turbulent Viscous Flows, Martina Franca, Italy, 2003. Editors: M. Cannone, T. Miyakawa (2006)

Vol. 1872: A. Friedman (Ed.), Tutorials in Mathematical Biosciences III. Cell Cycle, Proliferation, and Cancer (2006)

Vol. 1873: R. Mansuy, M. Yor, Random Times and Enlargements of Filtrations in a Brownian Setting (2006)

Vol. 1874: M. Yor, M. Émery (Eds.), In Memoriam Paul-André Meyer - Séminaire de Probabilités XXXIX (2006)
Vol. 1875: J. Pitman, Combinatorial Stochastic Processes. Ecole d'Eté de Probabilités de Saint-Flour XXXII-2002. Editor: J. Picard (2006)
Vol. 1876: H. Herrlich, Axiom of Choice (2006)
Vol. 1877: J. Steuding, Value Distributions of *L*-Functions (2007)
Vol. 1878: R. Cerf, The Wulff Crystal in Ising and Percolation Models, Ecole d'Eté de Probabilités de Saint-Flour XXXIV-2004. Editor: Jean Picard (2006)
Vol. 1879: G. Slade, The Lace Expansion and its Applications, Ecole d'Eté de Probabilités de Saint-Flour XXXIV-2004. Editor: Jean Picard (2006)
Vol. 1880: S. Attal, A. Joye, C.-A. Pillet, Open Quantum Systems I, The Hamiltonian Approach (2006)
Vol. 1881: S. Attal, A. Joye, C.-A. Pillet, Open Quantum Systems II, The Markovian Approach (2006)
Vol. 1882: S. Attal, A. Joye, C.-A. Pillet, Open Quantum Systems III, Recent Developments (2006)
Vol. 1883: W. Van Assche, F. Marcellàn (Eds.), Orthogonal Polynomials and Special Functions, Computation and Application (2006)
Vol. 1884: N. Hayashi, E.I. Kaikina, P.I. Naumkin, I.A. Shishmarev, Asymptotics for Dissipative Nonlinear Equations (2006)
Vol. 1885: A. Telcs, The Art of Random Walks (2006)
Vol. 1886: S. Takamura, Splitting Deformations of Degenerations of Complex Curves (2006)
Vol. 1887: K. Habermann, L. Habermann, Introduction to Symplectic Dirac Operators (2006)
Vol. 1888: J. van der Hoeven, Transseries and Real Differential Algebra (2006)
Vol. 1889: G. Osipenko, Dynamical Systems, Graphs, and Algorithms (2006)
Vol. 1890: M. Bunge, J. Funk, Singular Coverings of Toposes (2006)
Vol. 1891: J.B. Friedlander, D.R. Heath-Brown, H. Iwaniec, J. Kaczorowski, Analytic Number Theory, Cetraro, Italy, 2002. Editors: A. Perelli, C. Viola (2006)
Vol. 1892: A. Baddeley, I. Bárány, R. Schneider, W. Weil, Stochastic Geometry, Martina Franca, Italy, 2004. Editor: W. Weil (2007)
Vol. 1893: H. Hanßmann, Local and Semi-Local Bifurcations in Hamiltonian Dynamical Systems, Results and Examples (2007)
Vol. 1894: C.W. Groetsch, Stable Approximate Evaluation of Unbounded Operators (2007)
Vol. 1895: L. Molnár, Selected Preserver Problems on Algebraic Structures of Linear Operators and on Function Spaces (2007)
Vol. 1896: P. Massart, Concentration Inequalities and Model Selection, Ecole d'Été de Probabilités de Saint-Flour XXXIII-2003. Editor: J. Picard (2007)
Vol. 1897: R. Doney, Fluctuation Theory for Lévy Processes, Ecole d'Été de Probabilités de Saint-Flour XXXV-2005. Editor: J. Picard (2007)
Vol. 1898: H.R. Beyer, Beyond Partial Differential Equations, On linear and Quasi-Linear Abstract Hyperbolic Evolution Equations (2007)
Vol. 1899: Séminaire de Probabilités XL. Editors: C. Donati-Martin, M. Émery, A. Rouault, C. Stricker (2007)
Vol. 1900: E. Bolthausen, A. Bovier (Eds.), Spin Glasses (2007)
Vol. 1901: O. Wittenberg, Intersections de deux quadriques et pinceaux de courbes de genre 1, Intersections of Two Quadrics and Pencils of Curves of Genus 1 (2007)
Vol. 1902: A. Isaev, Lectures on the Automorphism Groups of Kobayashi-Hyperbolic Manifolds (2007)
Vol. 1903: G. Kresin, V. Maz'ya, Sharp Real-Part Theorems (2007)
Vol. 1904: P. Giesl, Construction of Global Lyapunov Functions Using Radial Basis Functions (2007)
Vol. 1905: C. Prévôt, M. Röckner, A Concise Course on Stochastic Partial Differential Equations (2007)
Vol. 1906: T. Schuster, The Method of Approximate Inverse: Theory and Applications (2007)
Vol. 1907: M. Rasmussen, Attractivity and Bifurcation for Nonautonomous Dynamical Systems (2007)
Vol. 1908: T.J. Lyons, M. Caruana, T. Lévy, Differential Equations Driven by Rough Paths, Ecole d'Été de Probabilités de Saint-Flour XXXIV-2004 (2007)
Vol. 1909: H. Akiyoshi, M. Sakuma, M. Wada, Y. Yamashita, Punctured Torus Groups and 2-Bridge Knot Groups (I) (2007)
Vol. 1910: V.D. Milman, G. Schechtman (Eds.), Geometric Aspects of Functional Analysis. Israel Seminar 2004-2005 (2007)
Vol. 1911: A. Bressan, D. Serre, M. Williams, K. Zumbrun, Hyperbolic Systems of Balance Laws. Cetraro, Italy 2003. Editor: P. Marcati (2007)
Vol. 1912: V. Berinde, Iterative Approximation of Fixed Points (2007)
Vol. 1913: J.E. Marsden, G. Misiołek, J.-P. Ortega, M. Perlmutter, T.S. Ratiu, Hamiltonian Reduction by Stages (2007)
Vol. 1914: G. Kutyniok, Affine Density in Wavelet Analysis (2007)
Vol. 1915: T. Bıyıkoğlu, J. Leydold, P.F. Stadler, Laplacian Eigenvectors of Graphs. Perron-Frobenius and Faber-Krahn Type Theorems (2007)
Vol. 1916: C. Villani, F. Rezakhanlou, Entropy Methods for the Boltzmann Equation. Editors: F. Golse, S. Olla (2008)
Vol. 1917: I. Veselić, Existence and Regularity Properties of the Integrated Density of States of Random Schrödinger (2008)
Vol. 1918: B. Roberts, R. Schmidt, Local Newforms for GSp(4) (2007)
Vol. 1919: R.A. Carmona, I. Ekeland, A. Kohatsu-Higa, J.-M. Lasry, P.-L. Lions, H. Pham, E. Taflin, Paris-Princeton Lectures on Mathematical Finance 2004. Editors: R.A. Carmona, E. Çinlar, I. Ekeland, E. Jouini, J.A. Scheinkman, N. Touzi (2007)
Vol. 1920: S.N. Evans, Probability and Real Trees. Ecole d'Été de Probabilités de Saint-Flour XXXV-2005 (2008)
Vol. 1921: J.P. Tian, Evolution Algebras and their Applications (2008)
Vol. 1922: A. Friedman (Ed.), Tutorials in Mathematical BioSciences IV. Evolution and Ecology (2008)
Vol. 1923: J.P.N. Bishwal, Parameter Estimation in Stochastic Differential Equations (2008)
Vol. 1924: M. Wilson, Littlewood-Paley Theory and Exponential-Square Integrability (2008)
Vol. 1925: M. du Sautoy, L. Woodward, Zeta Functions of Groups and Rings (2008)
Vol. 1926: L. Barreira, V. Claudia, Stability of Nonautonomous Differential Equations (2008)
Vol. 1927: L. Ambrosio, L. Caffarelli, M.G. Crandall, L.C. Evans, N. Fusco, Calculus of Variations and Non-Linear Partial Differential Equations. Cetraro, Italy 2005. Editors: B. Dacorogna, P. Marcellini (2008)

Vol. 1928: J. Jonsson, Simplicial Complexes of Graphs (2008)
Vol. 1929: Y. Mishura, Stochastic Calculus for Fractional Brownian Motion and Related Processes (2008)
Vol. 1930: J.M. Urbano, The Method of Intrinsic Scaling. A Systematic Approach to Regularity for Degenerate and Singular PDEs (2008)
Vol. 1931: M. Cowling, E. Frenkel, M. Kashiwara, A. Valette, D.A. Vogan, Jr., N.R. Wallach, Representation Theory and Complex Analysis. Venice, Italy 2004. Editors: E.C. Tarabusi, A. D'Agnolo, M. Picardello (2008)
Vol. 1932: A.A. Agrachev, A.S. Morse, E.D. Sontag, H.J. Sussmann, V.I. Utkin, Nonlinear and Optimal Control Theory. Cetraro, Italy 2004. Editors: P. Nistri, G. Stefani (2008)
Vol. 1933: M. Petkovic, Point Estimation of Root Finding Methods (2008)
Vol. 1934: C. Donati-Martin, M. Émery, A. Rouault, C. Stricker (Eds.), Séminaire de Probabilités XLI (2008)
Vol. 1935: A. Unterberger, Alternative Pseudodifferential Analysis (2008)
Vol. 1936: P. Magal, S. Ruan (Eds.), Structured Population Models in Biology and Epidemiology (2008)
Vol. 1937: G. Capriz, P. Giovine, P.M. Mariano (Eds.), Mathematical Models of Granular Matter (2008)
Vol. 1938: D. Auroux, F. Catanese, M. Manetti, P. Seidel, B. Siebert, I. Smith, G. Tian, Symplectic 4-Manifolds and Algebraic Surfaces. Cetraro, Italy 2003. Editors: F. Catanese, G. Tian (2008)
Vol. 1939: D. Boffi, F. Brezzi, L. Demkowicz, R.G. Durán, R.S. Falk, M. Fortin, Mixed Finite Elements, Compatibility Conditions, and Applications. Cetraro, Italy 2006. Editors: D. Boffi, L. Gastaldi (2008)
Vol. 1940: J. Banasiak, V. Capasso, M.A.J. Chaplain, M. Lachowicz, J. Miękisz, Multiscale Problems in the Life Sciences. From Microscopic to Macroscopic. Będlewo, Poland 2006. Editors: V. Capasso, M. Lachowicz (2008)
Vol. 1941: S.M.J. Haran, Arithmetical Investigations. Representation Theory, Orthogonal Polynomials, and Quantum Interpolations (2008)
Vol. 1942: S. Albeverio, F. Flandoli, Y.G. Sinai, SPDE in Hydrodynamic. Recent Progress and Prospects. Cetraro, Italy 2005. Editors: G. Da Prato, M. Röckner (2008)
Vol. 1943: L.L. Bonilla (Ed.), Inverse Problems and Imaging. Martina Franca, Italy 2002 (2008)
Vol. 1944: A. Di Bartolo, G. Falcone, P. Plaumann, K. Strambach, Algebraic Groups and Lie Groups with Few Factors (2008)
Vol. 1945: F. Brauer, P. van den Driessche, J. Wu (Eds.), Mathematical Epidemiology (2008)
Vol. 1946: G. Allaire, A. Arnold, P. Degond, T.Y. Hou, Quantum Transport. Modelling, Analysis and Asymptotics. Cetraro, Italy 2006. Editors: N.B. Abdallah, G. Frosali (2008)
Vol. 1947: D. Abramovich, M. Mariño, M. Thaddeus, R. Vakil, Enumerative Invariants in Algebraic Geometry and String Theory. Cetraro, Italy 2005. Editors: K. Behrend, M. Manetti (2008)
Vol. 1948: F. Cao, J-L. Lisani, J-M. Morel, P. Musé, F. Sur, A Theory of Shape Identification (2008)
Vol. 1949: H.G. Feichtinger, B. Helffer, M.P. Lamoureux, N. Lerner, J. Toft, Pseudo-Differential Operators. Quantization and Signals. Cetraro, Italy 2006. Editors: L. Rodino, M.W. Wong (2008)
Vol. 1950: M. Bramson, Stability of Queueing Networks, Ecole d'Eté de Probabilités de Saint-Flour XXXVI-2006 (2008)
Vol. 1951: A. Moltó, J. Orihuela, S. Troyanski, M. Valdivia, A Non Linear Transfer Technique for Renorming (2009)
Vol. 1952: R. Mikhailov, I.B.S. Passi, Lower Central and Dimension Series of Groups (2009)
Vol. 1953: K. Arwini, C.T.J. Dodson, Information Geometry (2008)
Vol. 1954: P. Biane, L. Bouten, F. Cipriani, N. Konno, N. Privault, Q. Xu, Quantum Potential Theory. Editors: U. Franz, M. Schuermann (2008)
Vol. 1955: M. Bernot, V. Caselles, J.-M. Morel, Optimal Transportation Networks (2009)
Vol. 1956: C.H. Chu, Matrix Convolution Operators on Groups (2008)
Vol. 1957: A. Guionnet, On Random Matrices: Macroscopic Asymptotics, Ecole d'Eté de Probabilités de Saint-Flour XXXVI-2006 (2009)
Vol. 1958: M.C. Olsson, Compactifying Moduli Spaces for Abelian Varieties (2008)
Vol. 1959: Y. Nakkajima, A. Shiho, Weight Filtrations on Log Crystalline Cohomologies of Families of Open Smooth Varieties (2008)
Vol. 1960: J. Lipman, M. Hashimoto, Foundations of Grothendieck Duality for Diagrams of Schemes (2009)
Vol. 1961: G. Buttazzo, A. Pratelli, S. Solimini, E. Stepanov, Optimal Urban Networks via Mass Transportation (2009)
Vol. 1962: R. Dalang, D. Khoshnevisan, C. Mueller, D. Nualart, Y. Xiao, A Minicourse on Stochastic Partial Differential Equations (2009)
Vol. 1963: W. Siegert, Local Lyapunov Exponents (2009)
Vol. 1964: W. Roth, Operator-Valued Measures and Integrals for Cone-Valued Functions (2009)
Vol. 1965: C. Chidume, Geometric Properties of Banach Spaces and Nonlinear Iterations (2009)
Vol. 1966: D. Deng, Y. Han, Harmonic Analysis on Spaces of Homogeneous Type (2009)

## Recent Reprints and New Editions

Vol. 1702: J. Ma, J. Yong, Forward-Backward Stochastic Differential Equations and their Applications. 1999 – Corr. 3rd printing (2007)
Vol. 830: J.A. Green, Polynomial Representations of $GL_n$, with an Appendix on Schensted Correspondence and Littelmann Paths by K. Erdmann, J.A. Green and M. Schoker 1980 – 2nd corr. and augmented edition (2007)
Vol. 1693: S. Simons, From Hahn-Banach to Monotonicity (Minimax and Monotonicity 1998) – 2nd exp. edition (2008)
Vol. 470: R.E. Bowen, Equilibrium States and the Ergodic Theory of Anosov Diffeomorphisms. With a preface by D. Ruelle. Edited by J.-R. Chazottes. 1975 – 2nd rev. edition (2008)
Vol. 523: S.A. Albeverio, R.J. Høegh-Krohn, S. Mazzucchi, Mathematical Theory of Feynman Path Integral. 1976 – 2nd corr. and enlarged edition (2008)
Vol. 1764: A. Cannas da Silva, Lectures on Symplectic Geometry 2001 – Corr. 2nd printing (2008)

*LECTURE NOTES IN MATHEMATICS*

Springer

Edited by J.-M. Morel, F. Takens, B. Teissier, P.K. Maini

**Editorial Policy** (for the publication of monographs)

1. Lecture Notes aim to report new developments in all areas of mathematics and their applications - quickly, informally and at a high level. Mathematical texts analysing new developments in modelling and numerical simulation are welcome.

   Monograph manuscripts should be reasonably self-contained and rounded off. Thus they may, and often will, present not only results of the author but also related work by other people. They may be based on specialised lecture courses. Furthermore, the manuscripts should provide sufficient motivation, examples and applications. This clearly distinguishes Lecture Notes from journal articles or technical reports which normally are very concise. Articles intended for a journal but too long to be accepted by most journals, usually do not have this "lecture notes" character. For similar reasons it is unusual for doctoral theses to be accepted for the Lecture Notes series, though habilitation theses may be appropriate.

2. Manuscripts should be submitted either to Springer's mathematics editorial in Heidelberg, or to one of the series editors. In general, manuscripts will be sent out to 2 external referees for evaluation. If a decision cannot yet be reached on the basis of the first 2 reports, further referees may be contacted: The author will be informed of this. A final decision to publish can be made only on the basis of the complete manuscript, however a refereeing process leading to a preliminary decision can be based on a pre-final or incomplete manuscript. The strict minimum amount of material that will be considered should include a detailed outline describing the planned contents of each chapter, a bibliography and several sample chapters.

   Authors should be aware that incomplete or insufficiently close to final manuscripts almost always result in longer refereeing times and nevertheless unclear referees' recommendations, making further refereeing of a final draft necessary.

   Authors should also be aware that parallel submission of their manuscript to another publisher while under consideration for LNM will in general lead to immediate rejection.

3. Manuscripts should in general be submitted in English. Final manuscripts should contain at least 100 pages of mathematical text and should always include

   - a table of contents;
   - an informative introduction, with adequate motivation and perhaps some historical remarks: it should be accessible to a reader not intimately familiar with the topic treated;
   - a subject index: as a rule this is genuinely helpful for the reader.

   For evaluation purposes, manuscripts may be submitted in print or electronic form, in the latter case preferably as pdf- or zipped ps-files. Lecture Notes volumes are, as a rule, printed digitally from the authors' files. To ensure best results, authors are asked to use the LaTeX2e style files available from Springer's web-server at:

   ftp://ftp.springer.de/pub/tex/latex/svmonot1/ (for monographs).

17 R 10-2 A

Additional technical instructions, if necessary, are available on request from: lnm@springer.com.

4. Careful preparation of the manuscripts will help keep production time short besides ensuring satisfactory appearance of the finished book in print and online. After acceptance of the manuscript authors will be asked to prepare the final LaTeX source files (and also the corresponding dvi-, pdf- or zipped ps-file) together with the final printout made from these files. The LaTeX source files are essential for producing the full-text online version of the book (see www.springerlink.com/content/110312 for the existing online volumes of LNM).

   The actual production of a Lecture Notes volume takes approximately 12 weeks.

5. Authors receive a total of 50 free copies of their volume, but no royalties. They are entitled to a discount of 33.3% on the price of Springer books purchased for their personal use, if ordering directly from Springer.

6. Commitment to publish is made by letter of intent rather than by signing a formal contract. Springer-Verlag secures the copyright for each volume. Authors are free to reuse material contained in their LNM volumes in later publications: a brief written (or e-mail) request for formal permission is sufficient.

**Addresses:**

Professor J.-M. Morel, CMLA,
École Normale Supérieure de Cachan,
61 Avenue du Président Wilson, 94235 Cachan Cedex, France
E-mail: Jean-Michel.Morel@cmla.ens-cachan.fr

Professor F. Takens, Mathematisch Instituut,
Rijksuniversiteit Groningen, Postbus 800,
9700 AV Groningen, The Netherlands
E-mail: F.Takens@math.rug.nl

Professor B. Teissier, Institut Mathématique de Jussieu,
UMR 7586 du CNRS, Équipe "Géométrie et Dynamique",
175 rue du Chevaleret
75013 Paris, France
E-mail: teissier@math.jussieu.fr

*For the "Mathematical Biosciences Subseries" of LNM:*

Professor P.K. Maini, Center for Mathematical Biology,
Mathematical Institute, 24-29 St Giles,
Oxford OX1 3LP, UK
E-mail: maini@maths.ox.ac.uk

Springer, Mathematics Editorial I, Tiergartenstr. 17
69121 Heidelberg, Germany,
Tel.: +49 (6221) 487-8259
Fax: +49 (6221) 4876-8259
E-mail: lnm@springer.com